Lecture Notes on Data Engineering and Communications Technologies

Volume 62

Series Editor

Fatos Xhafa, Technical University of Catalonia, Barcelona, Spain

The aim of the book series is to present cutting edge engineering approaches to data technologies and communications. It will publish latest advances on the engineering task of building and deploying distributed, scalable and reliable data infrastructures and communication systems.

The series will have a prominent applied focus on data technologies and communications with aim to promote the bridging from fundamental research on data science and networking to data engineering and communications that lead to industry products, business knowledge and standardisation.

Indexed by SCOPUS, INSPEC, EI Compendex.

All books published in the series are submitted for consideration in Web of Science.

More information about this series at http://www.springer.com/series/15362

Valentina E. Balas · Aboul Ella Hassanien ·
Satyajit Chakrabarti · Lopa Mandal
Editors

Proceedings of International Conference on Computational Intelligence, Data Science and Cloud Computing

IEM-ICDC 2020

Springer

Editors
Valentina E. Balas
Department of Automation and Applied Informatics and Department of International Relations Programs and Projects
Intelligent Systems Research Centre
Aurel Vlaicu University of Arad
Arad, Romania

Aboul Ella Hassanien
Department of Information Technology
Faculty of Computer and Artificial Intelligence
Cairo University
Giza, Egypt

Satyajit Chakrabarti
Institute of Engineering and Management
Kolkata, West Bengal, India

Lopa Mandal
Department of Information Technology
Institute of Engineering and Management
Kolkata, West Bengal, India

ISSN 2367-4512 ISSN 2367-4520 (electronic)
Lecture Notes on Data Engineering and Communications Technologies
ISBN 978-981-33-4967-4 ISBN 978-981-33-4968-1 (eBook)
https://doi.org/10.1007/978-981-33-4968-1

This Springer imprint is published by the registered company Springer Nature Singapore Pte Ltd.
The registered company address is: 152 Beach Road, #21-01/04 Gateway East, Singapore 189721, Singapore

Program Committee

Patron

Dr. Satyajit Chakrabarti, President, Institute of Engineering & Management, Kolkata, India
Dr. Amlan Kusum Nayak, Principal, Institute of Engineering & Management, Kolkata, India

General Chair

Dr. Satyajit Chakrabarti, Director, Institute of Engineering & Management, Kolkata, India

Convener

Dr. Lopa Mandal, Institute of Engineering & Management, Kolkata, India

Co-conveners

Dr. Baisakhi Das, Institute of Engineering & Management, Kolkata, India
Dr. Maumita Chakraborty, Institute of Engineering & Management, Kolkata, India
Dr. Sanchita Ghosh, Institute of Engineering & Management, Kolkata, India
Dr. Moutushi Singh, Institute of Engineering & Management, Kolkata, India
Mr. Subhabrata Sengupta, Institute of Engineering & Management, Kolkata, India

Editorial Board

Dr. Valentina E. Balas, Aurel Vlaicu University of Arad, Romania
Dr. Aboul Ella Hassanien, Cairo University, Egypt
Dr. Satyajit Chakrabarti, Institute of Engineering & Management, Kolkata, India
Dr. Lopa Mandal, IEM, Institute of Engineering & Management, Kolkata, India

Corporate/Research Keynote Speakers

Dr. Rajkumar Buyya, University of Melbourne, Australia
Dr. Arnab Bose, University of Chicago, USA
Dr. Joao Manuel R. S. Tavares, University of Porto, Portugal
Dr. Kouichi Sakurai, Kyushu University, Japan

Dr. Feng Hao, Warwick University, UK
Dr. Michael David, National Intelligence University, Washington D.C., USA
Dr. Parth Gupta, Amazon USA
Dr. Narang N. Kishor, Narnix Technolabs Pvt. Ltd, Delhi, India
Mr. Diptiman Dasgupta, IBM, India

Advisory Committee

Dr. Valentina E. Balas, Aurel Vlaicu University of Arad, Romania
Dr. Aboul Ella Hassanien, Cairo University, Egypt
Dr. Joao Manuel R. S. Tavares, University of Porto, Portugal
Dr. Kouichi Sakurai, Kyushu University, Japan
Dr. Michael David, National Intelligence University, Washington D.C., USA
Dr. Feng Hao, Warwick University, UK
Dr. Arnab Bose, University of Chicago, USA
Dr. Michael Sheng, Macquarie University, Sydney, Australia
Dr. Eric Filiol, Universite Bretagne Sud, France
Dr. Lalit Garg, University of Malta
Dr. Sudipta Roy, Washington University, Saint Louis, Missouri, USA
Dr. Biplab Sikdar, National University of Singapore
Dr. Franklin Jayasim, Communication Technology, Arizona, USA
Dr. Kaushik Deb, CUET, Bangladesh
Dr. Parth Gupta, Applied Scientist, Amazon USA
Dr. Muhammad Usman, University of South Wales, UK
Dr. Partha Pratim Das, IIT Kharagpur, India
Dr. Pramatha Nath Basu, Jadavpur University, India
Dr. Nabarun Bhattacharya, CDAC, India
Dr. Mita Nasipuri, Jadavpur University, Kolkata, India
Dr. Subir Kumar Sarkar, Jadavpur University, Kolkata, India
Dr. Ujjwal Maulik, Jadavpur University, Kolkata, India
Dr. Paramartha Dutta, Visva-Bharati University, India
Dr. Debasish Chakraborty, Indian Space Research Organisation, Kolkata, India
Dr. Samiran Chattopadhyay, Jadavpur University, India
Dr. Amlan Chakrabarti, Calcutta University, India
Dr. Aniruddha Nag, Computer Society of India, Kolkata, India
Mr. Subimal Kundu, Computer Society of India, Kolkata, India
Dr. Abhishek Das, Aliah University, India
Dr. Indraneel Mukhopadhyay, Institute of Engineering & Management, Kolkata, India
Ms. Banani Chakrabarti, Institute of Engineering & Management, Kolkata, India
Ms. Gopa Goswami, Institute of Engineering & Management, Kolkata, India

Technical Committee

Dr. Sudipta Roy, Washington University, Saint Louis, Missouri, USA
Dr. Biplab Sikdar, National University of Singapore

Dr. Siddhartha Bhattacharyya, VSB—Technical University of Ostrava, Czech Republic
Dr. Ranjan Mehera, Subex Inc., Colorado, USA
Dr. Kaushik Deb, CUET, Bangladesh
Dr. Nabendu Chaki, University of Calcutta, India
Dr. Sankhayan Choudhury, University of Calcutta, India
Dr. S. Sivasathya, Pondicherry University, India
Dr. Koushik Ghosh, Burdwan University, India
Dr. Sourav De, Coochbehar Government Engineering College, India
Dr. Matangini Chattopadhyay, Jadavpur University, India
Dr. Debasish Jana, TEOCO Software Pvt. Ltd., India.
Dr. Parama Bhowmik, Jadavpur University, Kolkata, India
Dr. Amitava Nag, CIT, Kokrajhar, India
Dr. Saprarshi Goswami, Bangabasi Morning College, Kolkata, India
Dr. Koushik Majumder, MAKAUT, India
Dr. Sourav Saha, IBM, India
Dr. Lopa Mandal, IEM, Kolkata
Dr. Pilli Emmanuel Shubhakar, MNIT Jaipur, India
Dr. Gaurav Trivedi, IIT Guwahati, India
Dr. Sujata Dash, North Orissa University, India
Dr. Aprajita Ojha, CSE, IIITDM Jabalpur, India
Dr. Sanchita Ghosh, IEM, Kolkata
Dr. Gaurav Trivedi, IIT Guwahati, India
Dr. Baisakhi Das, IEM, Kolkata
Dr. Maumita Chakraborty, IEM, Kolkata
Dr. Prabhat Kumar, NIT Patna, India
Dr. Abhishek Sribastava, IIT Indore, India
Dr. Jyoti Prakash Singh, NIT Patna, India
Dr. Raja Chakrabarti, NDMA Delhi, India
Dr. Rik Das, Xavier Institute of Social Service, Jharkhand, India
Mr. Amitava Chakraborty, IBM, India

Organizing Chairs

Dr. Lopa Mandal, Institute of Engineering & Management, Kolkata, India
Dr. Baisakhi Das, Institute of Engineering & Management, Kolkata, India
Dr. Maumita Chakraborty, Institute of Engineering & Management, Kolkata, India
Dr. Sanchita Ghosh, Institute of Engineering & Management, Kolkata, India

Organizing Committee

Mr. Partha Sarathi Paul, Institute of Engineering & Management, Kolkata, India
Mr. Avijit Bose, Institute of Engineering & Management, Kolkata, India
Mr. Rabi Narayan Behera, Institute of Engineering & Management, Kolkata, India
Dr. Satyasaran Changdar, Institute of Engineering & Management, Kolkata, India
Mr. Arup Kumar Chattopadhyay, Institute of Engineering & Management, Kolkata, India

Preface

As the world is moving toward Industry 4.0, computational intelligence, data science and cloud computing are gaining huge importance and relevance in our society in all possible ways. To address this need of the hour, International Conference on Computational Intelligence, Data Science and Cloud Computing (IEM-ICDC) 2020 was organized by the Department of Information Technology of Institute of Engineering & Management, Kolkata, India, held online using virtual platform during September 25–27, 2020.

The three-day conference provided a platform where researchers from all over the world, from academia and industry, took part in discussion and exchanged their views to make an intelligent, smart and secured society. Participation was from twenty-two different countries around the world. IEM-ICDC 2020 has been made possible with the generous support of the sponsors: Springer, Computer Society of India (CSI) and Centre for Development of Advanced Computing (CDAC).

Most significant new findings about AI and robotics, image processing and NLP, cloud computing and big data analytics as well as cyber security, blockchain and IoT and various allied fields were highlighted through insightful invited talks and innovative technical sessions.

Every detail of the conference has been reported in the proceedings with the hope that this will be helpful and be a motivating force for future research in the practical designs of cutting-edge systems in related fields.

The book is organized into six parts. Part one discusses research findings on "computational intelligence." Part two and part three present "computer vision" and "NLP and ICT," respectively. Part four highlights research works on "cloud and security," whereas Part five elaborates "data and security." The sixth part concludes the book with research papers on "IoT and blockchain."

We thank all the sponsors, authors, supporters and organizing team of the conference for the grand success of this event. In the current dire situation of the world, we hope the conference has come out with important research contributions to the technological aspects to combat the crisis and help the society at large.

Arad, Romania
Cairo, Egypt
Kolkata, India
Kolkata, India

Valentina E. Balas
Aboul Ella Hassanien
Satyajit Chakrabarti
Lopa Mandal

Contents

About the Editors

Dr. Valentina E. Balas is currently Full Professor in the Department of Automatics and Applied Software at the Faculty of Engineering, "Aurel Vlaicu" University of Arad, Romania. She holds a Ph.D. in Applied Electronics and Telecommunications from Polytechnic University of Timisoara. Dr. Balas is the author of more than 250 research papers in refereed journals and international conferences. Her research interests are in intelligent systems, fuzzy control, soft computing, smart sensors, information fusion, modeling, and simulation. She is Editor-in-Chief to International Journal of Advanced Intelligence Paradigms (IJAIP) and to International Journal of Computational Systems Engineering (IJCSysE), an Editorial Board member of several national and international journals, and Director of Intelligent Systems Research Centre in Aurel Vlaicu University of Arad. She is a member of EUSFLAT and SIAM and a senior member of IEEE, a member in TC – Fuzzy Systems (IEEE CIS), a member in TC – Emergent Technologies (IEEE CIS), and a member in TC – Soft Computing (IEEE SMCS).

Prof. Aboul Ella Hassanein is Founder and Chairman of the Scientific Research Group in Egypt, Cairo University (SRGE), and Professor of Information Technology at the Faculty of Computer and Artificial Intelligence, Cairo University. Professor Hassanien is Ex-Dean of the faculty of computers and information, Beni Suef University. Professor Hassanien has more than 1000 scientific research papers published in prestigious international journals and over 45 edited books/textbooks covering such diverse topics as data mining, medical images, intelligent systems, bioinformatics, social networks, aerospace engineering, and smart environment. His other research areas include computational intelligence, medical image analysis, security, animal identification, space sciences and telemetry mining and multimedia data mining. Prof. Hassanien won several awards including the Best Researcher of the Youth Award of Astronomy and Geophysics of the National Research Institute, Academy of Scientific Research (Egypt, 1990). He also received the State Award for Excellence in Engineering Sciences 2015. He holds the Medal of Sciences and Arts first-class from President of Egypt (2017).

Dr. Satyajit Chakrabarti is Pro-Vice Chancellor, University of Engineering & Management, Kolkata and Jaipur Campus, India and Director of Institute of Engineering & Management, IEM. As the Director of one of the most reputed organizations in Engineering & Management in Eastern India, he launched a PGDM Program to run AICTE approved Management courses, Toppers Academy to train students for certificate courses, and Software Development in the field of ERP solutions. Dr. Chakrabarti was Project Manager in TELUS, Vancouver, Canada from February 2006 to September 2009, where he was intensively involved in planning, execution, monitoring, communicating with stakeholders, negotiating with vendors and cross-functional teams, and motivating members. His research interest includes Data Mining, Nano Electronics, Nanotechnology. He has got many journal and conference publications and edited books to his credit.

Dr. Lopa Mandal is currently Professor of Department of Information Technology, Institute of Engineering & Management, Kolkata, India. She has completed her M.Tech/IT and Ph.D (Engineering) from Jadavpur University, Kolkata, India. Her research interest includes Artificial Intelligence, Data analytics, Distributed & cloud computing, Mobile computing and Learning technologies. She has got many National and International journal and conference publications and book chapters to her credit. She has been a Technical Program Committee member of different International Conferences and editorial/review board member of International Journals. She has been the convener or organizing chair and also part of the organizing committee of different International Conferences. She is also a member of Computer Society of India. She has got around 20 years of work experience in industry and academia.

Part I
Computational Intelligence

Deep Learning, Predictive Modelling and Nano/Bio-Sensing Technologies for Mitigation of the COVID-19 Pandemic

Asim Kar and Anuradha Kar

Abstract As the world battles the COVID-19 pandemic, artificial intelligence, epidemiological analysis and novel sensing technologies can play key roles in mitigating the impacts of this unprecedented global crisis. This paper reviews how the domains of deep learning, numerical modelling and bio-sensing have rapidly responded with solutions to this grim pandemic scenario through the development of new methods for early detection and forecasting of the impacts of the COVID-19 virus. The recent applications in the area of deep learning-based computer vision and image analysis tools are discussed which have emerged with promising early COVID-19 detection methods using clinical chest imagery. In addition, a variety of predictive models developed in the current year for estimating the transmission, infection and mortality rates due the novel coronavirus are described which provide crucial information for determining social and governmental measures for containing the viral outbreak. Finally, state-of-the-art clinical test methodologies using nano-level sensors for point-of-care testing practices are highlighted which may enable rapid and accurate diagnosis of the COVID-19 infections. Overall, this paper brings forward some of the most advanced technologies available today and analyses how these may help mankind in defence of the survival challenges posed by the COVID-19 virus. In order to allow readers to further explore the concepts presented in this paper, an open repository named Coro-Lib containing implementations of the computational methods reviewed in this paper is created and described at the end.

Keywords Deep learning · Modelling · Nanotechnology · Biosensor · COVID-19

A. Kar
Institute of Engineering and Management, EP Block, Sector V, Kolkata 700091, India
e-mail: asim.kar@iemcal.com

A. Kar (✉)
École Normale Supérieure de Lyon, 46 allée d'Italie, 69007 Lyon, France
e-mail: anuradha.kar@ens-lyon.fr

V. E. Balas et al. (eds.), *Proceedings of International Conference on Computational Intelligence, Data Science and Cloud Computing*, Lecture Notes on Data Engineering and Communications Technologies 62, https://doi.org/10.1007/978-981-33-4968-1_1

1 Introduction

As of August, 2020, it has been justified via scientific investigations that COVID-19 is a global pandemic caused by the acute infection of the novel coronavirus 2 (SARS-CoV-2) [1]. The pandemic virus has spread throughout the world within 2–3 months (from December 2019 to March 2020) infecting millions of people of all ages and taking away nearly a million human lives. The rates of infection and mortality are still rising in different parts of the world and some countries in Europe are on the verge of experiencing a second wave of the disease. The World Health Organization (WHO) working round the clock has become helpless because their bulletin declared that none of the existing vaccines would respond for the cure of this disease.

It has been observed since the COVID-19 virus outbreak that the major threat to life from this disease arises because the typical characteristics of the virus infection in patients are absolutely difficult to detect, understand or predict [2]. The virus can overtake within a very short span of time leading to collapse of the vital organs of human body. Besides the established social and clinical measures like imposition of lockdowns and mandatory use of masks, the best way to battle the impacts of this disease is to arrange for early detection of the infections, which only can increase the probability of survival of the patients. It is therefore an apt time for leveraging the power of new and advanced nano and bio-sensing technologies for non-contact sensing of COVID-19 virus and that of artificial intelligence (AI) and computational methods to assist in early detection and prediction of the disease occurrences.

In this paper, a detailed review of the latest and significant research developments aiming towards fast and accurate diagnosis of COVID-19 infections and forecasting the trends of the pandemic evolution is presented. These promising concepts and developments are discussed here to analyse their capabilities, limits and future prospects towards dealing with the COVID-19 pandemic in a systematic manner.

2 Deep Learning-Based Chest Image Analysis for Early Diagnosis of COVID-19 Infections

An extensive body of research has been done in the past 6 months towards using deep learning (DL) models for identification of COVID-19 features and distinguishing it from community-acquired pneumonia (CAP) and no-infection cases from chest imaging techniques such as computed tomography (CT), X-ray and ultrasound. Important results from these research works are summarized here.

2.1 Deep Learning with CT Images

The study in [2] proposes a joint algorithm using both CT images and clinical history data of patients such as symptoms, history of exposure and laboratory testing results to identify COVID-19 occurrence in 279 cases. The authors used a deep convolutional neural network (CNN) for learning the COVID-19 related features from the CT scan and support vector machine (SVM), random forest and multilayer perceptron (MLP) classifiers to classify patients with COVID-19 according to clinical information such as age, exposure to COVID-19 infection cases, fever, cough, cough with sputum and white blood cell counts. They achieved nearly 85% sensitivity in COVID-19 identification while that from a human radiologist is close to 75%. The authors in [3] present a strategy for detecting COVID-19 occurrence from CT images which minimizes the requirements of manual image labelling while achieving high classification accuracy of 96.2%. CT images from severely affected patients were chosen for representation learning on multiple spatial levels of features, such as large patches and small lesions, to locate the infections on the lungs. The authors used a fully convolutional model comprising of five convolutional blocks to perform a three-class classification between CAP, COVID-19 and non-pneumonia cases. The results indicate that while their model gets significant accuracy in the binary COVID-19 identification tasks, the classification performance declines while distinguishing COVID-19 from community-acquired pneumonia. Another weakly supervised deep learning method (named 3D DeCoVNet) is developed in [4] which uses 3D CT volumetric images for classification of COVID-19 and localizing lesions. The segmentation is done by 3D U-Net followed by another 3D classification network to predict the probability of the disease. The localization of the lesions is done in the classification network with the full pipeline achieving detection accuracy of 90%. Also, the method can discover lesions without the need for lesion region annotations during training.

2.2 Deep Learning with Chest X-Ray (CXR) Images

One of the drawbacks of CT scans is that it is a costly imaging procedure and may not be available in all health clinics or hospitals. In this respect, chest X-rays can be excellent alternatives for early detection of COVID-19 infections. They are available in most clinics, medical laboratory or hospitals and are also cheaper and faster for testing patients. A range of studies therefore reports the use of deep learning on CXR images such as [5]. In this, CXR data was gathered from open online repositories like Kaggle and randomly sampled. A CNN-based classification model is proposed with two convolutional and two dense layers which achieved 94.03% accuracy in distinguishing between COVID-19, viral pneumonia and no-infection cases. A binary classification model is presented in [6] which achieves 98% accuracy in distinguishing between COVID-19 versus no-infection classes and 87% in a multi-class classification problem between COVID-19 versus no-infection versus pneumonia cases. A new DL model named DarkCovidNet was developed based on the pre-existing Darknet-19 architecture which uses 17 convolution layers and multiple layer-wise filtering approaches.

The COVID-Net model in [7] is also a new deep CNN model which was trained with a large open dataset of CXR images that contains data from 13,870 patients. The model performs three-class classification between normal, non-COVID infection and COVID-19 cases and achieves an accuracy of 93.3%. The COVID-Net architecture is built using a residual projection-expansion-projection-extension (or PEPX) design pattern and has advantages such as lesser architectural and computational complexity than the VGG-19 and ResNet-50. Another new DL model named nCOVnet is proposed in [8] comprising of 24 layers with 18 convolutional layers and is based on transfer learning, with a VGG-16 model trained on ImageNet dataset. The study claims the proposed model to be suitable for small X-ray dataset sizes as it achieves 97.62% true positive rate in a binary classification task between COVID and no-COVID detection. A study addressing the factor of limited training data is [9] which uses a patch-based procedure with CNNs and also requires fewer trainable parameters. In this, small but numerous patches are created from each CXR image which are then input to a classification network built using ResNet-18 architecture. It achieves an accuracy of 91.9% in distinguishing between normal, bacterial, viral pneumonia and tuberculosis. A new model CoroNet is proposed in [9] which uses an Xception architecture pre-trained on ImageNet data. The network is further trained on a mixed CXR dataset containing images from patients with COVID-19, bacterial, viral pneumonia and no-pneumonia cases. It achieves 98.2% for four-class and 95% accuracy in the three-class classification task with COVID-19, pneumonia and no-pneumonia cases.

Several works focus on comparing the performance of existing deep learning CNN models on CXR images for detecting COVID-19 cases. In [10], a DL framework called COVIDX-Net is presented which comprises of seven different deep CNN models such as modified DenseNet201, InceptionV3, ResNetV2, Inception-ResNetV2, Xception, and MobileNetV for automatic COVID-19 detection from CXR images. The training dataset is short and comprises of 50 CXR images. The best performance was obtained from VGG19 and DenseNet201 models, although the MobileNetV2 was found to be fast and suitable for use on smart healthcare systems. Also, in [11], the performance of three different deep CNNs (ResNet50, InceptionV3 and InceptionResNetV2) were compared based on chest X-rays. This study reports the best classification accuracy from the pre-trained ResNet50 model (98%) followed by InceptionV3 and Inception-ResNetV2. The major issue faced by the models is limited amount of available training data containing signatures of COVID-19 infection.

2.3 Deep Learning with Ultrasound Images

Ultrasound chest imaging has several advantages such as being widely available, low-cost and real-time imaging and is often claimed to have better sensitivity than X-ray in detecting pneumonia. Several works have used lung ultrasound or LUS images with DL models for detecting COVID-19. In [12], a large and fully annotated LUS database is used that contains frame, video and pixel level annotations which can be used for segmenting the LUS frames. The study uses spatial transformer networks, which predicts disease severity score as well as locates pathological artefacts which are aggregated on a video scale. In [13], a large LUS dataset is assembled with 1103 LUS images with more than 50% COVID-19 instances. A CNN model named POCOVID-NET is trained on this dataset to distinguish between COVID-19, bacterial pneumonia and normal cases. The POCOVID-NET is based on the VGG-16 architecture pre-trained on the ImageNet data and achieves an accuracy of 89% on the three-class classification problem. This model is also integrated into a web-service which uses the trained model to perform predictions on LUS uploaded by any user.

3 Predictive Modelling of COVID-19 Epidemiological Patterns

Significant amount of research has been directed towards developing mathematical and machine learning models to predict the mortality count, rate of infections and understanding of the evolution of the COVID-19 pandemic which will allow the Governments to issue effective control and mitigation strategies for quantifying the COVID-19 infections. Below, several classes of the prediction models are described.

3.1 Mathematical Approaches—Compartmental Models

Mathematical models for infectious disease transmission using differential equations represent dynamic characteristics of the diseases, taking into consideration population growth, susceptible, symptomatic and asymptomatic individuals, transmission of the disease within the population and geo-spatial/age/behavioural characteristics of the population. These may use ordinary or partial or delay differential systems and are called compartmental models such as susceptible, infected, recovered (SIR) and susceptible, exposed, infected, removed (SEIR) models (Fig. 1).

The study in [14] used a modified SIR model to compare the epidemiological characteristics in Italy with that in China. It uses the daily count of confirmed cases, recovered cases and deaths from late January 2020 onwards. The model is a modified SIR model with a time varying transmission rate parameter ($\pi(t)$) as described in Eqs. (1)–(4). In the model S, I and R are the numbers of susceptible, infected and

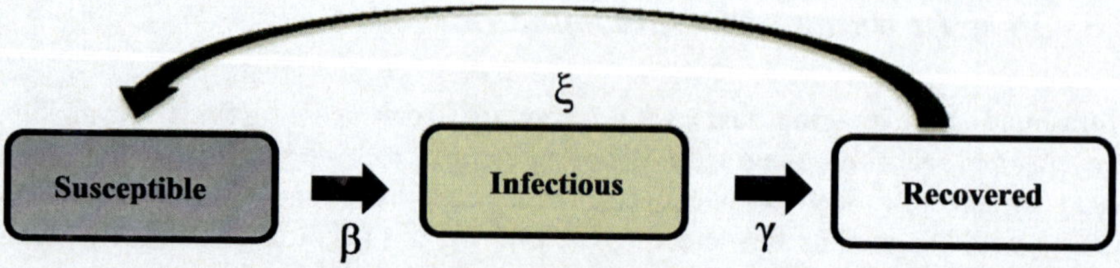

Fig. 1 Schematic of an SIR model where β is the infection rate, γ is the recovery rate, and ξ is the rate at which recovered individuals return to the susceptible state [15]

recovered individuals, N is the sum of these three. β is the average number of contacts per person per time and γ is the mean recovery rate. R_0 (reproductive number) represents the average number of cases arising from a primary case in a susceptible population.

$$\frac{dS}{dt} = -\pi(t)\frac{\beta SI}{N} \tag{1}$$

$$\frac{dI}{dt} = \pi(t)\frac{\beta SI}{N} - \gamma I \tag{2}$$

$$\frac{dR}{dt} = \gamma I \tag{3}$$

$$R_0 = \frac{\beta}{\gamma} \tag{4}$$

The study predicts a rapid spread of COVID-19 throughout Italy after February 21 due to a deficit in preventive measures and under-reporting of earlier infections. It also predicts that strict measures such as detecting and isolating individuals with symptoms, traffic restrictions, medical tracking and entry/exit screening in Italy should still be maintained until August 2020 to prevent further spread of the disease. In [15], a new epidemiological class is added to the existing SEIR model which comprises of super-spreaders individuals along with a fatality compartment that includes death due to the COVID-19 infection. The model is described as:

$$\frac{dS}{dt} = -\beta\frac{1}{N}S - I\beta\frac{H}{N}S - \beta'\frac{P}{N}S \tag{5}$$

$$\frac{dE}{dt} = \beta\frac{1}{N}S + I\beta\frac{H}{N}S + \beta'\frac{P}{N}S - \kappa E \tag{6}$$

$$\frac{dI}{dt} = \kappa\rho_1 E - (\gamma_a + \gamma_i)I - \delta_i I \tag{7}$$

$$\frac{dP}{dt} = \kappa\rho_2 E - (\gamma_a + \gamma_i)P - \delta_p P \quad (8)$$

$$\frac{dA}{dt} = \kappa(1 - \rho_1 - \rho_2)E \quad (9)$$

$$\frac{dH}{dt} = \gamma_a(I + P) - \gamma_r H - \delta_h H \quad (10)$$

$$\frac{dR}{dt} = \gamma_i(I + P) + \gamma_r H \quad (11)$$

$$\frac{dF}{dt} = \delta_i I + \delta_p P + \delta_h H \quad (12)$$

where S represents the susceptible class, E the exposed class, I the symptomatic and infectious class, P the super-spreaders class, the infectious but asymptomatic class A, the hospitalized H, recovery class R and fatality class F. Using this model, the number of deaths at each instant of time ($D(t)$) is given by Eq. (12), where δ_i, δ_p, δ_h are the death rates due to infected, super-spreaders and hospitalized individuals, respectively. It is reported that this new model although developed during the early phases of the pandemic was seen to adapt to the scenario in Wuhan, China and predicted a reduction on the daily number of confirmed COVID-19 cases over time.

A more inclusive SEIRA model in proposed in [16] which involves impacts from sub-groups of a population in predicting the spread of the COVID-19. A typical heterogeneous population is split into several sub-populations which share common characteristics, such as a common geographical zone, behavioural groups, social classes, and age groups, and there are interactions between the different classes. According to this model, the rate of variation of the susceptible population (S) varies as:

$$\frac{dS}{dT} = \text{Natural net growth} - \text{Intra-class infections} - \text{Interclass infections} \quad (13)$$

The advantages of this model are that it includes interaction between different population sub-groups which may help in better understanding of the pandemic evolution and also that the structure of the model can be modified to include population-specific factors. A further detailed compartment model is presented in [17] named SQUIDER which adds detected (D) and undetected (U) infected populations, quarantine (Q) and release from isolation and possible reinfection classes to the existing SIR model. This model is observed to fit accurately on data from eight states in USA such as Arizona, California, Florida, Illinois, Louisiana, New Jersey, New York State and Texas possibly because of the use of more compartments in the model. The model indicates that severe cases of the COVID-19 pandemic may be avoided with higher levels of isolation and social distancing.

3.2 Mathematical Approaches—Logistical Models

In the compartmental models of epidemic transmission and evolution that use classical differential equations, there are shortcomings arising from the assumption that the total number of people in a certain area is constant, because the population may undergo time variations including random interactions and connections between individuals. The study in [18], therefore, uses Logistic, Bertalanffy and Gompertz models to fit with SARS (2003) and COVID-19 data to predict the development trend of the virus infections, such as number of confirmed cases and death toll. The logistical model which best fits to the data from Wuhan is given by:

$$Q_t = \frac{a}{1 + e^{b - c(t - t_0)}} \tag{14}$$

where Q_t is the cumulative confirmed cases (deaths); a is the predicted maximum of confirmed cases (deaths). b and c are fitting coefficients. t is the number of days since the first case. t_0 is the time when the first case occurred. The logistic model predicts that the total number of confirmed cases will reach highest in the Wuhan province and China by April end and that also corresponds to the date when the COVID-19 epidemic will end in these regions. Three different growth models are used in [19] which are generalized logistic growth (Eq. 15), logistic growth (Eq. 16) and generalized growth model (Eq. 17) in order to predict the growth and increase in the confirmed positive COVID-19 cases in India. The growth parameters are r (rate of growth at an early stage), K (final epidemic size), C (number of aggregate cases at time t) and p the profile of distinct growth (0–1). The models are given as:

$$\frac{dC(t)}{dt} = r\left(1 - \frac{C(t)}{K}\right)C(t) \tag{15}$$

$$\frac{dC(t)}{dt} = r\left(1 - \frac{C(t)}{K}\right)C^{P}(t) \tag{16}$$

$$\frac{dC(t)}{dt} = rC^{P}(t) \tag{17}$$

Out of the models, the generalized linear model predicts an exponentially rising growth rate for the epidemic requiring major preventive measures at the earliest, while the logistic model depicts an optimistic scenario in terms of a much slower growth rate of infections. Another study predicting the growth rate of the COVID-19 infection is [20] which is modelled using an exponential decay function given below, where N is the number of infected people at time t, λ is the infection growth rate.

$$\frac{dN}{dt} = \lambda N, \quad \lambda_{t_i} = \frac{N_{t_i} - N_{t_{i-2}}}{2\Delta t N_{t_i}} \tag{18}$$

Using data from countries like and China and South Korea, this model predicts that number of total infections in Italy, Germany, Spain and Sweden have reached their peak by June 2020, whereas cumulative number of infections in India would be around 2.4 million in the beginning of August 2020. This study also reflects that India could have a slower decay of the infection growth rate if strict lockdown is imposed.

3.3 Mathematical Approaches—Time Series Prediction Models

ARIMA stands for auto-regressive integrated moving average—which are a class of mathematical models suited for prediction of natural adversities such as COVID-19. The model is defined as ARIMA (p, d, q) in which p is the order of the auto-regression (AR) component, d is the order of the differencing process to form a stationary times series, and q is the order of the moving average (MA) process. In an ARIMA model, the value of y at time t is estimated as:

$$y_t = \mu + \varphi_1 y_{t-1} + \cdots + \varphi_p y_{t-p} - \theta_1 e_{t-1} - \cdots - \theta_q e_{t-q} \tag{19}$$

where y_t is the value of a variable at time t, φ is the AR parameter, and θ is the MA parameter. These models have been used in several studies for forecasting COVID-19 epidemiological features such as number of daily confirmed infection cases and death rates. In [21], the ARIMA model is used to predict the infection and death rates of COVID-19 in Italy, Spain and Turkey, and the model results show that the number of infections in Italy and Spain would decrease in July and in Turkey in September 2020. Lowest death rate in Italy and Spain would occur in July but it would be highest in Turkey. In [22], the ARIMA model is used for forecasting the number the cases over a period of upcoming 20 days in India. The model reveals exponential increase in the number of cases in the 20 days owing to reasons such as large social gatherings and lack of isolation of individuals with recent travel history, some of whom were asymptomatic. In [23], the ARIMA model is fitted on data from 15 countries to estimate the COVID-19 disease spread parameters such as cumulative number of cases, deaths and recoveries over a sub-sequent period of 2 months (April–July, 2020). In this study, the ARIMA model was seen to predict a fast disease spread in the USA, the UK, Turkey, China and Russia, but fast recovery ratio in China, Switzerland, Germany, Iran and Brazil.

3.4 Machine Learning-Based Prediction Models

Machine learning models like recurrent neural networks (RNN) and long short-term memory (LSTM) have been applied for time series-based predictions of the trends of the COVID-19 for various countries. An LSTM network memorizes former patterns in data sequences and comprises of memory blocks and gates, and the former memorizes network parameters for long durations. The gates process the information with an activation function and have outputs either 0 or 1. The functional form of the three LSTM gates (input, forget, output) are as below.

$$J_t = \text{sigmoid}\left(w_J\left[h_{t-1}, k_t\right] + b_J\right) \tag{20}$$

$$G_t = \text{sigmoid}\left(w_G\left[h_{t-1}, k_t\right] + b_G\right) \tag{21}$$

$$P_t = \text{sigmoid}\left(w_P\left[h_{t-1}, k_t\right] + b_P\right) \tag{22}$$

where J_t is the function of the input gate, G_t and P_t the functions of forget and output gates, w_x the neuron coefficients at the gates, H_{t-1} the result from previous time step, k_t the input to the current function at time step t and b_x is the neurons at gate x.

In [24], the LSTM-based model was used for predicting number of positive and recovered cases in India over an upcoming period of 30 days while the country undergoes social isolation. Data from 30 January 2020 till 4 April 2020 is used and the study reveals that while the patient recovery rate in India is high, the time for recovery is large. The effect of lockdown is modelled by variation of the transmission parameter and it is seen that with lockdown (low transmission) the increase rate of total number of positive cases flattens significantly. In [25], data from several Indian states are grouped based on the daily growth rate and spread of COVID-19 for predictions with less than 3% errors for daily and less 8% errors for weekly predictions. Three variants of LSTM are used: stacked LSTM or deep LSTM which contains multiple hidden layers and memory units, convolutional LSTM where the input vector, output vector, hidden state vector and the gates form 3D tensors and Bi-LSTM which extracts temporal features using both past and future information. The bi-LSTM achieves lowest error (~3%) in predicting for all 32 Indian states on a short time scale.

4 Nano/Bio-Sensing Technologies for Point-of-Care Testing of COVID-19 Symptoms

The most common clinical tests for COVID-19 patients are laboratory-based methods using genetic analysis or by micro-neutralization assay (MNA) [26]. The genetic analysis method (RT-PCR) is performed on infected samples (swabs of sputum, saliva, nasal secretions) through RNA extraction, while the MNA test does virus neutralization on the culture cells with specific antibodies. These methods although accurate have drawbacks such as time delay and scarcity of test kits. In this respect, the concept of point-of-care testing (POC) testing is appropriate which aims to provide immediate, high sensitivity COVID test results to healthcare specialists (~within 10 min–2 h) and possibility to perform the tests at the site of the patient (thereby catering to a large patient population). A new and simple POC test method called STOPCovid is developed in [27] which can detect COVID-19 DNA or RNA signatures using lateral flow and fluorescence methods in 40–70 min. It has minimal sample processing steps and no need for sophisticated instrumentation which allows it to be deployed in general clinics, pharmacies and even homes.

A range of biosensor-based POC testing methods has been discussed in [28] such as: chip, paper, film, thread and graphene based ones whose characteristics are summarized in Table 1. Very recent works presenting developments in nano-sensors for COVID-19 detection include [29, 30]. In [29], a novel biosensor using field-effect transistor (FET) is proposed which is developed from graphene sheets coated with antibody for COVID-19 spike protein. It can detect the COVID-19 virus from antigen protein, cultured virus and patient clinical specimens without sample pre-treatment even in ultra-low concentrations nearing 1 femto-gram/mL. Another new biosensor for COVID-19 is reported in [30] which uses the plasmonic photo-thermal (PPT) effect and localized surface plasmon resonance (LSPR) sensing transduction. It includes 2D gold nano-islands with DNA receptors which can detect COVID-19 sequences through nucleic acid hybridization. It can detect sequences in extremely low concentrations (0.22 pM) and also distinguish COVID-19 from a mixture of genes.

Table 1 Comparing characteristics of POC bio-sensors (CFU: colony-forming unit)

Sensor type	Bio-analysis target	Test time (min)	Limit of detection
Chip based	Antibodies/nucleic acid	40–45	30–1000 CFU/mL
Paper based	Antibodies/nucleic acid	2–50	1.9×10^3 CFU/mL
Thread based	Antibodies	0.5	5000 CFU/mL
STOPCovid	Viral RNA	40–70	100 genome copies/reaction
Film based	Genes	50 s	30 CFU/mL
Graphene	DNA	<1	0.5 pg/mL protein
FET based	COVID-S-protein	3–10	242 genome copies/mL
Plasmonics/nano-particle	COVID-N-protein	60	0.18 ng/μL

5 Coro-Lib: Computational Resources for COVID-19 Mitigation

An open-source repository named Coro-Lib is created which contains implementations of the deep learning and predictive modelling concepts reviewed in this paper. This repository is hosted on GitHub under the following link:

https://github.com/anuradhakar49/Coro-Lib

In this repository, readers will find datasets and Jupyter notebooks with which they can implement several of the deep learning-based COVID image classification methods and ARIMA and SEIR-based prediction models. This repository will be updated periodically to include the latest computational resources for COVID-19 mitigation.

6 Discussions and Conclusion

The COVID-19 crisis is a global health emergency as of September 2020 and the pandemic is yet to show signs of receding in all the continents. The ongoing research and developments on sensing and computational techniques for early detection and prediction of COVID-19 outbreak are some of the best resources to tackle the global pandemic scenario until a vaccine becomes available. The research works summarized here to throw light on how global healthcare infrastructure and medical practices needs to change and adapt in order to reduce the viral transmission and risk of mortality as well as take pre-emptive measures based on future pandemic trends. On the other hand, open and transparent data sharing by health centres and governmental organizations is equally important since limited data is a major challenge towards building robust computational techniques for COVID-19 detection and forecasting. Overall, a synergy of operations between scientific research, technological developments and Governmental policies on a global scale is essential at this time to construct an effective and sustainable pandemic control system.

References

1. E.J. Williamson, A.J. Walker, K. Bhaskaran et al., Factors associated with COVID-19-related death using OpenSAFELY. Nature **584**, 430–436 (2020)
2. X. Mei, H. Lee, K. Diao et al., Artificial intelligence–enabled rapid diagnosis of patients with COVID-19. Nat. Med. (2020). https://doi.org/10.1038/s41591-020-0931-3
3. S. Hu et al., Weakly supervised deep learning for COVID-19 infection detection and classification from CT images. IEEE Access **8**, 118869–118883 (2020)

4. X. Wang, X. Deng, Q. Fu et al., A weakly-supervised framework for COVID-19 classification and lesion localization from chest CT. IEEE Trans. Med. Imaging **39**(8), 2615–2625 (2020)
5. K. Ahammed, M.S. Satu, M.Z. Abedin et al., Early detection of coronavirus cases using chest X-ray images employing machine learning and deep learning approaches. medRxiv, p. 2020.06.07.20124594 (2020)
6. T. Ozturk, M. Talo, E.A. Yildirim et al., Automated detection of COVID-19 cases using deep neural networks with X-ray images. Comput. Biol. Med. 121 (2020)
7. L. Wang, Z.Q. Lin, A. Wong, COVID-Net: a tailored deep convolutional neural network design for detection of COVID-19 cases from chest X-ray images. arXiv:2003.09871
8. H. Panwar, P.K. Gupta, M. Khubeb, R. Morales-Menendez, Application of deep learning for fast detection of COVID-19 in X-rays using nCOVnet. Chaos Solitons Fract. **138**, 109944 (2020)
9. Y. Oh, S. Park, J.C. Ye, Deep learning COVID-19 features on CXR using limited training data sets. IEEE Trans. Med. Imging **39**(8), 2688–2700 (2020)
10. A. Iqbal, J. Latief, M. Mudasir, CoroNet: a deep neural network for detection and diagnosis of COVID-19 from chest X-ray images. Comput. Methods Programs Biomed. **196**, 105581 (2020)
11. E. El-Din Hemdan et al., COVIDX-Net: a framework of deep learning classifiers to diagnose COVID-19 in X-ray images. ArXiv abs/2003.11055 (2020)
12. A. Narin et al., Automatic detection of coronavirus disease (COVID-19) using X-ray images and deep convolutional neural networks. ArXiv abs/2003.10849 (2020)
13. S. Roy et al., Deep learning for classification and localization of COVID-19 markers in point-of-care lung ultrasound. IEEE TMI **39**(8), 2676–2687 (2020)
14. J. Born, G. Brändle, M. Cossio et al., POCOVID-Net: automatic detection of COVID-19 from a new lung ultrasound imaging dataset (POCUS). CoRR abs/2004.12084 (2020)
15. J. Wangping, H. Ke, S. Yang et al., Extended SIR prediction of the epidemics trend of COVID-19 in Italy and compared with Hunan, China. Front. Med. **7**, 169 (2020)
16. F. Ndaïrou et al., Mathematical modeling of COVID-19 transmission dynamics with a case study of Wuhan. Chaos Solitons Fract. **135**, 109846 (2020)
17. Z.S. Khan, F. Van Bussel, A predictive model for Covid-19 spread applied to eight US states. arXiv:2006.05955 (2020), pp. 1–21
18. L. Jia et al., Prediction and analysis of coronavirus disease. arXiv:2003.05447 (2019)
19. M. Jain et al., Modelling logistic growth model for COVID-19 pandemic in India, in *2020 (ICCES)*, Coimbatore (2020), pp. 784–789
20. R. Bhardwaj, A predictive model for the evolution of COVID-19. Trans. Indian Natl. Acad. Eng. ID: covidwho-610593 (2020)
21. L. Bayyurt, B. Bayyurt, Forecasting of COVID-19 cases and deaths using ARIMA models. medRxiv, p. 2020.04.17.20069237 (2020)
22. H. Tandon, P. Ranjan, T. Chakraborty, V. Suhag, Coronavirus (COVID-19): ARIMA based time-series analysis to forecast near future. arXiv:2004.07859, pp. 1–11
23. R.K. Singh et al., Prediction of the COVID-19 pandemic for the top 15 affected countries: advanced autoregressive integrated moving average (ARIMA) model. JMIR Public Health Surv. **6**(2), e19115 (2020)
24. A. Tomar, N. Gupta, Prediction for the spread of COVID-19 in India and effectiveness of preventive measures. Sci. Total Environ. **728**, 138762 (2020)
25. P. Arora, H. Kumar, B. Ketan, Prediction and analysis of COVID-19 positive cases using deep learning models: a descriptive case study of India. Chaos Solitons Fract. **139**, 110017 (2020)
26. H.A. Hussein, R.Y.A. Hassan, M. Chino, F. Febbraio, Point-of-care diagnostics of COVID-19: from current work to future perspectives. Sensors **20**, 4289 (2020)
27. K.R. Jerome et al., Point-of-care testing for COVID-19 using SHERLOCK diagnostics. medRxiv. https://doi.org/10.1101/2020.05.04.20091231
28. J.R. Choi, Development of point-of-care biosensors for COVID-19. Front. Chem. **8**, 517 (2020)

29. G. Seo, G. Lee, M.J. Kim et al., Rapid detection of COVID-19 causative virus (SARS-CoV-2) in human nasopharyngeal swab specimens using field-effect transistor-based biosensor. ACS Nano **14**(4), 5135–5142 (2020)
30. G. Qiu, Z. Gai, Y. Tao, J. Schmitt, G.A. Kullak-Ublick, J. Wang, Dual-functional plasmonic photothermal biosensors for highly accurate severe acute respiratory syndrome coronavirus 2 detection. ACS Nano **14**(5), 5268–5277 (2020)

User Authentication Using Password and Hand Gesture with Leap Motion Sensor

Md. Adib Hassan, Qazi Shadman, Mishkat Haider Chowdhury, Sakib Al Hasan, and Jia Uddin

Abstract User authentication is becoming a significant factor in the field of modern technology. It is a process that permits a device to confirm the recognition of somebody who interfaces with a system asset. In the world of AI, machine learning is currently one of the leading research fields which is looking into practical implementation. In this paper, we propose a method where the user will enter the given password while the leap motion sensor will compare the behavioral data of the user with an existing dataset. Leap motion controller is a sensor which can recognize 3D movement of hands, fingers, and finger-like object with no contact. For our project, we chose to use dynamic time warping and Naive Bayes classifier algorithm. For the experiment, we captured the leap motion data of eight people and stored them in the database for user authentication and another eight people were used as third-party users. Moreover, as the password is being typed into the keyboard, it is important to restrict the intrusion of malicious third-party users. The proposed system demonstrates about 91% accuracy which rises to 93% in the best-case scenario.

Keywords Dynamic time warping (DTW) · Naive Bayes classifier · Leap motion sensor · Password authentication · Physiological biometrics · Behavioral biometrics

---

Md. A. Hassan · Q. Shadman · M. H. Chowdhury · S. Al Hasan
Department of Computer Science and Engineering, BRAC University, Dhaka, Bangladesh
e-mail: adib2k16@gmail.com

Q. Shadman
e-mail: qshadman2015@gmail.com

M. H. Chowdhury
e-mail: mhc.1996@gmail.com

S. Al Hasan
e-mail: hasansakib920@gmail.com

J. Uddin (✉)
Technology Studies Department, Endicott College, Woosong University, Daejeon, South Korea
e-mail: jia.uddin@wsu.ac.kr

V. E. Balas et al. (eds.), *Proceedings of International Conference on Computational Intelligence, Data Science and Cloud Computing*, Lecture Notes on Data Engineering and Communications Technologies 62, https://doi.org/10.1007/978-981-33-4968-1_2

1 Introduction

In the modern era, digital devices have come a long way in their uses. With IoT on the rise from improvement of technology, data security has become an essential [1]. There are multiple ways to get this resolved which include information based, token based and biometric based security systems. Abdulkader et al. [2] mentioned that information based consists of text passwords but has become unreliable as password can be hacked. Token-based rely on authentication through RFID keys [3], smart cards, etc., but they are vulnerable to token theft as even fingerprint and iris scan can be copied nowadays. According to experts, two-factor authentication could be the answer. However, it comes with limitations of being slow process and inefficient as well as being prone to token theft. Cmarsden [4] suggested tackling the complexity of IoT landscape, and biometric security is the way forward.

We have discussed the demerits of information and token-based authentication system on how easily it can be hacked. Therefore, we will focus more on biometric-based authentication as the solution. Biometric security cannot be forged; only genuine user has access to data. By any chance, if there is unfortunate event of someone forcefully forged in, then they can be easily identified with data confirmation as proof. This biometric authentication system enhances the security of the device or system and also makes it simpler and increasingly productive to manage key functions. One of the major factors of using this system is convenience as password does not need to be reset from time to time. Once the biometric authentication system is activated, no more costing or investment is needed for security, making it easier to operate with negligible chance of failures. Ghadawala [5] suggested moving forward with biometric authentication system as it is on the way to become a key to various factors of authentication. Overall, it can be said that with the immense advancement of science and technology, biometric-based is the ideal path to thread on for user authentication.

Rest of the paper is organized as follows. Section 2 includes literature review. The proposed model is in Sect. 3 and experimental results are in Sect. 4. Finally, conclude the paper in Sect. 5.

2 Literature Review

User authentication has been a reoccurring topic for research with different studies showing a variety of results and methods used to reach that decision. In this section, we will discuss about some related studies about user authentication. Chan et al. [6] have proposed a biometric user verification system using hand gestures during login. The experiment was carried out by 16 testers and used random forest algorithm, achieving accuracy of 99.97%. Tian et al. [7] presented challenge-response authentication (MoCRA) using in-air handwriting. It showed 98.82% authentication accuracy for 24 testers. Regarding the use of biometric hand gesture, Xiao et al. [8]

proposed an authentication system which accomplished an average of EER of 3.75% from 10 testers. The system had limitation, so verification time period is high. At the same time, Kamaishi and Uda [9] suggested a behavioral system simpler than the previous but has accuracy of 86.57%. The same can be said with Imura and Hosobe [10] and Maruyama et al. [11] who each have described similar methods for identification and have shown 96% (approximately) and 84.65% accuracy rate, respectively. Keep in mind that the studies mentioned so far all have the same drawback of being time consuming in terms of authentication, therefore, they are inefficient.

There is one paper with mechanism alike ours suggested by Manabe and Yamana [12], it introduces a two-factor authentication system using both input password and biometric of user's hands. The experimental results for 21 participants exhibit accuracy of 94.98% using random forest classifier. Another approach made by Atas [13] explicitly discuss about biometric recognition system based on hand tremor. The experiment was carried out using Naive Bayes classifier and achieved 95% accuracy rate. All the studies above indicate that hand structure is crucial in order to get optimum result.

3 Methodology

Biometric-based authentication is of various types. One of the most efficient types is by using leap motion sensor. Figure 1 shows the mechanism of our method through a flowchart. The way it will work is that a user will enter a password on a device.

While typing the password, the leap motion device will scan and analyze the movement of the hands of the user. After the password is entered, the system will perform a two-step authentication. Firstly, it will verify if the password is correct or not. If it is correct, then secondly it will compare the leap motion data of the user with its pre-stored dataset. If it finds a match, then the user will be authenticated. The user must pass both steps of authentication to be verified. In this way, the system becomes more secure even if the password gets leaked.

The proposed system will work in the following way. First, the user will enter their user data such as username and password. After entering that, the system will ask the user to enter motion data by placing their hands on the leap motion device for about 10–15 s. Once the system collects and processes the motion data, it will store it in the database, as illustrated in Fig. 2.

After the user data is registered into the database, they are secured for authentication. During the authentication phase, the user will attempt to login to their desired workplace by entering their user name and password. During the process, a leap motion device will capture the hand motion data of the user, as shown in Fig. 3. This is the first security layer of user authentication for the system. If the user enters the correct username and password they will be qualified for the second security layer of user authentication. Otherwise, they will be rejected.

In the second security layer, the system will pull up the stored motion data for the entered username and temporarily store it in memory. Afterward, the system

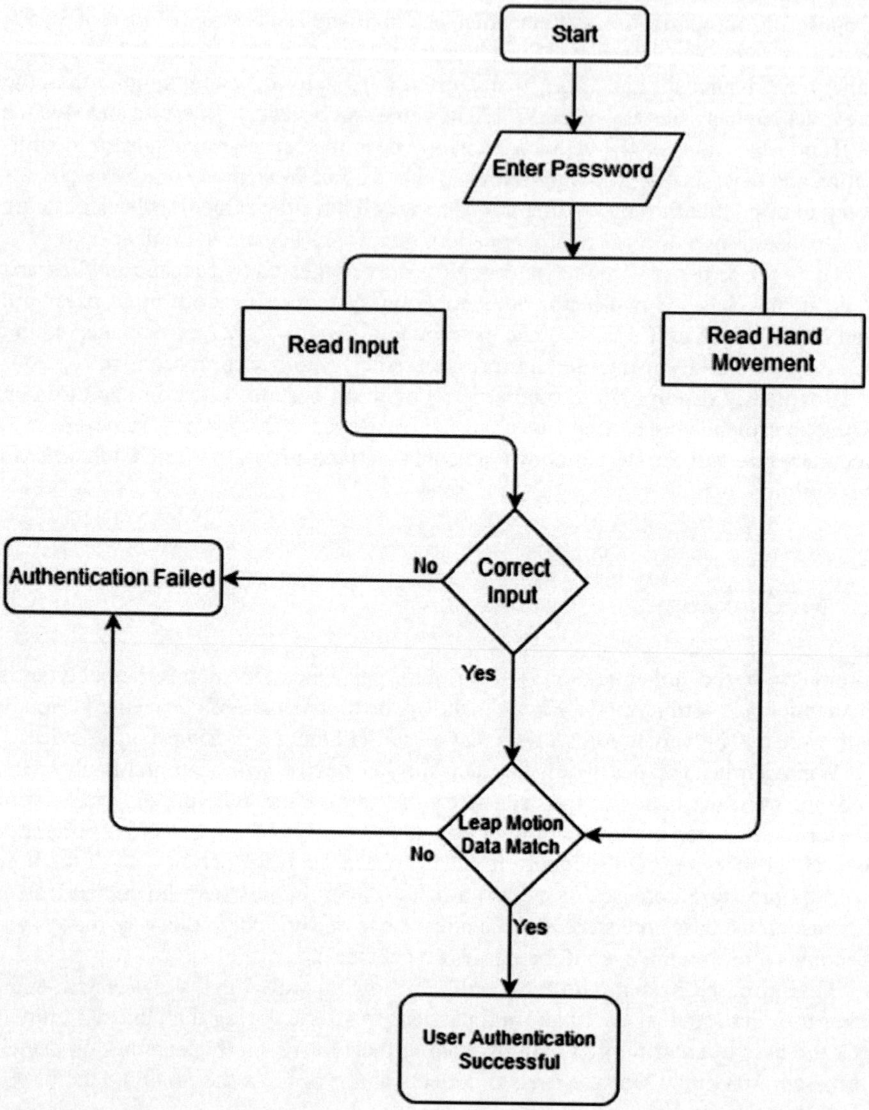

Fig. 1 Methodology of our proposed method

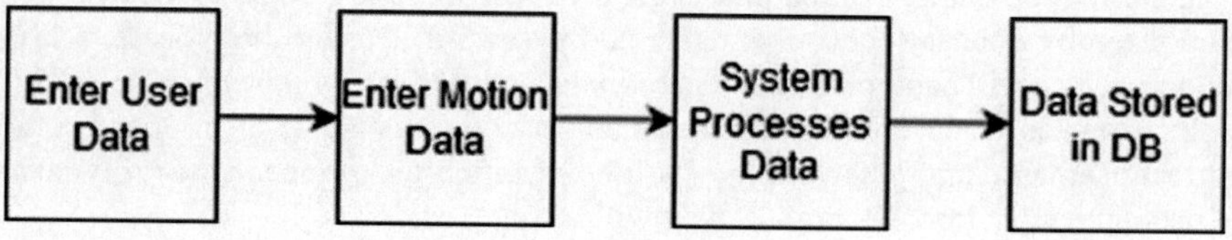

Fig. 2 Block diagram of registering user in the system

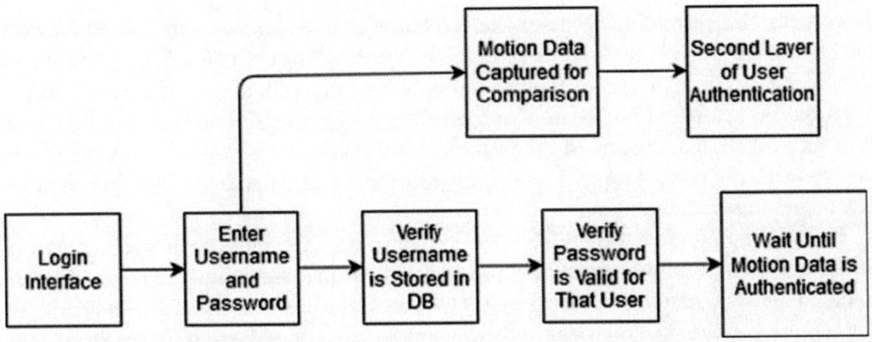

Fig. 3 First layer of user authentication

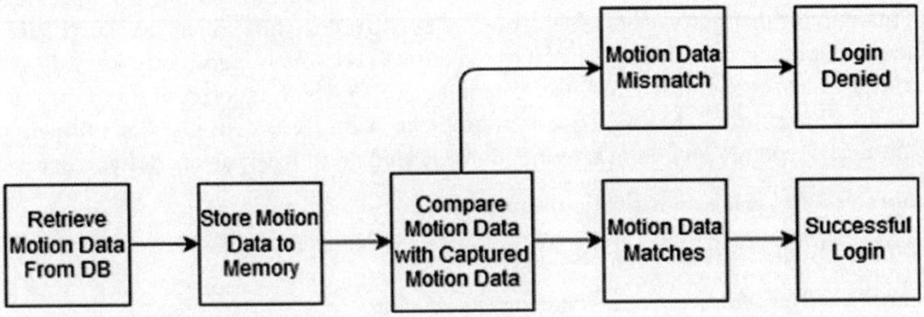

Fig. 4 Second layer of user authentication

will compare the captured motion data during user input earlier with the motion data stored in the database. If the data matches, then the user will be authenticated as a valid user and they will be successfully able to login, as illustrated in Fig. 4. Otherwise, the user will be rejected.

4 Experimental Analysis

The data collection from the leap motion device is a continuous process. So, we cannot immediately store it on a database for authentication use due to risk of system performance declining. Which is why, our system will first collect up to the most recent 1000 frames worth of user data which will then be converted to a sequence of bytes through serialization. To save several frames to the same file, we first create a way to differ between one frame from the other. The data size can vary from frame to frame. However, our system uses the following method by storing a 4-byte integer which carries the data size immediately before the data itself. This saves the serialized frame data in a file called "frames.data". After having a finite number of frames to work with, the system then passes the sequence of bytes to a deserializer to convert the bytes back into frame data. To read the frame data from "frames.data," the system then reads the first 4 bytes of the file to know how much data to read through to get a whole frame. After that, it simply repeats the process until reaching the end of the file. As a result, a new copy of the frame dataset is created which we can store in the database. Each frame data will be stored separately in the database with their unique ID, hand, and finger data.

Our main goal is to authenticate the user through leap motion data as accurately and quickly as possible, while they are giving their password input. This is why we decided to use dynamic time warping (DTW) and Naive Bayes algorithm to check for the accuracy of the system as well as how quickly it responded. For the experiment, we captured the leap motion data of eight people and stored them in the database for user authentication and another eight people were used as third-party users. Moreover, as the password is being typed into the keyboard, it is important to restrict the intrusion of malicious third-party users. Therefore, we aim to reduce false rejection rate (FRR), false acceptance rate (FAR), and false rejection (FR) which incorrectly identifies a valid user as invalid user and false acceptance (FA) which incorrectly identifies an invalid user as a valid user. To calculate these parameters, we use the following parameters, specifically, numT, numF, numFR, and numFA which are defined below.

numT = Total test data for valid users

numF = Total test data for invalid users

numFR = Total false rejections during experiment

numFA = Total false acceptances during experiment.

Furthermore, FRR and FAR are defined by the following equations.

FAR = numFR/numT

FRR = numFA/numF.

4.1 Applying DTW Algorithm Results

DTW or dynamic time array is an algorithm which allows us to compare or calculate the distance between two arrays of different length. In our case, this algorithm can be used to compare the frame data some user is providing while typing the password through the leap motion device with the user frame data stored in the database. Which is why during authentication phase, the system generates a DTW matrix by using a two-dimensional array which contains the average frame data computed from the database which we will call the source matrix. Additionally, the frame data being captured during the authentication phase are also converted into a two-dimensional array which we will call the converted matrices. These are temporarily stored until comparison phase ends. The length of the converted matrices may or may not be the same as the source matrix. The comparison process starts once the user hits the enter key, which will prompt the system to start comparing the converted matrices with the source matrix one by one. As mentioned above, we allowed for a deviation of up to 4% to be considered as acceptable. If all converted matrices are within 4% deviation of the source matrix, then the user is successfully authenticated. Otherwise, the user will be rejected even if they entered the correct password. To test the effectiveness of the system, each user was told to try to authenticate five times, for a total of 80 authentication attempts from 16 users. Additionally, in order to test the behavioral properties of the system, we ran three trials on separate days at different times. Table 1 shows the accuracy as well as the FAR and FRR of each trial.

Table 1 shows that if the system applies DTW algorithm, then we get an average of 80.9% accuracy which is not satisfactory for reliable security. In addition, the false acceptance rate (FAR) was also abnormally high at an average of 2.87. We speculate that the proposed system might not be compatible with DTW in this regard. The false rejection rate (FRR), however, had slowly declined with each trial which was one of the main goals of our system, as illustrated in Fig. 5. The average response time was noted to be around 2.93 s.

Table 1 Experimental result for DTW algorithm authentication

Experimental trial	Accuracy (%)	FAR (%)	FRR (%)
First trial	79.64	3.46	17.79
Second trial	82.13	2.66	16.10
Third trial	80.92	2.48	16.05

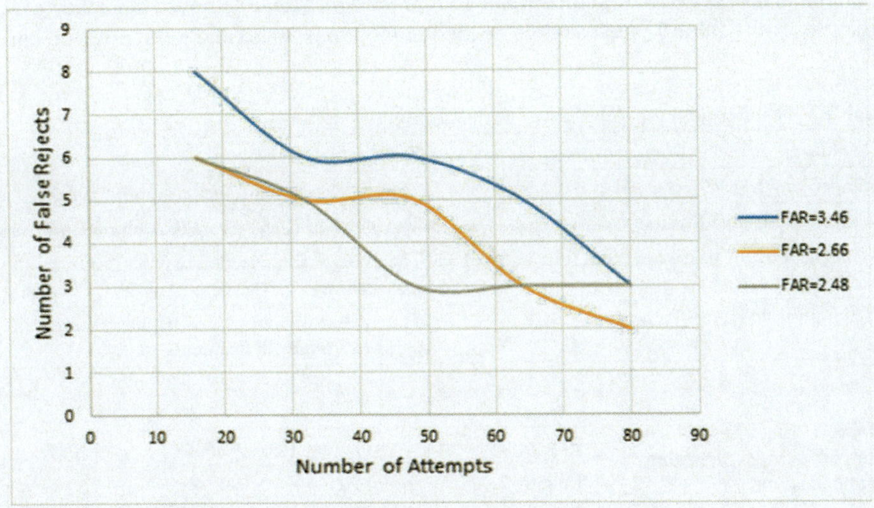

Fig. 5 Relationship between number of inputs and number of false rejections

4.2 Applying Naïve Bayes Algorithm Results

As Naive Bayes algorithm is very adaptable and a quick learning algorithm when assigned many features, we allowed the classifiers to learn as much as the user data as possible through 14 features (Table 2).

During the authentication phase, when the user is giving password input the system will capture the leap motion data until they hit the enter key. Afterward, the system will give a binary input of 0 or 1. We guess that the user needs x number of frames to input the password into the system, while the leap motion device also computes the same features x number of times as the password is being entered by the user. For every one of the x datasets, the system then judges whether or not the dataset is that of a valid user. If the dataset where the user is verified to be genuine becomes equal or more than some value z is when the system decides that the user is genuine and returns output 1. Here, we determine the parameter z by basing it on the dataset used for parameter adjustment. On the other hand, if the system decides that the user is an intruder or invalid, then it returns output 0.

Similar to testing the accuracy of DTW algorithm, each user attempted for authentication five times, for a total of 80 authentication attempts from 16 users. To test the behavioral aspect of the system, we ran three trials on separate days at different times. Table 3 shows the accuracy as well as the FAR and FRR of each trial. Table 3 shows that by applying Naive Bayes algorithm, the system gave an average of 90.63% of accuracy. Additionally, it is evident that FRR while having high percentage in the first and second trial, and it was greatly reduced by the third trial.

We speculate this is due to the fact that the users had gotten used to using the system.

It must also be noted that the FRR would be even less if we removed the behavioral features from Table 1, such as time needed to input password and palm position and

Table 2 Features used for Naive Bayes authentication

Feature	Explanation
Finger length (five features)	The length of each fingers (in millimeters)
Finger width (five features)	The width of each fingers (in millimeters)
Time needed for password input (one feature)	The average time needed to enter password (in seconds)
Palm position (three features)	The pitch, roll, and yaw of the hand during password input (in degrees)

Table 3 Experimental result for Naive Bayes algorithm authentication

Experimental trial	Accuracy (%)	FAR (%)	FRR (%)
First trial	92.22	1.78	19.32
Second trial	89.74	2.79	18.38
Third trial	89.94	1.40	12.53

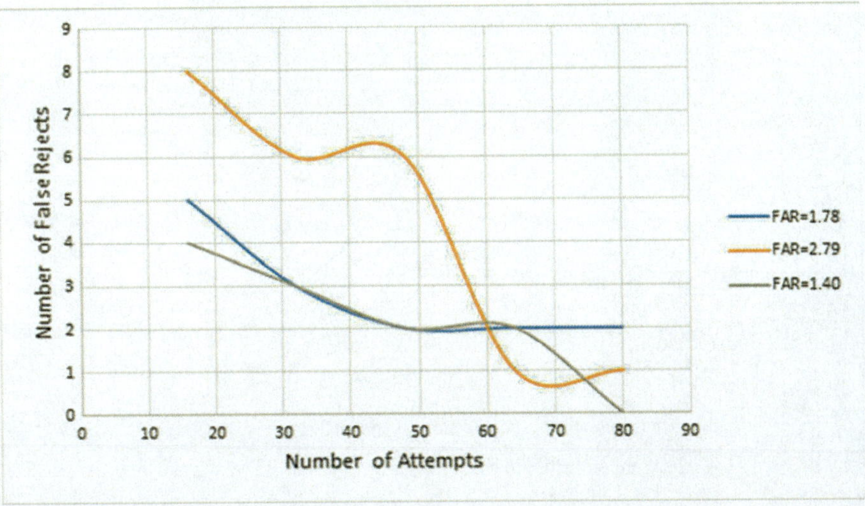

Fig. 6 Relationship between number of inputs and number of false rejections

only used physical features like finger length and width. This is because behavioral features are less likely to be accurate than physical features when users are not used to typing their password input. It gradually becomes more effective as users get used to typing their password, as shown in Fig. 6.

As evident in Table 4, the FRR is significantly reduced, with an average of 12.53% which is a 4.21% reduction from before. The accuracy also rises, with an average of 2.27% increase. Even though the FAR is much higher compared to before (averaging at 2.91%), as we also need to consider easy user interaction with the system we have determined it to be within acceptable values, as shown in Fig. 7. Furthermore, the response time also decreased from an average of 3.43 s to around 3 s.

Based on the results we can see that compared to DTW algorithm, Naive Bayes algorithm returned more accurate results. Even though, DTW had less false rejection rate than Naive Bayes, when comparing the average, the FRR was only at a 0.31% reduction. In addition, as mentioned above, the FRR of Naive Bayes is reduced more if we removed the behavioral features and only used the physical features. Lastly, the system also adapted and responded much quicker when Naive Bayes was applied. So, considering the accuracy, the negligible difference in FRR and the response time, Naive Bayes algorithm is better suited for our system. Figure 8 gives us a better understanding of the comparison between the two algorithms, including the remove the behavioral features from the Naive Bayes classifier.

Table 4 Experimental result for Naive Bayes algorithm authentication (without behavioral features)

Experimental trial	Accuracy (%)	FAR (%)	FRR (%)
First trial	92.22	3.56	14.23
Second trial	92.54	3.79	13.89
Third trial	93.94	1.39	9.53

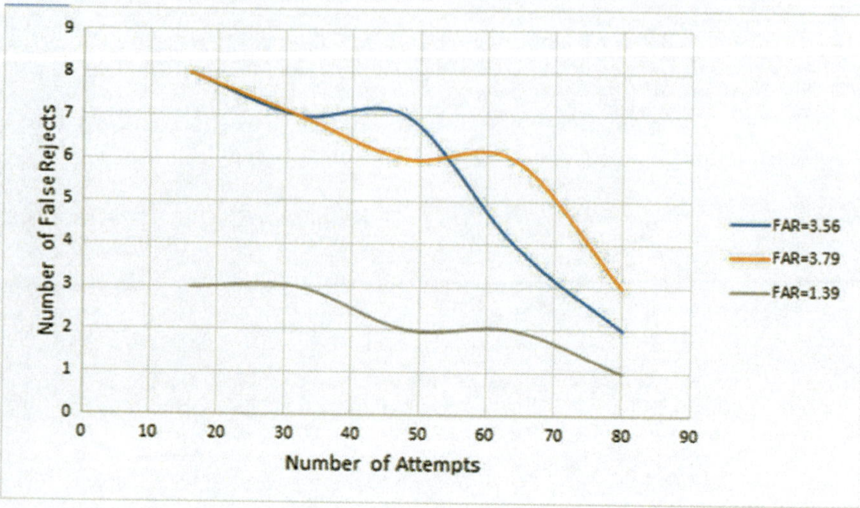

Fig. 7 Relationship between number of inputs and number of false rejections

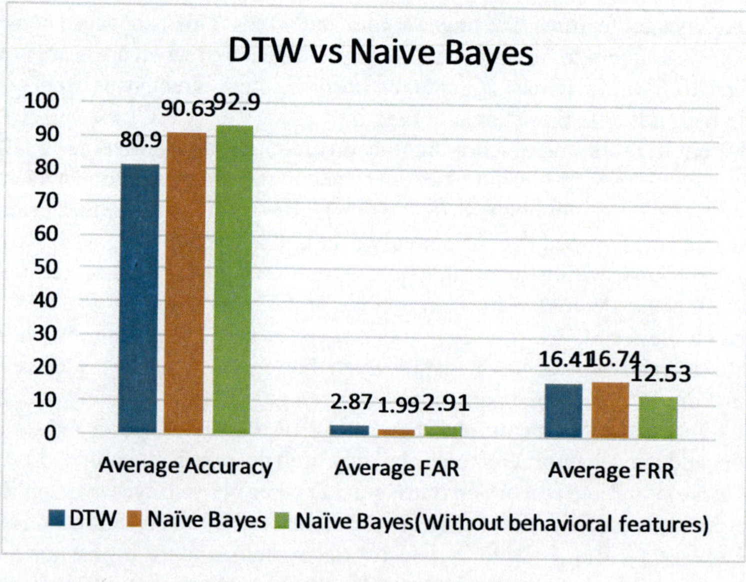

Fig. 8 Comparison chart between DTW and Naïve Bayes

5 Conclusion

In the proposed model, we have implemented a system where we are taking behavioral hand data and password from a user instead of only physiological data like fingerprint, FaceID, iris scanner, etc., which can be easily cloned by using artificial intelligence.

An end user's behavioral hand data is unique and varies from time to time. So, we are storing data of a user's hand gesture from a different time period on a database from where we can verify a particular user which is very difficult to duplicate by the hackers. Experimental results demonstrated that Naive Bayes algorithm provided the best accuracy rate of 93% compared to dynamic time warping algorithm. In contrast to other existing methods, our method is easy to implement because of its quick learning capability and quick response time.

References

1. A. Rahman, T. Rahman, N.H. Ghani, S. Hossain, J. Uddin, IoT based patient monitoring system using ECG sensor, in *2019 International Conference on Robotics, Electrical and Signal Processing Techniques (ICREST)*, Dhaka (2019), pp. 378–382
2. S. Abdulkader, A. Atia, M. Mostafa, Authentication systems: principles and threats. Comput. Inf. Sci. **8**(3) (2015)
3. S. Ahmed, T.M. Tan, A.M. Mondol, Z. Alam, N. Nawal, J. Uddin, Automated toll collection system based on RFID sensor, in *International Carnahan Conference on Security Technology*, India (2019), pp. 1–3
4. Cmarsden, 4 reasons why biometric security is the way forward [Internet]. Digitus Biometrics (2020). Cited 10 July 2020. [Online]. https://digitus-biometrics.com/blog/4-reasons-why-biometric-security-is-the-way-forward/
5. M. Ghadawala, 7 key benefits of security with biometrics. GlobalSign GMO Internet, Inc. (2020). [Online]. https://www.globalsign.com/en/blog/7-benefits-security-with-biometrics. Accessed 10/07/2020
6. A. Chan, T. Halevi, N. Memon, Leap motion controller for authentication via hand geometry and gestures, in *Lecture Notes in Computer Science* (2015), pp. 13–22
7. J. Tian, W. Xu, Y. Cao, S. Wang, Challenge-response authentication using in-air handwriting style verification. IEEE Trans. Depend. Secur. Comput. **17**(1), 51–64 (2020)
8. G. Xiao, M. Milanova, M. Xie, Secure behavioral biometric authentication with leap motion, in *4th International Symposium on Digital Forensic and Security (ISDFS)* (2016)
9. S. Kamaishi, R. Uda, Biometric authentication by handwriting with single direction using self-organizing maps, in *11th International Conference on Ubiquitous Information Management and Communication* (2017)
10. S. Imura, H. Hosobe, A hand gesture-based method for biometric authentication, in *Human-Computer Interaction Theories, Methods, and Human Issues* (2018), pp. 554–566
11. K. Maruyama, J. Shin, C. Kim, C. Chen, User authentication using leap motion, in *International Conference on Research in Adaptive and Convergent Systems* (2017)
12. T. Manabe, H. Yamana, Two-factor authentication using leap motion and numeric keypad, in *HCI for Cybersecurity, Privacy and Trust* (2019), pp. 38–51
13. M. Atas, Hand tremor based biometric recognition using leap motion device. IEEE Access **5**, 23320–23326 (2017)

Exploiting Transfer Learning Ensemble for Visual Sentiment Analysis in Social Media

Etu Chowdhury, Abu Nowshed Chy, and N. M. Istiak Chowdhury

Abstract With the growing popularity of social media platforms, everyday millions of users are sharing their emotions, views, and information about notable events through visual contents especially images. Analyzing the sentiments of these user-generated images provide prevalence opportunities to various organizations for decision making. Most of the prior studies focused on hand-crafted features and traditional deep learning-based approaches. Conversely, a little effort has been employed to exploit the state-of-the-art transfer learning methods. In this paper, we propose a visual sentiment analysis system adopting the strengths of the ensemble of transfer learning models. We employ three pre-trained deep convolutional neural network (CNN)-based feature extractors including VGG16, Xception, and MobileNet to train the corresponding base classifier. Finally, a majority voting-based fusion approach is employed to determine the sentiment label. Experiments on a benchmark Twitter image dataset indicate that our method achieves top-notch improvement compared to other state-of-the-art methods.

Keywords Visual sentiment analysis · Transfer learning · CNN · VGG16 · Xception · MobileNet · Majority voting

E. Chowdhury (✉) · A. N. Chy · N. M. I. Chowdhury
Department of Computer Science & Engineering, University of Chittagong, Chattogram 4331, Bangladesh
e-mail: etuchy.cu@gmail.com

A. N. Chy
e-mail: nowshed@cu.ac.bd

N. M. I. Chowdhury
e-mail: istiak@cu.ac.bd

V. E. Balas et al. (eds.), *Proceedings of International Conference on Computational Intelligence, Data Science and Cloud Computing*, Lecture Notes on Data Engineering and Communications Technologies 62, https://doi.org/10.1007/978-981-33-4968-1_3

1 Introduction

A cumulative number of social media (e.g., Twitter, Instagram, Facebook, and Flickr) provide various services to make it easier for users to communicate with others. Everyday a huge number of users are sharing their daily-life events and evolve their opinions on various issues using these services. These informal user-generated contents contain user's thoughts and emotions toward various entities including individuals, issues, events, products, services, organizations, and their attributes [32].

Nowadays, visual contents have gained much popularity in social media than textual contents due to its ability to express user emotions intuitively and efficiently. As a maxim goes, *A picture is worth a thousand words*". For example, Fig. 1 presents two images that are shared on social media. From the left-side image, we can easily deduce that the author of the image is in a happy mood and enjoying cherry blossoms, whereas from the right-side image we see that the author is in a sad mood due to the hazardous scenario of a natural disaster.

Visual sentiment analysis on social media refers to the computational study of determining user's sentiment from their shared images, which in turn helps to determine their inherent opinion, emotional tendency, and psychological states. It has become one of the most prominent research areas and practicable in many areas including behavior targeting, recommendation system, product promotion, and E-commerce with potential economic value [18]. A lot of researches has already been completed to explore sentiment from textual contents shared on social media [1]. But there are a very limited research works [14, 28] done on visual sentiment in computer vision.

In textual sentiment analysis, sentiment can be estimated easily utilizing textual, contextual, and semantic information. In contrast, visual sentiment analysis is more subjective which makes it challenging to explore effective visual features with strong discrimination ability of sentiment orientation. In this case, deep learning architecture is mostly used for its remarkable success in computer vision [15]. It provides accurate and robust feature representation of images. Among several neural network models, convolutional neural network (CNN) is widely used [4, 14, 28, 30] for

Fig. 1 Sample of positive (left) and negative (right) sentiment bearing images

image classification due to its ability to extract the spatio-temporal features in an image effectively. However, the informal visual images shared on social media have some unique characteristics and depend on other factors including people, regions, and time. Therefore, it is cumbersome to prepare a robust training sample. To overcome these limitations, one plausible solution is to transfer knowledge from existing pre-trained models.

The key contribution of this research includes: we propose a method for visual sentiment analysis that adopts the benefits of the ensemble of transfer learning models including VGG16 [23], Xception [7], and MobileNet [13]. Experimental results on the benchmark Twitter image dataset prove the efficacy of our method over the state-of-the-art methods.

We organize the rest of the paper as follows: Section 2 includes the discussion of prior researches, and we present our proposed framework in detail in Sect. 3. In Sect. 4, we analyze our experimental results and perform the comparative evaluation of our method against the related methods. Finally, we conclude our proposal with some plausible future notions in Sect. 5.

2 Related Work

Visual sentiment analysis studies the inherent emotional contents of an image to determine its sentiment orientation and can be formulated as an image classification task. Prior researches mainly focused on extracting the objective visual content features of images including color, texture, shape, and contour [10, 17]. Besides, Siersdorfer et al. [22] selected the discriminative sentiment-related features through the feature selection method and employed the machine learning approach to identify the image sentiment.

Borth et al. [3] presented a data-driven approach based on psychological theories and web information to generate a large-scale visual sentiment ontology (VSO). The ontology is later used to form a concept detector library, SentiBank for bridging the affective gap. Besides, Chen et al. [6] presented a sentiment detector of visual images named as DeepSentiBank where a deep CNN model is trained on using 1 million adjective–noun pairs (ANPs) tagged Flickr images, whereas Wang et al. [27] introduced a visual sentiment analysis framework using the deeply coupled adjective and noun neural networks. To address the problem of manual annotation, You et al. [30] presented a progressive CNN (PCNN) model to reduce the noise of weakly labeled image data. Vadicamo et al. [26] proposed a cross-media learning approach where they collected tweets with images and leveraged the tweet sentiment to annotate the image.

Besides, some approaches employed the transfer learning techniques and utilized the pre-trained models through fine-tuning for visual sentiment analysis. Campos et al. [4] conducted fine-tuning to train an AlexNet model on sentiment dataset collected from Twitter. Islam and Zhang [14] adopted the transfer learning approach for image sentiment analysis. They used hyper-parameters learned from a DCNN to initialize their model for reducing over-fitting.

A number of researches studied the merged model that incorporates both the local and global information of images as features in a unified architecture. These studies usually utilized the hand-crafted features to extract the local visual information and deep learning models to capture the global visual information of images. Among these works, Fengjiao and Aono [10] considered both hand-craft and CNN features in a merged model to explore visual sentiment. He et al. [11] employed a multi-attentive pyramidal model (MAP) where pyramidal segmentation is used to extract the local visual features from the global visual features obtained from CNN with an attention mechanism.

Besides social images, Tan et al. [25] proposed a method to learn the painting features and label emotions of painting images. They mainly utilized the transfer learning method and built a classification model based on VGG16 and ResNet50. Moreover, Paolanti et al. [19] compared the performance of various pre-trained CNN models including VGG16, ResNet, and Inception to analyze the sentiment of social images related to cultural heritage.

However, in our proposed method, we utilize the ensemble of transfer learning models without fine-tuning to extract the generic visual features and incorporate their strengths in a unified model. It helps our model to overcome the limitations of the small dataset and analyze the sentiment of social images effectively.

3 Proposed Framework

This section presents a detailed description of our visual sentiment analysis system. Figure 2 depicts the overall architecture of our proposed framework.

Given an image, at first, we perform the minimal preprocessing. To obtain the effective transfer learning features, we use the three deep convolutional neural net-

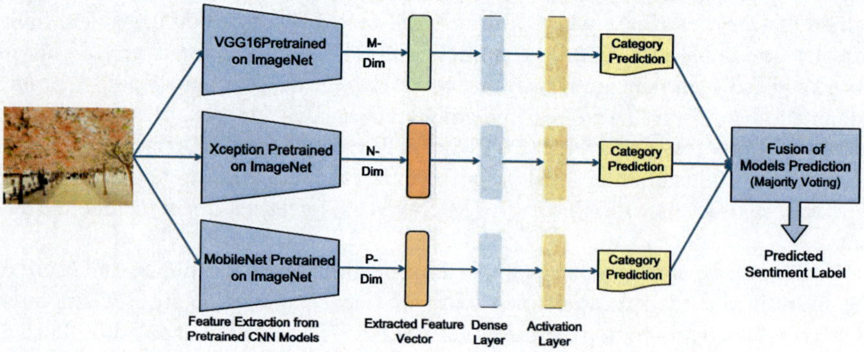

Fig. 2 Overall architecture of our visual sentiment analysis system

work (DCNN) models pre-trained on the ImageNet [8] dataset. We employ the VGG16 [23], Xception [7], and MobileNet [13] models to encode the image into the deep semantic feature vectors. The extracted features are then fed into the corresponding fully connected sentiment prediction module. Based on the prediction of each module, a majority voting scheme is employed to determine the final sentiment label of the image. For the simplicity of the discussion, we define the name of our proposed transfer learning ensemble-based method as TLEnsemble. Also, we elaborate on the description of each component in the subsequent sections.

3.1 Image Preprocessing

We employ a minimal image preprocessing stage initiated with resizing the image in a fixed shape. In social media, peoples are free to upload any kind of images. Therefore, it is inevitable to convert all the images into a uniform shape. Besides, social media applies the compression technique to store the vast amount of uploaded images. Considering this, to get the effective pixel value and maintain the original aspect ratio, we resize all the input images into a 160×160 shape.

3.2 Transfer Learning Using Pre-trained CNN Models

Transfer learning is a technique of overcoming the diverse learning paradigm and taking advantage of knowledge gained from the previous task to solve the related ones. Therefore, the learning process can be faster, more accurate, and need less training data. In our method, we employ the transfer learning to mitigate the limitations of the small dataset and extract effective features from various pre-trained models. However, we can ramify the prior work on utilizing transfer learning models into two types: i) models that employed fine-tuning according to the downstream task [29] and ii) models that did not employ fine-tuning the encoder parameters and utilize the generic transfer learning features [5, 9]. We follow the second type and transfer the learned features from the three pre-trained DCNN models including VGG16, Xception, and MobileNet without fine-tuning. Thus, our model gains a stable and reliable initialization state. The overall architectures of these three models are described as follows:

Architecture of VGG16 The visual geometry group of Oxford University proposed the model of VGG16 [23]. It composes of 13 convolutional layers and 3 fully-connected (FC) layers. The convolutional layers employ multiple 3×3 kernel-sized filters with a stride size of 1 and the pooling layers employ 2×2 filters with a stride size of 2. Upon removing the fully connected layer from the pre-trained VGG16 model, it can be employed to generate the deep semantic features of the images. We use the VGG16 model pre-trained on ImageNet [8] dataset to encode each image into M-dimensional feature vectors. We choose VGG16 because it uses very small receptive fields that help it to have a large number of weight layers and build the decision function more nonlinear.

Architecture of Xception François Chollet proposed the Xception [7] model in 2017. It has a simple and efficacious architecture with linear residual connections and a modified depthwise separable convolution inspired from InceptionV3 architecture. The core idea behind the modified depthwise separable convolution is that a pointwise convolution is followed by a depthwise convolution. Therefore, a 1×1 convolution is executed first and then a channel-wise spatial convolution is performed. It utilizes 36 convolutional layers to form the feature extraction module and three major partitions including entry flow, middle flow, and exit flow. We use the Xception model pre-trained on the ImageNet [8] dataset to encode each image into N-dimensional feature vectors.

Architecture of MobileNet MobileNet [13] architecture is widely used as it is a lightweight deep convolutional neural network. It composed of 28 layers and applies the depthwise and pointwise convolutions as a separate layer. It utilizes depthwise separable convolution which links a single filter to each input channel. Besides, a 1×1 convolution is applied by pointwise convolution to combine the output of depthwise convolution. In this architecture, a depthwise separable convolution is applied to minimize the model size and complexity. Also, a pointwise convolution is used to change the dimension. After each convolution, batch normalization and rectified linear units are employed. We use the MobileNet model pre-trained on the ImageNet [8] dataset to encode each image into P-dimensional feature vectors. We choose MobileNet because it is a low latency model, faster in performance, and reduces the number of parameters.

3.3 Model Training and Prediction

After getting the deep semantic feature vectors from VGG16, Xception, and MobileNet models, we pass them to a corresponding fully-connected dense layer and an activation layer, respectively to obtain the visual sentiment prediction results from the individual model.

The stochastic gradient descent (SGD) algorithm and Adam optimizer [16] are used to learn the model parameters. The SGD algorithm estimates the gradient of loss function utilizing a small batch of data and updates the model weights frequently to reduce the loss. We use the sparse categorical cross-entropy loss function and train the model as follows:

$$\text{Cross Entropy} = -(y^i \log(y^{\sim i}) + (1 - y^i)\log(1 - y^{\sim i}))$$

From the above-mentioned equation, we can observe that when the actual label y^i is 1, the second half of function disappears, whereas the first half becomes invisible in case the actual label is 0 ($y^i = 0$). We also apply dropout[12] to address the problem of over-fitting.

3.4 Fusion of Models Prediction

Upon getting the predicted category from VGG16, Xception, and MobileNet models, we employ an ensemble approach to incorporate them. We use the majority voting count-based scheme to obtain the final prediction label. The core idea behind this type of classification is that the final output class is selected based on the highest number of votes [20]. In prior research, this classification scheme was used to overcome the limitations of a single classifier [21]. Mathematically, it can be shown as follows:

$$\text{Class Label}(X) = \underset{c_i \in dom(y)}{\text{argmax}} \left(\sum_k f\left(y_k(x), c_i\right) \right)$$

where the prediction of the kth classifier is denoted by $y_k(x)$. Also, $f(y, c)$ implies an indicator function that can be shown as:

$$f(y, c) = \begin{cases} 1, \text{ if } & y = c \\ 0, \text{ if } & y \neq c \end{cases}$$

4 Experiments and Evaluation

4.1 Dataset Collection

We evaluate the performance of our visual sentiment analysis system on a benchmark Twitter image dataset[1] provided by You et al. [30]. The dataset consists of 1269 images having either of two sentiment labels including positive and negative. The authors used the crowd intelligence platform Amazon Mechanical Turk (AMT) and employed five AMT employees to annotate the sentiment label of the images. The dataset statistics are presented in Table 1, where "Five Agree" means all of the five AMT employees labeled the same sentiment for the stated number of images. Similarly, "At Least Four Agree" and "At Least Three Agree" imply that among five employees, at least four and at least three employees had the same label for the images, respectively.

We split the above-mentioned dataset between the training and test sets according to 80% and 20% of the total images from each category, respectively. Following this splitting ratio, we employ a fivefold cross-validation scheme for performance evaluation of our model.

[1] https://www.cs.rochester.edu/u/qyou/DeepSent/deepsentiment.html.

Table 1 Statistics of the Twitter image dataset [30]

Category	Five(5)-agree	≥ Four(4)-agree	≥ Three(3)-agree
Positive images	581	689	769
Negative images	301	427	500
Number of total images	882	1116	1269

4.2 Model Configuration

Our model is implemented on the Google Colab cloud system that provides 12 GB RAM, Intel Xeon CPU clocked at 2.20 GHz, and a powerful NVIDIA Tesla K80 GPU for parallel computation of the tensors.

In the Twitter image dataset, the resolution of each image was not the same. So, the images are resized to 160×160 resolution. Also, the extracted feature values are normalized and scaled in between 0.0 and 1.0. As described in Sect. 3.2, we used three models pre-trained on ImageNet [8] dataset for transfer learning features. For an input image with 160×160 along 3 input channels, the pre-trained VGG16 model provides, $M = (5 \times 5 \times 512) = 12{,}800$ transfer learning features. Similarly, the pre-trained Xception and MobileNet models provide, $N = (5 \times 5 \times 2048) = 51{,}200$ and $P = (5 \times 5 \times 1024) = 25{,}600$ transfer learning features, respectively. The output of the layer is dropped out with a probability of 0.5 before it is passed to the activation layer where softmax is applied to determine the final sentiment label of the images. We trained our network with 600 epochs and set the input batch size as 32. The default settings are used for the other parameters.

4.3 Experimental Results and Evaluation

We now evaluate the performance of our proposed method for visual sentiment classification. Four standard evaluation metrics including precision, recall, F1 score, and accuracy are used to estimate the classification performance. However, we consider the F1 score as the major evaluation measure for comparative performance analysis. The summarized results are presented in Table 2.

At first, we report the results of the individual model employing transfer learning features from pre-trained VGG16, Xception, and MobileNet models. It shows that when considering individual model Xception feature-based model obtained the best performance for the 3-agree dataset and MobileNet feature-based model obtained the best performance for the 4-agree and 5-agree dataset regarding the primary evaluation measure F1 score. Next, we report the performance of our proposed ensemble model, TLEnsemble. As shown in Table 2, we can observe that our proposed TLEnsemble method obtained the improved performance over the other three models. Our method

Table 2 Performance on individual model and our TLEnsemble model. Best results are highlighted in boldface

Model	At Least 3-agree				At Least 4-agree				5-agree			
	Prec	Rec	F1 Sc.	Acc	Prec	Rec	F1 Sc.	Acc	Prec	Rec	F1 Sc.	Acc
VGG16	**0.861**	0.682	0.761	0.740	0.828	0.732	0.777	0.740	0.904	0.810	0.854	0.818
Xception	0.792	**0.792**	0.792	0.748	0.820	0.790	0.805	0.762	0.862	0.862	0.862	0.818
MobileNet	0.797	0.766	0.781	0.740	0.841	**0.841**	0.841	0.803	0.897	**0.905**	0.901	0.869
TLEnsemble	0.836	**0.792**	**0.813**	**0.780**	**0.883**	0.819	**0.850**	**0.820**	**0.936**	0.888	**0.911**	**0.886**

Table 3 Comparative evaluation of our TLEnsemble method against other related methods [10, 14]

Method	At least 3-agree				At Least 4-agree				5-agree			
	Prec	Rec	F1 Sc.	Acc	Prec	Rec	F1 Sc.	Acc	Prec	Rec	F1 Sc.	Acc
GCH[22]	0.678	**0.836**	0.749	0.66	0.687	0.840	0.756	0.665	0.708	0.888	0.787	0.684
LCH[22]	0.716	0.737	0.726	0.664	0.725	0.753	0.739	0.671	0.764	0.809	0.786	0.710
GCH+BoVW[22]	0.683	0.835	0.751	0.665	0.703	0.849	0.769	0.685	0.724	0.904	0.804	0.710
LCH+BoVW[22]	0.722	0.726	0.723	0.664	0.751	0.762	0.756	0.697	0.771	0.811	0.790	0.717
SentiBank[2]	0.720	0.723	0.721	0.662	0.742	0.727	0.734	0.675	0.785	0.768	0.776	0.709
Sentribute[31]	0.733	0.783	0.757	0.696	0.750	0.792	0.771	0.709	0.789	0.823	0.805	0.738
CNN[30]	0.734	0.832	0.779	0.715	0.773	**0.855**	0.811	0.755	0.795	**0.905**	0.846	0.783
PCNN[30]	0.755	0.805	0.778	0.723	0.786	0.842	0.811	0.759	0.797	0.881	0.836	0.773
Islam et al.[14]	**0.851**	0.784	**0.816**	**0.787**	0.849	0.830	0.839	0.807	0.926	0.869	0.897	0.861
TLEnsemble	0.836	0.792	0.813	0.780	**0.883**	0.819	**0.850**	**0.820**	**0.936**	0.888	**0.911**	**0.886**

outperforms these three comparative models at least 2.1% and at best 5.2% for the 3-agree dataset, at least 0.9% and at best 7.3% for the 4-agree dataset, and at least 1.0% and at best 5.7% for the 5-agree dataset regarding the F1 score. Our method also obtained the best and competitive performance regarding other evaluation measures. This deduces the efficacy of our method for predicting the sentiment of visual images.

4.4 Comparison to Different Approaches

To further analyze the efficacy of our method, we compare the performance against the state-of-the-art-related methods. The comparative results are presented in Table 3. It shows that our proposed TLEnsemble method has surpassed other methods regarding the major evaluation metric F1 score for the 4-agree and 5-agree datasets. However, we obtained the competitive performance for the 3-agree dataset.

Siersdorfer et al. [22] utilized the global color histogram (GCH), the local color histogram (LCH), and the SIFT-based bag of visual words (BoVW) feature for estimating the sentiment of social images. In contrast, we didn't employ any hand-crafted features. We utilized the effective transfer learning features from pre-trained state-of-the-art models that made our feature extraction process easier. You et al. [30] employed a progressive CNN (PCNN) model where a subset of training samples is selected progressively to reduce noise. In contrast, we employed the ensemble of transfer learning feature-based models that helped to reduce the effect of noisy data. Islam and Zhang [14] used a deep CNN model having 22 layers where hyperparameters from pre-trained GoogLeNet [24] models were used for model initialization. In contrast, we utilized the feature from three transfer learning models and we didn't employ any downstream fine-tuning approach during feature extraction to keep the generalizability of the features.

In brief, we employed the ensemble of pre-trained models including VGG16, Xception, and MobileNet to extract effective and generic transfer learning features. Our model adopted the strengths of these models in a unified architecture which in turn helped our model to achieve better performance compared to other related methods.

4.5 *Discussion and Error Analysis*

To further analyze the performance of our TLEnsemble method, we provide the confusion matrix based on the Twitter image dataset [30] in Fig. 3. It schematizes the prediction results of our system in percentage form against the actual and the predicted classification ratio. It shows that most of the misclassification occur at the lower-left corner for all the three datasets. Since in our used dataset, the number of positive images is high compared to the negative images and we perform the category wise train–test split ratio; therefore, it is usual that most of the correct classification and misclassification occur in the positive image dataset. Since the dataset is imbalanced across classes, effective data augmentation might help our model to further improve the performance.

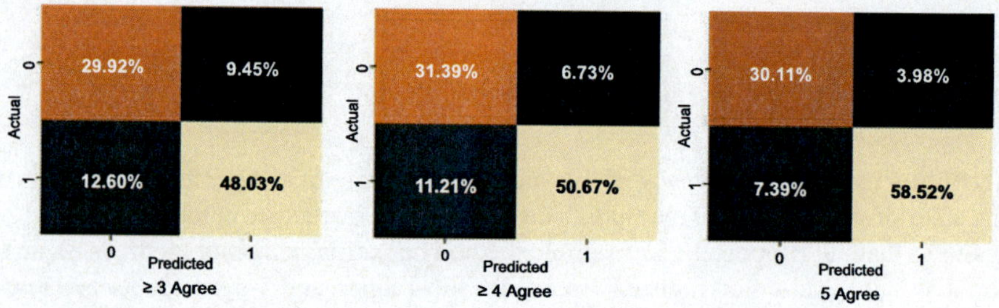

Fig. 3 Confusion matrix for at least three, at least four, and five agree datasets

Fig. 4 (Actual–Positive, Predicted–Negative) images

Fig. 5 (Actual–negative, predicted–Positive) images

Besides, to analyze the reason for misclassification, we present some misclassified sample images from the used dataset in Figs. 4 and 5, respectively. It shows that the ground truth of the images available in Fig. 4 is positive. But there is no clear positive indication available in these images. Consequently, our proposed method failed to classify them correctly. A similar scenario happened for the images presented in Fig. 5, where our model predicted the negative sentiment bearing images as positive.

5 Conclusion and Future Directions

In this paper, we introduce a very effective method for visual sentiment prediction which incorporates the strengths of various transfer learning models in a unified architecture. Employing different transfer learning models helps our method to learn the context of social media images from different levels of abstraction, which in turn improves the classification accuracy. Experiments on a benchmark Twitter image dataset deduce the proficiency of our method.

However, in the future, we have a plan to evaluate our method for the multilevel multi-class sentiment analysis, since an image may contain the sentiment of more than one type. We also have a plan to generalize our model for the other domain of interest.

References

1. J. Bollen, H. Mao, A. Pepe, Modeling public mood and emotion: Twitter sentiment and socio-economic phenomena, in *Fifth International AAAI Conference on Weblogs and Social Media* (2011)
2. D. Borth, T. Chen, R. Ji, S.F. Chang, Sentibank: large-scale ontology and classifiers for detecting sentiment and emotions in visual content, in *Proceedings of the 21st ACM International Conference on Multimedia* (2013), pp. 459–460
3. D. Borth, R. Ji, T. Chen, T. Breuel, S.F. Chang, Large-scale visual sentiment ontology and detectors using adjective noun pairs, in *Proceedings of the 21st ACM International Conference on Multimedia* (2013), pp. 223–232
4. V. Campos, B. Jou, X. Giro-i Nieto, From pixels to sentiment: Fine-tuning CNNS for visual sentiment prediction. Image Vision Comput. **65**, 15–22 (2017)
5. H. Cao, S. Bernard, L. Heutte, R. Sabourin, Improve the performance of transfer learning without fine-tuning using dissimilarity-based multi-view learning for breast cancer histology images, in *International Conference Image Analysis and Recognition* (Springer, Berlin, 2018)
6. T. Chen, D. Borth, T. Darrell, S.F. Chang, Deepsentibank: visual sentiment concept classification with deep convolutional neural networks. arXiv preprint arXiv:1410.8586 (2014)
7. F. Chollet, Xception: deep learning with depthwise separable convolutions, in *Proceedings of the IEEE Conference on Computer Vision and Pattern Recognition (CVPR)* (2017), pp. 1251–1258
8. J. Deng, W. Dong, R. Socher, L.J. Li, K. Li, L. Fei-Fei, Imagenet: a large-scale hierarchical image database, in *2009 IEEE Conference on Computer Vision and Pattern Recognition*. IEEE (2009), pp. 248–255
9. L. Fangxin, Z. Wenbo, W. Yanzhi, D. Changzhi, J. Li, Ausn: approximately uniform quantization by adaptively superimposing non-uniform distribution for deep neural networks. arXiv preprint arXiv:2007.03903 (2020)
10. W. Feng-jiao, M. Aono, Visual sentiment prediction by merging hand-craft and cnn features, in *2018 5th International Conference on Advanced Informatics: Concept Theory and Applications (ICAICTA)* (2018), pp. 66–71
11. X. He, H. Zhang, N. Li, L. Feng, F. Zheng, A multi-attentive pyramidal model for visual sentiment analysis, in *2019 International Joint Conference on Neural Networks (IJCNN)* (2019), pp. 1–8
12. G.E. Hinton, N. Srivastava, A. Krizhevsky, I. Sutskever, R.R. Salakhutdinov, Improving neural networks by preventing co-adaptation of feature detectors. arXiv preprint arXiv:1207.0580 (2012)
13. A.G. Howard, M. Zhu, B. Chen, D. Kalenichenko, W. Wang, T. Weyand, M. Andreetto, H. Adam, Mobilenets: efficient convolutional neural networks for mobile vision applications. arXiv preprint arXiv:1704.04861 (2017)
14. J. Islam, Y. Zhang, Visual sentiment analysis for social images using transfer learning approach, in *2016 IEEE International Conferences on Big Data and Cloud Computing (BDCloud), Social Computing and Networking (SocialCom), Sustainable Computing and Communications (SustainCom)(BDCloud-SocialCom-SustainCom)*. IEEE (2016), pp. 124–130
15. A. Karpathy, G. Toderici, S. Shetty, T. Leung, R. Sukthankar, L. Fei-Fei, Large-scale video classification with convolutional neural networks, in *Proceedings of the IEEE conference on Computer Vision and Pattern Recognition* (2014), pp. 1725–1732
16. D. Kingma, J. Ba, Adam: a method for stochastic optimization. arXiv preprint arXiv:1412.6980 (2014)
17. J. Machajdik, A. Hanbury, Affective image classification using features inspired by psychology and art theory, in *Proceedings of the 18th ACM international conference on Multimedia* (2010), pp. 83–92
18. D. McDuff, R. El Kaliouby, J.F. Cohn, R.W. Picard, Predicting ad liking and purchase intent: large-scale analysis of facial responses to ads. IEEE Trans. Affect. Comput. **6**(3), 223–235 (2014)

19. M. Paolanti, R. Pierdicca, M. Martini, A. Felicetti, E. Malinverni, E. Frontoni, P. Zingaretti, *Deep Convolutional Neural Networks for Sentiment Analysis of Cultural Heritage* (Remote Sensing & Spatial Information Sciences, International Archives of the Photogrammetry, 2019)
20. L. Rokach, Ensemble-based classifiers. Artif. Intell. Rev. **33**(1–2), 1–39 (2010)
21. U.A. Siddiqua, T. Ahsan, A.N. Chy, Combining a rule-based classifier with ensemble of feature sets and machine learning techniques for sentiment analysis on microblog, in *2016 19th International Conference on Computer and Information Technology (ICCIT)*. (IEEE, 2016), pp. 304–309
22. S. Siersdorfer, E. Minack, F. Deng, J. Hare, Analyzing and predicting sentiment of images on the social web, in *Proceedings of the 18th ACM International Conference on Multimedia* (2010), pp. 715–718
23. K. Simonyan, A. Zisserman, Very deep convolutional networks for large-scale image recognition. arXiv preprint arXiv:1409.1556 (2014)
24. C. Szegedy, W. Liu, Y. Jia, P. Sermanet, S. Reed, D. Anguelov, D. Erhan, V. Vanhoucke, A. Rabinovich, Going deeper with convolutions, in *The IEEE Conference on Computer Vision and Pattern Recognition (CVPR)* (2015)
25. W. Tan, J. Wang, Y., Wang, M. Lewis, W. Jarrold, CNN models for classifying emotions evoked by paintings
26. L. Vadicamo, F. Carrara, A. Cimino, S. Cresci, F. Dell'Orletta, F. Falchi, M. Tesconi, Cross-media learning for image sentiment analysis in the wild, in *Proceedings of the IEEE International Conference on Computer Vision Workshops* (2017), pp. 308–317
27. J. Wang, J. Fu, Y. Xu, T. Mei, Beyond object recognition: Visual sentiment analysis with deep coupled adjective and noun neural networks, in *IJCAI* (2016), pp. 3484–3490
28. C. Xu, S. Cetintas, K. chih Lee, L.J. Li, Visual sentiment prediction with deep convolutional neural networks. ArXiv abs/1411.5731 (2014)
29. Z. Yang, J. Yue, Z. Li, L. Zhu, Vegetable image retrieval with fine-tuning VGG model and image hash. IFAC-PapersOnLine **51**(17), 280–285 (2018)
30. Q. You, J. Luo, H. Jin, J. Yang, Robust image sentiment analysis using progressively trained and domain transferred deep networks, in *Twenty-Ninth AAAI Conference on Artificial Intelligence* (2015)
31. J. Yuan, S. Mcdonough, Q. You, J. Luo, Sentribute: image sentiment analysis from a mid-level perspective, in *WISDOM '13* (2013)
32. J. Zhao, K. Liu, L. Xu, Sentiment analysis: mining opinions, sentiments, and emotions (2016)

Automated Sorting of Rotten or Defective Fruits and Vegetables Using Convolutional Neural Network

Susovan Jana, Ranjan Parekh, and Bijan Sarkar

Abstract Modern peoples are very concerned about food safety. It is known to all that the rotten or defective fruits and vegetables are harmful to health. The rotten and defective fruits and vegetables may degrade the quality of food in the food processing industries. It may damage the fresh fruits and vegetables, which are in the surface contact with the rotten fruits or vegetables in the inventory of food processing industries as well as in the store of supermarkets. The rotten or defective fruits and vegetables should be detected and sorted as early as possible. The automated detection and sorting with the help of image analysis and deep learning can be very effective in industries. This paper proposes a convolutional neural network architecture for sorting the rotten or defective fruits and vegetables from a lot of fruits and vegetables. The proposed technique will classify the fresh and non-fresh (rotten or defective) fruits and vegetables using the trained convolutional neural network. The proposed convolutional neural network model has been tested on two datasets. The classification accuracy on the first and second datasets is 98.53% and 98.46%, respectively. The F1 scores are 99% and 99.18% on dataset 1 and dataset 2, respectively.

Keywords Deep learning · Convolutional neural network · Automated sorting · Fruits · Vegetables

S. Jana (✉) · R. Parekh
School of Education Technology, Jadavpur University, Kolkata 700032, India
e-mail: jana.susovan2@gmail.com

R. Parekh
e-mail: rparekh.edutech@jadavpuruniversity.in

B. Sarkar
Department of Production Engineering, Jadavpur University, Kolkata 700032, India
e-mail: bijan.sarkar@jadavpuruniversity.in

V. E. Balas et al. (eds.), *Proceedings of International Conference on Computational Intelligence, Data Science and Cloud Computing*, Lecture Notes on Data Engineering and Communications Technologies 62, https://doi.org/10.1007/978-981-33-4968-1_4

1 Introduction

Maslow's hierarchy of needs states that food is one of the physiological needs of human beings. The food is very essential for our daily life. The fruits and vegetables are majorly consumed by humans as food either directly or by processing. There are different categories of consumable fruits and vegetables. Nowadays, everybody is very much concerned about food safety and quality. The fresh fruits and vegetables are always in high demand in the open market as well as in the supermarket. The fresh fruits and vegetables are not only delicious to eat but also a good source of important nutrition constraints. The fruits and vegetables prevent from various seasonal diseases. Fresh fruits and vegetables are also used in food processing industries to make delicious food items. The rotten or defective fruits and vegetables belong to the non-fresh category. The fruits and vegetables become rotten after infected by harmful germs, bacteria, fungus, etc. The defect or damage may cause during transportation [1] or may be included in the lot by mistake. There are many other factors for the rot of fruits and vegetables, i.e., temperature, air, light, moisture, and microorganisms, etc. The health science suggests that eating the rotten or partially rotten fruits and vegetables are harmful to health, even though the damaged part is removed by the visual appearance. The quality of food may degrade if it has been processed with rotten or defective fruits and vegetables.

India is the second-largest fruit and vegetable producing country in the world just after China. The industries based on fruits and vegetables are growing rapidly in India. A rotten fruit or vegetable may cause damage to multiple fresh fruits and vegetables in the inventory because of surface contact. The rotten or defective fruits and vegetables need to be removed as early as possible from the inventory in the industries. The early removal of rotten fruits and vegetables reduces the chances of damage in the inventory. The human resources do this sorting based on the visual appearance of fruits and vegetables or by smelling. The manual sorting needs a large number of expert resources as well as a long time. The industry needs an automated application for sorting fresh and non-fresh fruits and vegetables.

Computer vision and machine learning-based automation are very popular in fruits and vegetables processing [2]. Computer vision mainly relies on the surface appearance of fruits and vegetables [3]. The changes are visible in surface color and texture compared with the fresh one when the fruits and vegetables become rotten or defective. Sometimes, the shapes also get changed due to defects. The intra-class dissimilarity increases the challenge. This paper proposed a convolutional neural network (CNN) architecture for sorting rotten or defective fruits and vegetables from the fresh one. The proposed CNN is designed and trained from scratch to classify fresh and non-fresh fruits and vegetables. The paper is structured with multiple sections. Section 2 describes the state of the art survey on this problem. Section 3 outlines a detailed description of the proposed CNN architecture. Section 4 brings details about the datasets. Section 5 presents the experimentation, results, and analysis of the work. Section 6 concludes the work with future scope.

2 Previous Works

The computer vision had already kept its footsteps for solving different automation challenges in the agricultural sector. This literature review has been done to find the prior works, which are dealing with the problem of automated sorting of non-fresh fruits and vegetables. Chandini and Maheswari proposed a technique for sorting the fresh and defective apple [4]. They have focused on two types of defects, i.e., rot and scab. At first, they converted the input color image into the HIS color space. K-means clustering was applied on HIS color space to segment the defective region. Contrast, correlation, energy, and homogeneity features were extracted from the gray level co-occurrence matrix (GLCM). A multiclass support vector machine (SVM) was trained with those features. The classifier predicts among three classes, i.e., fresh, rot, and scab. They were able to achieve 85.64% accuracy of classification. Karakaya et al. proposed an approach for the classification of rotten and fresh fruit [5]. At first, the Otsu segmentation technique was applied to the input images. The features, which were extracted from the segmented image for the classification, are histogram, GLCM, and bag of features. The proposed method was tested on 1200 images which were collected from an open-source dataset. They have trained the SVM classifier with RBF kernel and tested with 10-fold cross-validation. Yogesh et al. proposed a system to detect defective and non-defective fruit [6] using computer vision. They have also classified different stages of the defect. They had also collected 1200 samples in RGB color format. The images were pre-processed and segmented from the background. The extracted features were no of objects, connectivity, major axis, minor axis, eccentricity, area, perimeter, convex area, diameter, filled area, solidity, and Euler number. SVM performed more accurately than KNN and AlexNet in this context. Gómez-Sanchis et al. proposed an approach to detect the rotten citrus fruit caused by Penicillium fungi [7] using machine learning techniques. A hyperspectral image dataset was created. The extracted features from those images are citriculture, 114 spatio-spectral features, and 57 spectral features. The classification accuracy is maximum with the artificial neural networks (ANN). Kamalakannan and Rajamanickam proposed a system to detect and classify the defects of mandarin fruits using image analysis [8]. They had done a fuzzy segmentation. The feature for classification was binary wavelet transform (BWT). A rule-based linear classifier was applied to this feature for the task of classification. Capizzi et al. proposed a technique for defect detection and classification of defects of orange fruits using surface features [9]. They extracted HSV histogram and GLCM features for the classification of defects from orange. The radial basis probabilistic neural network was used for doing the task of classification. Ranjit et al. proposed a classification system for diseased and non-diseased fruits [10]. At first, k-means was applied to segment the defective region. Then, shape, texture, with color features with SVM were used for classification. A mixture of visual and non-visual features was also used to measure the freshness index of eggplant [11].

Identification and segmentation of the rotten spot are important to analyze the source of rot. It is very challenging to segment the rotten spot from the fruit and vegetable surface. Roy et al. surveyed the segmentation methods which had been already used for segmenting the rotten region of the fruits and vegetable surface [12]. The popular techniques are color-based, marker-based, and edge-based. A new segmentation technique for this problem was proposed by them after studying the prior techniques [13]. They had proposed a two-stage technique for segmenting the rotten region. In the first stage, they had segmented the fruit or vegetable object from the background in the YCbCr color space. In the second stage, k-means clustering was applied to divide the foreground into five clusters. The cluster, which contains the rotten region, was set like a mask. The masked region was segmented from the original image in Lab color space.

The previous approaches are mostly based on the histogram, surface texture, and color features. Those feature selections are specific for a particular type of fruit or vegetable. Hence, an automated framework for sorting is needed to detect rotten or defective items for any type of fruit or vegetable.

3 Proposed Approach

This paper addresses the challenges of sorting rotten or defective fruits and vegetables by proposing a CNN architecture. Figure 1 shows the architecture of the proposed CNN model. The CNN model will be trained with the labeled images of both fresh and non-fresh fruits and vegetables. The trained CNN model does the task of sorting. The trained network will predict the label between fresh and non-fresh for an unknown test sample.

The network contains 19 layers in total. The input layer is formed with 64×64 RGB images with zero center normalization. The next layer, which is just after the input layer, is a convolution layer. The convolution layer contains eight convolution filters of size 3×3, stride [1 1], and zero paddings. The padding size should be adjusted in such a way so that the output layer will have the size as the input layer. It identifies the most discriminating features between fresh and non-fresh fruits and vegetables. The surface color and texture of non-fresh fruits and vegetables are not the same as the fresh fruits and vegetables. The convolution layer continues the extraction of features as long as the training progresses. The convolution layer is followed by a batch normalization layer containing eight channels. The batch normalization layer normalizes the extracted features from the input layer. Next to the batch normalization layer is a ReLU layer, the ReLU layer adds nonlinearity with a nonlinear activation function as shown in Eq. (1). The reason for adding a batch normalization layer between convolution and ReLU is that the batch normalization layer reduces susceptibility to initial parameters as well as speed up the training process. Next to ReLU layer, a 2×2 max pooling layer is added to the net with stride [2 2] and padding [0 0 0 0]. A block is made with those four consecutive layers, i.e., convolution layer, batch normalization layer, ReLU layer, and the max pooling layer. Another three similar types of blocks are added in sequence just after the first block. The number of convolution filters is doubled in the blocks as the sequence progresses, i.e., 16, 32,

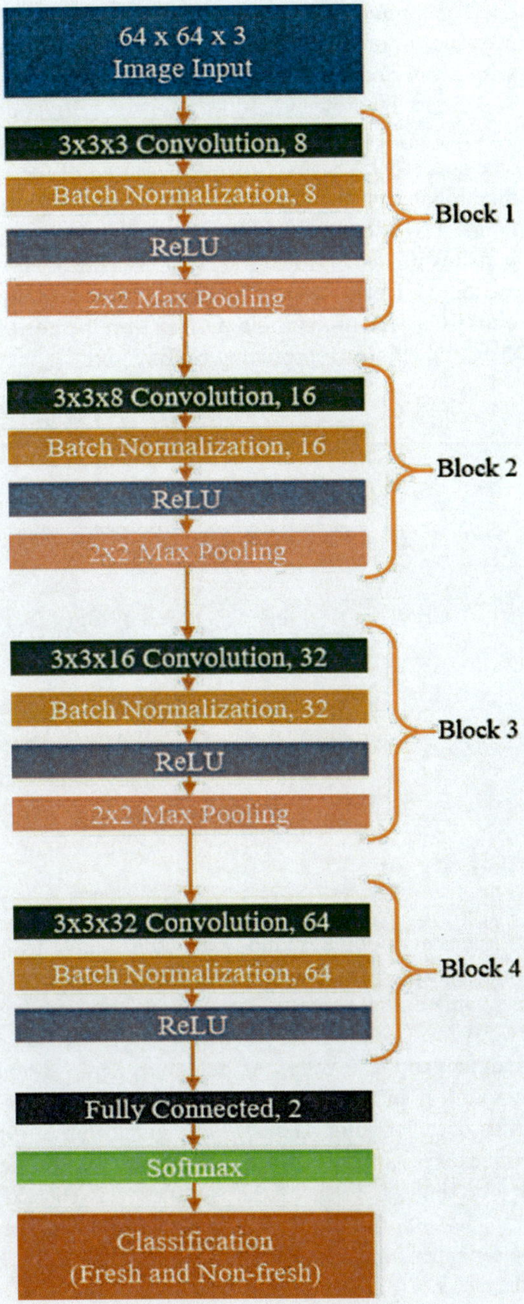

Fig. 1 Architecture of the proposed CNN model

64. The number of channels in the batch normalization layer is also doubled in the sequence of progression of blocks, i.e., 16, 32, 64. The max pooling layer is replaced with a fully connected layer in the final block, i.e., the 4th block in the sequence. A softmax layer is added just after the fully connected layer. Refer to Eq. (2). The softmax layer is followed by the classification layer, which is the final layer of the net. The softmax layer normalizes the output of the fully connected layer and produces the probability of prediction for an unknown sample. The prediction probability will be used by the classification layer. The final output is the class label, i.e., fresh or non-fresh. The classification layer uses the binary cross-entropy loss function. Refer to Eq. (3). Here, i stands for the number of classes. There are two classes as it is a binary classification problem, $t_1 = 1$ for the positive class and $t_1 = 0$ for the negative class. The loss is represented by Eq. (4).

$$f(x) = \begin{Bmatrix} x, \ x \geq 0 \\ 0, \ x < 0 \end{Bmatrix} \tag{1}$$

$$\sigma(z)_j = \frac{e^{z_j}}{\sum_{k=1}^{K} e^{z_k}} \tag{2}$$

$$\text{CE} = -\sum_{i=1}^{c=2} t_i \log(f(s_i)) = -t_1 \log(f(s_1)) - (1 - t_1)\log(1 - f(s_1)) \tag{3}$$

$$\text{CE} = \begin{cases} -\log(f(s_1)) & \text{if } t_1 = 1 \\ -\log(1 - f(s_1)) & \text{if } t_1 = 0 \end{cases} \tag{4}$$

4 Dataset

Apple is very popular among the consumable fruits and vegetables. Hence, we have selected apple fruit for experimentation. The datasets are downloaded from freely available online sources. The dataset 1 [14] contains 5031 images in total. There are two types of apple fruits, i.e., fresh apple and non-fresh apple. It has 232 fresh apple samples and 327 rotten or defective apple samples. The dataset contains a good amount of intra-class variations. The number of images is very less for working with deep learning. Hence, the eight different types of image augmentation techniques were applied to increase the number of images as well as introduce diversity in the dataset. Each of the samples was rotated in five different directions, i.e., 15°, 30°, 45°, 60°, and 75°. The other three augmentation techniques are adding salt and pepper noise, translation, and a vertical flip. Figure 2 shows the samples from dataset 1.

The dataset 2 is formed with the images collected from freely available online sources. The dataset 2 contains the images of rotten or defective apples (244) as well as fresh apples (15). The images of fresh apples are collected from different online

Fig. 2 Samples from dataset 1 **a, b** fresh and **c, d** non-fresh

Fig. 3 Samples from dataset 2 **a, b** fresh and **c, d** non-fresh

sources. The images of non-fresh apples are collected from a GitHub source [15]. In total, it contains 259 images. Figure 3 shows the samples from dataset 2. Finally, the images of both datasets are resized into *64* × *64* for uniform size and faster processing.

5 Experimentation, Results, and Analysis

The CNN model is proposed here for any type of fruits and vegetables. Apple is a very popular fruit and it has a lot of variety. Hence, we have experimented with apple to test the performance of the proposed CNN model. The proposed network has been trained with 3200 images from dataset 1. Table 1 depicts the total number of images and the distribution of training and testing. Dataset 2 will be used for only testing purpose. Hence, it does not contain any training image. The training images and testing images have been chosen randomly. The training images are shuffled in every epoch. The selection of hyperparameters is very important to build a robust model. The initial rate of learning is 0.01. The maximum number of the epoch is set as 25. Figures 4 and 5 show the plots of training versus validation accuracy and loss, respectively. It has been observed that the validation accuracy increases and loss

Table 1 Distribution of images on datasets

Dataset	Class	Total	Training	Testing
Dataset 1	Fresh	2088	1600	488
	Non-fresh	2943	1600	1343
Dataset 2	Fresh	15	–	15
	Non-fresh	244	–	244

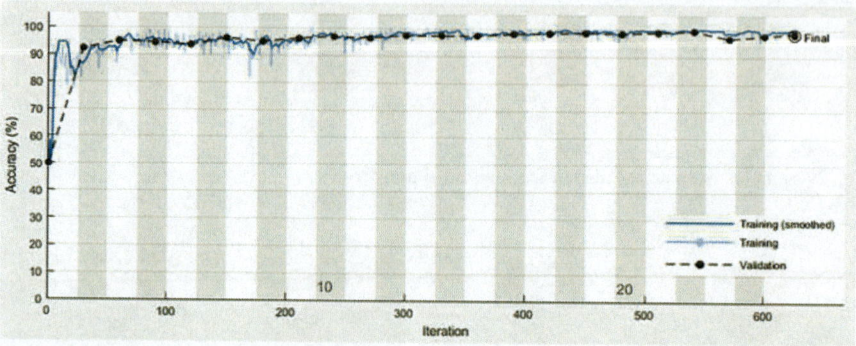

Fig. 4 Training versus validation accuracy

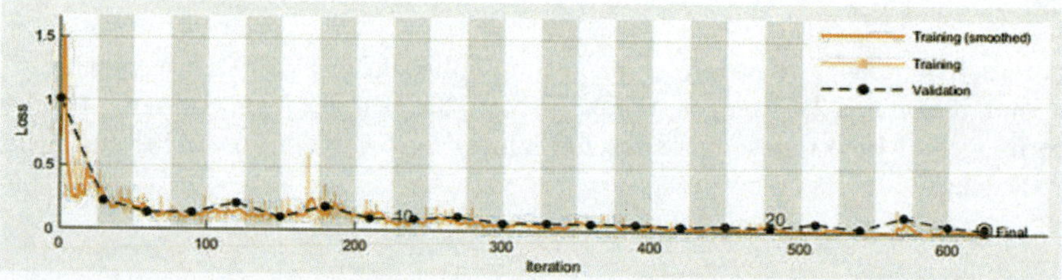

Fig. 5 Training versus validation loss

decreases as long as the iteration progresses. The low-level features are learned by the convolutional layers at the beginning of the network. The convolutional layers toward the end of the sequence learn the high-level features from the images. Figure 6 shows the high-level features learned by the final convolution layer. Figure 7 visualizes the fully connected layer.

The fruits and vegetables belong to the perishable category of products. The probability of getting a rotten or defective sample is very less in a recently delivered lot of fruits and vegetables. The number of damage may increase due to transportation delay or decrease in the sale. The damaged samples need to be removed as early as possible for the maintenance of the inventory. The detection and removal of non-fresh samples are more important than the fresh one. Hence, the non-fresh class is considered a positive class in this classification and fresh as a negative class. The samples, which are actually positive and also predicted as positive, are true positive (TP). The samples, which are negative but predicted as positive, are false positive (FP). The samples, which are negative and also predicted as negative, are true negative (TN). The samples, which are positive but predicted as negative, are false negative (FN). Those metrics vary depending on the distribution of training and testing samples. Hence, the overall accuracy is computed as performance metrics from those metrics using Eq. (5). The classification accuracy is not enough for this kind of class imbalance problem. Hence, some more performance metrics are computed, i.e., recall,

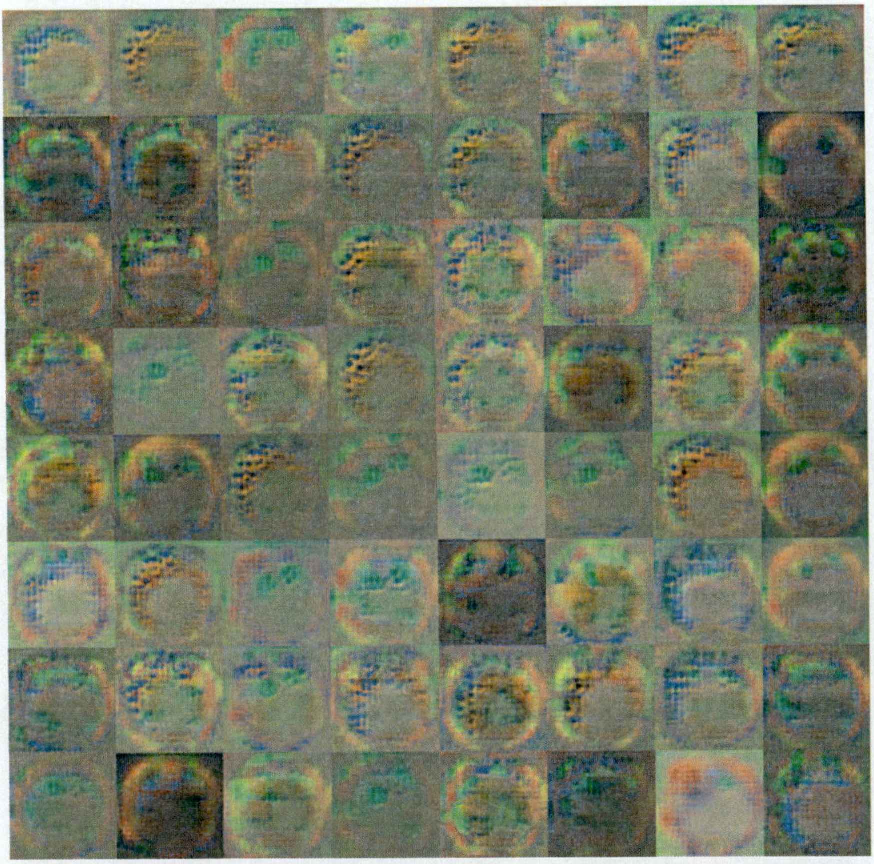

Fig. 6 Features learned by the final convolutional layer

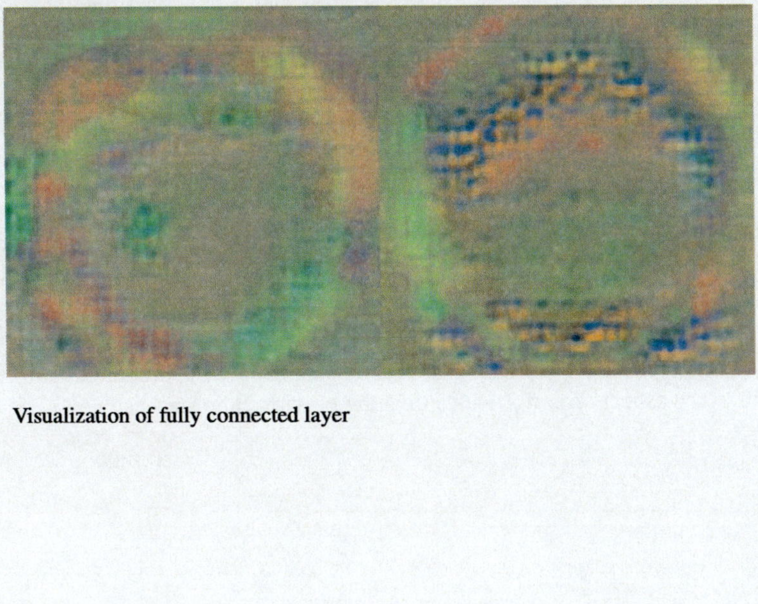

Fig. 7 Visualization of fully connected layer

Table 2 Results of the proposed approach on different datasets

Performance metrics	Dataset 1	Dataset 2
Fresh and predicted as fresh (%)	97.13	93.33
Fresh but predicted as non-fresh (%)	2.87	6.67
Non-fresh but predicted as fresh (%)	0.97	1.23
Non-fresh and predicted as non-fresh (%)	99.03	98.77
Overall accuracy (%)	98.53	98.46
Recall (%)	99.03	98.77
Precision (%)	98.96	99.59
F1 score (%)	99.00	99.18

precision, and F1 score. Refer to Eqs. (6)–(8). Dataset 2 is used for only testing the robustness of the network, which is trained by the training images of dataset 1. Table 2 presents the results using the proposed approach on two datasets.

$$\text{Accuracy} = \frac{\text{TP} + \text{TN}}{\text{TP} + \text{FP} + \text{FN} + \text{TN}} \tag{5}$$

$$\text{Recall} = \frac{\text{TP}}{\text{TP} + \text{FN}} \tag{6}$$

$$\text{Precision} = \frac{\text{TP}}{\text{TP} + \text{FP}} \tag{7}$$

$$\text{F1 Score} = \frac{2 * \text{Recall} * \text{Precision}}{\text{Recall} + \text{Precision}} \tag{8}$$

We have also implemented some previous approaches for sorting the rotten or defective fruits and vegetables. Capizzi et al. [9] and Karakaya et al. [5] used histogram features for the classification of fresh and non-fresh fruits and vegetables. The statistical features from GLCM were used by Karakaya et al. [5], Chandini and Maheswari [4], and Capizzi et al. [9]. Bag of features was used by Karakaya et al. [5]. A classification algorithm needs to be trained by those features. SVM classifier has been used by most of the researchers. Hence, the SVM classifier is trained and tested with these features separately on dataset 1, which was used during the experimentation of the proposed approach. Table 3 shows the experimental results of the previous approaches on dataset 1.

Figure 8 depicts the comparison of the proposed approach with the previous approaches on dataset 1. The x-axis of the plot represents the different performance metrics, i.e., overall accuracy, recall, precision, and F1 score. The y-axis represents values in percentage for the corresponding metrics. The red line is far away from the other lines for all the metrics. That means the proposed approach is performing better in every aspect of performance than the previous approaches.

Table 3 Result of previous approaches on dataset 1

Performance metrics	Grayscale histogram + SVM	GLCM features + SVM	Bag of features + SVM
Fresh and predicted as fresh (%)	42.42	72.54	78.48
Fresh but predicted as non-fresh (%)	57.58	27.46	21.52
Non-fresh but predicted as fresh (%)	37.01	35.67	19.14
Non-fresh and predicted as non-fresh (%)	62.99	64.33	80.86
Overall accuracy (%)	57.51	66.52	79.67
Recall (%)	62.99	64.33	80.86
Precision (%)	75.07	86.57	91.18
F1 score (%)	68.50	73.81	85.71

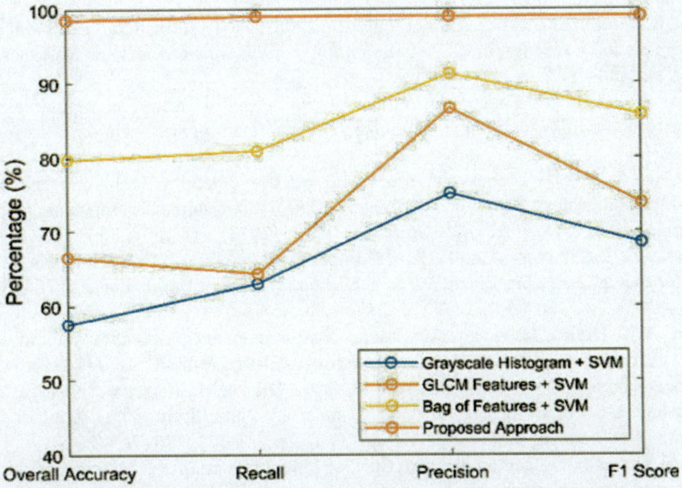

Fig. 8 Comparison of the proposed approach with different previous approaches based on the performance metrics, i.e., overall accuracy, recall, precision, and F1 score

6 Conclusion

The rotten or defective fruits and vegetables are very harmful to human health as well as for an inventory. The detection and removal of non-fresh fruits and vegetables are needed to maintain the quality of the store as well as reduction of loss due to damage. This paper proposes a convolutional neural network-based framework for sorting the rotten or defective fruits and vegetables. The proposed convolutional neural network model has been designed and trained from scratch to classify fresh and non-fresh fruits and vegetables. The training images of dataset 1 were used to train the network. The trained network has been tested on two different datasets for prediction. The accuracy and F1 score on the test images of dataset 1 are 98.53% and 99%, respectively. The dataset 2 is considered only for testing the robustness of the net, which is trained by the training images of dataset 1. The accuracy and F1 score on the test images of dataset 2 are 98.46% and 99.18%, respectively. The proposed approach outperforms previous approaches. The proposed approach can be applied to other types of fruits and vegetables. The measurement of the freshness level could be a good direction for future research.

References

1. J. Cao, C. Wang, S. Xu, Y. Chen, Y. Wang, X. Li, C. Sun, The effects of transportation temperature on the decay rate and quality of postharvest Ponkan (*Citrus reticulata* Blanco) fruit in different storage periods. Sci. Hortic. **247**, 42–48 (2019)
2. S. Jana, R. Parekh, Shape-based fruit recognition and classification, in *International Conference on Computational Intelligence, Communications, and Business Analytics*. Springer, Singapore (2017), pp. 184–196
3. S. Jana, R. Parekh, B. Sarkar, Automatic classification of fruits and vegetables: a texture-based approach, in *Algorithms in Machine Learning Paradigms*. Springer, Singapore (2020), pp. 71–89
4. A.A. Chandini, B.U. Maheswari, Improved quality detection technique for fruits using GLCM and multiclass SVM, in *2018 International Conference on Advances in Computing, Communications and Informatics (ICACCI)*. IEEE (2018), pp. 150–155
5. D. Karakaya, O. Ulucan, M.A. Turkan, Comparative analysis on fruit freshness classification, in *2019 Innovations in Intelligent Systems and Applications Conference (ASYU)*. IEEE (2019), pp. 1–4
6. Yogesh, A.K. Dubey, R. Ratan, A. Rocha, Computer vision based analysis and detection of defects in fruits causes due to nutrients deficiency. Clust. Comput. 1–10 (2019)
7. J. Gómez-Sanchis, J.D. Martín-Guerrero, E. Soria-Olivas, M. Martínez-Sober, R. Magdalena-Benedito, J. Blasco, Detecting rottenness caused by Penicillium genus fungi in citrus fruits using machine learning techniques. Expert Syst. Appl. **39**(1), 780–785 (2012)
8. A. Kamalakannan, G. Rajamanickam, Surface defect detection and classification in mandarin fruits using fuzzy image thresholding, binary wavelet transform and linear classifier model, in *2012 Fourth International Conference on Advanced Computing (ICoAC)*. IEEE (2012), pp. 1–6
9. G. Capizzi, G.L. Sciuto, C. Napoli, E. Tramontana, M. Woźniak, Automatic classification of fruit defects based on co-occurrence matrix and neural networks, in *2015 Federated Conference on Computer Science and Information Systems (FedCSIS)*. IEEE (2015), pp. 861–867
10. K.N. Ranjit, C. Naveen, H.K. Chethan, Fruit disease categorization based on color, texture and shape features. Int. J. Comput. Appl. **178**(49), 16–19 (2019)
11. S.N. Jha, T. Matsuoka, Development of freshness index of eggplant. Appl. Eng. Agric. **18**(5), 555 (2002)
12. K. Roy, S.S. Chaudhuri, S. Bhattacharjee, S. Manna, T. Chakraborty, Segmentation techniques for rotten fruit detection, in *2019 International Conference on Opto-Electronics and Applied Optics (Optronix)*. IEEE (2019), pp. 1–4
13. K. Roy, A. Ghosh, D. Saha, J. Chatterjee, S. Sarkar, S.S. Chaudhuri, Masking based segmentation of rotten fruits, in *2019 International Conference on Opto-Electronics and Applied Optics (Optronix)*. IEEE (2019), pp. 1–4

14. S.R. Kalluri, Fruits fresh and rotten for classification, https://www.kaggle.com/sriramr/fruits-fresh-and-rotten-for-classification. Accessed 14 May 2020
15. R. Siddiqi, Apple-defect-detection, https://github.com/raheelsiddiqi2013/apple-defect-detection. Accessed 28 June 2020

Fuzzy Time Series Forecasting of COVID-2019 Outbreak: A Case Study of U.S. Population

Arijit Chakraborty, Sajal Mitra, Dipankar Das, Debashis De, and Anindya J. Pal

Abstract The novel coronavirus (nCoV-2019) was first apparent in Wuhan city in China, which impacted the world and its peoples. This epidemic severely influenced the global equilibrium of humankind, including the USA, where the number of affected cases reached more than 4,323,160 by the end of July 2020. Therefore, the COVID-2019 outbreak scenario warrants a sound forecasting model to accurately predict the catastrophe in human lives that resulted from this pandemic. In this study, the Fuzzy Time Series (FTS) forecasting model for COVID-19 employed to analyze and predict the number of cumulative infected cases of the USA by employing the Abbasov and Mamedova model. Our experiment used 145 days of infected cases of the USA rendered from the World Health Organization (WHO). The optimized model achieved through tuning three hyper parameters of the Abbasov and Mamedova model. To estimate the model performance, we evaluated the forecast accuracy through the lenses of Mean Absolute Percentage Error (MAPE) and Theil *U* statistics, followed by a comparison between the forecasted with actual observations. We observed that the recommended FTS model's forecasting is reliable and acceptable up to 35 days ahead of forecasting.

A. Chakraborty (✉) · D. Das
Bachelor of Computer Application Department, The Heritage Academy, Kolkata, India
e-mail: arijit.chakraborty@heritageit.edu

D. Das
e-mail: dipankr.das@gmail.com

S. Mitra
Department of Computer Science and Engineering, Heritage Institute of Technology, Kolkata, India
e-mail: mitra.sajal@gmail.com

D. De
Department of Computer Science and Engineering, Maulana Abul Kalam Azad University of Technology, Kolkata, India
e-mail: dr.debashis.de@gmail.com

A. J. Pal
University of Burdwan, Burdwan, India
e-mail: anindyajp@gmail.com

V. E. Balas et al. (eds.), *Proceedings of International Conference on Computational Intelligence, Data Science and Cloud Computing*, Lecture Notes on Data Engineering and Communications Technologies 62, https://doi.org/10.1007/978-981-33-4968-1_5

Keywords COVID-2019 · Fuzzy time series · Mean absolute percentage error

1 Introduction

The novel Coronavirus-2019 (nCoV-2019) micro-organism creates disorders in humans is an extension to the Coronavirus family. The earlier versions of CoV, such as Middle East Respiratory Syndrome (MERS-CoV) and Severe Acute Respiratory Syndrome (SARS-CoV), witnessed in the Middle Eastern and Chinese provinces a few years back. The adversity of this virus is limited to humans and many avian organisms [1, 2]. The international body responsible for the taxonomy of viruses inferred the pathogenic nature of CoV with some notable exceptions in SARS-CoV; fundamentally, a β-CoV reported in China in 2002, and MERS-CoV reported in Saudi Arabia in 2012 [3]. A nCoV-2019 (COVID-2019) positive patient potentially infects more than one person with symptomatic pieces of evidence that appears within a fortnight. Additionally, many asymptomatic cases, i.e., COVID-2019 cases with no or mild symptoms bypass the standard precautionary measures necessary for breaking the COVID-2019 chains. This nCoV-2019 outbreak in a society categorized into several stages ranging from the localized transmission to more fatal stages such as neighborhood transmission and pandemic outbreak, wherein epidemic has a lengthened resurfacing in a particular nation. The gravity of the situation worldwide warrants the need to design robust and state-of-the-art methods to estimate the spread of contagious 2019-nCoV to support administration and government bodies to formulate a centralized logistic core to meet the challenges induced by COVID-2019. Time series based forecasting of any pandemic situation comprises a study of a systematic chain of values. Timely, estimates have a substantial impact on decision-making to arrest the outbreak chain of infectious nCoV-2019. In [4] proposed the concept of a fuzzy set to quantify the imprecision and vagueness involved in a system in terms of the degree of membership. Fuzzy based forecasting is a prudent stretch in analyzing the adversities of the COVID-2019 because of the fuzzy system's ability to handle uncertainty, imprecision associated with the unfortunate outbreak lease note that the first paragraph of a section or subsection is not indented. The first paragraphs that follow a table, figure, equation, etc. does not have an indent, either.

In this context, [5] proposed using Auto-Regressive Integrated Moving Average (ARIMA) and FTS models to depict the pandemic trajectory of COVID-2019 in African countries with a reasonable small Root Mean Square Error (RMSE). In their work, [6] considered ARIMA and FTS methodology for forecasting COVID-19 adversities in India and the most affected state of India, i.e., Maharashtra. Verma et al. [6] achieved a reasonable match between their estimates and real-time values.

Presently, scarce experimentation in adapting the FTS model done for estimating epidemiological casualties. This paper elucidates the use of FTS forecasting model in estimating the number of cumulative infections in the USA by employing the Abbasov and Mamedova model.

1.1 Motivations and Contributions

FTS approach can be considered as an unconventional approach for forecasting the COVID-19 outbreak. In this context, [7] showed comparable outcomes in the analysis of infectious disease surveillance. Tricahya and Rustam [8] presented precise estimates in forecasting the number of pneumonia patients in Jakarta using FTS. These investigations motivated us to incorporate the FTS technique's use to design a robust forecasting model to estimate the COVID-2019 adversities in the USA, where the proportion of COVID tests per million persons offers high-quality dataset for better assessment of the nCoV-2019 outbreak. In this context, major offerings of this publication as follow:

- Tested the applicability of the Abbasov and Mamedova FTS model to forecast the number of cumulative infections of COVID-19 in the USA.
- In the model testing phase, the Abbasov and Mamedova FTS model exhibited low MAPE (<10) and Theil U (close to zero) which indicates a reliable and acceptable forecast accuracy in performing 45 days ahead forecasting of cumulative infections of COVID19 in the USA.
- In the forecasting phase, the model forecast is reliable and acceptable up to 35 days ahead forecasting with low MAPE (<10) and Theil U (close to zero).

1.2 Dataset

The USA's COVID-19 cumulative infected person data from January 21st to July 28th 2020, i.e., a total of 190 no. of days, accumulated from the World Health Organization (WHO) [9] to create our dataset for examination.

Out of 190 days data, the first 100 days data, i.e., from January 21st to April 20th 2020, was treated as a training set, and next 45 days data, i.e., from April 30th to June 13th 2020 was treated as the test set. The training set also divided into two subsets. The first 80 days data considered a training subset, and the rest 20 days data considered for the validation set.

1.3 Context Diagram of the Proposed Study

The context diagram of our proposed forecasting model shown in Fig. 1.

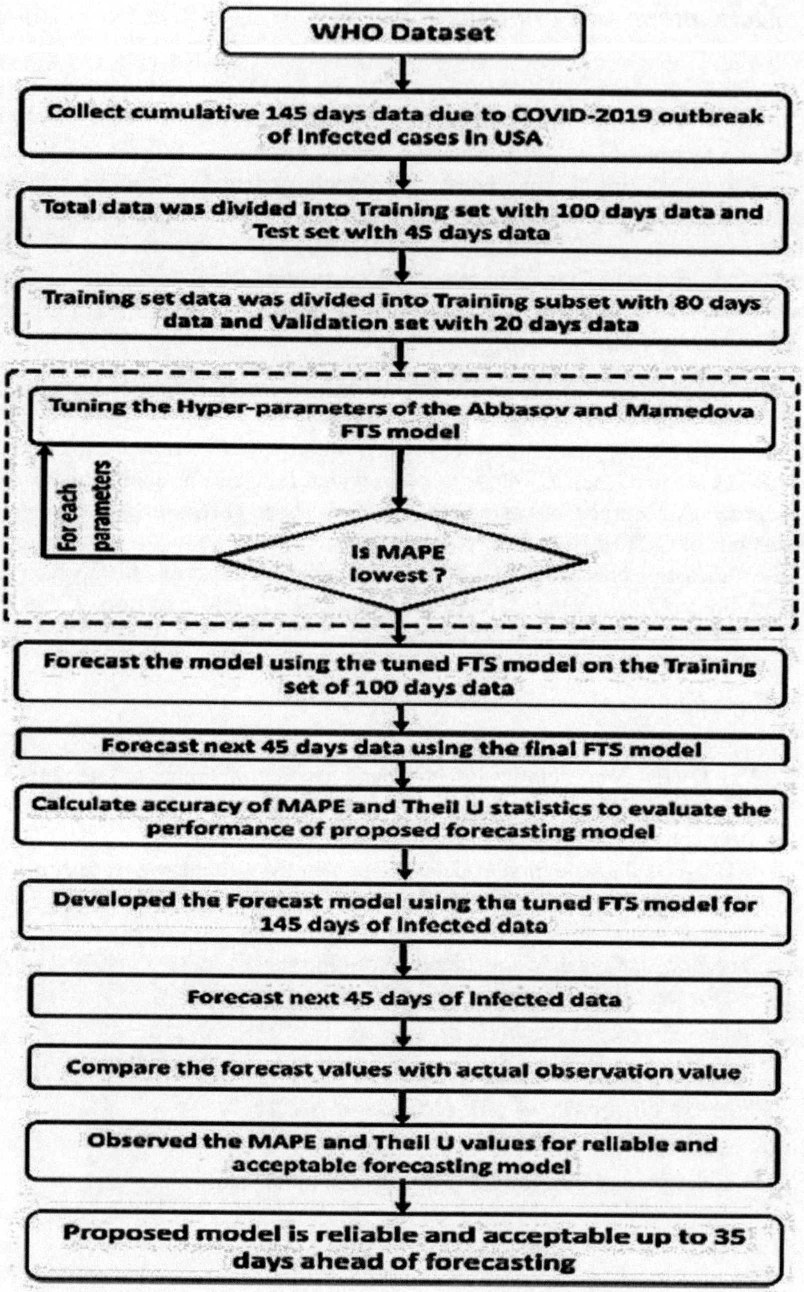

Fig. 1 Context diagram of the proposed model

2 Background Study

Authors of [10] proposed using the FPASSA-Adaptive Neuro-Fuzzy Inference System (ANFIS) optimization model in estimating adversities of the COVID-19 in Chinese regions using the WHO dataset between January 21st 2020 and February 18th 2020. They determined the outbreak pattern of ten days up the line using the ANFIS in conjunction with the Flower Pollination Algorithm (FPA) and the Salp Swarm Algorithm (SSA). Al-Qaness et al. [10] optimized the performance by fine-tuning ANFIS-parameters by employing FPASSA and subsequently analyzed their model with eight other models based on Coefficient of Determination (R^2), RMSE, MAPE, and Mean Absolute Error (MAE). They observed that the performance of FPASSA model outperformed other existing model with $R^2 = 0.9645$, RMSE = 5779, MAPE = 4.79, RMSRE = 0.07, MAE = 4271, and computing time = 23.30.

In another work, authors of [11] employed Moving Average (MA), Weighted Moving Average (WMA), and Single Exponential Smoothing (SES) time series model to generate a daily estimation of the COVID-19 outbreak using the dataset compromising of the WHO, the National Health Commission of China, and Johns Hopkins University. The authors measure their forecasting performance of forecasting based on RMSE, MAPE, and Mean Absolute Deviation (MAD) and concluded that the SES model is the best-fit candidate with a confidence interval of 95% and forecasted up to more than 200,000 no. of confirmed cases, 140,000 no. of recovered cases, i.e., a ratio of 95.6%, and 7000 no. of death cases with 3.5% ratio globally by April 9th 2020. This model also suggested that the no. of COVID-19 affected cases rises exponentially with more than 25% per day in the world.

Authors of [12] predicted the COVID-19 outbreak in India based on the prediction pattern of COVID-19 in China using machine learning approach. The authors considered the Kaggle dataset [13] of COVID-2019 misfortunes between January 2020 and April 3rd 2020, and analyzed the performance of time series forecasting models based on RMSE and MAE. They predicted that during the last week of April, the outbreak rise would be at the peak in India, which can be neutralized by the end of May. In another study, [14] offered a forecasting model of confirmed COVID-19 cases of seven cities of China using the Convolutional Neural Network (CNN) model. The authors considered the dataset between 23rd January 2020 and 2nd March 2020 of Surging News Network (SNN) and WHO. The accuracy and reliability of the CNN model's performance contrasted with the other three deep learning prediction techniques based on the values of RMSE and MAE. The CNN model exceeded the other models with the lowest RMSE and MAE values in a range between 109.439–325.857 and 102.943–284.573, respectively. In Table 1, a summary of the use of Goodness of Fit statistics in forecasting the COVID-2019 outbreak listed.

Table 1 Goodness of fit statistics in pandemic

Authors	Forecasting method		Forecast disease	Performance of best fit candidate based on			
	Statistical	Time series	COVID-19	RMSE	MAPE	MAD	MAE
[10]	✔	✔	✔	✔	✔	✔	✔
[11]	✔		✔	✔	✔	✔	
[12]	✔		✔	✔			✔
[14]	✔		✔	✔			✔
Our work		✔	✔		✔		

3 Methodology

The schematic representation of the data partition is shown in Fig. 2.

The Abbasov and Mamedova model [15, 16] have the following steps:

Step 1: Define the universal set U
Step 2: Divide U into n equal length intervals
Step 3: Define the fuzzy set using the formula

$$\mu_{A_i} = \frac{1}{1 + \left[C \times \left(u - u_m^i\right)\right]^2}$$

where u: A generic element of U
u_m^i: Middle point of the corresponding interval u_i, $(i = 1, 2, 3, 4, \ldots, n)$
C: A constant
Step 4: Fuzzyfication of the input data
Step 5: Select the value of w parameter $(1 < w < l)$

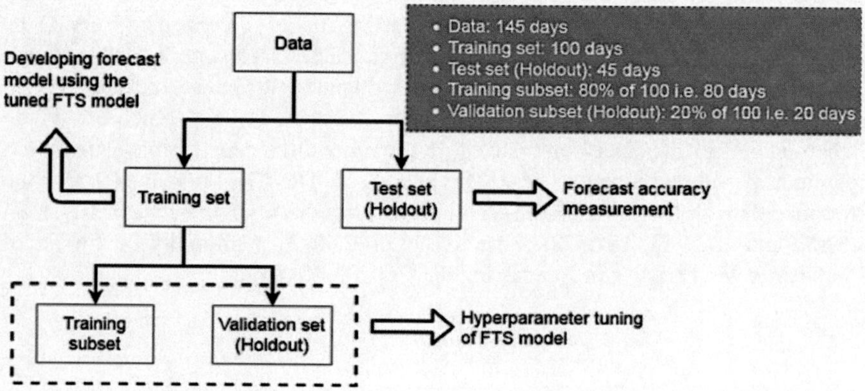

Fig. 2 Data partitioning scheme

where l is the number of days prior to the current day
Step 6: Defuzzyfication of the result.

In the Abbasov and Mamedova model, the following three hyper parameters are needed to be tuned to obtain the suitable values:

- Constant C
- Number of interval n
- Parameter w.

The search spaces of the hyper parameters are:

- $C = \{0.0010, 0.0009, 0.0008, 0.0007, 0.0006, 0.0005, 0.0004, 0.0003, 0.0002, 0.0001\}$
- $n = \{3, 4, 5, 6, 7, 8, 9, 10, 11, 12\}$
- $w = \{2, 3, 4, 5, 6, 7, 8, 9, 10\}$.

Abbreviations used in Fig. 3 are as follows: MAPE = Mean Absolute Percentage Error, C_1 = first value of C from its search space, n_1 = first value of n from its search space, n_2 = second value of n from its search space, w_1 = first value of w from its search space, w_2 = second value of w from its search space, C_{id} = identified C value having the lowest MAPE, n_{id} = identified n value having the lowest MAPE, w_{id} = identified w value having the lowest MAPE.

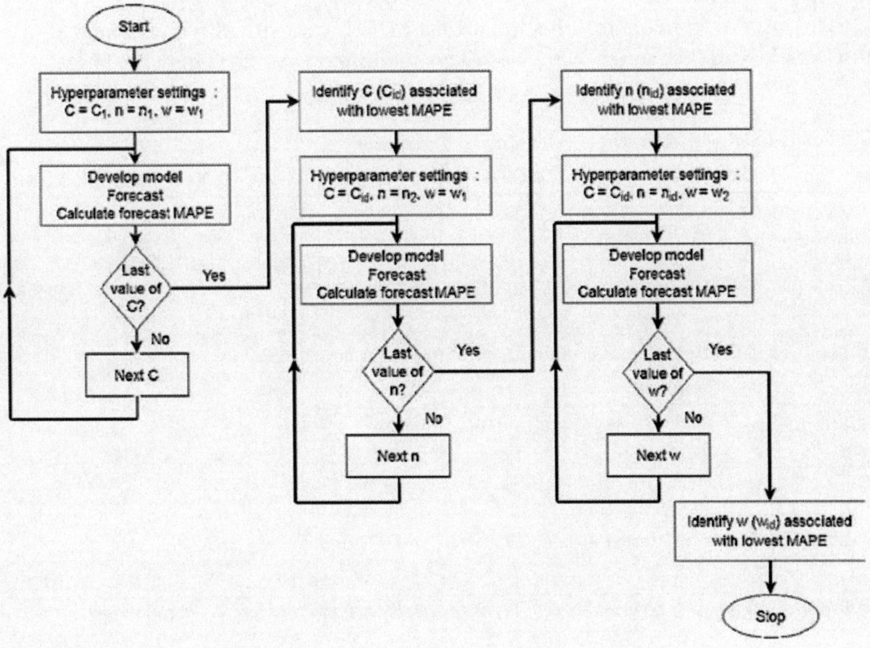

Fig. 3 Overview of hyper parameter tuning

In the hyper parameter tuning stage, we developed the model using the training subset data, and the forecast MAPE calculated using the out of sample forecast values and holdout validation set data. After the model is tuned, we developed the model using the tuned hyper parameter values on the training set data, and out of sample, forecasting is performed. The forecast MAPE and Theil U statistics are calculated using the forecast values and holdout test set data. We used the AnalyzeTS package [17] of the R software to develop the proposed FTS model.

4 Result and Discussion

4.1 Abbasov and Mamedova FTS Model Development

The tuning of the Abbasov and Mamedova FTS model's hyper parameter done using the training set of the first 100 days, i.e., from January 21st to April 29th 2020. The first 80 days of training subset data trained and rest 20 days' data used to validate the model. The results are given in Tables 2, 3, and 4 and visualized in Fig. 4.

We observe from Table 2 that the lowest MAPE value (3.329) is obtained at $C = 0.001$, $n = 3$, and $w = 2$. This identified C value is indicated as C_{id}.

We observe from Table 3 that the lowest MAPE value (0.8) is obtained at $n = 5$, $C = C_{id} = 0.001$, and $w = 2$. This identified n value is indicated as n_{id}.

We observe from Table 4 that the lowest MAPE value (0.780) is obtained at $w = 4$, $C = C_{id} = 0.001$, and $n = n_{id} = 4$. This identified w value is indicated as w_{id}.

Table 2 Model with different C values, $n = 3$ and $w = 2$

C	0.001	0.0009	0.0008	0.0007	0.0006	0.0005	0.0004	0.0003	0.0002	0.0001
20 step ahead MAPE	3.329	3.375	3.439	3.534	3.684	3.946	4.476	6.141	13.531	18.116

Table 3 Model with different n values, $C = C_{id} = 0.001$ and $w = 2$

n	3	4	5	6	7	8	9	10	11	12
20 step ahead MAPE	3.329	1.383	0.80	0.966	2.602	3.266	3.283	3.758	4.30	4.63

Table 4 Model with different w values, $C = C_{id} = 0.001$ and $n = n_{id} = 5$

w	2	3	4	5	6	7	8	9	10
20 step ahead MAPE	0.800	0.782	0.780	0.780	0.780	0.780	0.780	0.780	0.780

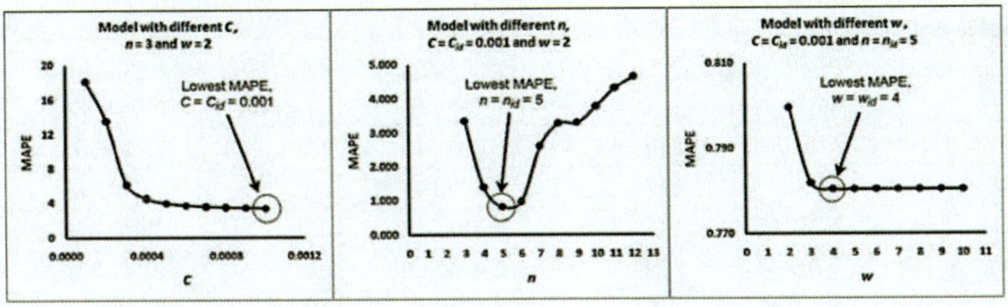

Fig. 4 Hyper parameter tuning of FTS model

From Tables 2, 3, and 4, we observed that the tuned hyper parameter values indicated as C_{id}, n_{id}, and w_{id} are 0.001, 5, and 4, respectively, for the first 100 days cumulative infected cases.

4.2 Model Evaluation

The forecast MAPE and Theil *U* statistics of the tuned Abbasov and Mamedova FTS model using different forecast horizons are given in Table 5. If we observe that the forecast MAPE is less than 10% and Theil *U* value is close to 0, we consider the forecast is reliable and acceptable [18, 19].

We observed from Table 5 that the forecasts of the proposed FTS model are reliable and acceptable up to 45 days ahead of forecasting, i.e., April 30th to June 13th 2020.

4.3 Forecast of Next 45 days

We used the hyper parameters of the tuned FTS model, developed the FTS model using the full dataset—i.e., 145 days data, and performed forecast of the number of cumulative infections in the USA for the next 45 days, i.e., June 14th to July 28th 2020. The results are given in Table 6.

Table 5 Forecast accuracy of the tuned FTS model

Model	Steps ahead						
	7	14	21	28	35	42	45
MAPE	0.6	1.04	1.91	2.66	3.39	3.98	4.24
Theil *U*	0.007	0.015	0.027	0.038	0.048	0.056	0.06

Table 6 Forecast results

Day ahead	Cumulative infections	Day ahead	Cumulative infections	Day ahead	Cumulative infections
1	2,032,721	16	2,402,318	31	2,773,560
2	2,056,372	17	2,427,067	32	2,798,310
3	2,080,667	18	2,451,817	33	2,823,059
4	2,105,341	19	2,476,566	34	2,847,809
5	2,130,076	20	2,501,316	35	2,872,558
6	2,154,823	21	2,526,065	36	2,897,308
7	2,179,572	22	2,550,815	37	2,922,057
8	2,204,322	23	2,575,564	38	2,946,807
9	2,229,071	24	2,600,314	39	2,971,556
10	2,253,821	25	2,625,063	40	2,996,306
11	2,278,570	26	2,649,813	41	3,021,055
12	2,303,320	27	2,674,562	42	3,045,805
13	2,328,069	28	2,699,312	43	3,070,554
14	2,352,819	29	2,724,061	44	3,095,304
15	2,377,568	30	2,748,811	45	3,120,053

4.4 Evaluation of Forecast Versus Observed Data

We observed from Table 7 that the forecast MAPE is less than 10%, and Theil U value is close to 0 conditions matched with the first 35 days data, i.e., the forecasts of the proposed FTS model are reliable and acceptable up to 35 days ahead of forecasting. Thus, our proposed FTS forecasting model can forecast reliable and acceptable cumulative infected data of the USA from June 14th to July 18th 2020 ahead of forecasting.

4.5 Discussion

We have tuned the following hyper parameters (C, n, w) of the Abbasov and Mamedova FTS model to obtain the final model with $C = C_{\text{id}} = 0.001$, $n = n_{\text{id}} = 5$, and

Table 7 Forecast accuracy of the tuned FTS model

Model	Steps ahead						
	7	14	21	28	35	42	45
MAPE	0.18	0.61	2.16	4.53	7.35	10.99	12.53
Theil U	0.002	0.009	0.034	0.068	0.108	0.15	0.166

$w = w_{id} = 4$. These fine-tuned hyper parameter used to develop an Abbasov and Mamedova FTS model considering data of 100 days followed by testing the results obtained on the next 45 days data to monitor the efficacy of the proposed approach and perceived that the model exhibited reliable and acceptable forecasting up to 45 days with both MAPE and Theil U well within set limits. In the next phase, we have considered the 145 days data, developed an Abbasov and Mamedova FTS model using the hyper parameter settings obtained from the tuned model and performed out of sample forecasting for the next 45 days. Upon comparison with the observed values, we noticed that the proposed FTS model worked reliably and effectively up to 35 days ahead of forecasting. Therefore, the proposed model used as an effective and reliable approach for forecasting the number of cumulative infections by COVID-19 in the USA.

5 Recommendation for Forecasting the COVID-19 Outbreak

The outcomes achieved by the FTS method to describe the COVID-2019 situation in the USA are satisfactory. Thus, the analysis of our proposed model using Abbasov and Mamedova FTS forecasting results is valid and reliable to estimate the pandemic outbreak pattern. In this context, authors of [20] used four time series forecasting, methods, namely ARIMA, the Early Aberration Reporting System (EARS), the Cumulative SUM (CUSUM), the Holt-Winters algorithm, for early detection of infectious epidemics in Korea. EARS with C3 syndromic surveillance techniques outperformed the four time series forecasting methods in terms of RMSE, symmetric MAPE (sMAPE), and Mean Absolute Deviation (MAD) value. In [21], authors investigated the performance of six different statistical time series modeling, such as ARIMA, the Holt-Winters additive model (HWAAS), TBAT, Facebook's Prophet, DeepAR, and N-Beats, to investigate the COVID-19 outbreak pattern in ten different countries and estimated seven days ahead confirmed cases data up to May 4th 2020. They compared the six different time series models for these ten countries and observed that ML-based time series methods offer more accurate estimates for a small portion of the population. In another work, authors of [22] employed time series analysis prediction models of the COVID-19 pandemic constructed on genetic programming (GP) for confirmed cases and death cases of Maharashtra, Gujarat, and Delhi in India. Based on the results, they observed that the proposed gene expression programming (GEP) based models are incredibly consistent and can be used as a benchmark for the COVID-19 cases in time series prediction methods in India. A summary of various estimation techniques, i.e., both statistical and time series, to portray the COVID-2019 outbreak pattern implemented by authors of [20–22] listed in Table 8.

Table 8 Disease outbreak estimation methods

Authors	Forecasting method		Dataset collection countries	Forecasting disease	
	Statistical	Time series		Pandemic	COVID-19
[20]	✔	✔	Korea	✔	
[21]	✔	✔	Ten countries	✔	✔
[22]		✔	India	✔	✔
Our work		✔	USA	✔	✔

6 Conclusion

The COVID-2019 outbreak has a global outreach, including the USA, where this unfortunate outbreak witnessed the height of infected cases affecting its day to day activities. Therefore, a proper forecasting model for perceiving reliable and accurate estimates of the USA's COVID-19 misfortunes is potentially substantial in reducing its negative impact on people, planning, and policy, including formulation and implementation of social and health-related policies critical for future planning and support authorities in combating this outbreak. In this study, we considered the 145 days' data of cumulative infected cases in the USA collected from WHO and the FTS approach employed to investigate and forecast the infected cases in the USA.

The model optimization is done by fine-tuning three hyper parameters, i.e., C, n, and w of Abbasov and Mamedova model. The fine-tuned model then used to forecast the next 45 days' infected data. We evaluated the forecast accuracy in terms of MAPE and Theil U statistics by comparing the forecasted data with original data. This new tuned FTS model used to forecast the next another 45 days infected data and again compared between the forecasted and actual values to obtain the new values of MAPE and Theil U. We observed that the forecasted MAPE is less than 10%, and Theil U value is close to 0, and on this basis, we conclude that our forecast model is reliable and acceptable. Furthermore, based on experimental results obtained, we observed that the forecasting performance of the proposed FTS model is reliable and acceptable for up to 35 days ahead of 45 days.

The proposed model is found to be a robust and reliable forecasting approach, potentially not limited to a country like the USA only, but also applicable for other countries, where this pandemic becomes uncontrolled. In conclusion, future modifications are possible to improve the FTS forecasting model's predictive accuracy by considering other aspects of pandemic outbreaks, which are indirectly or directly related to time series information.

References

1. D. Cavanagh, Coronavirus avian infectious bronchitis virus. Vet. Res. **38**(2), 281–297 (2007)
2. M.M. Ismail, A.Y. Tang, Y.M. Saif, Pathogenicity of turkey coronavirus in turkeys and chickens. Avian Dis. **47**(3), 515–522 (2003)
3. A.M. Zaki, S.V. Boheemen, T.M. Bestebroer, A.D. Osterhaus, R.A. Fouchier, Isolation of a novel coronavirus from a man with pneumonia in Saudi Arabia. N. Engl. J. Med. **367**(19), 1814–1820 (2012)
4. L.A. Zadeh, Fuzzy sets. Inf. Control **8**, 338–353 (1965)
5. C. Fatih, A. Hamimes, P. Mishra, Covid-19 statistics, strange trend and forecasting of total cases in the most infected African countries: an ARIMA and fuzzy time series approaches (2020)
6. P. Verma, S. Dwivedi, S. Dixit, Forecasting the COVID-19 outbreak: an application of ARIMA and fuzzy time series models (2020)
7. T. Zhang, X. Zhang, Y. Liu, Y. Luo, T. Zhou, X. Li, The analysis of infectious disease surveillance data based on fuzzy time series method. Int. J. Infect. Dis. **45**, 309–310 (2016)
8. S. Tricahya, Z. Rustam, Forecasting the amount of pneumonia patients in Jakarta with weighted high order fuzzy time series. IOP Conf. Ser. Mater. Sci. Eng. **546**(5), 052–080 (2019)
9. WHO I Novel Coronavirus—China, Situation report archived from WHO. https://www.who.int/emergencies/diseases/novel-coronavirus-2019/situation-reports. Accessed 2020/07/30
10. M.A.A. Al-Qaness, A.A. Ewees, H. Fan, M.A.E. Aziz, Optimization method for forecasting confirmed cases of COVID-19 in China. J. Clin. Med. **9**(674), 1–15 (2020)
11. H.H. Elmousalami, A.E. Hassanien, Day level forecasting for coronavirus disease (COVID-19) spread: analysis, modeling and recommendations. Quant. Biol. (2020)
12. S. Tiwari, S. Kumar, K. Guleria, Outbreak trends of coronavirus (COVID-19) in India: a prediction. Disaster Med. Public Health Prep. 1–6 (2020)
13. Kaggle, Novel corona virus 2019 dataset. https://www.kaggle.com/sudalairajkumar/novel-corona-virus-2019-dataset. Accessed 2020/04/03
14. C. Huang, Y. Chen, Y. Ma, P. Kuo, Multiple-input deep convolutional neural network model for COVID-19 forecasting in China (2020)
15. A.M. Abbasov, M.H. Mamedova, Application of fuzzy time series to population forecasting. Vienna Univ. Technol. **12**, 545–552 (2003)
16. H. Che-Ngoc, T. Vo-Van, Q.C. Huynh-Le, V. Ho, T. Nguyen-Trang, M.T. Chu-Thi, An improved fuzzy time series forecasting model, in *Econometrics for Financial Applications, ECONVN 2018. Studies in Computational Intelligence* (Springer, Vietnam, 2018)
17. T.T.N. Han, D.H. Nghi, M.T.H. Diem, N.T.D. My, H.V. Minh, V.V. Tai, P.M. Truc, *AnalyzeTS: Analyze Fuzzy Time Series*. R Package Version 2.3 (2019)
18. O.S. Oshodi, O.A. Ejohwomu, I.O. Famakin, P. Cortez, Comparing univariate techniques for tender price index forecasting: Box-Jenkins and neural network model. Constr. Econ. Build. **17**(3), 109–123 (2017)
19. R.Y. Fan, S.T. Ng, J.M. Wong, Reliability of the Box-Jenkins model for forecasting construction demand covering times of economic austerity. Constr. Manag. Econ. **28**(3), 241–254 (2010)
20. E. Yang, H.W. Park, Y.H. Choi, J. Kim, L. Munkhdalai, I. Musa, K.H. Ryu, A simulation-based study on the comparison of statistical and time series forecasting methods for early detection of infectious disease outbreaks. Int. J. Environ. Res. Public Health **15**, 1–18 (2018)
21. V. Papastefanopoulos, P. Linardatos, S. Kotsiantis, COVID-19: a comparison of time series methods to forecast percentage of active cases per population. Appl. Sci. **10**, 1–15 (2020)
22. R. Salgotra, M. Gandomi, A.H. Gandomi, Time series analysis and forecast of the COVID-19 pandemic in India using genetic programming. Chaos Solitons Fract. **138** (2020)

Electrical Power Demand Forecasting of Smart Buildings: A Deep Learning Approach

Samit Bhanja and Abhishek Das

Abstract In recent times, due to urbanization, the electrical power demand has increased rapidly, and that increases the greenhouse gases. So, there is a demand for optimum utilization of electrical power and that enforces to build a smart energy-efficient building. And to utilize the electrical power more optimally and efficiently, a more accurate and intelligent demand forecasting technique is also essential. In this work, a smart building energy management system (SBEMS) architecture is proposed. In this context, we also propose a deep learning model to forecast the electrical power demand based on weather conditions. The two most popular deep learning architectures such as 1D convolutional neural network (1D-CNN) and bidirectional long short-term memory (Bi-LSTM) recurrent neural network are merged to build the proposed forecasting model. The experimental results show that the proposed model produced more accurate results in comparison with other established forecasting models.

Keywords Deep learning · LSTM · CNN · Smart building · Electrical load forecasting

1 Introduction

In recent times, the use of the Internet and smart devices has increased very rapidly. Now, the Internet is everywhere, and Internetworking technology allows us to connect every object to the Internet. "Internet of Things (IoT)", [1, 2] allows the physical devices to communicate in real time without the intervention of human beings. Due to this IoT technology, billions of data are produced every second. If this large volume

S. Bhanja
Government General Degree College, Singur 712409, Hooghly, India
e-mail: samitbhanja@gmail.com

A. Das (✉)
Aliah University, New Town, Kolkata 700156, India
e-mail: adas@aliah.ac.in

V. E. Balas et al. (eds.), *Proceedings of International Conference on Computational Intelligence, Data Science and Cloud Computing*, Lecture Notes on Data Engineering and Communications Technologies 62, https://doi.org/10.1007/978-981-33-4968-1_6

of data is analyzed, then it will help to smoothen the daily livelihood of human beings and also helps to optimally utilize the available resources.

With the advancement of the smart grid system, it provides a footstep toward a trustworthy and preservable green energy system, [3, 4]. In recent times, most modern cities are adopted smart city, [5, 6] infrastructure to provide better livelihood concerning the social, environmental, and economic benefits to its citizens. One of the pillier of smart cities is the smart electrical energy management system. It will not only reduce the energy consumption of the end users but also helps to increase the uses of green energy. The home energy management system (HEMS), [7] provides a strong backbone and infrastructure for the smart electrical energy management system for the smart building.

Nowadays, especially in the cities, a large number of families live in a multi-storage building, and one of the primary resources required by them is electricity. If the resident of the building will not optimally utilize the electrical resource, then it leads to a wastage of resources. So, there is a need for an intelligent building management system that can automatically and intelligently utilize energy resources.

If the demand-side electrical load is known well in advance, then the electrical equipment can be managed in a better way to utilize the energy resources more optimally. So, the successful prediction of the electrical power demand plays a vital role in the smart building energy management system. The frequency of uses of the different electrical gadgets is varied with the weather conditions. Therefore, based on weather conditions, the electrical power consumption can be predicted.

In recent times, deep learning has become one of the popular tools for time-series prediction, [8]. There are numerous research work such as [9–13] has been done for electrical load forecasting. The author in [14] forecasts the electrical power demand of a building. Here, the author compares the performance of autoregressive integrated moving average (ARIMA) model with the autoregressive with exogenous output (ARX) model and shows that the ARX model overshadows the ARIMA model especially for the high order lagging. In recent times, deep learning-based hybrid models are developed for the time-series forecasting. In [15], the authors developed a hybrid model by combining two deep learning architectures, such as convolutional neural network (CNN) and long short-term memory (LSTM) recurrent neural networks to forecast the power demand. They show that the hybrid model performs better compared to the shallow deep learning models.

In this paper, a smart building energy management system (SBEMS) is proposed to intelligently use electrical energy. That will help to optimally utilize the energy resources and to support the increase of green energy resources. Here, as part of the SBEMS, a deep learning-based electrical power demand forecasting method has been proposed to forecast the electrical power load more accurately.

Rest of the paper is organized as follows: In Sect. 2, architecture of the proposed SBEMS is described. Proposed prediction module is presented in Sect. 3. The description about the dataset and the correlation analysis is done in Sect. 4. In Sect. 5, the detail experimental process and results are analyzed. The conclusion is presented in Sect. 6.

2 Proposed SBEMS Architecture

Figure 1 represents the architecture of the proposed smart building energy management system (SBEMS). This SBEMS consists of SBEMS center, smart meter, building energy storage system (BESS), communication system, and various smart devices. In this figure, the red color lines represent the communication line for the flow of control and data between the different components of the SBEMS, and the black color directed line represents the electrical power flow of SBEMS.

2.1 SBEMS Center

The SBEMS center is the brain of the proposed framework. It consists of three modules, viz. prediction module, monitor module, and management and control module.

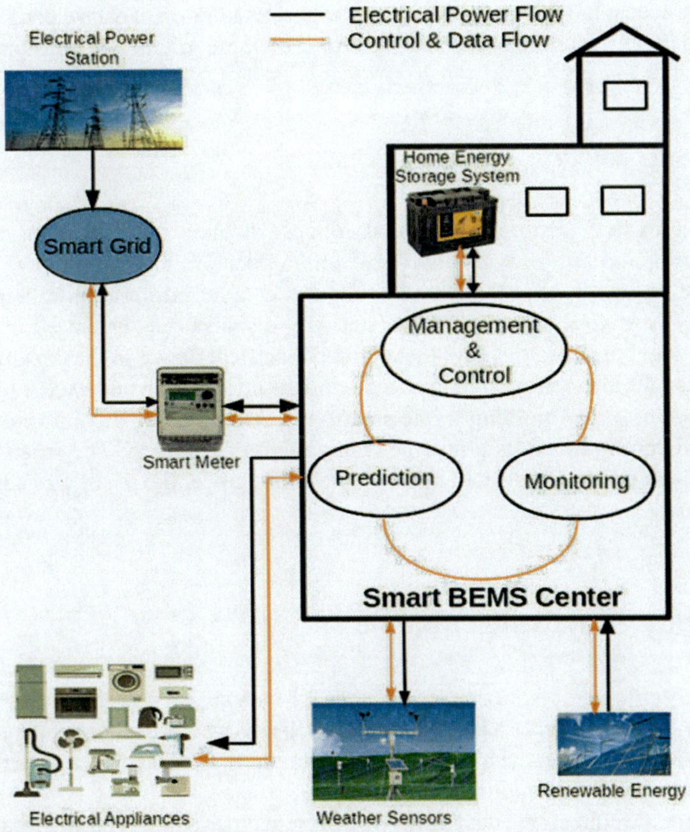

Fig. 1 The architecture of the SBEMS

The prediction module accepts the power consumption of various electrical equipment of the building, different weather conditions from various weather sensors, and renewable electrical power energy generated in the building for each time intervals. Now, based on these data, this module predicts the electrical power demands for the next time interval based on the weather conditions. It also predicts, how much renewable electrical energy can be generated in the next time interval.

The management and control module accepts information about how much energy is stored in the home energy storage system (HESS), [16] in the current time interval and also determines how much more energy can be stored in the HESS in the next time interval. After that, it sends information about energy requirements, and the excess renewable electrical energy will be generated in the building in the next time interval to the smart grid through the smart meter. This module also accepts the information from the smart grid through the smart meter regarding how much amount of energy the power grid can send to the building in the next time interval. This module is also optimized electrical power uses of the different electrical gadgets of the building based on the information received from the smart grid.

The monitoring module provides the display features. This display feature allows the user to access the real-time electrical power consumption of each electrical equipment. This feature will help the user to reduce their electric power consumption.

2.2 *Smart Meter*

Smart meter is an electronic device that not only displays the total electrical power consumption and also provides two-way communication to send the data and also to receive various types of instruction about real-time information quires. In this proposed framework, the smart meter not only displays the current electric power consumption, but it also displays the required electrical power in the next time interval. This meter also sends the power requirement information and excess renewable electric power of the building to the smart grid. After getting this information, the smart grid can optimize its power generation and distribution. The smart meter is also accepted information about how much electric power the smart grid can send to the building.

3 Proposed Prediction Module

In this section, we propose a prediction module for electrical power demand forecasting. This module is built as a part of our proposed smart building energy management system (SBEMS). Here, the module forecasts the short-term electrical power demand based on the weather conditions.

This module collects the data from two different types of sensors, viz. weather sensors and the electrical power consumption sensors. These sensors data are collected for a fixed time intervals. So, these data are the multivariate time-series data. If it is collected for T number of time intervals, then it is represented as $(X_1, X_2, \ldots, X_T)$, and each data X_i is also a vector of length k since they are collected from the k number of sensors.

This proposed work consists of two parts: data preprocessing and deep learning model construction.

3.1 Data Preprocessing

Since the data have been collected from two different types of sensors for each time interval, so in this work, two different datasets have been constructed, one for power consumption data collected from different electrical gadgets and others for weather-related data collected from the different weather sensors. Data are collected from different sensors, and there is a possibility of missing data. Here, all these missing values are replaced by the mean value of the data just before and after the missing value. Then for each time interval, the overall load of the building is calculated by summing up all the gadgets' power consumption of that time interval. This overall load dataset is treated as the target dataset of the proposed forecasting model.

There are a large number of weather sensors are available to sense the different parameters of the weather, viz. temperature, humidity, wind speed and direction, pressure, visibility, etc. Here the sensor optimization technique is adopted by reducing the number of sensors without affecting the overall weather condition. Here the correlation coefficient is calculated among all the weather parameters and based on these correlation coefficients, one of the parameters is removed from each pair of highly correlated parameters. Thus reduce the cost of the sensors. In this way, reduced weather dataset is generated. This reduced weather dataset is considered as the input dataset of the proposed forecasting model.

These time-series data can vary over a large range. So, to generate quality data by scaling down it to a small range, data normalization [18] is done.

3.2 Forecasting Model Construction

Here, the proposed forecasting model is described in detail. The proposed forecasting model is constructed by combining the two most popular deep learning architectures, viz. 1D convolutional neural network (1D-CNN) and bidirectional long short-term memory (Bi-LSTM) recurrent neural network as presented in Fig. 2. The first part of the model is the 1D-CNN block, the second part is the Bi-LSTM block, and the third part is the fully connected block.

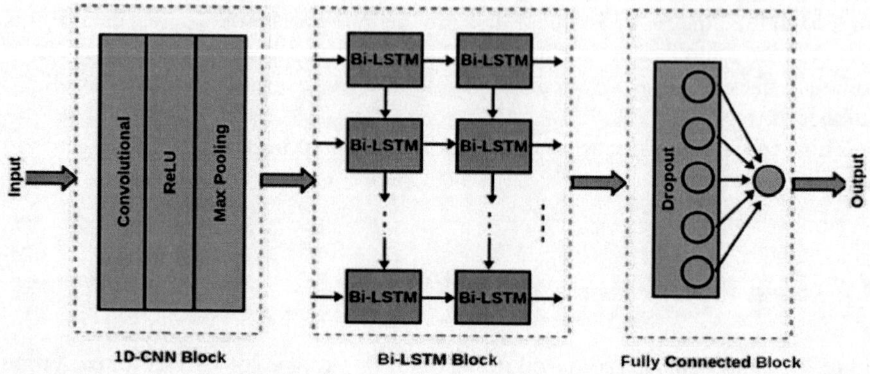

Fig. 2 Proposed forecasting model

3.2.1 1D-CNN Block

This 1D-CNN block consists of three different layers, viz. convolutional layer, rectifier linear units (ReLUs), and pooling layer. The convolutional layer performs the convolutional operation by striding the feature detector or filter. The output of the convolutional layers the output feature map. Next layer is the ReLU [17], and here, the nonlinear activation function $ReLU$ is applied on the feature map. The $ReLU$ function is as follows:

$$f(x) = \begin{cases} x, & \text{if } x > 0 \\ 0, & \text{if } x \leq 0 \end{cases} \tag{1}$$

In the pooling layer, the max-pooling function is used to pool-up the maximum from each sequence of output. Thus, the 1D-CNN block plucks the high-level features from each window of the input features matrix.

Bi-LSTM Block In this block, two LSTM layers have been created, and each LSTM layer can be expanded to the n number of bidirectional LSTM units along the x-axis. The main reason for selecting the Bi-LSTM units over the RNN is that LSTM can extract the features from a long sequence of historical time-series data in both forward and backward directions. It also overcomes the two major problems suffered by the standard RNN such as vanishing gradient problem and the exploding gradient problem. Bi-LSTM block takes the complex feature vector produced by the 1D-CNN block as the input, and it extracts the spatial and temporal features from this input feature vector.

Fully Connected Block In this block, all the output of the second layer of the Bi-LSTM block is connected with a single unit along with the dropout layer [19]. This dropout layer is added to overcome the overfitting problem. This overfitting problem is the most common problem suffered by almost all the machine learning and deep learning models. The output of the single unit of this module is the forecasted value.

4 Dataset Description and Correlation Analysis

For the simulation purpose of our proposed forecasting model, we use the "Smart Home Dataset with Weather Information" dataset, which is publicly available on the Kaggle Web site. This dataset contains the value of both the power consumption of different electrical gadgets and different weather parameters for each minute starting from 01-01-2016 12:00 AM. The dataset contains a total of 32 columns and 503910 rows. The column "house overall [kW]" represents the per minute overall power consumption of the house, and it is taken as the target vector of our forecasting model. This dataset also contains the weather information, and the weather parameters which we consider are as follows: "temperature", "icon", "humidity", "visibility", "summary", "pressure", "windSpeed", "cloudCover", "windBearing", and "dewPoint". Here, we generate two different parameters named "Clear" and "Rain_Snow" by manipulating the parameters "icon", "sum- mary", and "cloudCover". If the sky is clear, the value of the parameter "Clear" is set to 1 otherwise 0, on the other hand, the value of "Rain_Snow" is set to 1 for rainfall or snowfall otherwise 0. This weather-related data are considered as the input of the forecasting model.

Now, we convert the per minute dataset to an hourly dataset. Here, to generate the hourly dataset, we calculate per hour electrical power consumption by summing up the "house overall [kW]" power consumption for every 60 min. On the other hand, the hourly weather parameter's values are calculated by the mean value of each weather parameters for every 60 min. In this way, we generate the hourly dataset of 8398 h.

In the next step, to optimize the number of weather sensors, we analyze the correlation among all the weather parameters by using the Pearson correlation coefficient. The result of the correlation analysis is shown in Fig. 3.

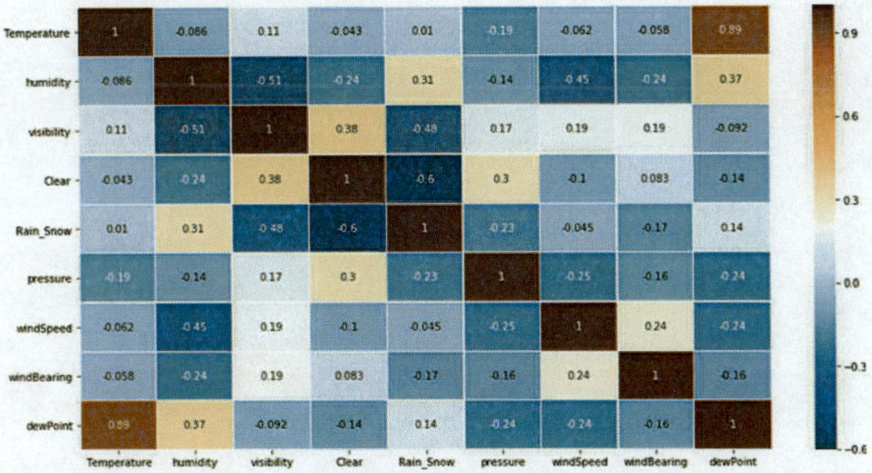

Fig. 3 Correlation coefficients among all the weather parameters

Here, the "Temperature" and "dewPoint" are highly positively correlated. Whereas "Clear" and "Rain_Snow", and "Humidity" and "visibility" are highly negatively correlated. So, to generate the optimized dataset, we remove the parameters "dewPoint", "visibility", and "Rain_Snow".

5 Experiments and Results

In this section, we describe in detail about the experiment and the results.

5.1 Experimental Setup

Here, the Python programming language and the Keras deep learning library are used for all the simulations.

Here, we find the autocorrelation coefficient of overall electrical power consumption, and the result is presented in Fig. 4. The figure shows that the same electrical power consumption pattern is repeated every 24 h. So here, the next hour's overall power consumption is predicted based on the previous 24 h of weather information. Here, the Tanh Estimator [18] data normalization technique is used. In the 1D-CNN block, ten filters are used with the filter size of 8. The striding length is set to 1, and the max-pooling window size is set to 2. In the fully connected layer, 20% of dropout is used to overcome the overfitting problem. Here, we used the ADAM optimization algorithm for the learning process of our model. We used 70% of the data (5878 h) as the training purpose, 15% of the data (1260 h) the validation purpose,

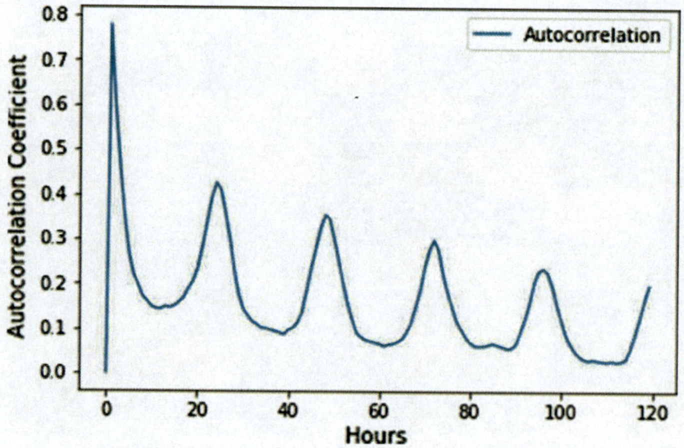

Fig. 4 Autocorrelation coefficient of hourly electrical power consumption

and the remaining 15% of the data (1260 h) for testing purposes of our model. In the validation phase, we depleted the k-fold cross-validation technique.

5.2 *Performance Metric*

To analyze the performance of our proposed work, we compare it with the other well-known models such as:

- **ARIMA**: Autoregressive Integrated Moving Average model.
- **ARX** [14]: Autoregressive with Exogenous Output model.
- **CNN-LSTM** [15]: The combination of convolutional neural network(CNN) and long short-term memory(LSTM) recurrent neural network model.

As the performance evaluation metric, here we use the root mean squared error (RMSE) and mean absolute percentage error (MAPE). The formula for calculating the errors is as follows:

$$\text{RMSE} = \sqrt{\frac{1}{n}\sum_{t=1}^{n}(\text{actual}_t - \text{predicted}_t)^2} \tag{2}$$

$$\text{MAPE} = \frac{1}{n}\sum_{t=1}^{n}\frac{|\text{actual}_t - \text{predicted}_t|}{\text{actual}_t} \times 100 \tag{3}$$

where actual_t and predicted_t are, respectively, the actual value and the predicted value at time-stamp t.

5.3 *Results*

In Fig. 5a–d, we plotted the actual electrical power consumption(kW) versus predicted electrical power consumption(kW), and the plot is zoomed for 150 h. From these figures, we can observe that our proposed model produces better results compared to the ARIMA, ARX, and CNN-LSTM models, particularly for the time-periods of wave peaks and wave valley. We also observe that the CNN-LSTM is superior compared to the ARIMA and ARX model, especially for the wave peak time-periods. In Table 1, we represent the testing errors of various models and the table depicted that our proposed model outperforms other models in terms of prediction errors. Our proposed model not only exhibits the lowest prediction error in terms of RMSE and MAPE, but it also produces a stable forecasting result even on the sudden rise and fall of the electrical demands.

Fig. 5 Forecasted results of different models **a** our proposed, **b** CNN-LSTM, **c** ARX, and **d** ARIMA

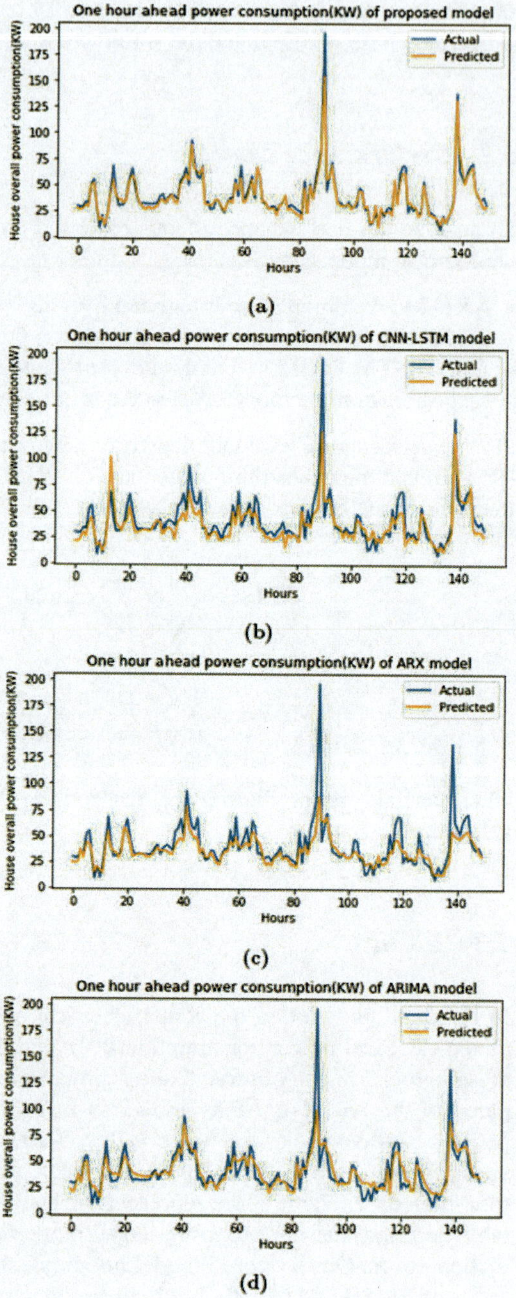

Table 1 Forecasting errors of various models

Forecasting model	RMSE	MAPE
ARIMA	1.1549	17.6383
ARX	0.9572	20.5206
Proposed model	**0.4401**	**9.3414**
CNN-LSTM	0.7880	13.3760

Bold values are used to highlight the results of our proposed work

6 Conclusion

In this study, a smart building architecture SBEMS is proposed to efficiently control and optimize the electrical power consumption of a building. Here also, in the context of this framework, an electrical load forecasting model is proposed based on the weather conditions.

The proposed smart building with SBEMS will play a vital role to intelligently use electricity with the demand response. This proposed architecture will also change the general concept of renewable energy utilization.

The proposed forecasting model has produced an excellent forecasting result and also outshined other well-established models. So, it can be concluded that the proposed smart building architecture (SBEMS) will definitely provide an improved way of electricity utilization and also proper utilization and distribution of the building's renewable energy.

As the future work, other parts of the proposed framework, such as management and control, monitoring, etc., will be developed.

References

1. J. Gubbi, R. Buyya, S. Marusic, M. Palaniswami, Internet of things (IoT): a vision, architectural elements, and future directions. Future Generation Comput. Syst. **29**(7), 1645–1660 (2013)
2. O. Hamdan, H. Shanableh, I. Zaki, A.R. Al-Ali, T. Shanableh, IoT-based interactive dual mode smart home automation, in *IEEE International Conference on Consumer Electronics (ICCE)* (2019), pp. 1–2
3. P. Haase, Intelli grid: a smart network of power. EPRI J. 17–25 (2005)
4. B.K. Bose, *Power Electronics in Renewable Energy Systems and Smart Grid: Technology and Applications* (Wiley, New York, 2019)
5. N. Komninos, C. Bratsas, C. Kakderi, P. Tsarchopoulos, Smart city ontologies: improving the effectiveness of smart city applications. J. Smart Cities **1**(1), 31–46 (2019)
6. S.P. Caird, S.H. Hallett, Towards evaluation design for smart city development. J. Urban Design **24**(2), 188–209 (2019)
7. Z.Y. Liu, *Basic Knowledge of Smart Grid* (China Electric Power Press, Beijing, 2010)
8. S. Bhanja, A. Das, Deep learning-based integrated stacked model for the stock market prediction. Int. J. Eng. Adv. Technol. **9**(1), 5167–5174 (2019)

9. Z. Yu, Z. Niu, W. Tang, Q. Wu, Deep learning for daily peak load forecasting-a novel gated recurrent neural network combining dynamic time warping. IEEE Access **7**, 17184–17194 (2019)
10. Y. Chen, P.B. Luh, C. Guan, Y. Zhao, L.D. Michel, M.A. Coolbeth, M. A., P.B. Friedland, S.J. Rourke, Short-term load forecasting: similar day-based wavelet neural networks. IEEE Trans. Power Syst. **25**(1), 322–330 (2009)
11. F.M. Bianchi, E. De Santis, A. Rizzi, A. Sadeghian, Short-term electric load forecasting using echo state networks and PCA decomposition. IEEE Access **3**, 1931–1943 (2015)
12. W. Kong, Z.Y. Dong, D.J. Hill, F. Luo, Y. Xu, Short-term residential load forecasting based on resident behaviour learning. IEEE Trans. Power Syst. **33**(1), 1087–1088 (2017)
13. R. Singhal, N.K. Choudhary, N. Singh, *Short-Term Load Forecasting Using Hybrid ARIMA and Artificial Neural Network Model, Advances in VLSI, Communication, and Signal Processing* (Springer, Singapore, 2020), pp. 935–947
14. K. Kandananond, Electricity demand forecasting in buildings based on ARIMA and ARX models, in Proceedings of the 8th International Conference on Informatics, Environment, Energy and Applications, vol.16 (2019), pp. 268–271
15. M. Kim, W. Choi, Y. Jeon, L. Liu, A hybrid neural network model for power demand forecasting. Energies **12**(5), 931
16. A.R. Al-Ali, A. El-Hag, M. Bahadiri, M. Harbaji, Y.A. El Haj, Smart home renewable energy management system. Energy Procedia **12**, 120–126 (2011)
17. V. Nair, G.E. Hinton, Rectified linear units improve restricted boltzmann machines, in *Proceedings of the 27th International Conference on Machine Learning (ICML-10)* (2010), pp. 807–814
18. S. Bhanja, S., A. Das, Impact of data normalization on deep neural network for time series forecasting (2018). arXiv preprint arXiv:1812.05519
19. N. Srivastava, G. Hinton, A. Krizhevsky, I. Sutskever, R. Salakhutdinov, Dropout: a simple way to prevent neural networks from overfitting. J. Mach. Learn. Res. **15**(1), 1929–1958 (2014)

Improved Performance of Recommender System Based on Demographic Attributes

Sushma Malik, Anamika Rana, and Mamta Bansal

Abstract In today's Internet era, human beings depend on the Internet for their daily work. With the increase of volatile growth of binary data and the number of visitors on the Internet, it has created data overload challenges and also increase the access time of information/data or products of interest on the Internet. Numbers of business opportunities are created with the implementation of the Internet and easily reach users. The Internet is employed to discover almost everything and anything meaningful information or product. Its 24 * 7 online service has provided the number of choices to the users due to which users faced the information overload problem. Internet users need some tools to search and fast access to appropriate information or products over the Internet. One such tool is called the recommendation system. Nowadays, major e-commerce Web sites are using RS to provide relevant suggestions to their customers. RS is generally used to suggest the most appropriate products to the customers. The functioning of the RS can be based on several parameters like item popularity on the Web site; characteristics of customers like the geographical location; or past buying behavior of customers. The RS is one of the tools that is used to increase the profit of Web sites and also retain the customers. The book RS is used by the Web sites to recommend the books to the customers on their interest. This paper presents the hybrid book recommendation system based on the collective character of *collaborative filtering (CF)* and collaborative-based (CB) filtering technique.

Keywords Recommendation system (RS) · Collaborative filtering (CF) · Content-based filtering (CB) · Hybrid technique · Knowledge-based (KB)

S. Malik (✉) · M. Bansal
Shobhit Institute of Engineering and Technology (Deemed-to-be-University), Meerut 250110, India
e-mail: sushmamalikiitm@gmail.com

M. Bansal
e-mail: mamta.bansal@shobhituniversity.ac.in

A. Rana
Maharaja Surajmal Institute of Technology, Janakpuri, New Delhi, India
e-mail: anamica.rana@gmail.com

V. E. Balas et al. (eds.), *Proceedings of International Conference on Computational Intelligence, Data Science and Cloud Computing*, Lecture Notes on Data Engineering and Communications Technologies 62, https://doi.org/10.1007/978-981-33-4968-1_7

1 Introduction

With the expansion of the Internet, human beings are using it to fulfill their daily needs. The Internet also acts as the best place for the grooming of e-commerce business globally. E-commerce Web sites have piles of products independent of physical boundaries. The entire products have their information over the Web sites and daily new products come into the virtual market. Identifying useful products from a thousand numbers of products is becoming a very complex task for the customers. So the customer is a strain with the product's information [1]. To overcome this information overload problem, many e-commerce sites are using the RS. It is very difficult for a single customer to use every available product on Web sites. This problem is solved by the RS because these systems recommend the products to customers concerning interest. RS is a software tool that helps the customers to find the right product for them by using their interest and using the experience of other customers [2]. RS can be the combination of a user; user interfaces data set and recommendation algorithms. In other words, RS can be said—for the user, to the user, and by the user [3]. This is the main motive of RS acceptance by e-commerce businesses. Customers articulate their opinion about online services and products in the form of ratings. The RS recommends the goods and services to them and other users by using these ratings. The main motive of the RS is not to retain the existing customers but also facilitate in attracting the new royal customers. It also assists to increase the revenue of online businesses and increase sales over their competitors. Recommender systems play a vital role in various domains by minimizing information overload problems like e-commerce, movies, e-learning, music, research articles, news, and books [4].

2 Literature Review

RS opens the innovative prospect of recover information on the Internet and also facilitates to resolve overload problems of information in the digital era and enables the users to access the items and services. CB, CF, and KB are the classic techniques of recommendation but they have some problems when used individually like cold start, diversity, data sparsity, etc. A hybrid technique is more used because it solves the problems of individual techniques by combining classic techniques [5]. The sparsity problem of the library book RS can be solved with the probabilistic-keyword CF method [6]. Personalized RS is designed to improve the utilization of book resources with a hybrid algorithm which is a grouping of CF and CB technique [7]. To develop the hybrid recommendation system, CF and association rule are used and also minimize the sparsity and cold start problem [8]. RS helps to retrieve useful information from an ocean of data with minimum efforts and time. CB and CF are mostly used techniques to design any kind of RS, but these techniques have several individual problems like cold start and synonym. These problems are overcome by the hybrid technique [3]. Proposed a book RS which is based on the identification of user opinion and social network by using the interest related to user like preferences, age, and also online accessible attributes on social networks. This approach helps to sort out the cold start problem by recommending the books to the new user in the nonexistence or user purchase history and this approach is also used for sorting out the item cold start problem when no one user had rated the items till now [9]. The proposed hybrid RS investigates the user behavior by using the features of various recommendation techniques like as CB, CF, and demographic [10, 11]. RS has become more popular in current times due to the overload problem of data on digital platforms and its wide area of applications. DBSCAN algorithm is proposed to recommend books based on user nature. The topic model-based (TMB) book RS used to improve the existing RS which suggests the books to users without a history of rating. In this RS, a profile of a user is created that stores all the subjects that are discussed on social platforms. Profile of user is matched with book descriptions and the most similar ones are suggested to the user [12]. Designed a hybrid book RS which combines the CF approach and association rule of data mining to suggest more precise and appropriate books to the reader on their needs [13]. A number of data mining techniques are used to develop the hybrid RS. The system will provide the books based on user interest and it analyzes the behavior of users by implementing the various recommendation algorithms like content-based, collaborative, and demographic. Hybrid book RS is designed with collective features of collaborative filtering, content-based, and association rule of mining. The content of the book, the quality of the book, and ratings given by other users are the main parameters that are used to design the system. The designed system is used offline that is why the system does not have performance problems [14]. To increase the performance and quality of the recommender system, NOVA merges the features of CF, CB, and many data mining techniques to find interesting useful suggestions from a huge amount of data sets. NOVA shows a better performance and quality results as compared to other techniques [15]. The book RS mines strong association rules based on improved Apriori algorithm and this improved algorithm help to mining the strong associations between books and also uses the mining result to recommend the books to the user based on their interest [16]. The book RS recommends the books to the user on their interest and store recommendations in the user Web profile. With the help of cloud computing, information and recourses are shared among the nodes. Cloud computing is the collection of nodes that are connected through the Internet. This emerging technique provides parallel and distributed computing with virtualization features. Nowadays, most of the portals and e-commerce sites are using cloud computing techniques for the implantation [17]. Designed the book RS with a number of parameters like the content of the book and quality of the book and recommend the book to the user by doing CF of ratings given by other users. Associative rules of mining are also used to design the RS who provides a stronger recommendation to the user [18].

3 Problem Statement

Recommender systems will assist the customers who do not have adequate knowledge about the products or online services on Web sites. It helps the customers to make decisions as to which item or services to purchase. Available book RS is based on either CB or CF for recommending to their customers. The proposed system uses the combination of CF and CB techniques to design the hybrid recommendation technique to propose the books to the customer on their interest. With the usage of these techniques, the cold start and data sparsity problem of RS can be minimized.

3.1 Methodology

This book recommendation model is based on the hybrid recommendation technique which is a combination of CF and CB technique along with demographic attributes like age, gender, education, etc., of the customer.

Collaborative Filtering Technique
Human beings always take the view of friends and family members before the purchase of any product due to their human nature. Collaborative filtering techniques recommend the product to the customer as a result of the views of other customers who have the same kind of interest. The main point to use this technique is that it is completely independent of any machine-readable representation of products being recommended. It also performs well on complex objects like music, movies, and books where taste or interest vary from customer to customer [19]. CF is the most popular technique and famous approach linked with RS. This approach recommends the products on the basis of the user needs [20].

Slop One Algorithm
Slop one algorithm uses two types of information; firstly, the other customers who rated the same book (like the adjusted cosine item based), and secondly, from the other books rated by the same customer (like the per usage average) [21]. It is used for the prediction of ratings and analyzes the average difference between the different books and their ratings along with the number of ratings for each book. The predictions are united by a weighted average if the customer has rated lots of books [2].

Content-Based Technique
The content-based technique recommends the goods to the customer concerning the buyer's past product buying history. The buying history of the customer provides the content of the goods in which the user is interested and selects the product from the pool of products. With this history, the taste of customers is easily identified. In this technique, products are recommended to the customer on the basis of customer characteristics [4]. It is based on customer likes and dislikes and hashing is used to map similar products to the same bucket with high probability [22]. To fulfill the

requirement, the MinHash function is used to find the similarity among the customers. The main motive of this system is to find similar customers and recommend the book rated by those similar customers.

Hybrid Technique

The hybrid technique combines the CF and CB techniques to provide a more effective recommendation to the customer. The hybrid technique uses the collaborative technique for the recommendation of books and finds out the customers who rated those books. Customers are differentiated based on the demographic features of customers like age, gender, etc. Now, the content-based technique is used to compare the filtered customers to find out the similarity with current customers. One more filter is used to find categories of books rated and viewed by selected similar customers in the previous step. With this technique, lots of comparisons of customers are reduced and the system provides the relevant recommendation to the customer.

Parameters Used for Analysis

Accuracy, precision, and recall are the three matrices that are used to evaluate the recommendation technique and also measure the performance of the system. Accuracy can be a workout as:

$$\text{Accuracy} = \frac{\text{Relevant Recommendations}}{\text{Total Possible Recommendations}} \tag{1}$$

Equation (1) is showing the ratio of relevant recommendations over the total possible recommendations and accuracy is the most intuitive performance measure in the recommendation. Precision can be a workout as:

$$\text{Precision} = \frac{\text{Relevant Recommendations}}{\text{Total Recommendations}} \tag{2}$$

Equation (2) is showing the ratio of correctly predicted positive recommendations to the total predicted positive recommendations. Recall can be a workout as:

$$\text{Recall} = \frac{\text{Relevant Recommendation}}{\text{Total Relevant Recommendations}} \tag{3}$$

Equation (3) is showing the ratio of correctly predicted positive recommendations to all recommendations in the actual class.

Comparative analysis of collaborative filtering, content-based, and hybrid techniques of recommendation has been done on account of accuracy, precision, and recall parameters.

Data Set

For the experiment, the used data set consists of files like ratings.csv, books.csv, book_tgs.csv, and tags.csv which has been taken from the Kaggle.com. The Rating.csv file contains the ratings which are given by the users. Book_id, user_id, and ratings are the attributes of the ratings.csv file. Books.csv contains information about books. Book_id, best_book_id, authors, publications, ISBN, title are the main attributes of the books.csv file. The Book_tags file contains the tag_id users have assigned to the books. Goodread_book_id, tag_id, and count are the attributes of book_tags.csv files while the tags.csv file contains the tag_names corresponding to the tag_ids.

Recommendation Techniques Are Compared Based on Accuracy
For the comparative analysis of techniques, we have executed each technique separately on the same data sets. Figure 1 represents that the hybrid technique performs well in comparison with the other two techniques when the number of customers is increasing. The result of the hybrid technique is best compared to other techniques because of the recommendation of products to the user is based on demographic attributes and customer preference (Table 1).

Recommendation Techniques Are Compared Based on Precision
The hybrid technique of recommendation has higher precision as compared to CF and CB techniques. Figure 2 represents the performance of the hybrid technique

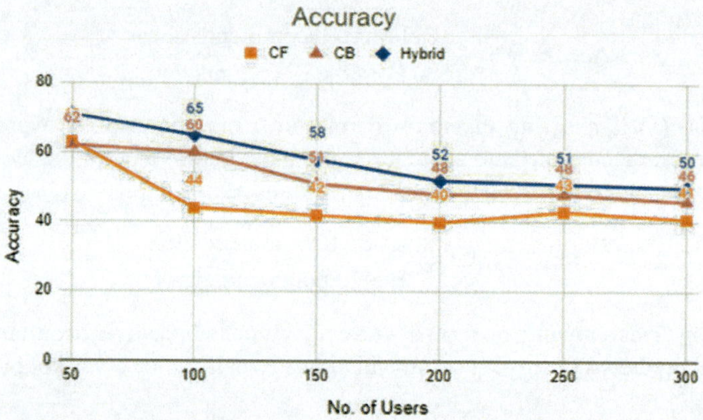

Fig. 1 Comparison of recommendation techniques based on accuracy

Table 1 Comparison of recommendation techniques based on accuracy

Accuracy			
No. of users	CF	CB	Hybrid
50	63	62	71
100	44	60	65
150	42	51	58
200	40	48	52
250	43	48	51
300	41	46	50

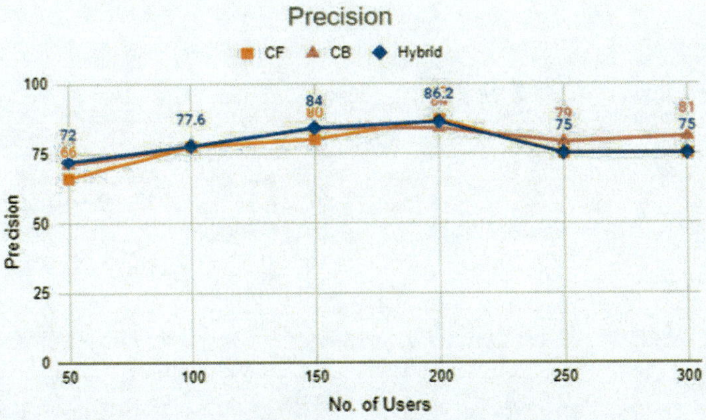

Fig. 2 Comparison of recommendation techniques based on precision

as compare to other techniques when the customer's numbers are increased. This result can happen because the hybrid technique overcomes the negative aspect of the other two techniques. CF recommends the books based on ratings given by the customers and CB technique recommend based on book attributes or features, but hybrid technique merges the features of both techniques and also enhance the accurate recommendation of products to the user on their needs. That is why the precision of the hybrid technique is more as compared to other techniques (Table 2).

Recommendation Techniques Are Compared Based on Recall
To calculate the recall, the relevant recommendations are used. Figure 3 represents the hybrid technique of recommendation has high recall as compared to CF and CB technique because the hybrid technique overcomes the negative aspects of the other two techniques by providing more relevant recommendations. When the number of customers is increased, the performance of the hybrid technique is much better as compared to the other two techniques.

The hybrid technique gave a more appropriate recommendation to the user by sorting the recommendation based on demographic attributes and also on the basis of customer preferences. With that recall of hybrid, the technique is increased

Table 2 Comparison of recommendation techniques based on precision

Precision			
No. of users	CF	CB	Hybrid
50	66	71	72
100	77.5	77.4	77.6
150	80	84	84
200	87	84	86.2
250	75	79	75
300	75	81	75

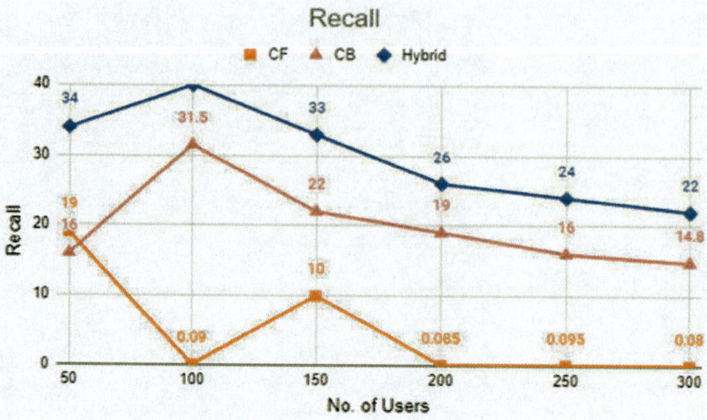

Fig. 3 Comparison of recommendation techniques based on recall

Table 3 Comparison of recommendation techniques based on recall

Recall			
No. of users	CF	CB	Hybrid
50	19	16	34
100	0.09	31.5	40
150	10	22	33
200	0.085	19	26
250	0.095	16	24
300	0.08	14.8	22

with this result, it observed that filter of recommendations based on demographic attributes and customer preferences; the hybrid technique has given the more relevant recommendations, thereby increasing the recall of hybrid technique (Table 3).

4 Conclusion

In the Internet era, e-commerce sites become an essential part of the human being in the fast going life. The functioning of e-commerce sites is based on the recommendation technique used to design the site. The RS is an integral part of the e-commerce sites because these sites need fast and relevant retrieval of desired information in very less time. E-commerce sites have huge data and RS are needed to retrieve the relevant data after processing. The main motive of RS is to predict customer interest concerning data and recommend the product accordingly. Now, e-commerce sites also provide lots of books to the customer but it is very difficult for the customer to select appropriate books from an ocean of books, but the RS recommends the books to the customer concerning needs and interest. Aiming to help the customer to defeat the problem, a hybrid technique is proposed which is a grouping of CF and CB technique along with demographic attributes to provide the superlative commendation of the books to the customer. Demographic attributes are applied to find the correlations between the customers to get recommendations and also solve the cold start problem for new customers and new books. The research demonstrates that a hybrid technique performs well as compared to traditional techniques like CF and CB technique.

References

1. A.S. Tewari, Generating items recommendations by fusing content and user-item based collaborative filtering, in *International Conference on Computational Intelligence and Data Science* (2020)
2. M. Chandak, S. Girase, D. Mukhopadhyay, Introducing hybrid technique for optimization of book recommender system, in *International Conference on Advanced Computing Technologies and Applications* (2015)
3. S. Malik, M. Bansal, Recommendation system: techniques and issues. Int. J. Recent Technol. Eng. (IJRTE) **8**(3), 2821–2824 (2019)
4. T. Anwar, V. Uma, CD-SPM: cross-domain book recommendation using sequential pattern mining and rule mining. J. King Saud Univ. Comput. Inf. Sci. (2019)
5. S. Malik, A. Rana, M. Bansal, A survey of recommendation systems. Inf. Resour. Manag. J. **33**(4) (2020)
6. N. Ifada, I. Syachrudin, M.K. Sophan, S. Wahyuni, Enhancing the performance of library book recommendation system by employing the probabilistic-keyword model on a collaborative filtering approach, in *4th International Conference on Computer Science and Computational Intelligence 2019* (2019)
7. Y. Tian, B. Zheng, Y. Wang, Y. Zhang, Q. Wu, College library personalized recommendation system based on hybrid recommendation algorithm, in *11th CIRP Conference on Industrial Product Service System* (2019)
8. A.E. Patil, S. Patil, K. Singh, P. Saraiya, A. Sheregar, Online book recommendation system using association rule mining and collaborative filtering. Int. J. Comput. Sci. Mob. Comput. **8**(11), 83–87 (2019)
9. H. Pasricha, S. Solanki, A new approach for book recommendation using opinion leader mining, in *Emerging Research in Electronics, Computer Science and Technology. Lecture Notes in Electrical Engineering* (Springer Nature, Singapore, 2019), pp. 501–515
10. S. Patil, V. Namdeo, A proposed hybrid book recommender system. Int. J. Comput. Appl. **6**(6), 14–18 (2016)
11. S. Bhosale, P. Nimse, S. Wadgaonkar, A. Yeole, *SuggestABook: A Book Recommender Engine with Personality Based Mapping*, vol. 159, no. 9 (2017)
12. H. Alharthi, D. Inkpen, S. Szpakowicz, Unsupervised topic modelling in a book recommender system for new users, in *SIGIR* (2017)
13. Z. Ali, S. Khusro, I. Ullah, A hybrid book recommender system based on table of contents (ToC) and association rule mining, in *INFOS* (2016)
14. S. Rajpurkar, D. Bhatt, P. Malhotra, Book recommendation system. Int. J. Innov. Res. Sci. Technol. **1**(11), 314–316 (2015)
15. D. Pathak, S. Matharia, C.N. Murthy, NOVA: hybrid book recommendation engine, in *3rd International IEEE Conference on Advance Computing Conference (IACC)* (2013)
16. Y. Zhou, Design and implementation of book recommendation management system based on improved apriori algorithm. Intell. Inf. Manag. **12**(03), 75–87 (2020). https://doi.org/10.4236/iim.2020.123006

17. D. Patel, J. Dangra, A book recommendation system for cloud computing: a survey. Int. J. Comput. Appl. **167**(5), 33–36 (2017)
18. A.S. Tewari, A. Kumar, A.G. Barman, Book recommendation system based on combine features of content based filtering, collaborative filtering and association rule mining, in *International Advance Computing Conference* (2014)
19. A. Sase, K. Varun, S. Rathod, D. Patil, A proposed book recommender system. Int. J. Adv. Res. Comput. Commun. Eng. **4**(2), 481–483 (2015)
20. K. Davagdorj, K.H. Park, K.H. Ryu, *A Collaborative Filtering Recommendation System for Rating Prediction* (2020), pp. 265–271
21. D. Lemire, A. Maclachlan, Slope one predictors for online rating-based collaborative filtering, in *SIAM Data Mining* (2005)
22. A.B. Bagwan, P. Singh, M. Sarkar, S. Satpute, S. Shirsth, Digital library using hybrid book recommendation engine. Int. J. Adv. Electron. Comput. Sci. **4**(3), 131–132 (2017)

A Novel Method for Hostility Management

Sai Tiger Raina, Amol Mathur, Charu Goyal, and Vedika Gupta

Abstract With the rise of social networking websites, there is a wave of user opinions expressed in text, image or different content. Different words and images are used to express one's personal feeling, but sometimes words and sentences are not used properly which results in an offensive or inappropriate language. Therefore, there is a need to look into the problem. There are many techniques to mine the sentiment from the most well-known social media, Twitter, where users post real-time reactions and opinions relating to everything. This is done by first analyzing the words properly by semantic analysis technique. Throughout this paper, we've a tendency to create a hybrid approach to exploitation of every word primarily based and lexicon-based types in which to figure out the linguistics orientation of the opinion words in tweets. Our aim is to optimize the classification of sentiments and prevent increasing hostility on social media platforms. Here, we used the machine learning technique to classify the text. Hostility management is finding out if a text is hostile or not. It can occur anytime or at any place. For cleaning text, we use lemmatization and stemming. To extract the features, we have different techniques in consideration that are bag of words, tfidfvectorizer, Word2Vec, Doc2vec and TF-IDF. We have developed an original technique that mixes existing approaches, providing the simplest coverage results and competitive agreement.

Keywords Bayesian algorithm · Hybrid approach · Hostility management · Word embedding

---

S. T. Raina · A. Mathur · C. Goyal · V. Gupta (✉)
Department of Computer Science and Engineering, Bharati Vidyapeeth's College of Engineering, Delhi, India
e-mail: vedika.gupta@bharatividyapeeth.edu

S. T. Raina
e-mail: sai.raina@gmail.com

A. Mathur
e-mail: amolmathur4869@gmail.com

C. Goyal
e-mail: charugoyal180@gmail.com

V. E. Balas et al. (eds.), *Proceedings of International Conference on Computational Intelligence, Data Science and Cloud Computing*, Lecture Notes on Data Engineering and Communications Technologies 62, https://doi.org/10.1007/978-981-33-4968-1_8

1 Introduction

Machine learning is a growing field that is tremendously being used nowadays to solve various tedious problems. Also, with the data being unprecedently growing and adding to big data [1], human intervention is cumbersome and time-taking process. The machine can go in the past and learn a behavior from its past and can perform in the same way to make predictions in the future with good precision. Human intervention is kept to the least as the machine does automatically what it has to do. In this data that is observed or accessed from the past experience and then used to find patterns that may help us in finding a solution to the problem.

Hostility management is a way of analyzing the text and extracting the sentiment to understand the emotion of the text. It can be done by analyzing the text and identifying a pattern and repetition of words that are usually associated with hostile behavior. It helps in identifying if a text is in any way hostile which is useful to keep in check cyber bullying, bullying at the workplace or even in educational institutes [2]. By hostile here we mean showing or feeling opposition or dislike; unfriendly. Hostility can occur any time whether inside or outside the workplace. Threat assessment professionals and forensic psychologists can use a variety of methods to predict who has violent nature, by determining the time, place but "if it will happen" remains far from certain [3].

To do hostility management, the semantic keywords play a crucial role in getting the information and for natural language processing. But measuring the similarities and differences between two words is a tedious task as similarities between words and emoticons change during time and across domains [4].

This can be done by using lemmatization as it is used for cleaning the text data. Another approach to do so is stemming which follows an algorithm which makes the process faster. It is a method in which all words with the same stem or root are reduced to a common form done by stripping each word in which stop words are removed. To remove these, we use the NLTK library which contains a set of stop words for the English language. Another thing we need to do is reduce the words to their basic format like love, loves and loved all become love as all of them convey the same sentiment. On comparing the two, we found out that lemmatizing gives a better output as compared to stemming as stemming sometimes does crude reduction of words [5, 6].

For extracting the features, we have different approaches like bag of words, TF-IDF, word2vec and doc2vec. For achieving the output, we used data from different source domains which helps in increasing the performance in the target domain. The common approach to do the task uses a bag-of-words feature set, but we choose TF-IDF because it is a general-purpose technique as it efficiently weighs word scores before the classification occurs as well as it works efficiently on the big dataset as compared to others feature extractions techniques. And we also used doc2vec and word2vec models on data from both source and target domains combined., then using them to calculate the average word vectors of all the words and then use it as input as a feature to classify our training.

Hostility management finds a plethora of applications in business since the Internet is an "Opinion Minefield" with unlimited potential for their use [7, 8]. A big one is brand monitoring, wherein whenever a customer mentions your brand name, you should know whether it is positive or not [9, 10]. To put this into perspective, when NFL quarterback Colin Kaepernick was announced to be a part of Nike's new marketing campaign, the sudden public response was disturbingly negative. A research indicated that more than 25% of the customers discontinue a product after a single bad customer experience. It should be no surprise that sentiment analysis plays a key role to prevent customers leaving. Monitoring product analytics would assist each company to decide what's working and what's worth changing. For a new startup, it is essential to observe market trends to launch successful products and for bigger companies to keep a track of dissatisfaction toward existing technologies and practices to be able to adapt to newer trends. Also monitoring user generated content would let you navigate through nascent stages of product development [11]. Apart from the aforementioned points, it should not hard to understand why it plays an important role to keep an eye out on the competitors since it ties in with the customer's perception of your brand with respect to your competition and gives an accurate representation as to how your marketing strategies stack up against theirs. And last, but certainly not least, is crisis management; there's nothing worse than discovering a problem too late. It's possible to detect when there's a sudden deluge of negativity surrounding your product since hostility management can work in real time. Promptly tackling the issues shows the customers how brisk the brand is, at resolving issues and injecting some positivity back into the influx of negativity. A concept as versatile as this is, by no means, limited to the miniscule number of applications mentioned in this paper.

2 Related Works

Back-propagation neural networks have some disadvantages as it trains the model slowly so another algorithm proposed a modified back propagation neural network which helps to increase the training speed. Earlier the model used for text categorization was a vector space model which needs high space to represent the data, and it also does not include the relationship between the words or terms which leads to poor accuracy so to avoid this another technique was used known as latent semantic analysis. As this technique instead of working on individual words, it creates a vector space for every word which helps for better understanding of the relationship between words [12].

From the next research paper, we get to know about the word embedding technique which helps to represent document vocabulary, and to capture context of words. According to Mikolov, the quality of words is calculated in a word similarity task, and after that the outcomes are compared to the previous outcome which were supposed to be the best based on the various types of neural network techniques. It also refers to that doc2vec is a more advanced version of word2vec. The doc2vec performs zealously when we train the model on large external corpora and then it can be improved by pre-trained word embeddings. From this paper, we get to know that the main problem is that information is not sufficient and the sparse format is not structured properly. To overcome this, semi-supervised learning and multiple document representation can be used. To increase the various types of feature set for classification, we can have three document representation methods: TF-IDF (term frequency-inverse document frequency) which follows bag-of-words scheme, another is Latent Dirichlet Allocation (LDA), and last Doc2vec (Document to vector) [13–15].

The next research papers explore long short-term memory (LSTM). Both these papers covered how LSTM is useful from the previous algorithm for recurrent neural networks (RNNs). Long short-term memory (LSTM) networks are a special kind of RNN, capable of learning long-term dependencies. LSTMs are explicitly designed to avoid long-term dependency problems. Remembering information for long periods of time is practically their default behavior [14]. The main aim of sentimental analysis is to detect numerical values and then classify as a binary form that is positive and negative. LSTM consists of both the target-level attention and the sentence-level attention. We also get to know about the disadvantages of this algorithm. In this paper, we also infer about the techniques of feature extraction technique, one of these is XGBoost. It processes only the input streams and does not mark sequence ends [16, 17].

In the next research paper, the author looked at the most popular sentimental analysis of Twitter and many bore social networking sites [18]. Authors in [4, 19] classified sentiment into positive, negative, and neutral classes. They also tried it with a model called unigram, feature-based model and last a tree kernel-based model. They tried it by combining two models with their features that were unigram and tree kernel. This was done with the help of two methods. Firstly, they classified the reply or messages into two categories: subjective and objective. Doing this they do not use manually annotated data to compose the training data as regularly supervised approaches. For better results, it is crucial to verify the potential value of using and then combining them. In a more efficient way, the author also proposed a feature set which captures an abstract overview of tweets and then it is composed by meta-information associated with words and analysis the specific characteristics of the pattern of how tweets are written. After this the author tried to evaluate the usefulness of existing lexical resources. Now, for building the training data they used the supervised learning approach, but for the use of hashtags in Twitter data, they used different corpora, and they used hashtag dataset (HASH) and the emoticon dataset (EMOT). They have done this by using three steps: tokenization, normalization and part of speech (POS), and with it they used various other features too like POS, but POS features were not as much useful for sentiment analysis in micro-blogging domain. Then after this, the author classifies the sentiments in positive and negative by using a novel approach with respect to query term. So this helps them a lot and they build a framework that treats classifiers and feature extractions as two different components. So, they were able to try different combinations of classifiers and feature extraction.

3 Experimental Analysis

In a language, a word can exist in various forms that actually carry the same meanings, usually being the derivatives of the same word. For example, "car" and "cars" fundamentally have the same meaning. Reducing this inflection in words is called Text Normalization or Word Normalization. It can be done in two ways, namely Stemming and Lemmatization, both differing in the way they accomplish the task. Stemming is the method of inflection reduction that reduces a plethora of words to a single stem, irrespective of whether or not that stem is comprehensible in the language. Here, stem refers to the part of a word to which inflectional suffixes are attached. For the English language, we can use Porter Stemmer or Lancaster Stemmer, which is a more aggressive form of Porter Stemmer. Lemmatization, unlike Stemming, reduces the inflections ensuring that the root is a comprehensible part of the language. A lemma is the canonical form, dictionary form, or citation form of a set of words. For sentiment analysis, it is better for the root we obtain after Text Normalization to have meaning in the language [5, 6, 14].

In branches of Computer Science that require high processing power like Machine Learning, Pattern Recognition and Image Processing, Feature Extraction must be done because the algorithms, especially the Machine Learning algorithms prefer well-defined fixed-length inputs and outputs. These algorithms cannot be fed raw text, it must be morphed into, more specific vectors of numbers. A method of doing the aforementioned is by using a bag-of-words model. A bag of words is a representation of text that lays down a vocabulary of known words and a measure of their presence. An increase in the vocabulary size means an increase in the vector representation. For a collection of books, the number of vectors may reach an unexaggerated figure in millions. It is of the essence to manage the vocabulary to avoid the formation of sparse vectors that require large memory and computational resources. Pattern features classifies the words into three classes—High-Frequency Words (HFW), Content Words (CW) and Regular Words (RW). Calculation of Word Frequency is done for the words in the dataset, not from an external corpus. HFWs typically consist of a sequence of punctuation marks along with URL, REF, TAG and RT meta-words for pattern extraction since they have a key role in pattern detection.

Punctuation feature divides the last feature type to make five generic features—word length of tweet, frequency of exclamation points, frequency of question markers, frequency of quotation points and frequency of uppercase words. The weight of the Punctuation Feature, denoted by wp, can be expressed as:

$\mathrm{wp} = (3 * \mathrm{Np})/(\mathrm{Mp} * (\mathrm{Mw} + \mathrm{Mng} + \mathrm{Mpa}))$, where Mw, Mng and Mpa represent maximal values of word, n-gram and pattern feature groups (Fig. 1).

We used a list containing 18,000 words, for key-based features, having sentiment strength falling in range $(1, -1)$ and we calculate key-based feature weight according to these word strengths. TF-IDF stands for term frequency inverse document frequency is a numerical statistic designed to demonstrate how important a word is to a document in a corpus. It can be implemented using one of the two

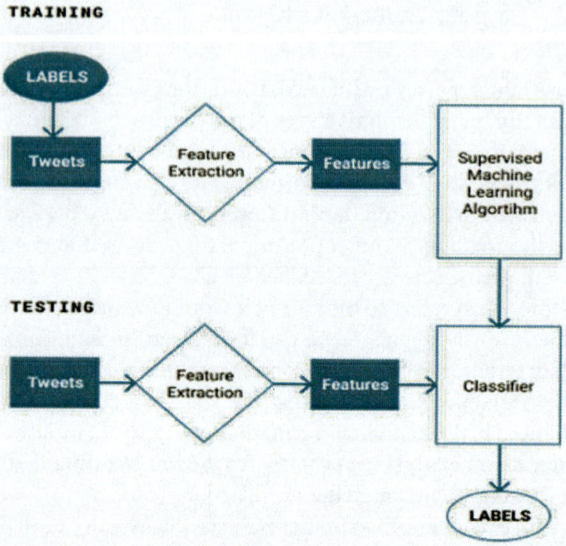

Fig. 1 Feature extraction

modules of Python—tfidftransformer and tfidfvectorizer. Tfidftransformer chronologically calculates word count using CountVectorizer, Inverse Document Frequency and only after that the tf idf scores. Tfidfvectorizer executes all three steps at once making it the better choice for our project [20, 21].

Word Embedding is yet another popular technique for representation for document vocabulary, possessing the ability to capture context of a word-syntactic and semantic similarity, relation with words and so on. Word Embedding is the representation of words into vectors, but the point to note is how it is done. Word2Vec is one of the most popular techniques to learn Word Embedding using a shallow neural network and can be obtained by skip gram and continuous bag of words (CBOW). The CBOW model is fed the context of a word as input and predicts the word corresponding to the context. It takes a C number of context words as input to predict the next word. The alternative approach to this is to use the context that we obtain as the output in CBOW, as the input, we get the skip gram model. Both have their pros and cons but which is the better choice, according to our research and experiments, the skip gram model works brilliantly for small data and appears to be able to represent rare words as well, whereas CBOW is faster and has more efficient representations for more frequent words [15].

It was a tough decision to choose the metric among precision, recall, F-Score and Accuracy, to measure and compare the performance of the various models. Precision, which is also known as positive predictive value, is the ratio of retrieved relevant instances where recall, also called sensitivity, is the ratio of retrieved relevant instances. Both, are therefore, an understanding and measure of relevance and together, are used to calculate F-Score. Considering the result, we obtain a delineation of performance for the algorithms applied on different subset sizes of the dataset.

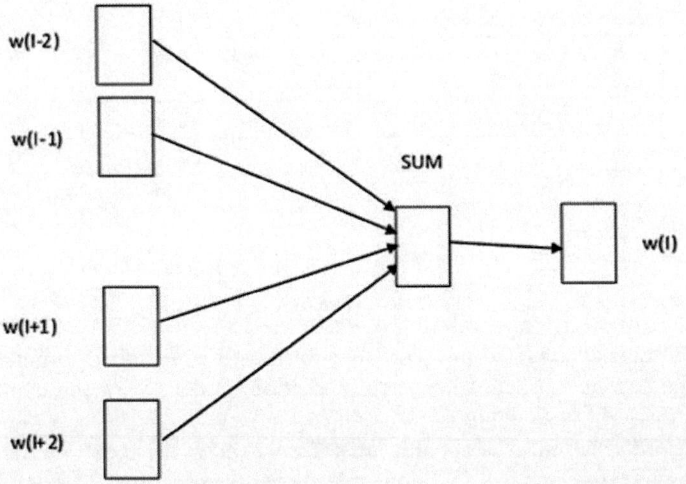

Fig. 2 Architecture of CBOW model of Word2Vec

$$\begin{aligned}\text{argmin}(-\log(\text{likelihood})) = \text{argmin}(&-\text{code}[t][i]\log(1-F)\\ &-(1-\text{code}[t][i])\log(F))\end{aligned} \quad (1)$$

where $f = \mathrm{e}^{-\sum_{t*} x(t*)P[T_i]}$; $F = \frac{f}{1+f}$; and likelihood $= (1-F)^{1-\text{code}[t][i]}$; $x(t^*)$ is the word vector (Fig. 2).

4 Result

In this work, in order to measure and analyze the performance we used the $F1$ score as the performance metric, the calculation of which makes use of Precision and Recall [16].

Precision is calculated by using the following formula:

$$\text{Precision} = \frac{\text{True Positive}}{\text{True Positive} + \text{False Positive}} \quad (2)$$

Recall is calculated using the formula:

$$\text{Recall} = \frac{\text{True Positive}}{\text{True Positive} + \text{False Negative}} \quad (3)$$

Once the Precision and Recall values are calculated, they are used for the calculation of the $F1$ score using the formula given below:

Table 1 $F1$ score with respect to various models

	Logistic regression	Support vector machine	Random forest	XGBoost
Bag-of-words	0.54	0.51	0.56	0.52
TF-IDF vectorizer	0.55	0.52	0.57	0.52
Word2Vec	0.63	0.62	0.51	0.66

$$F1\ \text{Score} = 2 \times \frac{\text{Precision} \times \text{Recall}}{\text{Precision} + \text{Recall}} \tag{4}$$

The $F1$ scores calculated in accordance to the aforementioned formulae for the each of the feature extraction for each prediction model have been mentioned in Table 1.

On close inspection, it is apparent that Word2Vec is the most accurate feature extraction procedure and has a prominent lead over the other two. The bag-of-words model consistently works the most inefficiently. The $F1$ score of bag of words and TF-IDF vectorizer has hardly much difference, dismissing them as potential options for the finished product. Word2Vec, irrespective of the model it works with, outperforms the other feature extraction techniques. Word2Vec along with XGBoost gives the best result.

5 Conclusion

For sentiment analysis and hostility detection, we used feature extraction techniques like bag of words, tfidfvectorizer and Word2Vec, out of which Word2Vec is working the most efficiently with any of the models.

Using Word2Vec along with XGBoost gives the best result, and it outperforms all other models with a score of over 65% which is good but not very high. This is due to the fact that Word2Vec doesn't allow conserving context over long periods of time, which is a definite shortcoming. This drawback can be overcome by using the long short-term memory (LSTM) or gated recurrent units (GRUs). In the future, we would like to refine our framework by incorporating the attention mechanism for better stability and accuracy. LSTM and GRU ensure this by having feedback networks.

References

1. V. Gupta, V.K. Singh, U. Ghose, P. Mukhija, A quantitative and text-based characterization of big data research. J. Intell. Fuzzy Syst. **36**(5), 4659–4675 (2019)
2. A. Alsaeedi, M.Z. Khan, A study on sentiment analysis techniques of twitter data. IJACSA **10**(2) (2019)
3. A. Binns, M. Bateman, Preliminary findings of hostility towards MPs on twitter using sentiment analysis (2017)
4. E. Kouloumpis, T. Wilson, J. Moore, Twitter sentiment analysis: the good the bad and the omg!, in *Fifth International AAAI Conference on Weblogs and Social Media*, July 2011
5. J. Plisson, N. Lavrac, D. Mladenic, A rule based approach to word lemmatization. Proc. IS **3**, 83–86 (2004)
6. J.B. Lovins, Development of a stemming algorithm. Mech. Translat. Comp. Linguistics **11**(1–2), 22–31 (1968)
7. R. Piryani, V. Gupta, V.K. Singh, U. Ghose, A linguistic rule-based approach for aspect-level sentiment analysis of movie reviews, in *Advances in Computer and Computational Sciences* (Springer, Singapore, 2017), pp. 201–209
8. V. Gupta, V.K. Singh, P. Mukhija, U. Ghose, Aspect-based sentiment analysis of mobile reviews. J. Intell. Fuzzy Syst. **36**(5), 4721–4730 (2019)
9. R. Piryani, V. Gupta, V.K. Singh, Generating aspect-based extractive opinion summary: drawing inferences from social media texts. Comput. Sist. **22**(1), 83–91 (2018)
10. R. Piryani, V. Gupta, V.K. Singh, Movie prism: a novel system for aspect level sentiment profiling of movies. J. Intell. Fuzzy Syst. **32**(5), 3297–3311 (2017)
11. V. Gupta, N. Rathore, Deriving business intelligence from unstructured data. Int. J. Inf. Comput. Technol. **3**(9), 971–976 (2013)
12. P. Gonçalves, M. Araújo, F. Benevenuto, M. Cha, Comparing and combining sentiment analysis methods, in *Proceedings of the First ACM Conference on Online Social Networks*, Oct 2013, pp. 27–38
13. J. Ramos, Using TF-IDF to determine word relevance in document queries, in *Proceedings of the First Instructional Conference on Machine Learning*, Dec 2003, Vol. 242, pp. 133–142
14. T. Mikolov, W.T. Yih, G. Zweig, Linguistic regularities in continuous space word representations, in *Proceedings of the 2013 Conference of the North American Chapter of the Association for Computational Linguistics: Human Language Technologies*, June 2013, pp. 746–751
15. T. Mikolov, K. Chen, G. Corrado, J. Dean, Efficient estimation of word representations in vector space (2013). arXiv preprint arXiv:1301.3781
16. C. Goutte, E. Gaussier, A probabilistic interpretation of precision, recall and F-score, with implication for evaluation, in *European Conference on Information Retrieval*, Mar 2005 (Springer, Berlin, Heidelberg), pp. 345–359
17. A. Joshi, K. Mehta, N. Gupta, V.K. Valloli, Indian language transliteration using deep learning, in *2018 IEEE Recent Advances in Intelligent Computational Systems (RAICS)*, Dec 2018 (IEEE), pp. 103–107
18. V. Gupta, R. Piryani, V.K. Singh, U. Ghose, An analytical review of sentiment analysis on twitter. Adv. Comput. Control Commun. Technol. **1**, 219 (2016)
19. M.R. Huq, A. Ali, A. Rahman, Sentiment analysis on twitter data using KNN and SVM. Int. J. Adv. Comput. Sci. Appl. (IJACSA) **8**(6), 19–25 (2017)
20. A. Agarwal, B. Xie, I. Vovsha, O. Rambow, R.J. Passonneau, Sentiment analysis of twitter data, in *Proceedings of the Workshop on Language in Social Media (LSM 2011)*, June 2011, pp. 30–38
21. T. Wilson, J. Wiebe, P. Hoffmann, Recognizing contextual polarity in phrase-level sentiment analysis, in *Proceedings of Human Language Technology Conference and Conference on Empirical Methods in Natural Language Processing*, Oct 2005, pp. 347–354

Detection of Invasive Ductal Carcinoma from Breast Histopathology Image Using Deep Ensemble Neural Networks

Sourodip Ghosh, Richik Ghosh, Shreya Sahay, and Suprava Patnaik

Abstract This paper presents a Deep Learning Ensemble framework for automatic detection of invasive ductal carcinoma from breast cancer histopathology images for diagnostic assistance. Pathological means of detecting IDC are predominantly arduous and tend to consume significant amount of time since the consulting pathologist must examine large specimens of benign samples before being able to ascertain the malignant ones. In the recent scenario, AI algorithms have meticulously outperformed some of the experts in analyzing medical images. We present an algorithm for breast cancer tissue detection on one of the largest datasets containing histopathological images of breast tissues samples. The proposed network has achieved a state-of-the-art performance through a balanced accuracy of 96.16% along with Matthews correlation coefficient of 0.923 and kappa score of 0.9231. Furthermore, using extended convolutional layers and the various techniques of feature mapping, the IDC-affected regions were methodically isolated from the test slides including the probable regions which might also be vulnerable. The training directory was set to 80%, followed by the test and validation directories, each of size 10%. The ensemble model is proven to be diverse and has significantly less chances of over-fitting without employing methods like cross-validation.

Keywords Invasive Ductal Carcinoma · Deep NN · Ensemble learning

S. Ghosh (✉) · R. Ghosh · S. Sahay · S. Patnaik
Department of Electronics Engineering, KIIT University, Bhubaneswar, India
e-mail: sourodip.ghosh02@gmail.com

R. Ghosh
e-mail: richikghosh98@gmail.com

S. Sahay
e-mail: sh10findit@gmail.com

S. Patnaik
e-mail: suprava.patnaikfet@kiit.ac.in

V. E. Balas et al. (eds.), *Proceedings of International Conference on Computational Intelligence, Data Science and Cloud Computing*, Lecture Notes on Data Engineering and Communications Technologies 62, https://doi.org/10.1007/978-981-33-4968-1_9

1 Introduction

1.1 Context

After successful deployment of Deep Neural Network (DNN) for many real-world image analysis applications, it is also gaining popularity for medical image processing. Cell structure and texture analysis play a crucial role in breast cancer pathology. Reliable automotive investigation of histopathological tissues can be a great support for treatment planning. Invasive ductal carcinoma (IDC) is the most prevailing type of breast cancer, encompassing about 80% of its totality. Unlike ductal carcinoma in situ (DCIS), which is confined within ducts, IDC has the tendency to rapidly infiltrate the duct as well as other surrounding areas [1]. The intensity of the disease can be assessed by a procedure called tumor grading which, however, is limited solely to areas particularly affected by invasive cancer [2]. Using a variety of methods, researches have laid the foundation to and inspected the detection of breast cancer in histopathological images. Images were isolated into histologic primitives (such as mitosis, epithelium). These quantifications were then set as the substratum for feature extraction (entities described in various scales) for differentiating between benign and malignant areas and further for other types of classification [3]. As discussed in [4], the aforementioned approach also appears to hold certain drawbacks—firstly, such specific approaches demand intensive research (in case of nuclei segmentation, it is vital to have a thorough perception of all possible incongruity in texture, morphology and anatomical structures) and secondly, the application of optimal parameters often becomes cumbersome for external observers, because the ability to analyze and adjust such parameters is often intrinsic to the developers of the algorithm. A large segment of the past methodologies included consolidating an enormous number of complex handcrafted features to create a visual substance as a representation of the BCa (breast cancer) histopathological images. The results of classification of these images depended considerably on the pre-processing done on the same, constituting of detection and segmentation. In more recent researches, feature learning inculcation and representation that do not assimilate domain proficiency have been integrated in complex learning procedures [5]. Deep learning (DL) [6] is part of a broader paradigm of feature learning, focused on improving upon learned representations with several instances of abstraction and yielding better representations with each iteration. DL approaches have recently gained reputation in the field of cellular feature analysis, pattern recognition and in situations where there is lack of domain understanding for feature introspection. It is also uniquely suited for processing of big data repositories and large datasets. Recent works on breast cancer detection have employed various CNN architectures since they have proven to be very efficient in analysis of image data. In [7], Han and team proposed the class structure-based deep convolutional neural network (CSDCNN) for multi-classification of different types on the BreakHis dataset. The algorithm leveraged high-end hierarchical feature representation. The main advantage of CNN over its predecessors is weight sharing and a significantly stronger feature extraction ability. Traditional machine learning methods have been extensively used in image analysis procedures, particularly in breast histopathology image analysis, such as mitosis detection, nuclei segmentation and cancer detection.

1.2 Literature Survey

A novel diagnosis architecture of breast cancer using whole slide images was introduced in [8]. It utilizes both histopathological image classification and content-based image classification image retrieval (CBHIR). The malignant patches in the processed images were detected through a probability map. He et al. [9] used a stacked sparse auto-encoder for rank-level fusion of local and integrated features for enhanced breast cancer diagnosis using histopathological images. Even though the classical machine learning architectures have managed to perform well and have achieved significantly good scores, the diagnostic performance still remains unreliable and to some extent, devoid of potential scope. To further enhance the performance capabilities, deep learning has been widely incorporated in many spheres of image and pattern analysis researches. Its techniques have the power to carry out automated image extraction and information retrieval from the data and thereby learn improved and abridged representations of the same. Hence, many researches have been conducted that constituted the use of deep learning for breast cancer classification and detection [10–17].

In [15], Celik et al. proposed transfer learning through DenseNet-161 and ResNet-50 models to distinguish between IDC negative and IDC positive histopathology images to achieve best balanced accuracy of 91.57% through DenseNet-161 and best F-score performance of 94.11% through ResNet-50 while also laying out insightful comparison with results of other previous works. Han et al. [7] laid out a proposition of the class structure-based deep convolutional neural network (CSDCNN), a new comprehensive recognition framework for multi-classification of subordinate classes of breast cancer—ductal carcinoma, fibroadenoma, lobular carcinoma and so on, reaching an average accuracy of 93.2%. Kassani et al. [18] used an ensemble network consisting of VGG19, MobileNetV2 and DenseNet201 on four different open-source breast histopathology image datasets to showcase the exceeding performance of the ensemble network with respect to that of the three aforementioned CNN architectures. The resultant accuracy scores were then compared against other state-of-the-art classifiers and different machine learning models. Performance as previously achieved by classical DCNNs, such as DenseNet, ResNet, InceptionV3 and VGG-16 individually, is seen to be boosted further in [19] where Wang et al. devise multi-network feature extraction model involving multi-level Inception V3 and multi-level VGG-16 which are thereby further combined with DenseNet-121 and ResNet-50. They pointed out that existing models fail to acknowledge contemporary relation among the features of two different DCNNs; to tackle this, they went on to propose the dual-network orthogonal low-rank learning (DOLL) method which utilizes various feature relations to remove redundant features. They also replaced the softmax layer in DCNN with ensemble-SVM classifiers. This experimental setup was then shown to improve performance and thereby proved its superiority over other contemporary methods. Gandomkar et al. [20] proposed the MuDeRN framework that used residual learning for classification of subclasses of benign and malignant breast cancer. Deep learning has evidently proven to be much more efficient and reliable than machine learning models, but often the former is prone to overfitting due to the added layers of abstraction and restricted training samples. Visual analysis and automatic detection of invasive ductal carcinoma (IDC) tissue segments of breast cancer using deep learning techniques were carried out by Cruz-Roa et al. in their paper [21], where CNN yielded the highest balanced accuracy (BAC) of 84.23% and F-measure of 71.80%. In [22], Araújo et al. designed an architecture for information retrieval at different scales, including nuclei and whole tissue structure. The features thus extracted were used to train a SVM network, generating optimum results over that of contemporary methods. The results obtained by the initial model evaluation of the individual architectures are followed by the final model training, which yields state-of-the-art results on the Breast Histopathology dataset.

2 Proposed Ensemble DNN Model

This research paper presents a deeply convolutional ensemble network comprising two pre-trained architectures, namely VGG16 and DenseNet, a lightweight CNN architecture called ShuffleNetV2 and a self-designed Custom CNN architecture. The proposed model is a framework of deep convolutional ensemble network. Model training involves two major steps. As shown in Fig. 1 (workflow), the initial classification employs the pre-trained architectures available at our disposal, namely VGG16, DenseNet and ShuffleNetV2. The motivation for selecting these particular architectures was because of their exemplary performance in terms of standard error metrics with respect to other architectures when operated on this particular dataset. The second step involves considering the performance of individual architectures and preparing the final model usage of deep ensemble neural network architecture.

2.1 *Histopathology Database*

The Breast Histopathology Images dataset has been used for the evaluation of the proposed model. The original publication can be found in [4]. This data contains 162 whole mount slide images of breast cancer, scanned in 40x magnification. This has been processed to extract patches of size 50×50, resulting in a total range of 277,524. The number of patches we used in our model is 160,000. A ground truth control setup was designated as the 0 labels, indicating non-IDC class. The data cohort consisted of two sub-folders of each patient labeled as '0' and '1' – '0' for non-IDC and '1' for IDC. The experiments were conducted, and the data was

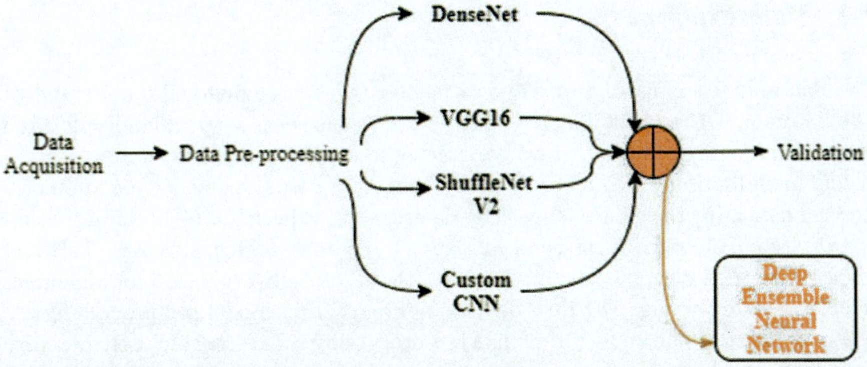

Fig. 1 Workflow (ensemble network)

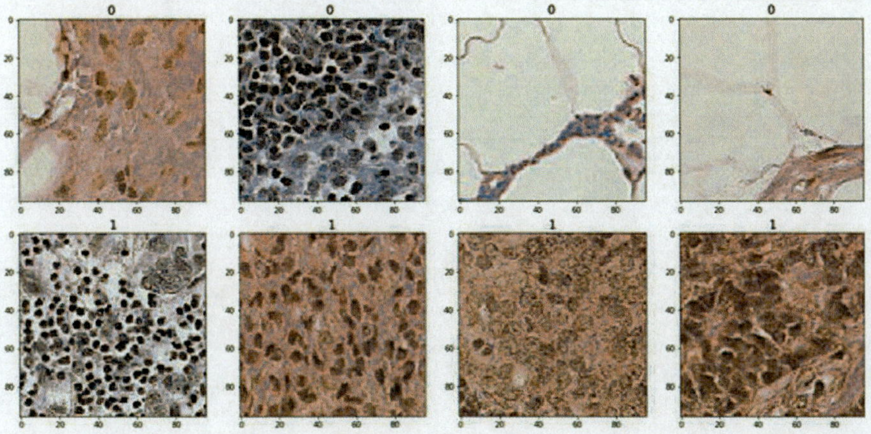

Fig. 2 Representation of labels in original dataset (0: Non-IDC, 1: IDC)

analyzed using TPU as a hardware assist. It is a custom-developed ASIC used to accelerate workload processing power. It facilitates advanced performance of linear computation and is thus used extensively in deep learning model training. Compared to GPU, it increases train computation time by 10x. The Breast Histopathology image data is significantly large; thus, it requires more potent processing power for fast and efficient computation. As the number of patches provided is 160,000, batches of size 10 each were arranged which subsequently proved to be helpful in running the model with the assistance of TPU. These run faster than GPUs if the appropriate amount of batched data is passed to it. This helped the authors to reinforce the scope of experimentation, as well as favoured the incorporation of high-end feature models (Fig. 2).

2.2 Data Pre-processing

The data initially consisted of 277,524 patches, with an individual class count of 198,738 non-IDC and 78,786 IDC patches. The dataset is very refined with high feature-containing patches that were extracted to make best-suited features for individual model training. No features were changed or color channels were altered to prevent damaging of feature maps. The dataset was reduced to 160,000 images and was dissected into training, test and validation directories, each of size 80%, 10% and 10% of the original, respectively. The layout was carefully organized to implement hyperparameter tuning. The final dataset highlights after model pre-processing are shown in Table 1. The images after data pre-processing with respect to each directory are shown in Fig. 3. The images depict the void of any interference with the original data. Some patches had disrupted features which were not suitable for model building; they were subsequently removed manually from the dataset.

Table 1 Number of images after data pre-processing

	Training set	Testing set	Validation set
No. of images	144,000	16,000	16,000

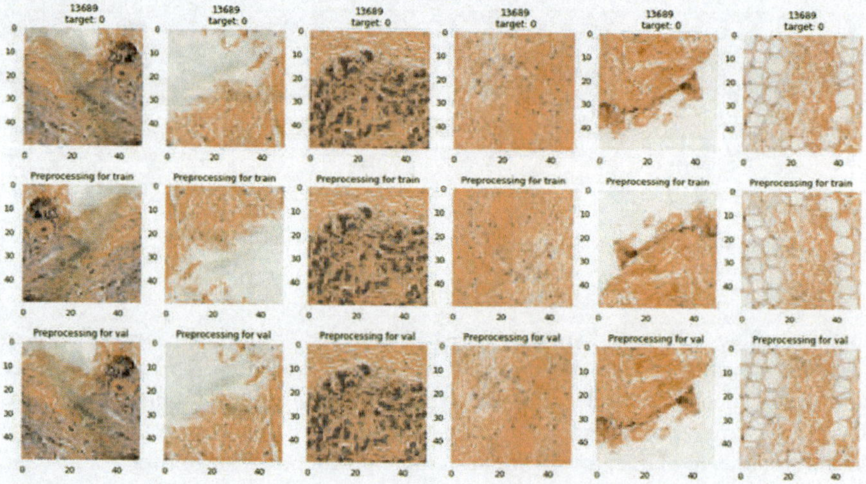

Fig. 3 Representation of slices in the respective directories after data pre-processing

2.3 Architecture Details

The proposed model is an ensemble network built using model architectures, namely

1. Pre-trained Architectures:
   (a) DenseNet
   (b) VGG16
   (c) ShuffleNet V2
2. Custom CNN Architecture

Pre-trained Architectures: The individual pre-trained DCNN models as discussed below were validated through 10 epochs, keeping the batch size fixed at 10. The batch size was kept small so that the hardware accelerator, TPU, performs to its optimal ability. Also, TPU demands lesser batch size to create tensor buckets for processing information and accessing features from images manually. The size of the images was kept at 94 × 94. No layers were frozen during the computation process.

(a) DenseNet: The DenseNet [23] model contains three blocks convoluted together. Each independent block contains 12 layers, supported with BN-ReLU conv layers. The architecture uses feature maps to transfer information, layer by layer. The initial weights were updated with Adam optimizer. This optimizer is suitable for situations where parameters are high in number. It is invariant to diagonal rescale of gradients. Also, the hyperparameters typically require very less tuning. Adam [24] was used to optimize the model parameters. This specifically converges the loss functions and updates weights. It combines the optimized performances of AdaGrad and RMSProp algorithms to regulate sparse gradients in channel noise. The number of epochs was set keeping in mind the fluctuations in validation and training accuracy and error, therefore avoiding chances of model overfitting.

(b) VGG16: VGG16 [25] is a huge network with approximately 138 million parameters. The 16 in VGG16 betokens that it has 16 layers that have weights. This follows a convolution neural network architecture with convolution layers of 3 × 3 filter having stride of 1. The padding method used here is same, along with which maxpool layers of 2 × 2 filter with a stride of 2 are being employed. The ending of this architecture consists of two fully connected layers that terminate with a block comprising softmax activation function for multiclass classification. Adam optimizer was used to evaluate loss function by utilizing updated weights.

(c) ShuffleNet V2: The ShuffleNet V2 architecture [26] was custom designed with added conv layers and FC layers in addition to dropouts. ShuffleNet is a very efficient CNN architecture for mobile devices. Here, depth-wise separable convolutions are used, which deal with the spatial patterns along with the depth dimensions. This splits the kernel in two segments for depth-wise and grouped convolutions on 1 × 1 conv layers which are point-wise grouped convolutions, to retrieve the channel dimensions. Once these operations are over, the channel is shuffled to make efficient computations. This architecture is guided by indirect metrics of computation complexity which are referred to as FLOPs. The direct metric, speed, is mainly dependable on factors like memory access expense and platform components. The model is optimized by the Adam optimizer which converges the mean square error with the increasing number of epochs.

Custom CNN Architecture: The input dimensions of the patches that are passed into the custom CNN model are $94 \times 94 \times 3$. This custom CNN model consists of three groups of three convolutional layers, each including ReLU activation function. Each of those groups further has one 3×3 max-pool layer and one dropout layer to avoid overfitting. This is succeeded by an FC layer of 256 output classes from which input is passed via a dropout layer to another FC layer consisting of two classes as output with a softmax activation function. This model is trained with NADAM optimizer having a learning rate set to 0.0001 and using binary cross-entropy as a loss function. The learning rate was intentionally kept less to make the overall model more sensitive. This model is trained by 10 epochs with batch size set to 10 in accordance to the accuracy and loss graph which was found to flatten after the mentioned number of epochs. NADAM rule is used for weight updation with a learning rate set to 0.0001 and employment of binary cross-entropy as loss function.

2.4 *Ensemble Learning*

At this stage, the organization of the considerable number of results accomplished from the individual architectures was done, and the evaluation highlights of the models on 16,000 test pictures were developed. Nonlinear weights were allocated based on the performance of the selected models. Negative correlation (NC) learning addresses mis-class diversity in neural network ensembles. As the ensemble size manifolds, the upper and lower bounds converge, thereby minimising the error rate. The FC and dense layers of the ensemble model were kept intact, and the remaining layers were at first frozen. The deep ensemble strategy guarantees the advancement of the model in terms of diversity, boosting accuracy and alleviating the issue of over-fitting. Along these lines, this implies the exclusivity of the proposed technique from every other study recently conducted on this dataset. The workflow on the ensemble model is showcased in Fig. 4.

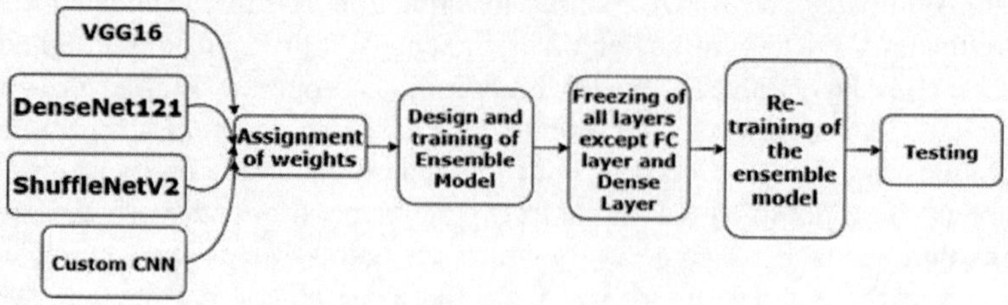

Fig. 4 Workflow (proposed ensemble model)

3 Results and Conclusion

3.1 Model Evaluation

The assigned weights from the individual architectures are passed to the ensemble model during design implementation. All layers, apart from FC layer and dense layer, were frozen, and the model was re-trained through 10 epochs. This ensured better optimization capability of the final model with respect to individual parameters (and thus the ensemble model was ready). Table 2 shows performance of individual architectures after first training, in terms of Accuracy, Precision, Recall and F1-score.

3.2 Result Analysis

The confusion matrix deduced from the ensemble framework is depicted in Fig. 6a. The balanced test accuracy (BAC) settled at 0.962 with an error rate of 0.038. The class-specific scores in terms of Sensitivity, Precision and F1-score are listed in Table 3. The overall sensitivity and precision was valued to be 0.973 and 0.949, respectively. Using values from sensitivity and precision, F1-score was calculated to be 0.961. The ensemble model correctly classifies 15,385 patches out of 16,000 test patches. This results to the final balanced accuracy (BAC) of 0.9616. Furthermore, Matthews correlation coefficient (MCC) was calculated to be 0.923. Cohen's kappa score stood at 0.923125, which indicates almost perfect agreement between the true labels and predicted labels during meta-analysis. The performance on training and validation directories per epoch-wise is shown in Fig. 5. The graph shows that the model has very low chances of over-fitting and thus guarantees an unbiased model. Correct classification functioning of the model is monitored using MCC parameter.

A total of 615 patches out of 16,000 test patches are mis-classified. Firstly, this is due to the evident fact that the model was unable to classify all the images with a perfect scope of precision. Secondly, there might be a possibility of error from the domain expert itself, in labeling all 277,524 patches with perfect precision, since the task was tedious and time-consuming. The model might have predicted the images correctly, while the medical expert had some ground truths, false.

Table 2 Performance evaluation of individual architectures

Metrics used	DenseNet	Custom CNN	VGG16	ShuffleNet V2
Accuracy	0.93	0.943	0.95	0.95
Precision	0.94	0.95	0.95	0.95
Recall	0.92	0.94	0.94	0.96
F1-Score	0.93	0.94	0.95	0.95

Table 3 Performance analysis using evaluation metrics

Metrics Used	Score
Precision (PPV)	0.949
Sensitivity (TPR)	0.973
Specificity (TNR)	0.950
Negative predictive value (NPV)	0.974
False positive rate (FPR)	0.049
F1-score	0.961
False discovery rate (FDR)	0.051
False negative rate (FNR)	0.027
Matthews correlation coefficient (MCC)	0.923
Balanced accuracy (BAC)	0.962
Cohen's kappa [27]	0.923

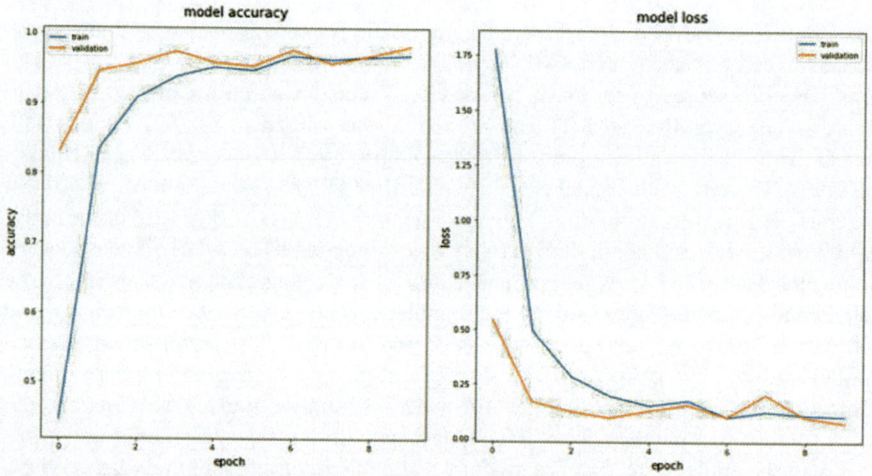

Fig. 5 Accuracy and loss graph versus epochs (ensemble model)

Figure 6b shows the breast histopathology tissue image of Patient ID: 10295, holding a ground truth of IDC positive. The tissue slice is visualized, and the possible zone containing cancer tissue is highlighted in red. The third part shows the probability of cancer affected region in the tissue in colour gradients of red (low $\rightarrow$ high : light red $\rightarrow$ dark red). These visualizations could further help doctors and researchers to correctly identify IDC-affected tissues, as well as the cancer affected region in that tissue.

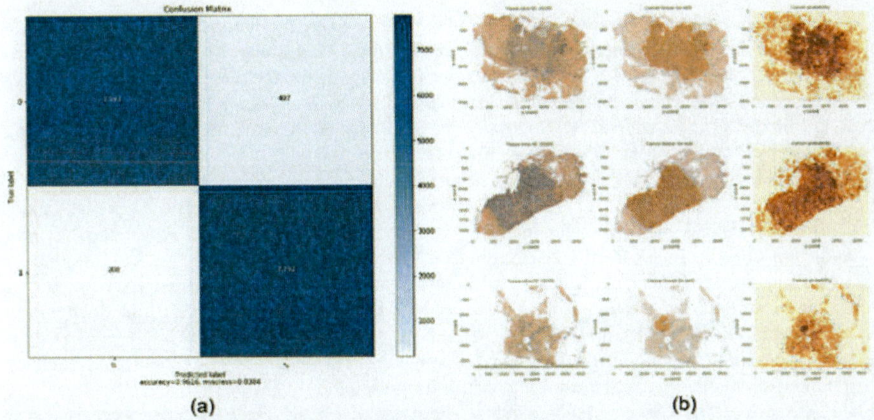

Fig. 6 **a** Confusion matrix of ensemble model **b** Sample predicted region

4 Conclusion and Scope for Future Work

This paper introduces an ensemble framework generated using four distinct architectures which were selected for their exceptional performance in the binary classification of IDC from normal patches across 277,524 patches. The overall performance was projected on the basis of various distinct evaluation metrics such as precision, sensitivity, Matthews correlation coefficient (MCC), Cohen's kappa score and so on, as shown in Table 3. The authors further encourage research in the domain of deep ensemble networks to predict more ground truth to address this vulnerable disease.

References

1. C. DeSantis, R. Siegel, P. Bandi, A. Jemal, Breast cancer statistics. Cancer J. Clinicians **61**(6), 408–418 (2011)
2. C.W. Elston, I.O. Ellis, Pathological prognostic factors in breast cancer. I. The value of histological grade in breast cancer: experience from a large study with long-term follow-up. Histopathology **19**(5), 403–410 (1991)
3. S. Naik, S. Doyle, S. Agner, A. Madabhushi, M. Feldman, J. Tomaszewski, Automated gland and nuclei segmentation for grading of prostate and breast cancer histopathology, in *2008 5th IEEE International Symposium on Biomedical Imaging: From Nano to Macro*, pp. 284–287 (2008)
4. A. Janowczyk, A. Madabhushi, Deep learning for digital pathology image analysis: a comprehensive tutorial with selected use cases. J. Pathol. Informat. **7** (2016)
5. Y. Bengio, A. Courville, P. Vincent, Representation learning: a review and new perspectives. IEEE Trans. Pattern Anal. Mach. Intell. **35**(8), 1798–1828 (2013)
6. Y. LeCun, Y. Bengio, G. Hinton, Deep learning. Nature **521**(7553), 436–444 (2015)
7. Z. Han, B. Wei, Y. Zheng, Y. Yin, K. Li, S. Li, Breast cancer multi-classification from histopathological images with structured deep learning model. Sci. Reports **7**(1), 1–10 (2017)

8. Y. Zheng, Z. Jiang, H. Zhang, F. Xie, Y. Ma, H. Shi, Y. Zhao, Histopathological whole slide image analysis using context-based CBIR. IEEE Trans. Med. Imag. **37**(7), 1641–1652 (2018)
9. K. He, X. Zhang, S. Ren, J. Sun, Deep residual learning for image recognition, in *Proceedings of the IEEE Conference on Computer Vision and Pattern Recognition*, pp. 770–778 (2016)
10. A. Pimkin, G. Makarchuk, V. Kondratenko, M. Pisov, E. Krivov, M. Belyaev, Ensembling neural networks for digital pathology images classification and segmentation, in *International Conference Image Analysis and Recognition* (Springer, Berlin, 2018), pp. 877–886
11. S. Vesal, N. Ravikumar, A. Davari, S. Ellmann, A. Maier, Classification of breast cancer histology images using transfer learning, in *International Conference Image Analysis and Recognition* (Springer, Berlin, 2018), pp. 812–819
12. N. Brancati, M. Frucci, D. Riccio, Multi-classification of breast cancer histology images by using a fine-tuning strategy, in *International Conference Image Analysis and Recognition* (Springer, Berlin, 2018), pp. 771–778
13. A.-A. Nahid, Y. Kong, Histopathological breast-image classification using local and frequency domains by convolutional neural network. Information **9**(1), 19 (2018)
14. F.A. Spanhol, L.S. Oliveira, P.R. Cavalin, C. Petitjean, L. Heutte, Deep features for breast cancer histopathological image classification, in *2017 IEEE International Conference on Systems, Man, and Cybernetics (SMC)* (IEEE, Berlin, 2017), pp. 1868–1873
15. Y. Celik, M. Talo, O. Yildirim, M. Karabatak, U.R. Acharya, Automated invasive ductal carcinoma detection based using deep transfer learning with whole-slide images. Pattern Recogn. Lett. (2020)
16. G.J.S. Litjens, C.I. Sánchez, N. Timofeeva, M. Hermsen, I.D. Nagtegaal, I. Kovacs, C.H. van de Kaa, P. Bult, B. van Ginneken, J. van der Laak, Deep learning as a tool for increased accuracy and efficiency of histopathological diagnosis. Sci. Rep. **6** (2016)
17. D. Wang, A. Khosla, R. Gargeya, H. Irshad, A.H. Beck, Deep learning for identifying metastatic breast cancer. ArXiv vol. abs/1606.05718 (2016)
18. S.H. Kassani, P.H. Kassani, M.J. Wesolowski, K.A. Schneider, R. Deters, Classification of histopathological biopsy images using ensemble of deep learning networks. arXiv preprint arXiv:1909.11870 (2019)
19. Y. Wang, B. Lei, A. Elazab, E.-L. Tan, W. Wang, F. Huang, X. Gong, T. Wang, Breast cancer image classification via multi-network features and dual-network orthogonal low-rank learning. IEEE Access **8**, 27779–27792 (2020)
20. Z. Gandomkar, P.C. Brennan, C. Mello-Thoms, Mudern: multi-category classification of breast histopathological image using deep residual networks. Artif. Intell. Med. **88**, 14–24 (2018)
21. A. Cruz-Roa, A. Basavanhally, F. González, H. Gilmore, M. Feldman, S. Ganesan, N. Shih, J. Tomaszewski, A. Madabhushi, Automatic detection of invasive ductal carcinoma in whole slide images with convolutional neural networks, in *Medical Imaging 2014: Digital Pathology*, vol. 9041 (International Society for Optics and Photonics, 2014), p. 904103
22. T. Araújo, G. Aresta, E.V. de Castro, J.F. Rouco, P. Aguiar, C. Eloy, A. Polónia, A. Campilho, Classification of breast cancer histology images using convolutional neural networks. PLoS ONE **12** (2017)
23. G. Huang, Z. Liu, L. Van Der Maaten, K.Q. Weinberger, Densely connected convolutional networks, in *Proceedings of the IEEE Conference on Computer Vision and Pattern Recognition*, pp. 4700–4708 (2017)
24. D.P. Kingma, J. Ba, Adam: a method for stochastic optimization. arXiv preprint arXiv:1412.6980 (2014)
25. K. Simonyan, A. Zisserman, Very deep convolutional networks for large-scale image recognition. arXiv preprint arXiv:1409.1556 (2014)
26. N. Ma, X. Zhang, H.-T. Zheng, J. Sun, Shufflenet v2: practical guidelines for efficient CNN architecture design, in *Proceedings of the European Conference on Computer Vision (ECCV)*, pp. 116–131 (2018)
27. J. Cohen, Introduces kappa as a way of calculating inter rater agreement between two raters. Educ. Psychol. Measurement (1960)

Neural Network-Based Surface Corrosion Classification on Metal Articles

Sanjay Kumar Ahuja, Manoj Kumar Shukla, and Kiran Kumar Ravulakollu

Abstract Generally, stainless steel has noticeable resistance to corrosion; however, it can still have corrosion because of gases, especially oxygen, and a high saline environment. The visual approach to detecting damage depends on the quality of the images and the knowledge and experience of the experts. Various destructive and non-destructive methods are available to detect damage in steel. Digital image processing techniques are widely accepted as a non-destructive approach for detecting the surface corrosion on iron articles. The selection of the algorithm depends on the environmental parameters, the type of the damage, and the lighting conditions and should also be cost-effective, fast, and with reasonable accuracy. Various image processing techniques in different combinations have been used by researchers to identify risks. This paper presents an adaptive neural network-based approach for the classification of different types of damage in metals. In the present paper, research in this field has been extensively reviewed to identify challenges and successes in this area. For the present work, convolution neural network with a custom kernel has been used to classify different types of corrosion. Pre-trained deep learning algorithms such as VGG, ResNet, and DensNet were used for classification. Finally, a customized VGG19 network with serially fused deep features and optimized hyper-parameters were used to improve the classification accuracy.

Keywords Computer vision · Image processing · Corrosion · Classification · Machine learning · Deep learning · Neural network · Convolutional network

---

S. K. Ahuja (✉)
AIIT, Amity University, Noida, Uttar Pradesh, India
e-mail: hi.sanjay.ahuja@gmail.com

M. K. Shukla
ASET, Amity University, Noida, Uttar Pradesh, India
e-mail: MKShukla@Amity.Edu

K. K. Ravulakollu
School of Computer Science, University of Petroleum and Energy Studies (UPES), Dehradun, India
e-mail: kiran007.r@gmail.com

V. E. Balas et al. (eds.), *Proceedings of International Conference on Computational Intelligence, Data Science and Cloud Computing*, Lecture Notes on Data Engineering and Communications Technologies 62, https://doi.org/10.1007/978-981-33-4968-1_10

Abbreviations

ANN	Artificial Neural Network
CNN	Convolutional Neural Network
FCN	Fully Convolutional Networks
GBM	Gradient Boosting Machine
HOG	Histogram of Oriented Gradients
LBP	Local Binary Pattern
R-CNN	Region-proposed Convolutional Neural Network
RF	Random Forest
SVM	Support Vector Machine
SSCN	Subspace Clustered Neural Network

1 Introduction

Metallic articles can have corrosion on the iron/steel surface when these are exposed to moisture. Corrosion is a chemical process that destroys metal surfaces. This is primarily because of chemical and electrochemical reactions caused by environmental conditions. As a result, there are many metallic components that lead to the low efficiency of the final use of metal parts. This leads to a shorter life of the objects, thus increasing the maintenance cost. Studying increased risk can help prevent such damage and prevent remediation.

Steel is an alloy of carbon and chromium in various percentages. The characteristics of steel, such as hardness, softness, and tensile strength, vary with different percentages of elements. It is used in construction and many other applications because of its high pressure bearing capability, improved corrosion resistance, and reduced maintenance costs.

The different types of surface corrosion are summarized in Table 1.

The general appearance for different types of corrosion is shown in Fig. 1.

The development of digital image processing algorithms opens up a wide variety of real-life possibilities for understanding different environments. Although physical and chemical tests are very effective at detecting damage, they are difficult to perform in large surface areas. Visual algorithms systems can be economical to perform non-destructive processes and can be used in large areas such as ships, metal bridges, and power poles.

Digital image processing techniques provide faster, logical, reasonably accurate, qualitative, and quantitative results. There are various image processing techniques already used by researchers and industries. Intensive studies have been done to detect and classify damage in metallic articles. This paper compares a number of identification and classification methods that researchers have adopted to detect damage.

Table 1 Types of corrosion

Type	Brief description	Cause	Appearance
General corrosion	A uniform corrosion that builds on the metal surface	Exposure to corrosive environment	Rough texture on the surface
Pitting corrosion	A localized corrosion leads to the cavities/holes on the metal surface	Chloride and other aggressive chemicals in the environment	Localized pits or holes on the surface
Crevice corrosion	A localized corrosion in the gaps/crevice at the joints contact of two metal/non-metal pieces	Contacts with metals and metals or non-metals	Similar to pitting corrosion
Intergranular corrosion	A corrosion at the grains/crystals of the material boundary due to impurities	Segregation of impurities at the metal surface boundary	Grains on the boundary
Galvanic corrosion	A corrosion caused by electrodes on two different metals contacts	Small electric current causes more anodic metal to corrode	Surface fogging or tarnishing with uniform surface corrosion

(a) (b) (c) (d) (e) (f)

(a) No Corrosion; (b) General Corrosion; (c) Pitting Corrosion; (e) Crevice Corrosion; (e) Galvanic Corrosion; and, (e) Intergranular Corrosion;

Fig. 1 Types of Corrosion (images adapted from various sources on Internet, n.d.)

2 Related Work

Learning-based approaches rely on supervised and unsupervised machine learning-based solutions. Researchers mainly use deep learning, neural network, support vector machine (SVM) [1], decision tree, and AdaBoost. Artificial neural network (ANN) is similar to human brain cells, a set of neurons where synaptic weights interconnect to form neural networks. Neurons work in a similar way as the human brain and acquire knowledge during the learning phase, which is used for validation during the test phase. The number of neurons and synaptic bursts may vary depending on the needs.

The adaptability of corrosion detection systems based on the computer vision has recently increased. In recent years, because of faster and affordable computing power, CNN has become a favorable choice to address complex issues related to computer vision. ILSVRCs challenge encourages users to perform 1000 class evaluations for class object recognition and image classification algorithms/architectures. To address these challenges, researchers have developed advanced computer vision-based algorithms that are widely being accepted in the industry for detection and classification of defects.

AlexaNet architecture was the first architecture to win the ILSVRC challenge. It includes five convention levels, maximum pooling levels, dropout levels, three fully connected levels, and ReLU as an activation function [13]. Another CNN architecture, VGGNet [14], has better performance than AlexaNet architecture even with a reduced number of parameters and increased network depth. AlexaNet error rate is 11.24%; however, the VGGNet error rate is 7.3%. Architecturally, there are more layers than AlexaNet, with smaller filter size, leading to a reduction in dimensions, but higher levels of features can be learned compared to previous CNNs.

Convolution neural network (CNN) object classification [4, 14–19] has achieved significant improvements in image segmentation [20] and object detection [21, 22]. The performance of CNN is better than manual approaches such as the HOG, and LBP [23]. Researchers are creating custom models for CNN to achieve better and faster results. The five layers of CNN proposed for detection of cracks and other damages in metal sheets, and [24] analyzed the structures to detect corrosion from the films. Their technique works the same way: First, large input images are sliced into small images of fixed size, and then, CNN is applied to detect and classify small images as crack or the corrosion. The different machine learning approaches adopted by different researchers have been summarized in Table 2.

Table 2 Summary—machine learning approaches

Technique	Sub category	References
Deep learning		[2, 3]
Neural network	Convolutional Neural Network	[4–6]
	Backward Propagation Neural Network	[7]
	LVQ Neural Network	[8]
	Gradient Boosting Machine (GBM)	[2]
	Subspace Clustered Neural Network (SSCN)	[2]
Support vector machine		[1, 7, 9–11]
Random Forest (RF)		[2]
K-Nearest Neighbors		[7]
AdaBoost		[12]
Decision Tree		[7]

3 Experiment

3.1 Training Dataset Preparation

Corrosion images were acquired from various publicly available images over the Internet for training the proposed deep learning model. A total of 563 images from four different corrosion grades were collected from the public domain. The images were sliced to 512 × 512 pixels for an in-depth study dataset. Subsequently, data augmentation techniques have been applied [17, 25–27] to improve efficiency and model visibility. Horizontal flip, vertical flip, a combination of both horizontal and vertical flip are used to increase the alignment in combination. This gave four times more information than the original dataset.

3.2 Model Setup and Transfer Learning

One major impediment for training deep learning models is the need for a large dataset; however, the transfer of model learning and fine-tuning techniques can significantly reduce the amount of data needed to train a model. For the present study, we have used an advanced custom kernel for faster-CNN model with custom layers and a custom dataset with transfer learning [28–31]. We trained the model on our custom dataset by starting a weighted block of correlated functions in a pre-trained weighted block in the MS COCO dataset. With the benefits of transfer learning, we inherited the model architecture with transfer learning, image data attributes, model-trained classes, and most importantly, the weight and bias of the final model. The CNN model is implemented using the open-source Tensor Flow library, with transfer learning of the pre-trained weights for other images. Based on the results of the model's practice and test information, a few more model hyperparameters are adopted and selected.

In our experiment, we used pre-trained deep learning algorithms such as VGG, ResNet, and DensNet for classification. Finally, a customized VGG19 network with serially fused deep features and optimized hyperparameters were used to improve the classification accuracy (Tables 3 and 4).

Table 3 Hardware setup

CPU	AMD Ryzen 3750H 2.30 GHz
RAM	16 GB
GPU	Geforce RTX 2060 6 GB DDR5
CUDA	1920 cuda cores
Tensor	240 tensor cores

Table 4 Software setup

Operating System	Ubuntu 20.04
Docker	19.03.8
Python	Python 3.6
CUDA	NVIDIA CUDA toolkit 10.2
cuDNN	cuDNN v8.0.1 for CUDA 10.2

3.3 Model Evaluation

Once the model is trained, it needs to be evaluated for the accuracy and performance of the classification results for True Positive (TrPo), True Negative (TrNe), False Positive (FaPo), and False Negative (FaNe). Based on these parameters, accuracy and performance can be calculated as

$$\text{Accuracy (ACC)} = \frac{\text{TrPo} + \text{TrNe}}{\text{TrPo} + \text{TrNe} + \text{FaPo} + \text{FaNe}'} \tag{1}$$

$$\text{Precision (PRE)} = \frac{\text{TrPo}}{\text{TrPo} + \text{FaPo}'} \tag{2}$$

$$\text{Recall (REC)} = \frac{\text{TrPo}}{\text{TrPo} + \text{FaNe}'} \tag{3}$$

$$F\text{score (FSO)} = \frac{2 * \text{PRR} * \text{REC}}{\text{PRE} + \text{REC}} \tag{4}$$

$$\text{Sensitivity (SEN)} = \frac{\text{TrPo}}{\text{TrPo} + \text{TrNe}} \tag{5}$$

$$\text{Specificity (SPE)} = \frac{\text{TrNe}}{\text{TrNe} + \text{FaPo}} \tag{6}$$

Precision is a good metric to determine when False Positive costs are high. Recall measures how many actual positives are produced by marking it as positive. This would be the ideal metric to select the best model when there is a high False Negative cost associated with it. Accuracy is affected by a large number of real negatives. Finally, the $F1$ score is used to find the equilibrium between recall for accurate and precise asymmetric distributions.

Table 5 Model performance

Architecture	Precision %	Recall %	$F1$ Score	Accuracy
VGGNet	92.1	88.2	90.1	91.3
ResNet18	93.2	91.2	92.2	93.1
Custom-CNN	94.3	93.3	93.8	95.4

Table 6 Validation results

Category	Precision	Recall	$F1$ Score	Accuracy
No corrosion	1.0	0.89	0.94	1.00
General	0.99	0.97	0.98	0.99
Pitting	0.98	0.96	0.97	0.98
Crevice	0.94	0.97	0.95	0.97
Galvanic	0.90	0.98	0.94	0.95
Inter granular	0.86	1.0	0.92	0.94

4 Results and Conclusions

A modified kernel with additional layers based on VGG16 was used as faster convolution neural network. Transfer learning was used to train the kernel for pre-trained weights. Pytorch was used with batch size 8, input size 224, learning rate 0.01, adamax optimizers on 10 epocs. Results were compared with other kernels, and performance was found to be better for the classification of corrosion for six different classes (Tables 5 and 6). Model performance is a trade-off between precision and recall rate as they are contradictory performance measures.

Our experiment shows that the performance of the custom kernel is significantly better than classic kernels for corrosion classification for six different classes. Increasing the depth of the model increases the capacity and is more efficient than increasing the nodes in a single layer (Fig. 2).

5 Future Work

Future work can include

- **Automated Image capture**—Currently, images are not automatically captured for vision-based corrosion detection systems, resulting in poor image stabilization, heavy background noise, and even poor image quality. A drone or CCTV camera feed may be used to capture images automatically with proper lighting conditions.
- **Improved algorithms for complex structure metal articles**—Mostly, algorithms are developed for the flat surface of metals; however, much research needs to be

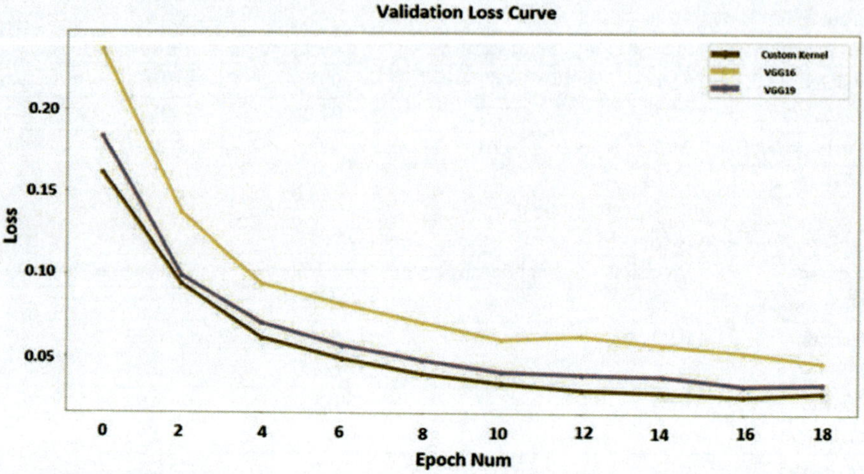

Fig. 2 Validation Loss

done for complex metal shapes and structures. Future work can be extended for complex metallic geometries that are actually used in many practical situations, such as transport, bridges, automotive, and heavy machinery.

- **Corrosion Growth Analysis**—Image-based risk detection techniques can be used for corrosion growth analysis and finding growth trends. This allows timely preventive measures to prevent metal surfaces from falling further.
- **Machine Learning**—When it comes to corrosion damage analysis over a large surface area, machine learning along with image processing techniques with knowledge of the past history of scan may lead to a false positive reduction. The algorithms are built to estimate the growth rate of damage, which can be analyzed for preventive maintenance.

References

1. S.K. Ahuja, M.K. Shukla, K.K. Ravulakollu, Surface corrosion detection and classification for steel alloy using image processing and machine learning, in *International Conference on Emerging Trends in Engineering & Technology on Information Security and Analytics*, Helix, 2018, pp. 3822–3827
2. C.I. Ossai, Corrosion defect modelling of aged pipelines with a feed-forward multi-layer neural network for leak and burst failure estimation. Eng. Fail. Anal. **110**, 104397 (2020)
3. L. Petricca, T. Moss, G. Figueroa, S. Broen, Corrosion detection using AI: a comparison of standard computer vision techniques and deep learning model (DMDB, MoWiN, CoSIT, CRIS, SIGL, ICBB, CNSA, CCSEIT, AIAP, 2016), pp. 91–99
4. S.K. Ahuja, M.K. Shukla, K.K. Ravulakollu, Surface corrosion grade classification using convolution neural network. Int. J. Recent Technol. Eng. **8**(3), 7645–7649 (2019)

5. S.A. Idris, F.A. Jafar, Image enhancement based on software filter optimization for corrosion inspection, in *2014 5th International Conference on Intelligent Systems, Modelling and Simulation (ISMS)* (IEEE, New York, 2014), pp. 345–350
6. A. Ortiz, F. Bonnin-Pascual, E. Garcia-Fidalgo et al., Visual inspection of vessels by means of a micro-aerial vehicle: An artificial neural network approach for corrosion detection, in *Robot 2015: Second Iberian Robotics Conference* (Springer, Berlin, 2016), pp. 223–234
7. H. Son, N. Hwang, C. Kim, C. Kim, Rapid and automated determination of rusted surface areas of a steel bridge for robotic maintenance systems. Autom. Constr. **42**, 13–24 (2014)
8. G. Vachtsevanos, K. Natarajan, R. Rajamani, P. Sandborn, *Corrosion Processes* (Springer, Berlin, 2020)
9. Y.-J. Lv, J.-W. Wang, J.J.-L. Wang, C. Xiong, L. Zou, L. Li, D.-W. Li, Steel corrosion prediction based on support vector machines. Chaos, Solitons & Fractals **136**, 109807 (2020)
10. Z. Liu, M. Genest, D. Krys, Processing thermography images for pitting corrosion quantification on small diameter ductile iron pipe. NDT & E Int. **47**, 105–115 (2012)
11. P.-H. Chen, H.-K. Shen, C.-Y. Lei, L.-M. Chang, Support-vector-machine-based method for automated steel bridge rust assessment. Autom. Constr. **23**, 9–19 (2012)
12. F. Bonnin-Pascual, A. Ortiz, Corrosion detection for automated visual inspection, in *Developments in Corrosion Protection* (InTech, 2014)
13. J. Alecci, Transfer learning for rain detection in images
14. K. Simonyan, A. Zisserman, Very deep convolutional networks for large-scale image recognition, arXiv preprint arXiv:1409.1556
15. C. Szegedy, V. Vanhoucke, S. Ioffe, J. Shlens, Z. Wojna, Rethinking the inception architecture for computer vision, in *Proceedings of the IEEE Conference on Computer Vision and Pattern Recognition*, 2016, pp. 2818–2826
16. K. He, X. Zhang, S. Ren, J. Sun, Deep residual learning for image recognition, in *Proceedings of the IEEE Conference on Computer Vision and Pattern Recognition*, 2016, pp. 770–778
17. A. Krizhevsky, I. Sutskever, G. E. Hinton, Imagenet classification with deep convolutional neural networks, in *Advances in Neural Information Processing Systems*, 2012, pp. 1097–1105
18. M.D. Zeiler, R. Fergus, Visualizing and understanding convolutional networks, in *European Conference on Computer Vision* (Springer, Berlin, 2014), pp. 818–833
19. O. Russakovsky, J. Deng, H. Su, J. Krause, S. Satheesh, S. Ma, Z. Huang, A. Karpathy, A. Khosla, M. Bernstein et al., Imagenet large scale visual recognition challenge. Int. J. Comput. Vision **115**(3), 211–252 (2015)
20. J. Long, E. Shelhamer, T. Darrell, Fully convolutional networks for semantic segmentation, in *Proceedings of the IEEE Conference on Computer Vision and Pattern Recognition*, 2015, pp. 3431–3440
21. J. Redmon, S. Divvala, R. Girshick, A. Farhadi, You only look once: Unified, real-time object detection, in *Proceedings of the IEEE Conference on Computer Vision and Pattern Recognition*, 2016, pp. 779–788
22. W. Liu, D. Anguelov, D. Erhan, C. Szegedy, S. Reed, C.-Y. Fu, A.C. Berg, SSD: Single shot multibox detector, in *European Conference on Computer Vision* (Springer, Berlin, 2016), pp. 21–37
23. Z. Guo, L. Zhang, D. Zhang, A completed modeling of local binary pattern operator for texture classification. IEEE Trans. Image Process. **19**(6), 1657–1663 (2010)
24. Y.-J. Cha, W. Choi, O. Büyüköztürk, Deep learning-based crack damage detection using convolutional neural networks. Comput.-Aided Civil Infrastruct. Eng. **32**(5), 361–378 (2017)
25. E. D. Cubuk, B. Zoph, D. Mane, V. Vasudevan, Q. V. Le, Autoaugment: Learning augmentation strategies from data, in *Proceedings of the IEEE Conference on Computer Vision and Pattern Recognition*, 2019, pp. 113–123
26. B. Zoph, E. D. Cubuk, G. Ghiasi, T.-Y. Lin, J. Shlens, Q. V. Le, Learning data augmentation strategies for object detection, arXiv preprint arXiv:1906.11172
27. J. Shlens, E. D. Cubuk, Q. Le, T.-Y. Lin, B. Zoph, G. Ghiasi, Learning data augmentation strategies for object detection, US Patent App. 16/416,848 (2019)

28. E. Holm, A. A. Transeth, O. Ø. Knudsen, A. Stahl, Classification of corrosion and coating damages on bridge constructions from images using convolutional neural networks, in *Twelfth International Conference on Machine Vision (ICMV 2019)*, Vol. 11433 (International Society for Optics and Photonics, 2020), p. 1143320
29. C. Ejimuda, C. Ejimuda, et al., Using deep learning and computer vision techniques to improve facility corrosion risk management systems 2.0, in *SPE Nigeria Annual International Conference and Exhibition* (Society of Petroleum Engineers, 2019)
30. Z. Shi, H. Hao, M. Zhao, Y. Feng, L. He, Y. Wang, K. Suzuki, A deep CNN based transfer learning method for false positive reduction. Multimedia Tools Appl. **78**(1), 1017–1033 (2019)
31. J. Yosinski, J. Clune, Y. Bengio, H. Lipson, How transferable are features in deep neural networks? in *Advances in Neural Information Processing Systems*, 2014, pp. 3320–3328

Automated Cockpit Voice Recorder Sound Classification Using MFCC Features and Deep Convolutional Neural Network

Rafeed Rahman, Mehfuz A. Rahman, and Jia Uddin

Abstract The main purpose of a cockpit voice recorder (CVR) is to record sounds in the cockpit of an aircraft, including the conversation of the pilots in order to facilitate the investigation of aircraft accidents and incidents. Although the audio recorded by the CVR is a crucial source of evidence for the aircraft crash investigation, it is often a strenuous and lengthy task to identify all kinds of sounds in that audio due to poor quality of sounds and current methodologies in practice. This paper proposes a model based on feature extraction and deep convolution neural network for automatic classification of CVRs recorded sounds. In the proposed model, MFCC vector of key sounds is extracted from the dataset which are then trained using a multiple layer-based convolution neural network. In the experimental evaluation, a sound dataset was utilized for training the models that consists of 5033 audio clips divided into several classes of major sounds for investigation. The data were extracted from online crash documentaries, videos and an existing spoken digit dataset. The experimental results demonstrate that the proposed CNN model exhibited classification accuracy of 97% which is much higher than the other state-of-art models—KNN, decision tree, and random forest classifier.

Keywords Cockpit voice recorder · Random forest · K-nearest neighbor · Transfer learning

R. Rahman (✉) · M. A. Rahman
Department of Computer Science and Engineering, Brac University, Mohakhali, Dhaka 1212, Bangladesh
e-mail: rafeedrahmansham2015@gmail.com

M. A. Rahman
e-mail: mehfuz.rahman097@gmail.com

J. Uddin
Technology Studies Department, Endicott College, Woosong University, Daejeon, Korea
e-mail: jia.uddin@wsu.ac.kr

V. E. Balas et al. (eds.), *Proceedings of International Conference on Computational Intelligence, Data Science and Cloud Computing*, Lecture Notes on Data Engineering and Communications Technologies 62, https://doi.org/10.1007/978-981-33-4968-1_11

1 Introduction

The main purpose of CVR is to keep audio recordings of aircraft cockpit in order to aid in aircraft accidents and incidents investigation. It is generally located in the tail cone section of an aircraft that is designed in such a way that it is fireproof and makes it easy to locate from crash debris. Unlike its earlier editions, modern CVR record data in digital form using nonvolatile solid-state memory replacing the earlier fireproof magnetic tape and can record a minimum of 2 h of latest cockpit audio. Modern CVR consists of four separate audio channels which includes the pilot's headset microphone, the co-pilot's headset microphone, and a cockpit area microphone (CAM) which is found on the overhead instrument panel between the two pilots and a fourth channel. All of these salient features of the CVR make it a vital source of information for detail investigation of aircraft accidents. For instance, CVR audio recording was the key evidence in solving the mystery of Germanwings Flight 9525 crash that killed 150 passengers and crew on March 2015 [1]. Despite this fact that cockpit voice recorder audio recording is an essential part of any aircraft disaster investigation, it takes audio forensic expert long hours of effort to successfully identify and analyze all the sounds in those audio recordings. Moreover, the complexity and quality of those audio signals recorded by the CVR can further complicate the task for the investigator [2] to timely scrutinize important information from this valuable source of evidence with accuracy. However, we believe that machine learning and its subset deep learning has the potential to contribute positively to this problem just like its contribution to many other problems.

From the very beginning of the last decade advancement in computer vision brought by deep learning, a subset of machine learning has attracted much of our attention. Earlier, deep learning is used in various applications such as large-scale image classification demonstrated by the transfer learning model AlexNet [3], medical diagnosis enhancement [4, 5], and object detection with unmanned area vehicle imagery [6, 7]. However, applying machine learning for audio signal processing is a field that is rapidly growing with many remarkable successes already achieved. For instance, environmental sound classification using deep convolution neural networks [8], classification of heart sound [9], and even sounds like birds [10] which is often difficult to classify. Further, some of the most prominent AI systems like virtual assistants Alexa introduced by Amazon, Siri by Apple, and Google Home are products of audio signal processing with machine learning [11].

Considering all the above-mentioned facts, this paper presents a model applying deep convolution neural network in order to automate the classification of various important sounds found on aircraft CVR recordings.

The rest of the paper is organized as follows: literature review in Sect. 2, detailed description of our proposed model in Sect. 3, experimental setup and results analysis in Sect. 4, and finally the conclusion in Sect. 5.

2 Literature Review

There are several researches that showed various ways of analyzing the cockpit voice recording audio. For instance, in [12], the author proposed a voice activity detection system for CVR that relies on iterative spectral subtraction and double thresholds. The results of their experiment indicate that spectral subtraction used to improve the signal to noise ratio can provide a practicable environment for the conventional double thresholds voice activity detection. Further, from the beginning of the twenty-first century, the non-speech sound of the cockpit voice recordings including the alarm sounds are also getting a lot of attention from the researchers [13–15]. Consequently, in [16], the author suggested an algorithm to detect alarm sounds in the CVR recording automatically. Their system was able to classify most of the sounds, however, they mentioned that it detects some other non-alarm sounds as alarms after testing their system on full CVR recordings. In [17], the authors conducted study to get the characteristic frequency of background sound of a cockpit from the CVR signal recordings. They initially introduced a cockpit voice decoding system in order to categorize different background signals and to get initial spectrum frequency and then used three different algorithms to find the characteristic frequency of background sounds. All their four approaches were reported to have similar results after verification using the LMS SC305 system. In [18], the paper suggested a new approach based on production rule and information fusion theory for analyzing and fault diagnosing of cockpit voice extracted from CVR. By and large, most of this works suggested various methodologies to analyze CVRs audio, however, they did not consider the enormous potential of various deep learning in their works.

On the other hand, there are many significant works done on signal processing using deep learning. In [8], the authors suggested a deep convolution neural network model in order to classify environmental sound and explained the importance of augmenting audio data to solve the problem of labeled data shortages. They recommend using class-conditional data augmentation for improvement in the performance of their models. Similarly, in [19], the authors suggested a different deep convolution neural network model for environmental sound classification which were tested on three different urban and environmental audio datasets. Their model was made of two convolutional layers with max-pooling and two fully connected layers. In [20], applied several CNN models for bird species classification based on bird sounds audio dataset. This work, however, had some limitations in terms of reducing dataset distortion. Furthermore, there was some good researches that applied CNN for various kinds of sound detection like speech recognition [21], heart sound classification [22], and musical instrument recognition [23]. Some of the recent research papers tried to give some good reviews on existing deep learning for signal processing techniques. For example, in [24–26], they analyzed the performance of existing CNN, LSTM, and other neural network architectures with respect to audio processing, whereas in [27], they surveyed different neural networks models for detection of acoustic events along with a suggestion for possible real-world scenarios.

3 Proposed Model

Figure 1 represents the CNN architecture that we used for our crash-related sound classification. This is a sequential model, i.e., a pile of layers. Initially, our model follows a pattern of the convolution layer and batch normalization for three consecutive steps. The first convolution layer contains 32 kernels of [4 × 4] dimension, the second convolution layer contains 48 kernels of dimension [3 × 3], and the final convolution layer contains 120 kernels of dimension [3 × 3].

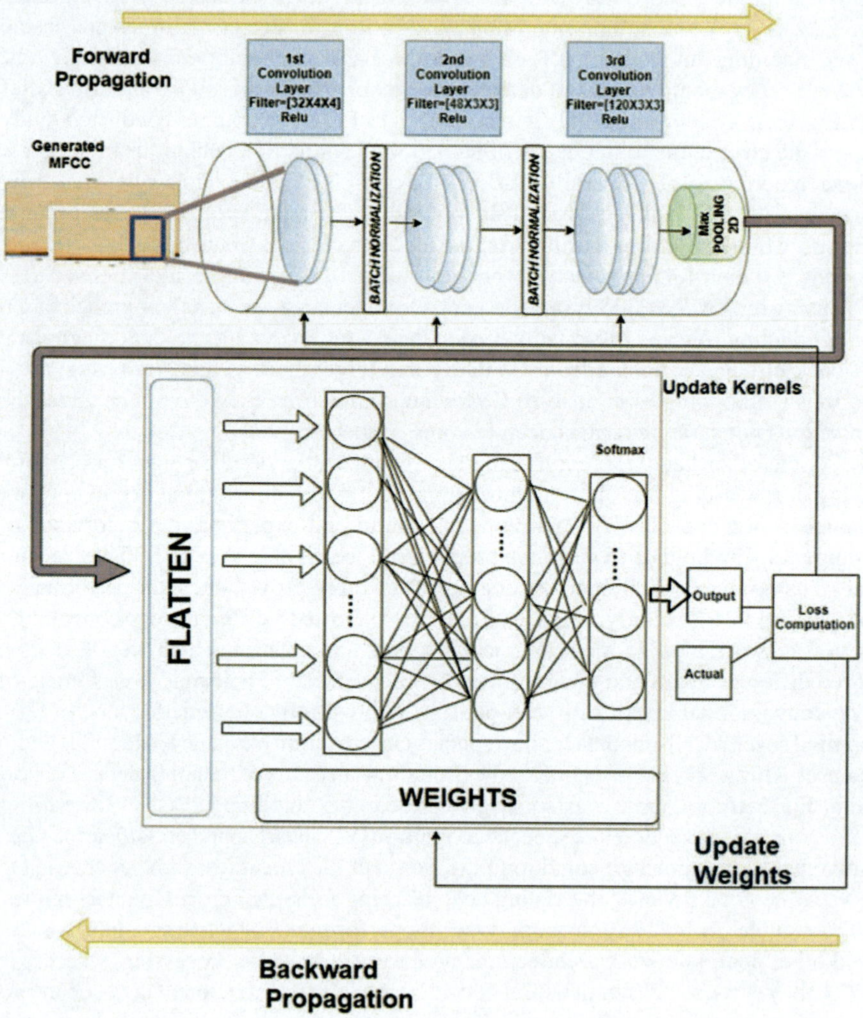

Fig. 1 CNN architecture

Batch normalization is done between the convolution operations. One of the notable factors that has caused deep learning to make an impact in the recent years is batch normalization. It greatly increases the speed of the model learning, i.e., learning rate can be kept higher. The model can train to high accuracy with 14 times lower iterations [28]. It is now a globally recognized technique as it provides stability along with speed in the training of the deep learning model [29]. This technique targets to enhance the training of DNN by controlling the input distributions. It also performs regularization [28].

The max-pooling layer of pool size [2 × 2] is used to collect the features generated from the convolution operation and forward those to next. It has the advantage of reducing the size of the input to be forwarded to the next layer and thus decreasing the computational complexity of the next layers.

We have used another regularization option which is dropout with rate assigned is 0.5. Thus, it ensures some randomly picked neurons are disengaged and prevented from having any impact on the activation of the next neuron in the forward computation of the neural network. Also, while propagating in the backward direction, it is ensured that any weight hanging computations are not done on these dropped out neurons. Dropout enhances the capability of neural networks in all sorts of supervised learning works such as visual classification, and audio classification [30].

Then, the output is flattened to be fed into the dense network of 128 neurons followed by 64 neurons and finally the output layer of 18 neurons due to the 18 classes that we have created for this classification of plane related sounds. The ReLU activation function is used in each layer. Equation (1) represents the ReLU activation function.

$$\text{Output} = \text{maximum}\{0, y\} \tag{1}$$

The output just retains any value other than negative. It enhances the training speed of the neural network due to its clipping of negative values and also swift computation of gradient. In the output layer, we have used the softmax activation function.

Categorical cross-entropy is used as the loss function

$$\text{Loss}(v, \hat{v}) = -\sum_{a=0}^{C}\sum_{b=0}^{D}\left(v_{ab} * \text{Log}(\hat{v}_{ab})\right) \tag{2}$$

where $\hat{v}$ is the predicted value.

The comparison between the actual distribution and the predictions are done by categorical cross-entropy where real class has probability one and the rest 0.

Mel-frequency cepstrum (MFC) is a display of a short-term log power spectrum that has undergone a linear cosine transformation on a nonlinear Mel scale of frequency [31].

MFCC is the most established tool in sound analysis and speech recognition as it is helpful in the case of modeling and audio contents in terms of frequency and sharpness [32]. MFCCs are a short vector of features which briefly represents the overall structure of spectral envelope.

The derivation of MFCCs is done by performing the Fourier transform of a signal, then mapping the power spectrum generated on Mel scale by triangular overlapping windows and then finally taking the cosine transformation of the log value of powers of each Mel frequencies. The MFCC vector is then the amplitudes that the resulting spectrum display [33]. A sample MFCC visualization is presented in Fig. 2.

We have used Librosa for loading the audio files. Librosa is a Python package for analyzing the audio files which provide all the necessary requirements that are needed for generating audio information.

We have written a method—wavtoMFCC (file path)—to create the mfcc vectors. Initially, the audio file is loaded with librosa.load function where parameters provided are the location of the audio file, sampling rate = 8000, and duration = 1.024 s. This function then returns the audio time series and the sampling rate.

The audio time series which is generated is then normalized using librosa.util.normalize function. Finally, mfcc vectors are generated using librosa.feature.mfcc function with parameters input is a normalized audio wave, sampling rate, and number of mfcc to return are 40, hop_length = 0.048 × sampling rate which is cast to an integer. Finally, wavtoMFCC (file path) return mfcc vector. Figure 3 illustrates the details conversion of WAV to MFCC conversion.

The MFCC vectors generated are then split into train and test sets by importing and using train_test_split function from sklearn.model_selection. We have assigned test size = 33% and random state = 1 as parameters of the train_test_split function.

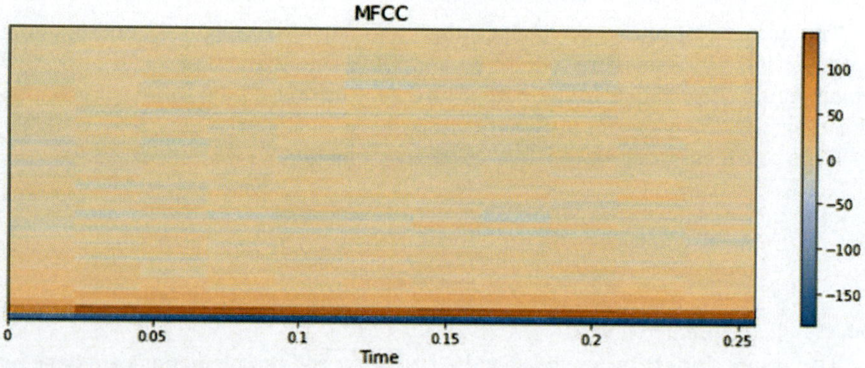

Fig. 2 MFCC visualization

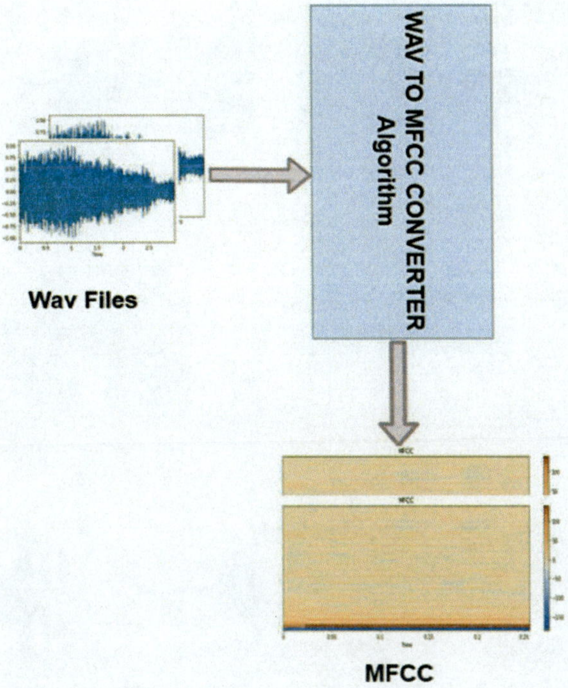

Fig. 3 WAV to mfcc converter

4 Experimental Setup and Results Analysis

We had gone through several crash videos which includes short documentaries and other videos from YouTube to collect the sounds for the classes. The documentaries were from Smithsonian Channel, National Geography, and many other channels such as The Flight Channel that contains videos based on the crashes that occurred in the past. Then, using those videos, we started generating sounds clips with Python. After we developed a complete dataset using the sound clips that we generated and the spoken digit dataset [34], we shifted our focus toward developing the MFCC vectors. Figure 4 illustrates the visualization of sample WAV files from the dataset.

The MFCC vectors that we prepared are used to create the training and testing dataset that were used for training our models. The outcome of training and classification is visualized at the end in terms of training and validation accuracies and classification.

Our main purpose is to detect all sorts of sounds that can be a clue for investigation of the crash [1]. So the classes that we have designed for our research contain all possible sounds. The dataset that we have prepared consisted of 5033 audio clips taken from documentary videos of plane crashes which also included the spoken digit data [32]. We have written a Python script to generate short audio clips in a WAV file format of 3 s each. The classes that we have chosen for our dataset are—panic conversation, emergency signals, explosion, crash, takeoff, engine running sounds,

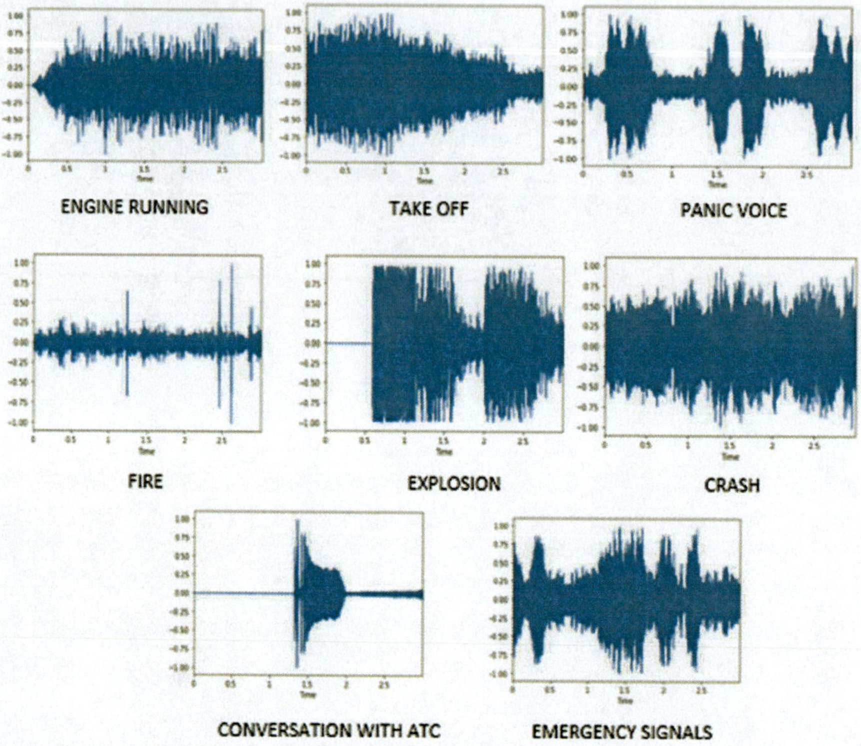

Fig. 4 Visualization of WAV files from dataset

and conversation with air traffic controller [ATC] sounds, respectively. Along with this, we have chosen a number [0–9] sound class and the audio clips that we use under these classes were taken from [32]. It contains 2000 audio clips of numbers [0–9].

For creating the classes of sounds, we have gone through some authentic crash articles and brief news about the investigations on the basis of which we designed our classes. The classes that we designed for sound detection can play a crucial role in aircraft crash investigation. The explosion sounds class that we have chosen among all the classes may contain clues that may lead to a direction in the crash investigations. Recently, we have seen the crash of Ukraine International Airlines Flight 752 which was traveling from Tehran to Keiv which is operated by Ukraine International Airlines (UIA) in January 2020 [35]. It crashed shortly after it took off from Tehran Imam Khomeini International Airport. All 176 passengers had died. Later an explosion video directs a new path to the investigation of what appeared to be a missile hit the plane. Our method of sound detection can automate the process by detection of any explosion sounds from CVR. The number [0–9] classes are considered as numbers play an important role in the conversation. A few years back, the plane of US Bangla Airlines [flight BS211] crashed while landing at Tribhuvan

International Airport in 2018 [36]. Later, the conversation of Pilot Abid Sultan with ATC was tracked. There was a clear confusion found between the ATC and the pilot about the 02 and 20 runway for landing [36–38]. It also showed the importance of our conversation with ATC class. The number of classes was primarily used in our sound detection system so that it can highlight this information and did not miss anything which can be huge contributions for investigation. If we consider the case of Nigeria Airways Flight 2120, the plane crashed on 11th July 1991 killing all 247 passengers and 14 crew members after taking off from Abdul Aziz International Airport. The crash reasons were later found to be the fire that was caused due to overheated tires from low inflated tires [39, 40]. So we have allocated a fire class so that our model can detect any fire sounds from audio of CVR. Alarm sounds can indicate after the crash the problems that may cause the incident. The Kenya airways Flight 431 crashed into the sea off the coast of Côte d'Ivoire after it took off from Félix-Houphouët-Boigny International Airport, Abidjan [41]. The investigation of this crash led to many sequences of events like overspeed warning sounds that were found from the CVR recordings. Also, the stall warning sounds that possess many clues for the investigation. So alarming sounds such as "bank angle," "fire alarm," "proximity warning," "cockpit stabilizer trim function indicator," etc., all sorts of sounds are present in our emergency signals class.

The Germanwings Flight 9525 took off from Barcelona–El Prat Airport in Spain and was heading toward Düsseldorf Airport in Germany [42]. But it crashed 100 km north-west of Nice in French Alps. The investigation later found that the crash was caused deliberately by the Co-Pilot Andreas Lubitz. The sounds from cockpit gave clue about the activity of the co-pilot and also the panicking shouts of the passengers and the Captain from behind was also heard in the sounds [43]. Thus, panic sounds and conversation from voice recorder sounds gave numerous clues of the aviation disasters that occurred.

After finalizing our classes, we used Python to generate audio clips of 3 s each from the audios of different crash documentaries and other audios from YouTube. Initially, we have taken those files as videos and then converted them to WAV audio clips. Then, we took short 3 s audio clips. For that purpose, we have installed Pydub in our Colab notebook that allows us to manipulate audio files with a simple and easy high-level interface. From Pydub, we imported AudioSegment and used it to take short clips from WAV files. All the models were trained and tested in Google Colab. The MFCC vector and the labels were separated into X_Train, X_Test, Y_Train, and Y_Test using train_test_split with a 33% test size. At first, our convolutional neural network model was trained with X_train and Y_train with validation state = 0.1 and batch size was 64. We have trained it for 50 epochs and observed variation in the training accuracies and validation accuracies per epochs. For visualizing the trend in training and testing, we used Python to generate the training and validation against Epochs curves. The curves in Fig. 5 represented both the training and validation accuracies and losses variation over 50 epochs. It is vivid that the training and validation accuracy has increased and the training and validation loss have decreased exponentially.

Figure 6 clearly displays the evaluation of our model for each of the convolution layers added to the model. It is vivid that there is a huge improvement in the evaluation

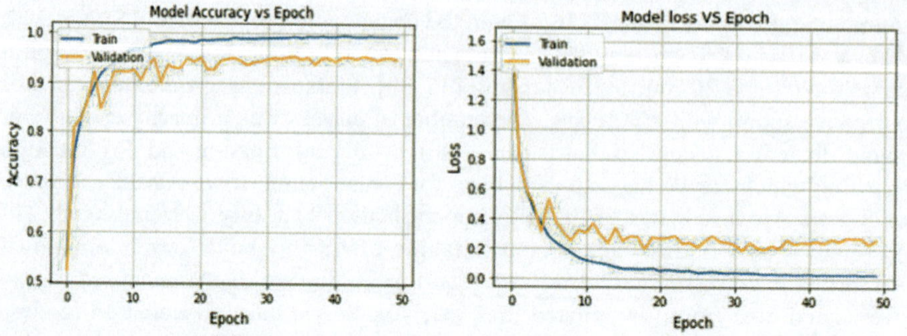

(a) (b)

Fig. 5 Training and validation accuracy and loss of CNN model

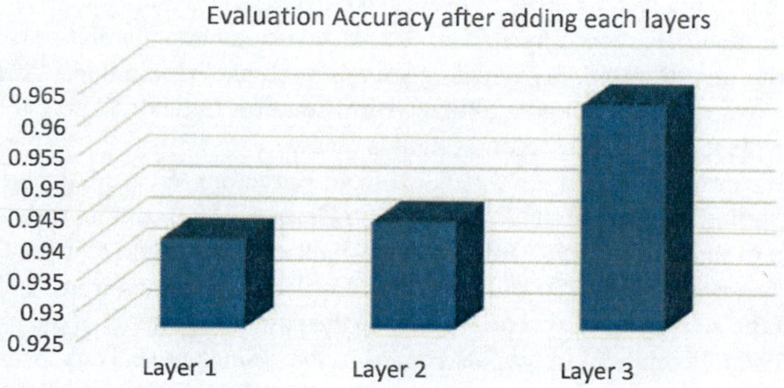

Fig. 6 Layer wise comparison of evaluation accuracies of the model

accuracy after the addition of third convolution layer. Figure 6 presents the layer wise comparison of the evaluation accuracies of the proposed model.

The X_Test was used for testing our model. For visualization of testing, the confusion matrix is a better technique as it summarizes the outcome of the prediction accuracy of the classification algorithm. The confusion matrix in Fig. 7 represents that our model has perfectly classified each sound.

The classification report generated in Fig. 8 represents the prediction quality of the classification algorithm. It represents precision, recall, and F1 score for each class. Recall indicated the capability of our CNN model to find all the true positive classes. The F1 score is the weighted harmonic mean of precision and recall. So the classification results of our CNN model again showed the accuracies in the prediction of our CNN architecture.

We have compared the outcome of CNN algorithm with three other machine learning algorithms—decision tree classifier, K-nearest neighbor, and random forest classifier in terms of classification accuracies.

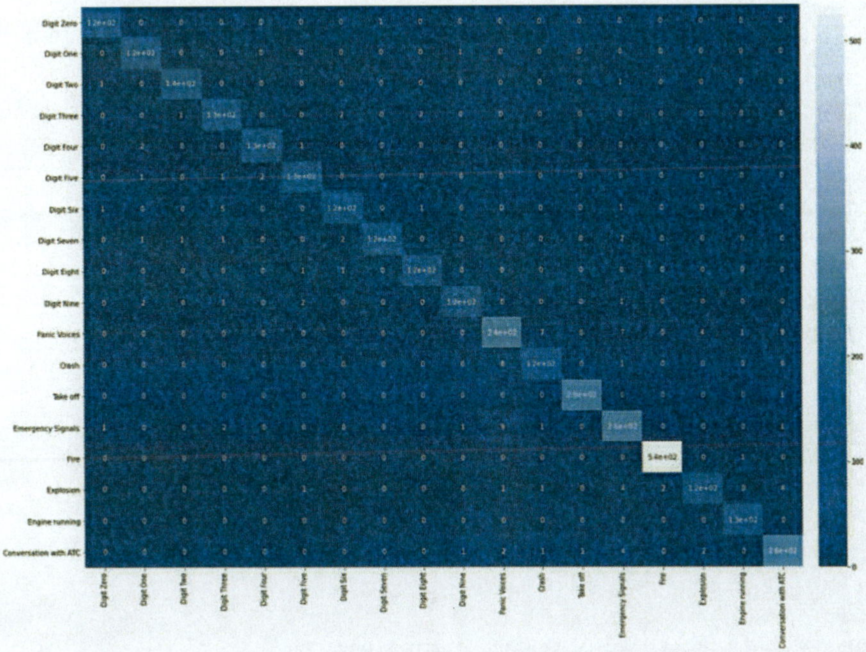

Fig. 7 Confusion matrix of our CNN model

Fig. 8 Classification report of our CNN model

	PRECISION	RECALL	F1-SCORE	SUPPORT
DIGIT ZERO	0.99	0.98	0.98	127
DIGIT ONE	0.89	0.98	0.93	129
DIGIT TWO	0.99	0.97	0.98	138
DIGIT THREE	0.94	0.90	0.92	147
DIGIT FOUR	0.95	0.95	0.95	121
DIGIT FIVE	0.98	0.97	0.98	142
DIGIT SIX	0.98	0.91	0.94	120
DIGIT SEVEN	0.97	0.98	0.98	125
DIGIT EIGHT	0.91	0.98	0.95	116
DIGIT NINE	0.98	0.94	0.96	139
PANIC VOICES	0.94	0.91	0.93	258
CRASH	0.9	0.92	0.91	142
TAKE OFF	1.00	1.00	1.00	279
EMERGENCY SIGNALS	0.94	0.90	0.92	275
FIRE	0.99	1.00	1.00	527
EXPLOSION	0.91	0.88	0.90	134
ENGINE RUNNING	0.97	0.99	0.98	132
CONVERSATION WITH ATC	0.89	0.94	0.92	266
ACCURACY			0.95	3317
MACRO AVG	0.95	0.95	0.95	3317
WEIGHTED AVG	0.96	0.95	0.95	3317

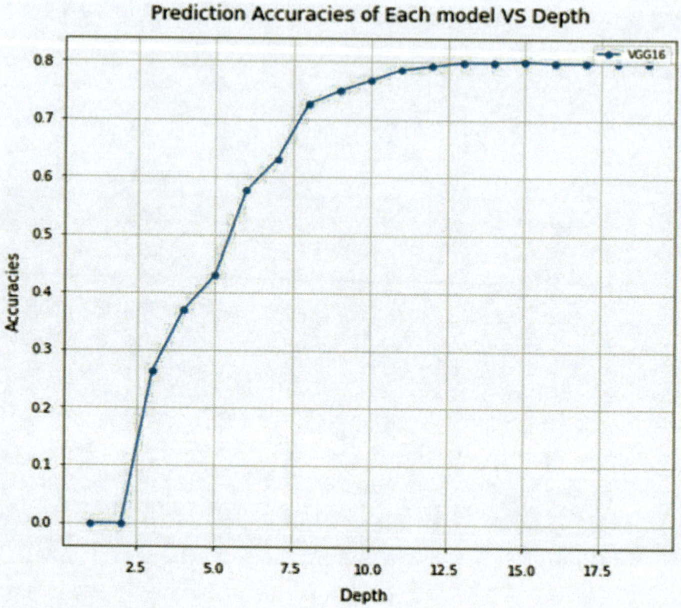

Fig. 9 Predict accuracy versus MAX-depth of decision tree

We have trained a decision tree classifier and used it to predict X_Test. Initially, we visualize how prediction accuracies vary of the decision tree classifier varies with its depth. The curve in Fig. 9 portraits the prediction accuracy score with the depth of the decision tree. It shows that the prediction accuracies gradually increased from depth = 3 and it reaches a maximum at epoch = 15 with a prediction score reaching 0.8.

To compare the classification outcome, the K-nearest neighbor algorithm is also incorporated. The value of *K* is varied and the prediction accuracy score noted. To observe the difference in prediction score with the value of *K*, a visualization in terms of bar chart was created in Fig. 10.

The pie-chart in Fig. 11 shows the maximum accuracies of the random forest classifier, decision tree classifier, and KNN algorithm. The classifiers were used to predict with X_Test and the output of the prediction were visualized in Table 1 classification report.

The outcomes had clearly portrayed that the proposed CNN model had outperformed these machine learning algorithms in classification of the sounds.

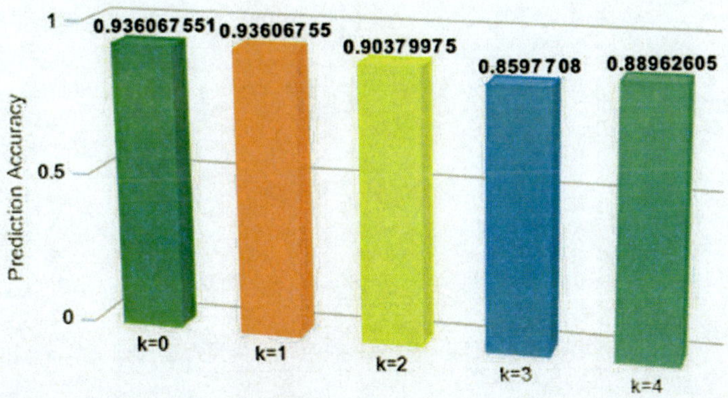

Fig. 10 Prediction accuracy per K_Neighbors

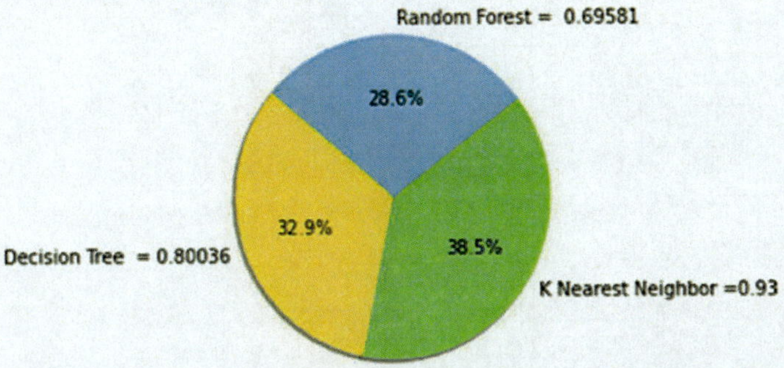

Fig. 11 Pie-chart indicating the prediction accuracies of random forest, KNN, and decision tree classifier

5 Conclusion

In this paper, we used our proposed CNN model and other existing machine learning model to analyses different kinds of sounds commonly present in CVR recording. We generated an audio dataset to train our models. The proposed CNN model gave the most accurate results compare to the machine learning algorithms reaching classification accuracy closer to 97%.

The future scope of this work includes but not limited to: audio forensic result verification in crime investigation, accident investigation via artificial intelligence, automated audio to text conversion, and noise pollution sound source classification.

Table 1 Classification report comparison of KNN, decision tree, and random forest, respectively

		Digit zero	Digit one	Digit two	Digit three	Digit four	Digit five	Digit six	Digit seven	Digit eight	Digit nine	Panic	Crash	Take off	Emergency signal	Fire	Explosion	Engine running	Conversation with ATC
KNN	Precision	0.99	0.90	0.98	0.89	0.98	0.90	0.92	0.91	0.90	0.94	0.91	0.84	0.98	0.92	0.98	0.92	0.93	0.94
	Recall	0.96	0.94	0.94	0.87	0.93	0.95	0.79	0.88	0.97	0.87	0.84	0.89	1.00	0.82	0.99	0.89	1.00	0.87
	F1_Score	0.98	0.92	0.96	0.88	0.95	0.92	0.85	0.89	0.93	0.90	0.87	0.87	0.99	0.87	0.99	0.90	0.96	0.89
	Support	125	125	142	135	130	135	125	130	119	129	265	140	261	275	536	138	134	272
Decision tree	Precision	0.74	0.72	0.70	0.61	0.8	0.68	0.68	0.57	0.71	0.76	0.78	0.85	0.98	0.73	0.74	0.74	0.98	0.80
	Recall	0.74	0.70	0.65	0.66	0.78	0.71	0.70	0.62	0.63	0.85	0.75	0.79	0.99	0.76	0.94	0.75	0.95	0.78
	F1_Score	0.75	0.71	0.67	0.63	0.79	0.69	0.69	0.60	0.67	0.80	0.76	0.82	0.98	0.75	0.94	0.75	0.97	0.79
	Support	125	125	142	135	130	135	125	130	119	129	265	140	261	275	536	138	134	272
Random forest	Precision	0.99	1.00	1.00	0.98	1.00	0.98	1.00	1.00	0.96	0.98	0.99	0.99	1.00	0.98	0.98	0.98	1.00	0.94
	Recall	0.71	0.61	0.43	0.45	0.78	0.42	0.37	0.37	0.66	0.48	0.70	0.66	0.98	0.59	0.98	0.68	1.00	0.94
	F1_Score	0.83	0.70	0.60	0.62	0.88	0.59	0.54	0.54	0.78	0.65	0.83	0.79	0.99	0.73	0.98	0.8	1.00	0.81
	Support	125	125	142	135	130	135	125	130	119	129	265	140	261	275	536	138	134	272

References

1. R.C. Maher, Application example 2: cockpit voice recorders, in *Modern Acoustics and Signal Processing Principles of Forensic Audio Analysis* (2018), pp. 137–142
2. D. Cheng, C. Yi, Z. Zhang, X. Xiao, H. Yao, Comparative analyses and experiment verification on cockpit background sound' characteristic frequency, in *Fourth International Conference on Innovative Computing, Information and Control* (2009), pp. 469–475
3. A. Krizhevsky, I. Sutskever, G.E. Hinton, Imagenet classification with deep convolutional neural networks, in *Advances in Neural Information Processing Systems* (2012), pp. 1097–1105
4. H. Chen, Y. Zhang, M.K. Kalra, F. Lin, Y. Chen, P. Liao, G. Wang, Low-dose CT with a residual encoder-decoder convolutional neural network. IEEE Trans. Med. Imaging **36**(12), 2524–2535 (2017)
5. M. Anthimopoulos, S. Christodoulidis, L. Ebner, A. Christe, S. Mougiakakou, Lung pattern classification for interstitial lung diseases using a deep convolutional neural network. IEEE Trans. Med. Imaging **35**(5), 1207–1216 (2016)
6. P. Chamoso, W. Raveane, V. Parra, A. González, UAVs applied to the counting and monitoring of animals, in *Ambient Intelligence-Software and Applications* (2014), pp. 71–80
7. Y. Xu, G. Yu, Y. Wang, X. Wu, Y. Ma, Car detection from low-altitude UAV imagery with the faster R-CNN. J. Adv. Transp. (2017)
8. J. Salamon, J.P. Bello, Deep convolutional neural networks and data augmentation for environmental sound classification. IEEE Signal Process. Lett. **24**(3), 279–283 (2017)
9. M. Tschannen, T. Kramer, G. Marti, M. Heinzmann, T. Wiatowski, Heart sound classification using deep structured features, in *Computing in Cardiology Conference* (2016), pp. 565–568
10. K. Qian, Z. Zhang, A. Baird, B. Schuller, Active learning for bird sound classification via a kernel-based extreme learning machine. J. Acoust. Soc. Am. **142**(4), 1796–1804 (2017)
11. M. Hoy, B. Alexa, S. Cortana, An introduction to voice assistants. Med. Ref. Serv. Q. **37**(1), 81–88 (2018)
12. X. Li, G. Li, X. Li, Improved voice activity detection based on iterative spectral subtraction and double thresholds for CVR, in *Workshop on Power Electronics and Intelligent Transportation System*, Guangzhou (2008), pp. 153–156
13. X.Y. Han, *The Study of Blind Signal Separation and Sound Identification for Cockpit Voice Recorder* (Department of Aeronautics and Astronautics, National Cheng Kung University, 2004)
14. M. Anderson, R. Diaz, J. Foxworth, *Wavelet Analysis of a Beech 1900 Cockpit Voice Recorder* (The University of Texas at Austin, Aerospace Engineering and Engineering Mechanics, 2002). [Online]. https://www.ae.utexas.edu/courses/ase463q/design_pages/fall02/wavelet/wavelethome.html. Accessed 11 June 2020
15. F. GosNIIAS, F.L. Gromov, Acoustics-background environment analysis in civil aviation vessel cockpit, in *XX Session of the Russian Acoustical Society*, Moscow (2008)
16. X. Xiao, H. Yao, C. Guo, Automatic detection of alarm sounds in cockpit voice recordings, in *IITA International Conference on Control, Automation and Systems Engineering*, Zhangjiajie (2009), pp. 599–602
17. D. Cheng, C. Yi, Z. Zhang, X. Xiao, H. Yao, Comparative analyses and experiment verification on cockpit background sound' characteristic frequency, in *4th International Conference on Innovative Computing, Information and Control*, Kaohsiung (2009), pp. 469–475
18. D. Cheng, S. Wan, L. Ji, Analysis & fault diagnosis of cockpit voice signals based on information fusion, in *4th IEEE International Conference on Information Science and Technology*, Shenzhen (2014), pp. 106–109
19. K.J. Piczak, Environmental sound classification with convolutional neural networks, in *25th International Workshop on Machine Learning for Signal Processing*, Boston (2015), pp. 1–6
20. K. Stefan, T. Wilhelm-Stein, H. Hussein, K. Holger, K. Danny, R. Marc, E. Maximilian, *Large-Scale Bird Sound Classification using Convolutional Neural Networks* (2017)

21. O. Abdel-Hamid, A. Mohamed, H. Jiang, G. Penn, Applying convolutional neural networks concepts to hybrid NN-HMM model for speech recognition, in *IEEE International Conference on Acoustics, Speech and Signal Processing*, Kyoto (2012), pp. 4277–4280
22. C. Potes, S. Parvaneh, A. Rahman, B. Conroy, Ensemble of feature-based and deep learning-based classifiers for detection of abnormal heart sounds, in *Computing in Cardiology Conference* (2016), pp. 621–624
23. Y. Han, J. Kim, K. Lee, Deep convolutional neural networks for predominant instrument recognition in polyphonic music. IEEE/ACM Trans. Audio Speech Lang. Process. **25**(1), 208–221 (2017)
24. H. Purwins, B. Li, T. Virtanen, J. Schlüter, S.Y. Chang, T. Sainath, Deep learning for audio signal processing. IEEE J. Sel. Top. Signal Process. **13**(2), 206–219 (2019)
25. Z. Ferdoush, B.N. Mahmud, A. Chakrabarty, J. Uddin, A short-term hybrid forecasting model for time series electrical-load data using random forest and bidirectional long short-term memory. Int. J. Electr. Comput. Eng. **11**(1), 763–771 (2021)
26. N. Sabab, A.R. Chowdhury, M.I. Nirjhor, J. Uddin, Bangla speech recognition using 1D-CNN and LSTM with different dimension reduction techniques, in *International Conference on Emerging Technologies in Computing* (2020), pp. 1–12
27. X. Xia, R. Togneri, F. Sohel, Y. Zhao, D. Huang, A survey: neural network-based deep learning for acoustic event detection. Circuits Syst. Signal Process. **38**(8), 3433–3453 (2019)
28. S. Ioffe, C. Szegedy, Batch normalization: accelerating deep network training by reducing internal covariate shift (2015)
29. S. Santurkar, D. Tsipras, A. Ilyas, A. Madry, How does batch normalization help optimization?, in *Advances in Neural Information Processing Systems* (2018), pp. 2483–2493
30. N. Srivastava, G. Hinton, A. Krizhevsky, I. Sutskever, R. Salakhutdinov, Dropout: a simple way to prevent neural networks from overfitting. J. Mach. Learn. Res. **15**(1), 1929–1958 (2014)
31. M. Xu, L.Y. Duan, J. Cai, L.T. Chia, C. Xu, Q. Tian, HMM-based audio keyword generation, in *Pacific-Rim Conference on Multimedia* (2004), pp. 566–574
32. K.V.K. Kishore, P.K. Satish, Emotion recognition in speech using MFCC and wavelet features, in *3rd IEEE International Advance Computing Conference (IACC)*, Ghaziabad (2013), pp. 842–847
33. M. Sahidullah, G. Saha, Design, analysis and experimental evaluation of block-based transformation in MFCC computation for speaker recognition. Speech Commun. **54**(4), 543–565 (2012)
34. [Online]. https://github.com/Jakobovski/free-spoken-digit-dataset#:~:text=A%20simple%20audio%2Fspeech%20dataset,time%20as%20data%20is%20contributed. Accessed 12 June 2020
35. [Online]. https://www.telegraph.co.uk/news/2020/01/08/iran-plane-crash-170-passengers-feared-dead-ukraine-boeing-737/. Accessed 12 June 2020
36. [Online]. https://www.thedailystar.net/frontpage/us-bangla-airlines-flight-crash-guilty-who-1547413. Accessed 12 June 2020
37. [Online]. https://www.firstpost.com/world/us-bangla-airlines-crash-in-kathmandu-nepal-airport-did-communication-gap-bring-down-aircraft-probe-on-role-of-pilot-atc-confusion-must-4388147.html. Accessed 13 June 2020
38. [Online]. https://www.thedailystar.net/world/south-asia/news/us-bangla-plane-crash-pilot-error-vs-tower-error-1694020. Accessed 13 June 2020
39. [Online]. https://news.google.com/newspapers?id=hAoxAAAAIBAJ&sjid=8eEFAAAAIBAJ&pg=1932%2C3182582. Accessed 13 June 2020
40. [Online]. https://lifeinsaudiarabia.net/nigeria-airways-flight-2120-crashed/. Accessed 13 June 2020
41. [Online]. https://www.bea.aero/docspa/2000/5y-n000130a/pdf/5y-n000130a.pdf. Accessed 15 June 2020

42. [Online]. https://www.nytimes.com/news-event/germanwings-flight-9525-crash. Accessed 15 June 2020
43. [Online]. https://www.thestar.com/news/world/2015/03/26/germanwings-co-pilot-intententionally-put-plane-in-dive-prosecutor-says.html. Accessed 16 June 2020

Bondhu Tank: An Automated Smart Water Management System

Md. Sazidur Rahman, Sazzad Alam Bhuiyan, Md. Arafat Hossain, Md. Jisan Bin Islam, and Jia Uddin

Abstract Water is an essential element in our daily life. Conventional water tanks in the households are unable to detect and regulate the level of the water. As a result, an overflow occurs which leads to a huge loss of water. For areas, especially with a shortage of water distribution, water wastage is not affordable. This paper demonstrates a digital water tank which can measure temperature, water level and change the water temperature controlled by an android application. The model not only reduces the amount of wasted water without manual intervention by controlling the water flow in the water tank but also saves a significant amount of power. Additionally, the user can monitor the water level and temperature of the water by the android application from anywhere which works through the Internet. The main focus of this project is to reduce the water wastage, energy consumption, and manpower involved in it which were noticeably addressed in the state-of-the-art models.

Keywords Arduino UNO · Water pump · Water chamber · Smart water system

---

Md. S. Rahman · S. A. Bhuiyan · Md. A. Hossain · Md. J. B. Islam
Department of Computer Science and Engineering, Brac University, Mohakhali, Dhaka 1212, Bangladesh
e-mail: mdsazidur02@gmail.com

S. A. Bhuiyan
e-mail: sazzadsabab@gmail.com

Md. A. Hossain
e-mail: thearft@gmail.com

Md. J. B. Islam
e-mail: jisanuday27@gmail.com

J. Uddin (✉)
Technology Studies Department, Endicott College, Woosong University, Daejeon, South Korea
e-mail: jia.uddin@wsu.ac.kr

V. E. Balas et al. (eds.), *Proceedings of International Conference on Computational Intelligence, Data Science and Cloud Computing*, Lecture Notes on Data Engineering and Communications Technologies 62, https://doi.org/10.1007/978-981-33-4968-1_12

1 Introduction

Water has always been an integral substance of everyday life. In the present environmental situation, lowering the wastage of water has become a demanding task. Water wastage has been a very common problem for underdeveloped and developing countries in the recent years. A large amount of water wastage occurs due to the overflow of water in the water tank in our households. Consequently, this affects families with middle to low income in third world countries like Bangladesh, India, etc., since they have to bear the extra expense for the water and power that is wasted.

To address this issue, researchers and developers have faced several constraints as the level of the water in the tank has to be monitored continuously; otherwise, overflow will occur. The major issue is to make the system automatic so no manual workforce is required to monitor and control the overflow in the tank. To make the water tank more user-friendly and automated, a number of researchers proposed different models for smart water tank [1–6].

Shah et al. [1], Agrawal and Singhal [2] and Theja [3] have proposed different models which use the ultrasonic sensor to measure the water level. The ultrasonic sensor sends ultrasonic ray to calculate the distance and measure the water level accordingly. They have also made an android-based application to control the water tank remotely which is connected to Google's firebase. Google's firebase system provides real-time database to the android application which is also connected to ESP8266 nodeMCU Wi-Fi module [7]. The android application directs the data to the Google's firebase. And then the firebase sends the real-time data to the Wi-Fi module. After that, the nodeMCU passes the data to the relay that controls the motor whether to turn it ON or OFF. Patil and Singh [4] made a system which contains two water tanks, and each of the pump contains three homemade water level sensors that detect the water level and send signals to the water pump through relays. This system is fully automatic, and it reduces the manpower involved in monitoring the system. Ntuli and Abu-Mahfouz [5] made a secured water management system to ensure secure booting, secure communication, and secure firmware updates. On the other hand, Shahanas and Sivakumar [8] offer a system, where they used raspberry pi for cloud server to make a water management system for smart city. This system can monitor water level and other factors through the cloud server created by raspberry pi.

Similarly, CC3200 Wi-Fi module, android application, and data cloud server have been used in the proposed model of Wadekar et al. [9]. This Wi-Fi module exactly works like ESP8266 nodeMCU module and can be connected to cloud server. The offered system by Tak et al. [10] sends the data retrieved by sensors to the web page, and the retrieved data can also be seen by the smartphone application. They have used here real-time database management system consisting of MySQL. They have created a database model to store the data more efficiently. Romano and Kapelan [11] built a water management system based on Artificial Neural network (ANN). In this system, they have developed various ANN models to develop the neural network and use it for managing water more efficiently and distribute it more conveniently than the before. Despite having all the models, there is no specific water tank and water management system used for our house, offices, etc.

The proposed model is aimed to solve this issue easily and efficiently. We have used a water level sensor that will be in the water, and it is more accurate than the sonar sensor, and it is completely automatic. This automated system helps the users to save a significant amount of water without any intrusion. The system turns on the water pump itself without any manual controlling as the water level falls low in the water tank and switches off the motor as the water level in the tank becomes full. The temperature of the water can also be controlled remotely by an android application named Bondhu Tank. This system will make the life of the users much easier as they do not have to worry about turning on and off the motor.

The major contributions of our project are outlined as follows:

- We propose an automated water management system that controls the water flow in the water tank without any touch.
- We develop a system, where the users can control and regulate the temperature of the water in the water tank so that they do not need any additional heater or cooler.
- An android application is proposed with which the users can control the temperature of the water remotely with the help of the internet connection from anywhere at ease.

Rest of the paper is structured as follows: Sect. 2 is mainly focused on the proposed model with detailed hardware descriptions. Section 3 includes the experimental setup and flowchart of the proposed system. Finally, concludes the paper in Sect. 4.

2 System Architecture

A prototype of the proposed system is presented in Fig. 1. The prototype can be divided into two sections: hardware and software.

2.1 Components Used in the Model

The list of components is given in the following sub-sections.

(1) *Arduino UNO*: In this project, Arduino is used for controlling the water pump. A 16 × 2 LCD screen is also connected to the Arduino UNO [12] which displays the condition of the water tank. A detail pin configuration of Arduino is presented in Table 1.
(2) *ESP8266 nodeMCU*: NodeMCU is an open-source IoT (Internet of Things) platform [7]. It has a firmware which runs on the ESP8266 Wi-Fi System on Chip (SoC). In this system, nodeMCU is used to control the temperature of

Fig. 1 Prototype of the Bondhu Tank system

Table 1 Pins configuration of Arduino

Pin number	Function	Type
7	Motor pin	Output
6	Motor pin	Output
A1	Sensor pin	Input
A0	Sensor pin	Input
0	LCD pin	Output
9	LCD pin	Output
Vin	Power from buck	Input
gnd	Ground	Input

the water. The nodeMCU is connected to TEC12706 through a relay which controls whether the water needs to get hot or cold. A detail pin specification is presented in Table 2.

(3) *LCD screen*: A 16 × 2 Liquid Crystal Display (LCD) is used to display characters. In this LCD screen, 32 characters can be displayed at a time. There are 2 rows and 16 characters can be placed at a time in one row. Here, the LCD

Table 2 Pins configuration of ESP8266

Pin number	Function	Type
D0	Heater pin	Output
D1	Cooler pin	Output
D2	RED LED for heater	Output
D3	Blue LED for cooler	Output

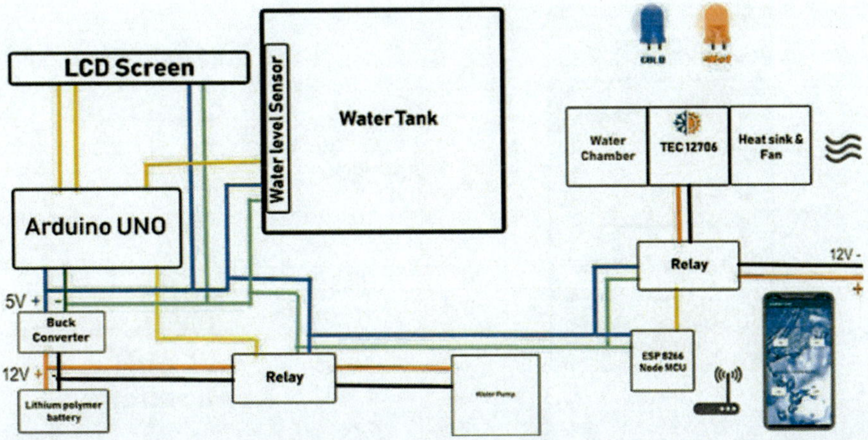

Fig. 2 Overview of the circuit configuration of Bondhu Tank

screen shows the level of water in the water tank. The LCD screen is connected to the Arduino from where it gets the status of the water tank.

(4) *Water pump*: In this proposed design, a 12 V DC water pump is used to move the water in the water tank from the main tank. This water pump can pass 100 L of water in just 1 min.

(5) *Buck converter*: Buck converter is used to step down the voltage. In this system, buck converter takes an input of 12 V power from lithium polymer battery and gives an output of 5 V power which supplies the Arduino UNO (Fig. 2).

(6) *Water chamber*: The water chamber here is for holding the water for some times to get hot or cold. This chamber is attached to the TEC12706 which controls the temperature. When the water enters the water chamber and remains there for some time. In the meantime, the temperature gets adjusted.

(7) *TEC12706*: TEC12706 is used in this model to control the temperature of the water. It is connected to a relay and attached to the water chamber. When the user needs hot water, he/she presses the button in the android application, the TEC12706 module becomes hot, and vice versa.

(8) *Android application*: The android application in this system is used to turn on/off the heater or cooler. The application is connected to Google's firebase database system which is connected to the nodeMCU module (shown in Figs. 3 and 4). When the user presses the "HEATER ON" button, the application sends the signal to Google's firebase, then from firebase to ESP nodeMCU. After that, the heater gets turned on. The same process is for turning off the heater and vice versa for cooler.

(9) *RELAY*: RELAY is used to turn on or turn off the current to the devices to which it is connected. In this system, 2 relays are used. Relay1 is associated to the water pump, and relay2 is connected to the TEC12706 module. When the

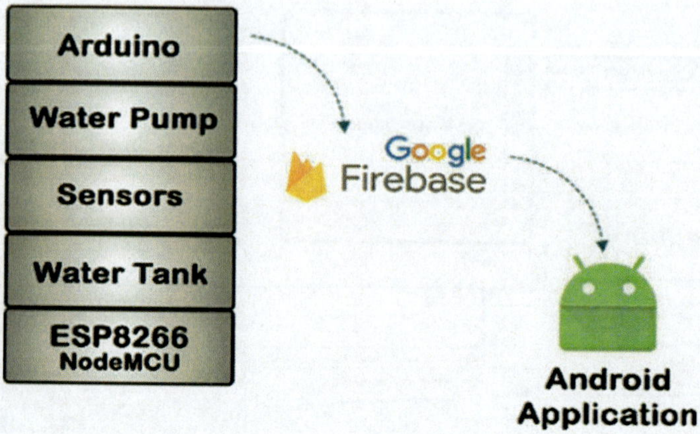

Fig. 3 Pipeline of the relation between the whole system and the android application through Google's firebase

Fig. 4 Android application user interface

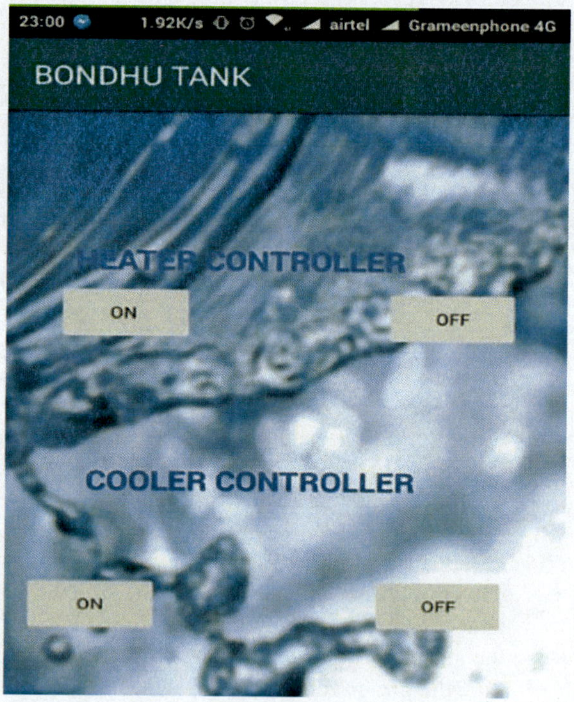

tank is full, the water pump is turned off automatically because the electricity flow is stopped by the relay. The same goes in the case of TEC12706.

(10) *Water level sensor*: The water level sensor measures the water level in the water tank. In this model, two of these sensors are implemented. The values from the sensors go to the Arduino and give output accordingly. When the tank is low, the motor is started automatically, and when the tank is full of water, the motor gets stopped according to the value of the water level sensors.

2.2 Flowchart of the Proposed Architecture

Figure 5 shows the mechanism of the Arduino UNO connected with the water level sensor. This sensor is theoretically output an analog signal between 0 and 1024. But in practical life, the range can be varied sensor to sensor. In this case, the range was in between 0 and 600. If the sensor value sent to the micro-controller is less than or equal to 100, then it will be displayed as "Empty" on the LCD screen.

Again, if the value is less than or equal to 200, then it will be printed "Low," and if the value obtained by the sensor is less than or equal to 300, then "Medium" will be printed, and if none of the above condition matches, then it will be displayed 'Full' on the 16×2 LCD screen.

(1) *Pin consideration*: In the case of ESP8266 nodeMCU in D0 and D1 pins, the heater and cooler are connected. The values of the pins are controlled by the Bondhu Tank android application. The values of those pins can either be 0 or 1. The D3 and D4 pins are connected to two LEDs. If the heater is on, then the red LED turns on, and when the cooler is on, the blue LED turns on as depicted in Fig. 6.

On the other hand, the water pump or the motor is connected to 6 and 7 no. pins of the Arduino. The water level sensors give output to the analog pins A0 and A1. The two LCD output pins are connected to 0 and 9 no pins of Arduino. The power and ground to the Arduino micro-controller come from Buck which are connected to two pins of Arduino which are Vin and gnd.

The details of the stages are as follows:

Algorithm 1

Stage 1: Checks whether the water level sensors are giving proper value or not. After the confirmation, the sensors' values are calibrated to measure the water level.

Stage 2: Motor responds according to the sensors' values.

Stage 3: Smartphone application sends data to Google's firebase properly and the values of Google's firebase are updated.

Stage 4: The firebase is connected to the heater and cooler via ESP8266 nodeMCU, and it is working properly.

Stage 5: All the components are synchronized to each other and are working properly.

The flow diagram illustrated in Fig. 6 shows the conditions that the heater and cooler follow which are connected to the ESP8266 nodeMCU. If the heater status is

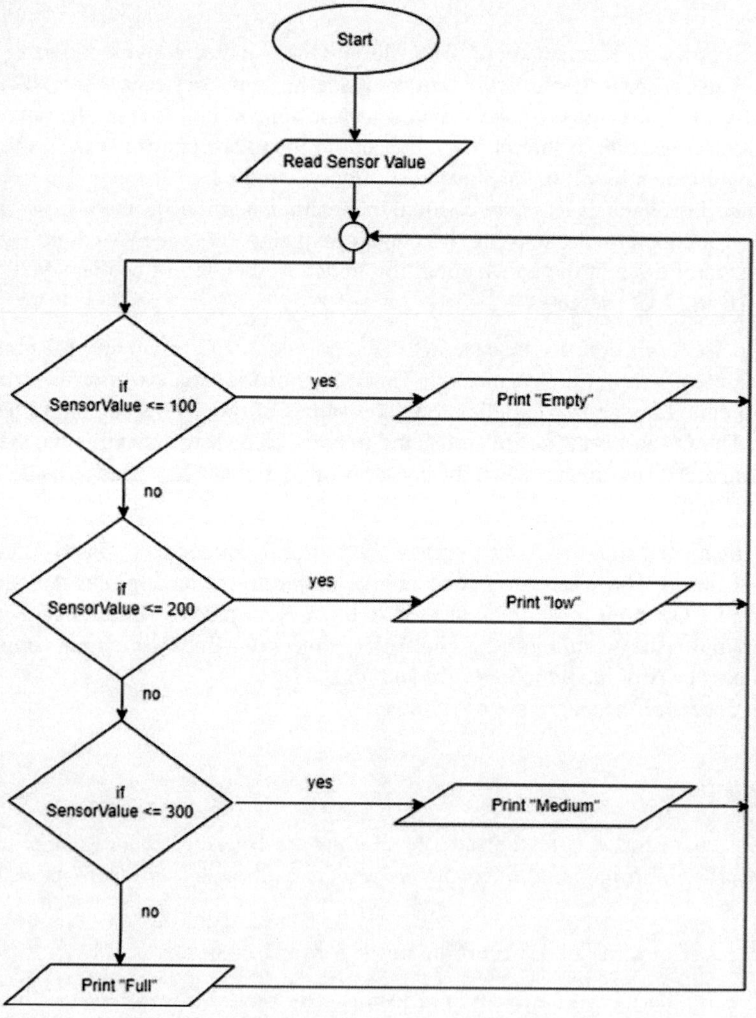

Fig. 5 Flowchart of the proposed architecture

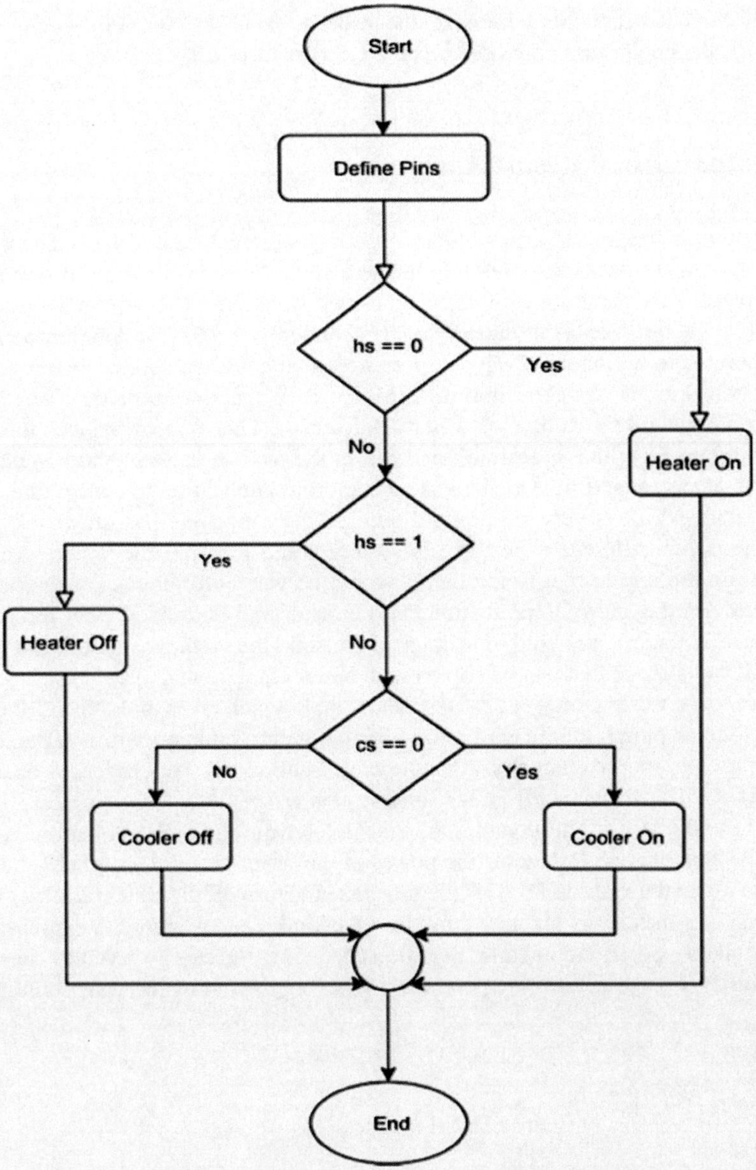

Fig. 6 Control flow of the ESP8266 nodeMCU with the heater and cooler

0, then the heater turns on; otherwise, the heater turns off. Again, if the cooler status is 0, then the cooler turns on; otherwise, the cooler turns off.

3 Experimental Result Analysis

This system is designed to save water from getting wasted. It passed several necessary tests to confirm that it runs smoothly through any unwanted situation. It was tested in various environments to avoid these unwanted situations. The system was running smoothly when it was taken outside in 37 °C and also in 16 °C indoor temperature. And there were no major or minor differences that could be spotted. After several calibration tests, it was clear that the Arduino code that was generated acted very responsive toward the values of water level sensor. This system reduces the need of waking up early in the morning and fill up the tank. It can save both water and our time at the same time. One can use the android application to control the water temperature which is very simple and quick. Every machine has drawbacks. Our machine is not different either. If this system could measure the water level and show it on the app in real-time, then it would be very informative for the people. Furthermore, the android application can be faster and optimized even further for better use. If there is any kind of leakage in the tank, the system cannot detect it. So, it is still a challenge that can be solved with better equipment and tweaks.

If we use a water pump of 1.5 BHP and a tank which takes an hour to fill up by this particular pump, then each time without Bondhu Tank system, it takes 1 min to turn off the motor manually after the tank is filled up. This process consumes 4095.54 KJ. But if we use Bondhu Tank system in this very same process, then it consumes 4028.4 KJ. These values are calculated from the below equations, where E denotes the energy, P denotes the power of the water pump, t_{full} defines the time taken to fill up the tank, and t_{extra} is the time taken to turn off the water tank manually. $E_{\text{w/o}\,B}$ defines the energy taken without Bondhu Tank system, and $E_{\text{w}/B}$ expresses the Energy taken using Bondhu Tank. Equation (2) defines the energy saved by the water tank. Lastly, Eq. (3) defines the percentage of energy saved by the water tank.

$$E = P(t_{\text{full}} + t_{\text{extra}}) \tag{1}$$

$$\text{Energy Saved} = E_{\text{without}\,B} - E_{\text{with}\,B} \tag{2}$$

$$\text{Energy Saved} = \frac{(E_{\text{without}\,B} - E_{\text{with}\,B})}{E_{\text{without}\,B}} * 100\% \tag{3}$$

If we compare these two processes, then we can save up to 1.66% of energy per process. If we continue this process with Bondhu Tank system 1 month or 30 days, then we can save up to 2014.2 KJ, and if we continue to use this system for 2 months, then we can save upto 4028.4 KJ which can save electricity of one day. Again, the

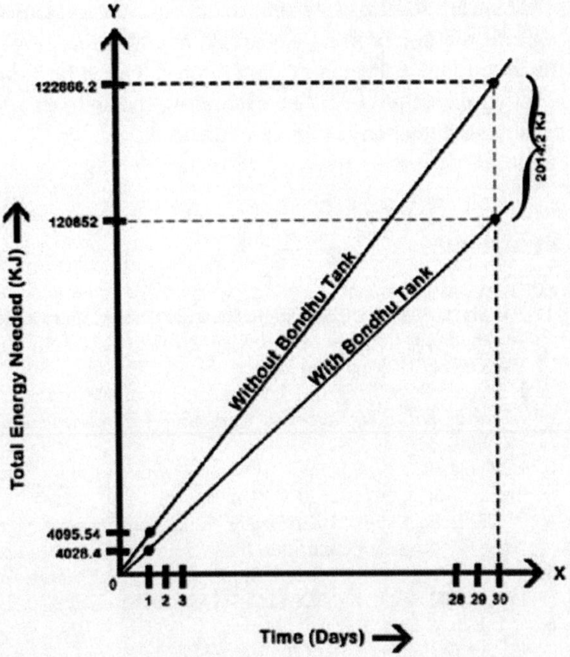

Fig. 7 Comparison of consumption of energy with and without Bondhu Tank system

water pump we have used of power 1.5 BHP can pump upto 10–120 L of water per minute. Without Bondhu Tank process, it takes 1 min extra to turn off the tank which causes the wastage of 10–120 L of water. But with Bondhu Tank system, we can save upto 10–120 L of water per process which is extremely efficient (Fig. 7).

4 Conclusion

Bondhu Tank addresses the features of solving the water wastage problems by avoiding excess usage of water and promises an efficient water management system as it controls the water levels automatically without any manual touch of the user. There are many other automated water level monitoring systems that are also present, but so far most of the systems have some major drawbacks as most of the systems use sonar sensors. It cannot give accurate readings when an unexpected object appears in between water and sonar sensor. Proposed Bondhu Tank is designed to overcome these drawbacks. Water level sensor was used in Bondhu Tank. As a result, it is giving more accurate values than the systems that use sonar sensors. We implemented an efficient automated water level monitoring and controlling system to give the perfect solution to this kind of daily difficulties that people face. This smart tank can be made in a wide range of different sizes to fulfill the needs of the water supply of apartments, buildings, bungalows, etc.

There are several directions to improve this system. One would be to use a machine learning method or a real-time reinforcement learning method to train the model from the individual water tank's data to produce the best output for the particular system. Another direction would be upgrading the android app and using leakage detecting sensors to detect any leakage in the tank.

References

1. P.P. Shah, A.A. Patil, S.S. Ingleshwar, IoT based smart water tank with android application, in *International Conference on I-SMAC (IoT in Social, Mobile, Analytics and Cloud) (I-SMAC)*, Palladam (2017), pp. 600–603
2. N. Agrawal, S. Singhal, Smart drip irrigation system using raspberry pi and Arduino, in *International Conference on Computing, Communication and Automation*, Noida (2015), pp. 928–932
3. B. Theja, IOT based smart water tank with android application. Int. J. Res. Appl. Sci. Eng. Technol. **6**(1), 2622–2627 (2018)
4. Y. Patil, R. Singh, Smart water tank management system for residential colonies using Atmega128A microcontroller. Int. J. Sci. Eng. Res. **5**(6) (2014)
5. N. Ntuli, A. Abu-Mahfouz, A simple security architecture for smart water management system. Procedia Comput. Sci. **83**, 1164–1169 (2016)
6. J. Uddin, S.M.T. Reza, Q. Newaz, J. Uddin, T. Islam, J. Kim, Automated irrigation system using solar power, in *7th International Conference on Electrical and Computer Engineering*, Dhaka (2012), pp. 228–231
7. B. Abid, M.R. Rumon, T. Sraboni, R. Hossain, F. Ahmed, J. Uddin, Design and implementation of an E-notice board using a NodeMCU, in *International Conference on Emerging Technologies in Computing 2020 (iCETiC'20)*. Online (London, 2020)
8. K.M. Shahanas, P.B. Sivakumar, Framework for a smart water management system in the context of smart city initiatives in India. Procedia Comput. Sci. **92**, 142–147 (2016)
9. S. Wadekar, V. Vakare, R. Prajapati, S. Yadav, V. Yadav, Smart water management using IOT, in *5th International Conference on Wireless Networks and Embedded Systems*, Rajpura (2016), pp. 1–4
10. H. Tak, D. Kwon, H.-G. Cho, Water tank monitoring and visualization system using smartphones. Int. J. Mach. Learn. Comput. 142–146 (2013)
11. M. Romano, Z. Kapelan, Adaptive water demand forecasting for near real-time management of smart water distribution systems. Environ. Model. Softw. **60**, 265–276 (2014)
12. A.M.M. Rahman, M.R. Hossain, M.Q. Mehdi, E.A. Nirob, J. Uddin, An automated zebra crossing using Arduino-UNO, in *International Conference on Computer, Communication, Chemical, Material and Electronic Engineering*, Bangladesh (2018), pp. 1–4

A Holistic Framework for Quality Evaluation of Fruits and Vegetables Suppliers

Susovan Jana, Bijan Sarkar, and Ranjan Parekh

Abstract The evaluation of the supplier is a very important task for the management of any industry in the modern-day supply chain. The supplier of fruits and vegetables is also coming under the same scanner. The fruits and vegetables are highly perishable products. The supplier selection of perishable products needs special care to reduce wastage inside the store. The evaluation of suppliers based on multiple conflicting criteria is a very challenging task. There may have many highly correlated criteria. This paper proposes a framework for supplier selection by integrating principal component analysis and technique for order of preference by similarity to ideal solution techniques. The principal component analysis is performed on the decision matrix to generate a reduced and uncorrelated set of criteria. The technique for order of preference by similarity to ideal solution is used to evaluate the suppliers of fruits and vegetables. The advantages of the proposed framework are that it reduces the dimensionality of criteria as well as the correlations among the criteria. A case study has been addressed to show the utility and effectiveness of the proposed framework.

Keywords Fruits · Vegetables · Supplier selection · Fuzzy set · PCA · TOPSIS

S. Jana (✉) · R. Parekh
School of Education Technology, Jadavpur University, Kolkata 700032, India
e-mail: jana.susovan2@gmail.com

R. Parekh
e-mail: rparekh.edutech@jadavpuruniversity.in

B. Sarkar
Department of Production Engineering, Jadavpur University, Kolkata 700032, India
e-mail: bijan.sarkar@jadavpuruniversity.in

V. E. Balas et al. (eds.), *Proceedings of International Conference on Computational Intelligence, Data Science and Cloud Computing*, Lecture Notes on Data Engineering and Communications Technologies 62, https://doi.org/10.1007/978-981-33-4968-1_13

1 Introduction

Fruits and vegetables have always been a profitable business in India. India is the second largest fresh fruit and vegetable producing country in the world just after China as per the National Horticulture Database in 2015–16. Fruits of worth 692.01 million USD and vegetable worth 777.25 million USD had been exported from India in 2018–19. The major importer of Indian fresh fruit and vegetables are UK, Netherland, UAE, Nepal, Sri Lanka, Bangladesh, Oman, Qatar, etc. The customer always wants the best quality fruit and vegetable as well as they are very much concerned about food safety. Maximizing the profit by providing better quality at lower prices was always challenging for the managerial person in the supermarket. Nearly, 30% of fruits and vegetables are wasted due to the poor supply chain management in the fruits and vegetable sector in India [1]. Figure 1 depicts the fruits and vegetables supply chain in India. The wastage of fruits and vegetables also happens in a store located in a supermarket. There are different reasons for wastage in the supermarket. The poor distribution network [2] for fruits and vegetables plays a lead role in wastage. The other reasons are bad supplier selection, poor storage facility, demand-supply gap, etc. The need is the reduction of wastage to assure a high profit. The need creates demand. The demand is an efficient and profitable supplier selection framework for the management of the supermarket.

Lin and Wu proposed a suppler selection technique for fruits and vegetables in supermarkets of Taiwan [3]. A fuzzy Delphi method was applied for screening the criteria at the initial level. Then, the fuzzy analytic hierarchy process (AHP) was used to identify relative priority among the criteria. Gupta et al. proposed a green supplier selection technique using fuzzy multi-criteria decision making (MCDM) theory [4]. Krishankumar et al. proposed a two-phase supplier selection technique for the automobile industry [5]. A linguistic-based aggregation is applied in the first phase. In the next phase, an extended preference ranking organization method for enrichment evaluation (PROMETHEE) was applied to the rank the alternatives in a fuzzy environment. A sustainable supplier selection of food was proposed for the airline industry [6] by integrating AHP, additive ratio assessment (ARAS), and multi-choice goal programming (MCGP).

This paper proposes a new framework for the supplier selection of fruits and vegetables. The set of experts analyzes the problem and identifies the conflicting criteria. Principal component analysis (PCA) is applied to reduce the number of

Fig. 1 Fruits and vegetable supply chain in India

criteria as well as correlations among the criteria. The technique for order of preference by similarity to ideal solution (TOPSIS) is used to rank the suppliers based on the closeness toward the positive ideal solution (PIS) and farness from the negative ideal solution (NIS). The paper has been organized as: Sect. 2 elucidates the background study on the techniques used in the proposed framework. Section 3 outlines the proposed framework. Section 4 describes the identification of criteria. Section 5 brings a case study with a numerical example. Section 6 concludes the work.

2 Preliminaries

This paper proposes a holistic approach to address the supplier selection problem for fruits and vegetables. The framework is a blend of PCA, fuzzy set theory, and TOPSIS technique. In this section, those techniques are described in brief.

2.1 Principal Component Analysis

The main goal of PCA is to reduce dimension and identify the hidden pattern of a dataset consisting of many variables correlated with each other. PCA was proposed by Pearson [7] and developed by Hotelling [8]. The steps for PCA:

Step 1: The raw decision matrix ($D_{m\times n}$) has been established with m number of alternatives and n number of criteria. There are two types of criteria, i.e., benefit (higher the better and indicated by "+" sign) and cost (lower the better and indicated by "−" sign).

Step 2: The benefit and cost criteria are normalized using Eqs. (1) and (2), respectively. The normalized matrix is $\mathrm{ND}_{m\times n}$.

$$d_i^+(j) = \frac{d_i(j) - [\min(d_i(j))]}{[\max(d_i(j))] - [\min(d_i(j))]} \tag{1}$$

$$d_i^-(j) = \frac{[\max(d_i(j))] - d_i(j)}{[\max(d_i(j))] - [\min(d_i(j))]} \tag{2}$$

Step 3: Compute the covariance matrix, which is an $n \times n$ symmetric matrix. Equation (3) is an example of an $n \times n$ covariance matrix.

$$C = \begin{bmatrix} \mathrm{cov}(1,1) & \mathrm{cov}(1,2) & \cdots & \mathrm{cov}(1,n) \\ \mathrm{cov}(2,1) & \mathrm{cov}(2,2) & \cdots & \mathrm{cov}(2,n) \\ \vdots & \vdots & \ddots & \vdots \\ \mathrm{cov}(n,1) & \mathrm{cov}(n,2) & \cdots & \mathrm{cov}(n,n) \end{bmatrix} \tag{3}$$

Step 4: Calculate the eigenvalues $(\lambda_1, \lambda_2, \ldots, \lambda_n)$ and the corresponding eigenvectors $(v_1, v_2, \ldots, v_n)$ from the covariance matrix. Refer to Eq. (4).

$$(C - \lambda I)v = 0 \tag{4}$$

Step 5: The number of dimensions is reduced from the scree plot of eigenvalues. The selected principal components should have the maximum variance.

Step 6: Assume, k number of eigenvectors are selected $(V_{n\times k})$ from n number of eigenvectors $(V_{n\times n})$, where $k < n$. The new decision matrix is computed by doing the steps mentioned below

- Transpose the selected eigenvectors, $\mathrm{EV}_{k\times n} = \left[V_{n\times k}\right]^{\mathrm{T}}$
- Transpose the normalized decision matrix, $\mathrm{TND}_{n\times m} = \left[\mathrm{ND}_{m\times n}\right]^{\mathrm{T}}$
- $\mathrm{TD}_{k\times m} = \mathrm{EV}_{k\times n} \times \mathrm{TND}_{n\times m}$
- Transpose $\mathrm{TD}_{k\times m}$ to get a new dataset $\mathrm{TD}_{m\times k} = \left[\mathrm{TD}_{k\times m}\right]^{\mathrm{T}}$ and rename columns.

2.2 *Fuzzy Set Theory*

A fuzzy set is a very popular mathematical tool to express vagueness and ambiguity. Prof. Lotfi Asker Zadeh introduced the fuzzy set [9] in 1965. In a fuzzy set, if anyone asks you "How good you are in Mathematics?," then the answer may be "Very Good," "Good," "Average," "Bad," and "Very Bad." Here comes the degree of membership. A fuzzy set $\tilde{A}$ can be defined as $\{(x, \mu_A(x))|x \in X\}$, where $\mu_{\tilde{A}}(x)$ is the membership function of x in universe of discourse X. Here, a triangular membership function is used to represent the expert's rating in the decision matrix. It is constructed with a triplet $\{a, b, c\}$. Refer to Eq. (5). Table 1 represents the linguistic ratings and corresponding triangular fuzzy numbers as well as defuzzified crisp values. The center of gravity (COG) method is used for the defuzzification of triangular fuzzy number.

Table 1 Linguistic rating, fuzzy number, and defuzzified crisp value

Linguistic terms	Abbreviation	Triangular fuzzy numbers	Defuzzified crisp number
Very low	VL	(1, 1, 3)	1.66
Low	L	(1, 3, 5)	3
Average	A	(3, 5, 7)	5
High	H	(5, 7, 9)	7
Very high	VH	(7, 9, 9)	8.33

$$\mu_{\tilde{A}}(x) = \begin{cases} 0 & \text{if } x \leq a \\ \frac{x-a}{b-a} & \text{if } a \leq x \leq b \\ \frac{c-x}{c-b} & \text{if } b \leq x \leq c \\ 0 & \text{if } c \leq x \end{cases} \quad (5)$$

2.3 Technique for Order of Preference by Similarity to Ideal Solution

The TOPSIS is a multi-criteria decision-making tool. It was originated by Hwang and Yoon [10]. Later, some modifications were made by Yoon [11]. The steps for TOPSIS:

Step 1: The computations of TOPSIS begin with the forming of a decision matrix (D). If there are m number alternatives and n number of criteria, then D will be a ($m \times n$) matrix. Refer to Eq. (6). The element d_{ij} in D will represent the rating of ith alternative and jth criteria.

$$D = \begin{array}{c} \\ A_1 \\ A_2 \\ \vdots \\ A_m \end{array} \begin{array}{c} \begin{array}{cccc} C_1 & C_2 & \cdots & C_n \end{array} \\ \begin{bmatrix} d_{11} & d_{12} & \cdots & d_{1n} \\ d_{21} & d_{22} & \cdots & d_{2n} \\ \vdots & \vdots & \ddots & \vdots \\ d_{m1} & d_{m2} & \cdots & d_{mn} \end{bmatrix} \end{array} \quad (6)$$

Step 2: The decision matrix is normalized by dividing with the root of squared summation of corresponding criteria. $R_{m \times n}$ denotes the normalized version of D.

Step 3: Construct a weighted normalized decision matrix ($S_{m \times n}$) by multiplying a particular weight with the corresponding criteria. Refer to Eq. (7).

$$s_{ij} = w_j * r_{ij} \quad (7)$$

Step 4: The highest score for benefit criteria and the lowest score for cost criteria are selected as *PIS*. The lowest score for benefit criteria and highest score for cost criteria are selected as *NIS*. Refer to Eqs. (8) and (9) for *PIS* and *NIS*, respectively.

$$A^+ = \left(s_1^+, s_2^+, \ldots, s_n^+\right) \text{ where } s_j^+ = \left\{\left(\max_i s_{ij} | j \in J\right) i = 1, 2, \ldots, m\right\} \quad (8)$$

$$A^- = \left(s_1^-, s_2^-, \ldots, s_n^-\right) \text{ where } s_j^- = \left\{\left(\min_i s_{ij} | j \in J\right) i = 1, 2, \ldots, m\right\} \quad (9)$$

Step 5: Calculate the *n*-dimensional Euclidean distance from *PIS* as well as from *NIS* using Eqs. (10) and (11), respectively.

$$E_i^+ = \sqrt{\sum_{j=1}^{n}\left(s_{ij} - s_j^+\right)^2}, \quad i = 1, 2, \ldots, m; \; j = 1, 2, \ldots, n \tag{10}$$

$$E_i^- = \sqrt{\sum_{j=1}^{n}\left(s_{ij} - s_j^-\right)^2}, \quad i = 1, 2, \ldots, m; \; j = 1, 2, \ldots, n \tag{11}$$

Step 6: Calculate the closeness coefficient (*CC*) for each alternative using Eq. (12). Rank the alternatives based on closeness coefficient in descending order.

$$CC_i = \frac{E_i^-}{E_i^+ + E_i^-}, \quad i = 1, 2, \ldots, m \tag{12}$$

3 Proposed Framework

Supplier selection for the supermarket is an MCDM problem. The supplier selection problem became more challenging for perishable items. Fruits and vegetables are highly perishable in nature though it will be stored in a freezer. In this paper, a holistic framework has been proposed for supplier selection of fruits and vegetables in a supermarket. Figure 2 depicts the complete overview of the proposed framework. The proposed framework has been divided into three phases.

Phase I: In the first phase, an expert committee is formed. The committee consists of three experts. Table 2 shows the detailed range of experts in the committee. The expert's committee will have a brainstorming session to find the criteria for supplier selection. The experts will give a rating for each of the suppliers based on their observations and expertise. The expertise comes through explicit knowledge as well as tacit knowledge, which can be compared with the tip of the iceberg. A decision matrix (*D*) is constructed with alternatives and criteria. The entries in the decision matrix are given in linguistic terms by experts following Table 1. The linguistic inputs are converted into triangular fuzzy numbers. The fuzzy numbers are defuzzified using the center of gravity method for further processing of the decision matrix.

Phase II: In the second phase, the principal component analysis is performed on the normalized decision matrix. The principal component analysis will reduce the dimension of criteria. It will generate a new set of criteria, which are uncorrelated principal components. The newly generated criteria will be benefit criteria. A new uncorrelated decision matrix is formed with principal components and alternatives. The weights of

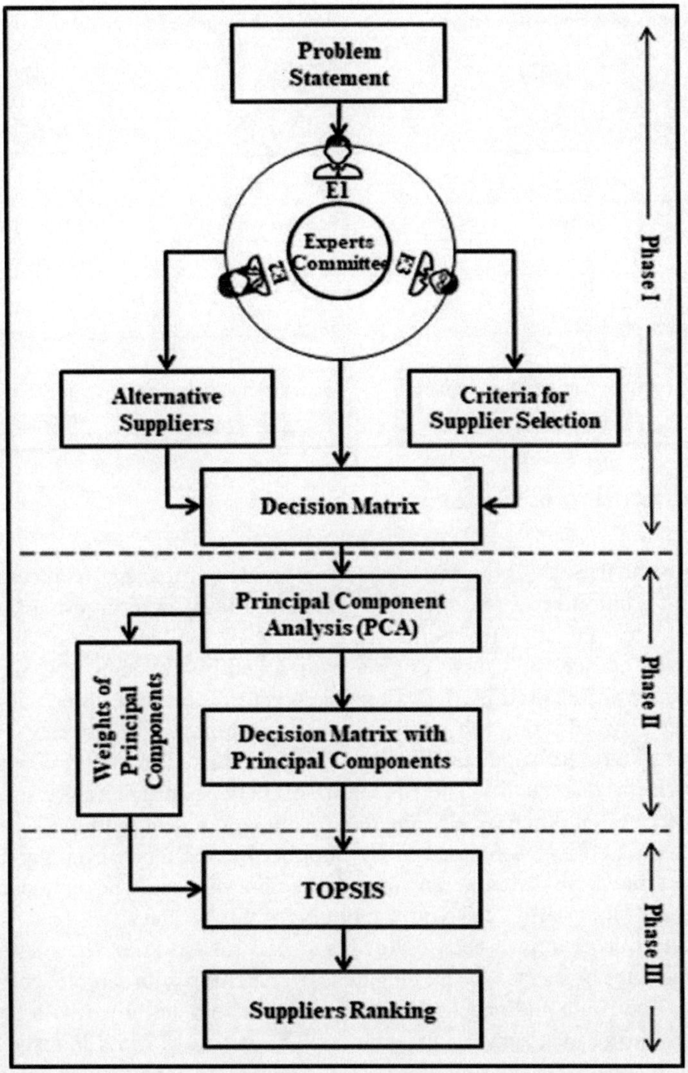

Fig. 2 Proposed framework for supplier selection of fruits and vegetable

the principal component will be determined by the normalized eigenvalues obtained from the PCA.

Phase III: In this phase, the weighted normalized matrix is obtained by multiplying weights with the corresponding criteria in the normalized decision matrix of principal components. Then, TOPSIS is applied to the weighted normalized matrix. The ranking or evaluation of the suppliers will be done by the descending order of the closeness coefficient value.

Table 2 Details of the experts in the committee (explicit knowledge)

Experts	 E1	 E2	 E3
Gender	Male	Female	Male
Age (years)	35	41	62
Qualification	BE	ME	Ph.D.
Experience (years)	12	18	32
Area of expertise	Agricultural engineer	Finance and strategic planning	Supply chain management

4 Identification of Criteria

Five criteria and their sub-criteria have been selected after managerial meetings with the experts' committee at the top level. These top-level criteria are cost, quality, service, delivery, and supplier profile. Table 3 depicts the hierarchy of the criteria. As mentioned earlier, the benefit criteria are marked by "+" sign, and non-benefit or cost criteria are marked by "–" sign. The cost criteria have three sub-criteria. i.e., product price, discount on purchase, payment policy. The price of a product is the major criteria for supplier selection of any product. The total product price is decided by the product quality. There is a trade-off between product price and product quality. Discount on purchase also matters for making a deal with the supplier. On spot, payment is always welcomed by the supplier but the managerial person of the supermarket may have different term-based payment policies. The payment method also matters for long-term deals with suppliers for this reason.

The quality criteria have been divided into five sub-criteria. These are product visual appearance of surface, packaging quality, matching with sample, consistency, and safety. The fruits and vegetable quality cannot be determined with any brand comparison like garments and other products. The quality of fruits and vegetables is determined by shape [12], size, volume [13], weight, etc. The surface color defines maturity. The price is directly depending on the quality of the fruits and vegetables. The packing quality also attracts retailers to buy from a specific supplier. Good packaging reduces the chances of damage. The delivered product should match the sample shown at the time of the deal. The consistency of the product quality is strongly desired by managerial personnel. The matching with sample and consistent quality build the trust in the supplier. The safety of the edible product is a concern for people all over the world. The farmers are using chemical fertilizer and pesticides for increasing production in a shorter time to follow market demand. Those products are not good for health. This is the reason that the retailers prefer organic products over the products cultivated using chemical fertilizer and pesticides.

Table 3 Hierarchical presentation of criteria

Serial no.	Criteria	Sub-criteria
1	Cost	Product price (−)
		Discount on purchase (+)
		Payment policy (+)
2	Quality	Surface appearance (+)
		Packaging quality (+)
		Matching with sample (+)
		Consistency (+)
		Safety (+)
3	Service	Service after the sale (+)
		Return and refund policy (+)
4	Delivery	Supplier's location (−)
		Damage during delivery (−)
		Delivery interval (−)
		Delivery efficiency (+)
		Timing flexibility (+)
		Quantity flexibility (+)
5	Supplier profile	Financial status (+)
		Cooperation (+)
		Storage condition (+)
		Storage capacity (+)
		Labor skill (+)
		Labor experience (+)

The service criteria have two sub-criteria, i.e., service after the sale and return and refund policy. The service after sale helps to build the relationship between supplier and managerial person. The supplier, who is more flexible in return and refund, gets more priority than other suppliers by the retailers.

The supplier's location, damage during delivery, delivery interval, delivery efficiency, timing flexibility, and quantity flexibility, fall under the delivery criteria. The supplier location is a very important criterion for the delivery of perishable items like fruits and vegetables. A large number of fruits and vegetable gets damaged during transportation. A local supplier also reduces the chances of damage as well as reduces transportation costs. The damage is the main concern of delivery. The damage can also be reduced by using the truck with a freezer. The availability of this kind of facilitated truck in India is very less. The interval of delivery and efficiency of delivery helps to find the trusted supplier for the long term. It satisfies the managerial team of the supermarket if the delivery is received in the time slot asked by them.

The flexibility of delivery quantity is the plus point for a supplier. The retailer will be pleased if the delivery quantity fulfills the current demand for this product.

The supplier profile has been divided into six sub-criteria. These are financial status, cooperation, storage condition, storage capacity, labor skill, and labor experience. The supplier profile is a very important criterion for long-term deals. The financial status of the supplier represents the stability of the supplier. The managerial personnel will prefer the supplier who is cooperative in every situation. The hygiene, capacity, and other conditions of the store will be looked at the time of supplier selection. The condition directly affects the safety of the food product. Skill and experience of labor add the value to the supplier profile.

5 A Case Study

The proposed framework has been validated with a case study and a numerical example. Six alternative suppliers have been identified, i.e., A1, A2, A3, A4, A5, and A6. The total number of criteria is 22 for supplier selection of fruits and vegetables. The members of the experts' committee have different levels of emotional intelligence. The experts' committee members are asked to give their rating for each criterion to all the alternative suppliers. They will come out with a single rating in linguistic terms. The linguistic terms are converted to triangular fuzzy numbers. Then, the defuzzification is done using the center of gravity method. The mapping has been already mentioned in Table 1. The defuzzified decision matrix is normalized to make it on the same scale. Table 4 depicts the defuzzified and normalized matrix. A covariance matrix is generated from the defuzzified and normalized decision matrix. Then, the eigenvalues and eigenvector are computed from the 22×22 covariance matrix. There will be 22 eigenvalues and one eigenvector for each of the eigenvalue. The eigenvalue represents the amount of variance of the corresponding eigenvector.

A scree plot is generated for the 22 eigenvalues. Figure 3 shows the scree plot. It is visible that from the sixth eigenvector, there is no variance because the eigenvalue is zero or nearly zero for all the eigenvectors. Hence, we have selected the first five eigenvectors. Then, *Step 6* of PCA is applied to get the reduced and uncorrelated decision matrix for supplier selection with five principal components as criteria. Table 5 shows the reduced and uncorrelated decision matrix. The eigenvalues of the corresponding principal components are normalized. These normalized eigenvalues are the weights of the corresponding principal components. Table 6 represents the normalized weights of principal components.

Now, we are ready to apply the TOPSIS method on the reduced uncorrelated decision matrix. Table 7 shows the weighted normalized decision matrix for TOPSIS. The positive ideal solution (*PIS*) and negative ideal solution (*NIS*) are identified from the weighted normalized decision matrix. Distance from positive ideal solution (*DPIS*) and distance from negative ideal solution (*DNIS*) is calculated by using Eqs. (10)

Table 4 Defuzzified and normalized decision matrix

	C1	C2	C3	C4	C5	C6	C7	C8	C9	C10	C11
A1	0.62	0.20	0.75	0.00	0.20	1.00	0.80	0.80	1.00	0.80	0.00
A2	1.00	0.50	1.00	0.38	0.50	0.38	0.20	0.00	0.75	0.50	0.00
A3	0.00	0.80	0.38	0.75	0.20	1.00	1.00	0.80	0.38	1.00	0.62
A4	0.00	1.00	0.00	0.75	0.80	0.38	0.50	0.20	0.00	0.00	0.25
A5	1.00	0.00	1.00	0.38	0.00	0.00	0.80	0.80	1.00	0.80	1.00
A6	0.25	0.50	0.38	1.00	1.00	0.38	0.00	1.00	0.38	0.80	0.62
	C12	C13	C14	C15	C16	C17	C18	C19	C20	C21	C22
A1	0.75	0.40	0.75	0.25	0.50	0.80	1.00	0.00	1.00	0.20	0.00
A2	0.75	1.00	1.00	1.00	0.00	0.50	0.38	1.00	1.00	0.00	0.63
A3	0.37	0.40	1.00	0.00	0.50	1.00	1.00	0.75	0.63	1.00	1.00
A4	1.00	0.00	0.75	1.00	1.00	0.80	0.38	0.38	1.00	0.50	1.00
A5	0.00	0.40	1.00	0.63	0.50	0.00	0.00	0.38	0.25	0.20	0.63
A6	0.75	1.00	0.00	0.63	1.00	0.00	0.38	0.75	0.00	0.50	0.63

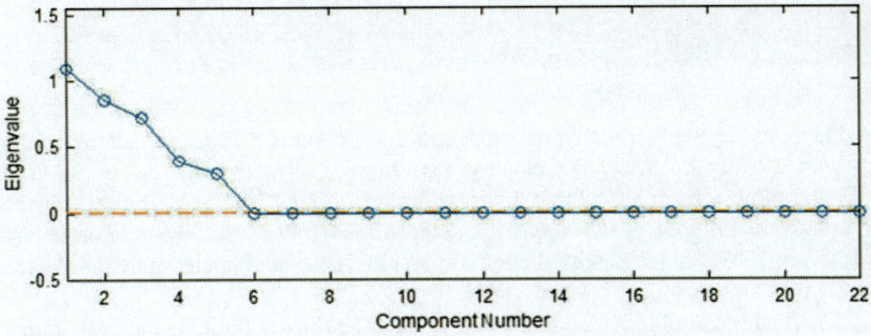

Fig. 3 Scree plot for all the principal components and corresponding eigenvalues

Table 5 Decision matrix with uncorrelated principal components

	PC1	PC2	PC3	PC4	PC5
A1	−0.49	1.59	0.00	0.73	0.42
A2	−0.46	0.17	−1.17	0.13	1.61
A3	0.79	1.45	0.95	−0.24	1.61
A4	1.63	0.44	−0.63	−0.49	0.45
A5	−1.15	−0.09	0.64	−0.74	0.67
A6	0.76	−0.82	0.74	0.76	0.97

Table 6 Normalized weights of principal components

PC1	PC2	PC3	PC4	PC5
0.32	0.26	0.21	0.12	0.09

Table 7 Weighted normalized decision matrix with principal components

	PC1	PC2	PC3	PC4	PC5
A1	−0.07	0.17	0.00	0.06	0.01
A2	−0.06	0.02	−0.13	0.01	0.05
A3	0.11	0.16	0.11	−0.02	0.05
A4	0.22	0.05	−0.07	−0.04	0.02
A5	−0.16	−0.01	0.07	−0.06	0.02
A6	0.10	−0.09	0.08	0.06	0.03

Table 8 DPIS, DNIS, CC, and rank

	DPIS	DNIS	CC	Rank
A1	0.31	0.33	0.52	4
A2	0.41	0.16	0.29	6
A3	0.14	0.44	0.75	1
A4	0.24	0.41	0.63	2
A5	0.44	0.22	0.33	5
A6	0.29	0.36	0.55	3

and (11), respectively. Finally, the closeness coefficient (*CC*) for each alternative is computed using Eq. (12). The closeness coefficient represents the closeness from *PIS* and farness from *NIS*. Table 8 shows the *DPIS*, *DNIS*, and *CC* for each of the alternatives. The rank of the supplier is determined based on the values of closeness coefficients. The supplier with the maximum closeness coefficient value is selected as the best among those six alternatives.

The final result of this case study depicts that the third supplier is the best among the alternatives. The final rank of the alternatives is **A3 > A4 > A6 > A1 > A5 > A2**. A sensitivity analysis is done to test the stability of the decision. The weights of the principal components are interchanged to generate different conditions. Then, TOPSIS will be applied for each of the conditions to check the stability of the decision. If there is n number of criteria, then we can generate C_2^n combination by exchanging the weights between two factors at a time. In this case study, we have five criteria after applying principal component analysis. Hence, 11 conditions are generated including the condition with actual weights for all the principal components. Figure 4 shows the radar diagram of sensitivity analysis. It depicts the variation of the closeness coefficient for different combinations of weights. Supplier 3 (A3) comes first and supplier 2 (A2) comes last for most of the combinations of weights. The score is not varying too much for the rest of the alternatives with respect to the different combinations of weights. This analysis signifies the proposed framework which is stable under varying conditions.

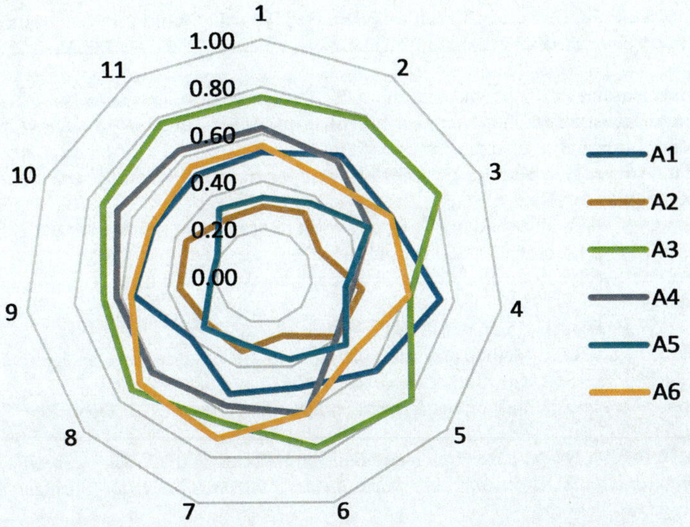

Fig. 4 Variation of closeness coefficients with a different combination of weights

6 Conclusion

This paper proposes a holistic framework for the supplier selection of fruits and vegetables in supermarkets. The previous frameworks for the supplier selection were mainly based on the combination of two techniques. These are the technique for computing weights of criteria and techniques for the ranking of suppliers. The previous frameworks suffer from lots of computations when the number of criteria is large. Those are dependent on the weights of experts' cognitive intelligence as well as emotional intelligence. This work blends PCA with TOPSIS to address the issues of the previous frameworks. This approach reduces the dimension of criteria as well as the correlation among the criteria. It also generates weights, which are not depending on the experts. This is the main contribution of this work. This framework can be validated for supplier selection problem in different sectors.

References

1. S. Negi, N. Anand, Supply chain efficiency: an insight from fruits and vegetables sector in India. J. Oper. Supply Chain Manag. **7**(2), 154–167 (2014)
2. F. Saeedi, E. Teimoury, A. Makui, Redesigning fruit and vegetable distribution network in Tehran using a city logistics model. Decis. Sci. Lett. **8**(1), 45–64 (2019)
3. P.C. Lin, L.S. Wu, How supermarket chains in Taiwan select suppliers of fresh fruit and vegetables via direct purchasing. Serv. Ind. J. **31**(8), 1237–1255 (2011)

4. S. Gupta, U. Soni, G. Kumar, Green supplier selection using multi-criterion decision making under fuzzy environment: a case study in automotive industry. Comput. Ind. Eng. **136**, 663–680 (2019)
5. R. Krishankumar, K.S. Ravichandran, A.B. Saeid, A new extension to PROMETHEE under intuitionistic fuzzy environment for solving supplier selection problem with linguistic preferences. Appl. Soft Comput. **60**, 564–576 (2017)
6. Y.K. Fu, An integrated approach to catering supplier selection using AHP-ARAS-MCGP methodology. J. Air Transp. Manag. **75**, 164–169 (2019)
7. K. Pearson, LIII. On lines and planes of closest fit to systems of points in space. Lond. Edinb. Dubl. Phil. Mag. J. Sci. **2**(11), 559–572 (1901)
8. H. Hotelling, Analysis of a complex of statistical variables into principal components. J. Educ. Psychol. **24**(6), 417 (1933)
9. L.A. Zadeh, Fuzzy sets. Inf. Control **8**(3), 338–353 (1965)
10. C.L. Hwang, K. Yoon, Methods for multiple attribute decision making, in *Multiple Attribute Decision Making* (Springer, Berlin, Heidelberg, 1981), pp. 58–191
11. K. Yoon, A reconciliation among discrete compromise solutions. J. Oper. Res. Soc. **38**(3), 277–286 (1987)
12. S. Jana, R. Parekh, Shape-based fruit recognition and classification, in *International Conference on Computational Intelligence, Communications, and Business Analytics* (Springer, Singapore, 2017), pp. 184–196
13. S. Jana, R. Parekh, B. Sarkar, Volume estimation of non-axisymmetric fruits and vegetables using image analysis, in *2019 International Conference on Computing, Power and Communication Technologies (GUCON)* (IEEE, 2019), pp. 628–633

Algorithm to Generate All Spanning Tree Structures of a Complete Graph

Suvrima Datta, Sumit Chakraborty, Maumita Chakraborty, and Rajat Kumar Pal

Abstract The objective of this paper is to propose an algorithm to generate all possible structures of spanning trees of an undirected complete graph of *n* vertices. The process starts with a star-tree (T) of the given complete graph and then replacing the edges of T one by one to generate different possible structures like chain, branch, etc. These spanning tree structures repeat themselves as we move from lower to higher values of *n*. The authors have attempted to find out some generalized expressions for different structures of spanning trees for a complete graph of order *n*.

Keywords Complete graph · Spanning tree structures · Star tree

1 Introduction

Let $G = (V, E)$ be an undirected complete graph where V is the set of vertices ($|V| = n$) and E is the set of edges. Spanning tree of G is a connected subgraph, $G' = (V, E')$ with minimum possible number of edges. Spanning trees have lots of applications in computer science, medicine, astronomy, biology, and chemistry. Generation of all possible spanning trees of a graph is a very popular problem since early 1960. Instead of finding out the trees explicitly, we have aimed to find out the

S. Datta (✉) · R. K. Pal
Department of Computer Science and Engineering, University of Calcutta, Block-JD-2, Sector III, Salt Lake, Kolkata 700106, India
e-mail: suvrimadatta2013@gmail.com

R. K. Pal
e-mail: pal.rajatk@gmail.com

S. Chakraborty
Department of Computer Science, Ramsaday College, Amta, Howrah 711401, India
e-mail: sm.chakraborty09@gmail.com

M. Chakraborty
Department of Information Technology, Institute of Engineering and Management, Y-12, Block-EP, Sector V, Salt Lake, Kolkata 700091, India
e-mail: maumita.chakraborty@gmail.com

V. E. Balas et al. (eds.), *Proceedings of International Conference on Computational Intelligence, Data Science and Cloud Computing*, Lecture Notes on Data Engineering and Communications Technologies 62, https://doi.org/10.1007/978-981-33-4968-1_14

different types of spanning tree structures, from where all the trees can be eventually formed. For the time being, we have restricted our work with only complete graphs. As the value of n (order of the input graph) increases, we found the structures do repeat themselves. This in time helps to formulate in obtaining the number of such structures, for different values of edges and vertices of different graphs.

2 Literature Survey

Several algorithms for generating spanning trees can generally be classified into the following three methods.

(i) *Test and Select Method*: In the first phase, all possible edge combinations are generated, such that no spanning tree combination is rejected. In the next phase, each combination is tested, and only unique tree combinations are selected using some tree testing algorithm. Char's algorithm [3], Sen Sarma's algorithm [8] are examples of this method.

(ii) *Elementary Tree Transformation Method*: Starting from an initial tree, the algorithms generate trees by replacing an edge of the tree by another suitable edge from the graph, such that no cycle is formed and all vertices remain connected. Hakimi's algorithm [4], Kapoor's algorithm [5], Naskar's algorithm [7] are examples of this method.

(iii) *Trees by Successive Reduction Method*: Original graph is divided into smaller subgraphs until the subgraphs become trivial like an edge. Spanning trees of the original graph are then obtained from the spanning trees of these trivial subgraphs. Winter's algorithm [10], Minty's algorithm [6] are examples of this method.

(iv) Besides the above methods for spanning tree generation, Chakraborty et al. [2] have proposed a completely different method of spanning tree generation using divide-and-conquer paradigm in their paper. In another paper by Chakraborty et al. [1], they have talked about two other types of spanning tree generation techniques, one of them with a complete graph.

3 Our Contribution

Here, we have considered an n-ordered complete graph (K_n) in which every pair of distinct vertices is connected by a unique edge. We know that the number of spanning trees in K_n is n^{n-2}, where n is the number of vertices in a complete graph, according to Cayley's theorem [9]. Let us consider the K_4 in Fig. 1.

The number of spanning trees of the K_4 graph of Fig. 1 is $4^{4-2} = 16$.

Let us consider a tree, namely a star-tree T (in Fig. 2) of the K_4. By replacing 1 through $n - 2$ number of edges from T, we can get different spanning tree structures for K_n. Let m denote the number of edges that can be replaced from T, where $m =$

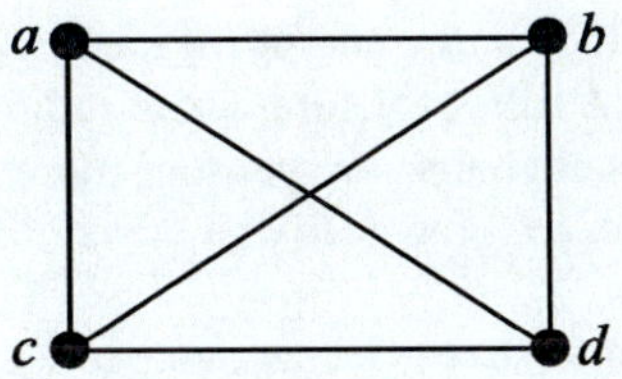

Fig. 1 K_4 graph

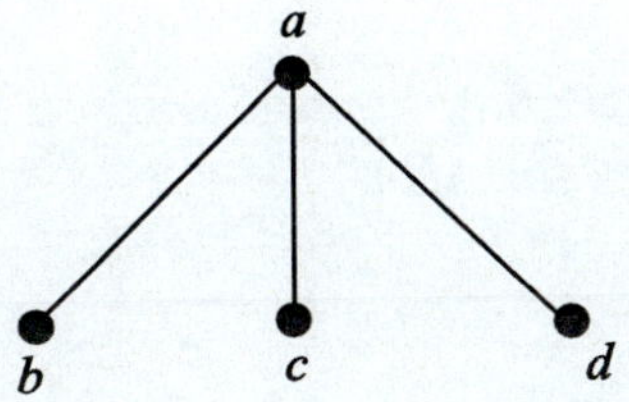

Fig. 2 Star-tree T of K_4

1, 2, 3,..., $n-2$. In spanning trees of Fig. 3(i), (ii) each, we have removed one edge from the above star-tree T, that is $m = 1$. We can choose any one edge from the edge set {*ab*, *ac*, *ad*} in $\binom{3}{1} = 3$ ways which can be added in two $(3 - 1 = 2)$ positions in $\binom{3}{1} \times 2 = 6$ ways. Thus, by replacing the edges *bd* and *cd* in Fig. 2, we can get the structures of Fig. 3(i), (ii) of diameter 3 each, respectively. Similarly, replacing edges *ab* and *ac*, one by one, from T, four more spanning tree structures similar to Fig. 3(i), (ii) can be obtained.

Now, we replace two edges. Two edges are selected from the edge set {*ab*, *ac*, *ad*} in $\binom{3}{2} = 3$ ways. We get two more spanning tree structures of diameter 2 by replacing {*ab*, *ad*} and {*ac*, *ad*} like the structure of Fig. 3(iii). In the structure of

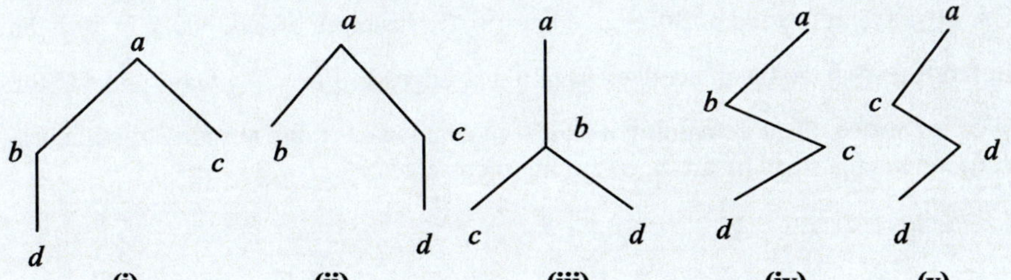

Fig. 3 Some of the spanning trees of K_4

Fig. 3(iii), two edges are added at the end of the same vertex, i.e., edges are added as branches. Now, we consider the structure where removed edges are added at the end of two different vertices as chains. We get four more spanning tree structures of diameter 3 by replacing $\{ac, ad\}$ and $\{ab, ad\}$ edges, like structures of Fig. 3(iv), (v).

The total number of spanning trees (K_4) = number of spanning trees by replacing one edge + number of spanning trees by replacing two edges + star tree = 6 + (3 + 6) + 1 = 16 = 4^2.

3.1 *Formula for Generating Some Common Structures of Spanning Trees for All* n

3.1.1 Formula 1

There are some common spanning tree structures of K_n for any value of n by replacing 1 through $n - 1$ edges. S_n represents the number of spanning trees obtained from the common structures by replacing one through $n - 2$ edges from a star-tree T.

$$
\begin{aligned}
S_n = {} & \left[\binom{n-1}{1} \times (n-2) + \binom{n-1}{2} \times (n-3) + \binom{n-1}{3} \times (n-4) + \cdots \right. \\
& \left. + \binom{n-1}{n-2} + 1\right] + \left[\binom{n-1}{2} \times 2! \times (n-1-2) + \binom{n-1}{3} \times 3! \right. \\
& \left. \times (n-1-3) + \cdots + \binom{n-1}{n-2} \times (n-2)!\right] = \sum_{m=1}^{n-2} \binom{n-1}{m} \\
& \times (n-1-m) + 1 + \sum_{m=2}^{n-2} \binom{n-1}{m} \times m! \times (n-1-m) \qquad (1)
\end{aligned}
$$

The above equation represents the total number of spanning trees obtained from the structures as shown in Fig. 4.

Replaced m edges have been added at the end of one vertex (in the structure of Fig. 4(i)) and at the end of different edges (in the structure of Fig. 4(ii)). In both the structures, we have taken m edges from $n - 1$ edges in $\binom{n-1}{m}$ ways and add the m edges at one of the remaining $n - 1 - m$ edges, and in the second structure, the replaced m edges can be arranged among themselves in $m!$ ways.

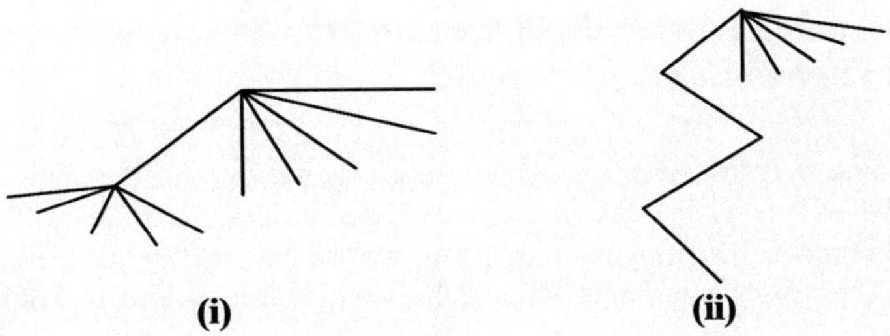

Fig. 4 Spanning tree structures for Eq. 1

3.1.2 Formula 2

In this section, we discuss about the structure which is formed only when $n \geq 5$ and $n - 1 = 2m$, where $m = 2$ through $n - 2$.

Expression (2) represents the total number of spanning trees obtained from the structures (i) and (ii) in Fig. 5, where the replaced m edges are added at the end of m different vertices.

$$\sum_{m=2}^{n-2} \binom{n-1}{m} \times m! \times \binom{n-m-1}{m} \tag{2}$$

Thus, if $n \geq 5$ and $n - 1 = 2m$,

$$S_n = \sum_{m=1}^{n-2} \binom{n-1}{m} \times (n-1-m) + 1 + \sum_{m=2}^{n-2} \binom{n-1}{m} \times m! \times (n-1-m) + \sum_{m=2}^{n-2} \binom{n-1}{m} \times m! \times \binom{n-m-1}{m} \tag{3}$$

If $n - 1 \neq 2m$, then Eq. 1 holds.

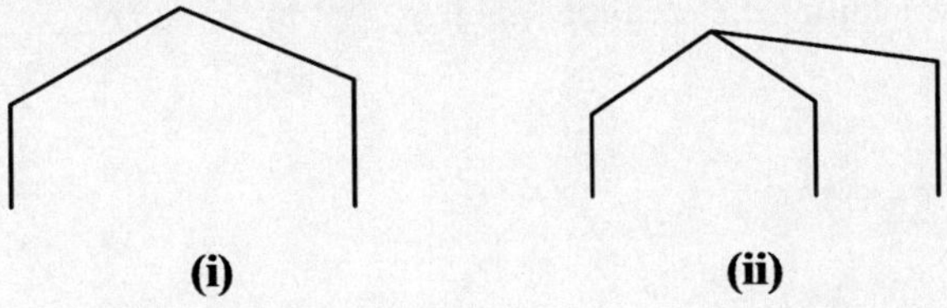

Fig. 5 Spanning tree structures for Expression (2)

3.2 Formula for Repetitive Spanning Tree Structures for Higher Values of n

Few spanning tree structures repeat themselves as we increase the value of n. For example when $n = 4$, by replacing an edge, we can get the structure like that in Fig. 6(i). For $n = 6$, by removing three edges, we can get similar structure like that in Fig. 6(ii). The structure of Fig. 6(i) is added in two positions in structure of Fig. 6(ii).

3.2.1 Recursion to Represent the Repeating Structures

To find out the number of spanning trees of n vertices, we need to find out the number of spanning trees for K_4 through K_{n-1}. In this section, we introduce a new term $F(n, m)$, where $F(n, m) = \binom{n-1}{m} \times (n - 1 - m)$, where n is the number of vertices and m is the number of replaced edges. The term $F(n, m)$ represents m edges being replaced from $n - 1$ edges and added at the end of $n - 1 - m$ different other edges. First we calculate the value of S_4 from Sect. 3.1.1. Then, $S_4 = \binom{3}{1} \times 2 + \binom{3}{2} + 1 + \binom{3}{2} \times 2! = 16 = N_4$. If S_5 denotes the number of common spanning tree structures by replacing one, two and three edges, and N_4 is the number of spanning trees of K_4, then $N_5 = S_5 + [N_4 - (\binom{3}{2} \times 2! + 1)] \times F(5, 3) = 89 + 36 = 125$.

Thus, the number of spanning trees of K_n is

$$N_n = S_n + \sum_{r=4}^{n-1} [N_r - \{(r-1)! + 1\}] \times F(n, (r-1)) + A \tag{4}$$

Here, S_n denotes the number of common spanning tree structures by replacing one through $n - 1$ edges from K_n. N_r is the number of spanning trees of K_r. We subtract $[(r - 1)! + 1]$ from N_r because the term is already included in S_n. Moreover, A represents the additional structures which are not in N_4 and N_5.

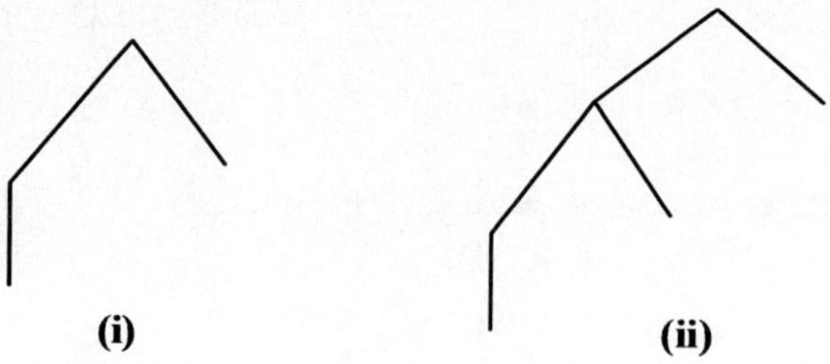

Fig. 6 Repetitive spanning tree structures for K_4 and K_6

3.3 *Partitioning*

As the value of n increases, the m number of removed edges is appended at the end of more than one edge instead of appending all the m edges at the end of one edge. In a star-tree, T, where $n = 5$, $m = 3$ number of edges can be added at the end of one edge only because the remaining number of edges is $n - 1 - m = 1$. Whereas, in a star-tree, T, where $n = 6$ and $m = 3$, the remaining number of edges is $n - 1 - m = 2$. So, we can add the three edges either at the end of one edge or three edges are partitioned as a $\langle 2 + 1 \rangle$ group and added at the end of two different edges and generate different structures. Structures (i), (ii), and (iii) in Fig. 7 represent m replaced edges being added at the end of one edge and generate different structures.

In Fig. 8, we removed three edges from a star-tree T of six vertices. These three edges are added at the end of one edge to generate different structures, shown in Fig. 8(i)–(iii). Then, they are partitioned as a $\langle 2 + 1 \rangle$ group and added at the end of two different edges, shown in Fig. 8(iv), (v).

3.3.1 Formulation of Partition

We denote the number of partitions by p.

- For any value of n, the minimum number of partitions is 2, i.e., $p_{\min} = 2$.
- For any value of n, the maximum number of partitions is $\lfloor (n-2)/2 \rfloor$, i.e., $p_{\max} = \lfloor (n-2)/2 \rfloor$. Thus, for any value of n, the number of possible partitions is $p =$ 2 through $\lfloor (n-2)/2 \rfloor$.

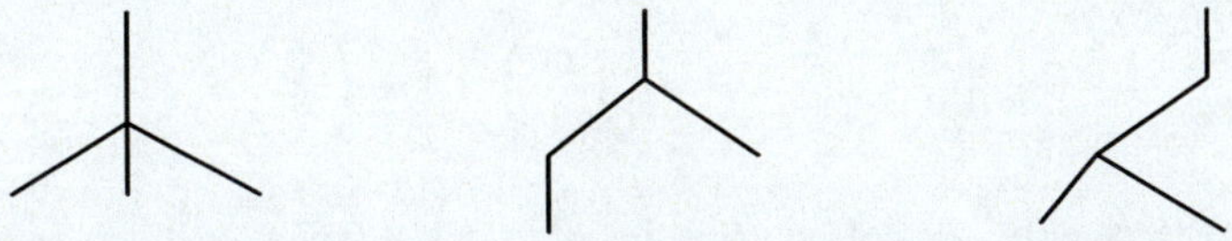

(i) **(ii)** **(iii)**

Fig. 7 Spanning tree structures where three removed edges are added at the end of one edge

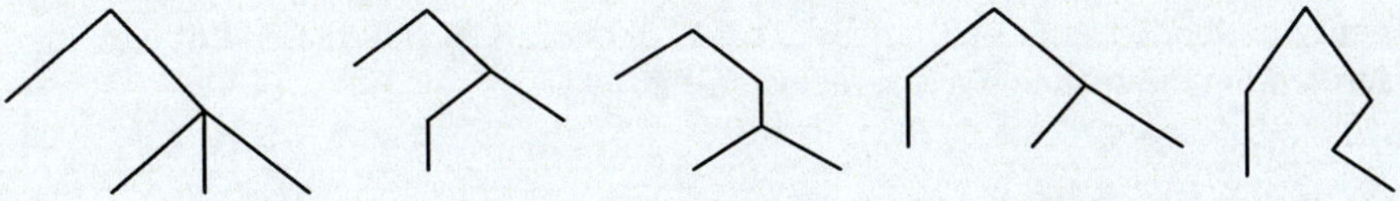

(i) **(ii)** **(iii)** **(iv)** **(v)**

Fig. 8 In (i), (ii), and (iii), the removed m edges from the star graph are added at the end of one edge, and in (iv), (v), the removed three edges are added at the end of two different edges

- For each partition p, the number of edges involved is $m = p + 1$ through $(n - (p + 1))$.
- Partitioning starts when $n = n_{\text{start}} = 2p + 1$.
- Thus, the number of partitions increases as we increase the value of n.

3.3.2 Partition Type

Depending on the number of vertices (i.e., the order of the graph) and the remaining number of edges $n - 1 - m$, we can divide m number of edges into different numbers of partitions.

Two Partitions

Now, we consider the case when m number of edges is divided into two parts and the parts are appended at the end of two different edges.

According to the formula, in Sect. 3.3.1, $n_{\text{start}} = ((2 \times p) + 1)$. When $p = 2$, $n_{\text{start}} = ((2 \times 2) + 1) = 5$.

Partition of $\langle z + (m - z)\rangle$ edge group having combination of branch and chain structures:

In this section, the removed m edges have been partitioned into $\langle z + (m - z)\rangle$ edges. If z edges are added as branch, then the remaining $m - z$ edges are added as chain at the end of two different edges and vice versa, where z may range from 1 to $m - 1$. Thus, this structure is a combination of branch and chain.

$$\sum_{m=3}^{n-3}\binom{n-1}{m} \times 2 \times \binom{n-1-m}{2} \times \left[\binom{m}{1} \times 1 + \binom{m}{2} \times 2! + \binom{m}{3} \times 3! + \binom{m}{4} \times 4! + \cdots + \binom{m}{(m-1)} \times (m-1)!\right] \tag{5}$$

When $z = 1$ and $(m - z) \geq 3$, $m - z$ edges represent all the spanning tree structures of K_n, where $n = m - z + 1$ and the structures are added at the end of one edge. In Fig. 9(i), (ii), we replace four edges from a star-tree T with $n = 7$, where four edges are divided as $\langle 1 + 3\rangle$ edges. Three edges are added at the end of one edge representing the spanning tree structures of K_4.

$$\sum_{m=4}^{n-3}\binom{n-1}{m} \times 2! \times \binom{n-1-m}{2} \times \binom{m}{m-1} \times [N_m - ((m-1)! + 1)] \tag{6}$$

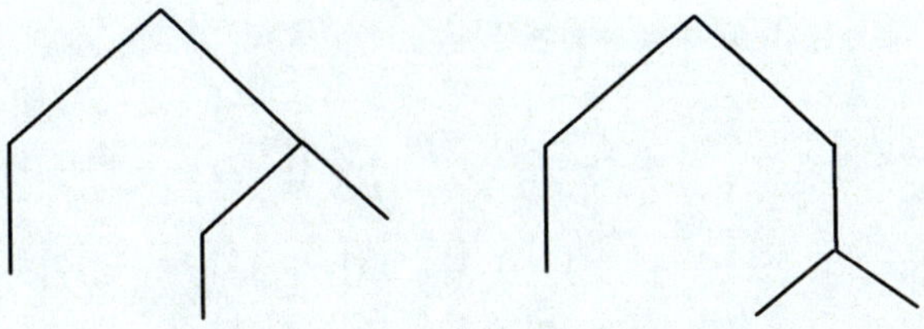

Fig. 9 Spanning tree structures of K_7 by removing four edges

Expression 6 represents the number of spanning trees obtained from the structures (i) and (ii) in Fig. 9, where N_m is the number of spanning trees of K_m.

Partition where both the parts contain more than two edges:

Here, we divide the removed m edges into a group of $\langle\lfloor(m/2)\rfloor + (m - \lfloor(m/2)\rfloor)\rangle$ edges.

If $r = m - r$, where $r = \lfloor m/2 \rfloor$, then

$$\sum_{m=4}^{n-3}\binom{n-1}{m} \times \binom{n-1-m}{2} \times \sum_{r=2}^{m/2}\left[\binom{m}{r} \times r! \times (m-r)! + \binom{m}{r}\right] \tag{7}$$

If $r \neq m - r$, then

$$\sum_{m=4}^{n-3}\binom{n-1}{m} \times 2! \times \binom{n-1-m}{2} \times \sum_{r=2}^{\lfloor m/2 \rfloor}\left[\binom{m}{r} \times r! \times (m-r)! + \binom{m}{r}\right] \tag{8}$$

Expressions 7 and 8 represent the number of spanning trees where r and $m - r$ edges are added as either chain structures or branch structures at the end of two different edges.

Case 1. When $r \neq m - r$, $r < 3$ and $m - r \geq 3$

$$\binom{n-1}{m} \times \binom{m}{r} \times 2! \times \binom{n-1-m}{2} \times \left[N_{m-r+1} - ((m-r)! + 1)\right] \times (1 + r!) \tag{9}$$

Figure 10(i), (ii) shows spanning tree structures where a group of $\langle 2 + 3 \rangle$ edges are added as branch in one part and the other part represents the spanning tree structures of K_4. The structures in Fig. 10(iii), (iv) represent the same spanning tree structures except two edges that are added as chain.

Case 2. $r = m - r, r \geq 3$, **and** $m - r \geq 3$

$$\binom{n-1}{m} \times \binom{m}{r} \times \binom{n-1-m}{2} \times ([N_{r+1} - (r! + 1)] \times [N_{r+1} - (r! + 1)] + ([N_{r+1} - (r! + 1)] \times 2) + ([N_{r+1} - (r! + 1)] \times 2 \times \mathbf{r}!)) \tag{10}$$

Expression 10 represents the number of spanning trees where the structures are same as shown in Fig. 10, including some additional structures shown in Fig. 11(i)–(iii). Figure 11 shows $\langle r + (m - r) \rangle$ edge group representing spanning tree structures of K_r.

Case 3. $r \neq m - r, r \geq 3$, **and** $m - r \geq 3$

$$\binom{n-1}{m} \times \binom{m}{r} \times 2! \times \binom{n-1-m}{2} \times (([N_{m-r+1} - ((m-r)! + 1)] \times [N_{m-r} - ((m-r+1)! + 1)]) + ([N_{m-r+1} - ((m-r)! + 1)]) + ([N_{m-r+1} - ((m-r)! + 1)] \times r!) + ([N_{m-r} - ((m-r+1)! + 1)] \times (m-r)!)) \tag{11}$$

Expression 11 represents the number of spanning trees where the structures are same as Figs. 10 and 11. As $r \neq m - r$ here, so r and $m - r$ represent spanning tree structures of K_r and $K_{(m-r)}$.

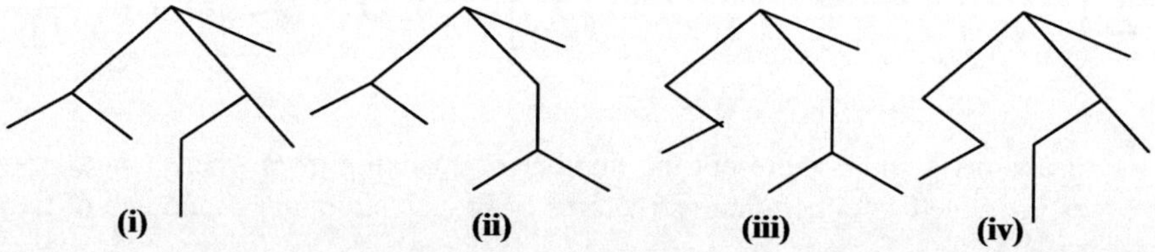

Fig. 10 Spanning tree structures represented by Expression 9

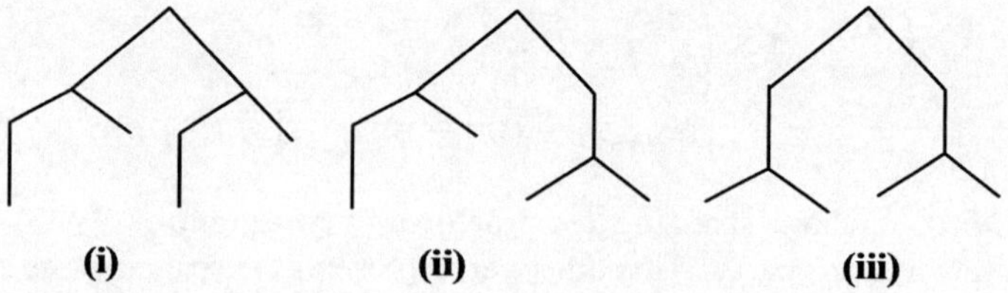

Fig. 11 Spanning tree structures which are represented by Expression 10

Three Partitions

Here, the replaced m edges are divided into a group of $\langle 1 + f + (m - f - 1)\rangle$ edges, where $f \geq 2$ and $m - f - 1 \geq 2$. There are two cases:

First Case: Expression 12 represents the number of spanning trees, where f and $m - f - 1$ edges are added as branch structure and $f = m - f - 1$.

$$\sum_{m=5}^{n-4} \binom{n-1}{m} \times \sum_{f=\lfloor \frac{m}{2} \rfloor}^{\lfloor \frac{m}{2} \rfloor + \left(\lfloor \frac{(m-1)}{2} \rfloor - 2\right)} \binom{m}{f} \times \binom{m-f}{m-f-1} \times \binom{n-1-m}{3} \times 3! \tag{12}$$

Second Case: Expressions 13 and 14 represent the number of spanning trees, where f and $m - f - 1$ edges are added as chain structure, or f number of edges are added as chain structure and $m - f - 1$ number of edges are appended as branch structure, where $f = m - f - 1$ and $f \neq m - f - 1$. f number of edges are added as branch structure, and $m - f - 1$ number of edges are appended as chain structure, where $f \neq m - f - 1$.

If $f = m - f - 1$,

$$\sum_{m=5}^{n-4} \binom{n-1}{m} \times \sum_{f=\lfloor \frac{m}{2} \rfloor}^{\lfloor \frac{m}{2} \rfloor + \left(\lfloor \frac{(m-1)}{2} \rfloor - 2\right)} \binom{m}{f} \times \binom{m-f}{m-f-1} \times \binom{n-1-m}{3} \times 3! \times f! \times (1 + (m - f - 1)!) \tag{13}$$

If $f \neq m - f - 1$,

$$\sum_{m=5}^{n-4} \binom{n-1}{m} \times \sum_{f=\lfloor \frac{m}{2} \rfloor}^{\lfloor \frac{m}{2} \rfloor + \left(\lfloor \frac{(m-1)}{2} \rfloor - 2\right)} \binom{m}{f} \times \binom{m-f}{m-f-1} \times \binom{n-1-m}{3} \times 3! \times (f! + (m - f - 1)! + (m - f - 1)! \times f!) \tag{14}$$

P Partitions

This section discusses about the structures in which each partition contains *s* number of edges either in branch or chain structure, where $s \geq 2$. The value of *s* in each partition is same. The number of trees from these structures is obtained from Expression 15.

If $s = 2$ through $\lfloor(m/p)\rfloor$ and if $s = m - (p-1) \times s$, then

$$\binom{n-1}{m} \times \binom{m}{s} \times \binom{m-s}{s} \times \binom{m-2s}{s} \times \binom{m-3s}{s} \times \cdots \times \binom{m-[(p-2)*s]}{s} \times \binom{n-1-m}{p} \times (1 + (s! \times s! \times s! \times \cdots p\,times)) \tag{15}$$

The value of *s* in each partition is same except one partition. For example, if there is *p* number of partitions, then $p-1$ number of partitions contains *s* number of edges, and one partition contains more than *s* number of edges.

Expression 16 represents the number of trees if $s \neq [m-(p-1)] \times s$,

$$\binom{n-1}{m} \times \binom{m}{s} \times \binom{m-s}{s} \times \binom{m-2s}{s} \times \binom{m-3s}{s} \times \cdots \times \binom{m-[(p-2) \times s]}{s} \times \binom{n-1-m}{p} \times p! \times \left(1 + (s! \times s! \times s! \times \cdots (p-1)\,times) \times \left[m-(p-1) \times s\right]!\right) \tag{16}$$

4 The Algorithm at a Glance

Algorithm for finding the number of spanning trees of a complete graph of order *n*

1. Take the value of *n* (number of vertices in the graph) as input.
2. Remove 1 through $n-2$ number of edges which is denoted by *m*.
3. For *p*: = 1 to $\lfloor(n-2)/2\rfloor$
4. If $p = 1$, determine the number of spanning trees from PARTITION1.
5. If $p = 2$, determine the number of spanning trees from PARTITION2.
6. If $p = 3$, determine the number of spanning trees from PARTITION3.
7. If $p > 3$, determine the number of spanning trees from PARTITION*N*.
8. Return PARTITION1 + PARTITION2 + PARTITION3 + PARTITION*N*.

Algorithm PARTITION1

1. If $n = 4$, return SP_n (SP_n is the number of common spanning tree structures of K_n).
2. If $n > 4$, calculate Eq. 4 in Sect. 3.2.1.

Algorithm for finding the value of SP_n

1. If $n - 1 - m \neq m$, calculate Eq. 1 in Sect. 3.1.1.
2. If ($n \geq 5$ and $n - 1 - m = m$), calculate Eq. 3 in Sect. 3.1.2.

Algorithm PARTITION2

1. For $m: = 3$ to $n - 3$
2. Calculate Expression 5 in Sect. 3.3.2.1.
3. For $m: = 4$ to $n - 3$
4. Calculate Expression 6 in Sect. 3.3.2.1.
5. For $m: = 4$ to $n - 3$

   5.1 For $r: = 2$ to $\lfloor m/2 \rfloor$
      5.1.1 If $r \neq m - r$ and ($r < 3$ and $m - r \geq 3$), calculate Expression 9 in Sect. 3.3.2.1.
      5.1.2 If $r = m - r$ and ($r \geq 3$ and $m - r \geq 3$), calculate Expression 10 in Sect. 3.3.2.1.
      5.1.3 If $r \neq m - r$ and ($r \geq 3$ and $m - r \geq 3$), calculate Expression 11 in Sect. 3.3.2.1.

   5.2 Calculate Expressions 7 and 8 in Sect. 3.3.2.1.

Algorithm PARTITION3

1. For $m: = 5$ to $n - 4$

   1.1 For $f: = \lfloor m/2 \rfloor$ to $(\lfloor m/2 \rfloor + (\lfloor (m-1)/2 \rfloor - 2))$
      1.1.1 Calculate Expressions 12, 13, and 14 in Sect. 3.3.2.2.

Algorithm PARTITION*N*

1. For $p: = 3$ to $\lfloor (n-2)/2 \rfloor$

   1.1 For $m: = p + 1$ to $v - (p + 1)$
   1.2 For $s: = 2$ to $\lfloor (m/p) \rfloor$
      1.2.1 If $s = [m - (p - 1)] \times s$, calculate Expression 15 in Sect. 3.3.2.3.
      1.2.2 If $s \neq [m - (p - 1)] \times s$, calculate Expression 16 in Sect. 3.3.2.3.

5 Data Structure and Complexity

In this algorithm, we are using array and stack data structures. Here, we represent the graph using adjacency matrix, which can be represented by a two-dimensional array. We may store the vertices of the graph in a stack and pop the vertices of the graph from the stack and generate different structures of spanning trees. In PARTITION1, we determine the number of trees in $O(n \times 2^n)$ time. Algorithm PARTITION2 takes $O(n \times 2^{2n-m})$ time. Algorithm PARTITION3 returns the number of trees in $O(2^{2n-m})$ time. Algorithm PARTITION*N* takes $O(n \times 2^{2n-m})$ time. So, the overall complexity of the algorithm is $O(n \times 2^n)$.

6 Experimental Results

We have implemented the above-mentioned algorithm on an Intel® core(TM) i3-4030U processor @1.90 GHz, with 4 GB RAM.

Table 1 shows the CPU time required to generate the trees for different number of vertices and the percentage of trees generated by the above algorithm. We have calculated the number of trees for four different diameters for $n = 4$ to $n = 11$. If $D3$ is the number of spanning trees of diameter 3, then $D3 = {}^{n-1}C_{n-3} \times 2 + {}^{n-1}C_{n-4} \times 3 + \ldots + {}^{n-1}C_1 \times (n-2) + {}^{n-1}C_{n-2} \times ({}^{n-2}C_1 \times (n-3) + {}^{n-2}C_2 \times (n-4) + \ldots + {}^{n-2}C_{n-3} \times 1)$. In Table 1, it is shown that when the number of partitions increases, time taken to find out the spanning trees also increases. Table 2 shows the number of spanning trees of different diameters for eight different complete graphs.

7 Conclusions and Future Work

In this paper, we have found out the number of spanning trees of some specific diameters in a complete graph. Here, we have attempted to generate a new approach for computing the spanning trees of a complete graph using recursion. First, we have taken a complete graph and generated a star-tree T. Then, we have replaced edges from T so that the different spanning tree structures are created. For n vertices, we have been able to generate diameter 2, diameter 3, diameter $n - 1$, and diameter $n - 2$ spanning trees. In future, we like to formulate expressions for all the structures of spanning trees of graphs with any order, so that eventually we will be able to evaluate the number of spanning trees for any given n.

Table 1 Table showing the number of trees with different diameters for $n = 4$–11

No. of vertices	Possible number of partition(s)	Number of trees generated	Percentage of trees generated (%)	CPU time to generate the trees (s)	Number of trees of diameter 2 (n)	Number of trees of diameter 3 ($D3$)	Number of trees of diameter $n - 1$ ($n!/2$)	Number of trees of diameter $n - 2$ ($n!/2$) × ($n - 4$)	CPU time to generate trees of a diameters (s)
4	1	16	100	1	4	12	12	4	0
5	1	125	100	1	5	60	60	60	0
6	1, 2	1296	100	2	6	210	360	720	0
7	1, 2	16,807	100	2	7	630	2520	7560	1
8	1, 2, 3	262,144	100	4	8	1736	20,160	80,640	1
9	1, 2, 3	4,782,969	100	4	9	4536	181,440	907,200	1
10	1, 2, 3, 4	100,000,000	100	6	11	11,430	1,814,400	10,886,400	1
11	1, 2, 3, 4	2,352,760,206	99.78	6	11	28,050	19,958,400	39,708,800	2

Table 2 Table showing spanning trees of different diameters for some complete graphs

Order of the complete graph	Diameter						Total
	2	3	4	5	6	7	
4	4	12	NA	NA	NA	NA	16
5	5	60	60	NA	NA	NA	125
6	6	210	720	360	NA	NA	1296
7	7	630	6090	7560	2520	NA	16,807
8	8	1736	70,144	89,456	80,640	20,160	262,144

References

1. M. Chakraborty, S. Chowdhury, R.K. Pal, Two algorithms for computing all spanning trees of a simple, undirected, and connected graph: once assuming a complete graph. IEEE Access **6**, 56290–56300 (2018)
2. M. Chakraborty, R. Mehera, R.K. Pal, A divide-and-conquer algorithm for all spanning tree generation, in *Proceedings of the 3rd International Doctoral Symposium on Applied Computation and Security Systems (ACSS), Proceedings Published as Book Chapter of the Series Advances in Intelligent Systems and Computing*, vol. 567 (Springer, 2016), pp. 19–36
3. J.P. Char, Generation of trees, two-trees and storage of master forests. IEEE Trans. Circuit Theory **CT-15**, 128–138 (1968)
4. S.L. Hakimi, On trees of a graph and their generation. J. Franklin Inst. **272**(5), 347–359 (1961)
5. S. Kapoor, H. Ramesh, Algorithms for enumerating all spanning trees of undirected and weighted graphs. SIAM J. Comput. **24**(2), 247–265 (1995)
6. G.J. Minty, A simple algorithm for listing all the trees of a graph. IEEE Trans. Circuit Theory **CT-12**, 120 (1965)
7. S. Naskar, K. Basuli, S. Sen Sarma, Generation of all spanning trees. Soc. Sci. Res. Netw. (2009)
8. S. Sen Sarma, A. Rakshit, R.K. Sen, A.K. Choudhury, An efficient tree generation algorithm. IETE J. Res. **27**(3), 105–109 (1981)
9. *Wikipedia*, Cayley's formula. https://en.wikipedia.org/wiki/Cayley%27s_formula
10. P. Winter, An algorithm for the enumeration of spanning trees. BIT Numer. Math. **26**(1), 44–62 (1986)

Size Invariant Ship Detection Using SAR Images

Atharva Potdar, Parth Kingrani, Rahul Motwani, Tanishqa Shetty, and Shalu Chopra

Abstract Synthetic aperture radar also is known as SAR imagery is considered a very essential source of data for overlooking maritime or naval activities, and its application towards oil spills and ship detection has been a crucial centre of many going on exploration works. Many object detection methods including conventional and deep learning have been contemplated. But, most of them have tremendous computational overloads and are vastly inaccurate. The proposed system is implemented by using a Python web framework Flask, Tensor flow for training the deep learning model, and a database called SQlite3 as a backend for downloading the saved images. The research work undertaken assists the end-user to monitor the activities of the ships, measuring their dimensions, and thus preventing potential mishaps.

Keywords Deep learning · SAR images · Tensor flow object detection · Neural network

A. Potdar (✉) · P. Kingrani · R. Motwani · T. Shetty · S. Chopra
Vivekanand Education Society Institute of Technology, Mumbai, India
e-mail: 2017.atharva.potdar@ves.ac.in

P. Kingrani
e-mail: 2017.parth.kingrani@ves.ac.in

R. Motwani
e-mail: 2017.rahul.motwani@ves.ac.in

T. Shetty
e-mail: 2017.tanishqa.shetty@ves.ac.in

S. Chopra
e-mail: shalu.chopra@ves.ac.in

V. E. Balas et al. (eds.), *Proceedings of International Conference on Computational Intelligence, Data Science and Cloud Computing*, Lecture Notes on Data Engineering and Communications Technologies 62, https://doi.org/10.1007/978-981-33-4968-1_15

1 Introduction

Real-time detection of objects in complex images such as SAR and radar is an essential but extremely challenging problem faced in maritime security. Previously it was difficult due to performance factors but recently algorithms like Faster Region-Based Convolutional Neural Networks (RCNN) have shown true potential for solving this problem. Synthetic-aperture radar (SAR) is used for the development of 2D or 3D images of varied objects. It is similar to how bats use echolocation to navigate inside of a cave. SAR uses radar antennas over a land area to provide a fine spatial resolution which has changed the game for accuracy in the depicted conventional radar, like analogue radars, images and hence is needed for designing a deep learning model. SAR is often mounted on a moving platform, such as an aircraft and uses echolocation to give precise information about the targeted landscape. The resolution generated by SAR depends upon the size of the aperture and not really on the aperture type. Hence they can have high resolution even with small antennas. A large amount of study has been done on detecting ships using SAR images. Traditional ship detection methods are usually very inaccurate and can lead a lot of false-positive outcomes which can lead to false alarms or discrepancies among the data [1–3].

2 Related Work

RCNN: It was proposed by R. Girshick in 2014 to tackle the issue of localization of an object in an efficient manner for object detection. It uses sliding windows on a different type of images to find proper regions which are also known as Exhaustive Search. To overcome the issues they came up with a Selective search algorithm which takes the benefit of currently existing Exhaustive Search and pairing up with segmentation and this algorithm gives nearly 2000 region proposals per image. Then they are fed to CNN Model which gives an output of 14,096 feature vectors from every individual region. The flaws if RCNN is it consumes more time for training and It nearly takes a minute to detect the object in the image even if you have a high-performance graphic processor.

Fast RCNN: R. Girshick solved the problems existing in the previous algorithm and came up with a better and faster algorithm and it's very much similar with RCNN but this time the input is being fed to CNN to extract a feature map of the neural network and which aids us in the identification of regions and we can reshape it with using pooling to make layer which is fully connected and we use softmax layer for estimating the region's class and gives the offset values of the region in form of bounding box.

Faster RCNN: Since selective search is slow and takes a longer time period to find out regions. It also takes a lot of memory and also affects the computational performance. Hence, Shaoqing came up with an algorithm which removes the algorithm of selective search and makes the network self-dependent to learn on its own. Faster

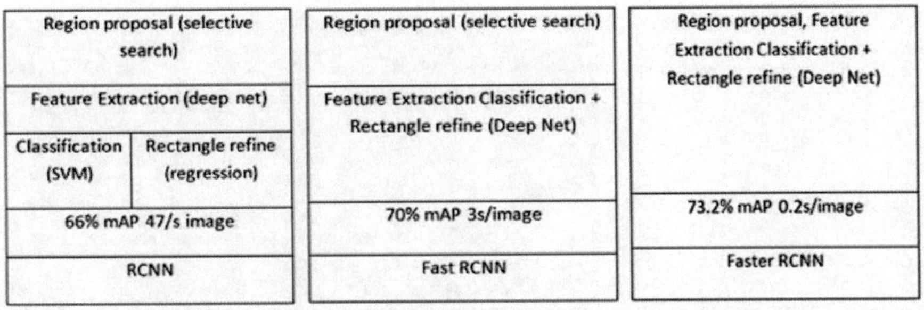

Fig. 1 The procedure and performance of RCNN, Fast RCNN and faster RCNN respectively

RCNN is extremely fast and it takes less than a second for detection. It is the fastest, efficient and robust RCNN model which is present. It can also be used for real-time (Fig. 1).

3 Dataset and Methodology

Dataset: We made a dataset of ships and vessels of SAR Images which is also known as SSDD and its renowned criterion for engineers and research fellows for gauging their models. One can acquire the dataset by sending a proposal for the dataset which is required for your research work. We labelled every ship with a bounding box through an application called LabelImg. We also obey a specific course of action as PASCAL VOC for making the customized dataset which is very much in use for object recognition and detection. We separate our dataset into two sections i.e. Training set and testing set having a ratio of 8:2. It is quite easier to make a dataset for construction over classification. The work of classification is invariant from translation compared from the work detection which is variant with classification. We formed a diverse dataset in order to cover various scenarios like different resolutions, different locations, different coordinates and many more. In order to make your model efficient, diversification of dataset is a must and it also helps us to achieve high performance. We have also presented a table below where it clearly describes Number of images (Num Img) and Number of ships (Num Ship).

Num Ship	1	2	3	4	5	6	7
Num Img	725	183	89	47	45	16	15
Num Ship	8	9	10	11	12	13	14
Num Img	8	4	11	5	3	3	0

It comprises 2456 ships in 1160 images. The SSDD is good enough to train our model, which will help us to obtain high accuracy (Fig. 2).

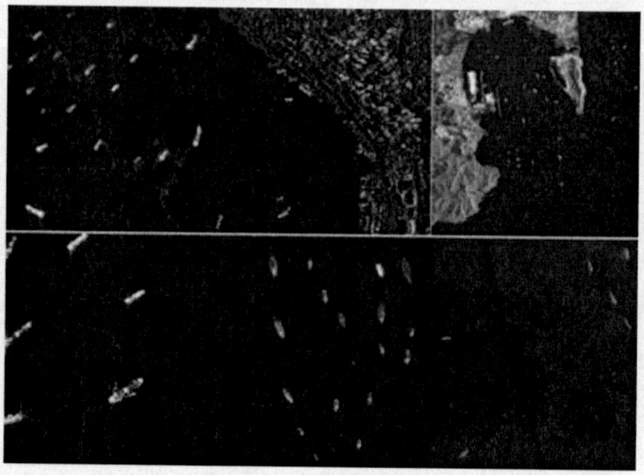

Fig. 2 Sample images from our dataset

Methodology: In this work, we have constructed a Tensor Flow Object Detection based end-to-end Faster RCNN to detect ships. First, the images of the training dataset are manually labelled using LabelImg under the class name "ship". Faster RCNN is a combination of 3 types of Neural Networks, namely—Feature Network, Region Proposal Network and Detection Network. The feature network generates good features from the images keeping the shape and structure of the image as it is. The RPN is a network with 3 layers—where a common layer is fed into classification and bounding box regression layers. Region of Interest, consisting of bounding boxes are generated, giving the count of the identified bounding boxes.

The RCNN network, referred to as the Detection Network takes input from the first two networks, generating the final class and bounding boxes.

Figure 3 explains complete working of the research work where a custom dataset was prepared by acquiring images from various sources and after collecting the dataset, it was then organized and every image was resized and refitted to a specific resolution. After matching every image's properties, they were separated into two parts i.e. Train and Test. The training data is used for training and the test data is untouched and used only for testing purposes. The ships present in the images were labelled using LabelImg using the procedure of PASCAL VOC for labelling under the class name "ship". The files obtained in the .xml format after labelling are converted into a single .csv file using a simple Python program, after which it is fed to the model for training using Tensor flow Object Detection module using Faster RCNN Inception v2. In parallel to the training of the model, User Interface has been developed using HTML, CSS and then the model was being integrated with the backend of UI using Flask with the latest weight file that has been obtained after training the model. The website is hosted on local host, and then Test images from the test dataset are used to obtain the desired result of our model. A vector file has also been generated with every result obtained and can be downloaded [4–7].

Figure 4 shows the complete working of Faster RCNN Inception v2 neural network, accustomed to our dataset.

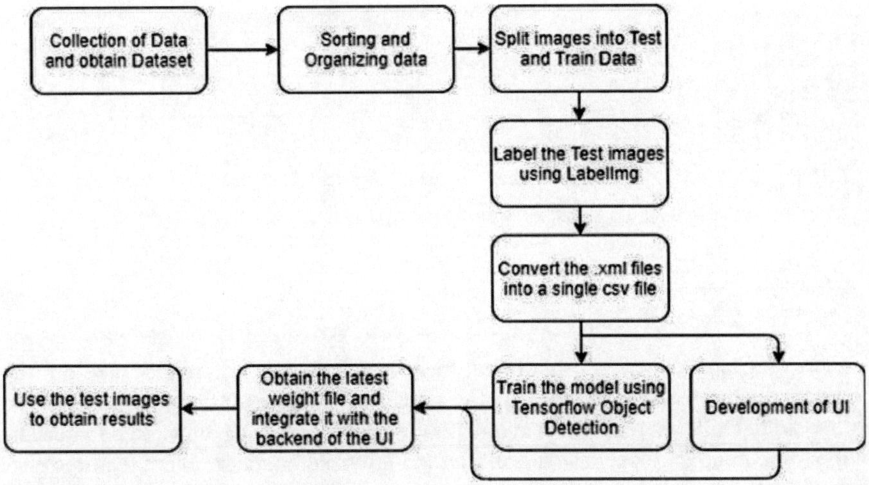

Fig. 3 Flowchart of the research work

Fig. 4 Faster RCNN inception v2 flowchart

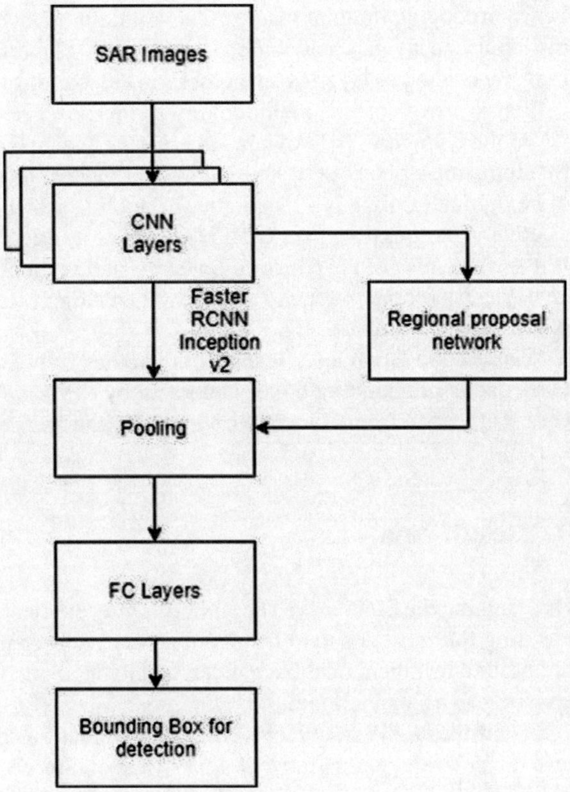

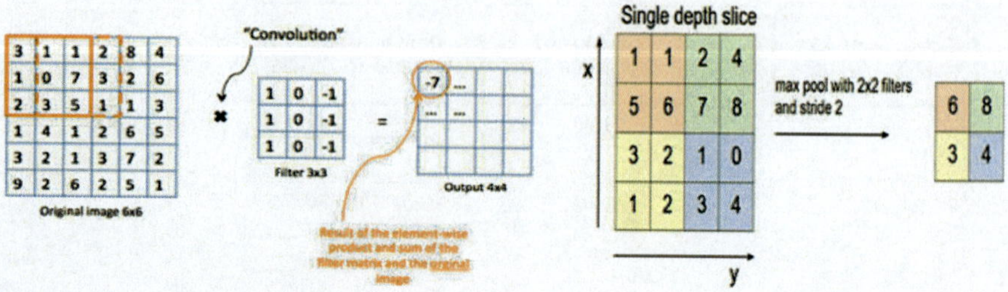

Fig. 5 Feature map and pooling in CNN

Convolutional Layers: In this layer, there are kernels or filters that take out the pertinent features of the images of the dataset. The convolutional layers can be compared to a filter in real life. Just like how a filter removes the unwanted materials, in the same manner, these layers don't let the unwanted data or object features pass through and only the desired objects are passed.

Training: To train the faster-RCNN architecture SGD is used to optimize convolution layer filters, RPN weights and the last fully connected layer weights combined, hence producing optimal results. Convolutional networks comprise 3 different and distinguished layers, Convolution layers, pooling layers and the last constituent which is a fully connected layer which helps us in the working of classification and detection.

Figure 5 explains the methodology followed for pooling in CNN.

Pooling: Pooling is known as decreasing the size of the feature map produced, by eliminating pixels with non-significant values, hence ensuring efficiency in the system. Max Pooling is a technique for pooling which is commonly used in CNN.

Region Proposal Network (RPN): RPN is a little neural network present on the last attribute map of the convolution layers and is simply used for predicting whether an object is present or not and if present then along with predicting the bounding box of the objects present.

Classes and Bounding Boxes: It is another fully connected neural network that takes pixels of bounding boxes generated by RPN and determines its class and also uses Regression for making the bounded box more precise.

4 Result Analysis

Determining the Efficiency: The work is highly efficient and complete working model ensuring the most accurate results in a very short span of time. The images which are used to test the model have given results in about 30 s along with its vector file, ensuring an efficient model.

Determining the Accuracy: When the work trained using the train dataset was tested, an average accuracy of 95% was obtained, when tested on the images belonging to the testing dataset (Fig. 6).

Fig. 6 Ships detected from SAR image near a port, image depicting the accuracy that the object detected is a ship and results of ships detected near shore area

We successfully trained our Faster RCNN Inception v2 model for detection of ships and vessels on our dataset. We used Tensor flow object detection module for object detection Evaluation of model was done using epoch metric which is a complete cycle through the full training dataset. It took more than 100 s of epochs to achieve a robust accuracy and the statement of the epoch can be different with respect to situations. The accuracy graph of our model is for measuring the accuracy with respect to the model parameters. We have also presented another graph for loss which is an addition of all type errors made during every iteration during the procedure of training and while testing. It is near to impossible to get an absolutely perfect model as it should have 0 losses. The aforementioned figures represent the accuracy and loss of our model and we achieved nearly 95% of accuracy while randomly selecting the results and since we nearly covered many types of cases, this can be used for commercial purposes isn't sporadic at all. It is being averaged across the recall values which lie from 0 to 1 (Fig. 7) [8–11].

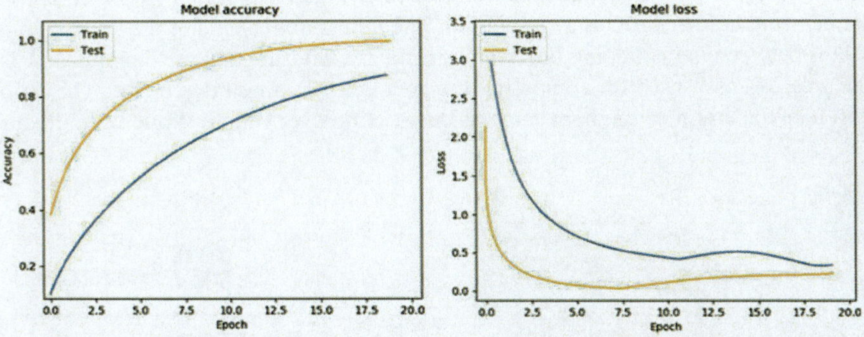

Fig. 7 Accuracy graph and loss graph of the model

5 Evaluation Measures

Average Precision: The Average precision is an evaluation metric which helps us in distinguishing various types of detectors. We figure out the area under the curve of the Precision x Recall curve. Average precisions tend to be zig-zag going in the direction of top and bottom, and it isn't an easy task to compare different curves present in the same plot because there are too many intersections between the curves and it isn't sporadic at all. It is being averaged across the recall values which lie from 0 isn't sporadic at all. It is being averaged across the recall values which lie from 0 to 1. Intersection over union is a beneficial metric which is widely used for evaluation for getting the precision of an object detector on a given dataset. This is extremely popular, and it is used widely in many challenges and Image processing competitions. Intersection over Union is used for quantifying the staging of Linear with Histogram oriented gradients and support vector machines object detectors and Neural Network detectors. It's applicable only for those algorithms where there is a prediction of bounding boxes. To use Intersection over Union for analyzing an (any) object detector:

1. The ground-truth (fundamental) bounding boxes.
2. The predicted bounding boxes through our model.

If we've got both of the aforementioned we qualify for applying Intersection over Union. We've stated an example of comparison of both of them (Fig. 8).

The box which is in red is a predicted bounding box and the other one which is in green is a bounding box which is the ground truth. This will aid in computation of Intersection over Union from these given bounding boxes. The above figure clearly states that we detected the presence of a stop sign in the picture. The bounding box, which is the ground truth, is in green and the other bounding box, which is the predicted one, is in red. Union divides the overlapping sector amongst the two aforementioned boxes by the area of union. After checking out the above equation we can conclude that IoU is a correlation. The numerator calculates the overlapping sector between the bounding box which is the ground truth and the denominator is the area of union used for computing the area taken by these two boxes. The ratio will help to determine the final value of Intersection over Union. While we train our

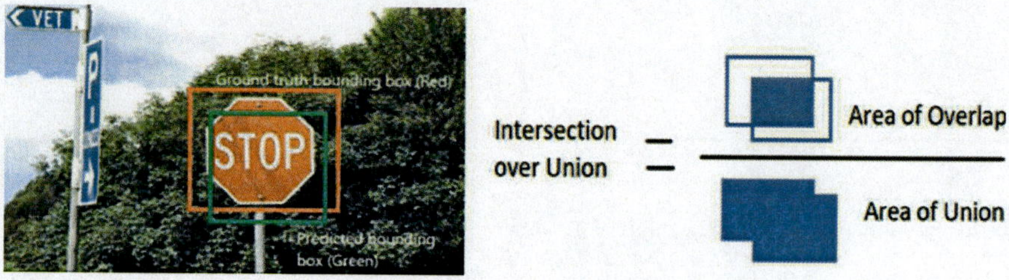

Fig. 8 Intersection over union actual representation and computing IoU

own object detection model, we'll require a dataset and which is being bifurcated into two groups:

1. A training set which is used for training our model.
2. Testing set for testing our model.

We can also have a verification set to refine the hyper parameters of the object detector.

The training and testing dataset will include:

1. The original pictures.
2. The ground truth boxes are related to the object(s) in the image. They're geometrical coordinates of the object which is there in the image.

Bounding boxes are the boxes which are being hand-labeled and it is the reason why it is known as 'ground truth'. Development of an object detector is done by adding training images+bounding boxes and then assessing its performance on the testing dataset. A good prediction is said when the IoU score is more than 0.5. In real life, there is a very minute chance that the geometrical coordinates of the box which is being predicted by the model will align perfectly with the geometrical coordinates of the other bounding box which is our ground truth. Due to the presence of some variations in the given parameters (box size, the scale of image pyramid and many more), there is extremely little to no chances of having a complete match of both the boxes. We define a metric that awards the predicted box for heavy overlapping with the ground truth (Fig. 9).

We have mentioned the nuances of poor and good enough Intersection over Union scores. The ones with the very high overlapping have a good score compared with the ones poor score in which the predicted bounding hasn't overlapped much with ground truth box which concludes us to say the Intersection over Union is a great metric for evaluation of object detectors. We don't focus for an exact match but we always have a goal of as much as possible of overlapping of the boxes. Intersection over the Union has been taken into consideration for this.

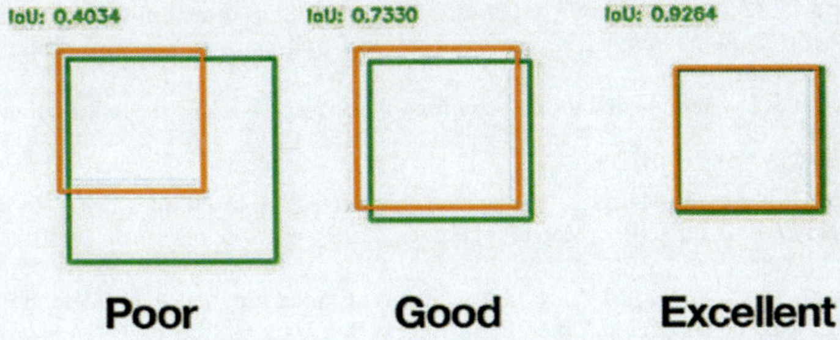

Fig. 9 A specimen of quantifying Intersection over unions for various bounding boxes

6 Conclusion

At the beginning of our work, our goal was to successfully develop a model which can detect ships and vessels which are present at the sea using SAR Images. We produced a high-quality dataset which is of SSDD benchmark. In the end, we've successfully worked into real-time detection having great robustness and performance which gives a result with an average accuracy of 95% for detection. We gauged the performance of the Tensor flow Object Detection module using Faster RCNN Inception v2 to detect the ship in different situations. The training outcomes on our dataset infers that Faster RCNN Inception v2 gives better results in real-time than the existing modules of deep learning. While training we faced some obstacles where the model can incorrectly detect docks, port, land instead of the ship but refining the labelling and with increment in the training dataset, one can overcome this situation and we achieved a great accuracy of detection. The module of RCNN which we selected is the backbone of our CNN whose accuracy and efficacy came at a pocket-friendly computational cost. The development of this research work can be used extensively for naval security and performing checks through the sea to detect any unusual activities. The methodology which we've implemented can be used for SAR Images having different resolutions and also we can distinguish between land and water and it also generates a vector file in which the estimated size of the ship is present [12–15].

References

1. R. Touzi, F.J. Charbonneau, R.K. Hawkins, P.W. Vachon, Ship detection and characterization using polarimetric SAR. Can. J. Remote. Sens. **30**(3), 552–559 (2004). https://doi.org/10.5589/m04-002
2. K. Ouchi, S. Tamaki, H. Yaguchi, M. Iehara, Ship detection based on coherence images derived from cross correlation of multilook SAR images. IEEE Geosci. Remote Sens. Lett. **1**(3), 184–187 (2004). https://doi.org/10.1109/LGRS.2004.827462
3. A. Marino, M.J. Sanjuan-Ferrer, I. Hajnsek, K. Ouchi, Ship detection with spectral analysis of synthetic aperture radar: a comparison of new and well-known algorithms. Remote Sens. **7**, 5416–5439 (2015)
4. X. Cui, Y. Su, S. Chen, A saliency detector for polarimetric SAR ship detection using similarity test. IEEE J. Sel. Top. Appl. Earth Obs. Remote Sens. **12**(9), 3423–3433 (2019). https://doi.org/10.1109/JSTARS.2019.2925833
5. F. Quina, J. Neves, P. Marques, "A look on ships detection techniques using SAR images," OCEANS 2019—Marseille, (Marseille, France, 2019), pp. 1–5, https://doi.org/10.1109/OCEANSE.2019.8867388
6. Y. Mao, Y. Yang, Z. Ma, M. Li, H. Su, J. Zhang, Efficient low-cost ship detection for SAR imagery based on simplified U-net. IEEE Access **8**, 69742–69753 (2020). https://doi.org/10.1109/ACCESS.2020.2985637
7. S. Kuttikkad and R. Chellappa, "Non-Gaussian CFAR techniques for target detection in high resolution SAR images," *Proceedings of 1st International Conference on Image Processing*, vol. 1 (Austin, TX, 1994), pp. 910–914. https://doi.org/10.1109/ICIP.1994.413444

8. A. Renga, M.D. Graziano, A. Moccia, Segmentation of marine SAR images by sub look analysis and application to sea traffic monitoring. IEEE Trans. Geosci. Remote Sens. **57**(3), 1463–1477 (2019). https://doi.org/10.1109/TGRS.2018.2866934
9. C. Brekke, S.N. Anfinsen, Y. Larsen, Sub band extraction strategies in ship detection with the sub aperture cross-correlation magnitude. IEEE Geosci. Remote Sens. Lett. **10**(4), 786–790 (2013). https://doi.org/10.1109/LGRS.2012.2223656
10. A. Renga, M. Daniela Graziano, A. Moccia, M. D'Errico, "Adaptive threshold and sub-look processing in ship detection by SAR," OCEANS 2015—Genova, Genoa. pp. 1–8 (2015). https://doi.org/10.1109/OCEANS-Genova.2015.7271538
11. D. Velotto, B. Tings, "Performance analysis of time-frequency technique for the detection of small ships in SAR imagery at large grazing angle and moderate metocean conditions," IGARSS 2018—2018 *IEEE International Geoscience and Remote Sensing Symposium*, Valencia. pp. 6071–6074 (2018). https://doi.org/10.1109/IGARSS.2018.8517472
12. C. Wang, S. Jiang, H. Zhang, F. Wu, B. Zhang, Ship Detection for high-resolution SAR images based on feature analysis. IEEE Geosci. Remote Sens. Lett. **11**(1), 119–123 (2014). https://doi.org/10.1109/LGRS.2013.2248118
13. R.L. Paes, J.A. Lorenzzetti, D.F.M. Gherardi, Ship detection using terra SAR-X images in the Campos basin (Brazil). IEEE Geosci. Remote Sens. Lett. **7**(3), 545–548 (2010). https://doi.org/10.1109/LGRS.2010.2041322
14. E.E. Kuruoglu, J. Zerubia, Modeling SAR images with a generalization of the rayleigh distribution. IEEE Trans. Image Process. **13**(4), 527–533 (2004). https://doi.org/10.1109/TIP.2003.818017
15. C. Corbane, L. Najman, E. Pecoul, L. Demagistri, M. Petit, A complete processing chain for ship detection using optical satellite imagery[J]. Int. J. Remote Sens. **31**(22), 5837–5854 (2010)

Generation of Simple Undirected Connected Random Graphs

Sumit Chakraborty, Maumita Chakraborty, and Rajat Kumar Pal

Abstract Random connected graphs are required to determine the performance of approximation algorithms, graph algorithms, algorithm for modeling and simulation, algorithms solving problems regarding database applications. Also, developing algorithms to generate all spanning trees required random connected graphs as input. In this paper, we have proposed an algorithm to generate random simple undirected graph instances that can be utilized to determine the performance and reliability of the above-mentioned algorithms. Our proposed algorithm creates random connected graphs with a given number of vertices and a distinct number of edges. We have been able to formulate the number of such generated graphs. The reliability and the time bound of the algorithm have also been established. We have included the computational results of the algorithm that are greatly encouraging.

Keywords Graph generation · Random graph · Simple connected graph

1 Introduction

A random graph is a graph in which the number of vertices, number of edges, and connections between them are determined in some random way [1, 2]. Random graphs are very useful in providing a test bed for researchers to determine the performance of the algorithms working with graphs. Random connected graphs are also

S. Chakraborty (✉)
Department of Computer Science, Ramsaday College, Amta, Howrah 711401, India
e-mail: sm.chakraborty09@gmail.com

M. Chakraborty
Department of Information Technology, Institute of Engineering and Management, Y-12, Block-EP, Sector V, Salt Lake, Kolkata 700091, India
e-mail: maumita.chakraborty@gmail.com

R. K. Pal
Department of Computer Science and Engineering, University of Calcutta, Block-JD-2, Sector III, Salt Lake, Kolkata 700106, India
e-mail: pal.rajatk@gmail.com

V. E. Balas et al. (eds.), *Proceedings of International Conference on Computational Intelligence, Data Science and Cloud Computing*, Lecture Notes on Data Engineering and Communications Technologies 62, https://doi.org/10.1007/978-981-33-4968-1_16

required to determine the performance of approximation algorithms, graph algorithms, algorithms for modeling and simulation, algorithms for solving problems related to database applications. Moreover, random connected graphs can be used to determine the reliability of the topological structure of a network.

In the context of the needs mentioned above, we have developed a simple and reliable algorithm to generate random connected simple graphs with same number of vertices and distinct number of edges. All spanning tree generation algorithms [3–5] require connected graphs as input. We have developed this algorithm to provide a test bed for the algorithms finding all spanning trees which need only connected graphs. That is why we focused only on the generation of some random connected graphs. We can formulate the number of connected graphs that can be generated by this algorithm when the number of vertices is given.

2 Literature Survey

Many researchers have provided with different algorithms to generate random graphs for the last few years [1, 6–12]. The input–output set and the purpose behind such generation of random graphs may vary for different algorithms. Some of them are given below:

a. Horn et al. [10] have proposed an algorithm to generate random graphs by considering the parameters like number of vertices, number of levels, degree of every vertex, and also the height and width of each graph. It is designed to provide a test set of graph instances for the development of an approximation algorithm.
b. Viger and Latapy [11] have proposed an algorithm to generate large graphs. It is very simple and efficient. Since it generates large graphs, it is especially useful in simulations for problems that have graph theoretic formulation.
c. Bayati et al. [6] have proposed an algorithm which uses a given degree sequence to generate random graphs. It can generate graphs with large girth by generating bipartite graphs. It is very useful in the design of high-performance Low-Density Parity Check (LDPC) codes.
d. A data parallel algorithm focuses on distributing the data across different vertices and operates concurrently on the data. Nobari et al. [1] have proposed a data parallel algorithm to generate random graphs. According to the authors, the generated graphs by the algorithm can fit into memory for further processing. The large graphs can be processed in parallel fashion. It can be used in database applications, simulations, and sampling.
e. The algorithm by Wang et al. [12] generates clustered random graphs in large scale. These large-scale graphs are suitable for real networks. The algorithm has the advantages in randomness evaluation and computational efficiency.
f. A parallel algorithm designed by Bhuiyan et al. [7] generates random graphs with a given degree sequence. For studying various structural properties and dynamics over a network, this algorithm is very useful.

g. Erdos et al. [9] provided a probabilistic approach to compute a random graph with n vertices and e edges. They mentioned that there are $m = \binom{n}{2}$ number of possible edges, and hence, e edges can be chosen in $\binom{m}{e}$ different ways. The authors also talked about several types of random graphs that may be generated, like connected and disconnected graphs, trees, graphs with isolated vertices, etc. The authors in the paper have also discussed about probability of a random graph being a connected one, probability that the greatest connected component of a random graph contains $n - r$ ($r = 0, 1, \ldots$) vertices, probability that a random graph should consist of exactly r (1, 2, …) connected components, and many others.
h. The algorithm by Chakraborty et al. [8] produces non-isomorphic connected graphs. At first, it generates the complete graph with a given number of vertices and then successively removes an edge in each iteration. It generates non-isomorphic connected graphs in its subsequent iterations and stops when a pendent vertex occurs. By considering the minimum degree of the vertices, it ensures that no disconnected graphs are being generated after removal of edges from the graph. Since they were interested only in generating spanning trees instead of generating spanning forests, they have not considered the disconnected graphs.

3 The Proposed Algorithm, *Random Connected Graph Generator* (RCGG)

In this work, the main goal is to develop a simple and efficient algorithm to obtain different simple, undirected, connected random graphs with a given number of vertices. We felt the requirements for a set of connected graphs, while studying and implementing different algorithms to generate all possible spanning trees [3, 4] of a simple, connected graph.

Our proposed algorithm, *Random Connected Graph Generator* (*RCGG*) given in Sect. 3.1 generates simple undirected connected graphs having the same number of vertices but containing different number of edges. It takes the number of vertices, n, as input and starts with the complete graph with n vertices. This complete graph contains $m = n(n - 1)/2$ number of edges. A complete graph contains m edges, and a spanning tree contains $n - 1$ edges. Thus, the algorithm successively removes edges from the complete graph $m - n + 1$ times, and in every iteration, it generates a connected graph, i.e., after the removal of the last edge, the graph comprises of $m - (m - n + 1) = n - 1$ edges only. So, the last or the final graph obtained is a spanning tree. In the context of selection of an edge, the randomness is incorporated in *RCGG*. It selects an edge, x randomly from E and deletes it from E (i.e., $E = E - \{x\}$) as well as the last modified graph we have and tests if the graph is still connected. If yes, then *RCGG* goes to the next iteration; otherwise, it selects a next edge, y from E (i.e., $E = E \cup \{x\} - \{y\}$), and checks whether the graph is still connected, and so on. Since *RCGG* removes exactly one edge in each iteration and produces one connected graph, it eventually generates $m - n + 2$ connected graphs with the same number of vertices including the complete graph. *RCGG* stops after $m - n + 1$ iterations because in the last iteration, it produces a spanning tree and removal of another edge undeniably makes it disconnected. Now, since the edges are selected randomly, the sets of connected graphs produced by *RCGG* are different in different executions of *RCGG*. The complete graph is the only graph which is common to the set of connected graphs.

3.1 *Algorithm Random Connected Graph Generator*RCGG*()*

Step1: *Begin*

Step 2: Read *n*. //*n* being an integer, same as the number of vertices in each generated graph.//

Step 3: Create a complete graph of *n* vertices, stores the edges in *E*, and print the graph.

Step 4: $m = n\times(n-1)/2$, $k = 1$.

Step 5: ***While*** $k \le m-n+1$ ***do***

Step 6: Select an edge $x = (u, v)$ randomly from *E* and $E \leftarrow E-\{x\}$.

Step 7: ***If*** the subgraph with the edges in *E* is connected, ***then***

Step 8: Print the subgraph with the edges in *E* and $k \leftarrow k+1$.

Step 9: ***Else***

Step 10: $E \leftarrow E\cup\{x\}$. Take another edge, *y* in *E*, next to *x*. $x \leftarrow y$ and go to Step 6, where either $y = (u, w)$, or $y = (t, v)$, or $y = (p, q)$.

Step 11: ***EndIf***

Step 12: ***EndWhile***

Step 13: ***End***

Theorem 3.1 *RCGG* generates exactly $m - n + 2$ number of connected graphs with *n* vertices and *e* edges, where $e = m, m - 1, m - 2, \ldots, n - 1$ and $m = n(n - 1)/2$.

Proof: Initially, algorithm *RCGG* presumes a complete graph with *n* vertices and *m* edges, where $m = n(n - 1)/2$ in Step 3 of the algorithm. It also keeps these *m* edges in set *E*, i.e., $|E| = m$. The loop in Step 5 iterates exactly $m - n + 1$ times.

In the *k*-th iteration of Step 5: $|E| = m - k + 1$. In Step 6, *RCGG* chooses an edge *x* randomly from the set of edges, *E*, i.e., after the *k*-th iteration, $|E| = m - k$. Now, if the tests for connectivity of the subgraph (V, E) in Step 7 results a success, i.e., the subgraph is still connected, then *RCGG* repeats Step 5 with $k = k + 1$ and the modified *E*. However, if the connectivity test fails, then a next edge of *x* from *E* is taken, and again the connectivity test is done in Step 7. Now, result of the test for connectivity will definitely be positive for at least one edge of *E*, because after removal of an edge from *E*, always $|E| \ge n - 1$ ($|E| = m - k$, $k \le m - n + 1 \Rightarrow |E|$

$\geq m - (m - n + 1) \Rightarrow |E| \geq n - 1$). So, in each iteration of Step 5, *RCGG* provides one connected graph with n vertices and $m - k$ edges.

Since Step 5 iterates $m - n + 1$ times, *RCGG* provides $m - n + 1$ connected graphs with n vertices and $(m - 1)$, $(m - 2)$, $(m - 3)$, …, and $(n - 1)$ edges, respectively. The last one is a spanning tree since it has exactly $n - 1$ edges, and at the beginning (in Step 3) of the algorithm, *RCGG* already assumed the complete graph (with n vertices and m edges), i.e., *RCGG* produces exactly $m - n + 2$ number of connected graphs. ◆

Theorem 3.2 Algorithm *RCGG* terminates after a finite number of iterations.

Proof: Step 5 of algorithm *RCGG* iterates at most $m - n + 1$ times since k is assigned the values 1, 2, 3, …, $m - n + 1$, i.e., the loop associated with Step 3 is countable. However, there is also an iteration in Step 7 if the connectivity test fails. For any value of k, one random edge x is chosen from E (where $|E| = m - k + 1$ before removal of x). If the subgraph (V, E) is connected, then the k-th iteration finishes and Step 5 iterates for $k = k + 1$. Otherwise, x is added to E, and a next edge of x from E is removed, and the subgraph (V, E) is tested for connectedness. This connectedness testing takes place at most for $m - k + 1$ number of edges in the k-th iteration. This test definitely becomes successful for some edge x in E since $|E| = m - k + 1, k \leq m - n + 1 \Rightarrow |E| \geq m - (m - n + 1) + 1 \Rightarrow |E| \geq n$. Then, the upper bound on the number of executions of Steps 7–10 is $m - k + 1$, i.e., the number of executions of Step 5 including the number of executions of Step 7 is bounded by $(m - n + 1) \times (m - k + 1)$. Hence, the algorithm terminates in a finite number of iterations. ◆

4 Data Structures and Complexity

Algorithm *RCGG* given in Sect. 3.1 creates a connected graph and stores it using an adjacency matrix. The adjacency matrix [13, 14] takes $O(n^2)$ space, where n being the number of vertices introduced into the graph. Initially, the assumed graph is a complete graph. From this complete graph, the algorithm gradually generates the different random connected graphs, by deleting edges one by one. Hence, no extra space is consumed for them. To perform the connectivity test, we take the help of *Breath First Search* (*BFS*) [15] algorithm. Two arrays of size n each are utilized to implement the *BFS* method, i.e., another $O(2n)$ space is required. So, *RCGG* requires $O(n^2 + n)$ space, or the space complexity of this algorithm is $O(n^2)$.

4.1 Time Complexity

The formation of the initial complete graph (in Step 3) of order n takes $n(n - 1)/2$ operations, i.e., $O(n^2)$ time. To select a random edge (in Steps 6 and 10) and to delete the same (in Step 8), the corresponding entries from the adjacency matrix, in order to generate another graph, take constant time. Step 5 iterates $m - n + 1$ times. The connectedness test in Step 7 can be performed using $O(n^2)$ operations, and Step 8 performs two operations only.

4.1.1 Worst Case

Steps 6 through 11 of the devised algorithm can be executed m times in the worst case. Since each vertex of a complete graph has its initial degree same as $n - 1$ and *RCGG* removes exactly one edge in each iteration in Step 5 for the values of k varying from 1 to $n - 2$, Step 8 executes just once for each k, as no disconnected graph is generated after any of the iterations. Incidentally, the connectivity test in Step 6 takes $O(n^2)$ time.

Now, for the values of $k > n - 2$, i.e., $k = (n - 1), n, (n + 1), \ldots, (m - n + 1)$, Step 7 can be executed at most $O(|E|)$ times to obtain the subsequent connected subgraphs, and $|E|$ becomes $m - k$ in the k-th iteration of Step 5. Thus, the maximum number of iterations of Step 10 for these values of k is

$$\sum_{n-1}^{m-n+1} (m-k) = (m-n+1) + (m-n) + \ldots + (n-1)$$

$$= \{1 + 2 + \ldots + (m - n + 1)\} - \{1 + 2 + \ldots + (n-2)\}$$
$$= (m-n+1)(m-n+2)/2 - (n-2)(n-1)/2$$
$$= (n^4 - 6n^3 + 13n^2 - 12n + 4)/4, \text{ where } m = n(n-1)/2.$$

Hence, Step 5 of the algorithm performs $(n^4 - 6n^3 + 13n^2 - 8n - 4)/4$ (obtained by adding $n - 2$ to the previous expression) operations or in $O(n^4)$ time. Now, combining the complexities of Steps 3, 5, and 6 (including the certification of connectedness), we conclude that the worst case complexity of *RCGG* is $O(n^2) + O(n^4) \times O(n^2)$, i.e., $O(n^6)$.

4.1.2 Best Case

Step 5 of the algorithm iterates at most $m - n + 1$ times. In the best case, the condition in Step 7 becomes true in the first attempt of each iteration of Step 5. Thus, Step 5 (together with the connectedness test) is completed in $O(n^2) \times O(n^2)$ time and also the operations executed for Step 3 when added to it takes $O(n^2)$ time. Hence, the best case complexity of *RCGG* is $O(n^4)$.

5 Illustration of *RCGG* with an Example

Let $n = 4$. *RCGG* computes the complete graph K4(V, E) in Step 3, and here, $E = \{e_1, e_2, e_3, e_4, e_5, e_6\}$, $V = \{v_1, v_2, v_3, v_4\}$, $m = 6$, as shown in Fig. 1.

K_4 is shown in the Fig. 1a.

$k = 1$: Here, the randomly selected edge, i.e., $x = e_4$ is deleted, and hence, E becomes $\{e_1, e_2, e_3, e_5, e_6\}$. The subgraph ($V$, E) in Fig. 1b is connected. So, we accept it and update $k = k + 1 = 2$.

$k = 2$: Here, the randomly selected edge, i.e., $x = e_1$ is deleted, and hence, E becomes $\{e_2, e_3, e_5, e_6\}$. The subgraph (V, E) in Fig. 1c is connected. So, we accept the deletion and update $k = k + 1 = 3$.

$k = 3$: Here, the randomly selected edge, i.e., $x = e_5$ is deleted, and hence, E becomes $\{e_2, e_3, e_6\}$. However, the subgraph (V, E), shown in Fig. 1d, is found to be disconnected. Hence, algorithm *RCGG* does not accept this deletion and rather, revert it back to the latest modified graph obtained from K_4 (of Fig. 1c), and a next edge, i.e., $y = e_6$ is considered for its deletion from edge set E. Now, E becomes $\{e_2, e_3, e_5\}$. The subgraph (V, E) after deletion of e_6 is depicted in Fig. 1e, which is found to be connected. So, we accept the deletion and update $k = k + 1 = 4$.

Here, the algorithm terminates since the condition $k \leq m - n + 1$ is dissatisfied for $k = 4$, $m = 6$, and $n = 4$. Hence, the algorithm outputs four preferred generated graphs shown in Figs. 1a–e with number of edges 6, 5, 4, and 3, respectively, and the last one in Fig. 1e is a spanning tree.

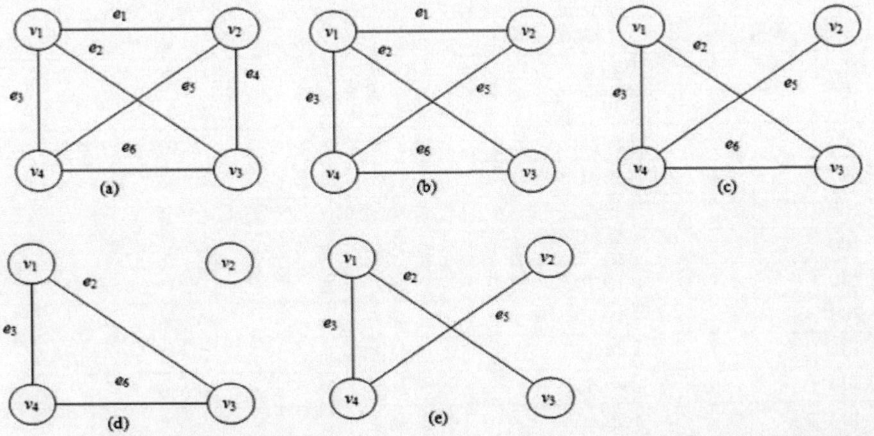

Fig. 1 **a** Complete graph of order 4, i.e., K_4, **b** One edge, e_4, is removed from K_4, **c** A second edge, e_1, is removed from the modified K_4, **d** A third edge, e_5, is removed from the latest modified K_4 that results in the graph no longer connected, **e** The last edge deleted (i.e., e_5) is added (or returned back) to the latest modified K_4, and then, a third edge, e_6, is removed from the newest modified K_4

6 Experimental Results

The algorithm, *RCGG*, developed in this paper, has been implemented on an Intel Dual Core processor running at 3.00 GHz, with 4 GB RAM capacity. Random graph instances have been generated for graphs of order (i.e., *n*) ranging from 10 to 600, *n* being the total number of vertices. The CPU time, taken by the algorithm to execute, for each such generation is shown in minutes and seconds in Table 1. As discussed in the earlier section, some selected edges are not removed to preserve the connectedness, as also has been shown in Table 1. Hence, for each such instance, i.e., the number of vertices (*n*), Table 1 includes the number of expected connected graphs (*CG*) generated by the algorithm, the number of (randomly) selected edges which are revert back to the previous subgraph and not to be removed (*NRE*), and the execution time. By considering the experimental results in Table 1, we observe that the CPU time needed to generate random connected graphs is very small. In every execution of *RCGG* with same number of vertices, it produces the same number of connected graphs but the CPU time and number of edges, which are selected but not removed, may differ. Because of the randomness property of *RCGG*, the sets of connected graphs produced by it, with the same number of vertices, may also be dissimilar or different.

Table 1 Results after execution of algorithm *RCGG*, where the order of the graph (i.e., the number of vertices belonging to the graph) varies from 10 to 600

Number of vertices (*n*)	Connected graphs generated by *RCGG* (*CG*)	Selected edges not removed (*NRE*)	CPU time (min: sec)
10	37	4	0: 0
10	37	5	0: 0
50	1177	271	0: 0
50	1177	172	0: 0
100	4852	1479	0: 0
100	4852	794	0: 0
200	19,702	6493	0: 7
200	19,702	4677	0: 6
300	44,552	11,874	0: 27
300	44,552	16,341	0: 30
400	79,402	38,136	1: 30
400	79,402	30,092	1: 27
500	124,252	55,866	3: 35
500	124,252	46,274	3: 28
600	179,102	76,798	7: 15
600	179,102	76,095	7: 15

7 Conclusions

Generation of random graphs is a topic of interest for researchers for different purposes like developing graph algorithms and approximation algorithms, simulation and sampling, scheduling algorithms [16], algorithms for database applications, etc. The generation of random graphs is often required to determine the performance of various other algorithms. In this paper, we have developed a random graph generation algorithm, *Random Connected Graph Generator* (*RCGG*), to provide a test bed for the above algorithms. *RCGG* has, thus, been able to generate only a set of simple, undirected, connected, graphs with any number of vertices. In the result section of this paper, we have generated such graphs with number of vertices ranging from 10 to 600. In terms of computation time and a large class of desired random graph generation, the efficiency of the devised algorithm has also been established.

In this paper, we have claimed to generate only a set of connected graphs with n vertices and e edges, where e varies from $n - 1$ to $m - n + 1$ and $m = n(n - 1)/2$. There may be many more connected graphs within the possible range of edges, which have not been generated since *RCGG* creates only one connected graph with a given number of edges (e). There are some future research possibilities to generate connected graphs of our desire with fixed number of vertices as well as fixed number of edges. So, there are scopes for doing a lot toward generating random graph instances. In that direction, this paper has attempted only to frame a basic guideline for the future researchers of this domain of interest.

References

1. S. Nobari, X. Lu, P. Karras, S. Bressan, Fast random graph generation, in *Proceedings of the 14th International Conference on Extending Database Technology* (2011), pp. 330–342
2. E.W. Weisstein, (RG) Random graph. *Wolfram MathWorld Built with Mathematica Technology* (2010)
3. J.P. Char, Generation of trees, two-trees, and storage of master forests. IEEE Trans. Circ. Theor. **15**(3), 228–238 (1968)
4. A. Rakshit, S. Sen Sarma, R.K. Sen, A.K. Choudhury, An efficient tree generation algorithm. IETE J. Res. Inst. Elect. Telecommun. Eng. (IETE), Taylor Francis **27**(3), 105–109, (1981)
5. Wikipedia, (ST) Spanning Tree. https://en.wikipedia.org/wiki/Spanning_tree, Last Accessed on 22 Sept 2018
6. M. Bayati, J.H. Kim, A. Saberi, A sequential algorithm for generating random graphs. Algorithmica, Springer **58**(4), 860–910 (2010)
7. H. Bhuiyan, M. Khan, M. Marathe, A parallel algorithm for generating a random graph with a prescribed degree sequence, in *Proceedings of the IEEE International Conference on Big Data* (2017), pp. 3312–3321
8. M. Chakraborty, S. Chowdhury, R.K. Pal, Generation of simple, connected, non-isomorphic random graphs, in *Proceedings of the Sixth International Doctoral Symposium on Applied Computation and Security Systems* (*ACSS*), *Proceedings Published as Book Chapters of the Series Advances in Intelligent Systems and Computing* (*AISC*) vol. 995, (Springer, 2019), pp. 69–77
9. P. Erdös, A. Rényi, On the evolution of random graphs. Publ. Inst. Math. Hung. Acad. Sci. **5**, 343–347 (1960)

10. M.V. Horn, A. Richter, D. Lopez, A random graph generator, in *Proceedings of 36th Annual Midwest Instruction and Computing Symposium* (2003), p. 6
11. F. Viger, M. Latapy, Efficient and simple generation of random simple connected graphs with prescribed degree sequence, in *Proceedings of the 11th Annual International Conference on Computing and Combinatorics* (*ICCC*) (Springer, 2005), pp. 440–449
12. C. Wang, O. Lizardo, D. Hachen, Algorithms for generating large-scale clustered random graphs. Netw. Sci. **2**(3), 403–415 (2014)
13. A.V. Aho, J.E. Hopcroft, J.D. Ullman, *Data Structures and Algorithms* (First Edition, Pearson, 1983).
14. E. Horowitz, S. Sahni, S. Anderson, *Fundamentals of Data Structures in C*, 2nd edn. (Universities Press Pvt. Ltd., Hyderabad, India, 2008).
15. N. Deo, *Graph Theory with Applications to Engineering and Computer Science* (Prentice Hall of India Pvt. Ltd., New Delhi, 2005).
16. D. Cordeiro, G. Mounié, S. Perarnau, D. Trystram, J.-M. Vincent, F. Wagner, Random graph generation for scheduling simulations, in *Proceedings of the Third International Conference on Simulation Tools and Techniques* (*SIMUTools*), Article no. 60 (2010), pp. 1–10

Capsnet-VGG16 Architecture for Cassava Plant Disease Detection

Avijit Bose, Debanjan Ghosh, Anish Banerjee, Debojyoti Saha, Pranita Ganguly, and Satyajit Chakrabarti

Abstract In the era advanced agricultural informatics and artificial intelligence, researchers and agronomist have tried to keep a check on the quality and yield of production as increasing demand for food in the future is something to worry about both in terms of quality and sustainability. CNN has been the preferred choice in the area due to its vast image processing abilities but, it misses out on few parameters like taking the spatial relationship into account, being non-robust towards affine transformation, etc. Here Capsule network comes into play which possesses these traits and might have a chance to produce better results, thus enhancing the food production and cultivation sector.

Keywords Agricultural informatics · Artificial intelligence · Production · Sustainability · CNN · Image processing · Spatial relationship · Non-robust · Capsule neural networks

A. Bose · D. Ghosh · A. Banerjee (✉) · D. Saha · P. Ganguly · S. Chakrabarti
Institute of Engineering and Management, Kolkata, India
e-mail: anish.banerjee1998@gmail.com

A. Bose
e-mail: avijit.bose@iemcal.com

D. Ghosh
e-mail: dghosh533@gmail.com

D. Saha
e-mail: debojyotisaha5@gmail.com

P. Ganguly
e-mail: pranitaganguly17@gmail.com

S. Chakrabarti
e-mail: styajit.chakrabarti@iemcal.com

V. E. Balas et al. (eds.), *Proceedings of International Conference on Computational Intelligence, Data Science and Cloud Computing*, Lecture Notes on Data Engineering and Communications Technologies 62, https://doi.org/10.1007/978-981-33-4968-1_17

1 Introduction

Increasing demand for food in the future is evident and is something to worry about in terms of both quality and availability. Agronomist and researchers have tried to keep a check on the quality and yield of production to keep the chances of economy and production losses minimum. Plant diseases spread at a rapid rate causing loss of production and pesticide implementation to translate to an accumulation of toxins within the crops resulting in severe repercussions and multifaceted depreciations. Detection and diagnosis of plant diseases visually through naked eyes is a very tedious task, but, deep learning has shown massive improvements in these segments and out-performed other predictive measures. CNN has been the most appropriate choice in this case, because of its extensive image processing abilities and has been very successful in detection and classification of plant leaves diseases. But it misses out on a few parameters like taking the spatial relationship into account, having large data demand, being non- robust towards affine transformation. In such cases, Capsule Neural Network is preferable over other networks.

2 Related Work

CNN architecture performed with 99% accuracy to detect plant diseases with a VGG architecture after extensive training on 25 classes [1]. AlexNet and GoogleNet CNN models were tested on PlantVillage dataset of a variety of coloured, black and white and segmented versions of the images [2], 3. Similar algorithms have been implemented on tomato leaf images for disease detection [4] with CNN architecture with 92% accuracy [5]. CNN models have been implemented on rice crop disease detection for 10 diseases upon naturally collected images [6]. Pest recognition on real-time images of tomato plant has been done with similar implementation [7]. After the boom in capsule networks, it has been applied in cervical image classification, brain tumour classification, traffic flow prediction, 3-dimensional image generation [8], emotion recognition, lung cancer screening and many more. But in terms of literature information, very less work has been conducted in the field of agriculture. It has been used in peanut leaf disease classification [9] and potato leaf disease classification [10]. Capsule network has been implemented in the identification of UAV Imagery [11] and hyperspectral image classification [12].

3 Capsule Network

In the classical convolutional neural network architecture, each neuron in a particular layer represents a pixel. Subsequently, it passes on this information forward to the corresponding layer. These corresponding convolutional layers consist of a bunch of

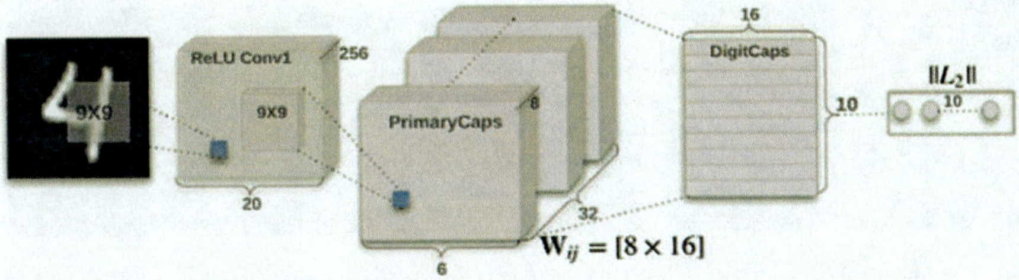

Fig. 1 Capsule network architecture

neurons, so that one particular neuron projects an entire bunch of neurons. Nevertheless, this framework is solely concerned about the existence of the object and its specific location. But, it does not hold itself accountable for the spatial relations and the direction of the object of concern.

Capsules are a relatively new concept, which is said to contain more information about each "object". We can describe a 'capsule' as a vector which specifies the features and prospect of an object. These features mainly include parameters like "pose" (position, size, orientation), velocity, hue, texture etc. capsule network can recognise unseen variations of a specified class without being trained on them (Fig. 1).

The architecture of capsule network is basically divided into two parts:

Encoders: It takes the image as an input, learns how to project it in the form of a 16-dimensional vector which holds all the information essentially required to render the image. It has three layers.

The Convolution layer, responsible for feature detection, which gets analysed in the later capsules.

Primary or lower capsule layer, containing 32 different capsuled and producing 4D vector output, Higher or digits capsule layer, using dynamic routing primary capsule would route to this particular layer.

The output of this layer is a 16D vector which contains all the instantiation parameters essential for restructuring the object of concern.

Decoders: The decoder obtains the 16D vector from the last digit capsule and decodes the instantiation parameters it holds (Fig. 2).

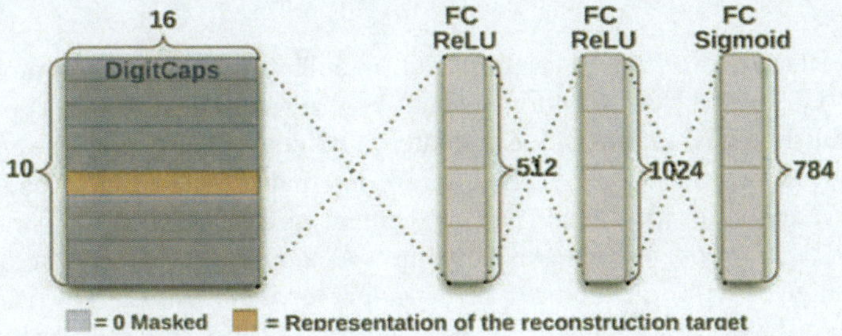

Fig. 2 Decoders in capsule networks

Fig. 3 Sample images in CBB, CBSD, CGM, CMD and healthy leave images

The decoder, along with the Euclidean distance loss function, determines how similar the restructured feature is in comparison to the original one, from which it is getting trained. The decoder is a simple feed-forward neural network which consists of three fully connected dense layers with a final output with the number of classes we intend to classify.

4 Dataset Overview

Cassava, a root vegetable, is a cellulose-rich insoluble fiber. It is a type of dietary fiber that aids in digestion and helps prevent constipation and diverticular diseases. Also, it is considered to be a prebiotic, a type of fiber that promotes the growth of probiotic bacteria as it ferments in the intestines. Rich in carbohydrates, this plant is of great importance for developing nations and is helpful to fight diabetes and diarrhoea. Cassava is prone to bacterial attacks of which Cassava Bacterial Blight (CBB) is the most common. Others are Cassava Mosaic Disease (CMD), Cassava Brown Streak Disease (CBSD) and Cassava Green Mite (CGM). Images of these diseases with a set of healthy plant images to test the accuracy of the capsule networks to classify the disease. A total of 5656 images was divided into a group of these five classes with almost equal distribution (Fig. 3).

5 Proposed Method

Many models tested on the dataset created, including the established models like VGG16, VGG19, ResNet, XceptionNet etc. showed varied results. In comparison to these established CNN models, we tested capsule neural networks. The variations of the Capsnet we used were with different internal architectures. Changing the layers, the number of capsules and pairing them up with established CNN models for getting the best of both worlds, we came up with a hybrid architecture comprising of a CapsuleNet with, the number of capsules 14 of dimension 16, paired with VGG16 (last_layer = block3_conv3), and our custom DNN, collectively referred to as CassavaNet in this paper (Fig. 4).

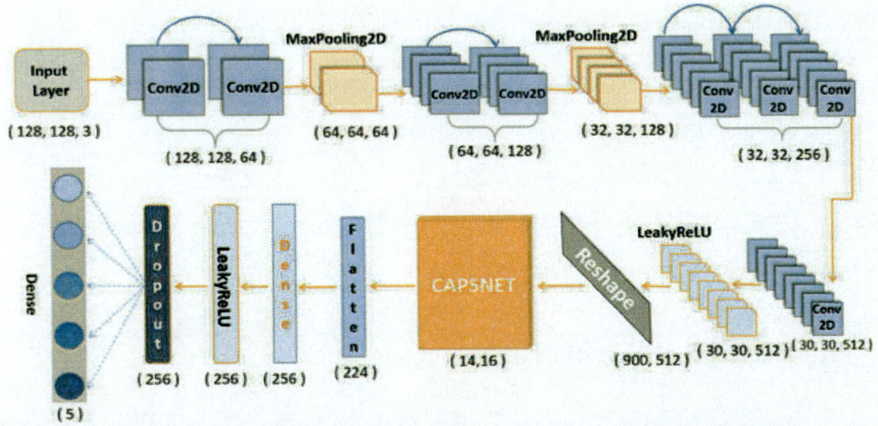

Fig. 4 CassavaNet architecture (Capsule network (512 layers) paired up with VGG16)

The complete architecture consists of the Input layer, Conv2D layers of increasing depth with MaxPooling2D and LeakyRelu in between. The diagram in Fig. 4 represents the detailed architecture with their names and dimensions.

6 Experimental Setup

The whole cassava dataset of 5,656 images belonging to 5 classes (CBB, CBSD, CGM, CMD, healthy), is used for this experiment. The original dimension of them were 500 × 660, which was reduced down to 128 × 128 to reduce the training time and decrease the internal parameters. The images were converted to MNIST format as capsule networks gave the best results when tested over MNIST dataset and for Keras implementation. Block-wise DCT scaling method is used to enhance the colored images. The input layer of the original Capsnet architecture, which was fed the images had layer (256) and filter dimension of (9 × 9) with stride (1). This layer is succeeded by the primary capsule with capsule dimension 8, 32 channels, and kernel size 9. The final layer has the same number of 'capsules' as the input labels, 5 in this case (CBB, CBSD, CDM, CMD and Healthy). The decoder consists of four fully connected layers. It was implemented with 128 and 512 filters in the first convolution layer, i.e. Conv1. Three CNN architectures (ResNet18, VGG19, VGG16) were trained on the same data, for comparison purpose along with VGG and ResNet paired with capsule networks. The training was conducted on RTX 2080 GPU using Python 3.7.

Table 1 Model accuracy comparison

Model	Training accuracy (%)	Validation accuracy (%)
Capsule network 256 layers	76.20	75.14
Capsule network 128 layers	56.30	51.7
Capsule network 512 layers	63.64	60.86
VGG16	90.3	87.92
VGG19	83.34	81.22
CapsNet (512) + VGG16 – [**CassavaNet**]	93.61	92.91

7 Results and Discussion

The results of training the mentioned algorithms yielded astonishing results with each model. The Cassava dataset showed the best results with capsule network paired with VGG16 model through transfer learning. Results from pure capsule networks with 216 and 512 layers were not satisfactory at all. However, pure CNN architectures showed some varying results as well. The VGG16 and VGG19 individually performed well, whereas ResNet and XceptionNet showed severe overfitting at a very early stage.

On the implementation of the selected models on the prepared dataset, diverse results could be seen. Both the training and testing accuracy of the respective models is listed in the table above (Table 1). So, in this comparison table, it is seen that in comparison to the other models we tested, our hybrid model, 'CassavaNet' performed better on the said dataset.

The dataset was composed of raw images taken by people and had hue, saturation and even sunlight effects. There were mixed images of leaves and even overlapping images. Grainy and discolored images were also there which might have accounted into these varying results. It is an overview of how the models perform and classify (Figs. 5, 6, 7 and 8).

8 Conclusion

In the field of classification, CNN has been immensely successful. Capsule Network is however relatively new in this field of agricultural informatics or classification. In this paper, we have tried to compare the performance of the Capsule framework with few pre-trained CNN architectures, with the extensive use of transfer learning, for classification of images of a diseased cassava plant, with the application of Capsule Network producing some comparable output, achieving a decent accuracy. According to the work that we have put up, our CassavaNet model [Capsule Network (512 layers) paired up with VGG16] gives comparatively better training accuracy of 93.61% and

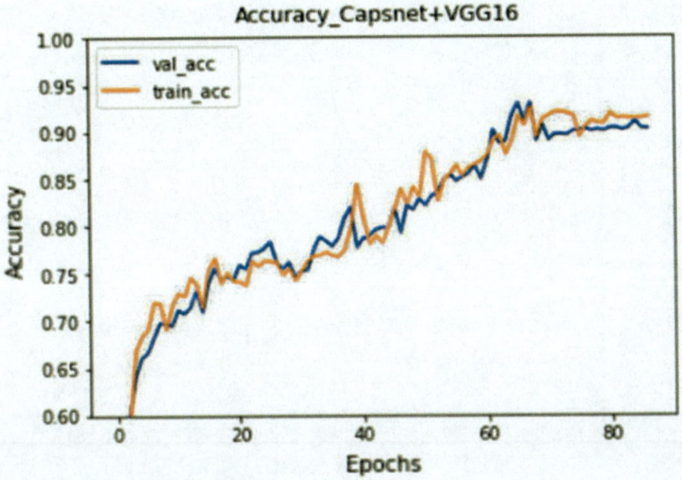

Fig. 5 Accuracy curve of CassavaNet model

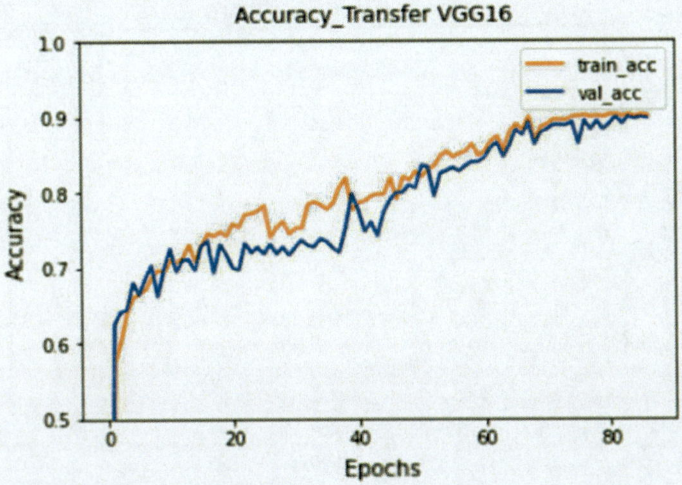

Fig. 6 Accuracy curve of VGG16 model

test accuracy of 91.26%, among the other models we implemented. Although the capsnet model with 128 layers was on the verge of overfitting, dense architectures like Inception V3, ResNet, XceptionNet showed early signs of high overfitting.

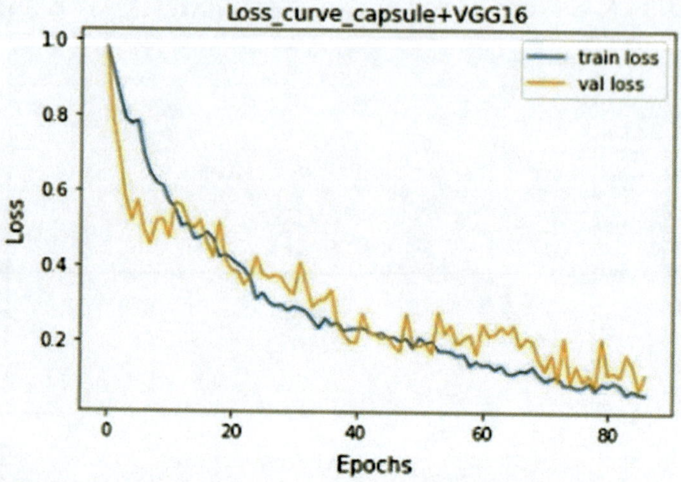

Fig. 7 Loss curve of CassavaNet model

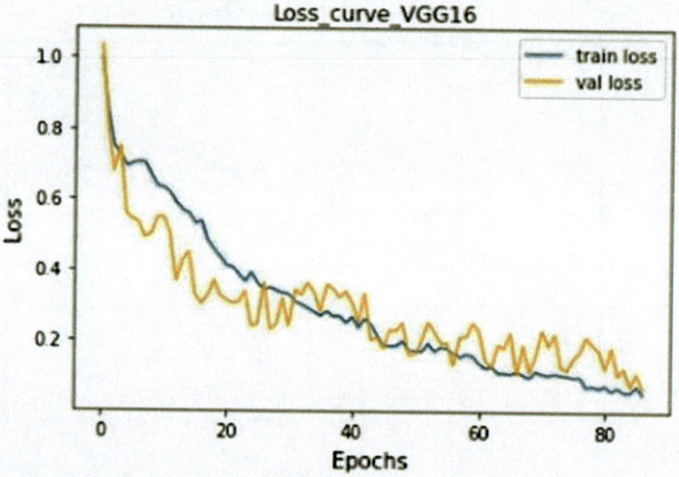

Fig. 8 Loss curve of VGG16 model

9 Future Scope

In future, we plan to explore the details of Capsule Network and implement them with various kind of images, like segmented images, over-lapping leaf images and also a few real-life images.

With the advanced digital and artificial intelligence efforts, we can imply this particular framework to enhance food production and cultivation in agricultural sectors. We need a hostile network to access agricultural sustainability and thereby reduce agriculture-related loss. The work in this paper can be further tested on other corresponding capsule networks with architectural changes as well for improvements.

References

1. K.P. Ferentinos, Deep learning models for plant disease detection and diagnosis. Comput. Electron. Agric. **145**, 311–318 (2018)
2. S.P. Mohanty, D.P. Hughes, M. Salathé, Using deep learning for image-based plant disease detection. Front. Plant Sci. **7**, 1419 (2016)
3. D. Hughes, M. Salathé, "An open access repository of images on plant health to enable the development of mobile disease diagnostics," arXiv preprint arXiv:1511.08060, (2015)
4. K. Zhang, Q. Wu, A. Liu, X. Meng, "*Can Deep Learning Identify Tomato Leaf Disease?*," Advances in multimedia, vol. 2018, Article ID 6710865, (2018)
5. M. Brahimi, K. Boukhalfa, A. Moussaoui, Deep learning for to-mato diseases: classification and symptoms visualization. Appl. Artif. Intell. **31**(4), 299–315 (2017)
6. G. Jaswal, R.C. Poonia, Selection of optimized features for fusion of palm print and finger knuckle? Based person authentication. Ex-pert Systems, e12523.
7. A. Fuentes, S. Yoon, S. Kim, D. Park, A robust deep-learning-based detector for real-time tomato plant diseases and pests recognition. Sensors **17**(9), 2022 (2017)
8. M.K. Patrick, A.F. Adekoya, A.A. Mighty, B.Y. Edward, "Capsule networks—a survey," J. King Saud Univ-Comput. Inform. Sci. (2019)
9. L. Xiao, H. Ouyang, C. Fan, T. Umer, R.C. Poonia, S. Wan, (2020). Gesture image segmentation with Otsu's method based on noise adaptive angle threshold. *Multimedia Tools and Applications*, pp. 1–22
10. S. Verma, A. Chug, A.P. Singh (2020) Exploring capsule networks for disease classification in plants. J. Stat. Manage. Syst.
11. Y. Li, M. Qian, P. Liu, Q. Cai, X. Li, J. Guo, H. Yan, F. Yu, K. Yuan, J. Yu, L. Qin, "The recognition of rice images by UAV based on capsule network". Cluster Comput. 1–10 (2018)
12. M.E. Paoletti, J.M. Haut, R. Fernandez-Beltran, J. Plaza, A. Plaza, J. Li, F. Pla, Capsule networks for hyperspectral image classification. IEEE Trans. Geosci. Remote Sens. **57**(4), 2145–2160 (2018)

Forecast Model Development of Some Selected Wholesale Price Index of India Using MLP

Dipankar Das and Satyajit Chakrabarti

Abstract In the war against hunger—timely, reliable, and acceptable forecast of food prices can provide huge leverage to society. The present WPI of India series (Base: 2011–12 = 100) lists sixty individual items in the manufacture of food products category. The study proposes forecast model development of some selected commodities of this category using the Multilayer Perceptron (MLP) approach. The researchers performed feature extraction of all the items, grouped them based on similarities, and developed the proposed forecast model for all the items of the largest group. The largest group consists of thirty-six indices—that is, sixty percent of the total (60 indices) under study. The study reveals that a promising result with the proposed forecast model outperformed regressions, ARIMA, and SES models in terms of average MAE and average MAPE. The model exhibited high accuracy (MAPE $< =$ 10) for approximately ninety-two percent of the indices of the largest group.

Keywords MLP model · WPI · Feature extraction · Forecast

1 Introduction

We know that the Wholesale Price Index (WPI), an important macroeconomic indicator, is used to measure the price fluctuations/changes in the wholesale stage. Timely and accurate forecast of these important indices help us in many areas—e.g., economic planning, financial measures, and monetary policy. We know that food is essential for life, and the importance of reliable and acceptable forecasts of food prices is paramount in the fight against hunger. To forecast time series, some of the classical approaches are the following: (i) ARIMA, (ii) Exponential smoothing

D. Das (✉) · S. Chakrabarti
University of Engineering and Management, Kolkata 700156, India
e-mail: dipankr.das@gmail.com

S. Chakrabarti
e-mail: satyajit.chakrabarti@iemcal.com

V. E. Balas et al. (eds.), *Proceedings of International Conference on Computational Intelligence, Data Science and Cloud Computing*, Lecture Notes on Data Engineering and Communications Technologies 62, https://doi.org/10.1007/978-981-33-4968-1_18

[1]. ANN is another important approach for time series forecasting and used by many [2, 3]. We used a Multilayer Perceptron (MLP) approach in the present study to develop the forecasting model of some selected WPI of the manufacture of food products of India.

2 Literature Review

Choudhary et al. employed an ANN approach with one hidden layer, developed a forecast model of the potato (daily wholesale price) in Delhi, India [4]. Liu and Yu compared the performances of ANN, long short-term memory, and multivariate linear regression models to forecast the raw grain prices and observed that the ANN model outperformed others [5]. Silva et al. developed an ANN model to forecast the sales of basic agricultural products [6]. Salazar Moreno and Zerecero Salazar analyzed the maize price in Mexico using an ANN approach [7]. Zaw et al. used a backpropagation neural network model to forecast the rice price in Myanmar [8]. Harris observed that the MLP approach (average MAPE 4.4 approx) is the best amongst the following machine learning approaches for forecasting the consumer price index of some food categories in Canada: Linear regression, multilayer perceptron, SMOreg, and M5PTree [9]. Gashaw developed an ANN model to forecast the daily prices of sesame in Ethiopia [10]. Anggraeni et al. applied the ANN approach for forecasting the chili price and achieved MAPE of 16.2% approximately [11].

Ohyver and Pudjihastuti developed an ARIMA-based forecast model for the price of rice [12]. Kumari et al. developed an ARIMA model to forecast the chili price in Khammam, India [13]. Mardianto et al. applied the ARIMA approach on two platforms—Amazon Sagemaker and Amazon Forecast to forecast thirty days ahead of rice prices and observed that the Amazon Sagemaker platform outperformed the other [14].

Lutfi et al. applied the simple exponential smoothing approach to analyze the food commodity price trends in Gowa, Indonesia [15]. Rosyid, Widiyaningtyas, and Hadinata compared three exponential smoothing approaches—single, double, and triple in the case of food price forecast in Java, Indonesia [16].

Volkov, Morkunas, and Skvarciany developed a retail food price forecast model for Lithuania using a multiple regression approach [17]. Ge and Wu used univariate non-linear regression to forecast the corn price in China [18].

Montero-Manso et al. used weighted combinations of different forecast methods, where the weights obtained from the extracted time series features [19]. Liew and Chou used the feature extraction approach with MLP and meta-regression to forecast stock performance in Taiwan [20]. Nguyen and Yoon proposed a feature engineering-based framework to predict stock prices [21].

3 Research Gap

We observed the following research gaps in the monthly WPI series (Base: 2011–12 = 100) in India: (i) feature extraction of the WPI of the individual items/commodities of manufacture of the food products category by analyzing the data from April 2012 to March 2016, (ii) grouping of the indices using extracted features, and (iii) development of forecast model of the individual indices of the largest identified group using the MLP approach.

4 Objectives of the Study

- Feature extraction of the indices of the individual items of manufacture of the food products category of the WPI series (Base: 2011–12 = 100) in India
- Grouping the indices having similar features and identification of the largest group
- Developing a forecast model for the WPIs belonging to the largest group using the MLP approach.

5 Methodology

5.1 Overview of Model Development

A diagrammatic representation of the steps of the model development is given in Fig. 1.

5.2 Data

We collected the data of the individual items/commodities of the manufacture of the food products category of the WPI (Base: 2011–12 = 100) of India from the data.gov.in, the Open Government Data (OGD) Platform India [22]. We collected sixty (60) month's data, starting from April 2012 to March 2017. There are sixty (60) items listed in this group. The data is monthly in nature.

Table 1 listed the indices (WPIs) under study. The item names sourced from data.gov.in, and their corresponding item codes designated by the researchers.

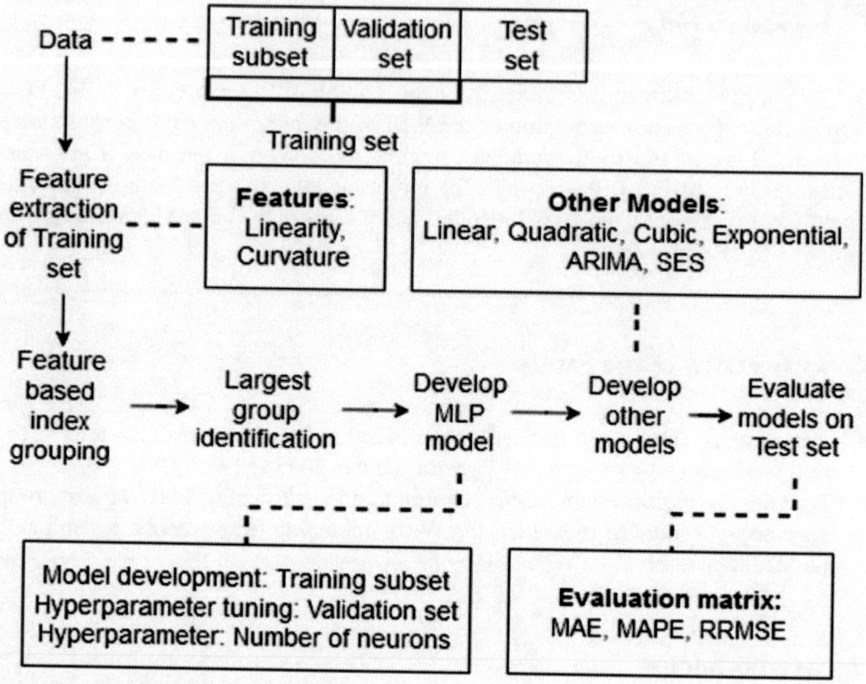

Fig. 1 Overview of model development

5.3 Data Division

We divided the data of each index as shown in Fig. 2. We used the training set data to extract the features of each index. We developed the MLP model using the training subset (1–36 months data) and used the validation set (37–48 months data) for hyperparameter tuning. We developed the final models using the training set (1–48 months data) and evaluated on the test set (49–60 months data). The training set consists of the forty-eight (48) month's data (from April 2012 to March 2016), and the holdout test set—the last twelve (12) month's data (from April 2016 to March 2017). We further sub-divide training set into a training subset (from April 2012 to March 2015) and a validation set (from April 2015 to March 2016).

5.4 Feature Extraction

We extracted linearity and curvature features of each index using the tsfeatures package of R [23]. We used the stl_features function of the package to extract the features of the univariate time series.

Table 1 List of items

Index code	Name of the index	Index code	Name of the index	Index Code	Name of the index
V1	Buffalo meat, fresh/frozen	V21	Cotton seed oil	V41	Molasses
V2	Meat of goat, fresh or chilled	V22	Condensed milk	V42	Bagasse
V3	Other meats, preserved/processed	V23	Ghee	V43	Gur
V4	Chicken/duck, dressed fresh/frozen	V24	Butter	V44	Honey
V5	Shrimps/prawns processed/frozen	V25	Ice cream	V45	Chocolate and cocoa powder
V6	Fish frozen/canned/processed	V26	Powder milk	V46	Sugar confectionary
V7	Fruit Juice including concentrates	V27	Maida	V47	Noodles and similar extruded products
V8	Fruit pulp	V28	Wheat flour (Atta)	V48	Processed tea
V9	Jams, jellies, marmalades, and puree	V29	Wheat bran	V49	Instant coffee
V10	Sauces of vegetables (Tomato, Chili, Soya, and others)	V30	Sooji (rawa)	V50	Coffee powder with chicory
V11	Vanaspati	V31	Flour of cereals other than rice, wheat, etc.	V51	Spices (including mixed spices)
V12	Mustard oil	V32	Gram powder (besan)	V52	Salt
V13	Soyabean oil	V33	Rice, non-basmati	V53	Instant Food/prepared meals based on vegetables
V14	Sunflower oil	V34	Basmati rice	V54	Corn flake
V15	Groundnut oil	V35	Rice products	V55	Whey powder
V16	Castor oil	V36	Vegetable starch	V56	Gola and similar cattle feed
V17	Rice bran oil	V37	Biscuit, cookies	V57	Rice bran extract
V18	Palm oil	V38	Bread, buns, and croissant	V58	Soya preparations excluding oil
V19	Rapeseed oil	V39	Cakes, pastries, and muffins	V59	Cotton seed oil cake
V20	Copra oil	V40	Sugar	V60	Mustard oil cake

Fig. 2 Overview of data division

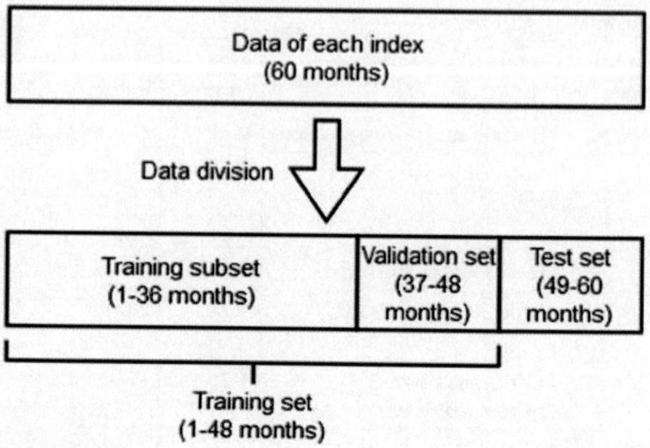

5.5 *Feature-Based Index Grouping*

Based on the extracted features of the indices, we created the following four (4) groups:

- Group 1: Linearity positive and curvature positive
- Group 2: Linearity positive and curvature negative
- Group 3: Linearity negative and curvature positive
- Group 4: Linearity negative and curvature negative.

5.6 *Largest Group Identification*

We identified the group which contains the maximum number of indices.

5.7 *Proposed Model*

We developed the MLP model of each WPI of the largest group. The proposed model architecture is as follows:

- Number of inputs: 2, i.e., Lag 1 and Lag 2 of the univariate time series
- Number of hidden layers: 1
- Number of output: 1.

We used the validation set data to tune the model. We performed hyperparameter tuning to obtain the number of neurons (N). The search space of N is as follows:

- Number of neurons (N) = {1, 2, 3, …, 8}

The diagrammatic representation of hyperparameter tuning is given in Fig. 3.

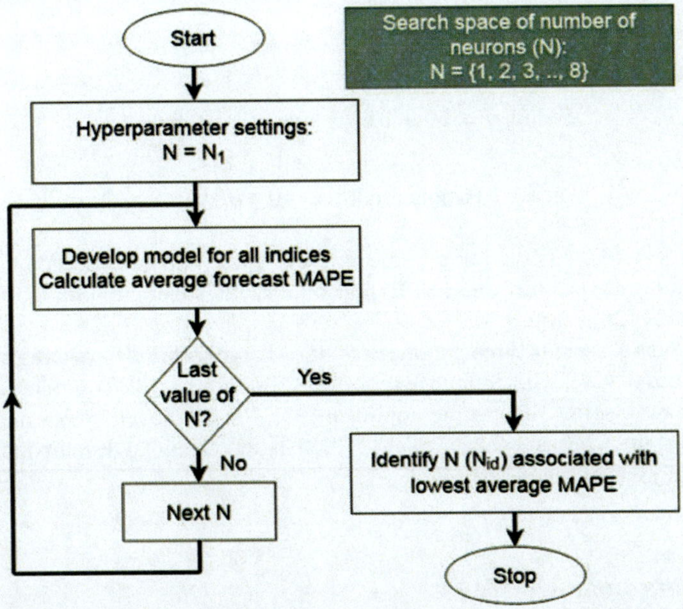

Fig. 3 Overview of hyperparameter tuning

For each index, we have trained twenty-five (25) networks to obtain the ensemble forecast. We combined the forecasts using the KDE estimation-based mode operator. We used the mlp function of the nnfor package of R to develop the proposed MLP model [24].

For each index, we developed an MLP model using a particular hyperparameter setting (i.e., using an N value from the search space), performed twelve (12) steps ahead forecast, and calculated forecast MAPE. We repeated this step for all the indices, calculated the average MAPE for the particular hyperparameter settings. We repeated these steps for all the N values from the search space. We identified the hyperparameter associated with the lowest average MAPE, which gave us the tuned model.

5.8 *Other Models*

We compared the forecast of the proposed MLP model with regression, ARIMA, and SES models. We used four (4) regression models: linear, quadratic, cubic, and exponential. The linear, quadratic, cubic, and exponential models are given as [25]:

$$\text{Linear: y} = \text{b0} + \text{b1} * x \tag{1}$$

$$\text{Quadratic: } y = b0 + b1 * x + b2 * x^2 \quad (2)$$

$$\text{Cubic: } y = b0 + b1 * x + b2 * x^2 + b3 * x^3 \quad (3)$$

$$\text{Exponential: } y = b0 * e^{b1x} \quad (4)$$

In Eqs. (1)–(4), y is the index value and x is time. We developed the (i) Linear, (ii) Quadratic, (iii) Cubic, and (iv) Exponential models using the lm () of the stats package of R [26].

The ARIMA accepts three parameters—(i) AR order, (ii) differencing order, and (iii) MA order [29]. The Simple Exponential Smoothing (SES) model is another univariate time series forecasting approach [30]. We employed auto.arima and ses functions of the forecast package of R [27, 28], respectively, to develop the ARIMA and SES models of the indices.

5.9 *Evaluation Matrix*

We used the Mean Absolute Error (MAE) [31], Mean Absolute Percentage Error (MAPE) [31], and Relative Root Mean Square Error (RRMSE) [32] to evaluate the forecast. We considered the model accuracy to be excellent if the RRMSE < ten (10) percent [32]. The model accuracy is high if the MAPE < = ten (10) percent [33].

6 Data Analysis and Findings

We performed the feature extraction of sixty (60) indices under study. Table 2 listed the linearity and curvature features of the indices.

The category wise grouping of the indices is shown in Fig. 4.

From Fig. 4, we observed that the largest group is Group 2 (positive linearity, and negative curvature features). It consists of sixty percent of the indices under study. Table 3 listed the indices of Group 2.

Hyperparameter tuning of the MLP model is shown in Fig. 5.

From Fig. 5, we observe that at N = 1 the average MAPE is the lowest. Therefore, the identified N value (Nid) is 1. The proposed final MLP model is given in Fig. 6.

We developed the forecast model of the WPIs of Group 2 using the MLP (proposed) approach, performed twelve (12) steps ahead forecast, and calculated the forecast performances. The evaluated performances of the model (proposed MLP model) are shown in Fig. 7.

The mean and median of MAE of the thirty-six (36) selected WPIs are 6.7 and 5.41, respectively. The mean and median of MAPE of these indices are 4.86 and

Table 2 Linearity and curvature features

Index code	Linearity	Curvature	Index code	Linearity	Curvature	Index code	Linearity	Curvature
V1	49.27	−25.22	V21	−25.30	5.40	V41	61.44	−42.76
V2	80.05	−21.85	V22	72.86	−12.62	V42	59.67	−29.15
V3	55.86	−8.31	V23	51.07	−10.67	V43	−7.31	−20.06
V4	50.38	−21.21	V24	61.72	−11.42	V44	122.01	−4.24
V5	51.35	−88.60	V25	51.17	−0.23	V45	49.48	−9.00
V6	56.00	−6.89	V26	47.62	−44.63	V46	25.32	0.78
V7	15.57	1.63	V27	37.93	−33.17	V47	42.96	−1.38
V8	44.94	2.18	V28	38.47	−23.21	V48	21.51	−16.02
V9	58.29	5.84	V29	62.94	−14.53	V49	24.69	−3.78
V10	32.33	4.83	V30	47.35	−29.67	V50	76.42	20.47
V11	−13.87	−5.39	V31	74.31	4.60	V51	75.80	−0.53
V12	20.61	37.50	V32	33.32	99.70	V52	11.64	−20.17
V13	−30.78	−1.68	V33	44.46	−29.36	V53	47.09	7.58
V14	−17.62	6.60	V34	−9.46	−53.13	V54	74.80	4.96
V15	−37.37	53.15	V35	16.90	−32.83	V55	70.18	−58.65
V16	15.85	−23.80	V36	12.76	−7.40	V56	61.43	−23.16
V17	−22.21	5.60	V37	39.34	−4.22	V57	8.62	−17.27
V18	−30.17	−7.66	V38	58.61	−16.01	V58	61.28	−43.28
V19	11.48	30.13	V39	50.11	−9.20	V59	80.28	11.76
V20	147.22	−54.67	V40	−34.71	−7.48	V60	151.67	121.23

Fig. 4 Groups of indices

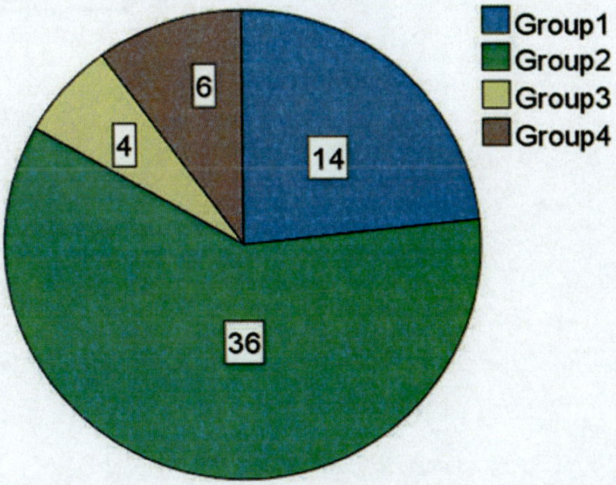

Table 3 Indices of group 2

Index code	Index code	Index code	Index code	Index code	Index code
V1	V16	V26	V35	V42	V51
V2	V20	V27	V36	V44	V52
V3	V22	V28	V37	V45	V55
V4	V23	V29	V38	V47	V56
V5	V24	V30	V39	V48	V57
V6	V25	V33	V41	V49	V58

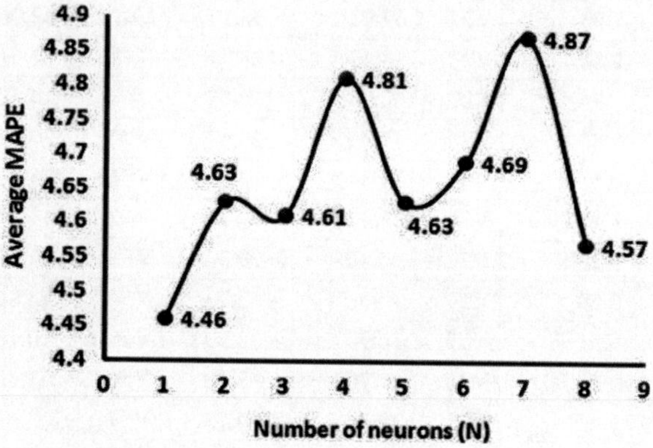

Fig. 5 Hyperparameter tuning of the MLP model

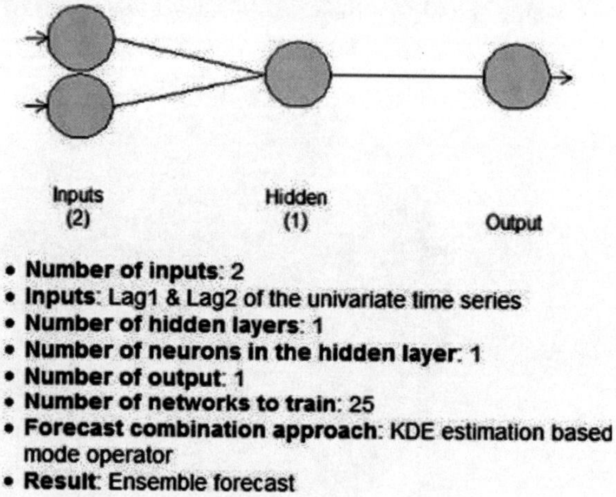

Fig. 6 Architecture/specifications of the MLP model

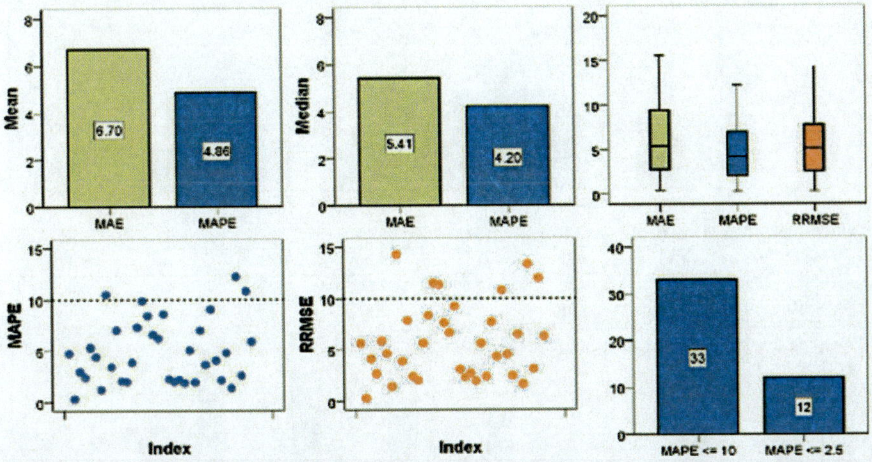

Fig. 7 Forecast performance of the MLP (proposed) model

4.2, respectively. There is no outlier in all three evaluation matrix. There are only three (3) indices (out of 36), where the MAPE is greater than ten (10). Therefore, the proposed MLP model exhibited high accuracy for 91.67% of the WPIs. Thirty (30) indices have the RRMSE values less than ten (10) which indicates excellent model accuracy (83.33%) for them.

We compared (Fig. 8) the proposed final MLP model with the following six (6) models: linear, quadratic, cubic, exponential, ARIMA, and SES.

From Fig. 8, we observed that the proposed MLP model has the minimum values of average MAE, average MAPE, median MAE, and median MAPE. In terms of these four (4) criteria, the proposed MLP model outperformed others. The proposed MLP model is of high accuracy for thirty-three (33) indices out of thirty-six (36), i.e., for 91.67% cases. This performance is on par with the SES model's in terms of MAPE $< = 10$. We also noticed that the proposed MLP model outperformed others for low values of MAPE (MAPE $< = 2.5$).

7 Conclusion

From the previous works of the literature, we noted some research gaps using the data of the present WPI series of India. The main objectives are to fill up some of these research gaps through performing feature extraction of the indices of the individual items of manufacture of the food products category of the WPI series (Base: 2011–12 = 100) in India, grouping the indices having similar features, identification largest group, and developing a forecast model for the indices (WPIs) of the largest group using the MLP approach.

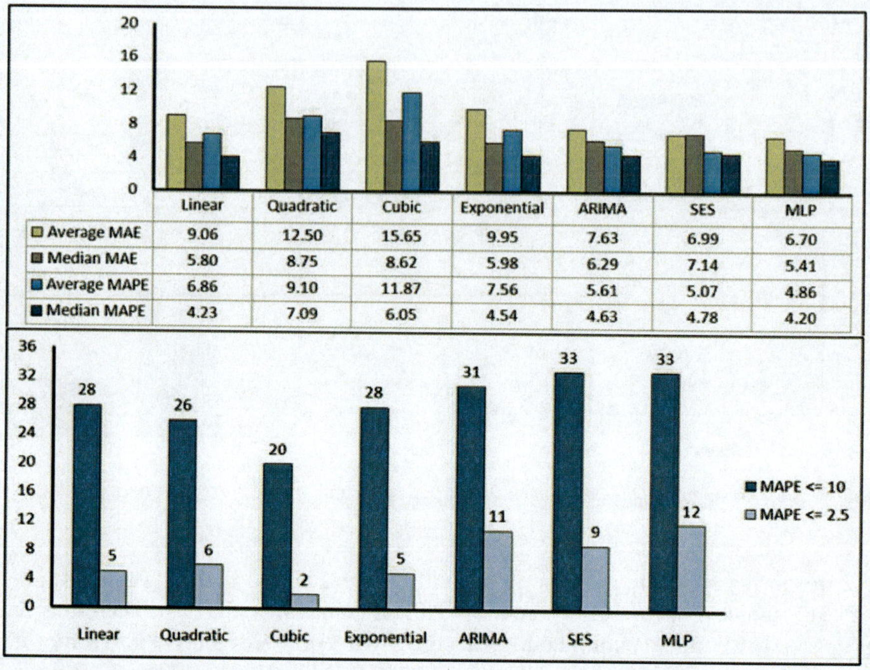

Fig. 8 Forecast performance comparison of MLP with others

The majority (i.e., 60%) of the indices of the individual items of manufacture of the food products category of the present WPI series (Base: 2011–12 = 100) in India simultaneously exhibited positive linearity and negative curvature features from April 2012 to March 2016. We developed an MLP model to forecast the indices with these features—identified as Group 2 in this work.

The architecture/specifications of the proposed MLP model is as follows: two inputs (Lag 1 and Lag 2 of the univariate time series), one hidden layer with one neuron, twenty-five networks to train, and results in an ensemble forecast. The proposed MLP model outperformed regressions, ARIMA, and SES models in terms of average MAE, average MAPE, median MAE, and median MAPE. The proposed MLP model exhibited high accuracy (MAPE $<=10$) for the majority (91.67%) of the indices from Group 2, i.e., positive linearity and negative curvature. In 33.33% of cases, the proposed MLP model exhibited MAPE $<=2.5$ and outperformed all others.

Therefore, this MLP model may be used to make twelve (12) steps ahead forecast of the WPIs having the said features. We have not applied the model either on the indices from the other identified groups or different sets of indices having the same set of features, studying this will be our future scope.

References

1. J. Brownlee, *11 Classical Time Series Forecasting Methods in Python (Cheat Sheet)*. Machine Learning Mastery. (2018, August 5). https://machinelearningmastery.com/time-series-forecasting-methods-in-python-cheat-sheet/
2. A. Tealab, Time series forecasting using artificial neural networks methodologies: a systematic review. Future Comput. Inform. J. **3**(2), 334–340 (2018). https://doi.org/10.1016/j.fcij.2018.10.003
3. A. Tealab, H. Hefny, A. Badr, Forecasting of nonlinear time series using ANN. Future Comput. Inform. J. **2**(1), 39–47 (2017). https://doi.org/10.1016/j.fcij.2017.05.001
4. K. Choudhary, G.K. Jha, P. Das, K.K. Chaturvedi, Forecasting potato price using ensemble artificial neural networks. Indian J. Extension Educ. **55**(1), 73–77 (2019)
5. N. Liu, J. Yu, Raw grain price forecasting with regression analysis. *Proceedings of the 2019 International Conference on Modeling, Simulation and Big Data Analysis (MSBDA 2019)*, (2019). https://doi.org/10.2991/msbda-19.2019.58
6. J. Silva, N. Varela, H. Martínez Caraballo, J. García Guiliany, L. Cabas Vásquez, J. Navarro Beltrán, N. León Castro, An early warning method for basic commodities price based on artificial neural networks. *Advances in Neural Networks—ISNN 2019*, (2019), pp. 359–369. https://doi.org/10.1007/978-3-030-22796-8_38
7. R. Salazar Moreno, O. Zerecero Salazar, An artificial neural network model to analyze maize price behavior in Mexico. Appl. Math. **09**(05), 473–487 (2018). https://doi.org/10.4236/am.2018.95034
8. T. Zaw, K.M.M. Tun, A.N. Oo, Price forecasting by back propagation neural network model. *2019 International Conference on Advanced Information Technologies (ICAIT)*, (2019). https://doi.org/10.1109/aitc.2019.8921396
9. J. Harris, (Jay), A machine learning approach to forecasting consumer food prices. *Dalspace.Library.Dal.Ca*, (2017). https://hdl.handle.net/10222/73170
10. E. Gashaw, Sesame price prediction using artificial neural network, (2019)
11. W. Anggraeni, F. Mahananto, M.A. Rofiq, K.B. Andri, Sumaryanto, Z. Zaini, A.P. Subriadi, Agricultural strategic commodity price forecasting using artificial neural network. *2018 International Seminar on Research of Information Technology and Intelligent Systems (ISRITI)*, (2018). https://doi.org/10.1109/isriti.2018.8864442
12. M. Ohyver, H. Pudjihastuti, Arima model for forecasting the price of medium quality rice to anticipate price fluctuations. Procedia Comput. Sci. **135**, 707–711 (2018). https://doi.org/10.1016/j.procs.2018.08.215
13. R.V. Kumari, P. Venkatesh, G. Ramakrishna, A. Sreenivas, Chilli price forecasting using auto regressive integrated moving average (ARIMA). Int. Res. J. Agric. Econ. Stat. **10**(2), 290–295 (2019)
14. I. Mardianto, M.I. Gunawan, D. Sugiarto, A. Rochman. (2020). Comparison of rice price forecasting using the ARIMA method on amazon forecast and sagemaker. *Jurnal RESTI (Rekayasa Sistem Dan Teknologi Informasi)*, *4*(3), 537–543. https://doi.org/10.29207/resti.v4i3.1902
15. M. Lutfi, H. Muttaqien, A. Apriliani, H. Zainuddin, Y. Yuyun, Application of the Naïve bayes algorithm and simple exponential smoothing for food commodity prices forecasting. *Proceedings of the 1st International Conference on Science and Technology, ICOST 2019, 2–3 May, Makassar, Indonesia*, (2019). https://doi.org/10.4108/eai.2-5-2019.2284613
16. H.A. Rosyid, T. Widiyaningtyas, N.F. Hadinata, Implementation of the exponential smoothing method for forecasting food prices at provincial levels on java Island. *2019 Fourth International Conference on Informatics and Computing (ICIC)*, (2019). https://doi.org/10.1109/icic47613.2019.8985872
17. A. Volkov, M. Morkūnas, V. Skvarciany, Is it possible to predict food retail prices? Evidence from Lithuanian market. *Proceedings of 6th International Scientific Conference Contemporary Issues in Business, Management and Economics Engineering '2019*, (2019). https://doi.org/10.3846/cibmee.2019.018

18. Y. Ge, H. Wu, Prediction of corn price fluctuation based on multiple linear regression analysis model under big data. *Neural Comput. Appl.* (2019). https://doi.org/10.1007/s00521-018-03970-4
19. P. Montero-Manso, G. Athanasopoulos, R.J. Hyndman, T.S. Talagala, FFORMA: feature-based forecast model averaging. Int. J. Forecast. **36**(1), 86–92 (2020). https://doi.org/10.1016/j.ijforecast.2019.02.011
20. C. Liew, T.N. Chou, The prediction of stock returns with regression approaches and feature extraction. J. Adm. Bus. Stud. **2**(3), 107–112 (2016)
21. T.-T. Nguyen, S. Yoon, A novel approach to short-term stock price movement prediction using transfer learning. Appl. Sci. **9**(22), 4745 (2019). https://doi.org/10.3390/app9224745
22. Data.gov.in, *Wholesale Price Index (Base Year 2011–12) Upto May 2017*. Open government data (OGD) platform India, (15 June 2017). https://data.gov.in/resources/wholesale-price-index-base-year-2011-12-upto-may-2017
23. R. Hyndman, E. Wang, Y. Kang, T. Talagala, (NA), tsfeatures: time series feature extraction. R package version 0.1. https://github.com/robjhyndman/tsfeatures/
24. N. Kourentzes, nnfor: time series forecasting with neural networks. R package version 0.9.2, (2017). https://CRAN.R-project.org/package=nnfor
25. IBM Knowledge Center, (n.d.). www.Ibm.Com. Retrieved July 1, 2020, from https://www.ibm.com/support/knowledgecenter/SSLVMB_23.0.0/spss/base/curve_estimation_models.html
26. R Core Team, *R: A Language and Environment for Statistical Computing*. R foundation for statistical computing, Vienna, Austria, (2019). https://www.R-project.org/
27. R. Hyndman, G. Athanasopoulos, C. Bergmeir, G. Caceres, L. Chhay, M. O'Hara-Wild, F. Petropoulos, S. Razbash, E. Wang, F. Yasmeen, *Forecast: forecasting functions for time series and linear models*. R package version 8.4, (2018). https://pkg.robjhyndman.com/forecast
28. R.J. Hyndman, Y. Khandakar, "Automatic time series forecasting: the forecast package for R." *J. Stat. Softw. 26*(3), 1–22 (2008). https://www.jstatsoft.org/article/view/v027i03
29. Introduction to ARIMA models, Duke.Edu. (2019). https://people.duke.edu/~rnau/411arim.htm
30. J. Brownlee, *A Gentle Introduction to Exponential Smoothing for Time Series Forecasting in Python*. Machine learning mastery, (24 April 2019). https://machinelearningmastery.com/exponential-smoothing-for-time-series-forecasting-in-python/
31. DescTools package | R Documentation. (n.d.). www.Rdocumentation.Org. Retrieved 10 July 2020, from https://www.rdocumentation.org/packages/DescTools/versions/0.99.36/topics/Measures%20of%20Accuracy
32. M. Despotovic, V. Nedic, D. Despotovic, S. Cvetanovic, Evaluation of empirical models for predicting monthly mean horizontal diffuse solar radiation. Renew. Sustain. Energy Rev. **56**, 246–260 (2016). https://doi.org/10.1016/j.rser.2015.11.058
33. V. Yadav, S. Nath, Novel hybrid model for daily prediction of PM10 using principal component analysis and artificial neural network. Int. J. Environ. Sci. Technol.c **16**(6), 2839–2848 (2018). https://doi.org/10.1007/s13762-018-1999-x

Educational Data Mining and Students' Academic Performance Prediction

Subhabrata Sengupta, Anish Banerjee, and Satyajit Chakrabarti

Abstract The progress of a country strongly depends on the education system of the country. There is a drastic change in the education system across the whole world. Data mining methods and techniques play an important role in the today's world, and it is used for decision making in education system to make decisions related to the students' academic status. Presently, dropping of students has increased with the higher education. This situation is directly related with the fame of the institute. Various students' information is being maintained with different values. As data insertion and retrieval processes are going on with this existing system, there are no scopes of intelligence related with the various existing system models. Data mining is used for sorting the educational problem by analyzing the students' performance using the techniques like linear regression, logistic regression and artificial neural network. In this research, the paper mainly focused in random forest algorithm to measure the performance of the students. Educational Data Mining (EDM) is used to obtain the new style to discover various intelligent paths to predict different semantic knowledge.

1 Introduction

Semantic extraction of knowledge in data mining is the latest development in the field of educational data mining. In different educational institutes, semantic level educational data mining is used to extract different distinct patterns and knowledge [1]. Data mining and knowledge discovery applications are the recent trends and are significant in decision making. In recent days, data mining and knowledge discovery

S. Sengupta · A. Banerjee (✉)
Department of Information Technology, Institute of Engineering and Management (IEM), Kolkata 700091, India
e-mail: anish.banerjee1998@gmail.com

S. Chakrabarti
Department of Computer Science and Engineering, Institute of Engineering and Management (IEM), Kolkata 700091, India

V. E. Balas et al. (eds.), *Proceedings of International Conference on Computational Intelligence, Data Science and Cloud Computing*, Lecture Notes on Data Engineering and Communications Technologies 62, https://doi.org/10.1007/978-981-33-4968-1_19

applications are the new trends which are very much significant in personalized decision making. To introduce various data mining techniques significance of machine learning, artificial intelligence is wide spread [2]. Mainly, descriptive and predictive, these two different data mining tasks are used. In the field of descriptive data mining, there is an extensive use of association rule mining, clustering, etc., to figure out the different trends from the extensive dataset.

To find the patterns hidden in large dataset, various techniques of association rule mining and clustering are used regarding the domain of descriptive data mining. SVM, decision tree, regression analysis, etc., and all these methods are used to predict a class from a large number of datasets, with the help of predictive data mining. EDM is also used to take smart decision making [3]. The paper also focuses on the algorithm-based performance prediction that that dynamically predicts the score based on the basis of questions answered, and also on the basis of processing other information, his future career track or preferred region of study or job can be predicted through career counselling.

2 Extensive Literature Review

El Gamal A F proposed about prediction of students' performance in programming language courses. In the publication, he stated about students' mathematical background, programming knowledge, problem solving knowledge, previous academic performance, etc. In his publication, he mentioned three phases, namely data pre-processing, attribute selection and rule extraction algorithm [4].

Suyal and Mohod [5] identify the students' performance by applying the association and classification rule. Their prime focus is to diminish the failure rate by classifying the students who need special attention [5].

Bresfelean proposed a novel approach to discover the students' education details with various decision tree algorithms in weka, ID3, J48, etc. [6].

There has been a study to evaluate the performance of students' by clustering the grades into various classes using CGPA. To identify the weak performance to improvement in performance various methods on neural network, regression analysis is being used [7].

Ramaswami and Bhaskaran proposed that the improvement in performance in higher education is the turning point of the students future. He applied predictive data mining model in his research work to identify slow learners. He applied CHAID prediction system to analyse the performance and he got the accuracy about 44.69% [8].

Brijesh et al. proposed that to achieve highest level of quality in higher education system, prediction about students' performance is one factor. In his research paper, he stated about the extraction of hidden knowledge from the education dataset with the help of data mining model. He also stated that to evaluate the students' performance and to classify the dataset, he used decision tree algorithm for his experiment [9].

Wook et al. proposed in their paper that they analyse two data mining mechanism like Artificial Neural Network (ANN) and the consolidation of clustering and decision tree classification techniques for the students' performance prediction. They find out the students having high chances of failing by considering parameters like encouragement in studies, interests, family encouragement in studies and past history in academic performance [10].

M. P. G. Martins et al., proposed in their research about random forest algorithm to predict the students' performance in a college. To get good academic prediction performance (R2 = 0.79), they implement 11 variables, starting from initial [11].

Rajni Jindal et al. interpreted that Educational Data Mining (EDM) plays an important role to analyse data in educational background by applying different data mining tools. They mainly highlighted on components, research scopes of EDM by stating its related tools, techniques and educational results [12].

P Nithya et al. contemplated the educational data using EDM. EDM uses data mining techniques, various tools and research works to find useful data from the datasets obtained. EDM still is on the rise and focuses on the unique trends of data from the educational datasets [13].

Saa [1] gives an important insight and inspect educational data, especially students' performance. EDM is the domain which deals with mining of educational data to make the educational organizations smarter in respect of student's performance, gaining knowledge, etc. [14].

P. Thakur et al. proposed about the future scope towards data mining. They stated that data mining is the most powerful weapon for analyse and prediction in students' performance, fraud detection, advertising, stock market prediction, etc. [15].

According to K. Parmar et al., the students' academic performance in the higher education gives a new direction for their excellence in their career. Different data are being increased regarding different educational sectors day by day. Classification and prediction are the major tools used in data mining, and they are in the trend for a long time. It is very much important to generalize the information contained in those models to support decision making in this domain. Specific classifier method is used to generalise information containing in the data mining insights [16].

S. Roy et al. stated that application of data mining methods on educational data helps in raising the quality of teaching as well as methods of teaching, identifying students weak in any particular domain and predicting students' academic performance. It helps in identifying the no. of students who are weak, drop out, needs improvements and obviously who are good in academics [17].

Md. Fahim Sikder et al. described student's educational results and performances depend on both academic background and family conditions. With the help of different machine learning algorithm, educational data mining helps to extract relevant data from a rich dataset and analyse these data. Using neural network, they evaluate their results in the form of Cumulative Grade Point Average (CGPA) and compare that with real CGPA [18].

S. E. Sorour et al. stated that prediction of students' academic achievement with high accuracy will bring a vast change in education system. This system will help to understand individual student more deeply by analysing the daily performance. Latent Semantic Analysis technique (LSA) is used to get the high accuracy in predicting student's grade. It uses semantic analysis from the thoughts of students with the use of statistically acquired abstraction instead of discrete words. Artificial Neural Network model (ANN) is used to examine students' thoughts to estimate the performances of the students. Prediction accuracy reaches 82.6 and 76.1% via this proposed method [19].

Studies uncover Educational Data Mining (EDM) and Learning Analytics (LA) investigate is a rising division, which is unfurling helpful information from instructive databases for some reasons, for example, anticipating understudies' prosperity. This can be useful for improvement in current instructive frameworks. Existing techniques have utilized highlights which are for the most part related to the understudies monetary and instructive area. Test results show that the proposed strategy essentially beats existing techniques because of the misuse of family consumptions and understudies' very own data highlight sets [20].

E-learning frameworks produce gigantic measures of information, whose examination may turn into an overwhelming assignment which makes it important to utilize computational investigative strategies and devices. Utilization of information disclosure strategies to break down verifiable understudy course grade information so as to foresee whether an understudy will drop out of a course has been proposed. Logistic regression models are utilized with the end goal of classification. Analyses directed with information on more than 100 understudies for several separations learning courses confirm the prescient intensity of our proposition [21].

The alarm device was intended to catch understudy logins, a culmination of online exercises, engagements and commitment. Information was caught from Moodle and measurable statistical models based on regression and was utilized to decide any connection between's understudy's online conduct and their exhibition. Understudy execution in this advanced education course could be anticipated dependent on their normal logins every week and the normal fruition paces of exercises. The exactness of the model was 60.8% [22].

An absence of work despite everything exists on utilizing EDM strategies for universal appraisal studies. This examination means to fill the hole in the present writing by applying a decision tree, a Bayesian network, a logistic regression and neural networks. During the development of models, 11 understudy based elements have been considered. The outcomes show that logical regression beats different calculations as far as estimating order execution [23].

Utilization of blended technique investigation and looking through three well-known libraries, including the Learning Analytics and Knowledge (LAK) meeting, the SpringerLink and the Web of Science (WOS) databases, has been conducted. Subsequently, there was an introduction of a review of the various strategies utilized by the investigations and their related undertakings. To pick up bits of knowledge into the pattern bearing of the various undertakings, the distributions into their partners were bunched. At long last, the restrictions of those investigations and talked about the most encouraging future lines and difficulties were handled. The consequences of this audit may help colleges to dispatch their own learning analytics extends or improve existing ones [24].

The current examination advances a relapse scientific model dependent on the random forest algorithm, created to foresee, at a beginning period, the worldwide scholarly presentation of the students of a polytechnic higher education establishment. The investigation focuses on the universe of an establishment made out of five schools as opposed to following the standard technique of wiping out the forecast to one single explicit degree course [25].

Fitting choices making for application or inversion of advancement is mostly founded on basic data gave to the Human Resource Management (HRM). This is to guarantee the coherence of greatness for scholarly staff in setting up and fortifying training towards the achievement of the colleges. Also, distributing is a centre segment in the assessment of scholastic ability that is influenced by the examination, oversight and meeting at the nearby college in Malaysia [26].

The utilization of AI with Educational Data mining (EDM) to anticipate student execution has consistently been a significant research territory. Foreseeing scholarly outcomes is one of the arrangements that plans to screen the advancement of understudies and envisions understudies in danger of bombing the scholastic pathways [27].

Foreseeing the understudies' presentation has become a difficult undertaking because of the expanding measure of information in instructive frameworks. With regards to this, recognizing the elements influencing the understudies' presentation in advanced education, particularly by utilizing prescient information mining strategies, is still hard to come by. This field of research is generally distinguished as instructive data mining. Subsequently, the fundamental point of this examination is to distinguish the most usually considered elements that influence the understudies' exhibition, just as, the most widely recognized information mining procedures applied to distinguish these components [28].

Data mining offers solid strategies for various areas including training. In the instruction field, the exploration is growing quickly expanding because of an enormous number of understudy's data which can be utilized to develop important example relating learning conduct of understudies. The organizations of training can use instructive information mining to look at the presentation of understudies which can bolster the establishment in perceiving the understudy's exhibition. In information, mining, arrangement is a natural procedure that has been actualized generally to discover the exhibition of understudies [29].

The capacity to misuse understudies' feelings utilizing distinctive AI procedures is viewed as a significant technique for arranging and manoeuvrings in a shared instructive condition. The headway of AI innovation is the sound development of large information advances. This helps the applications dependent on Sentiment Mining (SM) utilizing information to turn into a typical stage for information mining exercises. Be that as it may, almost nothing has been studied on the sentiment application utilizing a tremendous measure of accessible educational data [30].

Utilizing a mixed learning programme in a Chinese secondary school as the case, this investigation embraced information mining ways to deal with build-up prescient models utilizing character characteristics [31].

EDM investigations instructive information utilizing the turn of events and the use of Data Mining (DM) techniques and calculations to information put away in scholastic information storehouses. The aim is to survey which techniques and calculations of DM can be utilized in the investigation of instructive information to improve dynamic. Besides, it assesses these calculations utilizing a dataset made out of understudy information in the software engineering school of a private college. The centre of the investigation is to find patterns and examples of study in the graduation rate marker [32].

Learning analytics and instructive data mining have made some amazing progress in a brief timeframe. There has been a presentation of a dream for certain directions the field ought to go: towards greater interpretability, generalizeability, transferability, applicability and with clearer evidence for effectiveness [33].

3 Problem Statement

The way towards investigating designs in enormous informational indexes with the execution of AI, insights, and database frameworks is called data mining. By and large, understudies' exhibition is assessed towards the finish of an assessment. The proposed mechanism works in large dataset which are related with students' performance.

3.1 Preparation Process of Data

Sampling method will be used to collect different student related data from an educational institution.

3.2 Modification of Data

Selected states are needed for data mining in this step. The students' roll no. 10th and 12th marks, assignment marks, gender, parent's qualification and income were taken as indicators for prediction (Table 1).

Table 1 Student-related variables

Serial	Description	Possible values
A	Gender	Male/female
B	Class	General/schedule cast/schedule tribe/other backward classes
C	Medium of teaching	English/Bengali/Hindi
D	Family annual income	BPL/ Poor/ Medium/ High
E	10th grade	<40, 40–59, 60–79, 80–90, >90
F	12th grade	<40, 40–59, 60–79, 80–90, >90
G	Graduation (Semester grade)	10–O (Outstanding) 9–E (Excellent) 8–A (Very good) 7–B (Good) 6–C (Fair) 5–D (Bad) Less than 5–F (Fail)
H	Father's qualification	NA, elementary, secondary, higher secondary, graduate, post graduate, doctorate
I	Mother's qualification	NA, elementary, secondary, higher secondary, graduate, post graduate, doctorate
J	Profession of father	NA, Government job, private job, self employed
K	Profession of mother	NA, Government job, private job, self employed, house wife
L	Education loan	Yes/No
M	Location of residence	Rural, urban, others

4 Proposed System Architecture: (Basic Conceptual Architecture)

(See Figs. 1, 2 and 3).

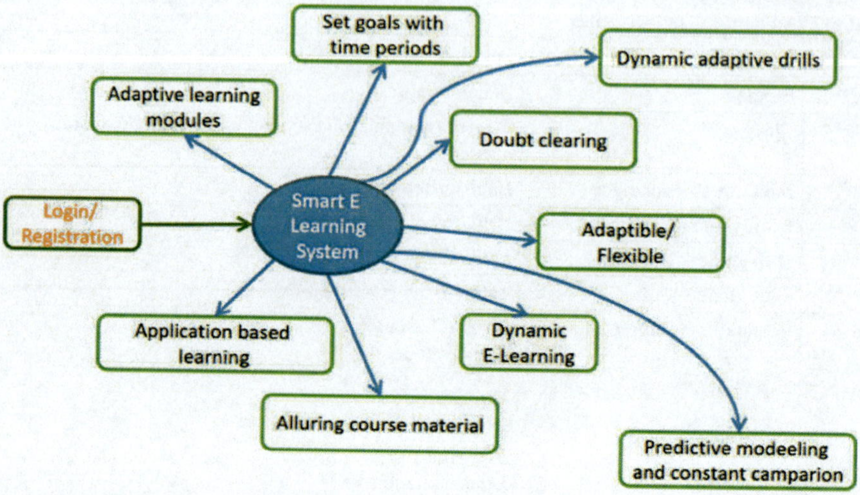

Fig. 1 Central Design of proposed smart learning system

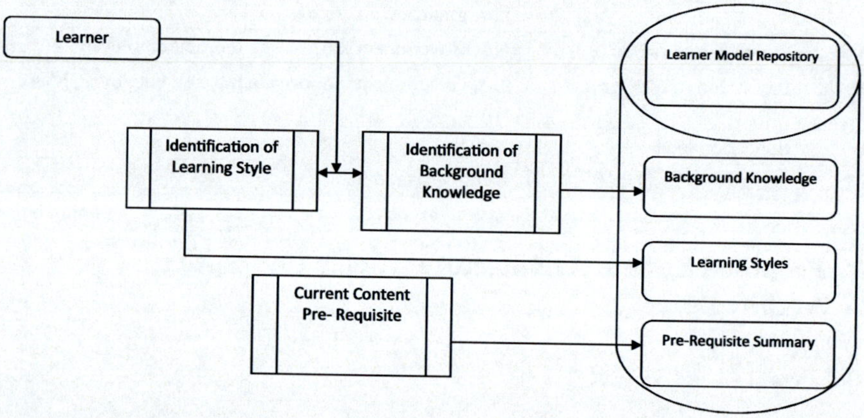

Fig. 2 Proposed learner model

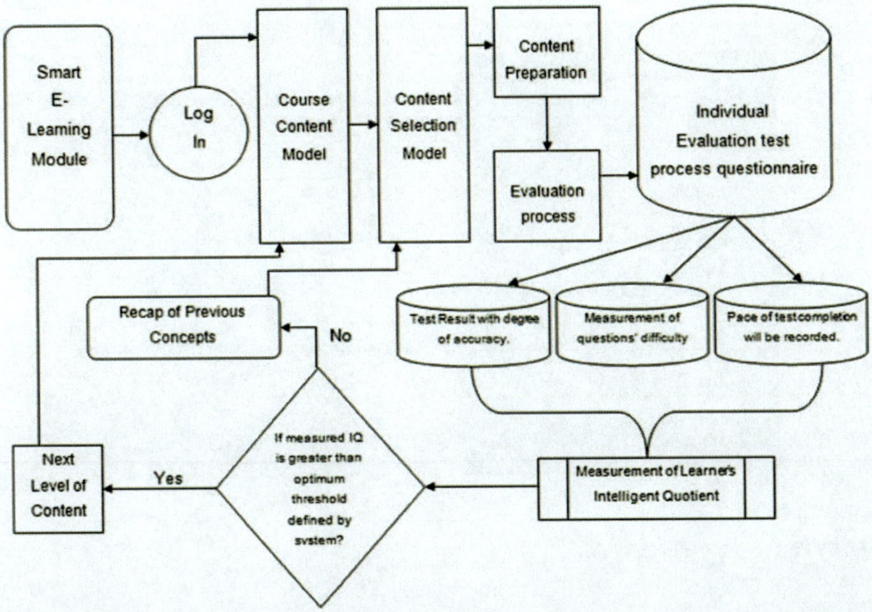

Fig. 3 Detailed architecture of proposed model

5 Results and Simulation

The following plots are based on the dataset of different universities. The visualizations are the representation in form of counts versus information with respect to different criteria.

Figure 4 beneath shows the quantity of male and female understudies whose information could be recovered. It very well may be seen that female understudies are more in number. The following one that is Fig. 5 shows the check of understudies having a place with different ethnic gatherings. For the absence of prejudice and test reason rather than genuine ethnic gatherings, it has been indicated as gathering A, bunch B, bunch C, bunch D and gathering E. Group A having the lowest number of students and Group C, the highest.

Parental background educational qualification has been held a ground for the students' interest in further studies in most of the cases. Figure 6 gives us an idea about the minimal qualification that most parents have. It was seen that most of the guardians have some professional education and a bunch finished till their master's (Figs. 7, 8, 9 and 10).

The dataset, when plotted additionally, gives us the sort of supper procured by the understudies during their residency in instructive foundations, where it very well may be seen from Fig. 4 that the vast majority of them decided on standard lunch, but rest had subsidized or free lunch. It can be inferred that the latter group belonged

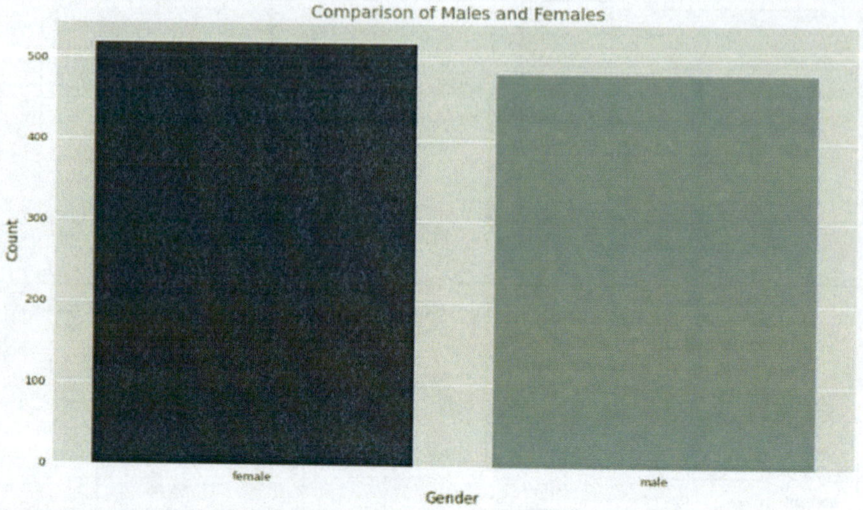

Fig. 4 Plot to show gender count

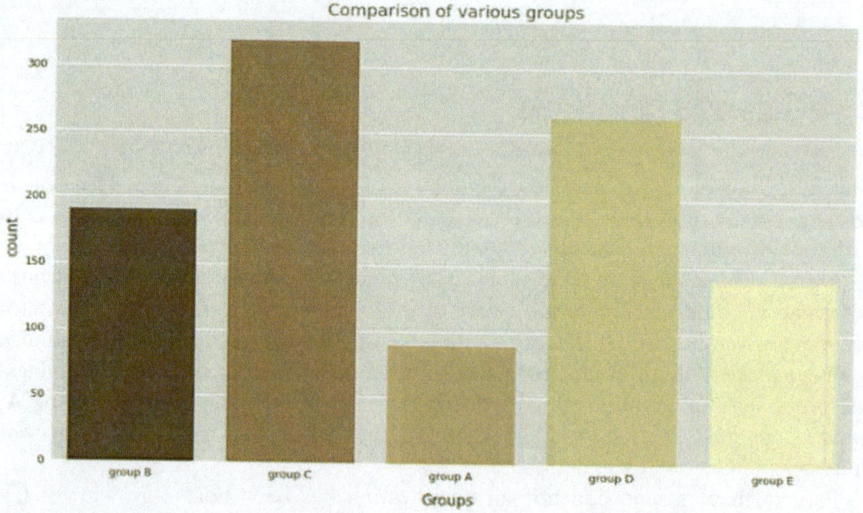

Fig. 5 Comparison of various ethnic groups

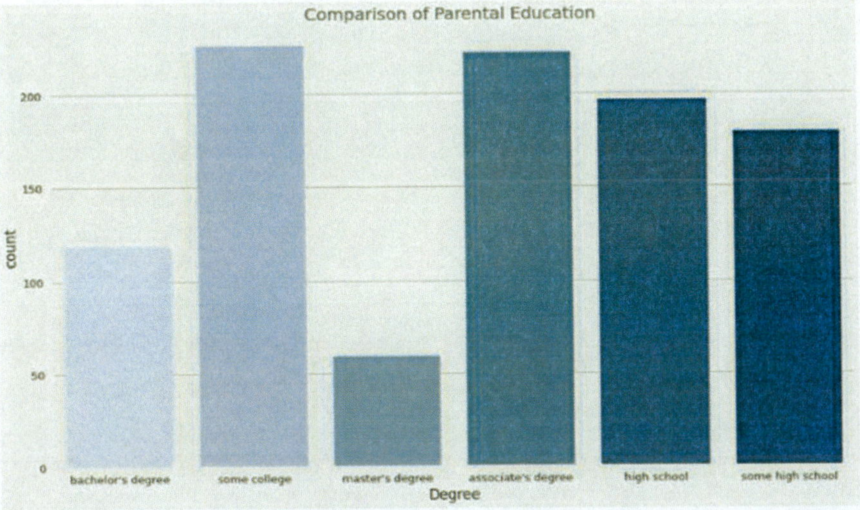

Fig. 6 Comparison of parental education

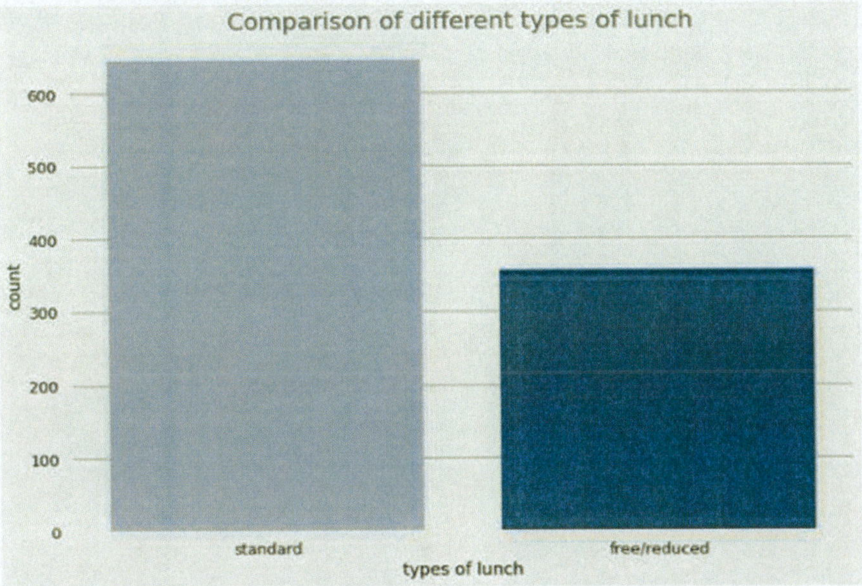

Fig. 7 Comparison of lunch types and their count

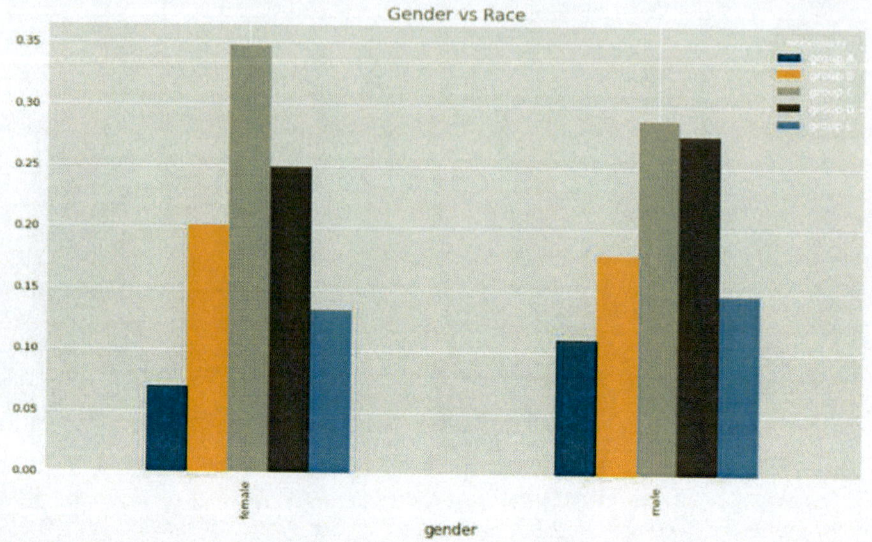

Fig. 8 Gender distribution of various ethnic groups

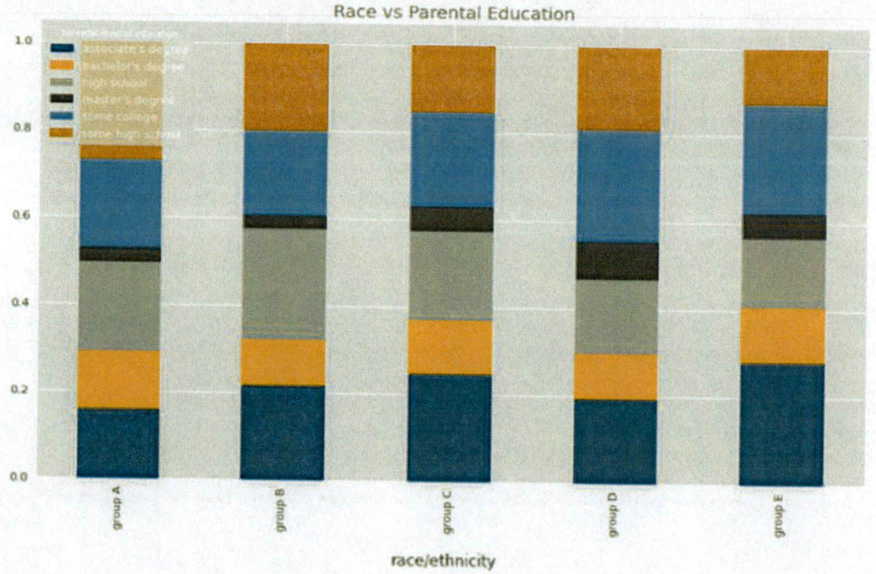

Fig. 9 Parents of various ethnicities on their education

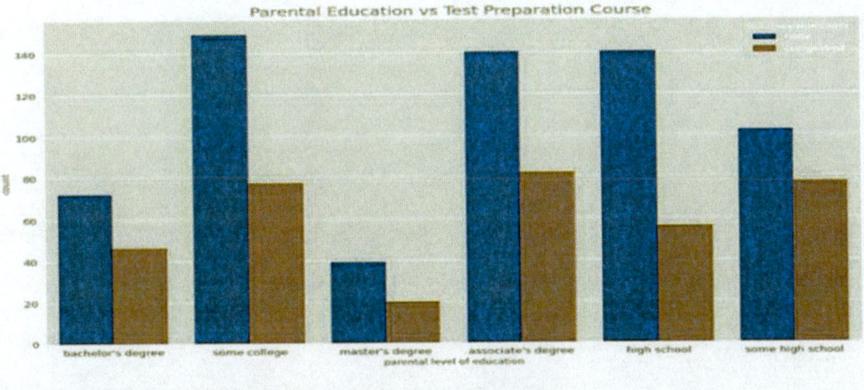

Fig. 10 Plot shows count of test preparation parental education

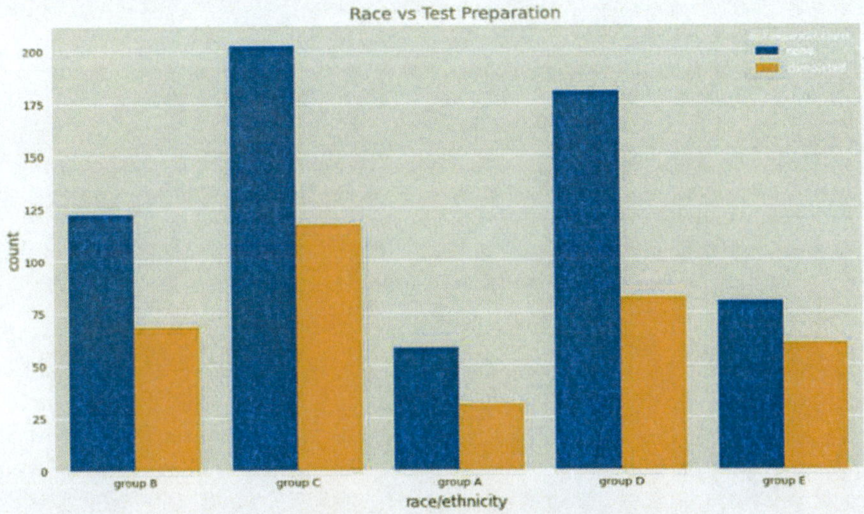

Fig. 11 Plot shows count of test preparation course to that of ethnicity

to a lower income group or can also be saving up their expenses to bear their tuition fees.

The students were assessed on the basis of their skills in various fields like math, reading and writing. It was observed that more than 35 students had scored 65 in their math skill test, and nearly 34 students have scored 72 in their reading skills.

Underneath in Fig. 11, we have the dissemination for the tally of male and female applicants concerning the ethnic gatherings—Group A (dark blue), Group B (ochre), Group C (dim), Group D (dark dim) and Group E (light blue). Figure 12 is about the count of parents of the ethnic groups and their educational qualification. The legend decodes the colour bars.

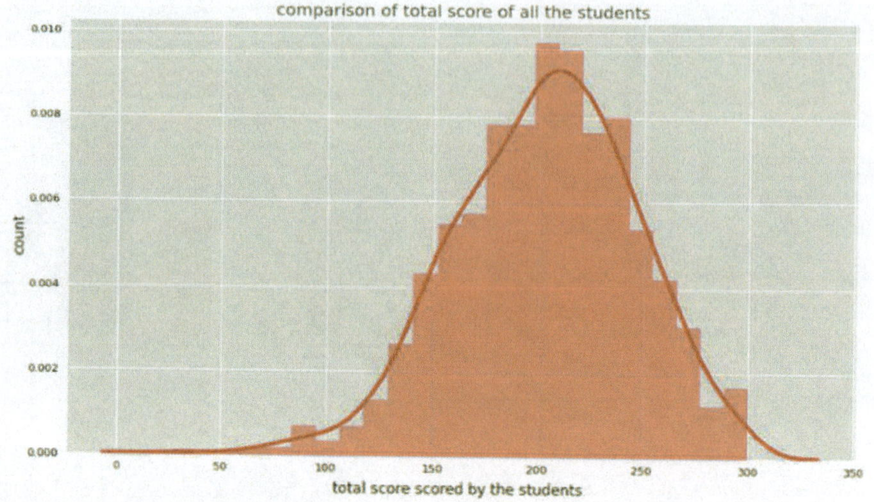

Fig. 12 Range of total score of students

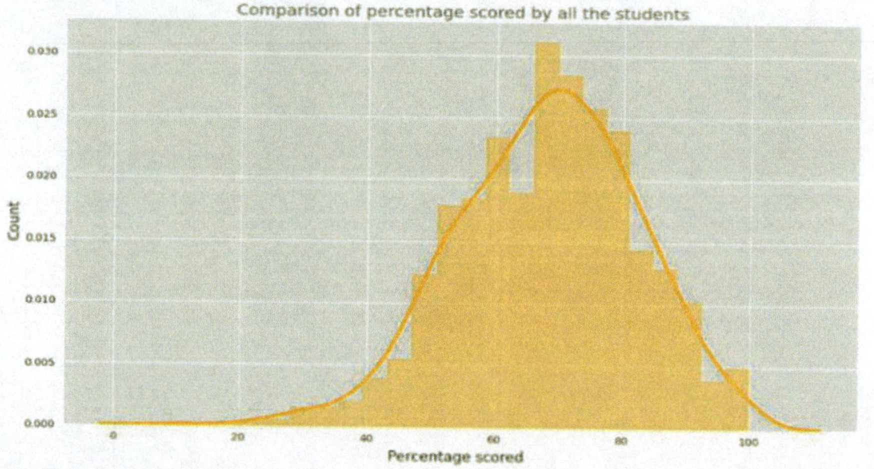

Fig. 13 Percentage scored range

Here, the numbers of students who have completed the test preparation (brown) and the ones who have not completed (blue) with respect to their parental qualification have been depicted in Fig. 13. Figure 15 is a similar plot, where instead of parental education, ethnicity is accounted (Fig. 14).

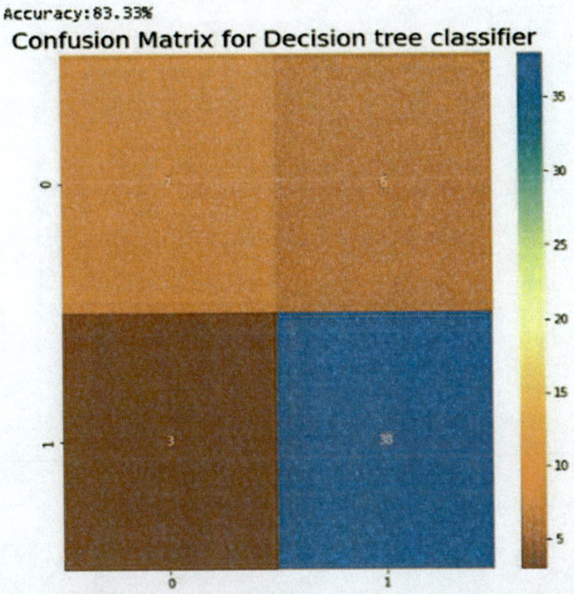

Fig. 14 Confusion matrix for decision tree classifier

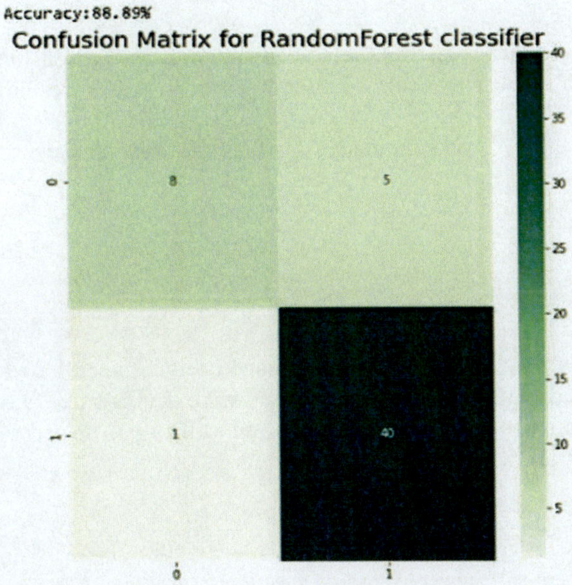

Fig. 15 Confusion matrix for random forest classifier

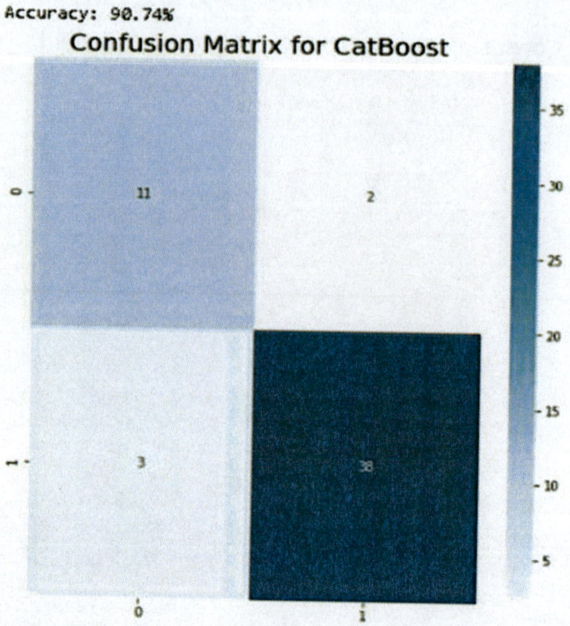

Fig. 16 Confusion matrix for catboost classifier

It was analysed that whether the student fraction has passed (major sector) or has failed with respect to skills mentioned earlier—maths, reading and writing, respectively:

Figures 16 and 17 shows the range in which the students have scored and their percentage, respectively.

5.1 *Model Results*

The confusion matrix and their accuracy have been obtained based on the model used on the cleaned dataset. The models used were decision tree classifier, random forest classifier, catboost, LGBM classifier and XGBoost. The accuracy was nearly 83%, 89%, 91%, 89%, 89%, respectively, (Fig. 18).

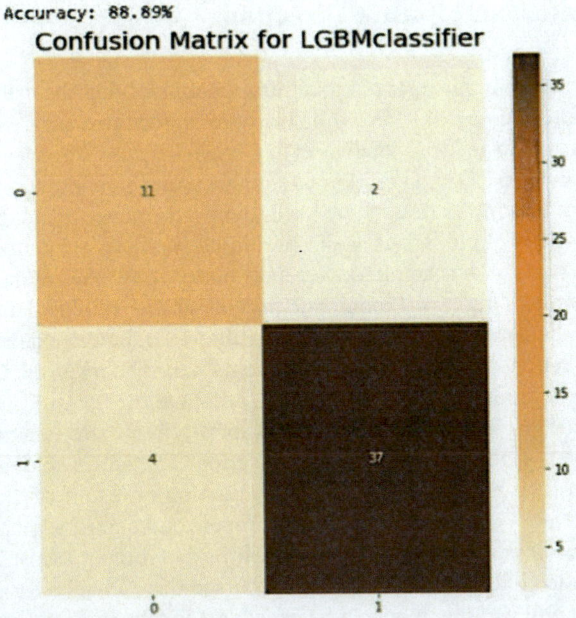

Fig. 17 Confusion matrix for LGBM classifier

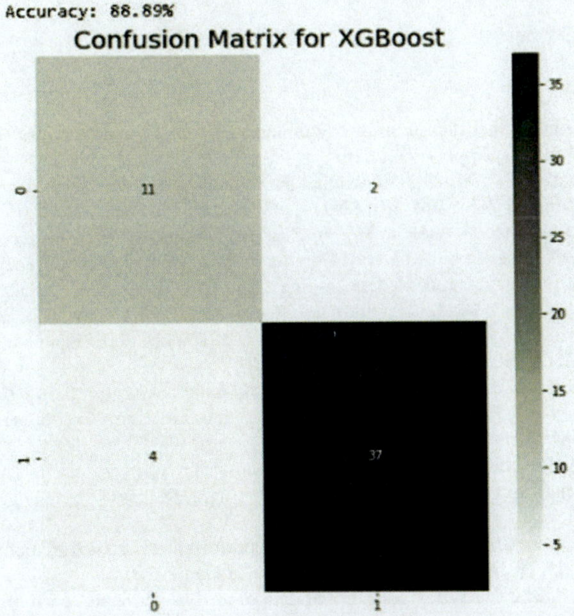

Fig. 18 Plot shows confusion matrix for XGBoost classifier

6 Conclusions and Future Direction

As could be seen from our research that our model involving the catboost is able to perform with an accuracy of 91%, while the other algorithms are close to 90%. The main challenge was to get a proper dataset to work in because we saw that despite a lot of attention is given to CGPA or marks of a student, it may not be always the case as we saw in the plots that in the dataset we used, the performance was highly correlated with the parental background as well. Additionally, there are numerous colleges over the world that follows an alternate report philosophy and getting a dataset that sums up the whole thought will require district insightful training dataset extraction. To predict the final result, various attributes like attendance, examination marks, seminar presentation will play vital role. Students' family background and economic condition help to predict whether he/she will continue higher education or not. If a student has good performance and excellent merit, then there is a probability that they can pursue a scholarship for higher education. The entire students and teacher communities will be benefited when better evaluation of results of the students will happen. The future work could be to give a proper roadmap of what a person needs to do if the person is not able to perform well by combining the output of several different scenarios like predicting whether a person has a chance for further studies or if the person will be able to open a business and suggesting a proper roadmap for the same. Also a text generation model could be used to use all the aforementioned points to generate a sentence to give the user keeping the sentiment of the statement as positive.

References

1. A.A. Saa, "Educational data mining & students' performance prediction". (IJACSA) Int. J. Advan. Comput. Sci. Appl. **7**(7) (2016)
2. B.K. Baradwaj, S. Pal, Mining educational data to analyze students' performance. (IJACSA) Int. J. Adv. Comput. Sci. Appl. **2**(6) (2011)
3. T. Mishra, D. Kumar, S. Gupta, *Mining Students" Data for Performance Prediction.* 4th International Conference on Advanced Computing and+Communication Technologies (2014)
4. A.F. ElGamal, "An educational data mining model for predicting student performance in programming course". Int. J. Comput. Appl. (0975–8887), **70**(17) (May 2013)
5. S.R. Suyal, M.M. Mohod, "Quality improvisation of student performance using data mining techniques". Int. J. Sci. Res. Publ. **4**(4) (April 2014)
6. V.P. Bresfelean, "*Analysis and Predictions on Students' Behavior Using Decision Trees in Weka Environment*", Proceedings of the ITI 2007 29th International Conference on Information Technology Interfaces, pp. 25–28, (June 2007)
7. O.F. Noah, B. Barida, T.O. Egerton, "Evaluation of student performance using data mining over a given data space". Int. J. Recent Technol. Eng. (IJRTE) ISSN: 2277–3878, **2**(4) (Sep 2013)
8. M. Ramaswami, R. Bhaskaran, "A CHAID based performance prediction model in educational data mining". IJCSI Int. J. Comput. Sci. Issues. **7**(1) (Jan 2010)
9. B.K. Baradwaj, S. Pal, Mining educational data to analyse students' performance. (IJACSA) Int. J. Adv. Comput. Sci. Appl. **2**(6) (2011)

10. M. Wook, Y.H. Yahaya, N. Wahab, M.R.M. Isa, N.F. Awang, H.Y. Seong, "*Predicting NDUM Student's Academic Performance Using Data Mining Techniques*, Paper presented at International Conference of Computer and Electrical Engineering, ICCEE. (28–30 Dec, 2009)
11. M.P.G. Martins, V.L. Migueis, D.S.B. Fonseca, "A data mining approach to predict undergraduate students' performance", *2018 13th Iberian Conference on Information Systems and Technologies (CISTI)*
12. R. Jindal, M.D. Borah, "A survey on educational data mining and research trends". Int. J. Database Manage. Syst. (IJDMS) **5**(3) (June 2013)
13. Dr. P. Nithya, B. Umamaheswari, A. Umadevi, "A survey on educational data mining in field of education". Int. J. Adv. Res. Comput. Eng. Technol. (IJARCET) **5**(1) (Jan 2016)
14. A.A. Saa, "Educational data mining and students' performance prediction". (IJACSA) Int. J. Adv. Comput. Sci. Appl. **7**(5) (2016)
15. P. Thakar, A.M. Manisha, "Performance analysis and prediction in educational data mining: a research travelogue". Int. J. Comput. Appl. (0975 – 8887) **110**(15) (Jan 2015)
16. K. Parmar, D. Vaghela, P. Sharma, "Performance prediction of students using distributed data mining". *IEEE Sponsored 2nd International Conference on Innovations in Information Embedded and Communication Systems ICIIECS'15*
17. S. Roy, A. Garg, "Predicting academic performance of student using classification techniques". *2017 4th IEEE Uttar Pradesh Section International Conference on Electrical, Computer and Electronics (UPCON) GLA University, Mathura*, Oct 26–28, 2017
18. F. Sikder, J. Uddin, S. Halder, "Predicting students yearly performance using neural network: a case study of BSMRSTU". *2016 5th International Conference on Informatics, Electronics and Vision (ICIEV)*
19. S.E. Sorour, T. Mine, K. Goda, S. Hirokawa, "Predicting students' grades based on free style comments data by artificial neural network". *2014 IEEE Frontiers in Education Conference (FIE) Proceedings*
20. Predicting student performance using advanced learning analytics by Ali Daud a,d a Faculty of Computing and Information Technology, King Abdulaziz University
21. Data mining for modeling students' performance: a tutoring action plan to prevent academic dropout
22. Early warning system as a predictor for student performance in higher education blended courses, (2018)
23. Identifying the classification performances of educational data mining methods: a case study for TIMSS, (2016)
24. Learning analytics in higher education—a literature review philipp leitner, (2017)
25. M.P.G. Martins et al, A data mining approach for predicting academic success—a case study, *ICITS 2019*, AISC 918, (2019), pp. 45–56
26. Z. Ali Othman et al. Publication model for academic talent based on apriori technique. (Springer Nature Singapore Pte Ltd., 2019)
27. A. Qazdar et al. A machine learning algorithm framework for predicting students performance: a case study of baccalaureate students in Morocco. (Springer Science+Business Media, LLC, part of Springer Nature 2019)
28. A.A. Saa et al, Factors affecting students' performance in higher education: a systematic review of predictive data mining techniques. (Springer Nature B.V. 2019)
29. B.K. Francis et al, Predicting academic performance of students using a hybrid data mining approach. (Springer Science+Business Media, LLC, part of Springer Nature 2019)
30. R.K. Jena, Sentiment mining in a collaborative learning environment: capitalising on big data. *Behaviour and Information Technology*, (2019)
31. F. Wu, S. Lai, Linking prediction with personality traits: a learning analytics approach. *Distance Education*, (2019)
32. O. Moscoso-Zea et al, Evaluation of algorithms to predict graduation rate in higher education institutions by applying educational data mining. Australas. J. Eng. Educ. (2019)
33. R.S. Baker, Challenges for the future of educational data mining: the baker learning analytics prizes. J. Educ. Data Min. **11**(1) (2019)

Deep Learning Approaches to Improve Effectiveness and Efficiency for Time Series Prediction

Dildar Ali, Nitin Tiwari, Baisakhi Das, Samit Bhanja, and Abhishek Das

Abstract The variation and dependency on diverse parameters of time series data make predictions a very complicated process. Also, the effectiveness and efficiency for predicting the values of a time series are important in a variety of areas such as stock exchanges, natural language processing, and sensor networks. Artificial neural networks have been demonstrated to be valuable in such cases to predict the time series data. As recurrent neural network (RNN) is more suitable for predictions of sequential data, we are considering long short-term memory (LSTM) and gated recurrent unit (GRU) in this paper. For comparing with different deep learning frameworks, we use Keras because Keras makes an environment to run new experiments easier and faster. In this paper, we provide a comparative analysis of different deep learning frameworks on the basis of training speed and accuracy of prediction using different metrics, namely mean absolute error (MAE), root mean square error (RMSE), and time taken by different frameworks. The proposed method is expected to be a promising method in the field of time series prediction to choose the suitable deep learning frameworks for a deep learning model.

Keywords Time series · Long short-term memory · Gated recurrent unit · Keras · TensorFlow · Theano · CNTK

D. Ali · A. Das (✉)
Aliah University, Kolkata, West Bengal, India
e-mail: adas@aliah.ac.in

D. Ali
e-mail: dildarskp1994@gmail.com

N. Tiwari · B. Das
Institute of Engineering and Management, Kolkata, West Bengal, India
e-mail: nitintiwry007@gmail.com

B. Das
e-mail: baisakhi.das@iemcal.com

S. Bhanja
Government General Degree College, Singur, Hooghly, West Bengal, India
e-mail: samitbhanja@gmail.com

V. E. Balas et al. (eds.), *Proceedings of International Conference on Computational Intelligence, Data Science and Cloud Computing*, Lecture Notes on Data Engineering and Communications Technologies 62, https://doi.org/10.1007/978-981-33-4968-1_20

1 Introduction

Time series forecasting is one of the important and challenging tasks in the field of data science and machine learning. Every business needs to predict some reasonable future and this prediction helps investors to make more accurate and effective decisions. Some common examples of time series are stock market forecasting, product selling forecasting, Internet traffic flow, and many more. The similarity between these problems is all are using their historical information to predict future values. The basic assumptions of different statistical models are time series which follows a particular distribution which can be represented by some linear equations. In general, time series are behaving like nonlinear, and it is very hard to apply these traditional models to get effective results.

In recent days, the deep learning algorithms are become widely popular in the field of time series prediction. Deep learning algorithms provides new approaches for predictions where the relationship between several variables are modelled deeply and uses layered hierarchies. The most noticeable quality of ANNs is their natural ability of modeling nonlinearities with no suspicions about the statistical distribution of the observations. Between all the different deep learning architectures, the most suitable for time series forecasting would be one that can process sequential data and maintain information about the past. The architecture that fulfills these requirements is the recurrent neural network (RNN) and in particular its variants like long short-term memory (LSTM) and gated recurrent unit (GRU). These have shown state-of-the-art results in many application areas such as machine translation, handwriting, and speech recognition [1–3].

Finally, with the advent of many successful use cases, the application of deep learning frameworks has been growing rapidly. Each framework is built in a different manner and for different purposes. Due to this variety, it is often necessary to decide which one is the most beneficial deep learning framework for a specific task. In this paper, we have compared the performance of TensorFlow, Theano, and CNTK as the back-end engine of Keras for time series prediction and later predictions are done using deep learning models with suitable deep learning framework.

2 Literature Survey

Two significant parts of time series data mining can be recognized as forecasting and classification. The time series forecasting problem has been performed overwhelmingly utilizing statistical-based strategies, for example, the linear autoregressive (AR) models because of their flexibility and adaptability to model many stationary processes. The autoregressive moving average (ARMA) model is used for time series prediction by Weron and Misiorek [4]. The model successfully provides the relationships among different variables but it only gives approximation in case of real-world complex problems. Also, the ARMA model performance generally degrades on the time series data with trends and seasonality. Later autoregressive integrated moving average (ARIMA) is used to reduce first-order non-stationarity in the time series data [5]. However, the different order of differencing amplifies the high-frequency noise in the time series, and different types of techniques are required to determine the order of an ARIMA model. Engle introduced the autoregressive conditional heteroscedasticity (ARCH) model to capture the second-order non-stationarity [6].

Compared to statistics-based forecasting models, neural network approaches have different unique characteristics which are nonlinear, data-driven, nonparametric, and being more flexible and universal so that it is applied to more complicated models. Later Zhang gives a conservable survey on different neural networks [7]. Feedforward neural network models (FNNs) parameterized with a back-propagation algorithm have been utilized for nonlinear time series forecasting [8]. They are known to outflank conventional statistical methods such as regression and Box–Jenkins approaches in functional approximation, but they assume the flow of time series is time-invariant. FNNs with recurrent feedback associations have endeavored for time series forecasting [9]. Such dynamic recurrent neural network (RNN) models permit the estimating of nonlinear time series happening in different areas [10]. A deep learning network-based framework on extreme learning machine using autoencoder (ELM-AE) has been introduced in [11]. Recurrent neural network the most successful deep learning architecture in case of sequential data is applied for time series forecasting [12–14]. Biao Zhang, Deyi Xiong, and Jinsong Su show that the RNN succeeds in powerfully identifying interpretation-related source tokens according to the partial target sequence [3]. In comparison with recurrent learning, feedforward neural network, back-propagation through time, and Elman nets LSTM give much more successful runs and learn faster from the sequences. Bhanja and Das develop a hybrid deep learning model by concatenating two deep learning architectures LSTM and 1D-CNN for stock market prediction. They show that the hybrid model produced more accurate results compared to the shallow deep learning architectures [15].

Therefore till now, we have been seen that there are several machine learning techniques are being used to improve the accuracy of the different forecasting models but in this paper instead of introducing a new architecture, we proposed a different approach to provide a comparative analysis of different deep learning frameworks on the basis of training speed and accuracy of prediction using different metrics.

3 Artificial Neural Networks

3.1 Recurrent Neural Networks (RNN)

A recurrent neural network (RNN) was introduced by Jeff Elman [16]. A recurrent neural network (RNN) is a type of deep neural network (DNN) that can handle the sequential data [7].

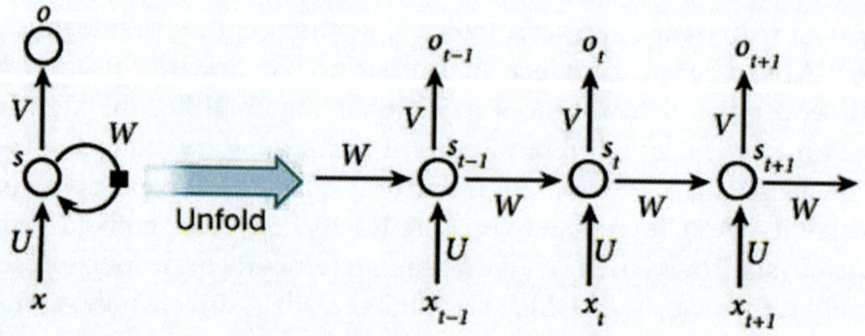

Fig. 1 Recurrent neural network and its unfolded version to show how the network's memory is built [17]

More specifically, this network has feedback loops that are in it, making it capable of preserving historical information from previous time steps (Fig. 1).

Normally, the RNN takes a sequence of vectors as input such as $x_t = (x_1; x_2 \ldots, x_T)$ computes the hidden states $s_t = (s_1, s_2, \ldots. s_T)$ and finally outputs $o_t = (o_1, o_2, \ldots. o_T)$ for every time step t, $t \in \{1, 2, \ldots T\}$. The hidden states represent the memory of the network and are computed by passing the weighted input at the current time step x_t with the weighted hidden state of the previous time step s_{t-1} through an activation function f.

3.2 *Long Short-Term Memory Networks (LSTM)*

Long short-term memory networks are a special kind of recurrent neural network that is competent of learning long-term dependencies. These networks present the memory cell that replaces the conventional artificial neurons in the hidden layers of the network. With the help of these memory cells, networks can effectively associate with memories. An RNN network feeds the output of the previous block as the input for the next iteration and uses its feedback connection to store recent events in the form of activation. Not at all like multilayer perceptron, LSTM uses different processing elements or blocks to build a more accurate recurrent network that is capable of learning long-term dependencies.

The memory in LSTM can be seen as a gated cell, and these gates are used to decide which part of the information from previous network needs to store and which part of the information is to be eliminated at every time steps [18]. It uses a sigmoid activation function in the neural network layer and weights to give importance to information. This helps the gates to protect and control the flow of information overcoming the vanishing gradient problem that the conventional RNN has. The stochastic gradient descent-based algorithm is used to utilize constant error propagation (neither exploding or vanishing) through its inner units. LSTM can be used for time series predictions [19].

3.3 Gated Recurrent Unit (GRU)

The gated recurrent unit was introduced by Kyunghyun Cho [20], and it is another variant of RNNs that follows closely the architecture of the LSTM. Both variants can track long-term dependencies while mitigating the problems of the training phase in RNNs [21]. Similarly, to the LSTM architecture, the GRU unit uses a mechanism to modulate the flow of information inside the unit. It mainly combines the input and forget gates into a single update gate which helps to determine how much information needs to pass from the current step to the next steps and also has an additional reset gate which decides how much past information to forget. Additionally, the GRU does not maintain a cell state and exposes its full content without any adjustment. That being said, the GRU unit is simpler than the LSTM and has fewer parameters which make it increasingly popular and more computationally efficient.

4 Methodology

4.1 Proposed Approach

This work aims to tackle a specific task with deep learning and explore the prediction efficiency and the speed of training of the popular deep learning frameworks and libraries. Subsequently, we want to present the state-of-the-art deep learning architectures for sequence-based tasks (LSTM, GRU), their internal workings as well as strategies that are broadly utilized in deep learning. Also, we need to show the steps of how to manipulate time series data and outline them in a frame that is reasonable for the LSTM model's training. Finally, these models were trained with the available deep learning frameworks that serve as the back-end engine of Keras, and comparisons were made in terms of prediction accuracy and training speed for different datasets and different network architectures such as LSTM, GRU, etc (Fig. 2).

4.2 Model Details

The same architecture is used for all the datasets, to determine the effectiveness of the frameworks in terms of the size of the different datasets. Also, after experimentation with the tuning strategy of the hyper-parameters, we select the ones that lead the model to produce the best prediction performance on all validation datasets. The predictor is a stateful multi-layered model that consists of two stacked LSTM layers with 128 and 256 neurons, respectively, and a dense layer with a single neuron to deliver a single prediction (one-step ahead forecasting). The primary LSTM layer returns as output of the whole sequence (time steps), while the second one returns only the last step, thus dropping the temporal dimension. In-between layers we include

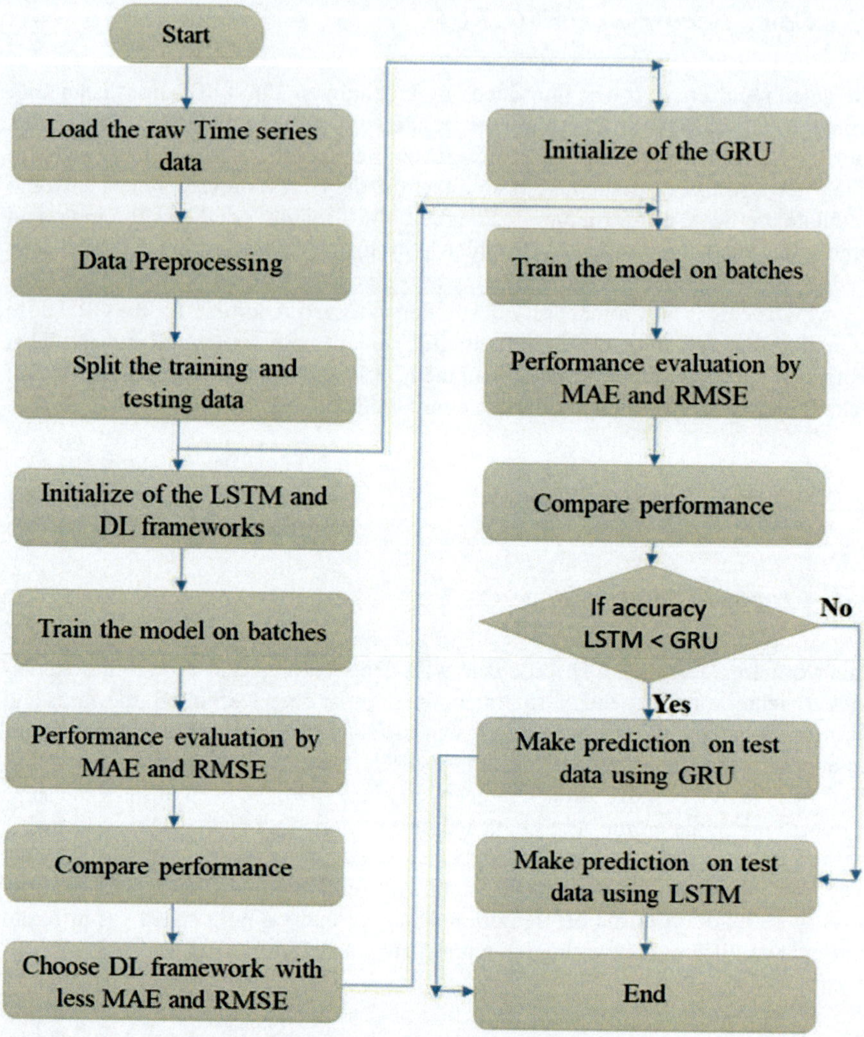

Fig. 2 Proposed approach flowchart

a dropout layer of 0.2, to prevent the model from overfitting. The model is trained for 50 number of epochs with early stopping and a batch size of 128. The objective loss function is MSE, and RMSprop [22] is used as the network's optimizer with a learning rate of 0.001. Later, the learning rate is monitored from the starting and is reduced by a factor of 0.5 in case there is no improvement in the validation loss for 5 epochs. The time series has been converted to input–output pairs with 24 past time steps and one output, with a process. We use 70% of data for training, 10% for validation, and the last 20%to test the prediction performance (Fig. 3).

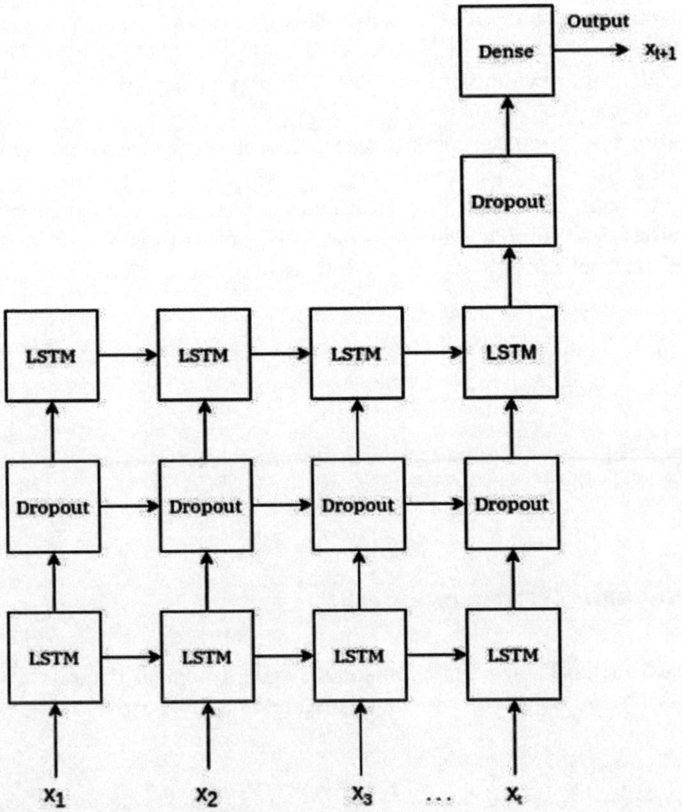

Fig. 3 Illustration of the LSTM model

5 Experimental Results and Discussion

5.1 Brief Outline of Datasets

The proposed method is evaluated on three datasets consisting of various complexities and challenges as they are belonging to time series data. The datasets used in the experiments are publicly available and their details are as follows.

Beijing PM2.5 dataset—This hourly dataset contains the PM2.5 data of the US Embassy in Beijing. This dataset contains 43,824 no. of rows and 13 no. of attributes. This dataset is available at the UCI Machine Learning Repository, and this dataset is open for all.

Internet traffic data—This dataset contains aggregated traffic in the UK. It was collected between November 19, 2004, at 09:30 h and January 27, 2005, at 11:11 h. Data is collected at five-minute intervals. This dataset contains 19,888 no. of rows and one attribute. This dataset is publicly available at Time Series Data Library.

Zurich monthly sunspot numbers—This dataset describes a monthly count of the number of observed sunspots for over 230 years (1749–1983). This dataset contains 2820 no. of rows and one attribute. This dataset is publicly available at Time Series Data Library.

The Beijing PM2.5 dataset contains also meteorological data, but we are interested in forecasting the PM2.5 concentration, thus we are going to drop all the other attributes. Furthermore, we can observe that the dataset contains some missing values, hence an additional preprocessing step has to be done to remove these rows. The Zurich and Internet traffic datasets are left as it is because there are no missing values.

5.1.1 Dataset Visualization

To get a visual grasp of the patterns that govern the time series, the datasets are plotted over their whole range (Figs. 4, 5 and 6).

5.2 *Evaluation Metrics and Results*

The proposed method is evaluated on some standard popular datasets, and all the experiments have been done on a system with Intel i5-8265U with 3.9 GHz processor

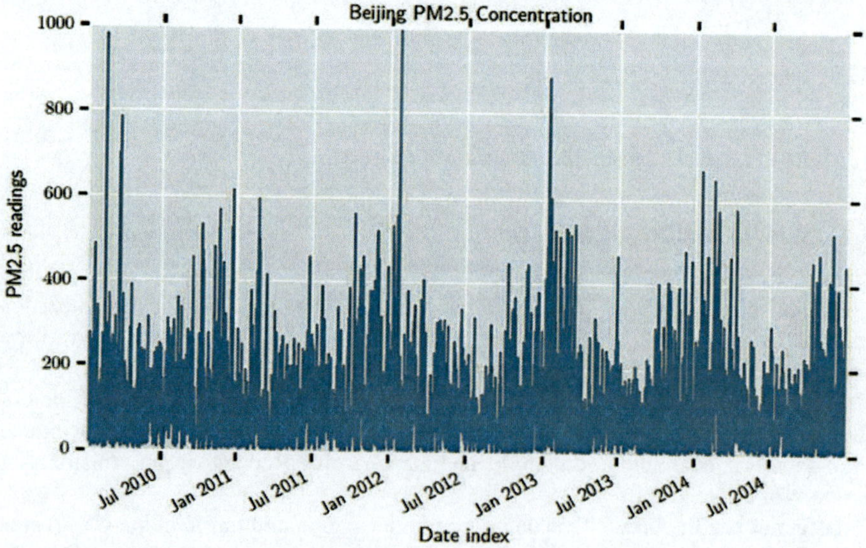

Fig. 4 Beijing PM2.5 dataset visualization

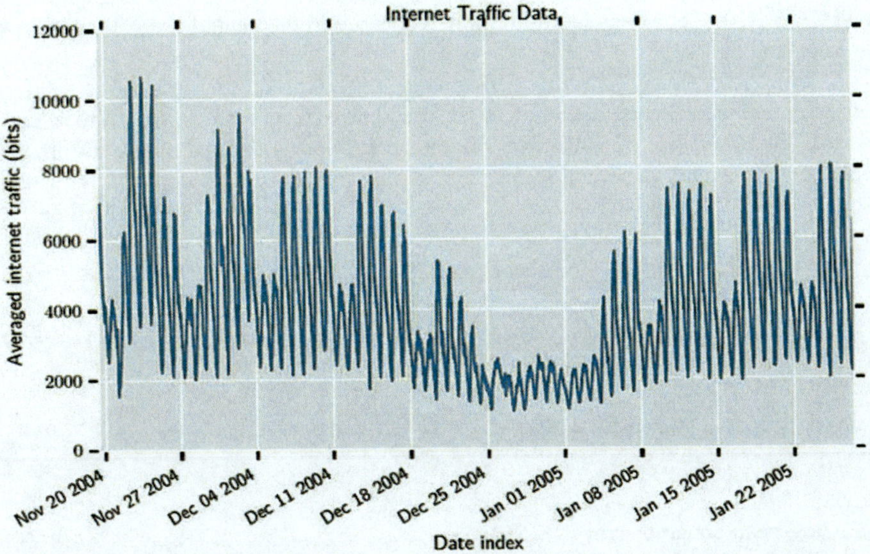

Fig. 5 Internet traffic dataset visualization

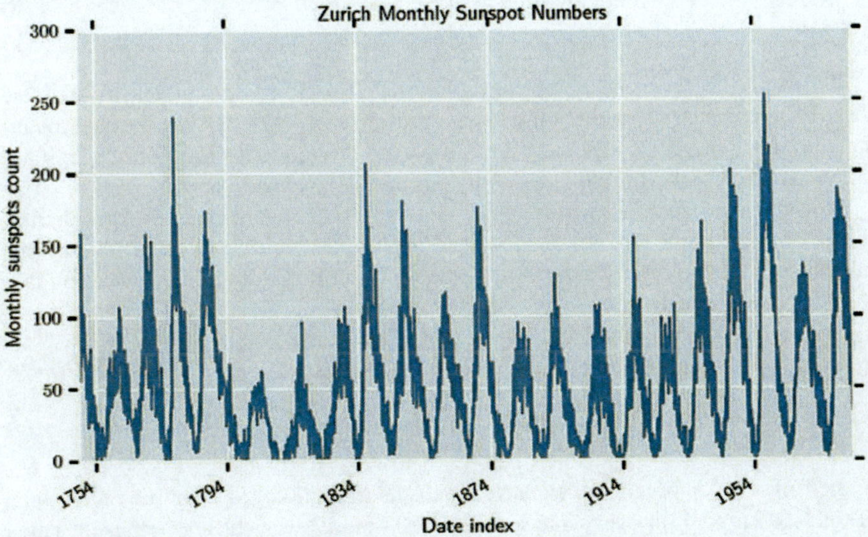

Fig. 6 Zurich sunspots dataset visualization

and 8 GB RAM. The deep learning frameworks are implemented using TensorFlow as back end [1].

In order to evaluate the actual performance of the forecasting model, we need to quantify and measure the deviation of the predictions from the help of real measurements. This deviation is called the forecast error and is defined as the difference between the actual value and the forecast value. If the error is denoted as E_t, the actual value as X_t, and the forecasted value as X_t, then the forecast error will be

$$\text{Forecast Error, } E_t = X_t - \widehat{X}_t \tag{1}$$

Some of the most popular forecasting errors measures are
Mean absolute error (MAE):

$$\text{MAE} = (\frac{1}{N})\sum_{i=1}^{N}|E_t| \tag{2}$$

Root mean-squared error (RMSE):

$$\text{RMSE} = \sqrt{(\frac{1}{N})\sum_{i=1}^{N}|E_t|^2} \tag{3}$$

To assess the performance of the frameworks for the LSTM model, we will use three different metrics, namely the mean absolute error (MAE), [22] the root mean-squared error (RMSE), and the time it took for the model to train on the training dataset for 50 epochs (Table 1).

Additionally, we trained with TensorFlow a GRU network as an exact replica of the LSTM, to compare their performance. From Table 2, it is clear that GRU achieves better score for the given datasets while being faster during training, indicating that it might be better candidate for modeling the time series.

5.3 *Predictions Visualization*

In the following subsections, we have given the first 500 out-of-sample predictions of the GRU model for every dataset. The GRU model was chosen because it gains the best MAE and RMSE scores, as depicted in Table 2.

5.3.1 Beijing PM2.5 Dataset

Figure 7 depicts the first 500 continuous plots of one-step ahead forecasted values and the actual values on the test dataset.

Table 1 Framework evaluation metrics of the LSTM for every dataset

Dataset	Beijing PM2.5			Internet traffic			Zurich sunspots		
Metrics	MAE	RMSE	Time	MAE	RMSE	Time	MAE	RMSE	Time
TensorFlow	11.25	20.75	18 m 54 s	62.15	84.56	9 m 53 s	14.06	19.00	2 m 5 s
Theano	11.30	21.23	29 m 26 s	62.17	84.16	17 m 37 s	14.18	20.23	4 m 13 s
CNTK	12.59	21.67	35 m 58 s	61.97	84.67	20 m 36 s	13.63	19.75	4 m 49 s

Table 2 LSTM vs GRU evaluation metrics for every dataset

Dataset	Beijing PM2.5			Internet traffic			Zurich sunspots		
Metrics	MAE	RMSE	Time	MAE	RMSE	Time	MAE	RMSE	Time
LSTM	11.25	20.75	18m54s	62.15	84.56	9m53s	14.06	19.00	2m5s
GRU	10.94	19.69	16.51 s	57.91	78.76	8m40s	13.41	18.39	1 m 12 s

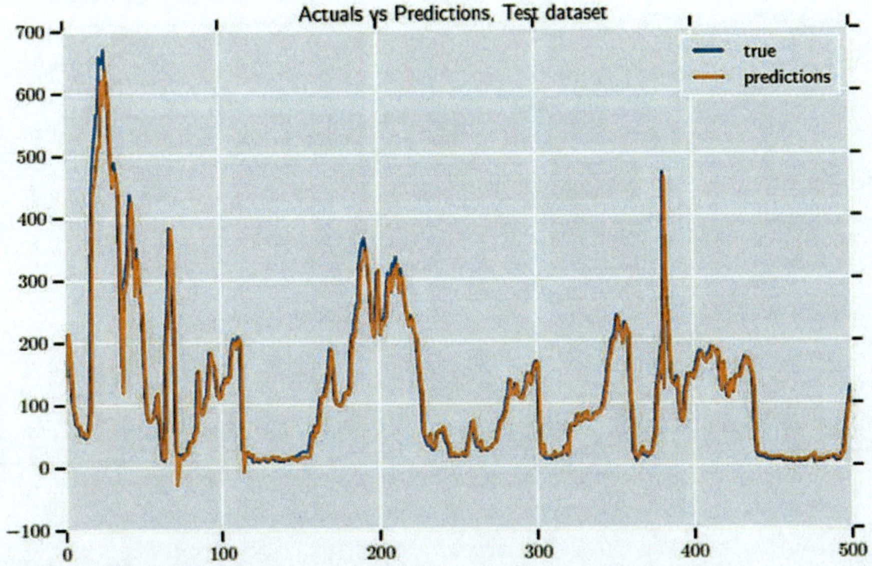

Fig. 7 One-step ahead predictions on the Beijing PM2.5 test dataset

5.3.2 Internet Traffic Dataset

Figure 8 depicts the first 500 continuous plots of one-step ahead forecasted values and the actual values on the test dataset.

5.3.3 Zurich Sunspots Dataset

Figure 9 depicts the first 500 continuous plots of one-step ahead forecasted values and the actual values on the test dataset.

6 Conclusion

In this work, we have presented the state-of-the-art deep learning framework for sequence-based works (LSTM, GRU), their internal workings as well as strategies that are broadly used in deep learning. Although different mathematical operations are used by different frameworks, the result of the experiments is almost the same except the speed of training. We can see that Theano being a little bit faster than the CNTK and TensorFlow being the fastest among all. As GRU takes fewer parameters, it is computationally expensive than LSTM, and as a result, GRU gives faster training speed. It is additionally demonstrated that the GRU can give the superior

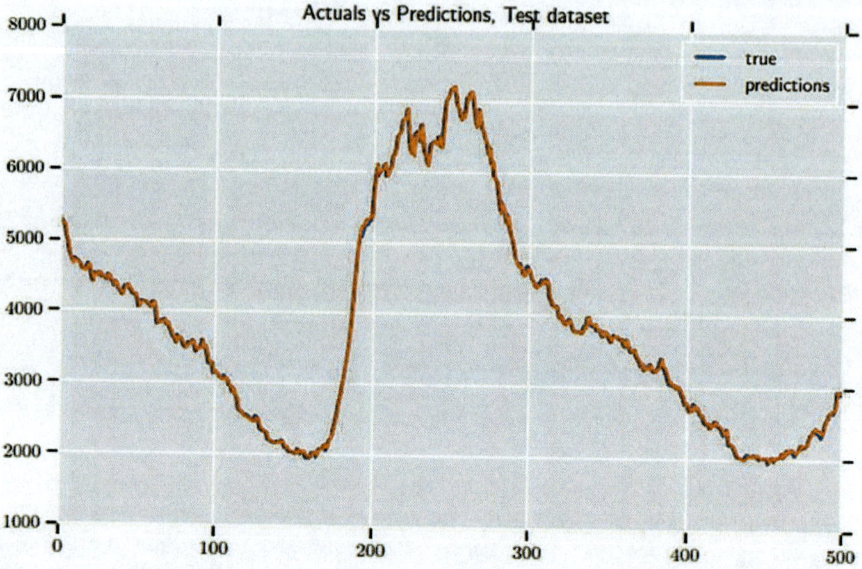

Fig. 8 One-step ahead predictions on the Internet traffic test

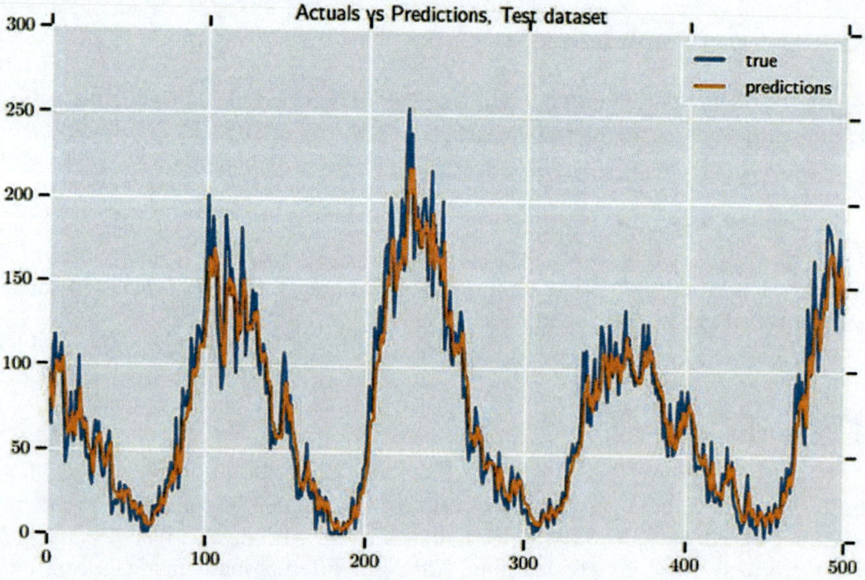

Fig. 9 One-step ahead predictions on the Zurich sunspots test dataset

expectation that comes about than the LSTM, having better evaluation scores as shown by the results presented in Table 2. To whole up, it appears that the victory of the framework's comparison is Google's TensorFlow because it accomplished roughly the same assessment score, but it is quicker than the others. Subsequently, to construct LSTM and GRU models for time series prediction, TensorFlow appears to be the foremost appropriate framework as the back-end engine of Keras in a CPU (non-GPU) setup, by utilizing the default arrangements given within the installation rules of each framework.

References

1. A. Graves, M. Liwicki, S. Fernández, R. Bertolami, H. Bunke, J. Schmidhuber, A novel connectionist system for unconstrained handwriting recognition. IEEE Trans. Pattern Anal. Mach. Intell. **31**(5), 855–868 (2008)
2. A. Graves, A.R. Mohamed, G. Hinton, Speech recognition with deep recurrent neural networks, in 2013 IEEE International Conference on Acoustics, Speech and Signal Processing 26 May 2013 (pp. 6645–6649). IEEE
3. J. Su, S. Wu, D. Xiong, Y. Lu, X. Han, B. Zhang, Variational recurrent neural machine translation, in Thirty-Second AAAI Conference on Artificial Intelligence (27 Apr 2018)
4. R. Weron, A. Misiorek, Forecasting spot electricity prices: a comparison of parametric and semiparametric time series models. Int. J. Forecast. **24**(4), 744–763 (2008)
5. J. Fan, Q. Yao, *Nonlinear Time Series: Nonparametric and Parametric Methods*. (Springer Science and Business Media, 11 Sep 2008)
6. R.F. Engle, Autoregressive conditional heteroscedasticity with estimates of the variance of United Kingdom inflation. Econometrica: J. Econometric Soc. 987–1007 (1 Jul 1982)
7. Q. Zhuge, L. Xu, G. Zhang, LSTM neural network with emotional analysis for prediction of stock price. Eng. Lett. **25**(2) (1 Apr 2017)
8. A. Lapedes, R. Farber, *Nonlinear Signal Processing Using Neural Networks: Prediction and System Modelling*. (1 Jun 1987)
9. C. de Groot, D. Würtz, Analysis of univariate time series with connectionist nets: a case study of two classical examples. Neurocomputing. **3**(4), 177–192 (1991)
10. C.M. Kuan, T. Liu, Forecasting exchange rates using feedforward and recurrent neural networks. J. Appl. Econometrics. **10**(4), 347–364 (1995)
11. A. Dagar, R. Bala, R.P. Singh, *Financial Time Series Forecasting Using Deep Learning Network*, in International Conference on Application of Computing and Communication Technologies 9 Mar 2018 (pp. 23–33). (Springer, Singapore)
12. F.A. Gers, D. Eck, J. Schmidhuber, Applying LSTM to time series predictable through time-window approaches, in *Neural Nets WIRN Vietri-01 2002* (pp. 193–200). (Springer, London)
13. S. Bhanja, A. Das, *Impact of Data Normalization on Deep Neural Network for Time Series Forecasting*. arXiv preprint arXiv:1812.05519. (13 Dec 2018)
14. S. Bhanja, A. Das, *Deep Neural Network for Multivariate Time Series Forecasting*. International Conference on Frontiers in Computing and Systems (COMSYS-2020). (2020)
15. S. Bhanja, A. Das, Deep learning-based integrated stacked model for the stock market prediction. Int. J. Eng. Adv. Technol. (9), 5167–5174 (Oct 2019)
16. J.L. Elman, Finding structure in time. Cogn. Sci. **14**(2), 179–211 (1990)
17. Y. LeCun, Y. Bengio, G. Hinton, Deep learning. Nature **521**(7553), 436–444 (2015)
18. C. Olah, Understanding lstm networks. (Aug 2015)
19. A. Moawad, The magic of lstm neural networks. (Feb 2018)

20. K. Cho, B. Van Merriënboer, C. Gulcehre, D. Bahdanau, F. Bougares, H. Schwenk, Y. Bengio, Learning phrase representations using RNN encoder-decoder for statistical machine translation. arXiv preprint arXiv:1406.1078. (3 Jun 2014)
21. Y. Bengio, P. Simard, P. Frasconi, Learning long-term dependencies with gradient descent is difficult. IEEE Trans. Neural Networks **5**(2), 157–166 (1994)
22. T. Tieleman, G. Hinton, Lecture 6.5-rmsprop: Divide the gradient by a running average of its recent magnitude. COURSERA: Neural Netw. Mach. Learn. **4**(2), 26–31 (Oct 2012)

Genetic Algorithm-Based Imperceptible Image Steganography Technique with Histogram Distortion Minimization

Pratik D. Shah and Rajankumar Bichkar

Abstract In image steganography, the confidential information is hidden in a cover image to camouflage the existence of covert communication. The steganography algorithm should be highly imperceptible to avoid being detected by a visual steganalysis attack. However, at the same time, the image steganography algorithm should also have the capability to resist statistical steganalysis attacks. These statistical steganalysis attacks can be histogram attack, RS analysis, chi-square attack, etc. A genetic algorithm (GA)-based imperceptible image steganography scheme that can resist histogram attack is presented in this paper. GA is used to find suitable locations in the cover image to insert the secret message so that the data embedding process causes very few distortions in the cover image. At the same time, GA has to ensure that these locations cause very minimum histogram distortion. The proposed technique can insert the secret data at 1 bit per pixel capacity (bpp). The proposed technique yields imperceptible stego images with an average PSNR value of 51.34 dB at 1 bpp data embedding capacity.

Keywords Genetic algorithm · Image steganography · Information hiding · LSB steganography

P. D. Shah (✉)
G H Raisoni College of Engineering and Management, Savitribai Phule Pune University, Pune, Maharashtra, India
e-mail: shahpratik219@gmail.com

R. Bichkar
Vidya Pratishthan's Kamalnayan Bajaj Institute of Engineering and Technology, Baramati, Maharashtra, India

V. E. Balas et al. (eds.), *Proceedings of International Conference on Computational Intelligence, Data Science and Cloud Computing*, Lecture Notes on Data Engineering and Communications Technologies 62, https://doi.org/10.1007/978-981-33-4968-1_21

1 Introduction

The Internet has enabled the data transfer quite conveniently at a very high speed. On the other hand, it has also caused many data security-related issues. To secure the data, cryptography is used but it reveals the existence of secure communication. To eliminate this problem, steganography was invented. The goal of steganography is to hide the presence of secret communication [1]. It can be attained by embedding the secret data in the cover image without generating noticeable modifications to it [2]. The insertion of the secret data in the cover image creates the stego image [3]. The prime objective of the image steganography is to reduce the variation between the cover and the stego image [4]. The major parameters used for evaluating any steganography algorithm are imperceptibility, payload capacity and security [5]. Developing a steganography technique that can excel in all three parameters is an extremely difficult task.

In the last decade, a vast number of researchers have worked on various steganography algorithms. But, mostly, their research was dedicated to improvement in imperceptibility. Few of them have even worked on the improvement of the imperceptibility and payload capacity together. But, limited studies have worked on increasing imperceptibility and enhancing security simultaneously. Some researchers have explored the use of evolutionary computation algorithms, to develop techniques that can resist steganalysis attacks and also produce stego images that have good imperceptibility.

A steganography technique having good visual quality and variable data embedding capacity using GA was proposed in [1]. This technique was developed in the spatial domain with a payload capacity of up to 4 bpp. The main focus of this research was to use GA for searching the best order and arrangements to insert the secret message in the image. Ranyiah Wazirali et al. [2] presented a high-payload steganographic scheme that focuses on increasing the imperceptibility by using various operations like optimized pixel scanning order, circular shifting, flipping secret bits and transposing secret data. Their technique used GA to find optimum solutions. A GA and particle swarm optimization (PSO)-based approach to increase the imperceptibility with a good payload was presented in [5]. Similarly, few other approaches [6–8] have explored PSO for the improvement of visual quality and payload capacity. Hemanth et al. developed a GA-based technique in Fresnelet transform and discrete ripplet transform [9]. The authors used traditional GA and modified GA approaches to enhance the performance of the technique. Few other techniques have also explored the use of GA in transform domain for improvement in the imperceptibility with a huge payload capacity [10–12]. These techniques used wavelet transform and discrete cosine transform.

In this paper, we propose an imperceptible steganography technique that can resist the histogram analysis. The data embedding capacity of this technique is 1 bit per pixel. In the proposed technique, GA is used to improve the imperceptibility of the stego image, and at the same time, it also helps in reducing the distortions caused in the stego image due to the secret data insertion. The GA is very useful in identifying the best locations in the cover image to embed the secret data. The GA fitness function ensures that the identified locations result in an imperceptible stego image with very slight degradation in the histogram. The proposed technique is explained with a detailed description in Sect. 2. The experimental results are discussed in Sect. 3. The key findings of the paper are presented in Sect. 4; in the same section, the future scope is also discussed.

2 Proposed Steganography Technique

To insert the secret message in the cover image, LSB replacement steganography is used. The cover image is converted in blocks of fixed size, and the secret message is inserted in LSB of each pixel in the block. However, the secret message is not inserted sequentially. Genetic algorithm is used to find the starting location and the direction for inserting the secret data. GA is also used to decide the polarity of the secret data for the specific block of the cover image. The polarity decides whether to complement the secret data or to keep it unchanged. These parameters are discussed in Sect. 2.1. To reduce the distortions in the histogram, the histogram distortion minimization technique developed by Yadav and Ojha is used [13]. The mechanism used to reduce the histogram distortion is explained in Sect. 2.2. The implementation details of GA and the chromosome structure are discussed in Sect. 2.3. Finally, the algorithm for data insertion and extraction is presented in separate sections.

2.1 Parameters for Secret Data Insertion

To improve the imperceptibility of the stego image, GA is used to find the proper locations in the cover image to hide the secret data. To achieve this task, we first convert the secret data in a binary bit array. To find the proper locations in the cover image to insert the secret data, we convert the cover image in 16 non-overlapping blocks. If the size of the cover image is 512×512 pixels, then the size of block will be 128×128. The normal LSB replacement steganography embeds the secret data row by row from the first pixel in the left to the last pixel in the right. But, in the proposed technique, we do not follow the traditional approach instead of that we search for the best starting point, i.e., x-offset and y-offset. In each block, we try to find a separate starting point, which can give the best matching between secret data and cover image LSB's. In this technique, along with searching for the best starting points, we also search for different embedding directions. To understand the combined effect of x-offset, y-offset and direction, we will consider a block of size 4×4. In Fig. 1, four different embedding directions are shown for different starting points.

In Fig. 1, we can observe various possible orders to insert the data in the cover image. The *x*-offset and *y*-offset will decide the starting point, and the direction parameter will be responsible for the order of embedding. The data can be inserted

12	13	14	15
16	1	2	3
4	5	6	7
8	9	10	11

(a) x-offset=1, y-offset=1 and direction=0

8	9	10	11
4	5	6	7
16	1	2	3
12	13	14	15

(b) x-offset=2, y-offset=1 and direction=1

3	2	1	16
7	6	5	4
11	10	9	8
15	14	13	12

(c) x-offset=0, y-offset=2 and direction=2

11	10	9	8
7	6	5	4
3	2	1	16
15	14	13	12

(d) x-offset=2, y-offset=2 and direction=3

Fig. 1 Effect of *x*-offset, *y*-offset and direction on the secret data embedding process

from the left to right direction or conversely. Similarly, after the end of a particular row is attained, we may go for the next bottom row or top row. These decisions are dependent on the direction parameter. In addition to the above parameters, one more parameter is used prior to inserting the payload. This parameter is data polarity (*dp*); the *dp* parameters decide the polarity of secret data, i.e., either to complement the secret data before insertion or keep it unchanged.

The above process of using *x*-offset, *y*-offset and direction to find the embedding order is used from the technique developed by Kanan and Nazeri in [1]. They have used several other parameters in their technique, but we have selected only the above-stated parameters in our technique. Kanan and Nazeri have not divided the image into blocks for inserting the data, and their technique had variable data hiding capacity.

2.2 *Histogram Distortion Minimization (HDM)*

Histogram of stego image after LSB replacement steganography undergoes huge modifications. These modifications drastically increase as the payload increases. Figure 2 depicts the histogram of the clock test image and corresponding stego images after 1 bit LSB steganography. In Fig. 2, it can be observed that after 1 bit LSB steganography, the histogram is severely changed, and we can easily detect the presence of steganography by visual histogram analysis.

To reduce the modifications in the histogram, Yadav and Ojha [13] developed histogram distortion minimization. Consider a cover image $C = \{c_1, c_2, \ldots, c_{m\times n}\}$, the value of c lies between 0 and 255. Due to LSB steganography, a pixel with intensity $2n$ may change to $2n + 1$ or pixel with intensity $2n + 1$ which may change to $2n$ where n is between 0 and 127. In some cases, the pixel may even remain unchanged if the secret data bit and LSB of the cover image are the same. To note the change in histogram an one-dimensional array named $q = (q_i)_{i=0}^{255}$ of size 256 is initialized with $q_i = 0$. The change in the histogram due to LSB steganography

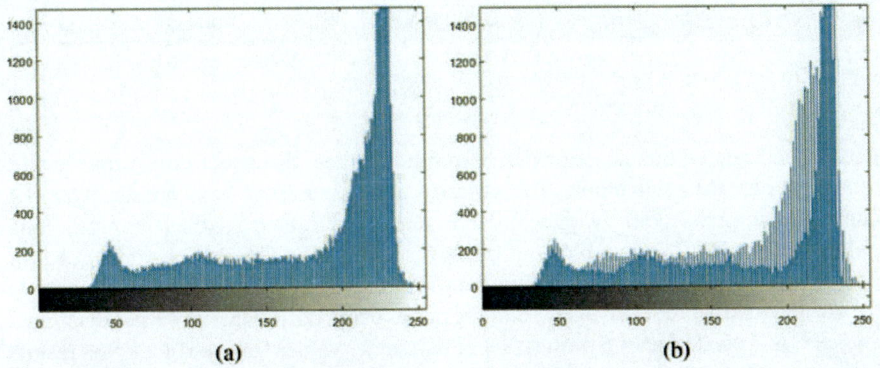

Fig. 2 Histogram of clock image, **a** Original cover image, **b** Stego images after 1 bit LSB steganography

is noted in q. If the pixel intensity $2n$ changes to $2n + 1$, then $q_{2n} = q_{2n} - 1$ and $q_{2n+1} = q_{2n+1} + 1$; if the pixel intensity $2n+1$ changes to $2n$, then $q_{2n+1} = q_{2n+1} - 1$ and $q_{2n} = q_{2n} + 1$. Defining the array weight $W(q)$ as the absolute sum of entries in q as follows:

$$W(q) = \sum_{i=0}^{255} |q_i|$$

To reduce the changes in the histogram, a steganography technique should have a minimum value of $W(q)$. In the proposed technique, our focus is on increasing the imperceptibility of the stego image and minimizing the value of $W(q)$. To achieve this challenging objective GA is employed.

2.3 Genetic Algorithm

GA is inspired by Darwin's theory of evolution, and it obeys the principle, 'survival of the fittest'. GA is a population-based technique, and each individual in the population is a possible solution for the problem being solved. These individuals contest one another for their survival and to produce new solutions for the next generation. The population evolves as the generation progress. To evaluate the efficiency of each individual in the population, a fitness function is defined, and it is used to assess their performance [14].

In the proposed technique, GA is used for finding the best values for the starting point and the direction of data insertion for each block of the cover image. Similarly, GA also helps in deciding the polarity of the secret data for the particular block. The block size of the cover image is 128×128; hence, for each block, the offset parameters will require 7 bits each. In the proposed technique, four different possible orders of

x-offset	y-offset	direction	polarity

Fig. 3 GA chromosome gene for each block of cover image

data embedding or the directions are explored. Hence, the direction parameter will require 2 bits for each block. The polarity parameter is of 1-bit length. The GA chromosome consists of 16 genes; each gene has x-offset, y-offset, direction and polarity parameter for each block of the cover image. The GA chromosome gene for one block with all parameters is shown in Fig. 3.

The payload is hidden in the LSB of the pixels using the parameter values discussed above. GA is used to search the best set of parameter values for each block that results in good imperceptibility and fewer distortions in the histogram. The length of each GA chromosome is 272 bits. The initial population will be produced randomly, and it consists of 100 chromosomes. In each iteration, all the chromosomes will contest one another to produce the next generation. The fitness function used to assess the competence of all the individuals in the population consists of two factors: The first one is the stego image PSNR value, and the second one is the histogram distortion array weight $W(q)$. The fitness function is shown in Eq. 1. The values of α and β are the coefficients, which allocate the specific weightage to the imperceptibility and the histogram distortion minimization. Suppose our priority is to get a more imperceptible stego, then the value of α should be kept higher as compared to the value of β and vice versa for the other case.

$$\text{Fitnessfunction} = \alpha \times PSNR + \beta \times \left(\frac{1}{W(q)}\right) \tag{1}$$

To produce new individuals for the next generation, various GA operators are used. To choose chromosomes from the current population, the roulette wheel selection strategy is employed. It selects two individuals from the populations for the next operations. Two-point crossover is performed on these chromosomes to produce two new individuals. A mutation is performed on these new solutions with a probability of 0.1. All the above steps are repeated until entire new population is not obtained. Later, the elitist strategy is used to retain the best individual in the population. The entire process is executed for 25,000 iterations in search of a near-optimal solution.

Consider a cover image that has m rows and n columns, the maximum amount of the secret data that can be inserted using the proposed technique is $m \times n$ bits. The proposed technique can also be used to insert fewer amounts of data. However, $m \times n$ bits will be the maximum capacity for data insertion. We hide the GA chromosome value as the secret key in the stego image. Two approaches can be employed for inserting the secret key in the stego image. The first approach is used in the case when the size of the bit array is greater than $(m \times n) - 272$ bits. In this case, the secret key is hidden in the second LSB of the border pixels of the image. The second approach is used when the size of the bit -array is less than $(m \times n) - 272$ bits. In this case, we start hiding the secret key in LSB of the last pixel in the bottom rightmost corner of the stego image and continue towards left.

2.4 Algorithm for Secret Message Insertion

Input: Cover image $I = \{i_1, i_2, \ldots, i_{m\times n}\}$,
Secret data (image) $D = \{d_1, d_2, \ldots, d_{(m/4)\times(n/2)}\}$
Output: Stego image $O = \{o_1, o_2, \ldots, o_{m\times n}\}$

1. Initialize an one dimensional bit arrays *sm* of length $len=m\times n$ as,
   $sm \leftarrow (D_8\ to\ D_1)\ \forall\ d_i$
2. Initialize
   $popmax \leftarrow population\ size\ (100)$
   $imax \leftarrow number\ of\ iterations\ (25000)$
3. Randomly generate *popmax* chromosomes for GA.
4. *for i = 1 to imax*
   for p = 1 to popmax
   Divide the cover image in non-overlapping blocks of size 128 × 128.
   Interpret the values of x-offset, y-offset, direction and polarity for each block from the GA chromosome as explained in sec.2.1.
   Obtain the stego image *O* by insert the data from the bit-array in the LSB's of the pixels in each block using the LSB steganography.
   For the stego image *O*, calculate the W(q) and the PSNR value.
   Store the fitness value of the chromosome in the array $fitness(p)$ using the fitness function defined in equation 1.
   end
   Select two individuals from the population using roulette wheel selection.
   Perform two-point crossover on these chromosome to get two new individuals.
   Perform mutation on these chromosomes with the mutation probability of 0.1.
   Repeat the above steps of selecting two individual and performing the crossover and mutation operators till we don't get 100 new chromosomes.
   Apply elitism.
   end
5. Choose the best solution and insert the chromosome as the secret key in predefined positions of the stego image.

2.5 Algorithm for Extracting Secret Data

Input: Stego image $O = \{o_1, o_2, \ldots, o_{m\times n}\}$
Output: Secret data (image) $D = \{d_1, d_2, \ldots, d_{(m/4)\times(n/2)}\}$

1. Obtain the secret key from the predetermined positions of the stego image.
2. Get the value of x-offset, y-offset, direction and polarity for all the blocks from the secret key.
3. Get all the LSB from the stego image and store them in the bit-array *se*.
4. Rearrange and modify the bits in the bit-array *se* as per the value of the secret key.
5. Recreate the secret data image *D* using *se*.

3 Experimental Results and Discussion

MATLAB 2018b software running on an Intel Core i5 processor containing a RAM of 8 GB was used for the implementation of the proposed algorithm. The research was performed using USC-SIPI image database [15]. The test images of size 512 × 512 were used as the cover images, and they are shown in Fig. 4. The cropped pepper image of size 128 × 256 was used as the secret data. To evaluate the imperceptibility of the stego images, the PSNR parameter was used. To perform histogram analysis, city block distance, Euclidean distance and Kullback-Liebler divergence were used along with visual histogram analysis. The equation of all these is given below. Let $H(I)$ and $H(O)$ be the histogram of cover and stego image, respectively. Let A and B be the probability density function (PDF) of $H(I)$ and $H(O)$, respectively.

$$A = \frac{H(I)}{len} \quad \text{and} \quad B = \frac{H(O)}{len} \quad \text{where} \quad len = M \times N$$

City block distance	Euclidean distance	Kullback-Liebler divergence
$d_{L1} = \sum_{i=1}^{len} \lvert A_i - B_i \rvert$	$d_{L2} = \sqrt{\sum_{i=1}^{len} \lvert A_i - B_i \rvert^2}$	$d_{KL} = \sum_{i=1}^{len} A_i ln \frac{A_i}{B_i}$

The comparison between the results obtained by the proposed technique and the LSB replacement steganography at 1 bpp payload capacity is shown in Table 1. The proposed technique produces the stego images that have an average PSNR value of 51.33 dB. The comparison between the splash test image and the corresponding stego image is shown in Fig. 5. There is an improvement in the PSNR value of the stego images for all images when compared to the traditional LSB approach which proves

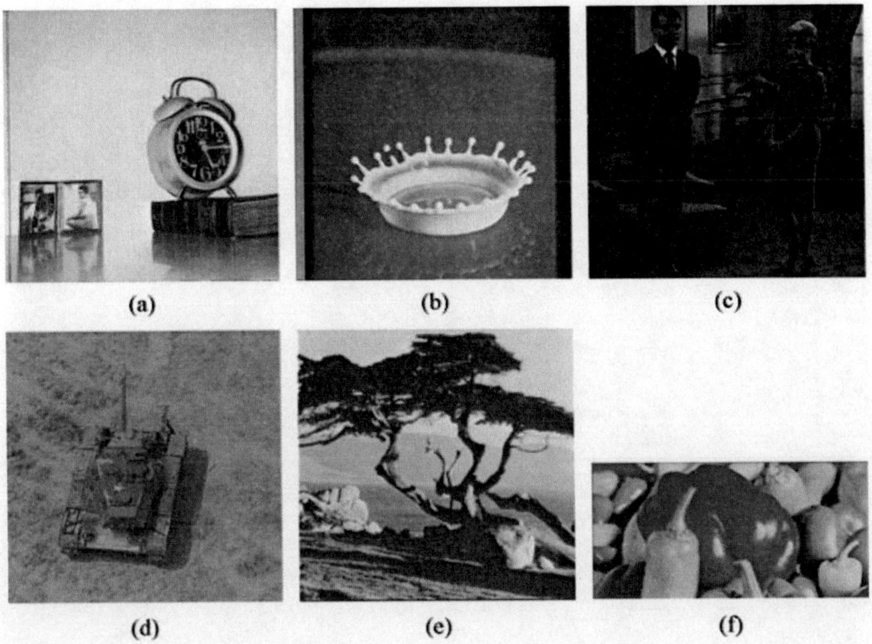

Fig. 4 Test images **a–e** Cover images (512 × 512) (clock, splash, couple, tank and tree) **f**–Secret message image (128 × 256) (pepper)

Table 1 PSNR value of stego images

Test image	LSB steganography	Proposed technique
Clock	51.08	51.34
Splash	51.12	51.33
Couple	51.17	51.38
Tank	51.15	51.35
Tree	51.14	51.33

the imperceptibility of the proposed technique. This improvement in the imperceptibility is also complemented by the reduction in the histogram distortions. The histogram of the splash test image and the corresponding stego images generated by the traditional LSB steganography and the proposed technique is presented in Fig. 6. In the same figure, the absolute difference between the histogram of the original image and the stego image obtained from proposed technique is also shown. Various distance measure values of the stego images are presented in Table 2. The value of the distances measured for the stego images obtained by the proposed technique is significantly lower as compared to the LSB steganography. This affirms the effectiveness of our algorithm in minimizing the distortions in the histogram. In the fitness

(a) (b)

Fig. 5 **a** Splash test image and **b** Splash stego image by proposed technique

function, the coefficient values used are $\alpha = 1$ and $\beta = 100$. These values can be varied as per the requirement, and this decision can also be dependent on the cover image.

4 Conclusion

An imperceptible image steganography scheme capable of resisting a histogram attack and having a moderate payload capacity is presented in the paper. GA was quite effective in finding the solutions that can satisfy the imperceptibility and security parameter simultaneously. The result of the proposed technique in each case outperformed the traditional LSB steganography. The stego image produced by the proposed technique had minute changes in its histogram as compared to the original cover image. One of the best parts of the proposed technique is its ability to control the imperceptibility and the security of the stego image. We can customize the security and imperceptibility aspect of the proposed algorithm by simply changing the coefficients in the GA fitness function. The future work will focus on using different evolutionary computation techniques for improving the results of the proposed technique.

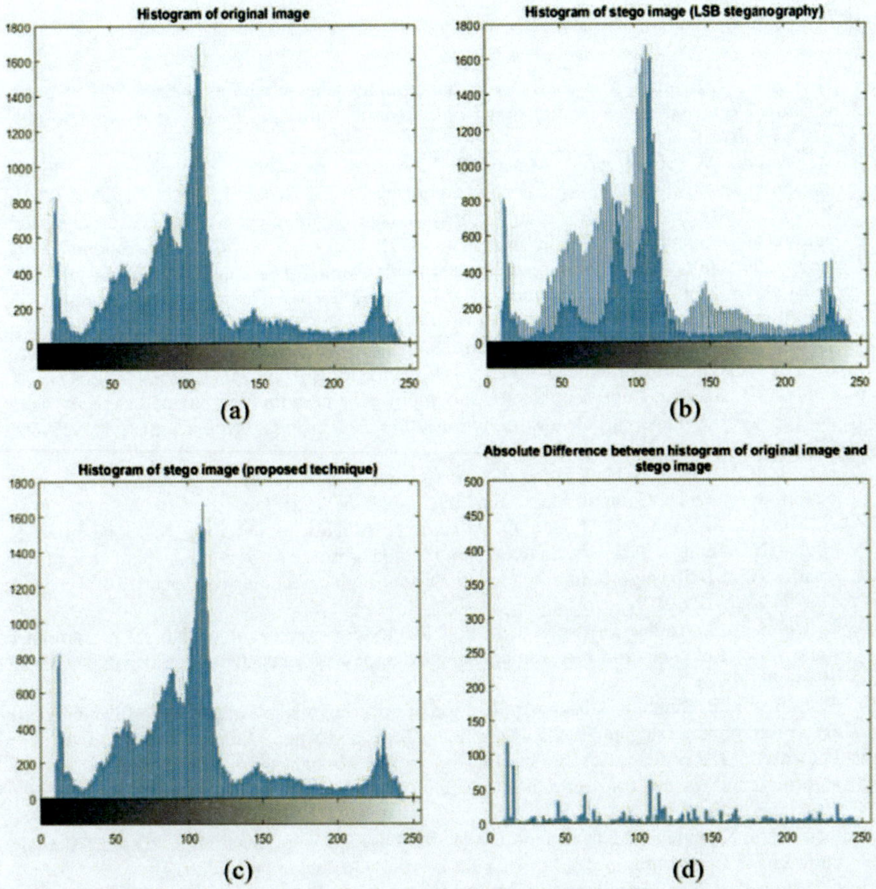

Fig. 6 Histogram of **a** splash test image, **b** stego image received by LSB steganography, **c** stego image obtained from proposed technique and **d** absolute difference between **a** and **c**

Table 2 Comparison of d_{L1} d_{L2} and d_{KL} for LSB steganography and the proposed technique

Test image	LSB steganography			Proposed technique		
	d_{L1}	d_{L2}	d_{KL}	d_{L1}	d_{L2}	d_{KL}
Clock	0.2989	0.0376	0.0832	0.0790	0.0168	0.0065
Splash	0.3670	0.0334	0.1121	0.0424	0.0047	0.0030
Couple	0.2756	0.0472	0.0667	0.1164	0.0437	0.0442
Tank	0.1905	0.0224	0.0281	0.0194	0.0028	0.0007
Tree	0.2589	0.02579	0.0626	0.0480	0.0058	0.0048

References

1. H. Kanan, N. Bahram, A novel image steganography scheme with high embedding capacity and tunable visual image quality based on a genetic algorithm. Expert Syst. Appl. **41**(14), 6123–6130 (2014)
2. R. Wazirali, W. Alasmary, M. Mahmoud, A. Alhindi, An optimized steganography hiding capacity and imperceptibly using genetic algorithms. IEEE Access **7**, 133496–133508 (2019)
3. P. Shah, R. Bichkar, A secure spatial domain image steganography using genetic algorithm and linear congruential generator, in *International Conference on Intelligent Computing and Applications, Advances in Intelligent Systems and Computing* (Springer, 2018), pp. 119–129
4. R. Biswas, I. Mukherjee, S. Bandyopadhyay, Image feature based high capacity steganographic algorithm. Multimedia Tools Appl. **78**(14), 20019–20036 (2019)
5. S. Maheswari, D. Hemanth, Performance enhanced image steganography systems using transforms and optimization techniques. Multimedia Tools Appl. **76**(1), 415–436 (2017)
6. P. Shah, R. Bichkar, Imperceptible steganography scheme with high payload capacity using genetic algorithm and particle swarm optimization. Int. J. Eng. Adv. Technol. **9**(1), 917–923 (2019)
7. P. Bedi, R. Bansal, P. Sehgal, Using PSO in a spatial domain based image hiding scheme with distortion tolerance. Comput. Electr. Eng. **39**(2), 640–654 (2013)
8. A. Mohsin, A. Zaidan, B. Zaidan, O. Albahri, A. Albahri, M. Alsalem, K. Mohammed, S. Nidhal, N. Jalood, A. Jasim, A. Shareef, New method of image steganography based on particle swarm optimization algorithm in spatial domain for high embedding capacity. IEEE Access **7**, 168994–169010 (2019)
9. D. Hemanth, J. Anitha, D. Popescu, L. Son, A modified genetic algorithm for performance improvement of transform based image steganography systems. J. Intel. Fuzzy Syst. **35**(1), 197–209 (2018)
10. S. Pramanik, R. Singh, R. Ghosh, Application of bi-orthogonal wavelet transform and genetic algorithm in image steganography. Multimedia Tools and Appl. (Article in press) (2020)
11. E. Ghasemi, J. Shanbehzadeh, N. Fassihi, High capacity image steganography based on genetic algorithm and wavelet transform, in *Intelligent Control and Innovative Computing* (Springer, 2012), pp. 395–404
12. R. Biswas, S. Bandyapadhay, Random selection based GA optimization in 2D-DCT domain color image steganography. Multimedia Tools Appl. (Article in press) (2019)
13. G. Yadav, A. Ojha, Hamiltonian path based image steganography scheme with improved imperceptibility and undetectability. Appl. Soft Comput. **73**, 497–507 (2018)
14. D. Goldberg, Genetic algorithms in search optimizations and machine learning, pearson education India, (2006)
15. USC-SIPI Image Database, https://sipi.usc.edu/database/

A Machine Learning Approach Toward Fast Track Decision in Judicial System

Snehangshu Ghosh, Md. Tabrez Nafis, and Safdar Tanweer

Abstract The advancement in artificial intelligence has led to the modernization of many fields including games, sports and shopping experiences but the judicial system has much more potential in overcoming some of the basic needs which is time and money. The modern techniques which can be used to solve some of the most complex problems like AlphaGo and other protein predicting algorithms to fight against cancer are not any more a fantasy but a reality. And with all these latest developments, the judicial system has more opportunity to grab and make a better society. But using the CNN and clustering, we can make a simpler approach to solve the large pending cases in the lower courts in Indian judicial system. In this paper, there has been an approach to make a way to use the techniques and find a way to solve the problem of pending cases in the lower court.

Keywords Machine learning · Convolution neural network (CNN) · Differential privacy

1 Introduction

In India, the primary sources of law are—statutes (established laws) and precedents (prior cases) [1]. Since both have their major role to play as a part of lawyer's career usually the statutes are learned in the law school and case studies/moot court are a part of precedents. But the complexity is not in the way people work with these things but the unfilled judges in the high court and subordinate courts. India has a population of 1.3 billion people, and as of August 2019, 3.5 crore cases have been

S. Ghosh · Md. T. Nafis (✉) · S. Tanweer
Department of Computer Science and Engineering, Jamia Hamdard, New Delhi, India
e-mail: tabrez.nafis@gmail.com

S. Ghosh
e-mail: snehangshughosh9@gmail.com

S. Tanweer
e-mail: safdartanweer@yahoo.com

V. E. Balas et al. (eds.), *Proceedings of International Conference on Computational Intelligence, Data Science and Cloud Computing*, Lecture Notes on Data Engineering and Communications Technologies 62, https://doi.org/10.1007/978-981-33-4968-1_22

pending due to lack of judges. It is estimated that around 63% vacant seats are there in the 21 high courts and 75% seats are unfilled for judges in subordinate courts. Which is a very staggering percentage for the largest democratic country in the world, but these cases could be solved using machine learning algorithms and more specifically using supervised learning as it turns out that a lot of the pending cases do belong to some similar kind of cases. Even the WHO talks about the sustainable development goal in which the 16th is Peace, Justice and Strong Institution. In this paper, there has been made an effort to solve the problem of a case where supervised learning is used.

2 Related Work

There are many counties who have started even implementation in this field for the staggering amount of cases such as in the case of European Union, USA and China. Even though there are constant changes and allegations regarding the system and its judgment, still it is a very efficient way as we move forward in future. It is not only that the machine learning algorithms are faster than the traditional system, but also they are more transparent and easily available. Even in comparison of five different algorithms, SVM is used to give more confidence as compared to others [2]. Even though the idea of using mathematics goes back to even sixteenth century, the real application of it could only be leveraged in the twentieth century as the availability of data is there and the development of the processing power is increased. The use of digital information and how it affects the life of everyone has changed. Artificial intelligence is considered to mimic the process of the human brain but using a lot of selective power in terms of GPUs and currently considered as narrow AI. The concept of narrow AI is that the systems today are only capable of dealing with specific problems and with a significant amount of data and time to learn about it as in the case of AlphaGo [3]. The history of using mathematics in law can be traced back to the 1600s when Gottfried Wilhelm Leibniz [4] introduced the propositional logic in the law and found out the outcome of it [5]. Then, after him the 1940s Lovinger came up with some of it [6]. There after the era of knowledge representation in the 1960s and 1970s had some major contributions due to mainly the development of diode and microprocessors after which the contributions Kashely's legal knowledge and information system came up and T. Mccarty on legal knowledge-based system also came up. Stanford University codex started in 2006 has some of the major work in this area. But with all that being said the real work has been done in countries like the USA, European Union and China. Even in China's Hangzhou court as part of a pilot project, the AI models are used for the hearing with a virtual court using virtual judges backed by powerful AI algorithms for solving large numbers of cases. They use the blockchain for the protection of the evidence; right now, it is used to program to solve cases such as Internet trade issues, copyright cases and disputes over online product sales. Even as use of the Internet is on the rise they can be used in a variety of ways [7]. European Union human rights have also used these techniques in different cases.

3 Proposed Model/Framework for Solution:

Human intelligence versus Artificial intelligence

Human intelligence is what we consider as the ability to learn from experience, adapt to new situations and handle them and use this knowledge to manipulate according to one's environment. But as we are moving forward in this world there is a larger need for us to consider the shortcoming of human and what the computer's are good at and make the best out of the both instead of seeing it from the standpoint of machine versus human there is a very large ground for the collaboration for both of them.

In the above model, people reach the lower court where they present the case in front of the lawyer where he works on the case file and presents the data about the chances of winning and all the laws that are in favor of it. After which, the case is filed in the courts and is presented before the judge where depending upon the pendency

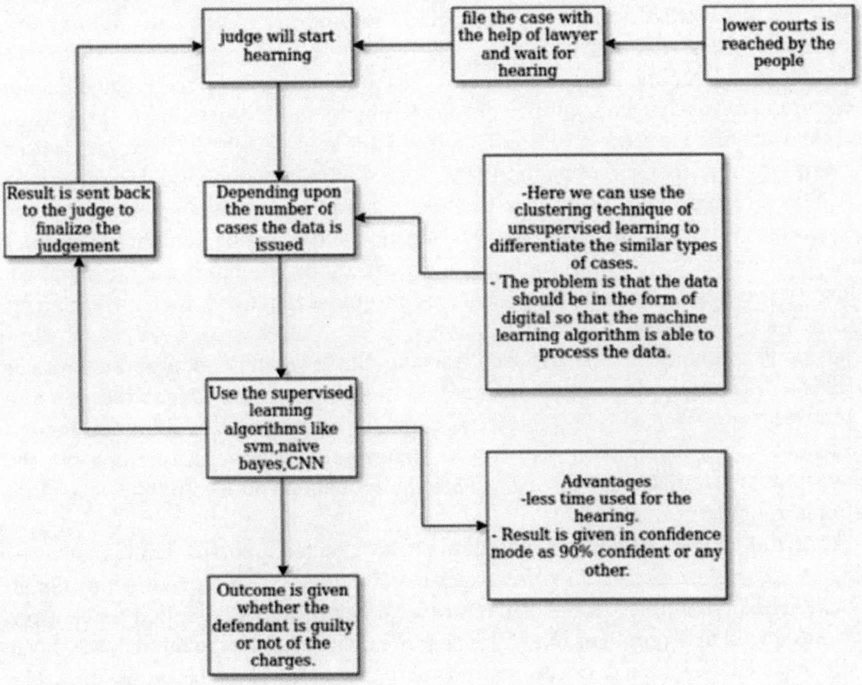

Fig. 1 Flowchart of how the system will work

of the cases a date is given for the hearing and the judgment is made based on the evidence present. Which can take from 2 to 20 years? Which is where the problem arises and to solve this problem we can use the unsupervised learning to differentiate the different number of cases and based on that then we can use those cases with any supervised learning algorithm to solve the cases. A number of algorithms can be used from using classification, decision tree and Bayesian techniques. In some previous experiments [2], SVM is considered to give the best result but the validity of it could be different for different cases and laws in different countries. In making these decisions, the use of the different parameters such as age, number of previous offends, place and sex needs to be considered for making any decision. There are a lot of benefits in this way as the cases will be solved with more training and data the accuracy can be increased to a very large level not only that there the use of digital data for the evidence can create a lot of comfort for the people which usually are protected by blockchain technology. Even with all the training and testing data when the cases are solved using these techniques, the output is given in the form of confidence such as 90 or 70% confident to be found as guilty which is sent to the judge for the final judgment of the case. Initially, the system may not be very efficient but with a lot of training this model can really solve the problem of the pendency of cases in Indian judicial courts.

Machine learning algorithms: Usually, the ML algorithms can be divided into supervised learning, unsupervised learning and reinforcement learning and within each category there are different techniques, and the best results are found using ensemble of these techniques.

Supervised learning: This can be defined as the learning from the labeled data or the learning form what we know as right and wrong by showing it to them. Regression and classification are some of the techniques in this area. The 90% of the data is used for training the model and remaining 10% is used for testing the model for accuracy.

Logistic regression: This is the regression technique in which when the number of the variables is binary they are called classification and when the number of variables increases then they are called multiclass classification. The interesting thing about logistic regression is that the function $h(x)$ is a sigmoid function which differentiates it from linear regression.

Using the techniques of supervised learning, the law can be incorporated into as there is a major area where the laws are all written and we have to load the data into it and then train the data on the previous cases in any particular field. Even though the techniques are simple enough for the training of the data and finding out the results that are available for the user to see the probability of finding the case close to punishment or not.

Artificial neural network (ANN) is an interconnected group of artificial neurons that has a natural property for storing experiential knowledge and making it available for use. The first simplest form of feed forward neural network, called perceptron, was introduced by Rosenblatt in 1957. These are connected via the input of any layer to the other via weighted connections and the output is fed at the last stage of it. The most interesting part is that sometimes the results are not what is expected as the hidden layer is considered as a black box which is a field in itself famously known as computer vision [8].

The combination of these layers is known as the multilayer perceptron where we have more than one layer and these small nodes or neurons [9]. As these neurons have weights we can think of this as the laws which are present in the Indian constitution and put the question as the input of the network and find what is the result at the output. These will make great results as these methods have proved to be very successful in the model in terms of gene modification, playing Go game and also the game of Shogi and chess by DeepMind.

Differential privacy: In the Fig. 2 there is a differentiation between male and female and the no. of offends that are committed over the years and the model in run using a logistic regression for the result and it is shown in the above (Fig. 2). It is a technique where the privacy of the user is protected via using Laplace mechanism where the actual data is also faked or not faked based on probability and then sent to the server and the process can be reversed accurately. Differential privacy addresses the paradox of learning nothing about an individual while learning useful information about a population [10]. This technique has a greater use to hide the real data and only use the data such as that which can be used to solve the problem without exposing

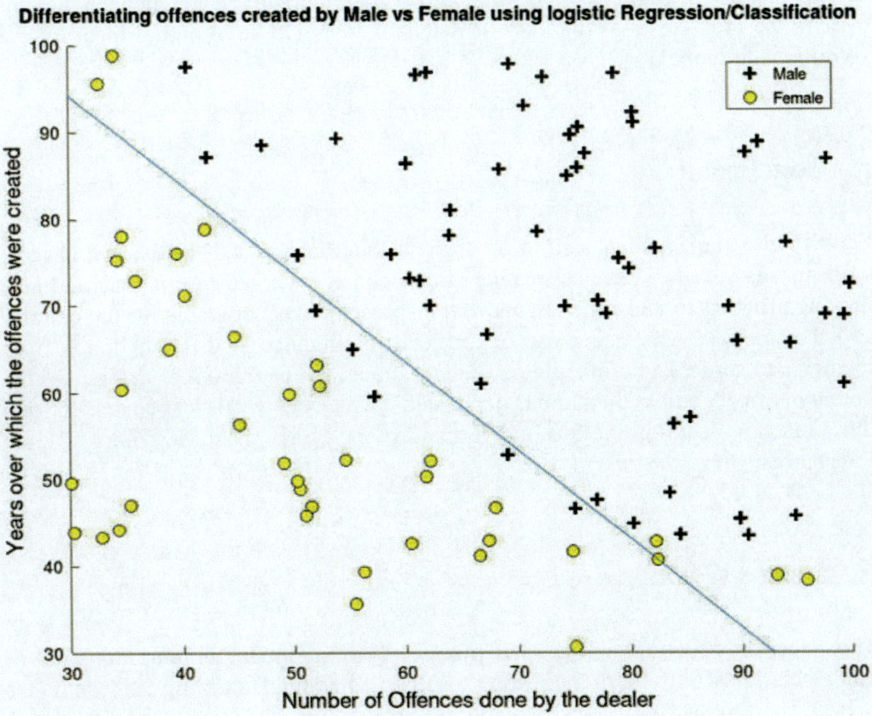

Fig. 2 Differentiation of the male offenders from female offenders over the property cases using logistic regression

the real data as the engineer will be working on the data which is already hidden and cannot be reversed.

4 Current Scenario

There has been a lot of work that has been done in training the model for the artificial judge but a lot of work can also be done in the field of privacy and security with the help of blockchain to protect the evidence and with which the databases can be secured for good reasons. But not only that there is a greater way of putting the different models that can work according to the data and cases which is possible only when the data is available for the models to run, a lot of major work has been done in the field of text summarization but the process is not yet efficient as it should be may be because of the language issue and the format in which the different high courts and subordinate courts does that. Hence, there needs to structure so the data can be fed into the model for which we can get the result out. Not only that there are ways to use the natural language processing for the defendants but also can be used for speech frequency etc. As China has made a pilot project using virtual court for defendants. But the main focus of this paper is that with all the technical challenges there is the problem of availability of data which should be in digital form for the algorithms to work [11].

5 Conclusion

Fourth industrialization is said to be artificial intelligence and for that we have to work in this area, we have seen other countries have worked in this area and also making progress to make the law more transparent and accountable for the country and their people. The rising use of artificial intelligence in different fields from medicine to surgery, pharmacy, etc., the law should not be left out as this makes the country stronger and as the famous quote said "justice delayed is justice denied" and these machine learning models once trained properly can make decisions fast and transparent.

6 Future Work

This paper only focused on the basic machine learning model without much use of other techniques but a lot of interesting results can be found out using the techniques of deep learning such as the LSTM network and also using the OpenMind differential privacy techniques for greater availability of the data to work on. Even though the early days of using machine learning techniques are very naive, there are many techniques which can be used such as that of using a convolution neural network.

References

1. Using Case Facts to Predict Accusation Based on Deep Learning. https://doi.org/10.1109/QRS-C.2019.00038
2. A Predictive Performance Comparison of Machine Learning Models for Judicial Cases. https://doi.org/10.1109/SSCI.2017.8285436
3. Mastering the game of Go with deep neural networks and tree search: https://www.nature.com/articles/nature16961
4. Leibniz as legal https://www.iep.utm.edu/leib-log/
5. Scholar: https://www.scielo.org.za/scielo.php?script=sci_arttext&pid=S1021-545X2014000100003
6. Machine Learning and Law By Harry Surden
7. AI judges via chat app and virtual court: https://learningenglish.voanews.com/a/robot-justice-the-rise-of-china-s-internet-courts-/5201677.html
8. Y. Kim, Convolutional neural networks for sentence classification. arXiv preprint arXiv:1408.5882 (2014)
9. G. Kurata, B. Xiang, B. Zhou, Improved neural network-based multi-label classification with better initialization leveraging label co-occurrence, in *Proceedings of the 2016 Conference of the North American Chapter of the Association for Computational Linguistics: Human Language Technologies* (2016), pp. 521–526. [17] G. Chen, D. Ye, Z. Xing, J. Chen, E. Cambria, Ensemble application of convolutional and recurrent neural networks for multi-label text categorization, in *2017 International Joint Conference on Neural Networks (IJCNN)* (IEEE, 2017), pp. 2377–2383
10. The Algorithmic Foundations of Differential Privacy: https://www.cis.upenn.edu/~aaroth/Papers/privacybook.pdf
11. A. Joulin, E. Grave, P. Bojanowski, T. Mikolov, Bag of tricks for efficient text classification. arXiv preprint arXiv:1607.01759 (2016)

Part II
Computer Vision

Detection of COVID-19 Using ResNet on CT Scan Image

Sourav Paul, Sweta Agarwal, and Ranjita Das

Abstract SARS-CoV-2, the main cause of COVID-19, is a new coronavirus identified as the cause of a disease outbreak in 2019 that originated in China. COVID-19 was declared a pandemic for the entire globe in March 2020 by the World Health Organization. The transmission rate of this virus is fast and spreads from person to person through discharge from the nose or droplets of saliva when an infected person sneezes or coughs. Even COVID-19 may spread from people who are not showing symptoms. An Antibody test might take 2–3 weeks to reveal the results of COVID-19. Presently, the Saliva Test for COVID-19, which takes 48 h to reveal the results, is most commonly used. In a study of more than 1000 patients published in the journal Radiology, Chest Computed Tomography surpassed lab testing in the diagnosis of COVID-19. Researchers concluded that CT scan images should be used as the primary screening tool for COVID-19. In remote village areas, we may not have any specialized doctor who can diagnose the CT scan image taken from a suspected COVID-19 patient. This proposed work can diagnose COVID-19 suspected CT scan images very fast, without the intervention of any specialized doctor. CT scan images have been taken and trained in the Residual Networks model, which gives a maximum of 95.97% accuracy.

Keywords COVID-19 · Residual networks · Transfer learning

S. Paul (✉) · S. Agarwal
Department of Computer Science and Engineering, Sikkim Manipal Institute of Technology, Sikkim Manipal University, Sikkim, India
e-mail: sourav.p@smit.smu.edu.in

S. Agarwal
e-mail: sweta.agarwal96.sa@gmail.com

S. Paul · R. Das
Department Computer Science and Engineering, National Institute of Technology, Aizawl, Mizoram, India
e-mail: rdas@nitmz.ac.in

V. E. Balas et al. (eds.), *Proceedings of International Conference on Computational Intelligence, Data Science and Cloud Computing*, Lecture Notes on Data Engineering and Communications Technologies 62, https://doi.org/10.1007/978-981-33-4968-1_23

1 Introduction

COVID-19 is a contagious disease caused by the newly discovered coronavirus, SARS-CoV-2. Coronaviruses cause diseases in mammals and birds. It is a group of related RNA viruses, which in humans causes respiratory tract infections. These respiratory tract infections might be mild (such as the common cold) or fatal, which causes COVID-19 [1], SARS [2], and MERS [3]. It has been declared a global health emergency by the World Health Organization. There is no effective vaccine against the coronavirus disease. Due to the contagious nature of the virus, it needs to be identified at an early stage to stop it from spreading, by immediately isolating the infected person.

Most COVID-19 patients experience mild to moderate illness. People of old age or with a medical history of diabetes, cardiovascular disease, chronic respiratory disease, and cancer have higher chances of developing a fatal illness when suffering from COVID-19.

Presently, the diagnosis of COVID-19 is done using RT-PCR (Reverse Transcription-Polymerase Chain Reaction) or blood specimens or gene sequencing for respiratory. Due to limitations in sample collection, transportation, and slow and high false-positive rate (30%) [4], along with kit performance, many COVID-19 patients suffer from inaccurate diagnosis. The recent growth of medical science led to more extensive use of medical digital imaging devices, like the CT scan. Chest CT may be considered as a primary tool for the detection of COVID-19 [5]. It is difficult for doctors to read a large number of CT scan images, making it time-consuming, and also leading to inaccurate diagnosis.

Several efforts have been made to research and develop a system of diagnosis based on deep learning [11]. ResNet-50 is a deep learning architecture based on convolutional neural networks, suitable for image recognition. It can automatically extract features from images and obtain information from them. Through thorough training of the model, using the available images, the diagnosis can be made fast and efficient.

A deeper network was thought to give better performance. Increasing the number of layers resulted in the degradation of the accuracy of the model. Residual Networks, with the help of skip connections that make use of shortcut paths, improves the performance of a deeper network.

2 Previous Works

Soares et al. [6] introduces a model using a baseline method, an eXplainable Deep Learning approach. The model delivers an F1 score of 97.31%. The approach used is based on recursive calculations and is non-iterative.

Sharma et al. [7] classified chest X-rays, using several deep convolutional neural network-based models like VGG, MobileNet, Xception, and many more. Unsatisfied with the outcomes of the above-mentioned models, A model using Residual Attention Network was introduced, with a testing accuracy of 98%.

Alqudah et al. [8] implements multiple machine learning algorithms and compares their performance. The algorithms include Convolutional Neural Networks, Random Forest, and Support Vector Machine. The comparison shows the performance of CNN to be better than the other algorithms with a testing accuracy of 95.2%. The CNN model was used for performing classification as well as extracting graphical features for hybrid system implantation.

The SARS-CoV-2 CT scan dataset contains 2482 CT scans in total, including images of infected and uninfected patients and is publicly available. The images are of real patients from Sao Paulo, Brazil hospitals. The aim of the dataset is to encourage research and development of Artificial Intelligence models that can classify the CT-scan images as infected or uninfected by SARS-CoV-2.

3 Methodologies Used

The work presented in the paper aims at detecting COVID-19 infected patients by classifying the CT scan images as infected or uninfected. ResNet-50 has been implemented for achieving the task of classification. ResNet-50 is a 50-layer deep Convolutional Neural Network. The images are preprocessed, before feeding them into the model, to enhance certain features further improving the performance.

3.1 Dataset

Data has been collected from Kaggle.com [9]. The name of the dataset is sarscov2-ctscan. Total 2482 CT scan images are there, out of which 1252 CT scan images are of infected patients and 1230 CT scan images are of uninfected patients. All the images in the dataset are real, collected from hospitals in Sao Paulo, Brazil.

3.2 Image Preprocessing

Images in the dataset are preprocessed before feeding them into the model for training. Image preprocessing includes resizing, labeling of the images, etc. It enhances certain features, which helps in improving the performance of the model [10].

- Resize: The dataset consists of different size images that are resized to a base size, which is the input image size of the model.
- Labeling: Labeling of images is done for the model to be trained accordingly and extract features based on the labels.

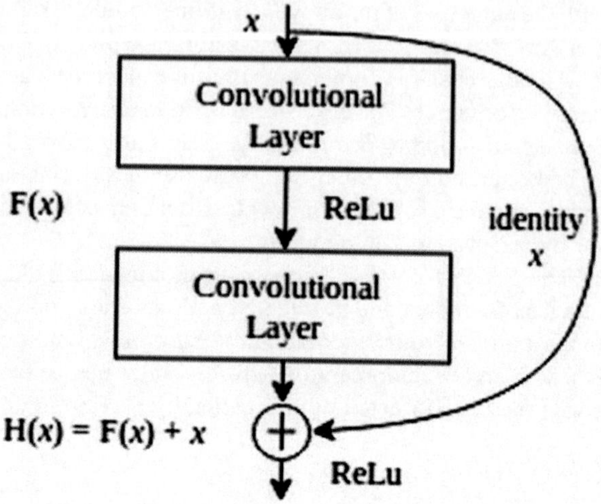

Fig. 1 Skip connection

3.3 Residual Networks

Residual Networks are also known as ResNet. ResNet is a Convolutional Neural Network having the advantage of adding a greater number of layers without the problem of vanishing gradient. ResNet uses a skip connection technique that combines the input of the previous layer with the input of the current layer using the ReLU activation function to avoid vanishing gradient problem function.

In residual learning, the network has been developed as, $H(x) = F(x) + x$, thus, the residual function can be written as, $F(x) = H(x) - x$, Now, if $F(x) = 0$, it forms an identity map $H(x) = x$, where x represents the pre-layer input, $F(x)$ is the network mapping before the sum, and $H(x)$ is the network mapping from the input to summation as shown in Fig. 1.

3.4 Convolutional Neural Networks

Convolutional Neural Network (CNN) is specially used for image processing, classification, segmentation, and for other auto correlated data. CNN acquires features from a raw dataset. Unlike other neural networks, it extracts the features automatically from the raw dataset. The main characteristic of CNN is to apply a non-linear function on a raw dataset and produce abstracted outcomes. The Backpropagation algorithm is one of the best learning algorithms, suited for CNN, due to its simple computation.

CNN is a stack of convolutional and pooling layers leading to a series of fully connected layers followed by the output layer. The aim of the convolutional layer is representation of features and the pooling layer reduces the size of the feature map. A

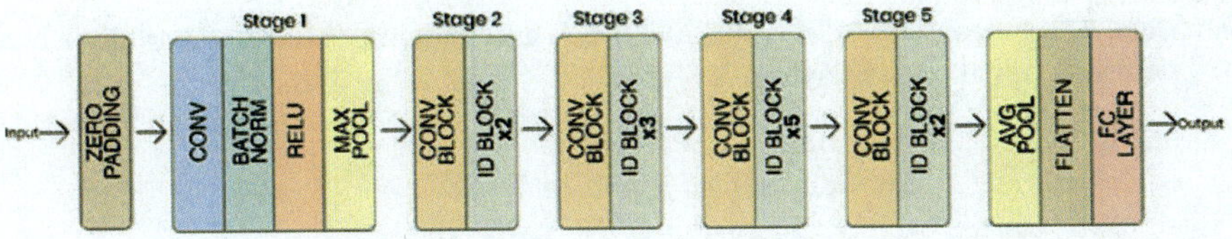

Fig. 2 Architecture of ResNet-50

fully connected layer does not preserve special information. SoftMax Regression is commonly used for classification purposes as it provides a good probability function. Flatten is the function that converts the pooling feature map to a single column and passes the outcome to the fully connected layer. CNN has many layers, all the layers have their own input and output, so there is a problem of getting the same distribution of input to all the layers. Batch normalization is used to ensure all the layer gets the Gaussian distribution.

3.5 Transfer Learning of the Residual Neural Network

The architecture of the model created using ResNet-50 as shown in Fig. 2 is explained below:

- Zero Padding: Zero-padding symmetrically adds zeroes to the input matrix. The size of the filter used is 3 × 3.
- Stage 1: It consists of a convolutional layer with 64 kernels, each of dimension 7 × 7, and a stride of 2 × 2. Batch Normalization, ReLU activation function, and max pooling with filter size 3 × 3 and a stride of 2 × 2 are applied. The total number of layers in this stage is 1.
- Stage 2: It consists of 1 convolutional block and 2 identity blocks. Each block has 3 convolutional layers having,
  - 64 kernels, each of dimension 1 × 1,
  - 64 kernels, each of dimension 3 × 3, and
  - 128 kernels, each of dimension 1 × 1, respectively.

  The 3 layers are repeated for all the 3 blocks and hence, the total number of layers in this stage is 9.
- Stage 3: It consists of 1 convolutional block and 3 identity blocks. Each block has 3 convolutional layers having,
  - 128 kernels, each of dimension 1 × 1,
  - 128 kernels, each of dimension 3 × 3, and
  - 512 kernels, each of dimension 1 × 1, respectively.
    The 3 layers are repeated for all the 4 blocks and hence, the total number of layers in this stage is 12.

- Stage 4: It consists of 1 convolutional block and 5 identity blocks. Each block has 3 convolutional layers having,
  - 256 kernels, each of dimension 1×1,
  - 256 kernels, each of dimension 3×3, and
  - 1024 kernels, each of dimension 1×1, respectively.

  The 3 layers are repeated for all the 6 blocks and hence, the total number of layers in this stage is 18.
- Stage 5: It consists of 1 convolutional block and 2 identity blocks. Each block has 3 convolutional layers having,
  - 512 kernels, each of dimension 1×1,
  - 512 kernels, each of dimension 3×3, and
  - 2048 kernels, each of dimension 1×1, respectively.

  The 3 layers are repeated for all the 3 blocks and hence, the total number of layers in this stage is 9.
- Output Layer: It consists of an average pool with a stride of 2×2, followed by a fully connected layer.

The dataset containing 2482 images was randomly divided into training and testing data with a distribution of 80% and 20% respectively, with randomly selected images. Hence, the training data had 1984 images and the testing data had 497 images.

$$\begin{aligned}\text{The total number of layers} : &\ 1(\text{stage } 1) + 9(\text{stage } 2) + 12(\text{stage } 3) + 18(\text{stage } 4)\\ &+ 9(\text{stage } 5) + 1(\text{output layer}) = 50\,\text{layers}.\end{aligned}$$

4 Evaluation Metric

Confusion Matrix, Recall, Precision, and $F1$ score have been used to evaluate the performance of the proposed model. The confusion matrix is the best way to evaluate a binary classification algorithm. In Fig. 5, the column shows the actual class and the row shows the predicted class. It also shows the true positive value and true negative value, that is, 233 and 249, which indicates that out of 497 testing data, the model classified 482 data correctly.

$$\text{Accuracy} = \frac{\text{TP} + \text{TN}}{\text{TP} + \text{FP} + \text{TN} + \text{FN}}$$

Recall, Precision, and $F1$ score are the matrices that are generated from the confusion matrix. Precision says how often the model predicts the correct positive cases.

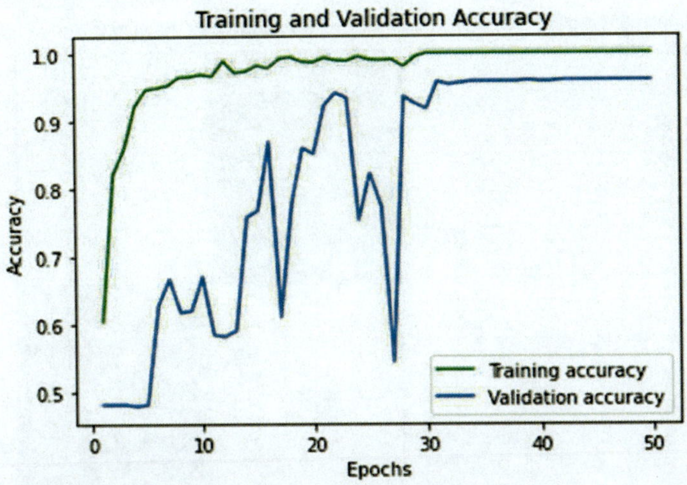

Fig. 3 A figure showing training and validation accuracy graph

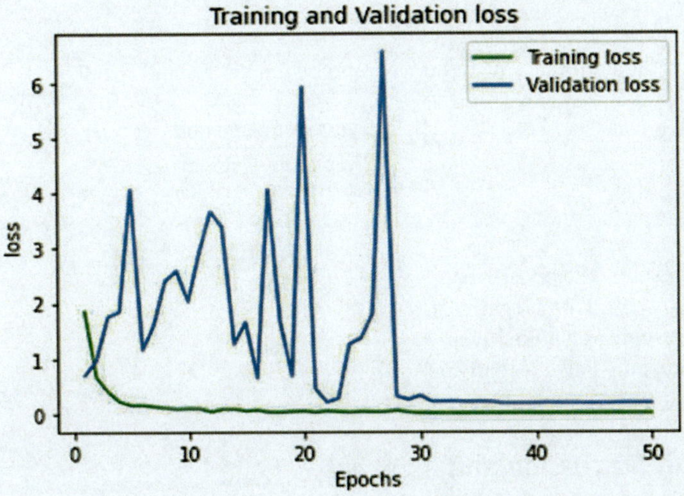

Fig. 4 A figure showing training and validation loss graph

$$\text{Precision} = \frac{\text{TP}}{\text{TP} + \text{FP}}$$

Recall measures the fraction of positive outcomes that are correctly classified.

$$\text{Recall}(r) = \frac{\text{TP}}{\text{TP} + \text{FN}}$$

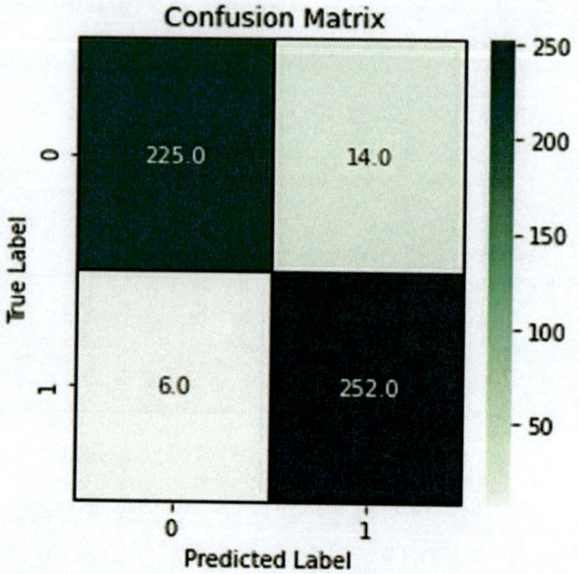

Fig. 5 Confusion matrix showing testing data classification

$F1$ score is the harmonic mean between precision values and recall.

$$F1\,\text{Score} = 2\frac{\text{Precision} \times \text{Recall}}{\text{Precision} + \text{Recall}}$$

where,

TP represents True Positives,
TN represents True Negatives,
FP represents False Positives,
FN represents False Negatives.

5 Result Discussion and Analysis

The model was trained using the training data and validated against the testing data and results were as follows:

- **100%** accuracy on training data.
- **95.97%** accuracy on testing/validation data.
- Precision: **0.414**.
- Recall: **0.9740**.
- F1 Score: **0.9574**.

Figure 3, shows the training and validation accuracy graph, where the green line represents the training accuracy and the blue line represents the validation accuracy. Figure 4, shows the training and validation loss graph, where the green line represents the training loss and the blue line represents the validation loss. Figure 5 shows the confusion matrix. All the above-mentioned metrics like Precision, Recall, and $F1$ Score are calculated using the generated confusion matrix.

6 Conclusion

COVID-19 is a disease caused by SARS-CoV-2, a type of coronavirus. Globally, 12,401,262 confirmed cases, including 559,047 deaths, have been reported to the World Health Organization [1], as of 8:15 am CEST, 12 July 2020. Diagnosis is done primarily using RT-PCR but due to limitations in sample collection and transportation, along with kit performance, many COVID-19 patients suffer from unnecessary or delayed diagnosis.

A Chest CT scan, a routine test for pneumonia diagnosis, helps in faster diagnosis by the identification of certain specific characteristics. The work presented in the paper uses a publicly available SARS-CoV-2 CT scan dataset, collected from real patients in Sao Paulo, Brazil. A deep learning method using ResNet-50 is introduced for classifying the images as infected and uninfected. ResNet-50 is an advanced form of CNN that uses shortcut paths for achieving state-of-the-art performance and improving optimization for the deeper networks. Through thorough training of the model, using the available images, it achieved a training accuracy of 100% and testing accuracy of 95.97%.

References

1. WHO Coronavirus Homepage, https://www.who.int/health-topics/coronavirus, last accessed 11 July 2020
2. D.J. Cleri, A.J. Ricketti, J.R. Vernaleo, Severe acute respiratory syndrome (SARS). Infect. Dis. Clin. **24**(1), 175–202 (2010)
3. WHO MERS-CoV Homepage, https://www.who.int/news-room/fact-sheets/detail/middle-east-respiratory-syndrome-coronavirus-(mers-cov), last accessed 12 July 2020
4. Y.H. Wu, S.H. Gao, J. Mei, J. Xu, D.P. Fan, C.W. Zhao, M.M. Cheng, JCS: An Explainable COVID-19 Diagnosis System by Joint Classification and Segmentation. arXiv preprint arXiv: 2004.07054 (2020)
5. T. Ai, Z. Yang, H. Hou, C. Zhan, C. Chen, W. Lv, Q. Tao, Z. Sun, L. Xia, Correlation of chest CT and RT-PCR testing in coronavirus disease 2019 (COVID-19) in China: a report of 1014 cases. Radiology, 200642 (2020)
6. E. Soares, P. Angelov, S. Biaso, M.H. Froes, D.K. Abe, SARS-CoV-2 CT-scan dataset: A large dataset of real patients CT scans for SARS-CoV-2 identification. medRxiv (2020)
7. V. Sharma, C. Dyreson, COVID-19 detection using Residual Attention Network an Artificial Intelligence approach. arXiv preprint arXiv:2006.16106 (2020)

8. A.M. Alqudah, S. Qazan, H. Alquran, I.A. Qasmieh, A. Alqudah, COVID-2019 Detection Using X-Ray Images and Aritificial Intelligence Hybrid Systems (2020)
9. Kaggle SARS-CoV-2 CT Scan Dataset Homepage, https://www.kaggle.com/plameneduardo/sarscov2-ctscan-dataset, last accessed 12 July 2020
10. Towards Data Science Image Pre-processing Homepage, https://towardsdatascience.com/image-pre-processing-c1aec0be3edf, last accessed 08 July 2020
11. W. Kong, P.P. Agarwal, Chest imaging appearance of COVID-19 infection. Radiol. Cardiothorac. Imag. 2.1, e200028 (2020)

On Mitigation of False Positive Problem in Singular Value Decomposition-Based Digital Image Watermarking

Subhadeep Koley

Abstract Due to the effortless sharing of multimedia data (mainly images) over the communication channels and social media platforms, copyright protection of such data has become the topic of research interest. This paper presents a robust and false positive free watermarking scheme for digital image copyright protection with singular value decomposition (SVD) in the Stockwell transform domain. Unfortunately, most of the earlier SVD-based watermarking algorithms suffer from false positive or watermark ambiguity problem. In this study, we showed an approach to mitigate the false positive problem. A human visual system (HVS)-based method has been incorporated via discrete Cosine Stockwell transform (DCST). After projecting the input image in the DCST domain, SVD has been performed on the DCST coefficients. Finally, a total insertion-based approach has been presented to embed the watermark logo into the singular matrix of the DCST coefficients. Furthermore, the proposed approach's security has been augmented by encrypting it with a novel cryptic algorithm called the chaotic logistics map. A detailed false positive analysis proves that the proposed method is free from all kinds of watermark ambiguity problems. Qualitative and quantitative comparisons with other state-of-the-art methods delineate the primacy of the proposed algorithm under most types of malicious signal processing and geometric impairments.

Keywords Digital watermarking · Cryptography · SVD · Discrete Cosine Stockwell Transform · False positive free · Chaotic Logistics Map

S. Koley (✉)
Department of Electronics and Communication Engineering, RCC Institute of Information Technology, Kolkata, West Bengal, India
e-mail: subhadeepkoley@gmail.com

V. E. Balas et al. (eds.), *Proceedings of International Conference on Computational Intelligence, Data Science and Cloud Computing*, Lecture Notes on Data Engineering and Communications Technologies 62, https://doi.org/10.1007/978-981-33-4968-1_24

1 Introduction

Digital watermarking is a critical component of any multimedia communication architecture, which has many applications like copyright protection, multimedia authentication, copy control, etc. It is the process of embedding supplementary information (image, text, etc.) inside multimedia files, which can be extracted later, on the basis of some user-given keys to prove ownership [1]. Nowadays, digital image watermarking has become a topical subject of research due to the effortless sharing of images over the communication channels and social media platforms. An efficient image watermarking approach is always a trade-off between imperceptibility and robustness [1]. Recently, SVD-based image watermarking schemes gained popularity for their extreme robustness [2, 3]. Usage of the singular matrix of a cover image provides great robustness, as the singular values are mostly unfazed by signal processing attacks. SVD-based watermarking methods also provide excellent imperceptibility, as minor changes in the singular matrix do not affect the perceptual quality of the cover image [2]. However, more often than not due to wrong embedding approach, SVD-based frameworks suffer from false positive problem (FPP) [2]. FPP occurs when the decoder extracts out a watermark from a watermarked image even when the watermark was never inserted [1]. In this way, an adversary can maliciously profess ownership violating the sole purpose of establishing a rightful ownership [4]. Despite FPP being a well-understood issue in the field of image watermarking, most of the SVD-based approaches mainly focus on the watermark imperceptibility and robustness rather than resolving the crucial issue of ensuring rightful ownership [2]. Therefore, a lot of the published SVD-based watermarking methods [5–13] suffer from the FPP or watermark ambiguity problem. In most of those methods, the authors used optimization, image transforms, etc., to increase robustness or imperceptibility but those techniques have no effect on the FPP issue. FPP issue occurs in most of the algorithms mentioned above because those algorithms only embed the S component of the watermark by modulating the S component of the cover image, and the U and V components of the watermark are kept safe with the owner for using as keys during extraction. Later, they extract the watermark from the watermarked image based upon user-provided U and V components. Also, as a matter of fact, the U and V components contribute to the major geometric and shape properties of an image, and the S component only determines the degree of contribution of U and V in the image formation [2]. Therefore, attackers can easily extract out their malicious watermark logo from the watermarked image simply by swapping the original U and V watermark component with that of theirs. In this way, they can extract a watermark from the watermarked image, which was never embedded in the beginning. We also confirmed this problem via an experiment where we have taken two watermark logos '$W1$' and '$W2$' and performed SVD individually on both to get the orthogonal and singular matrices. During reconstruction, we swapped the orthogonal matrices of $W1$ with those of W2 and vice versa. As predicted earlier, the reconstructed image strikingly resembles the image whose orthogonal matrices were used during reconstruction (Fig. 1). This proves that the orthogonal matrices hold the

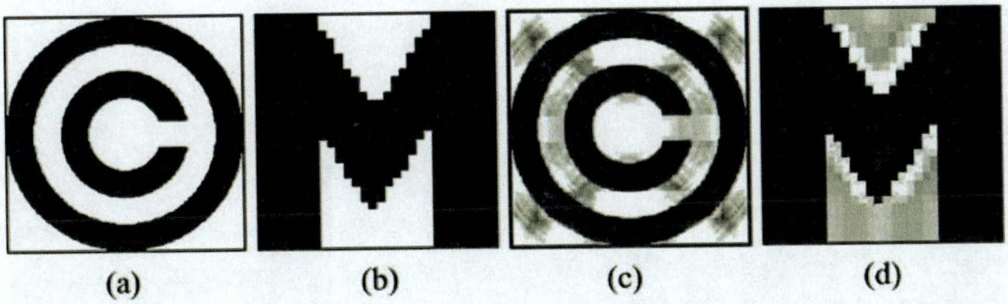

Fig. 1 a Watermark '$W1$' b Watermark '$W2$' c Image reconstructed with U_{W1}, V_{W1}, S_{W2} d Image reconstructed with U_{W2}, V_{W2}, S_{W1}

major information of an image, and a watermarking framework is not faithful where the end users provide those matrices.

To mitigate this problem, an effective solution is given by [2] where the authors embed the watermark's principal component into the S component of the cover image. Though this solution can handle the issue of FPP, but it still depends on the user-provided V component of the original watermark. To countermeasure this problem, in this paper, we proposed a total insertion of the watermark image into the singular values of the DCST coefficients of the cover image, which entirely eliminates (Detail in Sect. 2.4) the dependency of the user-provided U and V watermark components. In this way, we designed a trustworthy watermarking scheme, which ensures the rightful ownership. Furthermore, due to the obvious advantages of DCST over discrete Cosine transform (DCT) and discrete wavelet transform (DWT) [14], we have transformed the image into the DCST domain prior to the SVD for added robustness and imperceptibility. The rest of the paper contains other sections like proposed methodology and result analysis followed by conclusion.

2 Proposed Methodology

2.1 *Discrete Cosine Stockwell Transform (DCST)*

Stockwell transform (ST) given by Stockwell [15] is an efficient time-frequency domain decomposition, which generates absolute referenced phase and frequency information of a signal. Stockwell transform was designed to overcome the limitations of the time-frequency domain wavelet transform and frequency domain Fourier transform. Stockwell transform decomposes the image in various resolutions along with accurate time-frequency separation. To overcome the high redundancy of the discrete Stockwell transform (DST), an orthonormal version of the DST was introduced namely the discrete orthonormal Stockwell transform (DOST) [14]. Similar to DCT, DOST also provides significant energy concentration, which in turn provides

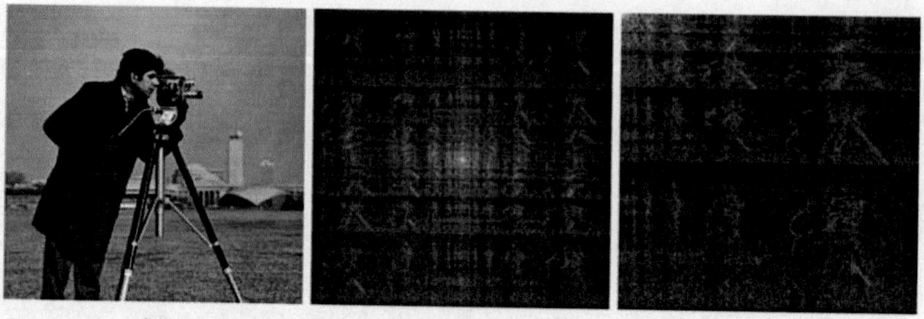

(a) (b) (c)

Fig. 2 a Original cover image **b** DOST coefficients **c** DCST coefficients

excellent imperceptibility and robustness. The ST basis function can be expressed as [15],

$$\mathcal{D}_{\nu,\omega,\gamma}[k] = \frac{exp\,(i\pi\gamma)}{\sqrt{\omega}} \sum_{n=\nu-\frac{\omega}{2}}^{\nu+\frac{\omega}{2}-1} exp\left(-i\frac{2\pi}{N}kn\right) exp\left(i\frac{2\pi}{N}\gamma n\right) \tag{1}$$

where ν is the center frequency of each band, ω is the bandwidth, γ is the spatial location. Also, DOST can be generated by the product of $\mathcal{D}_i$ matrices as,

$$\text{DOST} = \left(\bigoplus_{i=1}^{k} \mathcal{D}_i\right) \text{DFT} \tag{2}$$

where *DFT* is the Fourier transform operator. Similarly, we can express discrete Cosine Stockwell transform (DCST) simply by replacing the *DFT* operator with a DCT operator [14].

$$\text{DCST} = \left(\bigoplus_{i=1}^{k} \mathcal{D}_i\right) \text{DCT} \tag{3}$$

Fig. 2 has been presented to depict the original cover image along with the corresponding DOST and DCST coefficients.

2.2 *Singular Value Decomposition*

In the past, SVD has been used in many image processing and matrix manipulation tasks. SVD generates one diagonal and two orthogonal matrices from a two-dimensional matrix. Let $\mathcal{I} = \mathbb{R}^{n_1 \times n_2}$ be a two-dimensional image matrix, then using SVD, $\mathcal{I}$ can be decomposed as,

$$\mathcal{I} = \mathcal{U} * \mathcal{S} * \mathcal{V}^T \tag{4}$$

where $\mathcal{U} \in \mathbb{R}^{n_1 \times n_1}$ and $\mathcal{V} \in \mathbb{R}^{n_2 \times n_2}$ are the orthogonal matrices ($\mathcal{U}^T\mathcal{U} = 1$ and $\mathcal{V}^T\mathcal{V} = 1$), $\mathcal{S} \in \mathbb{R}^{n_1 \times n_2}$ denotes the diagonal matrix carrying the singular values λ_i sorted in descending manner where $1 \leq i \leq r$ and r represents the rank of the matrix $\mathcal{I}$. Here, the $\mathcal{U}$ and $\mathcal{V}$ components contribute to the significant geometric properties of an image, and the $\mathcal{S}$ component only governs the degree of participation of $\mathcal{U}$ and $\mathcal{V}$ in image formation.

2.3 *Chaotic Logistics Map (CLM)*

Logistics Map (LM) is a discrete and dynamic system, which depicts a controlled chaotic behavior [16]. Logistics map can mathematically be expressed as

$$x_{n+1} = \psi \cdot x_n\,(1 - x_n) \tag{5}$$

where ψ is the control parameter, n is the iteration number and x is the state variable of the LM. At $\psi > 3.57$, LM system delineates a controlled chaotic state with periodicity [16]. In this paper, we have used the LM attractors to create a random encryption matrix. The proposed encryption framework is described in Algorithm 1.

Algorithm 1: Proposed encryption algorithm

Input: $\mathcal{W}, key1, key2$
Output: $\mathcal{W}_{enc}$
$itr \leftarrow m \times n$
$for\ n = 1$ to itr
  $x_{n+1} \leftarrow key1 \cdot x_n\,(1 - x_n)$
end
$x \leftarrow$ Reshape x into a matrix of size $m \times n$
$randMat \leftarrow x \odot x$
$randInt \leftarrow$ Generate a random integer using $key2$ as a seed
$roundRandMat \leftarrow \lfloor (randMat \times randInt)\ \%256 \rfloor$
$\mathcal{W}_{enc} \leftarrow image \oplus roundRandMat$

The control parameter ψ has been used as one of the encryption keys in this scope. Another key is used as a seed to generate a random number. A random permutation matrix has been generated by taking hadamard product of x with itself. After that, we rounded off the permutation matrix with the randomly generated integer. Therefore, for every unique set of encryption keys, the rounded permutation matrix is unique. Encrypted image is generated by taking XOR between the rounded per-

Fig. 3 **a** Original watermark **b** Encrypted watermark (SSIM = 0.0036, NCC = 0.1264) **c** Decrypted with one wrong and one right key (SSIM = 0.0032, NCC = 0.1163) **d** Decrypted with two wrong keys (SSIM = 0.0028, NCC = 0.1099) **e** Decrypted with two correct keys (SSIM = 1.0, NCC = 1.0

mutation matrix and the original image. The decryption process is exactly same as the encryption process, as performing XOR again between the rounded permutation matrix and encrypted image will yield the original image. Figure 3 shows the original watermark and the corresponding encrypted and decrypted images for right as well as wrong keys. Form Fig. 3, it is clear that the encrypted watermark has no similarity with the original image. Encrypted image's Structural SIMilarity Index (SSIM) and Normalized Cross Correlation (NCC) values of 0.0036 and 0.1264 also indicate the same. Contrastingly, the image decrypted with correct keys is exactly same as the original watermark with SSIM and NCC values of 1.0 and 1.0, respectively. Moreover, in Fig. 3, we can see that the images decrypted with wrong set of keys have no similarity with the original one. This proves that the encryption scheme is sensitive enough to withstand counterfeit key attacks. We call this proposed encryption framework as *chaotic logistics map (CLM)*. Being an area retaining transform (i.e., size of the encrypted image is same as the original image), CLM does not affect the payload capacity.

2.4 Encoder and Decoder

Let $\mathcal{I}$ be a 2D gray scale image of size $p \times q$, and $\mathcal{W}$ be the gray scale watermark image of size $m \times n$. The images can be represented as

$$\mathcal{I} = \mathcal{P}\,(i, j)\,|0 \leq \mathcal{I} < p, 0 \leq \mathcal{I} < q, \mathcal{P}(i, j) \in \{0, 1, 2, \ldots, 255\} \tag{6}$$

$$\mathcal{W} = \mathcal{R}\,(i, j)\,|0 \leq \mathcal{W} < m, 0 \leq \mathcal{W} < n, \mathcal{R}(i, j) \in \{0, 1, 2, \ldots, 255\} \tag{7}$$

After CLM encryption, the original watermark image $\mathcal{W}$ is transformed into $\mathcal{W}_{enc}$ and can be expressed as

$$\mathcal{W}_{enc} = \text{CLM}(\mathcal{W}, key1, key2) \tag{8}$$

where CLM is the chaotic logistics map encryption operator, $key1$ and $key2$ are the user-provided encryption keys. $key1$ and $key2$ can be any integer upto 32 bit. The decrypted watermark can be expressed as

$$\mathcal{W}_{dec} = \text{ICLM}(\mathcal{W}_{enc}, key1, key2) \tag{9}$$

where ICLM is the inverse chaotic logistics map encryption operator. The watermark encoding and decoding strategies are depicted in Algorithm 2 and Algorithm 3.

Algorithm 2: Proposed encoding algorithm

Input: $\mathcal{I}, \mathcal{W}, key1, key2$
Output: $\mathcal{I}_{wmkd}$

$DCST_{\mathcal{I}} \leftarrow$ Apply DCST on $\mathcal{I}$
$[\mathcal{U}, \mathcal{S}, \mathcal{V}] \leftarrow$ Apply SVD on $DCST_{\mathcal{I}}$
$\mathcal{W}_{enc} \leftarrow$ Apply CLM on $\mathcal{W}$ using $key1$ and $key2$
$\mathcal{S}' \leftarrow$ Modify $\mathcal{S}$ to embed $\mathcal{W}_{enc}$ using Eq. (10)
$DCST'_{\mathcal{I}} \leftarrow$ Compute inverse SVD using $\mathcal{U}, \mathcal{S}'$, and $\mathcal{V}$
$\mathcal{I}_{wmkd} \leftarrow$ Compute inverse DCST with $DCST'_{\mathcal{I}}$

To exploit the supremacy of the HVS perception, we transformed the cover image into the DCST domain. After that, performing SVD, we get the two orthogonal and one diagonal matrices. After getting the singular matrix $\mathcal{S}$, the gray scale watermark is embedded inside $\mathcal{S}$ using the following equation,

$$\mathcal{S}' = \mathcal{S} + (\zeta \cdot \mathcal{W}_{enc}) \tag{10}$$

where $\mathcal{S}$ and $\mathcal{S}'$ are the original and modified singular matrices, respectively, $\mathcal{W}_{enc}$ is the encrypted watermark, and ζ is the watermark embedding factor. After getting the modified singular matrix, we perform inverse DCST with the modified singular matrix and other two unaltered orthogonal matrices to get the watermarked image $\mathcal{I}_{wmkd}$. At the time of decoding, firstly the DCST and SVD are applied on both the watermarked and the original cover image. The embedded watermark $\mathcal{W}_{enc}$ is extracted from the $\mathcal{S}'$ using the following equation,

$$\mathcal{W}_{enc} = (\mathcal{S}' - \mathcal{S})/\zeta \tag{11}$$

Algorithm 3: Proposed decoding algorithm

Input: $\mathcal{I}, \mathcal{I}_{wmkd}, key1, key2$
Output: $\mathcal{W}_{dec}$

$DCST_{\mathcal{I}} \leftarrow$ Apply DCST on $\mathcal{I}$
$[\mathcal{U}, \mathcal{S}, \mathcal{V}] \leftarrow$ Apply SVD on $DCST_{\mathcal{I}}$
$DCST'_{\mathcal{I}} \leftarrow$ Apply DCST on $\mathcal{I}_{wmkd}$
$[\mathcal{U}', \mathcal{S}', \mathcal{V}'] \leftarrow$ Apply SVD on $DCST'_{\mathcal{I}}$
$\mathcal{W}_{enc} \leftarrow$ Extract the watermark from $\mathcal{S}'$ and $\mathcal{S}$ using Eq. 11
$\mathcal{W}_{dec} \leftarrow$ Apply CLM encryption on $\mathcal{W}_{enc}$ using $key1$ and $key2$

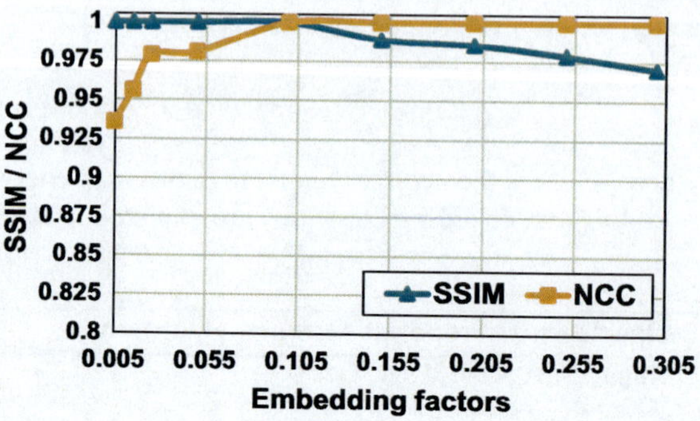

Fig. 4 SSIM/NCC values for different embedding factors

The decoder will generate the decrypted watermark image $\mathcal{W}_{dec}$ upon supplying the keys, which were used to encrypt the watermark. To determine the optimum value of the embedding factor ζ, Fig. 4 has been presented.

SSIM values for the 'cameraman' watermarked image (against the original cover image) and NCC values for the extracted '$W1$' logo (against the original logo) have been plotted in Fig. 4 for different embedding factors. As seen in Fig. 4, $\zeta = 0.104$ provides optimized trade-off between imperceptibility and robustness.

3 Results and Discussion

To judge the effectiveness of the proposed scheme, we have taken total 25 different gray scale images of size 256 × 256, and one gray scale image of size 128 × 128. The proposed method has been compared with other state-of-the-art methods viz. [5, 6, 8, 11, 17–21]. All experiments have been designed and performed on MATLAB® 2018a.

3.1 Imperceptibility Analysis

Imperceptibility or perceptual signal quality is one of the most important parameters to quantify when real-life digital image watermarking is being considered. Figure 5 has been presented with a few cover images along with the corresponding watermarked images for qualitative comparison. It is evident from Fig. 5 that there is no perceivable trace of the hidden watermark in the watermarked images.

Moreover, in Fig. 6, we have compared our proposed approach with other state-of-the-art methods in terms of peak signal-to-noise ratio (PSNR). As seen in Fig. 6, the proposed method offers highest PSNR among all.

Fig. 5 **Top row**: Cover images. **Bottom row**: Watermarked images. **a** Elaine **b** Boat **c** Pentagon **d** Barbara

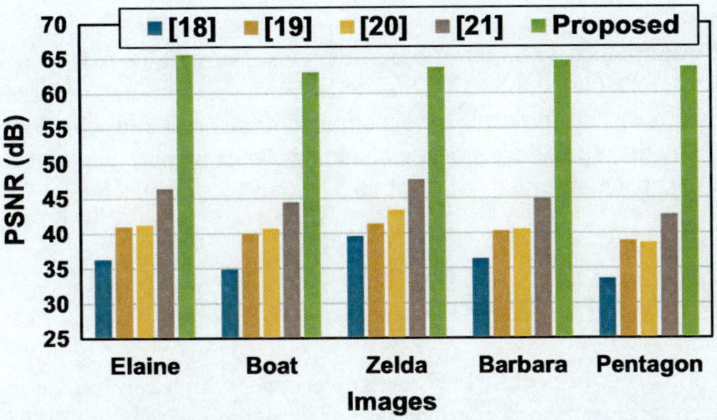

Fig. 6 PSNR comparison with other methods

3.2 *Robustness Analysis*

Robustness refers to the potentiality of the embedded watermark to resist common malicious signal processing and geometric attacks. To justify the proposed algorithm's robustness, we have performed attacks of various degrees on the watermarked image. Figure 7 has been presented with recovered watermarks for a few different attacks. As seen in Fig. 7, the embedded watermark is robust enough to resist common attacks.

To compare the proposed framework's robustness with other ultramodern methods, Table 1 has been presented where we have compared the NCC values of our

Fig. 7 **Top row (from left to right)**: Sharpening (3 × 3), Median filtering (3 × 3), Average filtering (3 × 3), Gaussian filtering (3 × 3). **Bottom row (from left to right):** JPEG (QF=60), Salt & Pepper noise (0.1), Cropping (Upper left), Gamma correction (0.9)

recovered watermarks with that of other methods. Form Table 1, it is evident that, for all reported attacks, the NCC values of our proposed scheme outperform other reported methods. This indicates the superiority of the proposed scheme in terms of robustness. On the more, the average embedding and execution time of the proposed framework has been observed to be only 0.87 s and 1.2 s, respectively.

3.3 *False Positive Error Analysis*

As discussed earlier in Sect. 1, FPP is a grave issue, which can hinder any sophisticated watermarking framework. The proposed method mitigates FPP problem by removing the dependency of user-given orthogonal matrices. To verify this experimentally, Fig. 8 has been presented where we have plotted the NCC values of 50 random watermarks. As seen in Fig. 8, only at position 15, the NCC value is greater than the detection threshold (0.7), which is the location of the original embedded watermark. Moreover, the detector response of all other random watermarks is way below the detection threshold of 0.7 and never crossed 0.1552. Therefore, it is evident that the proposed scheme is free of FPP, as the detector response of the original watermark is far greater than that of random counterfeit watermarks. Furthermore, a detail overall framework comparison with other methods (viz. [5, 6, 8, 11, 17]) has been shown in Table 2. In Table 2, it can be seen that the proposed framework outperforms other competing methods with an average PSNR, mean square error (MSE), and NCC values of 64.086 dB, 0.0259, and 0.9845, respectively. Also, unlike the proposed method, all other SVD-based methods presented in Table 2 suffer from

Table 1 NCC comparison with other state-of-the-art methods for different attacks

Attacks	[18]	[19]	[20]	[21]	Proposed
Sharpening(3 × 3)	0.9100	0.9800	0.9590	0.9890	0.9912
Median filter(3 × 3)	0.8930	0.9730	0.9160	0.9880	0.9951
Median filter(5 × 5)	0.7470	0.8490	0.6960	0.8840	0.9913
Average filter(3 × 3)	0.8740	0.9690	0.9080	0.9860	0.9887
Average filter(5 × 5)	0.7260	0.8230	0.6770	0.8630	0.9758
Gaussian filter(3 × 3)	0.9810	0.9980	0.9840	0.9980	0.9991
Gaussian filter(5 × 5)	0.9810	0.9980	0.9840	0.9990	0.9875
JPEG (QF = 40)	0.9100	0.9630	0.9380	0.9830	0.9857
JPEG (QF = 60)	0.9900	0.9850	0.9620	0.9980	0.9989
JPEG (QF = 80)	0.9950	0.9990	0.9880	0.9990	0.9999
Salt & Pepper noise (0.1)	0.5250	0.6450	0.5370	0.6730	0.9789
Rotation (0.5)	0.6390	0.6740	0.6510	0.6920	0.9915
Rotation (0.7)	0.5520	0.5580	0.6050	0.610	0.9867
Cropping (Upper left)	0.8230	0.8500	0.8970	0.9100	0.9785
Cropping (Lower right)	0.8280	0.8360	0.9020	0.9020	0.9655
Histogram equalization	0.7980	0.9800	0.9220	0.9980	0.9998
Resize (512 → 256 → 512)	0.9210	0.9850	0.9250	0.9990	1.0000
Resize (512 → 1024 → 512)	0.9930	0.9910	0.9880	0.9990	1.0000
Gamma correction($\gamma = 0.9$)	0.8360	0.9930	0.9760	0.9990	1.0000
AWGN($\sigma^2 = 0.001$)	0.6330	0.7390	0.5780	0.8850	0.9325
AWGN($\sigma^2 = 0.005$)	0.5870	0.6730	0.5590	0.8120	0.9158
Scaling (factor = 0.50)	0.9010	0.9450	0.9050	0.9580	0.9897
Scaling (factor = 0.75)	0.9130	0.9710	0.9180	0.9890	0.9925

FPP. Moreover, the proposal of the novel chaotic logistics map helps achieve added security.

4 Conclusion

In this paper, a DCST domain robust and invisible digital image watermarking has been presented. The proposed method impregnates the watermark inside the cover image by adjusting the singular values of the DCST coefficients of the cover image. FPP of SVD-based image watermarking has been mitigated via a total insertion-based method. Furthermore, a novel image encryption scheme has been proposed to further increase the security of the hidden watermark. Several experiments have

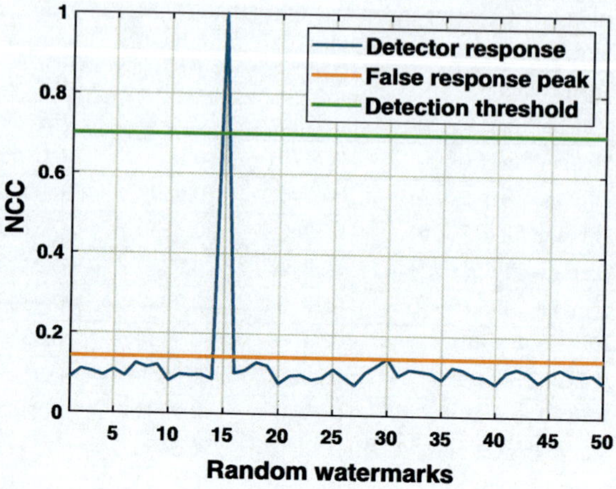

Fig. 8 Watermark detector responses for random watermarks

Table 2 Overall framework comparison with other state-of-the-art methods

Methods	[5]	[6]	[8]	[11]	[17]	Proposed
Host size	512 × 512	256 × 256	256 × 256	256 × 256	512 × 512	256 × 256
Watermark size	512 × 512	32 × 32	64 × 64	256 × 256	64 × 64	128 × 128
Watermark type	Gray scale	Binary	Gray scale	Gray scale	Binary	Gray scale
Domain	Wavelet	Wavelet	Wavelet	Wavelet	Wavelet	Discrete Stockwell
Technique used	SVD	SVD	SVD	SVD	SVD	SVD
Imperceptibility	++	++	+	+++	++	++++
Avg. PSNR (dB) ↑	54.722	52.490	40.283	62.890	55.217	64.086
Avg. MSE ↓	0.2190	0.3660	6.0916	0.0334	0.1956	0.0259
Robustness	++	++	+	++	+++	++++
Avg. NCC ↑	0.9640	0.9586	0.3044	0.9345	0.9830	0.9845
Capacity (bpp) ↑	8.0000	0.0156	0.5000	8.0000	0.0156	2.0000
FPP	Yes	Yes	Yes	Yes	Yes	No
Encryption	N.A.	N.A.	N.A.	N.A.	N.A.	Novel CLM

been performed to compare the proposed scheme with other state-of-the-art methods. Experimental results presented here prove the primacy of the proposed scheme in terms of imperceptibility, robustness, and security. Moreover, the detector response curve presented in Fig. 8 experimentally proves that the proposed scheme is free from FPP. In near future, we aim to make the framework faster for real-time hardware implementation.

References

1. S. Koley, A feature adaptive image watermarking framework based on phase congruency and symmetric key cryptography. J. King Saud Univ. - Comput. Inf. Sci. (2019). https://doi.org/10.1016/j.jksuci.2019.03.002
2. I.A. Ansari, M. Pant, C.W. Ahn, ABC optimized secured image watermarking scheme to find out the rightful ownership. Optik **127**(14), 5711–5721 (2016). https://doi.org/10.1016/j.ijleo.2016.03.070
3. M. Ali, C.W. Ahn, Comments on 'optimized gray-scale image watermarking using DWT-SVD and Firefly Algorithm'. Expert Syst. Appl. **42**(5), 2392–2394 (2015). https://doi.org/10.1016/j.eswa.2014.10.045
4. H.C. Ling, R.C.W. Phan, S.H. Heng, Comment on 'robust blind image watermarking scheme based on redundant discrete wavelet transform and singular value decomposition'. AEU - Int. J. Electron. Commun. **67**(10), 894–897 (2013). https://doi.org/10.1016/j.aeue.2013.04.013
5. N.M. Makbol, B.E. Khoo, Robust blind image watermarking scheme based on redundant discrete wavelet transform and singular value decomposition. AEU - Int. J. Electron. Commun. **67**(2), 102–112 (2013). https://doi.org/10.1016/j.aeue.2012.06.008
6. A. Mishra, C. Agarwal, A. Sharma, P. Bedi, Optimized gray-scale image watermarking using DWT-SVD and Firefly Algorithm. Expert Syst. Appl. **41**(17), 7858–7867 (2014). https://doi.org/10.1016/j.eswa.2014.06.011
7. G. Bhatnagar, A new facet in robust digital watermarking framework. AEU - Int. J. Electron. Commun. **66**(4), 275–285 (2012). https://doi.org/10.1016/j.aeue.2011.08.005
8. G. Bhatnagar, B. Raman, Q.M.J. Wu, Robust watermarking using fractional wavelet packet transform. IET Image Process. **6**(4), 386–397 (2012). https://doi.org/10.1049/iet-ipr.2010.0400
9. G. Bhatnagar, Q.M. Jonathan Wu, A new logo watermarking based on redundant fractional wavelet transform. Math. Comput. Model. **58**(1–2), 204–218 (2013). https://doi.org/10.1016/j.mcm.2012.06.002
10. G. Bhatnagar, Q.M. Jonathan Wu, Biometrics inspired watermarking based on a fractional dual tree complex wavelet transform. Futur. Gener. Comput. Syst. **29**(1), 182–195 (2013). https://doi.org/10.1016/j.future.2012.05.021
11. P. Pandey, S. Kumar, S.K. Singh, A robust logo watermarking technique in divisive normalization transform domain. Multimed. Tools Appl. **72**(3), 2653–2677 (2014). https://doi.org/10.1007/s11042-013-1577-7
12. R.A. Ghazy et al., Block-based SVD image watermarking in spatial and transform domains. Int. J. Electron. **102**(7), 1091–1113 (2015). https://doi.org/10.1080/00207217.2014.963892
13. S. Rastegar, F. Namazi, K. Yaghmaie, A. Aliabadian, Hybrid watermarking algorithm based on singular value decomposition and Radon transform. AEU - Int. J. Electron. Commun. **65**(7), 658–663 (2011). https://doi.org/10.1016/j.aeue.2010.09.008
14. A. Vaish, M. Kumar, Color image encryption using singular value decomposition in discrete cosine Stockwell transform domain. Opt. Appl. **48**(1), 25–38 (2018). https://doi.org/10.5277/oa180103
15. R.G. Stockwell, A basis for efficient representation of the S-transform. Digit. Signal Process. A Rev. J. **17**(1), 371–393 (2007). https://doi.org/10.1016/j.dsp.2006.04.006
16. R.L. Devaney, *An Introduction to Chaotic Dynamical Systems*, 2nd edn. (Addison-Wesley, 2018)
17. S. Maloo, M. Kumar, N. Lakshmi, A modified whale optimization algorithm based digital image watermarking approach. Sens. Imaging **21**(1), 26 (2020). https://doi.org/10.1007/s11220-020-00291-6
18. C. Das, S. Panigrahi, V.K. Sharma, K.K. Mahapatra, A novel blind robust image watermarking in DCT domain using inter-block coefficient correlation. AEU - Int. J. Electron. Commun. **68**(3), 244–253 (2014). https://doi.org/10.1016/j.aeue.2013.08.018
19. A.M. Abdelhakim, H.I. Saleh, A.M. Nassar, Quality metric-based fitness function for robust watermarking optimisation with Bees algorithm. IET Image Process. **10**(3), 247–252 (2016). https://doi.org/10.1049/iet-ipr.2015.0379

20. A.M. Abdelhakim, M. Abdelhakim, A time-efficient optimization for robust image watermarking using machine learning. Expert Syst. Appl. **100**, 197–210 (2018). https://doi.org/10.1016/j.eswa.2018.02.002
21. V. Sharma, and R. N. Mir, An enhanced time efficient technique for image watermarking using ant colony optimization and light gradient boosting algorithm. J. King Saud Univ. - Comput. Inf. Sci. (2019). https://doi.org/10.1016/j.jksuci.2019.03.009

Image Registration with a Comparative Feature Matching Approach

Biprojit Roy and Geetanjali Oishe

Abstract Image registration is the technique of aligning multiple images that are captured from diverse sources, dissimilar viewpoints, or different times. In general, the image registration algorithms construct a complete image by mapping the source images on the target image by rotation or translation. This mapping is executed by finding the relative transformation. There are several steps to obtain the desired goal. One of the most important steps is feature matching, as based on the obtained result of matched features, the relative transformation is found. In this paper, a comparison among multiple feature matching algorithms has been presented to provide a better understanding of the image registration technique. Additionally, the outcomes after applying those algorithms onto the images have also been presented to show which of them perform better in given circumstances. This can be useful in various fields of computer vision such as remote sensing, automatic target recognition, and medical imaging.

Keywords Image registration · Transformation · Feature matching

1 Introduction

The illustration of a scene can be different if the images are captured in diverse situations like different light conditions, different time periods, by using different sensors, etc. So, it becomes very difficult to gather essential information from those images. The high functioning human eyes may also find trouble to do these jobs. Here, the idea of image registration can be presented to find the key information and combine them as a whole [1]. Back in 2008 Yang Mingqiang and Kpalma Kidiuo proposed the essential characteristics of a shape to be a feature that also contains the idea of

B. Roy (✉)
SCT Bangla Limited, Dhaka 1219, Bangladesh
e-mail: bipro.roy@gmail.com

G. Oishe
Synesis IT Limited, Dhaka 1215, Bangladesh
e-mail: geetanjali.oishe@gmail.com

V. E. Balas et al. (eds.), *Proceedings of International Conference on Computational Intelligence, Data Science and Cloud Computing*, Lecture Notes on Data Engineering and Communications Technologies 62, https://doi.org/10.1007/978-981-33-4968-1_25

translation, rotation, scaling, noise resistance, and reliability [2]. Image registration [3] is very important in the field of image analysis by which the required information is extracted by combining gathered information from different data sources. The goal this research stands for is to provide a strong understanding of different stages of image registration by constructing a scene from the sub-scenes gathered from various sources. In the process of image registration, one of the most significant steps is feature matching. This step is important because the final image we obtain from image registration is constructed by mapping the source images. As a result, it requires the matched parts of the source images which is the result of feature matching. In this paper, more emphasis has been given to feature matching describing different algorithms such as SURF, BRISK, and MSER. The paper presents distinctive methodologies in their individual interest areas. There are two basic categories of image registration, area-based and feature-based image registration [4]. That is why there is a set of algorithms to perform image registration based on the input type and the required output.

2 Grayscale Conversion

As later in the process feature detection and extraction will be performed, first, the images need to be converted into grayscale images. The color images are multidimensional which is not supported by the used algorithms. So, to work with the images, grayscale conversion is required which will convert the images into binary images. Grayscale conversion is done by eliminating saturation and hue and it returns an intensity-based image (Fig. 1).

Fig. 1 **a** Input image 1, **b** Input image 2, **c** Grayscale image 1, **d** Grayscale image 2 [1]

3 Feature Detection Using SURF

In this approach, SURF [5] feature detection is used to detect the features from the images because of its speedy processing method. To speed up the process, integral images are used. The grayscale images have grayscale values that are summed up in a table which is contained by integral images. Gaussian approximation is applied which is basically applying a square shape filter on the integral image.

$$\mathrm{S(x, y)} = \sum_{i=0}^{x}\sum_{j=0}^{y} I(i, j) \tag{1}$$

SURF uses a blob-based detector to find the interest points of the images. The point with the maximal determinant is taken and to measure the change around that point the determinant of the matrix is used.

$$H(\mathrm{P}, \sigma) = \begin{pmatrix} L_{xx}(\mathrm{P}, \sigma) & L_{xy}(\mathrm{P}, \sigma) \\ L_{yx}(\mathrm{P}, \sigma) & L_{yy}(\mathrm{P}, \sigma) \end{pmatrix} \tag{2}$$

At the point x of image $I(x, y)$, the second-order Gaussian derivatives convolution are Lxx (P, σ). To get the upper stage image of the image pyramid, Gaussian filters are applied to smoothen the image. Initially, a 9×9 mask is applied to the image, and after that the filter size is increased to 15×15, 21×21. Then, the interpolation is performed which is a very essential method as the diversity between every octave is comparatively larger than the other (Fig. 2).

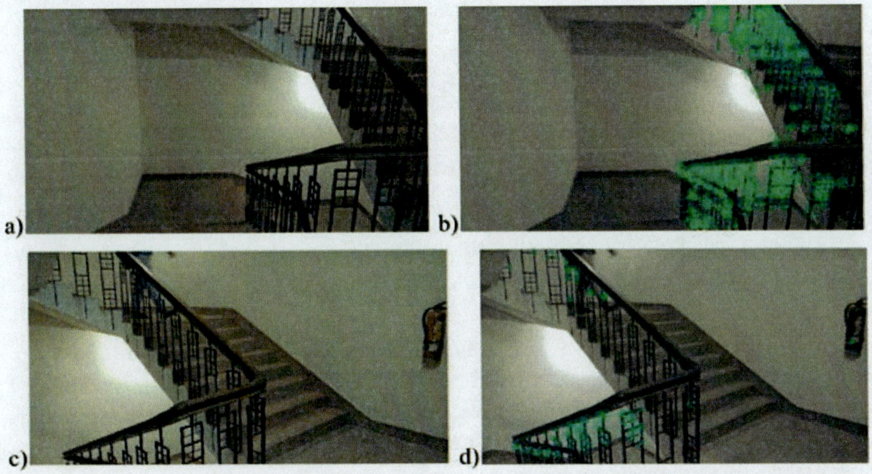

Fig. 2 **a** Input image scene 1, **b** All detected features using SURF, **c** Input image scene 2, **d** 150 strong detected features using SURF [1]

Fig. 3 Point feature (SURF) and corner feature (Harris corner detector) [1]

4 Feature Extraction

The n features of an image are encoded in an $n * 1$ array which is also known as feature vector [6]. The contents of the array can be symbolic, numerical or both. In mathematics, assuming $n =$ number of features and $T =$ transpose operation a numerical feature vector x is measured as:

$$x(x1, x2 \ldots xn)T \tag{3}$$

As the images have a minimum similarity, so the extracted features will help to reduce the interest points and focus only on the data that are necessary. The detected points (SURF algorithm) or corners (Harris corner detection algorithm) [7] are taken as inputs and resulted in feature vectors those are extracted from them, also known as descriptors (Fig. 3).

5 Feature Matching Using SURF, BRISK, and MSER

Recognizing the common features between two images is the main target in this part. The most popular feature matching [8] techniques are:

1. Unique intensity/color configuration of region (Fig. 4)
2. Geometry or topology of region (Figs. 5 and 6).

Due to the inconsistency of data, images have some similarity but are not absolutely same. As different sensors can be involved, so the quality or perspective of the images may differ. The results of the matching of a pair of images using SURF, BRISK, and MSER methods are shown below.

Fig. 4 Feature matching using intensity/color [9]

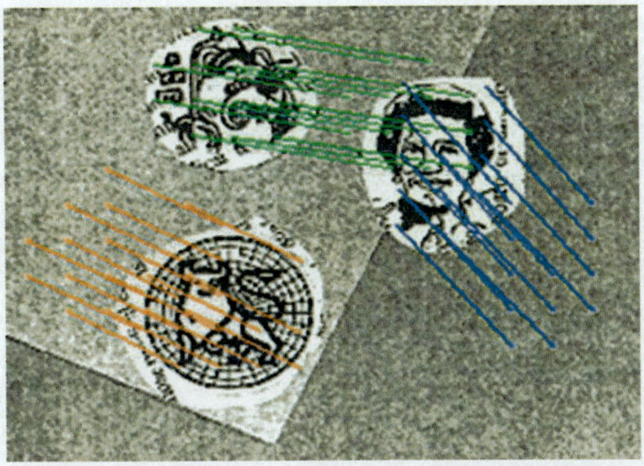

Fig. 5 Feature matching by topology of region (similar vectors are colored the same) [9]

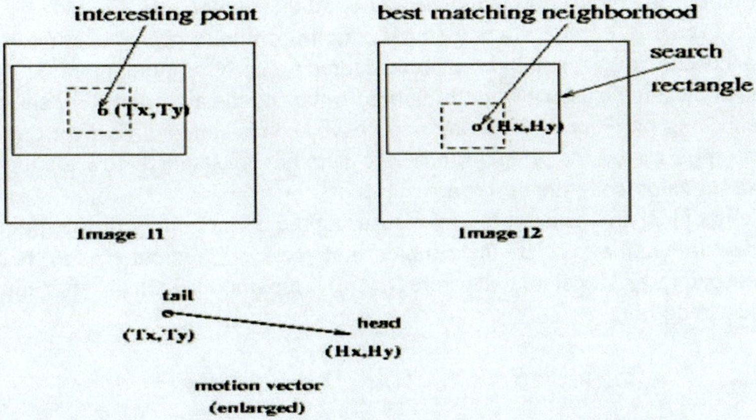

Fig. 6 Searching the best matching neighborhood for features [9]

Fig. 7 Matched features using SURF algorithm [1]

5.1 *SURF Feature Matching*

During feature matching, the sign of the Laplacian (i.e., the trace of the Hessian matrix) for the underlying interest points is included. The interest points are generally found at blob-type structures. The sign of the Laplacian distinguishes bright blobs on dark backgrounds from the reverse situation. As the feature was already computed during feature detection phase, no extra computational cost is needed. The matching stage only compares if the features have the same type of contrast. In SURF, rotational invariance is achieved from the orientation of the point of interest (Fig. 7).

5.2 *BRISK Feature Matching*

BRISK matches two descriptors by computing their Hamming distance. The variation between two descriptors is calculated by the difference of bits between them. Also, the operations are reduced to a bitwise XOR and then a bit count.

BRISK descriptor [10] has a predefined sampling outline where pixels are sampled over concentric rings. A small patch is considered for each sampling point. The patch is smoothed using Gaussian smoothing before starting the algorithm. For sampling, short and long pairs are used. Long pairs have the distance above a set threshold distmin and are used for orientation. Short pairs have distance below distmax and are used for calculating the descriptor.

Rotation in BRISK is achieved by measuring the orientation of the keypoint and using that orientation to rotate the sample. First, the local gradient $g(p_i, p_j)$ between sampling pair (p_i, p_j) is calculated where $I(p_i, p_j)$ is the smooth intensity after applying Gaussian smoothing.

$$g(p_i,\ p_j) = (p_i - p_j).\ I(p_j,\ j) - I(p_j,\ j)p_j - p_i^2 \tag{4}$$

After that, all local gradients between the long pairs are summed and from the sum the arctangent (qy/qx) between y and x components is taken as the angle of the

Fig. 8 Matched features using BRISK algorithm [1]

Fig. 9 Feature matching using MSER [1]

keypoint. For the descriptor to become more invariant of rotation, the short pairs need to be rotated by that angle (Fig. 8).

5.3 MSER Feature Matching

The intensity image (MSER+) and the inverted image (MSER−) are used to compute MSER regions [11]. Measurement regions can be selected at multiple scales: the size of the actual region, $1.5c$, $2x$ and $3x$. As the matching is performed robustly, so the increased distinctiveness of large regions is necessary (Fig. 9).

6 Transform Model Estimation

Estimating the geometric transform from the matched features is the next necessary step that results in a geometrically transformed object which reportedly maps the inliers of the first image in the inliers of the second image.

Then, the M-estimator sample consensus (MSAC) algorithm is used to exclude the outliers. MSAC is an arbitrary algorithm in nature so the results may vary each time. [12, 13].

Assuming w is the number of inliers/points in a particular data, it is not well-known beforehand most of the time, so an approximate value needs to be presumed. And assuming n points are needed to estimate a model, the probability of all the n points being inliers is $w2$ and all n points being outliers are $(1 - w2)$. Then, the chance of never selecting a set of n points would be the probability to the power of k which is same as $1 - p$.

$$1 - p = (1 - wn)k \tag{5}$$

This, after taking the logarithm of both sides, leads to:

$$K = \log(1 - p)/\log(1 - wn) \tag{6}$$

7 Geometric Transformation

This is the final part of image registration. Basically, this is the application of the previous step. Basic parts of transformation are translation, rotation, and scaling. Where translation is changing the position of the object, rotation is angling the object, and scaling is changing the size of the object (Fig. 10).

Geometric transformation has two types, projective and affine transformation. Projective transformation preserves all the lying points after transformation by preserving the co-linearity. However, it does not preserve parallelism. On the other hand, the affine transformation also preserves the parallelism with the same properties of projective transformation. In most cases, projective transformation brings better result as it can avoid the parallelism of affine transformation (Fig. 11).

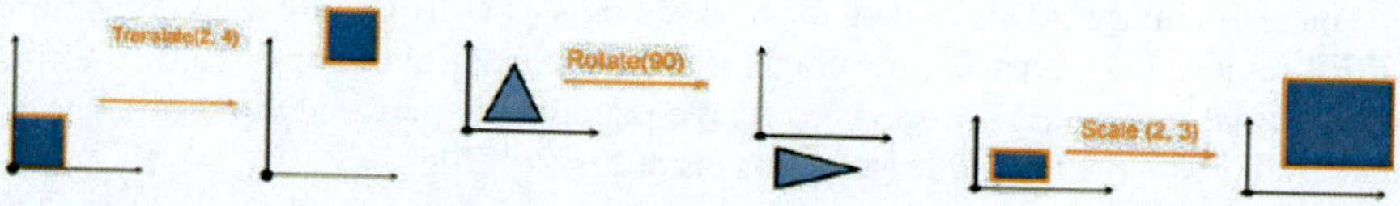

Fig. 10 Translation (left), rotation (middle), and scaling (right) [14]

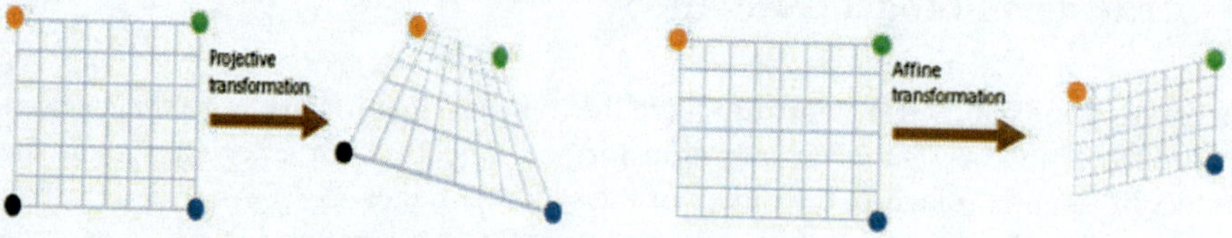

Fig. 11 Projective versus affine transformation [15]

Fig. 12 Geometric transformation (projective)

We have used the projective transform as we tested on multiple situations and the projective derives better result than affine. However, in some cases, where the camera is placed far from the captured image, affine transformation works better. According to Eq. 7, projective transformation contains eight DOF where last parameter is set to 1 so that the matrix can be normalized. Due to having eight DOF, minimum four points are required to complete this transformation [16].

$$H = \begin{bmatrix} h_{11} & h_{12} & h_{13} \\ h_{21} & h_{22} & h_{23} \\ h_{31} & h_{32} & 1 \end{bmatrix} \tag{7}$$

Next step is to calculate image coordinates which are obtained by Eqs. 8 and 9. After obtaining image coordinates, projective transformation is found from the source matched points applying RANSAC [16].

$$x_i = \frac{h_{11x} + h_{12y} + h_{13}}{h_{31x} + h_{32y} + 1} \tag{8}$$

$$x_i = \frac{h_{11x} + h_{12y} + h_{13}}{h_{31x} + h_{32y} + 1} \tag{9}$$

This result shows the transformation of the images (Fig. 12).

8 Result and Analysis

We have applied SURF, MSER, and BRISK feature matching algorithms on the input images along with both affine and projective methods. In order to get the best combination out of them, the execution time and accuracy are measured for each case. What we have found is, all of the mentioned algorithms are rotation invariant. SURF

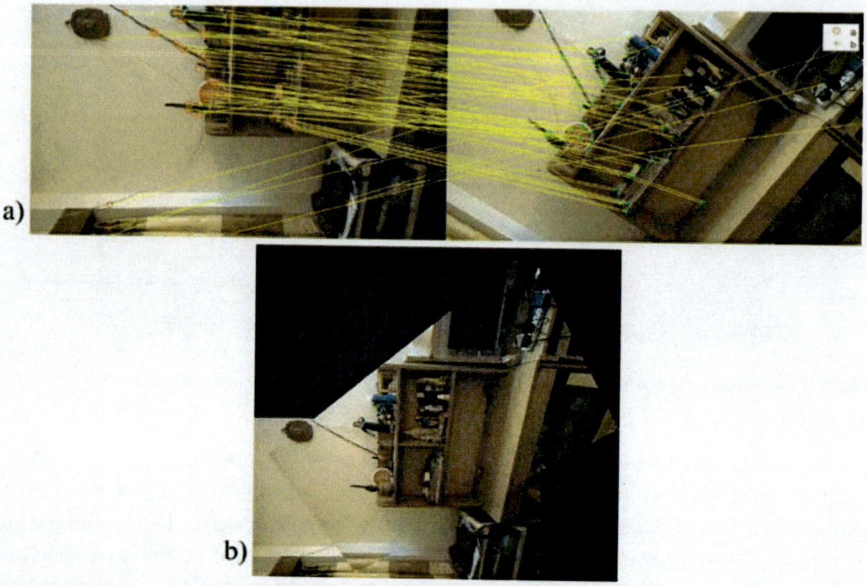

Fig. 13 **a** Matched features of images captured from different viewpoints, **b** Registered output image

feature detection combined with projective transformation gives almost perfect result in most of the cases. Though it takes relatively more time to execute, this combination is more universal than the others and works fine in almost every situation. However, observing the feature detected images, we can find that SURF detector spots a lot of interest points (corners, junctions, blob, etc.) due to noise fluctuation. In order to discard these superfluous interest points, a threshold can be used [17]. The application of MSER performs well in small regions and on images containing homogeneous regions with distinctive boundaries.

Results of SURF feature detection using projective transformation in diverse scenarios (Figs. 13, 14, 15, and 16).

9 Conclusion

This paper primarily demonstrates an elaborate overview of feature matching stage of image registration, applying various algorithms. Additionally, this paper also provides a descriptive overview of the whole image registration process. The main algorithms which were compared for feature matching are SURF, BRISK, and MSER. Based on the diversity of the source images, each algorithm has its own

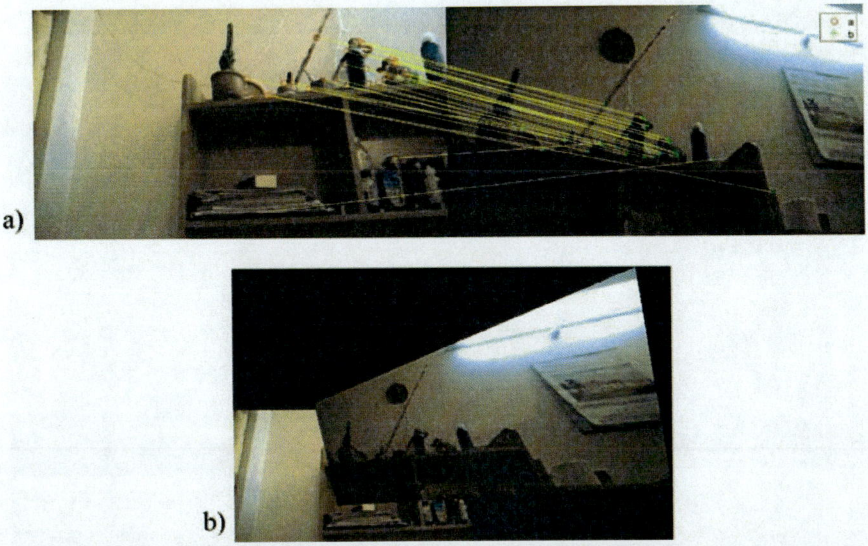

Fig. 14 **a** Matched features of images captured in different lighting conditions, **b** Registered output image [1]

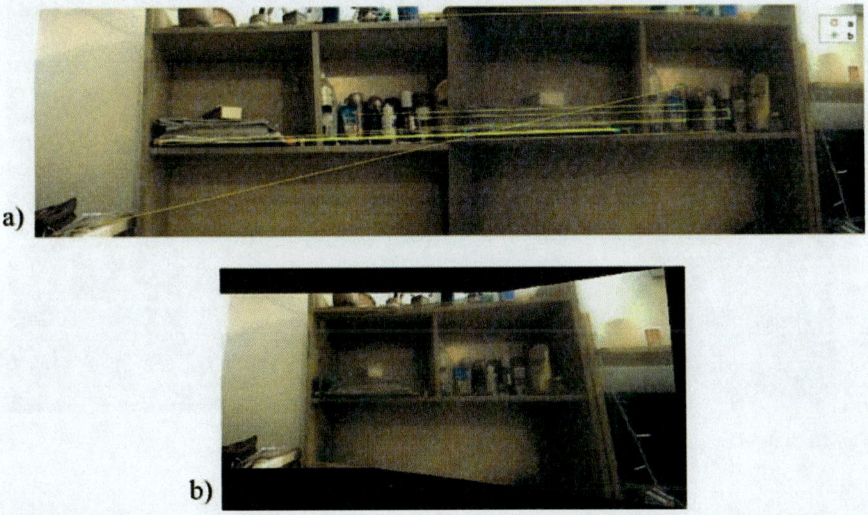

Fig. 15 **a** Matched features of images of different quality, **b** Registered output image [1]

a)
b)
c)
d)
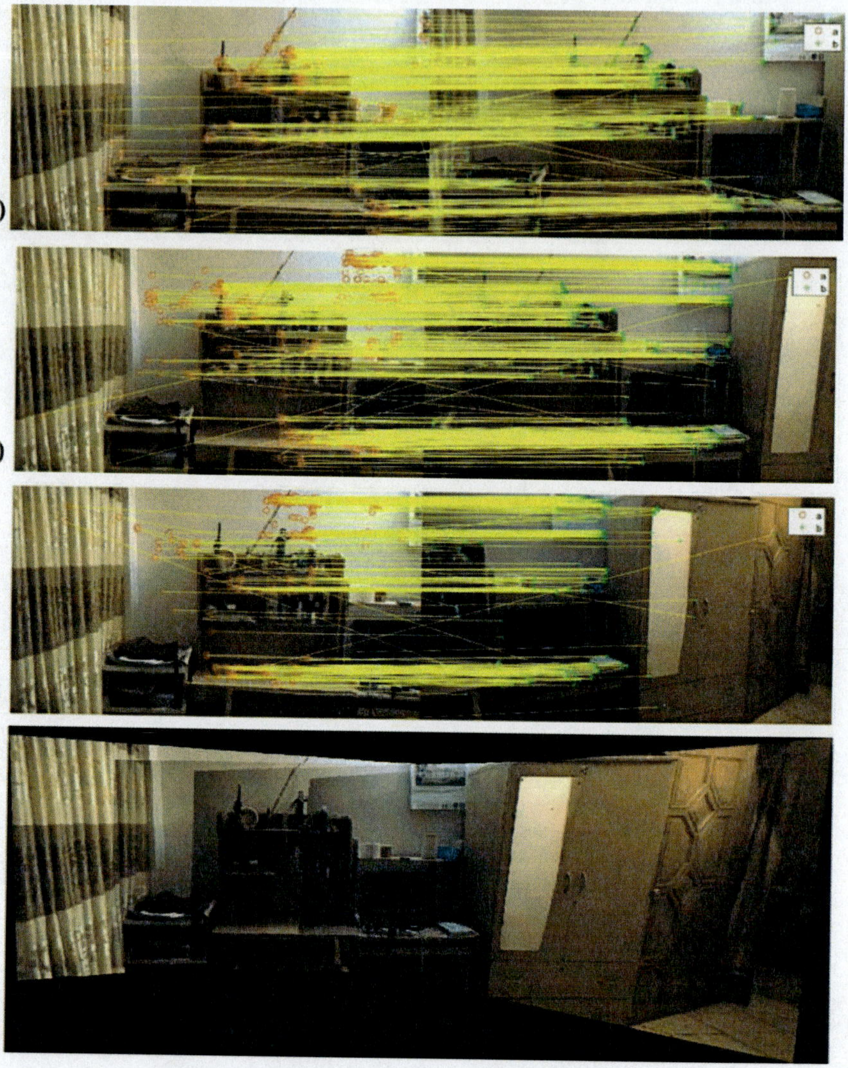

Fig. 16 **a**, **b**, **c** Matched features from four input images, **d** Registered output image [1]

benefits and drawbacks. The obtained output image was not always enhanced in quality as diverse sources of inputs were present and enhancement of an image is also over-looked as registering the image with different feature matching algorithms was the main objective.

References

1. G. Oishe, B. Roy, M.H. Ali, M. Mostakim, An improved image registration technique for shape reconstruction. Undergraduate Thesis, BRAC University, Dhaka Bangladesh. Retrieved from https://hdl.handle.net/10361/8244 (2017)
2. M.V. Wyawahare, P.M. Patil, H.K. Abhyankar, Image registration techniques: an overview. Int. J. Signal Process. Image Process. Pattern (2009)
3. M. Yang, K., Kpalma, J. Ronsin, A survey of shape feature extraction techniques, in *Pattern Recognition*, ed. by P.-Y. Yin (2008), pp. 43–90
4. A. Aichert, Feature extraction techniques. Camp Medical Seminar WS0708 (2008)
5. H. Bay, A. Ess, T. Tuytelaars, L.V. Gool, Speeded-up robust features. Comput. Vis. Image Underst. **110**, 346–359 (2008)
6. S. Lim, K. Lee, O. Byeon, T. Kim, Efficient iris recognition through improvement of feature vector and classifier. ETRI J. **23**(2) (2001)
7. C. Harris, M. Stephens, *A Combined Corner and Edge Detector* (Plessey Research Roke Manor, United Kingdom, 1988).
8. A. Baumberg, Reliable feature matching across widely separated views. IEEE Conf. Comput. Vision Pattern Recog. https://doi.org/10.1109/CVPR.2000.855899. Print ISSN: 1063-6919. INSPEC Accession Number: 6651714 (2000)
9. G. Stockman, Point/feature matching methods, https://www.cse.msu.edu/~stockman/RegTutorial/matchingPart1.ppt (2004)
10. S. Leutenegger, M. Chli, R. Siegwart, BRISK: binary robust invariant scalable keypoints
11. Region detectors, https://www.micc.unifi.it/delbimbo/wp-content/uploads/2011/03/slide_corso/A34%20MSER.pdf, last accessed 18 July 2020
12. R. Hartley, A. Zisserman, *Multiple View Geometry in Computer Vision* (Cambridge University Press, Science and Engineering Library, 2003).
13. P.H.S. Torr, A. Zisserman, MLESAC: a new robust estimator with application to estimating image geometry. Comput. Vis. Image Underst. (2000)
14. 2D Geometrical Transformations, https://www.cs.brandeis.edu/~cs155/Lecture_06.pdf, last accessed 17 May 2017
15. Affine and Projective Transformations, https://www.graphicsmill.com/docs/gm5/Transfor-mations.htm, last accessed 29 Mar 2017
16. R. Redzuwan, N.A.M. Radzi, N.M. Din, I.S. Mustafa, Affine versus projective transformation for SIFT and RANSAC image matching methods. IEEE Int. Conf. Signal Image Process. Appl. (2015). https://doi.org/10.1109/icsipa.2015.7412233
17. E. Oyallon, J. Rabin, An analysis of the SURF method. Image Process. Line **5**, 176–218 (2015)

An Improved Decision Tree Classification Approach for Expectation of Cardiotocogram

G. Ravi Kumar, K. Venkata Sheshanna, S. Rahamat Basha, and P. Kiran Kumar Reddy

Abstract Cardiotocography (CTG) is a synchronous account of fetal pulse (FHR) and uterine withdrawals (UC). It is one of the most widely recognized analytic strategies to assess maternal and fetal prosperity during pregnancy and before conveyance. By watching the Cardiotocography follow designs specialists can comprehend the condition of the embryo. The motivation behind this investigation is to assess the exhibition of decision tree classifier with the quality determination measures for cardiotocogram information expectation. Upheld measures are "Gini" for the Gini index and "entropy" for the information gain. Yields of a cardiotocogram are normally deciphered as having a place with one of three states: physiological, dubious and neurotic. Programmed order of these states dependent on cardiotocographic information is the objective of this paper. From the result examination we can reason that the execution of decision tree arrangement relies upon the trademark property determination decision measures. Exploratory outcomes have uncovered that the decision tree classifier with entropy delivered palatable outcomes, and accomplished better outcomes with an accuracy of 86.73%. An examination of the decision tree with the entropy and gini choice measures is likewise given.

Keywords CTG · UC · FHR and decision tree

G. R. Kumar (✉) · K. V. Sheshanna
Department of Computer Science, Rayalaseema University, Kurnool, India
e-mail: grkondaravi@gmail.com

S. R. Basha
Department of CSE, RGM College of Engineering and Technology, Nandyal, India

P. K. K. Reddy
Department of CSE, MRIET, Hyderabad, Telangana, India

V. E. Balas et al. (eds.), *Proceedings of International Conference on Computational Intelligence, Data Science and Cloud Computing*, Lecture Notes on Data Engineering and Communications Technologies 62, https://doi.org/10.1007/978-981-33-4968-1_26

1 Introduction

Lately, it has been discovered that huge measure of illness determination. Information mining procedures have been applied to extricate information from this clinical information with the goal that illness expectation turns out to be simple [1, 2]. Cardiotocography (CTG) the most well-known procedure to watch fetal wellbeing. Cardiotocography (CTG) is a synchronous account of fetal pulse (FHR) and uterine constrictions (UC) and it is one of the most widely recognized symptomatic methods to assess maternal and fetal prosperity during pregnancy and before conveyance [3, 4]. FHR designs are watched physically by obstetricians during the procedure of CTG examinations. Calculation and other information mining strategies can be utilized to examine and group the CTG information to stay away from human missteps and to help specialists to take a choice. There are a few signs preparing and Computer programming-based methods for deciphering average Cardiotocography information. The fundamental point of the current investigation is to decide the most productive machine learning method to arrange fetal pulse signals [5, 6]. This paper shows the significance of utilizing attribute evaluator techniques for decision tree to choose the characteristics with high effect on the dataset that give more commitment to the accuracy.

The paper is sorted out as follows; Sect. 2 presents data about CTG gadget and its job in diagnosing fetal status. Segment 3 gives the proposed model, calculation and its advantages. Area 4 presents an exploratory consequence of the model and investigation of other arrangement procedures in correlation. Toward the end, end and future work are recorded in segment 5.

2 Cardiotocography (CTG)

2.1 A Subsection Sample

Cardiotocography (CTG) the most well-known strategy to watch fetal wellbeing. It is a mix of two signs: Fetal Heart Rate (FHR) and Uterine Contractions (UC). It is one of the most widely recognized symptomatic methods to assess maternal and fetal prosperity during pregnancy and before conveyance. By watching the Cardiotocography follow designs specialists can comprehend the condition of the embryo. The CTG examination done by obstetricians during FHR design perception helps in perceiving fetal state, for example, physiological, suspect, and pathological [7, 8]. Thus, flourishing of undeveloped organism can be envisioned and taken consideration in advance. Cardiography is a successful and non-intrusive procedure for assessing the fetal wellbeing. The unborn baby pulse and the mother's uterine constrictions are recorded on paper like ECG. This procedure is practical and can prompt early recognizable proof of neurotic states, for example, intrinsic heart deformities, hypoxia or fetal pain. The hazard conditions can be distinguished in beginning phases with the goal that the obstetrician can mediate to give therapeutic measures before more harm is done to the creating baby. There are a few sign handling and PC programming-based strategies for deciphering a run of the mill Cardiotocography information.

It is utilized to decide the fetal state both during pregnancy and conveyance. It is a blend of two signs: Fetal Heart Rate (FHR) and Uterine Contractions (UC). CTG is utilized to decide those children who need oxygen during work. At that point choice is made by the obstetricians whether to convey the infant by caesarean segment or by common birth. Visual examination of CTG information frequently prompts inaccurate understandings and consequently Computer programming based helped frameworks are fundamental in characterizing the recorded CTG information for dynamic by the obstetricians [9].

3 Decision Tree

Decision trees are one of the most broadly utilized procedures for classification problems. The key favorable position of Decision trees over different strategies is that they are truly interpretable, and in numerous applications, for example, manufacturing, financial analysis, medicinal services, this interpretability is frequently favored over different techniques that may have higher exactness yet are generally un-interpretable [10–12].

They construct order models as a treelike structure with inner roots otherwise called possibility nodes speaking to qualities, branches demonstrating results and leaf nodes or end nodes giving the class marks. The primary principle node called root or decision node speaks to the entire example. The way from root to leaf gives the characterization rule. A decision tree makes a preparation model which predicts class or estimation of target factors by choice guidelines that are deduced from the preparation information gave.

The objective is to assemble a model from a lot of models related with the classes to discover a depiction for every one of the classes from the regular properties between the models [9, 13]. When this model has been constructed, we can remove a lot of grouping rules. In this model, the removed guidelines are then used to order new items whose class is obscure. The grouping is finished by voyaging away from the root to a leaf. The class returned is the one that is generally visited among the models on the sheet. At each inner hub of the tree, there is a test which compares to a characteristic in the learning base and a branch relating to every one of the potential estimations of the property. At each leaf node, there is class esteem. A way from the root to a node in this way compares to a progression of traits with their qualities. This flowchart-like structure with recursive dividing helps client in dynamic. It is this perception, which effectively imitates the human-level reasoning. That is the reason decision trees are straightforward and decipher.

The main test is to choose the right property as the root node and inner nodes. To discover the credit that is to turn into the root node the many property selectors can be utilized. Well known measures are information gain, gain ration and gin index. In

Information gain the trait with most elevated data gain is picked as root. The means associated with building a tree are splitting, stopping, and pruning. The roots are distinguished utilizing qualities, for example, Gina Index, Gina Ratio or Information increase starting from the root node. Parting is done at the root node and ensuing inward hubs are made and are preceded till a halting standard is met.

3.1 Information Gain or Entropy

Information gain is the estimation of changes in entropy after the division of a dataset dependent on a characteristic. It indicated by the estimation of information gain, we split the node and fabricate the decision tree. It computes how much data an element gives us about a class. A decision tree calculation consistently attempts to amplify the estimation of information gain, and a node having the most noteworthy information gain is part first [11]. One way to deal with measure degrees is using entropy:

$$\text{Entropy} = -\sum_{i=1}^{n} pi \ \log 2 \ pi$$

where *pi* is the non-zero likelihood that a subjective tuple in *D* has a place with class *Ci* and is evaluated by $|Ci, D|/|D|$. A log capacity to the base 2 is utilized, in light of the fact that the data is encoded in bits.

3.2 Gini Index

Gini index is a proportion of polluting influence or virtue utilized while making a decision tree in the CART (Classification and Regression Tree) calculation. A characteristic with the low Gini list ought to be favored when contrasted with the high Gini list [9, 11]. There is one progressively metric which can be used while building a decision tree is Gini Index. Gini record evaluates the contaminating impact of a data fragment *D*, formula for Gini Index can be recorded as:

$$Gini(D) = 1 - \sum_{a}^{b}$$

where *pi* is the likelihood that a tuple in *D* has a place with class *Ci* and is assessed by $|Ci, D|/|D|$. The aggregate is figured over m classes.

4 Experimental Results

4.1 Dataset

The CTG dataset comprises of estimations of FHR and UC for the embryo, the significant highlights of cardiotocograms grouped by an obstetricians' master, and the informational collection is accessible openly at the information mining store of University of California Irvine (UCI) [14]. Informational collection was part into preparing information and testing information with rates 70 and 30% separately. The informational index has 21 ascribes and characterized by the FHR design or fetal state class code. In this investigation, fetal state class code is utilized as the objective trait rather than FHR design class code and characterization into one of three gatherings Normal, Suspicious or Pathological (NSP) classes. The dataset incorporates an aggregate of 2126 examples. Obstetricians could group the condition of fetal as should be expected, pathologic or dubious state as indicated by estimations of its highlights.

4.2 Result and Discussion

In our test, we tuned the hyper boundaries for characteristic determination measure rules, including the greatest depth and number of least leaves. Depth of the decision tree is used to add most extraordinary significance to the decision tree after the tree is broadened, least number of leaves is limit is used to incorporate the base number of tests required to be accessible at a leaf node. Picking the correct target work for the correct issue is critical. Fitting the model will necessitate that the quantity of most extreme depth and least number of leaves to be determined. We experiment the decision tree model with most extreme depth is set to 12 and least number of leaves is set to 5. We utilized the two characteristic determination measure rules i.e., entropy and Gini rules of decision tree to assess the exhibition.

The outcomes are introduced utilizing the grouping precision in rate for inside train information and test information independently as appeared in Figs. 1 and 2. The order is done between two quality determination standards of information at once; the outcomes were assessed utilizing arrangement exactness for train and test information. We watch the order accuracy in Figs. 1 and 2 for both train and test where, accuracy improves as entropy strategy is ideal parting measures. The outcome shows the reliance of the assessment precision on the parting standards for both preparing and testing information. Order precision for train information under entropy is 84.65%, while Gini criteria has got 80.36% and the characterization exactness for test information under entropy is 86.73% while Gini rules has accomplished 82.54% exactness. So, in the two cases entropy parting models got most noteworthy exactness of the decision tree grouping.

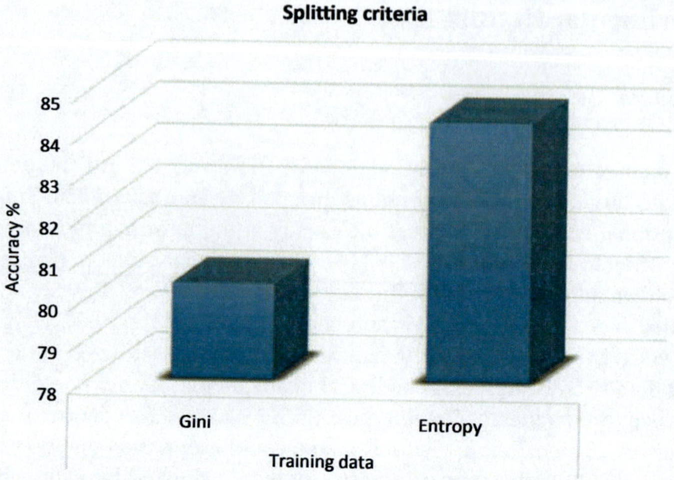

Fig. 1 Performance of training data

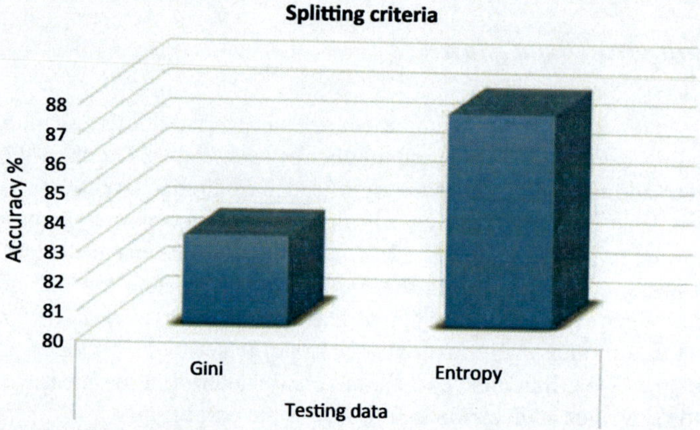

Fig. 2 Performance of testing data

5 Conclusion

In this paper, we actualize a model based CTG information arrangement framework utilizing a decision tree which can order the CTG information dependent on its preparation information with two standard characteristic choice proportions of Gini and entropy. Request precision is significantly destitute upon the Decision Tree is trademark assurance measure rules for smoothing out. The goal to get high accuracy of gauge is fulfilled by Decision Tree using Entropy measure. As per the showed-up results, the presentation of the decision tree with entropy boundary-based characterization approach gave noteworthy execution. It was discovered that, the decision tree-based classifier was fit for distinguishing Normal, Suspicious and Pathologic condition, from the idea of CTG information with generally excellent exactness.

References

1. S.B. Rao, S.R. Basha, G. Ravi Kumar, A comparative approach of text mining: classification, clustering and extraction techniques. J. Mech. Continua Math. Sci. Special Issue (5), 120–131
2. G.R. Kumar, S. R. Basha, S.B. Rao, A summarization on text mining techniques for information extracting from applications and issues. J. Mech. Continua Math. Sci. Special Issue (5), 324–332
3. Z. Alfirevic, D. Devane, G.M. Gyte, Continuous cardiotocography (CTG) as a form of electronic fetal monitoring (EFM) for fetal assessment during labour. Cochrane Database, 1:CD004116, Syst Rev (2006)
4. D. Ayres-de-Campos, J. Bernardes, A. Garrido, J. Marques-deSa, L. Pereira-Leite, Sis Porto 2.0: A program for automated analysis of cardiotocograms. J. Mater. Fetal Neonatal. Med. **9**, 311–318 (2000)
5. S.R. Basha, J.K. Rani, J.J.C. Prasad Yadav, A novel summarization-based approach for feature reduction, enhancing text classification accuracy. Eng. Technol. Appl. Sci. Res. **9**(6), 5001–5005 (2019). ISSN 1792-8036
6. S.R. Basha, J.K. Rani, J.J.C. Prasad Yadav, G.R. Kumar, Impact of feature selection techniques in text classification: an experimental study. J. Mech. Continua Math. Sci. Special Issue (3), 39–51 (2019). ISSN: 0973-8975
7. Y. Zhang, Z. Zhao, Fetal state assessment based on cardiotocography parameters using PCA and AdaBoost, in *2017 10th International Congress on Image and Signal Processing, Biomedical Engineering and Informatics (CISP-BMEI)*. IEEE
8. M.L. Huang, Y.Y. Hsu, Fetal distress prediction using discriminant analysis, decision tree, and artificial neural network. J. Biomed. Sci. Eng. **5**(9), 526–533 (2012)
9. P.N. Tan, M. Steinbach, V. Kumar, *Introduction to Data Mining* (Addision-Wesley, 2005)
10. G.R. Kumar, V.S. Kongara, Dr. G.A. Ramachandra, An efficient ensemble based classification techniques for medical diagnosis. Int. J. Latest Technol. Eng. Manage. Appl. Sci. **II**(VIII), 5–9 (2013). ISSN: 2278-2540
11. J. Han, M. Kamber, Data mining concepts and techniques, in *The Morgan Kaufmann series in Data Management Systems*, 2nd edn. (Morgan Kaufmann, San Mateo, CA, 2006)
12. I.H. Witten, E. Frank, *Data Mining: Practical Machine Learning Tools and Techniques*, 2nd edn. (Morgan Kaufmann, San Francisco, 2005).
13. S.R. Basha, J.K. Rani, A comparative approach of dimensionality reduction techniques in text classification. Eng. Technol. Appl. Sci. Res. **9**(6), 4974–4979 (2019). ISSN: 1792-8036
14. UCI Machine Learning Repository. https://archive.ics.uci.edu/ml/

Helper Gloves for Differently Abled Patients Using Arduino

Namrata Banerjee, Apurba Kumar Ghosh, and Darothi Sarkar

Abstract Helper Gloves assists patients with hearing and speech impairment to convey their medical needs to medical workers. This project consists of a mobile application which enables the patient to convey their needs to medical helpers and control home appliances from any part of the room. Also, the application helps medical workers to update health information of patients for doctors to see in real time. Altogether this project aims to build an user-friendly application which cuts off the hindrances in communication in aspect of medical requirements for disabled patients.

Keywords Deaf and dumb patient · Microcontroller-based system · Smart glove · Android application · Room automation · Health gadget

1 Introduction

The vast population which India has also constitutes of its 2.4 million deaf and dumb population [1]. People born with disabilities like hearing impairment eventually end up speechless due to failure of learning verbal communication. Most of the times, these people find fluid communication, a "luxury" which is a daily comfort for normal people like us. This trouble aggravates more when they are hospitalized or stricken by other health issues and in more need of assistance. Conveyance of needs becomes paramount. And inability of normal people to always understand their sign language is what creates an obstacle just like language barrier between the patient and medical workers. This

N. Banerjee (✉) · A. K. Ghosh · D. Sarkar
Computer Science and Engineering Department, Institute of Engineering and Mangaement, Kolkata, India
e-mail: banerjeen31@gmail.com

A. K. Ghosh
e-mail: basic6211@gmail.com

D. Sarkar
e-mail: darothisarkar@rediffmail.com

V. E. Balas et al. (eds.), *Proceedings of International Conference on Computational Intelligence, Data Science and Cloud Computing*, Lecture Notes on Data Engineering and Communications Technologies 62, https://doi.org/10.1007/978-981-33-4968-1_27

gloves along with the mobile application can succour such a person in need to call other people in charge of the patient and deliver essential requirements wanted by them. It also eases the patients' need to call for assistance to control the room appliances. Immediate health log update feature via Google sheet when nurses add any new information about the patient's health so that doctors can see the information in real time is also another aspect of the mobile application.

Our work involves wearing the gloves by the needy and contingent on the change in flow of electricity in the buttons as pressed by the patient, an integer value linked with each particular button is communicated to the Android application which in turn converts it logically into voice/message and in texts and visuals. The device senses the change in electricity flow carefully and executes the different predefined functions accordingly. At this stage, the predefined functions loaded in the device are limited but can be enhanced by adding more in the future development of the project.

Google Sheets have been used in this project for the information updation aspect where data entry takes place through the application interface and passed on through Google Forms to Google Sheet. For room automation, wireless communication between the embedded system and the mobile application is implemented.

1.1 Literature Review

Variety of models of smart gloves had been developed with the same intention to cater to the advantage of disabled people. Many wireless means have also been incorporated to the systems to accomplish the same. Slight impediment to all these systems is their complications and size. Devices pinpointing the problems faced by disabled patients solely are not abundant and more recognized for generalized purposes.

In the year 2001, Ryan Petterson was the first person to design Hand Talk glove. He used his own sign language for the purpose [2, 3]. His creation consisted of a golf leather glove attached with ten flexible sensors whose position could be measured when a change during movement of the hand was detected over the shift in placement of the fingers. The embedded microcontroller at the back of the hand converted the alterations in electrical current to digital value and channelled the value wirelessly to the computer [4]. The major drawback of that system was its mobility due to the computer.

For clearer communication between the speech and hearing impaired people in their mainstream sign language, a student from Goldsmith University developed a system which involved an expressive hand glove called "Sign Language Glove" that converted gestures made by the hand of the user into visual and audible dialogue [5].

A sensor-equipped gloves brought to light by some students named Bhargav Bhat, Hemant Sikaria, Jorge Meza and Wesley Jin of Carneige Mellon University. It demonstrated the conversion of hand gestures into voiced words [6]. This simplistic model made use of Bluetooth to transmit the data for different hand gestures with a limited vocabulary of 32 words and was the first of its kind. The output was made through a cellphone. This co-operative glove was able to bring down the communication barrier between deaf and speech impaired person and people.

All the above gloves are solely meant for communication purposes, but our project gives the fusion of Arduino-based room automation as well as communication in a much simpler form for patients centric problems.

Problems faced by the disable patients regarding this can be overcome by our method. Our work performs as a compact articulate microcontroller-based system and covers the following functions:

- Impart the requirements of the patients into voice, text and visuals.
- Help a person to control his room appliances if he could not walk to switchboard.
- Patient information updation through Android app in real time.
- Compact design.

1.2 Objectives

The project aims at the following results

To uproot the communication hesitation of speech impaired patient and cater to their medical requirements [7].

Help medical workers understand the requirement of patients and provide real-time patient information to doctor in remote location through Google Sheets.

To find suitable technique to implement a home control system [8] with affordable price.

Idea of having control over home environment from anywhere within the house and at any time. To design low-cost system with user-friendly functionality [9].

To benefit the ones with joint pains or conventional patients with disorders and also make the life of others more comfortable.

2 Methodology

2.1 Block Diagram

The block diagram depicts the technical architecture of the project. Hardware used are as follows—Arduino Nano, Arduino UNO R3 Bluetooth Module HC05433Mhz Rf Module (transmitter and receiver), Android Smartphone Buttons, Resistors (1 K), 1N4007 Diode, DC fan, Lamp (230 V).

Relay (12 V, 240 Coil Resistance), Alternator (AC Voltage Source, 230 V, 50 Hz), Silicon NPN Low Power High Frequency Bipolar Transistor (500 mW, 200 °C), DC Generator (12 V), Power Supply for Arduinos (5 V), Power Supply for home appliances.

Software used in this project are—Arduino IDE, MIT App Inventor, Proteus 8 Professional, Google Sheets.

Here, we are accomplishing our goal by diving the project into two modes of working—MODE 1 and MODE 2. Below are the block diagrams for the same (Figs. 1 and 2).

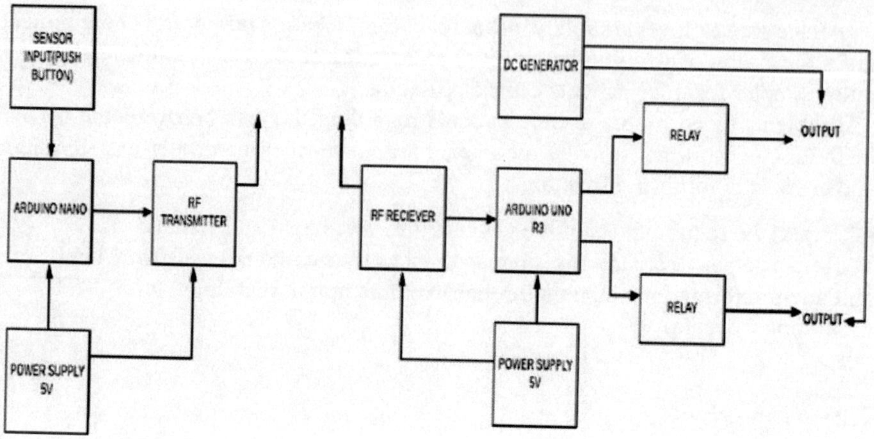

Fig. 1 Block diagram for MODE 1

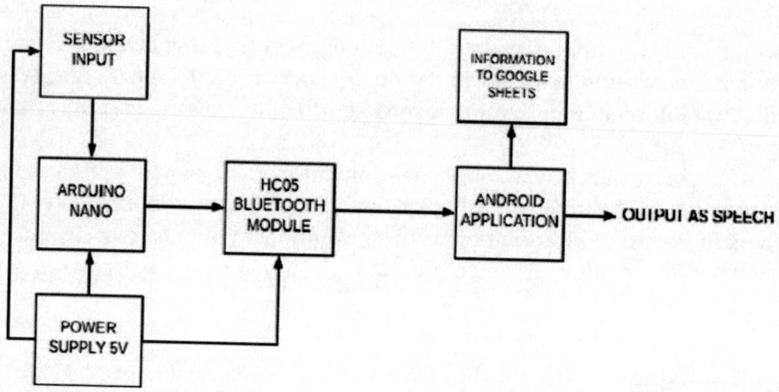

Fig. 2 Block diagram for MODE 2

2.2 *Proposed Strategy*

See Fig. 3.

2.3 *Flowchart*

See Figs. 4 and 5.

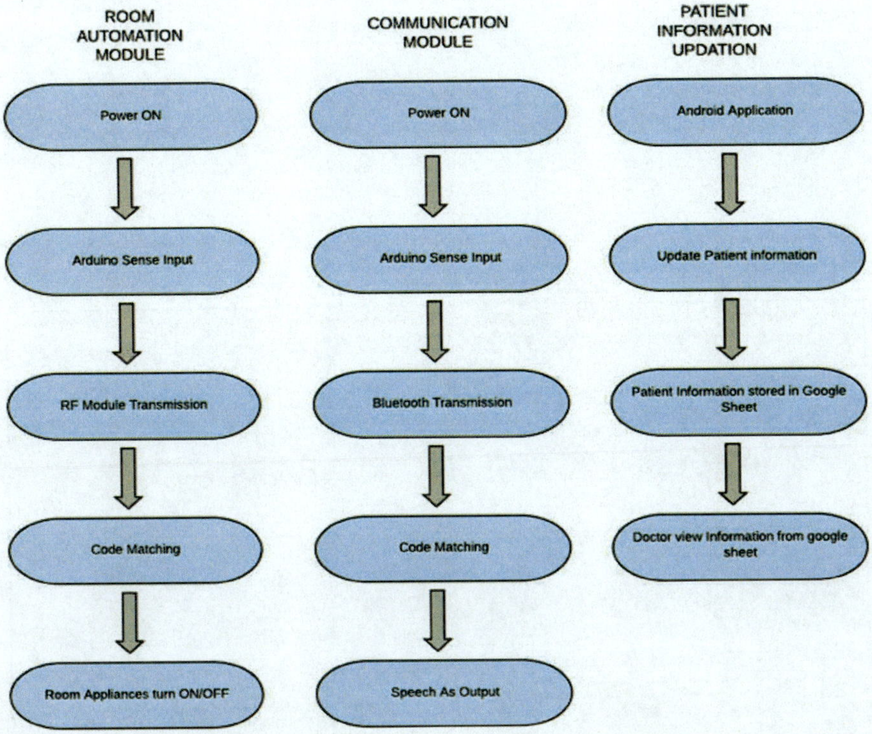

Fig. 3 Proposed strategy

2.4 *Working Principal*

The patient wears the glove and five push buttons acting as sensors are wired through the inside of the fabric of the gloves. Every push button is assigned INPUT_PULLUP value where it makes use of its internal resistance and remains as HIGH INPUT in normal unpressed condition. The push button which is assigned as the change mode button when pressed for the first time activates MODE 1, and when pressed the next time, shifts to MODE2. This algorithm based on alternative mode is implemented to change modes in the two-mode system.

The total processing and functionality of the project is divided into two modes:

MODE 1(room automation mode):

Arduino Nano is employed to assemble signals from the sensors placed on the glove. Then the processed output is send over to the Arduino UNO with the help of RF Module, and depending on the Arduino Nano output which acts as the input of Arduino UNO, room appliances are accordingly turned ON and OFF.

Working principle for MODE1—Once change mode button is assigned to MODE 1, Arduino Nano senses drop in voltage for the remaining pins and if that happens,

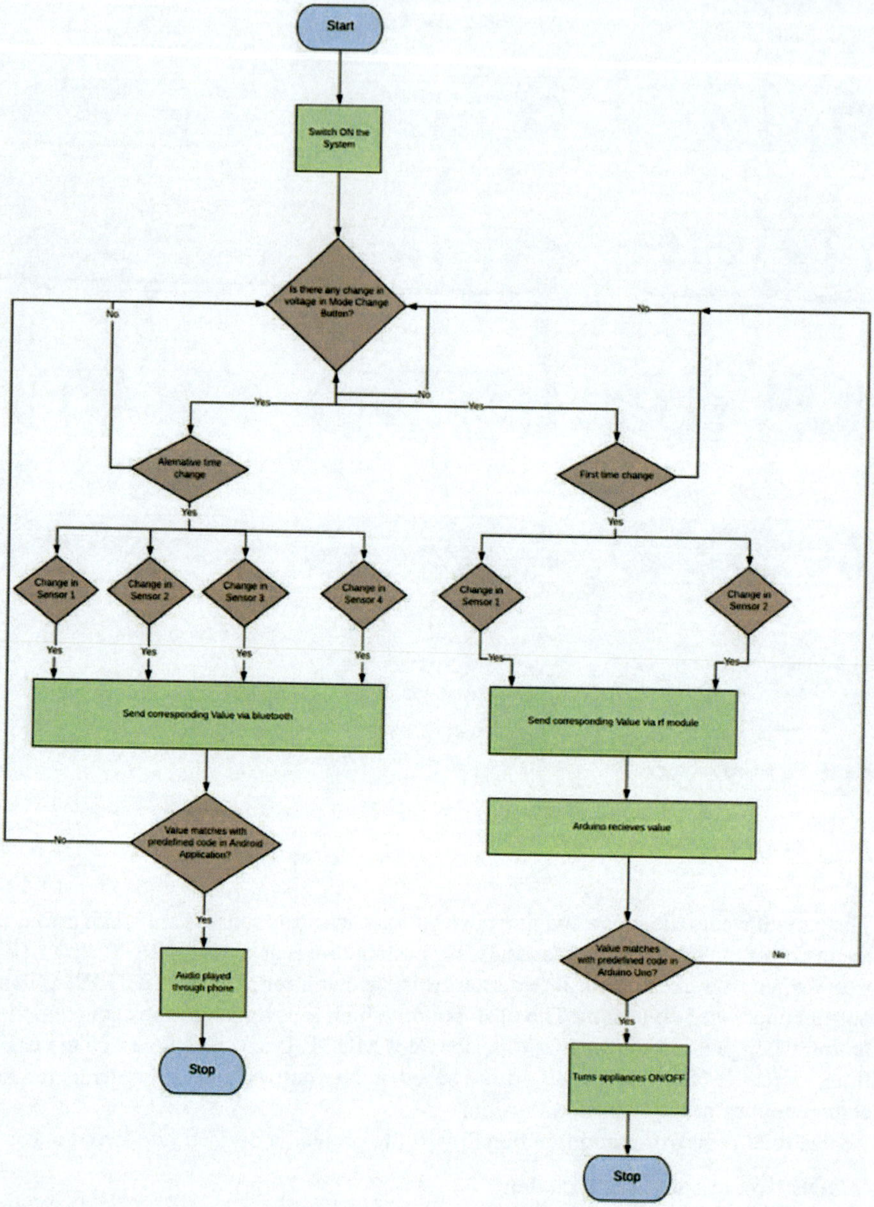

Fig. 4 Flowchart for the proposed system

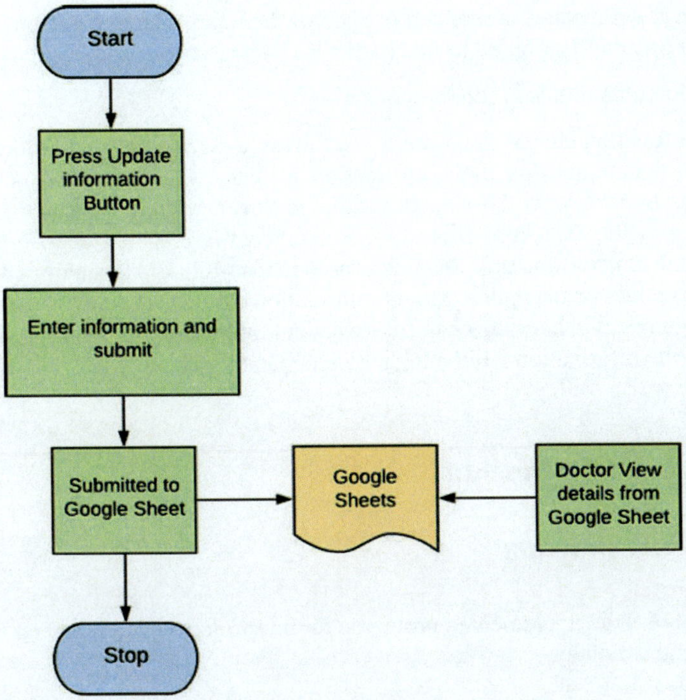

Fig. 5 Flowchart for the information updation functionality

sends predefined integer values as set in the code for each particular pin, and corresponding room automation functionality to RF transmitter which transmits it to the RF receiver attached to the Arduino UNO R3.On receiving integer values, Arduino UNO R3 matches the value to the functionalities set in the UNO R3 code and makes the digital output for preassigned pins to the integer HIGH. Hence switching ON the light and fan with help of relay, DC generator, alternator, diode, transistors and resistors. Same push buttons are used as toggle switch to switch off the previously ON appliance.

MODE 2(communication mode):

The processed output by the Arduino Nano to the MIT app to display the text and output associated conjointly via a Bluetooth link to an Android smartphone or a private laptop consisting of text to speech package (application) and speech output is obtained.

Working principal of MODE 2—Once the change mode button is assigned to MODE 2, Arduino Nano senses drop in voltage at the remaining pins, and if voltage drops at certain pin, send preassigned integer value to the Android application through Bluetooth HC05 Module as serial input. Android application on receiving the value, compared it with values predefined in code. If there is a match, it performs the text to speech operation assigned to that particular integer in the code.

Information updation in Android application

The third functionality of the system is allowing assigned medical workers of the patient to update and log their information so that doctor can view them in real time. This is done with the help of a Google Sheet, address of which is already predefined in the MIT App Inventor code. A form is prepared depending on the required information and prefilled link is used in the MIT App Inventor code as well. When user clicks on the update patient information button in the application, they are moved to a screen where they can fill up patient information and submit thereafter. Instantly, the information will be logged in the Google Sheet.

3 Practical Implementation

3.1 Circuit Diagram

Figure 6 shows the implemented prototype for our project which is run on Proteus 8 Professional simulator.

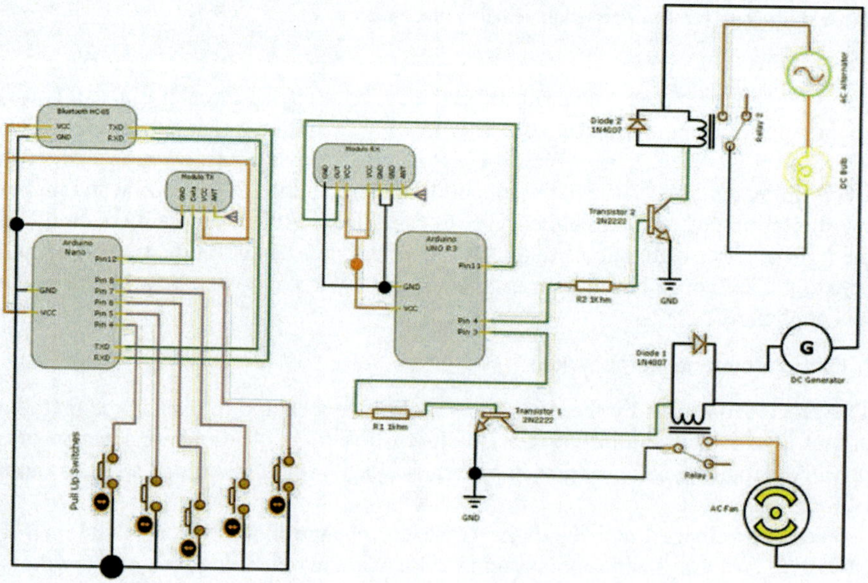

Fig. 6 Circuit diagram

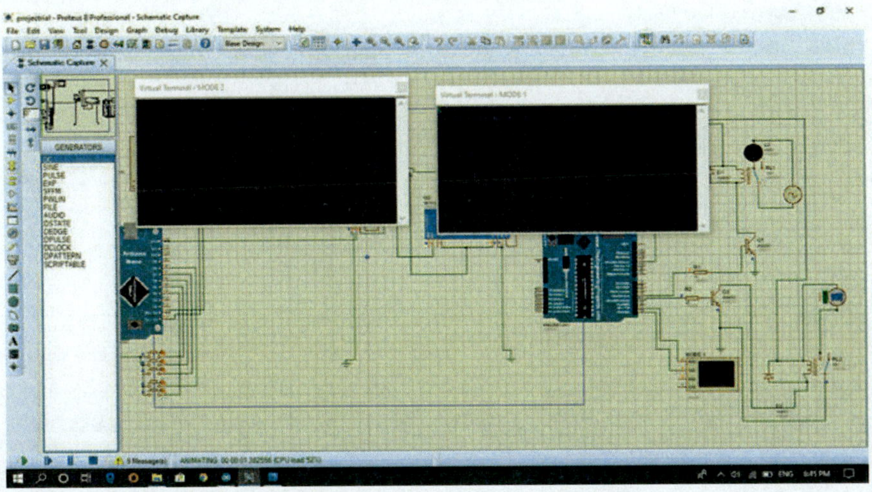

Fig. 7 Proteus light ON simulation

4 Results

The desired results which were sighted to solve the problems of a disabled patient were attained by running the project after assembling it.

Mode 1 Results (Room Automation)

Light ON—here in simulator, since R1 is showing RED, and Relay 1 is having a closed connection, light is ON.

Same switch is used to turn ON/OFF the light. Once the Arduino receives corresponding value through rf transmitter, it makes the output of pin 4 HIGH. Hence, light is turned ON (Fig. 7).

Fan ON—here in simulator, since R2 is showing RED, and Relay 2 is having closed connection, fan is ON.

Same switch is used to turn ON/OFF the fan. Once the Arduino receives corresponding value through rf transmitter, it makes the output of pin 3 HIGH. Hence, fan is turned ON (Fig. 8).

Light OFF—here in simulator, since R1 is showing BLUE, and Relay 1 is having open connection, light is OFF. The switch is used as a toggle switch to turn ON/OFF the light. Once the Arduino receives corresponding value through RF transmitter, it makes the output of pin 4 LOW. Consequently, the light gets turned OFF (Fig. 9).

Fan OFF—here in simulator, since R2 is showing BLUE, and Relay 2 is having open connection, fan is OFF (Fig. 10).

Same switch is used to turn ON/OFF the fan. Once the Arduino receives corresponding value through rf transmitter, it makes the output of pin 3 LOW. Hence, fan is turned off.

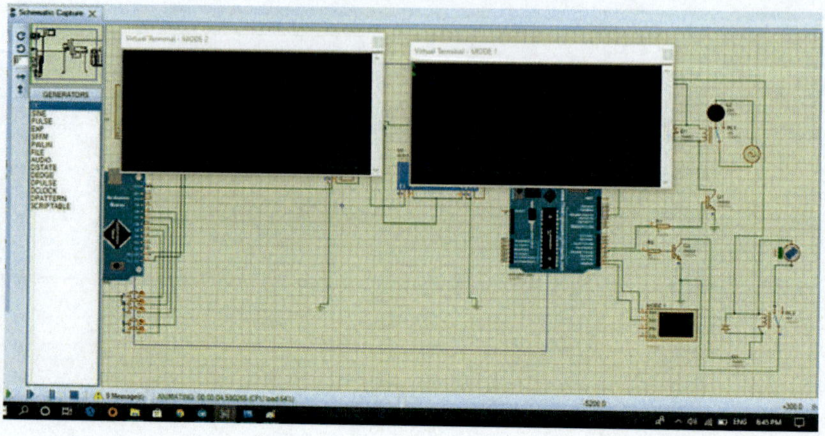

Fig. 8 Proteus fan ON simulation

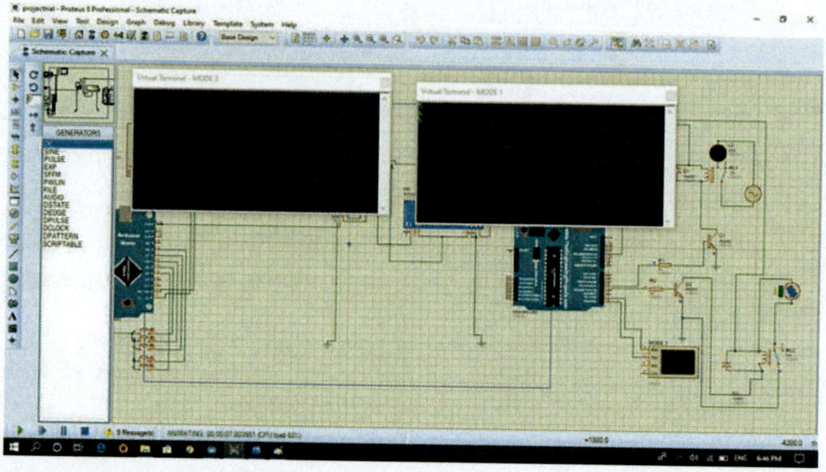

Fig. 9 Proteus light OFF simulation

Mode 2 Results (Android Application Run)

On being connected to Bluetooth, our Android app Helping Hand shows "Welcome" caption as a surety that our app is connected successfully to the glove (Fig. 11).

Information updation screen and submission—Real-time information updation can be done by medical workers by filling up the textboxes as shown in Fig. 12. Once submitted, submitted notifier is fired so that one can be sure of the submission. If wrong input type is inserted in the text box, the information will not be accepted by the application.

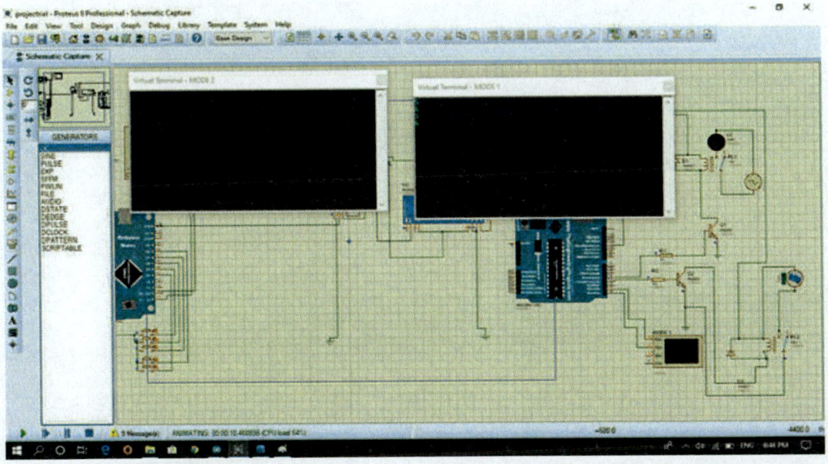

Fig. 10 Proteus fan OFF simulation

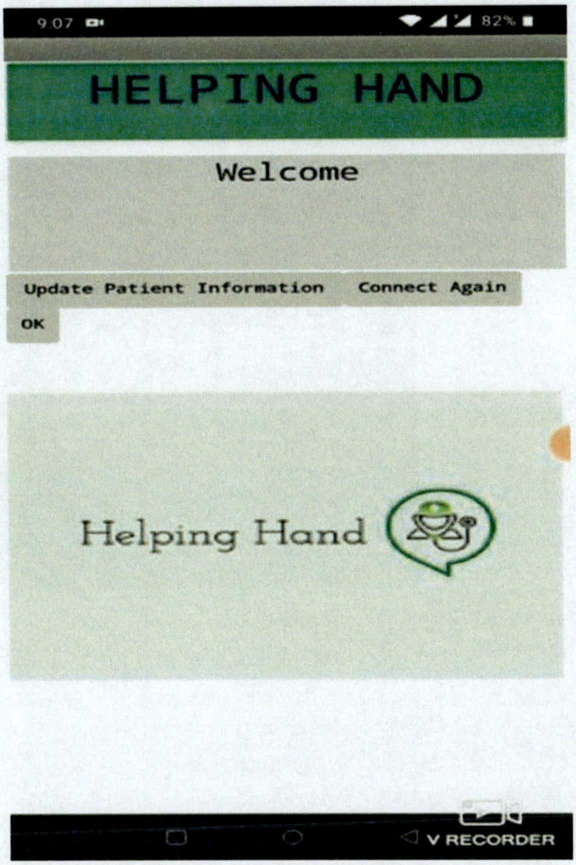

Fig. 11 MODE 2 Bluetooth connected

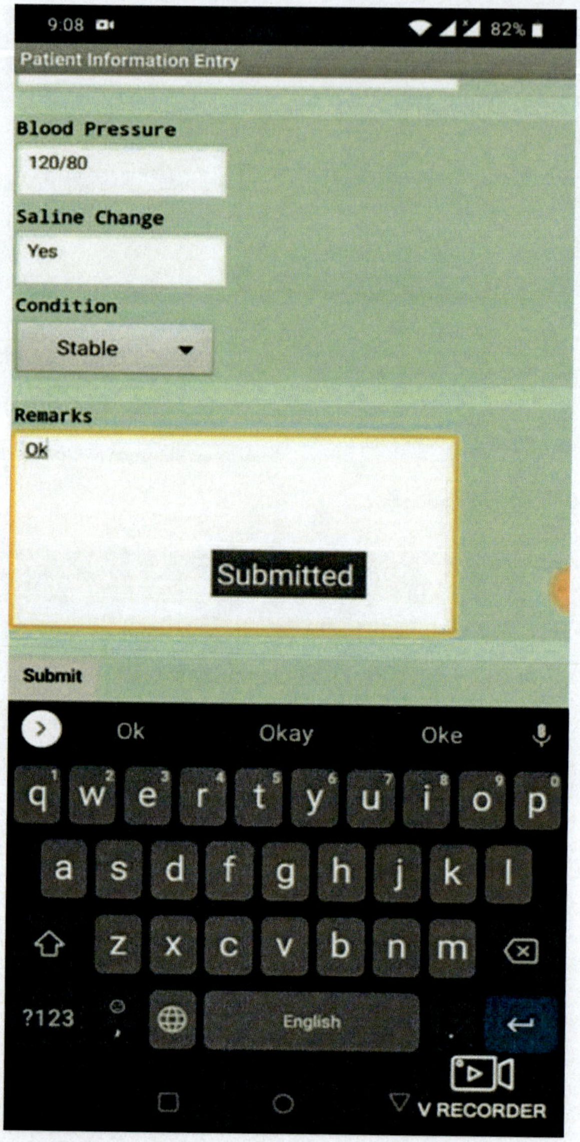

Fig. 12 Patient information submission

Google Sheet after updation—Once someone submits information through the app, instant reflection of the data is shown in Google Sheet (Fig. 13).

Connect again clicked-If the application needs to be disconnected from a particular patient and needs to be connected to other patient, then connect again button is provided for the purpose which disconnects the existing Bluetooth connection and enables the Bluetooth selection again (Fig. 14).

On button press 1—If a patient wants immediate medical assistance, then Button 1 in the gloves can be pressed by the patient and the following display of the app is given in Fig. 15.

On button press 2—If patient wants to have assistance for washroom purposes, Button 2 can be pressed (Fig. 16).

On button press 3 of the glove, patient can call for water if needed (Fig. 17).

On button press 4 of the glove, patient can call medical worker having the app for his Bluetooth device if he wants to go for a walk outside the room (Fig. 18).

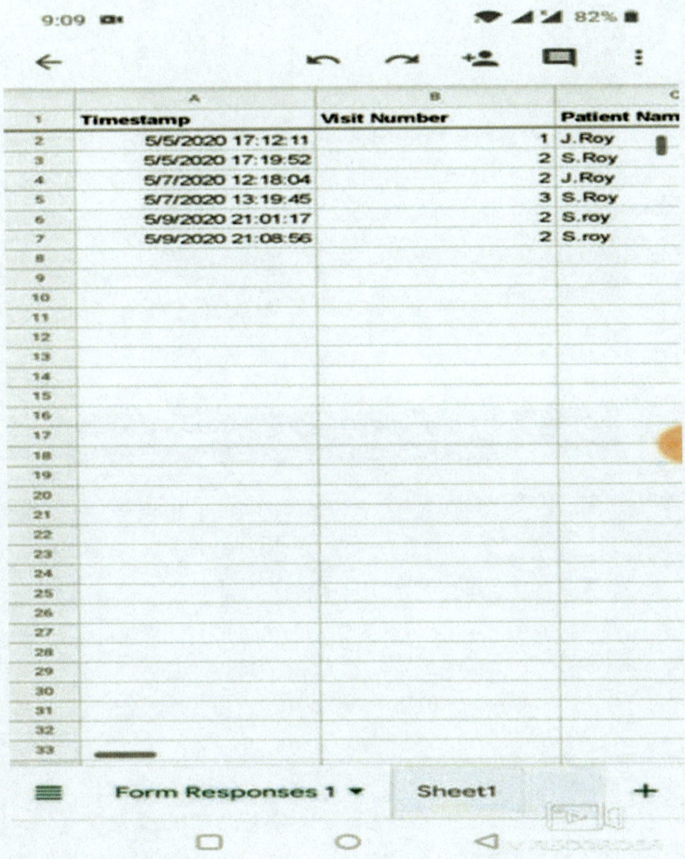

Fig. 13 Google Sheet updation

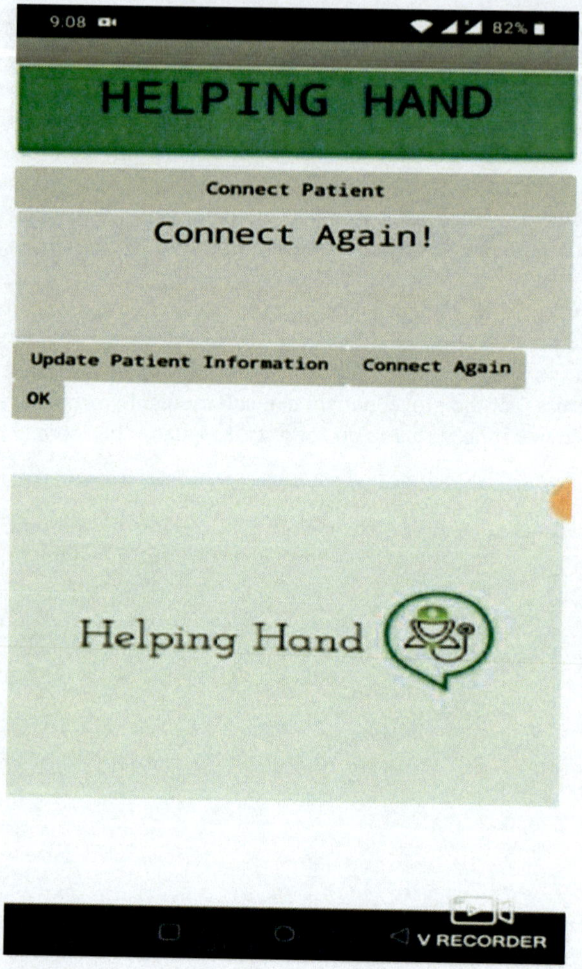

Fig. 14 Connect again screen

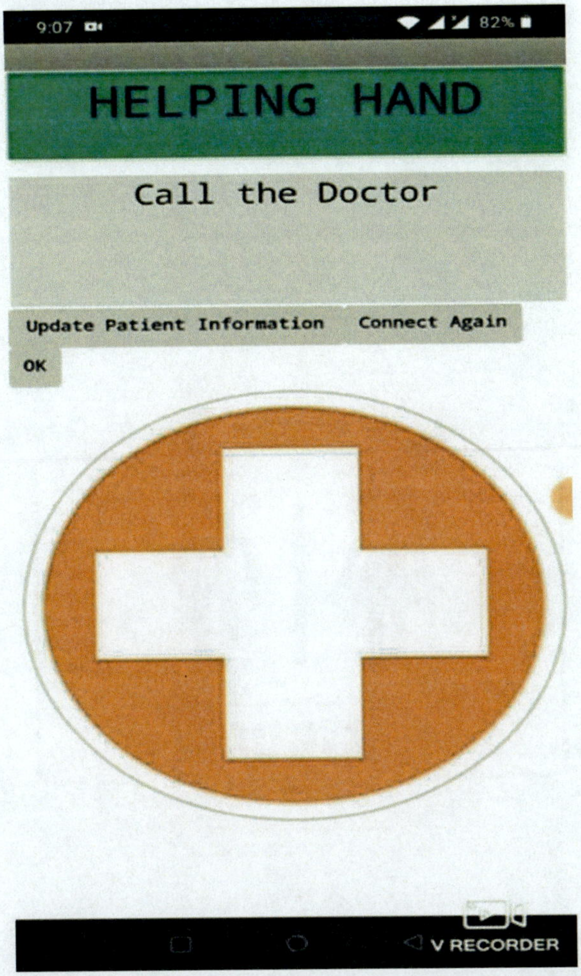

Fig. 15 Android app button press 1

Fig. 16 Android app button press 2

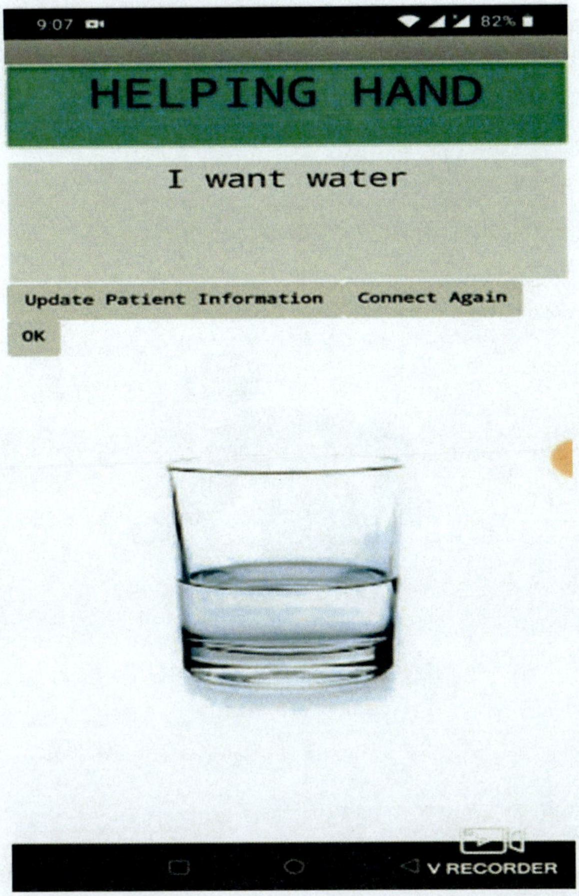

Fig. 17 Android app button press 3

Fig. 18 Android app button press 4

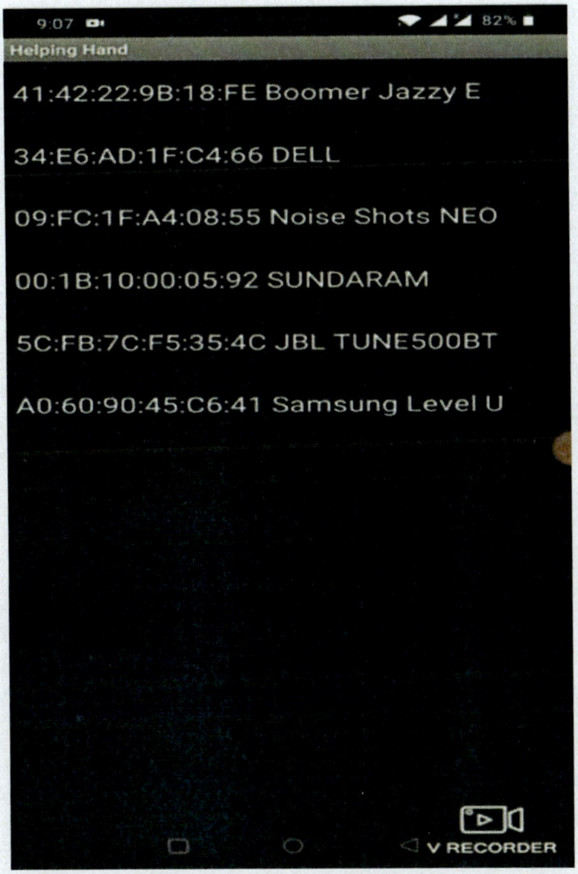

Fig. 19 Bluetooth selection list

Bluetooth selection list—Fig. 19 shows the list of available Bluetooth devices if connected.

Patient button is clicked in the app (Fig. 19).

5 Conclusions

Our system successfully reduces the communication barrier between medical helpers and differently abled patients as intended in the objective. This will specifically shed light on problems disabled patients generally face in hospitals and inability of medical helpers to understand their requirement until and unless they enquire about it.

It provides advantages of being cost effective, flexible to users, compact as we use Arduino Nano for gloves and allows instant communication as well as room automation and information updation.

Generally, WiFi connectivity has a range 50 m or less of while Bluetooth has coverage area of 10 m Adding WiFi module in place of Bluetooth, which demands data connectivity all the time unlike our system but definitely will have an edge over our system in respect of coverage area.

Our application allows input by all the medical workers into the same default Google Sheet we have set up; hence, all the doctors have to share the form and view according to their names and their particular patients. This will reduce user-friendly capability of our project upto some extent. Personalized Google Sheets for doctors can also be added in the future to extend the scope of the project in this respect.

Helping Hand development till now enables medical worker to connect to a particular patient, he/she has to disconnect first from an already connected patient and then establish Bluetooth connection with the next one. In reality, a medical worked deals with more than patient at a time and has to cater to all their requirements. Multiple Bluetooth device connection to one individual Android system will definitely make the project shine on practical grounds.

References

1. P. Verma, S.L. Shimi, S. Chatterji, *Design of Smart Gloves*, NITTTR Chandigarh
2. A.C. Chougule, S.S. Sannakki, V.S. Rajpurohit, *Smart Glove for Hearing-Impaired*
3. https://lemelson.mit.edu/winners/thomas-pryor-and-navidazodi
4. T. Jaya, V. Rajendran, *Hand-Talk Assistive Technology for the Dumb*, Department of Electronics and Communication Engineering School of Engineering, Vels Institute of Science Technology and Advanced Studies (VISTAS), Chennai, India
5. D.L. Patel, H.S. Tapase, P.A. Landge, P.P. More, A. P. Bagade, *Smart Hand Gloves For Disable People*, Department of Electronics and Telecommunication Engineering, BDCE, Sevagram, Wardha (MH)-442102, India
6. R. Shriwas, N. Vihirkar, K. Morey, N. Gulhane, *A Preview Paper on Hand Talk Glove*, Department of EXTC, J.D.I.E.T Yavatmal, Maharashtra, India
7. V. Chowdhry, M. Padhi, B.B. Mohanty, S. Biswal, *Perioperative Challenges in Management of a Deaf and Dumb Patient Posted for High-Risk Cardiac Surgery*
8. Design of a Home Automation System Using Arduino Nathan David University of Nigeria nathan.david@unn.edu.ng Abafor Chima University of Nigeria chrischuks56@gmail.com Aronu Ugochukwu University of Nigeria ugochukwuaronu@gmail.com Edoga Obinna University of Nigeria butler4ril@gmail.com
9. R. Singh, P. Singh, K. Yadav, *Wireless Communications Enlargement: A Review of Advancement in Technologies*, vol. 4, No. 4 School of Energy and Environmental Studies, Devi Ahilya University, Khandwa Road, Indore 452001, India Ḃ Solar Energy Centre, Ministry of New and Renewable Energy, Gurgaon 110003, India Accepted 17 July 2014, Available online 01 Aug 2014

A Brief Study and Overview of Image Fusion Methods

Vineeta Singh and Vandana Dixit Kaushik

Abstract Due to some camera limitations, it is not practically possible to click all the objects focused in a single scene, while human observer likes to see the clear images with all the fine details present in the scene. Similarly for diagnosis purpose, all the pertinent information is required in a single modality image, but due to high cost instrument limitations, it is not possible in practical, to shot at a single instance, all the pertinent and essential details in a solo modality. Thus, concept of image fusion has an essential role here. With the advancement in the technology and recent applications, image fusion model has become quite beneficial in different areas like military, medical, surveillance as well as defense, etc. Here, we will review different kind of image fusion models with advantages and disadvantages.

Keywords Multi-focus image fusion · DWT · PCA · SWT · A survey · Fusion rules

Abbreviations

CEIF	Contrast Enhancement and Image Fusion
MI	Mutual Information
SSIM	Structural Similarity Index Measurement
DCNN	Deep Convolution Neural Network
PCNN	Pulse-Coupled Neural Network
NSCT	Non-Subsampled Contourlet Transform
JSL	Joint Subband Learning
SR	Sparse Representation

V. Singh (✉) · V. Dixit Kaushik
Department of Computer Science and Engineering, Harcourt Butler Technical University Kanpur, Kanpur, India
e-mail: vineeta.singh.cs@gmail.com

V. Dixit Kaushik
e-mail: vdkaushik@hbtu.ac.in

V. E. Balas et al. (eds.), *Proceedings of International Conference on Computational Intelligence, Data Science and Cloud Computing*, Lecture Notes on Data Engineering and Communications Technologies 62, https://doi.org/10.1007/978-981-33-4968-1_28

1 Background of the Work

In the current era of image processing technical scientific world, the concept of image fusion with different methodologies is evolving. The process of image fusion involves different instances of same scene to combine together to get a newly formed image which possess all the pertinent information. The image fusion concept helps tremendously to human beings as well as machines for interpreting better results as per the requirement of the user. There are also many application areas such as medical research, forensic, remote sensing area, camera applications. recognition of objects, detection of targets has become easier with the advent of image fusion concept [1, 2].

The concept of image fusion involves two or more than two instances of a same picture with different focus areas, and finally, those images are fused in such a way so that resultant image includes all the hidden information pertinent to the observer. Image fusion methods may be classified on the basis of source of datasets and on the basis of objective of the fusion method. It can be image fusion based on multi-modality, multi-view, multi-focus, multi-temporal and for restoration purpose image fusion.

2 Literature Survey

Meher et al. [1] carried out a review on image fusion methods on region basis. Liu et al. [2] obtained a satisfactory fusion performance so they purposed a convolution neural network (CNN) model, compared with other previous existing fusion methods, and finally, it is concluded with the novel multi-focus image fusion models. Main part of this model is parallel computing which increases the practical usage. There are different color images experimental at various level as well as the infrared images and medical images (Brain image). They mapped between source image (main) and focus map. First time, they used CNN model in image fusion but the disadvantage is that there is a lot of research gap between these two (CNN and image fusion) concepts. Liu et al. [3] devised a new method based on a novel fusion scheme with cartoon-texture decomposition. Here, some fusion model was used for conversion of multi-focus image sets into two components, namely cartoon content as well as texture content. The simulated results depicted that the proposed fusion model has given better quality but the problem is to build up a potential image decomposition algorithm to boost up the process.

Daniel et al. [4] devised a new fusion model which was named as optimum spectrum mask fusion method for medical images. This activity was done with a gray wolf optimization algorithm. Existing (OSMF) scheme was used for MR-SPECT, MR-PET and MR-CT of simple brain images. All experimental results improved the best results and better than other pixel-based fusion techniques.

Koutouzi et al. [5] had proposed image fusion techniques for TEVAR technology. TEVAR is used for visualization with medical image fusion that really helps in medical imaging fields. Li et al. [6] proposed a new multi-focus image fusion-based scheme based on multi-scaling of image processing technique. A novel weighted fusion model scheme was examined by incorporating non-local means filtering (NLF). This proposed model has three different sections:—identification of focused and fused pixel, correction of identifying results and finally fusion map generation of original image. They used various types of images (Pepsi and Lab) with the all parameters in different segments. This study is incomplete with the creation of initial decision maps.

Manviya and Bharti [7] had illustrated in their research paper about different image fusion methods with their advantages and disadvantages. Li et al. [8] reviewed different image fusion models based on pixel level, which were classified based on sparse representation, multi-scale decomposition and different transforms-based hybrid methods. Ma et al. [9] reviewed different image fusion models for visible image with infrared image fusion, different application areas as well as performance metrics. Li and Dong [10] studied pixel level image fusion with different methodologies. Li et al. [11] devised a fusion model for multi-focus image fusion based on the block consistency verification and filtering techniques. Zhang et al. [12] reviewed fusion models for multi-sensor and multi-modality images based on sparse representation. Maqsood et al. [13] proposed a multi-focus image fusion model based on multi-scale image matting concept. Maqsood and Javed [14] proposed a image fusion scheme for multi-modality-based medical image fusion by sparse representation and two-scale image decomposition.

Cao et al. [15] devised a multi-focus image fusion model on low-cost camera pictures for Canola Phenotyping Sensors. Yang et al. [16] devised a fusion model for multi-focus images. It uses pulse-coupled neural network (PCNN) and use of robust sparse representation (RSR) was done. Firstly, here decomposition of every source image takes place with the help of RSR to get a residual matrix as well as sparse coefficients matrix. Further, the spatial frequency is computed of the residuals matrix and a salience map for the source image is created which acts as linking strength for the PCNN. At starting, decision map is created by comparing ignition frequency maps of the input pictures generated via PCNN. Further, at last, decision map is generated with the help of morphological operations such that opening operation and closing operation. At the end, combined output image incorporates a weighted fusion method.

Li et al. [17] devised a novel image fusion-based model for creation of an informative fused image via blending or combining many images. This fusion model relies on two-scale image decomposition technique. Here, two layers are generated, first one is base layer which has wider variations in intensity and second layer is detail layer which has smaller intensity variations. A new weighted average technique which uses guided filtering is devised to use spatial consistency to blend both of the base and detail layers.

Fig. 1 Multi-focus image fusion example: I image (near focus) II image (far focus) III image (fused image)

Liu et al. [2] introduced convolution neural networks, abbreviated as CNNs for image fusion purpose. Here, researcher discussed the suitability as well as importance of CNNs in context of image fusion model. Gonzalez et al. [18] devised a book titled "Digital Image Processing using MATLAB." Here, image processing fundamentals as well as software principles used in practical implementation were described adequately.

Levin et al. [19] devised a natural image matting fusion scheme based on closed-form solution. Here, it was stated that high-quality mattes of natural images may be produced by a mini amount from user input. Yin et al. [20] devised a novel image fusion scheme for multi-modal medical images in the field of non-subsampled shearlet transform abbreviated as NSST. Here initially, NSST decomposition is carried for input images to produce their multi-direction as well as multi-scale representations. Fusion of bands based on high-frequency is accompanied via a parameter-adaptive pulse-coupled neural network, in short, PA-PCNN model. Sample images for Fig. 1 are taken from Lytro multi-focus image dataset [21]. In [22] pulse-coupled neural network (PCNN), in [23] guided filtering (GF) algorithm, in [24] multi-scale weighted gradient-based fusion (MWG), in [25] sparse representation (SR), in [28] deep convolution neural network (DCNN), in [26] joint subband learning (JSL), in [27] NSCT have been proposed and described in detailed form with wider discussion.

3 Different Type and Levels of Fusion Methods

At the current state of the art, many image fusion schemes have been devised to combine multi-focus images. An example of multi-focus image fusion concept is illustrated in Fig. 1. An image fusion concept diagram is shown in Fig. 2. Generally, these proposed methods can be categorized into two broader categories: spatial domain and transformed domain. For spatial domain, fusion procedure is applied directly on the values of pixel. While for transformed domain, an input image is at

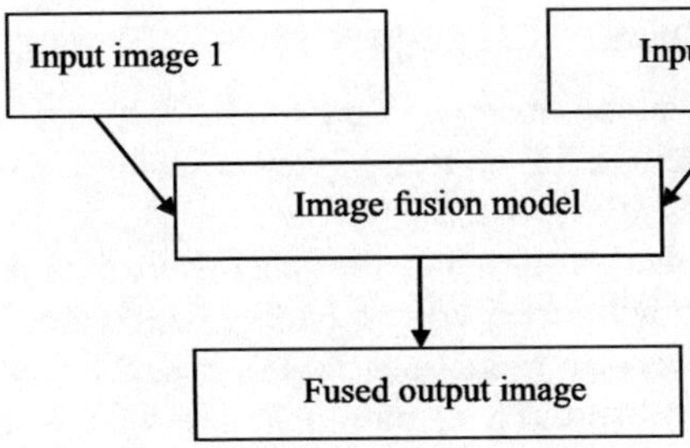

Fig. 2 Image fusion concept

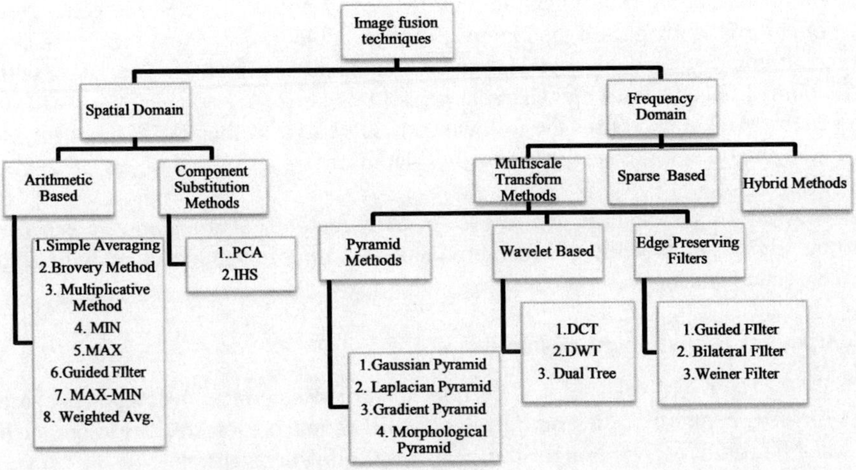

Fig. 3 Image fusion classification diagram

first, transformed to multi-resolution components. Process of image fusion is basically performed at four different levels: pixel level, feature level, signal level as well as decision level. Detailed diagram is present in Fig. 3.

3.1 Spatial and Frequency Domain

Spatial domain includes "simple fusion rules." These rules are directly applied on the pixel levels. Append the source image pixel to fused pixel. Easiest method yet encounters distortions in final fused output image. Let us consider transform domain, initially the source image is transformed into transform domain and further fusion rules are implemented over the transform domain image and afterward inverse transformation is applied to get the fused output image.

Arithmetic-Based Fusion Rules

1. **Simple Averaging**: Very simple method. Take the respective pixel values from the source image and calculate the average. Now combined output image will contain average pixel intensity values.
2. **Brovey Method**: Brovey method-based fusion scheme involves product of every multi-spectral image band with every corresponding panchromatic image band. Then normalization is done with multi-spectral image and repeatedly performs this step for every single band. Thus, output image with high-resolution is obtained via intensity modulation.
3. **Multiplicative Fusion Rule Method**: In this method, PAN and MS image are combined yielding color preservation. The arithmetic fusion rules such as addition, subtraction, multiplication and division methods are used for generation of chromatic image and then intensity image. Color preservation is provided by multiplicative method. It yields spectral bands possessing high correlation with source image features and characteristics.
4. **MIN**: MIN method uses the following strategy, source image with low contrast and pixel having lowest value is considered out of respective pixels of source images.
5. **MAX**: MAX method is just reverse of MIN method. This fusion rule extracts the highest pixel value and then produces an output image leveraged with high contrast features.

Component Replacement Fusion Rules:

1. **PCA**: Here with the help of correlated variables, principal components are formed. It mainly relies on datasets and does not possess fixed vectors such as FFT, DCT, etc. PCA algorithm may have following steps:
   (i) Produce column vectors from source image.
   (ii) Co-variance matrix formation takes place.
   (iii) Using the matrix in step (ii) Eigen values, Eigen vectors are calculated.
   (iv) Mean of Eigen vectors is calculated and further normalization is done.
   (v) With each pixel of input image, multiply the step (iv) Eigen value
   (vi) Now final summation is done for both of the matrices to calculate the fused output image.

2. **IHS Method**: IHS is an abbreviation of Intensity Hue Saturation. Intensity is the amount of reflected light that comes toward our eyes. Hue denoted saturation. Color wavelength is represented by Hue. It projects visual form of the scene by substituting any of the elements (I, H, S) by another element of the source image. Hence, it has spectral information requires to be controlled. Thus, high resolution is added in this manner. It is very old fusion scheme mainly used for sharpening of images.

3.2 *Multi-scale Transform Methods*

Pyramid Methods

At specific levels, pyramid structures are formed of source image by pyramid method in combination of selective approach. Initially pyramid structures are decomposed, further all the decomposed images are integrated, at last to get the combined output image, inverse PT is used. Gaussian pyramids are primary used pyramid structures. Gaussian pyramid is formed by convolution of an image accompanied with a Gaussian low-pass filter. Further down sampling is performed through a factor of 2.

Laplacian pyramid works on the principle of multi-scale image fusion scheme. Here, low resolution images are taken as an input and high-resolution output images are produced. This fusion rule has following four steps:

(i) Blurring with the help low-pass filter.
(ii) Subsamples are constructed.
(iii) Interpolation takes place.
(iv) Differencing scheme is adapted at all the existing pyramid levels. Take the source image at first level, and then divide it to the further levels.

3.3 *Methods Based on Wavelets*

DWT: DWT means discrete wavelet transformation. DWT produces outputs in time domain as well as in frequency domain. DWT has more benefits over the Fourier transformation. In DWT, input images are transformed into different frequency subbands LH, HL, HH, LL. DWT changes the input image in approximation part and detail part, i.e., into different frequency components. Then further these coefficients are formed at certain levels and further fusion rules are implied to fuse those coefficients. Then apply inverse DWT to get the final combined output image. This method lacks in directionality. It has shift variance problem, aliasing problem. Due to these drawbacks, duel-tree complex wavelet transformation was evolved but it suffers from improper curve and edges, though duel-tree complex wavelet transformation has lessen the spectral distortion in an image but it also has setbacks like low spatial resolution.

Discrete Cosine Transform (DCT): Here, source image is broken into n*n size blocks then discrete cosine transform (DCT) is calculated. At each block level, fusion schemes are accompanied and further combined coefficients are generated. At last, inverse DCT is imposed to get the resultant image. DCT has compactness properties.

3.4 Sparse Based

Sparse-based image fusion model works in the following manner. In this input, image signals are reused as a straight line integration of a "certain" elements out of a pre-trained dictionary, where the sparse coefficients demonstrate the input image characteristics. Steps can be summarized as:

(i) Input images are broken down into overlapping patch segments and every patch is rewritten as a vector.
(ii) Sparse representation operation is done on input image patches via trained dictionaries.
(iii) Apply some fusion method and fuse the sparse representation segments.
(iv) Produce the final fused output images with the help of their respective sparse notations.

This methodology is used by several fusion rules such as PCA, adaptive sparse representation as well as multi-scale dictionary learning, in context of image fusion research area.

3.5 Hybrid Methods

Hybrid method involves many kind of fusion schemes like combination of multi-scale transform with sparse representation. Combination of sparse methods and neural networks can be used in this category. Basic objective here is to integrate the characteristics from multi-scale transform with region detection to secure the detail part yielding more accurate and rectified outcomes.

4 Performance Metrics

To check the clarity and effectiveness of the combined output image as per required parameters, many performance metrics have been proposed. Some of them are mean, entropy, standard deviation, mutual information (MI), peak signal to noise ratio (PSNR), root mean square error (RMSE) as well as structural similarity index measurement (SSIM).

Entropy: How much information content is available in an image is known as entropy and it is calculated as following:

$$E_n = \sum_{j=0}^{G} P(i)\big(\log_2 p(i)\big) \tag{1}$$

where G denotes the number of gray levels of the image and $p(i)$ denote the normalized probability of occurrence of every single gray level.

Standard Deviation: To get the value of standard deviation, first calculate the mean or average of the data set. Further, minus the mean from every number present in the data set, and then calculate square for every difference. A combined output image with size $P \times Q$, standard deviation can be represented by following expression,

$$\sqrt{\frac{1}{PQ}\sum_{i=1}^{P}\sum_{j=1}^{Q}\left(I_{\text{fuse}}(i, j - m\prime)^2\right)} \tag{2}$$

where $I_{\text{fuse}}\ (i, j)$ denotes the intensity value of (i, j)th pixel and m' denotes the sample mean of all the present pixel values of the image.

MI: The mutual information of two images is a quantity of mutual dependence between two images, given by:

$$\text{MI} = \frac{1}{2}[\text{MI}[A(g, h), T(g, h)] + \text{MI}[B(g, h), T(g, h)]] \tag{3}$$

where $A(g, h)$ and $B(g, h)$ represent two input images and $T(g, h)$ shows the fused image.

Peak Signal to Noise Ratio (PSNR): The proportion of maximum signal power and the noise which affect the fidelity is calculated by PSNR and is given as:

$$\text{PSNR} = \frac{1}{2}[\text{PSNR}[A(g, h), T(g, h)] + \text{PSNR}[B(g, h), T(g, h)]] \tag{4}$$

Root Mean Square Error (RMSE): It is a measure for computing the error of fusion model to predict quantitative data and is given by,

$$\text{RMSE} = \frac{1}{2}[\text{RMSE}[A(g, h), T(g, h)] + \text{RMSE}[B(g, h), T(g, h)]] \tag{5}$$

Structural Similarity Index Measurement (SSIM): It is used to demonstrate the similarity of the structural information content present in the two images under comparison. The SSIM of two images r and s is defined as

$$\text{SSIM}(r, s) = \frac{(2\mu_r\mu_{s+}C_1)(2\sigma_{rs}\mu_{s+}C_2)}{\left(\mu_r^2 + \mu_s^2 + C_1\right)\left(\sigma_r^2 + \sigma_s^2 + C_2\right)} \tag{6}$$

where μ_r, mean intensity of image r and σ_r, standard deviation of image r utilized as an amount of image contrast. $C1$, $C2$ denotes the constant terms.

Average Gradient (AG): Depicts the clarity of the image and can be calculated as following:

$$\mathrm{CEIF} = \frac{1}{M} \sum_{(u,v)\in\Omega} S(u, v)O(u, v) \tag{7}$$

Here $S\ (u,\ v)$ shows the edge strength enhancement quantity and $O(u,\ v)$ depicts the coincidence of $(u,\ v)$ over the edge direction.

Contrast Enhancement and Image Fusion (CEIF): CEIF metric is utilized for the assessment of the contrast enhancement and image fusion.

5 Conclusion

Different image fusion schemes have been reviewed in this paper. A brief comprehensive summary is given in Table 1. DWT has better advantages as compared to other listed methods in this paper. DWT with hybrid model of other methods can provide more encouraging results.

5.1 *Comparative Results*

We can see from Table 2 that experimental results have been carried out for Lytro multi-focus image dataset and one table listed here in this paper for the "image of Fig. 1." We have included four metrics MI, CEIF, AG and SSIM. Bold highlighted results are showing results with better performances. As we can see, **JSL** has given better results for **Mutual information, CEIF and Average gradient** fusion metrics while MWG has given better results for **Structural Similarity Index Measurement** fusion metric as compared to other listed algorithms in Table 2. SSIM value closer to 1 depicts the higher similarity with the source image. Larger value of CEIF denotes the better contrast enhancement and better preservation of features of the source image.

Table 1 Fusion methods comparison table

Method names	Pros	Cons
Spatial Domain Methods		
Averaging method,	Simple and ease of implementation	Contrast not as desired
Brovery method,	Faster execution speed, less execution time, good contest output image	Distortion of color
Multiplicative method		
Max		Blurred image
MIN method		
MAX–MIN method		
Weighted average method	Detection reliability	High value of SNR
HIS method	High sharpening value, speedy processing	Only RGB bands used in process so color distortion issue arises
Laplacian pyramid and Gaussian pyramid,	Edges are preserved, no aliasing effect, self-inversion, invariant in rotation. Good visual effects	Decomposition levels influence fused output image
Gradient pyramid		
Morphological pyramid		
DCT method	Series of waveforms created out of input image, beneficial for real-time applications mainly	Not good quality output image
Discrete wavelet transform	Better SNR ratio in comparison of pixel-based fusion schemes	Directionality issue, self-alias creation, shift variance issue
Hybrid fusion models	Good quality fused output image	Slow + Time Taking
PCA + DWT hybrid method	Efficient, high spatial resolution	Complexity High
NSCT [27]	For image representation, it has property of shift-invariance, multi-dimensionality and multi-resolution	Research scope is still there for improvement
Sparse Representation (SR) [25]	Uses sparse-based coding	Spatial inconsistency, details tend to be smoothed
DCNN [28]	The generation of ground truth images in context of DCNN training can be fulfilled via an artificial way	Further research is still going on
Pulse-coupled neural network [22]	Synchronous pulsed output, controllable parameters and adjustable threshold	Linking coefficient from image statistics is a challenging task
GF (Guided Filtering) [23]	Fast and effective	–
JSL (joint subband learning) [26]	Robust and effective	Further improvement is in progress

(continued)

Table 1 (continued)

Method names	Pros	Cons
MWG (Multi-scale weighted gradient-based fusion) [24]	Basis of the method is multi-scale structure-basis focus measure	–

Table 2 Lytro multi-focus image dataset performance evaluation metrics results for "image of Fig. 1" for different algorithms

S. No	Performance Metrics	NSCT	SR	DCNN	PCNN	GF	JSL	MWG
1	MI	7.1463	7.4629	9.2636	9.2739	7.7835	**9.8154**	8.4194
2	CEIF	0.3028	0.3001	0.4030	0.3947	0.3255	**0.5006**	0.4937
3	AG	7.3756	7.3037	7.1980	7.7473	7.6255	**8.3385**	7.6438
4	SSIM	0.6629	0.5605	0.7875	0.6815	0.6154	0.7119	**0.8154**

References

1. B. Meher, S. Agrawal, R. Panda, A. Abraham, A survey on region based image fusion methods Inf. Fusion **48** https://doi.org/10.1016/j.inffus.2018.07.010
2. Y. Liu, X. Chen, H. Peng, Z. Wang, Multi-focus image fusion with a deep convolutional neural network. Inf. Fusion **36**, 191–207 (2017). https://doi.org/10.1016/j.inffus.2016.12.001
3. Z. Liu, Y. Chai et al., A novel multi-focus image fusion approach based on image decomposition. Inf. Fusion **35**, 102–116 (2017)
4. E. Daniel, J. Anitha et al., Optimum spectrum mask based medical image fusion using gray wolf optimization. Biomed. Signal Process. Control **34**, 36–43 (2017)
5. G. Koutouzi, C. Sandström, P. Skoog, H. Roos, M. Falkenberg, 3D image fusion to localise intercostal arteries during TEVAR. EJVES (2017). https://doi.org/10.1016/j.ejvssr.2017.03.001
6. H. Li, H. Qiu, Z. Yu, B. Li, Multifocus image fusion via fixed window technique of multiscale images and non-local means filtering. Signal Process. **138**, 71–85 (2017)
7. M. Manviya, J. Bharti, Image fusion survey: a comprehensive and detailed analysis of image fusion techniques, in *Social Networking and Computational Intelligence*, Lecture Notes in Networks and Systems 100. https://doi.org/10.1007/978-981-15-2071-6_53
8. S. Li, X. Kang, L. Fang, J. Hu, H. Yin, Pixel-level image fusion: a survey of the state of the art. Inf. Fusion **33**, 100–112. ISSN 1566-2535 (2016). https://doi.org/10.1016/j.inffus.2016.05.004
9. J. Ma, Y. Ma, C. Li, Infrared and visible image fusion methods and applications: a survey. Inf. Fusion. **45**, 153–178 (2019). https://doi.org/10.1016/j.inffus.2018.02.004
10. M. Li, Y. Dong, Review on technology of pixel-level image fusion, in *Proceedings of 2nd International Conference on Measurements, Information and Control (ICMIC)*, vol. 1, pp 341–344, Harbin, China (2013). https://doi.org/10.1109/mic.2013.6757979
11. X. Li, F. Zhou, J. Li, Multi-focus image fusion based on the filtering techniques and block consistency verification, pp 453–457, doi: https://doi.org/10.1109/icivc.2018.8492825.
12. QiangZhang et al., *Sparse representation based multi-sensor image fusion for multi-focus and multi-modality images: A review* (2018), pp 57–75. https://doi.org/10.1016/j.inffus.2017.05.006
13. S. Maqsood et al., Multiscale image matting based multi-focus image fusion technique. Electronics **9**, 472 (2020). https://doi.org/10.3390/electronics9030472
14. S. Maqsood, U. Javed, Multi-modal medical image fusion based on two-scale image decomposition and sparse representation. Biomed. Signal Process. Control **57**, 101810 (2020). https://doi.org/10.1016/j.bspc.2019.101810

15. T. Cao, A. Dinh, K.A. Wahid, K. Panjvani, S. Vail, Multi-focus fusion technique on low-cost camera images for canola phenotyping. Sensors (Basel, Switzerland) **18**(6) (2018).https://doi.org/10.3390/s18061887
16. Y. Yang, Yang et al., Robust sparse representation combined with adaptive PCNN for multifocus image fusion. IEEE Access 2018 **6**, 20138–20151 (2018). https://doi.org/10.1109/ACCESS.2018.2822688
17. S. Li, X. Kang, J. Hu, Image fusion with guided filtering. IEEE Trans. Image Process. Publ. IEEE Signal Process. Soc. **22** (2013). https://doi.org/10.1109/TIP.2013.2244222
18. R.C. Gonzalez, R.E. Woods, S. Eddins, *Digital Image Processing Using MATLAB* (Prentice Hall, New York, NY, USA, 2004)
19. A. Levin, D. Lischinski, Y. Weiss, A closed-form solution to natural image matting. IEEE Trans. Pattern Anal. Mach. Intell. **30**(2), 228–242 (2008). https://doi.org/10.1109/TPAMI.2007.1177
20. M. Yin, X. Liu, Y. Liu, X. Chen, Medical image fusion with parameter-adaptive pulse coupled neural network in nonsubsampled shearlet transform domain. IEEE Trans. Instrum. Meas. **68**, 49–64 (2019). https://doi.org/10.1109/TIM.2018.2838778
21. Lytro Multi-focus Image Dataset taken from, https://www.researchgate.net/publication/291522937_Lytro_Multi-focus_Image_Dataset, accessed on June 2020.
22. M. Monica Subashini, S.K. Sahoo, Pulse coupled neural networks and its applications. Expert Syst. Appl. **41**(8), 3965–3974 (2014)
23. S. Li, X. Kang, J. Hu, Image fusion with guided filtering. IEEE Trans. Image Process. **22**(7), 2864–2875 (2013)
24. Z. Zhou, S. Li, B. Wang, Multi-scale weighted gradient-based fusion for multi-focus images. Inf. Fusion **20**, 60–72 (2014)
25. B. Yang, S. Li, Multifocus image fusion and restoration with sparse representation. IEEE Trans. Instrum. Meas. **59**(4), 884–892 (2010). https://doi.org/10.1016/B978-0-12-816718-2.00010-5
26. Z. Zhong, T. Shen, Y. Yang, Z. Lin, C. Zhang, Joint sub-bands learning with clique structures for wavelet domain super-resolution. Adv. Neural Inf. Process. Syst. (2018). https://arxiv.org/abs/1809.04508
27. Q. Zhang, B.-L. Guo, Multifocus image fusion using the nonsubsampled contourlet transform. Signal Process. **89**(7), 1334–1346 (2009)
28. G. Sreeja, O. Saraniya, Chapter 3 - Image fusion through deep convolutional neural network, in *Deep Learning and Parallel Computing Environment for Bioengineering Systems*, ed. by Arun Kumar Sangaiah, (Academic Press, 2019), pp. 37–52, ISBN 9780128167182. https://doi.org/10.1016/B978-0-12-816718-2.00010-5.

Detection of Plant Leaf Disease Using Image Processing and Deep Learning Technique—A Review

Gagneet kaur, Priyanka Arora, and Jasdeep kaur

Abstract Agriculture employs 50% of liveware of the nation. The agricultural yield gets affected by fruit and plant leaf diseases. It is necessary to take vital steps against the deterioration of yield. Image processing techniques provide essential apparatus to look after these diseases in plants. Mostly designed systems work in four steps. Once a picture is captured of doubtful fruit or leaf, this is called image acquisition. The picture is sent to the system, and it goes under pre-processing, segmentation, feature extraction, and classification. The acquired image is processed to fit in the system by setting its resolution. Clusters are made as it becomes easy for the system to extract color features, boundary features, and histograms, etc. The final step classifies the input data into pre-defined categories of different diseases. This paper is a hub of some acceptable techniques and highlights the success ratio of variant methods used for disease detection.

Keywords Segmentation · Feature extraction · Clustering

1 Introduction

Agriculture plays a big role in the world's economy in providing food safety by improving the quality of yield and reducing the wastage of the yield. In the present era, plants get infected by different types of diseases at very first stage that is when a seed starts to strew into a small sapling. The reason behind the plague on the plant could be any climatic condition or pre-defined plant diseases. Climatic conditions like temperature, airflow, and solar radiation are far away from man's hands to control. Few other natural factors affecting the yield present in soil are clay content, pH value, and base saturation [1]. Also soil erosion and soil degradation are crucial strands that

G. kaur (✉)
Student, Guru Nanak Dev engineering college, Ludhian, Punjab, India
e-mail: jaggigagneet@gmail.com

P. Arora · J. kaur
Faculty of Guru Nanak Dev engineering college, Ludhian, Punjab, India

V. E. Balas et al. (eds.), *Proceedings of International Conference on Computational Intelligence, Data Science and Cloud Computing*, Lecture Notes on Data Engineering and Communications Technologies 62, https://doi.org/10.1007/978-981-33-4968-1_29

affect the propagation of plants. Greater minds in agricultural industry defined some plant and fruit diseases. The plants needed to be treated at the very initial stage of their life cycle so that they can be harvested properly [2]. Just like doctors, medicament human body, there exist experts for plants too. These adepts scan the plants with naked eyes and can reveal the diseased plant. But such proficient is very expensive to afford by layman. So a system is needed that can be used by anyone, anytime, and at any place. The system does not require human collaborations but work simultaneously. One way is using image processing techniques. Image processing helps to fabricate an apparatus that can perceive the flaw by capturing images of the plant leaves or fruit [3, 4]. In this paper, some already used techniques in image processing for detecting the fruit and plant leaf diseases are assessed. The different malady discussed in this paper is differentiated based on part of the plant it affects. Likewise some diseases affect the root of the plant, some attack the fruit, and some diseases attack the whole plant before it starts to swell and show the symptoms on the leaves. The root malady is caused by fungi or saprophytes present in the soil [5]. Experts also found that mainly these five crops get attacked via root, i.e., rice, maize, wheat, potato, and soya bean are affected by 137 pathogens and pests like bacterial wilt, fungi, viruses, parasites, vertebrates, arthropods, and mollusks [6]. Plant diseases like Alternaria leaf spot, brown spot, mosaic, gray spot, rust, scab, and rot, etc., attack the leaves and demolish the plant till the root. The most observable damage is the innumerable spots on the leaves which can assassinate the whole leaf. Mosaic is irregular mottling of the leaf, vein appears lighter than normal and leaf gets curled or puckered. Gray spot is caused by fungal pathogens. It attacks maize commonly. Rust directly damages the buds or small fruits of guava or apple plants. Scab appears on the fruit during wet summers.

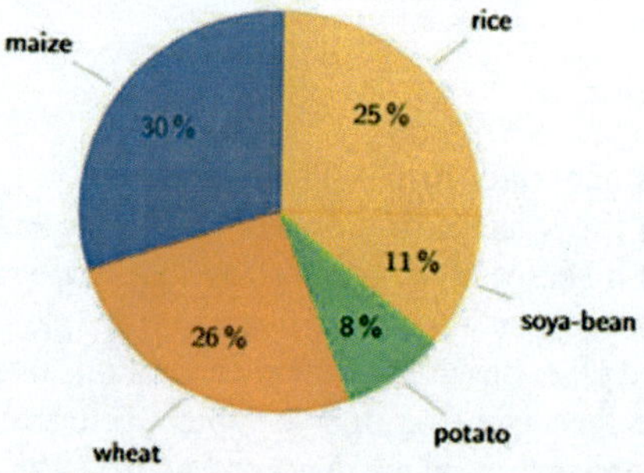

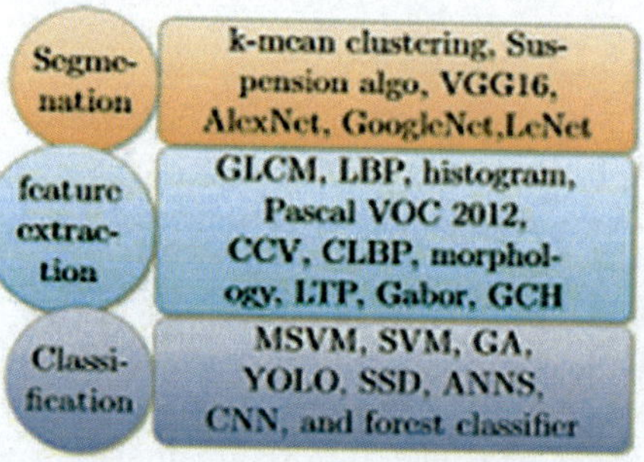

This paper revolves around four techniques that help in the processing of image and finding the flaws. It is very important to perform the techniques accurately to attain the best results. These steps are listed below.

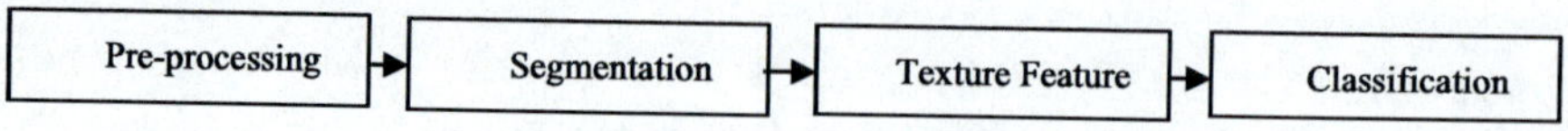

Before these four steps, there are two more steps: image acquisition and data collection. Data collection is a gathering of data as input of any system related to disease detection. This data consists of images of leaves, fruits, and roots, etc. Image acquisition is acquiring images of plant leaves, fruits, or roots. The images are captured via mobile cameras [7], professional cameras, DSLRs, and other photo capturing devices. These photos can be of any type, good or bad quality, high or poor resolution, blurred, noised, shady background, sun-kissed, etc. To make the acquired images fit for systems, these undergo pre-processing steps.

1.1 Pre-processing

It is referred to as the lowest level operation performed on images, where input and output images are intensity images. The agenda behind pre-processing is to improve the image data by enhancing the important features of the image. Pre-processing also works on the brightness in the three ways: 1. pixel transformation, 2. geometric transformation, and 3. restoration. Pre-processing of the image also requires prior knowledge of the image for pixel transformation, for example nature of degradation, image acquisition device properties, and surroundings in which images are captured and information about object is searched in the image. Geometric transformation eliminates the geometric distortion when the image is captured. The pre-processing can be performed using already trained data set, plant village data set, pathological images, etc. These images are then sent for segmentation.

1.2 Segmentation

It is the first step in any computer vision application. It partitions the pixels of an image into groups that correlate with the object. Segmentation is based on two intensity values, and these are similarity and discontinuity. Discontinuity is partitioning the image based on a change in intensity, and similarity is based on grouping of image intensity according to pre-defined criteria. Segmentation mainly includes segments or clusters, and it is easy to perform by k-mean clusters, VGG16 Net, AlexNet, GoogLe Net, CoNet, global thresholding techniques, suspension algorithms, and strong correlation-based methods, etc. K-means clustering is an unsupervised technique. VGG16 Net, CoNet, AlexNet, and GoogLeNet are multilayer neural network architectures used to recognize patterns from pixels of minimal pre-processing [8, 9].

1.3 Feature Extraction

The basic features high level and low level can be extracted from any image. Low-level feature of any image is the change in edge direction known as curvature. High-level features are region and patches. In feature extraction, analyzing texture features is an important thing. Texture is a pattern without an analytical description that helps in pattern classification. Feature extraction is technically done by GLCM, LBP, color histogram, Pascal VOC 2012, CCV, CLBP, morphology, LTP, Gabor feature, GCH, fuzzy, and NAG algorithms, etc. Gray-level co-occurrence matrix (GLCM) considers the spatial relationship of pixels for examining texture. Local binary pattern (LBP) is an operator that labels the pixel of an image by thresholding the neighborhood of each pixel and provides the result in binary form. Compound local binary pattern (CLBP) is the advancement of LBP as it labels 3 * 3 neighborhood, i.e., eight neighbor pixels around the central pixel and assigns 16-bit code to a pixel. Local ternary pattern (LTP) is also an extension to LBP. It uses a threshold constant to threshold pixels into three values. Color histogram is the distribution of colors in an image. Color coherence vector (CCV) is an image descriptor. It is used to extract color-related features from the image and represent a low-dimensional image.

1.4 Classification

It is a process to classify an image upon visual content. An example is to detect human figure present in an image or not. It is a trivial task for a human to judge for robust images. Classification is done using image classifiers. Some commonly used classifiers are MSVM, SVM, GA, YOLO, SSD, ANNS, and CNN in caffee framework and forest classifier, etc. Support vector model (SVM) and multi-support vector model (MSVM) are supervised models associated with algorithms to analyze data for classification and regression [10, 11]. Genetic algorithm (GA) is an unbiased optimization technique used for image enhancement and segmentation for large solution space. You only look once (YOLO) is a technique in which whole data is looked at once. Single-shot detector (SSD) is the fastest detector that works on a single shot. Artificial neural networks (ANN) and the convolution neural network (CNN) both are deep learning neural networks used for visual image recognition. These techniques work behind the scenes [12–15].

2 Related Work

This paper will provide a review from the year 2012 to the year 2019 on different plant-related diseases. These diseases affect the plant from roots to fruits. The fruits and vegetables that get attacked by such diseases are apple, potato, peanut, lemon, papaya, pomegranate, cardamom, pigeon pea, tomato, grapefruit, orange, lime, and other citrus fruit. The systems designed to detect the diseases from leaves or fruit images are very advance. Such systems made this task easy and non-expensive. This apparatus works in four procedural steps. Each step is performed with many different techniques. Every author proposed the best technique as per his need. Here we will compare such techniques and their percentage of success and consider which technique is best for each step.

2.1 Pre-processing

This step includes collecting infected as well as healthy data, managing the collected data, and transforming it into a suitable state to perform different detection methods. The data is collected from different places and in different ways. The challenges that let pre-processing take place are: fruit covering noisy background, best contrast between fruit covering and background, change in lighting and variations, input image low intensity value, distinct wavelength bands, and huge number redundancy [16]. Systems engaged in disease detection used plant village named data set from a Web site kaggle.com [17, 18], previously trained data [19, 20], pathological images collected from two apple experiment stations [21], and pathological collected images used PCA jittering proposed by AlexNet to reduce over-fitting. Some systems collected only the infected leaves, placed it into a ziplock bag to keep it moist, later captured the images of leaves in the laboratory [22]. An apparatus needed to perform data augmentation and image annotation for pre-processing [23]. Image annotation is selecting object spots from a diseased image. After annotation, an XML file is generated that contains classes of diseases and coordinate values of each bounding box. Data augmentation is used to solve the problem of over-fitting. Over-fitting occurs due to random noise and errors [24]. Other two different techniques specially designed for detection of apple fruit disease are as follows: one is YOLO and the second is SSD. Firstly for applying you only look once (YOLO), acquired images are re-sized to 448 * 448. Then this re-sized images are sent to CNN [25, 25], and future resulting detection is done using non-max suppression algorithm [27]. The concept of single-shot detector is almost similar to YOLO but SSD divides the output into a set of default boxes over aspect ratio and scale per feature map location. During training set, his default boxes are compared to ground-truth boxes [28]. Image clipping, i.e., cropping of leaf image to get interested image region and applying histogram equalization for intensity distribution is another combination for pre-processing [29]. In some papers, the pre-processed images needed to be re-sized. Further images are transformed from colored to black and white. Then resolution is improved, noise is removed, image is enhanced, and intensity values are distributed using different algorithms. These images are filtered using different hybrid methods like 3D box filtering, Gaussian filters, smoothing filter, top-hat filtering, median filters, blue filters, HSI, HSV, and many more.

2.2 Segmentation

After the collection of data and applying different filters on the images, next comes the segmentation step where the images are segmented into different segments based on various characteristics it shows. This operation can be applied to an image for grouping the images into different clusters based on the same features or some similarity shown in background, foreground, and boundary. Segmentation can be performed using K-means clustering [19–22, 30], Otsu's method, converting RGB to HSI model [29], global thresholding technique [31], strong correlation-based methods [17], suspension algorithm, and single-shots detection [28]. For boundary and spot detection, the RGB color of the image is converted to HSI for segmentation. This conversion is done by considering eight connecting pixels. K-means clustering algorithm: (1) Select the center of K clusters randomly. (2) Assign a cluster to each pixel of the image and decrease the distance between the pixel and center of the cluster. (3)Compute cluster center and average all pixels of the cluster. Repeat steps 2 and 3 until convergence is attained. Otsu's thresholding algorithm converts gray-level images to binary images by setting pixels in O and 1. The pixel below threshold level is labeled 0, and pixels above threshold level are labeled 1. [32] Min [21] also used k-means clustering and divided the clusters into three segments. Segmentation can also be performed with SSD. In SSD technique, object presence and its location are looked once from bounding box coordinates, and associated class probability is evaluated directly from the image pixel. It also computes all approaches at once and provided a framework for training as well as testing [33]. Jiang [23], Rangarajan et al. [18], and Alvaro Fuentes [34] used deep learning architectures based on CNN. Deep learning architecture named AlexNet and VGG16Net segmented the image creating mini-batches by modifying images and evaluating weight and learning rate. Other CNN models like GoogleNet, CoNet, and Lenet, where GoogleNet performed inspection and rainbow concatenation. Rainbow concatenation enhances multi-scale disease object and small-scale disease object detection performance. Cognitive ocean network (CoNet) is the mainstream of future ocean science and engineering developments [35], and LeNet is composed of two parts. The first part consists of a self-taught feature extraction model, and the second part consist of classification model [36]. Liu et al. [37, 38] applied AlexNet based on neural network for segmentation by partially removing connected layers and adding the pooling layer. Further GoogleNet structure is introduced for inspection, and the NAG algorithm for optimization of convolution neural network was implemented over GPU platform in caffee framework. AlexNet—a pre-trained model consists of five convolution layers and three fully connected layers. Every convolution layers consist of a variant number of filters of different dimensions, like convolution layer 1 with dimensions 11 * 11 * 3 and 96 filters, convolution layer 2 with dimensions 5 * 5 * 48 and 256 filters, convolution layer 3 with dimensions 3 * 3 * 256 and 384 filters, convolution layer 4 with dimensions 3 * 3 * 192 and 384 filters, and convolution layer 5 with dimensions 3 * 3 * 192 and 256 filters, which represent the height, width, and depth, respectively. These filters are applied over the input of dimension 227 * 227 * 3 and dot product of

Table 1 Crop loss estimates for some economically important plant viruses in developing countries [41]

Crop	Disease	Country	Losses (%)
Banana	Streak	Worldwide	82
Cassava	Mosaic	India	13–3
Cardamom	Mosaic	India	70–100
Cereals	Yellow dwarf	Worldwide	11–12
Cocoa	Swollen shoot	Africa	100
Peanut	Bud necrosis	India	89
Pigeon pea	Mosaic	India	30
Potato	Virus *X* and *Y*	Worldwide	10
Soyabean	Mosaic	Worldwide	8–25
Tomato	Leaf curl	Tropical and subtropical region	93–100
Sweet potato	feathery mottle and chlorotic stunt	Worldwide	50

filter matrix over the image is carried out. The VGG16 net pre-trained model is a stacked version of the AlexNet model. It consists of 13 layers. Each convolution layer consists of a filter of very small dimension 3 * 3 along with a fully connected ReLU layer. Small dimension filters reduce the number of parameters [39] (Table 1).

2.3 Feature Extraction

This step extracts the useful and detectable features from the segmented images. The main features that can be extracted are color, texture, and shape. Color features and texture features play a vital role. Color features can be extracted by color coherence vector (CCV) and global color histogram (GCH) [20, 40], GLCM using HSV color [17, 21], CCV [19], calculating chroma, hue angle, statistical feature [22], LBP, CLBP, LTP, Gabor [30] and global color histogram, and color coherence vector which tells about the probability of colors present in the image, and it is efficient and invariant to minor changes [30]. Fuzzy curves and fuzzy surfaces are used for texture as well as surface features. Texture features were extracted by complete local binary pattern (CLBP), local binary pattern (LBP) [20, 21], and genetic algorithm [17]. A.S.M. Shafi measured statistical, structural, and spectral of image using GLCM and stored the features like contrast, correlation, and homogeneity in feature database [31]. Just like A. S. M. Shafi, K. Naga Subramanium looked after spectral and spatial information from electromagnetic spectrum including visible and near-infrared region of hyperspectral images. Raheel Siddiqui used two different techniques: 1. SSD: as the name tells it to detect multiple objects in a single shot. It also computes all approaches at once and provides a framework for training as well as testing and 2. YOLO, i.e., your only look once. Both techniques used PASCAL VOC 2012 challenge for performance evolution [28].

2.4 Classification

Finally pre-processed, segmented, and feature extracted images are classified based on types of disease they are infected. This is a final step. Yin Min used the SVM algorithm. It builds a model that assigns new examples into one category on the other. The collected data builds a confusion matrix. This matrix classification is done in support vector machine (SVM) and *k*-nearest neighbors (KNN), where SVM provides 98% accuracy [21]. Manisha Bhange et al. (2016) also used SVM for training and classification. SVM is slow for training purposes but is still used as it always provides an accurate result. After SVM, the clusters are divided into two portions [19]. The advancement of SVM that is multi-support vector machine (M-SVM) is trained on the stored database. This leads to the detection of a single type of fruit disease from a single image. It provides high accuracy up to 93–94% [20, 31]. The other two proposed techniques are SSD and YOLO. Both techniques include precision–recall curve average precision. Both are good for apple disease detection, but whereas SSD shows better result than YOLO [28]. DSSD is the deconvolution single-shot detector. It is the augmentation of SSD with deconvolution layer for large-scale context in object detection and improving accuracy. Another hybrid technique is MSVM with PCA. Particular component analysis (PCA) is used for feature reduction [17], and it is proved to be best combination for classification in terms of accuracy and time consumption as it provided 4.3% of accuracy in just 3.13 min of time consumption [17]. Bhavini J. Samajpati told that the classification is done in a tree structure using random forest classifiers (Table 2).

Table 2 Comparison of classifier-based techniques [16]

Classifiers	Advantages	Disadvantages
SVM	It works best when training images have some noise. It is utilized for linear and nonlinear classification. Also it is easy to recognize	Algorithm is very complex and slow. It can work with only two classes
ANN	Efficient in managing data as it is self-adapting approach	It requires large amount of data for training, so it is very time consuming. Also it faces scalability issues
KNN	It is easy to implement	Very sensitive to irrelevant data and training is time consuming
Decision tree	Good to describe small size data. It compares the accuracies of other classifiers	It has been used for specific data set by noisy regression or classification task

Thus, result is found using a fusion of features Gabor + CLBP and Gabor + LTP. Linear ternary pattern (LTP) is the advancement of LBP, and the Gabor feature is used for texture analysis and segmentation. Thus, hybrid of Gabor + LTP gives good results than a combination of CLBP + Gabor [30]. Before feature extraction training, cross-validation and testing are performed by ANN SVR by Elham Omraniet al (2014). Two SVRs were investigated using radial basis function polynomial function, whereas SVR-RBF is superior to SVR poly. [22]. Aravind K. Rangarajan et al. (2018) performed classification on plant images using architecture AlexNet and VGG16 which gave accuracy to classification up to 97.29%. The technique proposed by Bin Liu provided 97.62% accuracy, more than the standard AlexNet model [37].

3 Conclusion

It is a great idea to detect and classify fruit and plant diseases through image processing as it can be considered savage for farmers. A fully integrated smart system that can be used by any person, anywhere to save the yield. The most comfortable thing such a system provides is that it can detect diseases from any kind of image in a natural way. Table 3 represents the results of various systems, and different techniques are used for all four steps along with the percentage of accuracy attained during these procedures. Although a combination of different techniques completes a system, but looking at a single step, there are best methods to be used. Images captured by DSLR in laboratories are best for image acquisition, the color transformation methods from RGB to BW, 3D box filtering, Gaussian, and median filter is accurate for pre-processing, k-means clustering method is considered to be accurate for segmentation, and the techniques like LBP, MSVM, GA, and CNN are mostly used for feature extraction and classification.

All the techniques studied in this survey are appropriate for detection, whereas few methods gave high accuracy than others. Like SSD is considered better than YOLO, ANNS gave higher accuracy than SVM. But MSVM with genetic algorithm gave the highest accuracy of 98.20%.

For future scope, although it is a vast topic as there are many fruits to discover, also this field is rarely touched by researchers in the past two years. Plants and fruits yield needs attention to overlook the condition in order to save the outcome before sapling gets affected.

Table 3 Overview of techniques

Author	Year	Pre-processing	Segmentation	Texture Feature	Classiffication	Accuracy
Shivram Dubey	2012	Previously trained data	K-means Segmentation	GCH, CCV, LBP, CLBP	M-SVM	93%
Elham Omrani	2014	Laboratory testing	K-means	Hue angle, statistical feature, and GLC	SVM and NN	
Manisha Bhange	2015	Already trained data set	K-means cluster	Morphology, color, and CCV	M-SVM	
Bhavini J. Samajpati	2016		K-means clustering	LBP, CLBP, LTP, and Gabor	Forest classifier	96
Bin Liu	2017	Pathological images from 2 apple experiment stations	CNN-based AlexNet and GoogleNet inception	Apply NAG algorithm to optimized network	CNN in caffee framework on GPU platform	97.62
Muhammad A. Khan	2018	Plant village database	Strong correlation method	Color, color histogram, and LBP	Selection, reduction using SVM	98.20
Aravind k. Rangarajan	2018	Plant village	Deep learning architecture mini-batch,	Weight and bias learning rate	AlexNet and VGG16net	96.19
Yin min	2018	Filtered and resize	Three segments K-means clustering	GLCM, LBP	SVM	98
Raheel siddiqi	2019	Trained data set	Non-max suspension algorithm	PASCAL VOC 2012	YOLO and SSD	77.5
Peng Jiang	2019	Data augmentation	Image annotation	CNN model for GoogleNet inspection and rainbow concatenation	INAR-SSD	78.80

References

1. K.P. Jiang, K.D. Thelen, Effect of Soil and Topographic Properties on Crop Yield in a North-Central Corn-Soybean cropping system. Agron. J. **96**(1), 252 (2004). https://doi.org/10.2134/agronj2004.0252
2. R. Zhou, S. Kaneko, F. Tanaka, M. Kayamori, M. Shimizu, Disease detection of Cercospora Leaf Spot in sugar beet by robust template matching. Comput. Electron. Agric. **108**, 58–70 (2014)
3. G. Ying, L. Miao, Y. Yuan, H. Zelin, A study on the method of image pre-processing for recognition of crop diseases, in *2009 International Conference on Advanced Computer Control*, Singapore, (2009), pp. 202–206. https://doi.org/10.1109/ICACC.2009.10
4. M. Jhuria, A. Kumar, R. Borse, Image processing for smart farming: detection of disease and fruit grading, in *2013 IEEE Second International Conference on Image Information Processing (ICIIP-2013)*, Shimla, (2013), pp. 521–526. https://doi.org/10.1109/ICIIP.2013.6707647
5. T.A. Tattar, Root Diseases. *Diseases of Shade Trees*, (Revised Edition), (Elsevier, 1989), pp. 152–167. https://doi.org/10.1016/B978-0-12-684351-4.50016-8
6. S. Trimble, *Detecting Plant Root Diseases Pests*, blog
7. A. Johannes, A. Picon, A. Alvarez-Gila, J. Echazarra, S. Rodriguez-Vaamonde, A.D. Navajas, A. Ortiz-Barredo, Automatic plant disease diagnosis using mobile capture devices, applied on a wheat use case. Comput. Electron. Agric. **138**, 200–209 (2017). https://doi.org/10.1016/j.compag.2017.04.013
8. S. Phadikar, J. Sil, Rice disease identification using pattern recognition techniques, in *2008 11th International Conference on Computer and Information Technology*, Khulna, (2008), pp. 420–423. https://doi.org/10.1109/ICCITECHN.2008.4803079
9. P.R. Rothe, R.V. Kshirsagar, Cotton leaf disease identification using pattern recognition techniques, in *2015 International Conference on Pervasive Computing (ICPC)*, Pune, (2015), pp. 1–6. https://doi.org/10.1109/PERVASIVE.2015.7086983
10. P. Rajan, B. Radhakrishnan, L.P. Suresh, Detection and classification of pests from crop images using support vector machine, in *2016 International Conference on Emerging Technological Trends (ICETT)*, Kollam, (2016), pp. 1–6. https://doi.org/10.1109/CCECE.2017.7946594
11. M. Islam, A. Dinh, K. Wahid, P. Bhowmik, Detection of potato diseases using image segmentation and multiclass support vector machine, in *2017 IEEE 30th Canadian Conference on Electrical and Computer Engineering (CCECE)*, Windsor, ON, (2017), pp. 1–4. https://doi.org/10.1109/CCECE.2017.7946594
12. A. Kamilaris, F.X. Prenafeta-Boldú, Deep learning in agriculture: a survey. Comput. Electron. Agric. **147**, 70–90 (2018). https://doi.org/10.1016/j.compag.2018.02.016
13. Y. Lu, S. Yi, N. Zeng, Y. Liu, Y. Zhnag, Identification of rice diseases using deep convolution neural networks, Neuro Comput. **267**, 378–384. https://doi.org/10.1016/j.neucom.2017.06.023
14. X. Zhang, Y. Qiao, F. Meng, C. Fan, M. Zhang, Identification of maize leaf diseases using improved deep convolution neural networks. IEEE Access **6**, 30370–30377 (2018). https://doi.org/10.1109/ACCESS.2018.2844405
15. K.P. Ferentinos, Deep learning models for plant disease detection and diagnosis. Comput. Electron. Agric. **145**, 311–318. https://doi.org/10.1016/j.compag.2018.01.009
16. Z. Iqbal, An automated detection and classification of citrus plant diseases using image processing techniques: a review. Elsevier (2018). https://doi.org/10.1016/j.compag.2018.07.032
17. M.A. Khan, An optimized method for segmentation and classification of apple diseases based on strong correlation and genetic algorithm based feature selection. IEEE (2018). https://doi.org/10.1109/ACCESS.2018
18. A.K. Rangarajan, Tomato crop disease classification using pre-trained deep learning algorithm. Elsevier (2018). https://doi.org/10.1016/j.procs.2018.07.070
19. M. Bhange, Identification of apple leaf diseases based on deep convolution neural networks. Symmetry MDPI. https://doi.org/10.3390/sym10010011

20. S. Dubey, Detection and classification of apple fruit diseases using complete local binary pattern, in *Third International Conference on Computer and Communication Technology* (2012). https://doi.org/10.1109/ICCCT.2012.76
21. Y. Min, Plant leaf detection and classification using image processing. Int. J. Res. Eng. (2018). https://doi.org/10.21276/ijre.2018.5.9.4
22. S.H.-N.-M.E. Omrani, B. Khoshnevisan, Potential of radial basis function based support vector regression for apple disease detection. Measurement (2014). https://doi.org/10.1016/j.measurement.2014.05.033
23. P. Jiang, Real-time detection of apple leaf diseases using deep learning approach based on improved convolution neural. IEEE **7**. https://doi.org/10.1007/s42452-019-1393-4
24. N. Srivastava, G. Hinton, A. Krizhevsky, I. Sutskever, R. Salakhutdinov, Dropout: a simple way to prevent neural networks from over-fitting, J. Mach. Learn. Res. **15**(1), 1929–1958 (2014)
25. K. Simonyan, A. Zisserman, Very deep convolution networks for large-scale image recognition, [Online]. Available: https://arxiv.org/abs/1409.1556
26. H.P.J. Jeong, N. Kwak., Enhancement of SSD by concatenating feature maps for object detection, [Online]. Available: https://arxiv.org/abs/1705.09587
27. A. Caglayan, A.B. Can, Volumetric object recognition using 3-d CNNS on depth data. IEEE Access **6**, 20058–20066 (2018). https://doi.org/10.1109/ACCESS.2018.2820840
28. R. Siddiqi, *Automated Apple Defect Detection Using State-of-the-Art Object Detection Techniques*. Springer Nature Switzerland (2015). https://doi.org/10.1007/s42452-019-1393-4
29. A.S.D. Khirade, Plant disease detection using image processing, in *2015 International Conference on Computing Communication Control and Automation* (2015). https://doi.org/10.1109/iccubea.2015.153
30. B.J. Samajpati, Hybrid approach for apple fruit diseases detection and classification using random forest classifier, in *International Conference on Communication and Signal Processing*. https://doi.org/10.1109/ICCSP.2016.7754302
31. A.S.M. Shafi, Fruit disease recognition and automatic classification using MSVM with multiple Feature. Int J Comput Appl. https://doi.org/10.5120/ijca2018916773
32. M.R. Badnakhe, P. Deshmukh, Infected leaf analysis and comparison by otsu threshold and k-means clustering (2012)
33. S. N.-W. M.E. Leibe, J. Matas, SSD: single shot multi box detector, in Computer vision|ECCV 2016. ECCV 2016. Lecture Notes in Computer Science, vol. 9905. (Springer, Cham, 2016). https://doi.org/10.1007/978-3-319-46448-02
34. A. Fuentes, S. Yoon, S. Kim, D. Park, A robust deep-learning-based detector for real-time tomato plant diseases and pests recognition. Sensors **17**(9), 2022 (2017). https://doi.org/10.3390/s17092022
35. H. Lu, D. Wang, Y. Li, J. Li, X. Li, H. Kim, ..., I. Humar, CONet: A Cognitive Ocean Network. IEEE Wireless Commun. **26**(3), 90–96 (2019). https://doi.org/10.1109/mwc.2019.1800325
36. A. Jihen, A deep learning approach for banana leaf disease classification. BTW Workshops (2018). https://doi.org/10.1016/j.compag.2018.07.032
37. B. Liu, Identification of apple leaf diseases based on deep convolution neural networks. Symmetry MDPI. https://doi.org/10.3390/sym10010011
38. H. Durmuş, E.O. Güneş, M. Kırcı, Disease detection on the leaves of the tomato plants by using deep learning, in *2017 6th International Conference on Agro-Geo informatics*, Fairfax, VA, (2017), pp. 1–5. https://doi.org/10.1109/Agro-Geoinformatics.2017.8047016
39. J. Shijie, J. Peiyi, H. Siping, S. Haibo, Automatic detection of tomato diseases and pests based on leaf images, in *2017 Chinese Automation Congress (CAC)*, Jinan, (2017), pp. 2537–2510. https://doi.org/10.1109/CAC.2017.8243388
40. J.Q.D.M.B. Dae Gwan Kim, T.F. Burks, Classification of grapefruit peel diseases using color texture feature analysis. Int. J. Agric. Biol. Eng. **2**(3), 4150 (2009). https://doi.org/10.3965/j.issn.1934-6344.2009.03.041-050
41. B.L. Patil, Plant viral diseases: economic implications. Ref. Module Life Sci. (2020). https://doi.org/10.1016/b978-0-12-809633-8.21307-1

Part III
NLP and ICT

Agitation on Social Media Amid Terrorist Attacks: A Case Study on Jammu and Kashmir

Megha Chaudhary and Divya Bansal

Abstract With the advent of the Internet, social media has penetrated into a major part of world's population. It has become a platform of opinion sharing and communication for plethora of users. In this paper, the prime motive is to study the variation of opinions on social media platform Twitter amid terroristic activities. The case study is done on Jammu and Kashmir region of India, which is one of the most vulnerable areas to terrorism in the country. Twitter data is collected for over a year, and temporal analysis of sentiments is done, to evaluate the effect of terroristic activities on public opinions. The data was collected using trending hashtags for two major terrorism attacks, i.e., Pulwama Attack and Jammu Bus Grenade Blast which occurred in year 2019. The variation of sentiments w.r.t. severity of a terroristic attack is analyzed. It is observed that a terroristic attack having high impact in terms of life loss and national security breach has prolonged agitation of negative opinions on Twitter.

Keywords Twitter · Sentiment analysis · Terrorism · Naïve Bayes classification

1 Introduction

In the current scenario, with the reachable access to Internet services across the globe, social media platforms have gained a wide popularity. People communicate and verbalize their opinions on these platforms. Twitter is one of the platforms where people can post images, text and videos. Currently, it has around 330 million monthly active users and 145 million daily active users [1]. This infers a huge content on public opinions being created on Internet on daily basis. People discuss over common topics on Twitter using trending hashtags on the topic.

M. Chaudhary (✉) · D. Bansal
Cyber Security Research Center, Punjab Engineering College (Deemed to Be University), Chandigarh, India
e-mail: meghaa91@gmail.com

D. Bansal
e-mail: divya@pec.edu.in

V. E. Balas et al. (eds.), *Proceedings of International Conference on Computational Intelligence, Data Science and Cloud Computing*, Lecture Notes on Data Engineering and Communications Technologies 62, https://doi.org/10.1007/978-981-33-4968-1_30

It has been observed by researchers, social media trends vary before and after a terroristic attack [2, 3]. There are many users who have radical thoughts and spread their propagandas by tweeting negative posts amid terrorist attacks. Twitter has also been actively working to delete such accounts [4]. In this paper, a case study is done on two major terrorist events in Jammu and Kashmir, India. Due to the dispute over the Kashmir area between India and Pakistan, there are many active militant groups making the state one of the most vulnerable states to terrorism [5]. The two major attacks focused here are Pulwama Attack on February 14, 2019, when a bomb was placed in an army bus killing 46 army personnel and Jammu Bus Grenade Blast on March 7, 2019, killing 3, injuring 28-35 civilians [6]. The tweets were collected for period—February 2019 to March 2020 using trending hashtags pertinent to the two given attacks.

Sentiment analysis is done on the tweets extracted on the aforementioned events. A temporal analysis of sentiments is done over the year. Apart from the two major attacks in the given year, due to prevalent terrorism activities in the state tweets on terrorism are often posted. This can be attributed to frequent cease-fires on the borders, firings, etc. Hence, the analysis is extended to comparison with tweets on just terrorism, not related to any of the two events, to evaluate the agitation and unrest among the masses w.r.t terrorism in Kashmir. The results are evaluated using visualization graphs on sentiments and wordclouds.

The paper is organized into five sections. The related work is presented in Sect. 2 followed by methodology in Sect. 3. Data visualizations and results are discussed in Sect. 4. In Sect. 5, conclusion is drawn from the observations marked in the analysis.

2 Related Work

With the advancement in Internet technologies, the social media platforms have paved a way for people to easily connect to the world. The popular platforms are Twitter, Facebook, YouTube, Instagram, etc. There are various users on these platforms who actively participate in spreading terroristic propagandas. Such users influence the candidates vulnerable to brainwash in context of terrorism. The terroristic organizations also carry on online recruitments by spreading such propagandas [7].

The authors in [8] have shown the presence of terroristic content on messenger applications and social media platforms using keyword-based search. Another platform for sharing intimidating content related to terrorism is weblogs social networks [9]. The content on such platforms can be extracted and analyzed using various data mining techniques such as link predictions, network analysis and sentiment analysis which can help the defense agencies to get valuable information, e.g., unrest in public and hidden radical communities on social networks.

Researchers have been carrying on work in area of social media analytics with respect to terroristic events. In context of Jammu and Kashmir, the authors in [10] did sentiment analysis on Uri terror attack in Kashmir which took place in September 2016. The author collected data using Twitter API for over a month using trending hashtags as search keywords. The author studied the survival of the tweets over time based on the count of retweets and favorite count. Sentiment analysis is done using Naïve Bayes–support vector machine algorithm. The observation was made that the negative tweets survive more than the positive tweets.

Another event analysis was done by authors in [11] where the revocation of special provisions of Article 370 of constitution of India was done. This article marked the abolition of limited autonomy granted to Jammu and Kashmir. The authors did opinion mining by applying sentiment analysis on tweets collected using keyword-based extraction. Python library TextBlob was used for sentiment analysis. It was observed that Pakistani users tweeted more on trade-related issues after the event and the Indians tweeted on terrorism in the state having optimistic opinions on the event.

Apart from Kashmir, Twitter analysis has also been done for other terroristic organizations and countries with different analytical parameters. Those can be geographical, demographical and link analysis on user profiles [12]. The authors in [13] collected tweets related to ISIS terrorist organization from eight different countries. The countries were equally distributed as four from east and four from west. Tweets were collected based on keyword search, and sentiment analysis was done using Opinion Lexicon dictionary in R language. The results showed that the opinions on ISIS on two opposite sides of the globe were similar. Countries all across the world perceive it as a notorious organization and extrapolate fear from the same.

3 Methodology

The methodology is divided into four steps. Figure 1 represents the flow of the steps.

3.1 Data Collection

Data collection is done from Twitter using Twitter Streaming API for the period February 2019 to March 2020. The tweets were extracted on the basis of trending hashtags for two terroristic events, i.e., Pulwama Attack and Jammu Bus Grenade Blast. A set of tweets was extracted which were related to terrorism apart from the two aforementioned events called "other related tweets". The trending hashtags used for data collection are shown in Table 1.

The "other related tweets" were extracted such that they do not contain any of the hashtags from the two terroristic events. Also, it is evident that other tweets are too generic to relate to terrorism from Jammu and Kashmir only. So, they may be containing tweets about terrorism all across the globe. So, to narrow down the tweets to Jammu and Kashmir, another set of mandatory hashtags is used, i.e., "#JammuKashmir, #Kashmir, #Jammu, #JammuAndKashmir, #SouthKashmir and #NorthKashmir". Hence, the tweets having hashtags from "other related tweets"

Fig. 1 Methodology

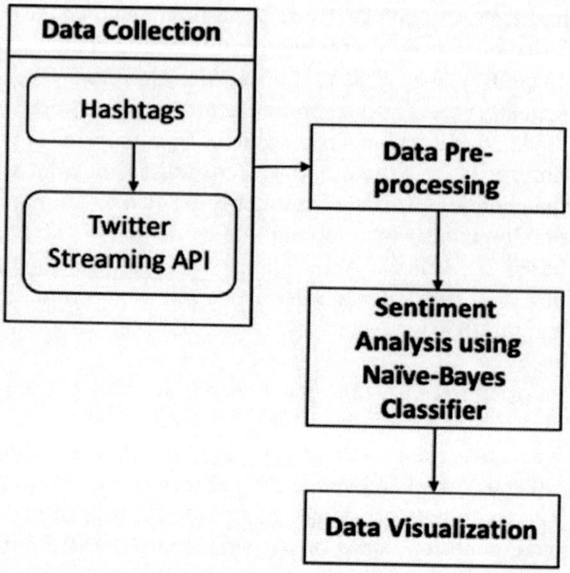

Table 1 List of hashtags for data collection

S.No	Terrorism Event	Trending Hashtags
1	Pulwama Attack (February 14, 2019)	#PulwamaTerrorAttacks, #PulwamaAttack, #PulwamaEncounter, #Pulwama
2	Jammu Bus Grenade Attack (March 7, 2019)	#JammuTerrorAttack, #jammubusblast, #jammubusstand, #JammuBlast, #busstopblast, #JammuBusStandBlast, #busstopbombblast
3	Other related tweets	#jihad, #MilitantsKilled, #Terrorists, #Encounter, #Militants, #Jihadist, #Jihadis, #JEM, #Islamist, #Islamists, #Islamophobia, #militantskilled, #terrorists, #encounter, #militants, #jihadist, #jihadis, #JEM, #islamist, #islamists, #islamophobia

should contain at least one hashtag from set of mandatory hashtags. The tweets were also narrowed down to extraction in English language only.

3.2 Data Preprocessing

The methodology adopted for sentiment analysis is Naïve Bayes classifier which is a supervised machine learning technique. Since the supervised machine learning technique needs training data, it was created manually by annotating 2000 tweets as positive and negative. After the annotation, for creation of the Naïve Bayes model based on training data, data preprocessing of the annotated tweets is done. The set of tasks in preprocessing included tokenization of tweets for each category, i.e., positive tweets and negative tweets. The tokenized tweets are cleaned by removing stopwords, punctuations and web links. The tokens are further processed through lemmatization. This makes the training dataset optimized for training the Naïve Bayes classifier for sentiment analysis.

3.3 Sentiment Analysis

The model used for sentiment analysis is the Naïve Bayes classifier. The classifier uses Bayes theorem of probability to classify a sentence as positive or negative. The Bayes rule is defined by two equations:

$$P(c|x) = P(x|c)P(c)/P(x) \tag{1}$$

$$P(c|X) = P(x_1|c)x\ P(x_2|c)x\ P(x_3|c)$$
$$x\ P(x_4|c)x \ldots\ldots\ldots x\,P(x_n|c)x\,P(c) \tag{2}$$

Here *P*(*c*|*x*) is the posterior probability of the class *c*(positive or negative in the context of sentiment analysis) given predictor *x*. *P*(*c*) is prior probability of class. *P*(*x*|*c*) is the likelihood, i.e., probability of predictor given the class and *P*(*x*) is prior probability of predictor. For each sentence, the posterior probability for each class is calculated, and the class with the highest posterior probability is the result of the classification. Since the model is supervised, the data annotated in the previous set is divided into training and testing data in 70:30 ratio. The model is then trained using NLTK sentiment analysis Naïve Bayes classifier. Natural Language Toolkit (NLTK) is a set of libraries in Python which implements various algorithms on natural language processing, e.g., sentiment analysis, topic modeling, etc.

3.4 Data Visualization

The temporal effect of sentiments of the twitter users is visualized with respect to three aforementioned categories of events. The visualization is done using histograms and wordclouds. The results are discussed in the next section.

4 Results and Discussions

A total of 38178 tweets were collected for all the categories. The temporal analysis of the tweets is done as follows.

4.1 Tweets Per Month Analysis

It can be observed from Fig. 2 that for Pulwama Attack which took place in Feb 2019, the tweets are highest in the month of occurrence followed by gradual decrease to the end of the year. It can be seen that the tweets abruptly rise for the Pulwama Attack in month of August, and this can be attributed to a major political decision for the state taken by the Government of India [11]. People standing against the decision condemn it by reminding the anger for the brutal Pulwama Attack. Another spike in tweets for Pulwama attack is seen around February 2020, which marks the completion of one year of attack. Since the attack was on Indian Army killing 46 army personnel, people remember the attacks by paying condolences and homage to the martyrs through their tweets.

For Jammu terror attack, the spike is seen only in the month in which the attack occurred. The attack is hardly discussed after that. This can be attributed to less human life loss as compared to the former terrorist attack. Tweets related to terrorism other than these two terroristic attacks are posted throughout the year. But, it can be

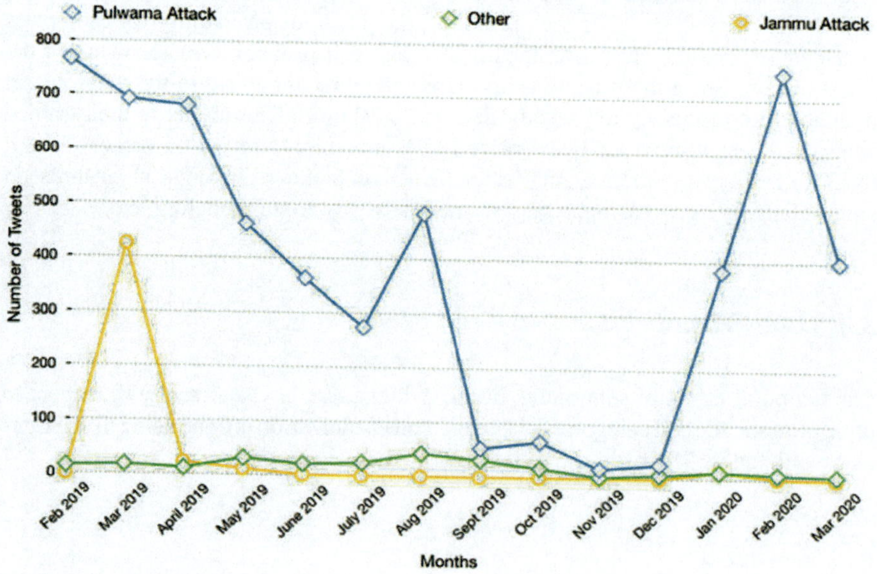

Fig. 2 Number of tweets per month

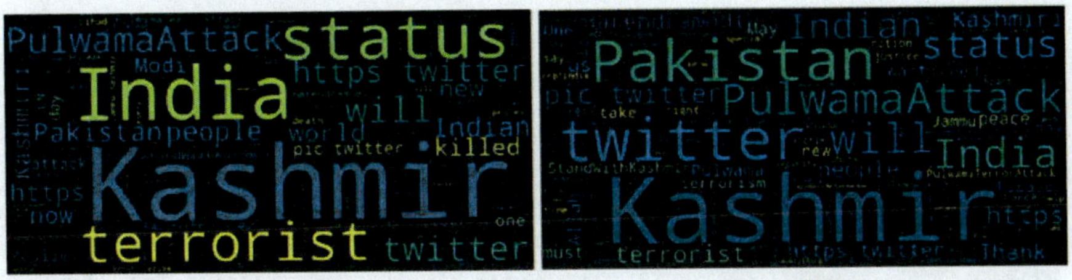

(a) (b)

Fig. 3 Wordclouds on negative and positive tweets **a** Negative tweets wordcloud **b** Positive tweets Wordcloud

observed that the majority tweets on terrorism contain hashtags on Pulwama Attack. It infers a huge prolonged unrest among the people of Kashmir.

4.2 WordClouds

The sentiment analysis classified the tweets as negative and positive based on training data. The wordclouds for the two are shown in Fig. 3. It can be observed that negative tweets have more weightage to words like “killed”, “terrorist”, “attack”, “death” and “Jihad”, whereas the positive has words “peace”, “martyr”, “support”, “Allah”, “thank” and “please”. The positive tweets are mostly thankful tweets to the martyrs for their support to the country and also to the prime minister of the country.

4.3 Temporal Visualizations of Sentiments

Distribution of number of tweets across the time w.r.t. sentiments for each terroristic event aforementioned is shown in Fig. 4. The X-axis represents the time in terms of consecutive combination of two months from February 2019 to March 2020, while the Y-axis represents the number of tweets. It can be observed that the negative tweets are always greater than the positive tweets all across the year, and the maximum tweets are in context of Pulwama Attack. The number of tweets gradually decreases with increasing time from the occurrence of the event.

The Jammu Blast has significant number of tweets only in the month of the attack, i.e., March 2019, with negative tweets more in number. The spike in tweets for Pulwama Attack is seen on completion of one year after the attack. The tweets after one year of the attack are still negative in majority. The positive tweets in case of Pulwama Attack are accounted to the thankful tweets for the martyrs in the attack. Also, in response to the attack, airstrikes by Indian Air Force were done on Pakistani militant groups which marked positive tweets by the Indian users on Twitter.

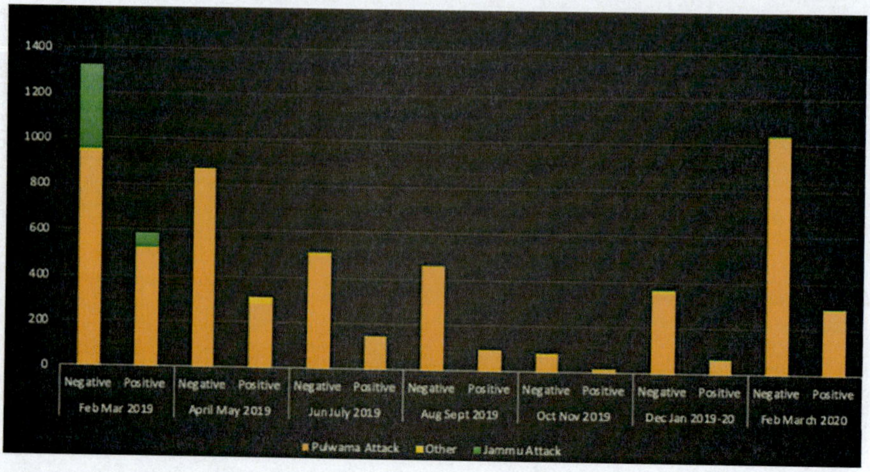

Fig. 4 Distribution of positive and negative tweets over time

5 Conclusion

With the advent of the Internet, social media platforms have become a means of opinion sharing among mass audiences. Twitter having a large base of users has given them a platform to influence a plethora of diverse public across the world. People can keep their viewpoint on any subject matter of their interest. From the analysis done in this paper, it can be concluded that a large number of users actively participate to share their opinions on terroristic events. It can be either hate posts, posts on condolences or posts on political impact of the terroristic events. The analysis infers that the agitation on social media increases at the time of the attack and gradually decreases over time. The severity of the attack also determines the prolonged existence of posts on Twitter. Severity can be defined in terms of life loss and loopholes in national security. This analysis can help defense and intelligence agencies determine the level of unrest among the public. They can also determine the users who are deliberately creating hate content on social media and radicalizing other users.

References

1. Twitter Statistics Every Marketer Should Know in 2020 [Infographic]. https://www.oberlo.in/blog/twitter-statistics, last accessed 09 Aug 2020
2. M. Almoqbel, S. Xu, Computational mining of social media to curb terrorism. ACM Comput. Surv. **52**(5), (2019)
3. I. Terziyska, S. Shah, X. Luo, Are recent terrorism trends reflected in social media? In *Proceedings—14th IEEE International Conference on Mobile Ad Hoc and Sensor Systems. MASS 2017*, (IEEE, Florida USA, 2017), pp 535–539

4. G. Kalpakis, T. Tsikrika, I. Gialampoukidis, S. Papadopoulos, S. Vrochidis, I. Kompatsiaris, Analysis of suspended terrorism-related content on social media, in *Community-Oriented Policing and Technological Innovations*, ed. by G. Leventakis, M. Haberfeld (SpringerBriefs in Criminology. Springer, Cham, 2018)
5. Insurgency in Jammu and Kashmir—Wikipedia, https://en.wikipedia.org/wiki/Insurgency_in_Jammu_and_Kashmir. Last accessed 09 Aug 2020
6. List of terrorist incidents in India—Wikipedia, https://en.wikipedia.org/wiki/List_of_terrorist_incidents_in_India. Last accessed: 09 Aug 2020
7. A. Saad, A. Alhumaid (2019) How terrorist groups use social networking media to attract and recruit new members, in *Proceedings 13th International Conference Innovation* Information Technology. *IIT* (IEEE, UAE, 2019), pp. 129–134
8. T. Kolajo, O. Daramola, Leveraging big data to combat terrorism in developing countries, in *Conference on Information Communication Technology and Society* ICTAS 2017—Proceeding (IEEE, Umhlanga, South Africa, 2017)
9. C.C. Yang, T.D. Ng, Terrorism and crime related weblog social network: Link, content analysis and information visualization, in *IEEE Intelligence and Security Informatics ISI 2007* (IEEE, New Brunswick, NJ, USA, 2007), pp. 55–58
10. P. Garg, H. Garg, V. Ranga, Sentiment analysis of the Uri terror attack using Twitter, in *Proceeding—IEEE International Conference Computer Communication and Automation ICCCA 2017* (IEEE, Noida, India, 2017), pp. 17–20
11. R. Patil, N. Gada, K. Gala, Twitter data visualization and sentiment analysis of article 370, in *6th International Conference on Advanced Communication Control and Computing ICAC3 2019*, (IEEE, Mumbai, India, 2019), pp. 1–4
12. M. Dawson, M. Leibel, A. Adeboje, Open source intelligence: performing data mining and link analysis to track terrorist activities. Inf. Technol. - New Gener. Adv. Intell. Syst. Comput., **558**, 159–163 (2018)
13. S. Mansour, Social media analysis of user's responses to terrorism using sentiment analysis and text mining. Procedia Comput. Sci. **140**, 95–103 (2018)

Bilingual Machine Translation: Bengali to English

Sauvik Bal, Supriyo Mahanta, and Lopa Mandal

Abstract The present work proposes a novel architecture of machine translation from Bengali language to English Language. After taking the Bengali sentences as input, syntactical analysis has been done. After that it is analyzed semantically, and the sentence is rearranged as per English grammar. A dictionary has been prepared and used for mapping from Bengali to English words. The proposed system is based on the concept of syntax-based and meaning-based approach. Proper grammatical rules are used for translating the sentences from Bengali to English language. In some cases, the proposed system performed better compared to the Google translator output.

Keywords Machine translation · Bengali to English translation · Parse tree · POS tagging

1 Introduction

In today's world, language translation is very much required for communication. Machine translation is one of the important tools by which language translation can be done very easily. Translation time will be reduced when machine translation techniques are applied. Also machine translation process is able to translate a sentence from one language to many different language at a time. Though the performance of machine translation is not as good as human being, many problems may occur in machine translation due to lack of resources, proper sentence structure analysis,

S. Bal (✉)
University of Engineering and Management, Jaipur, India
e-mail: sauvikbal@gmail.com

S. Mahanta
Kreeti Technologies Pvt. Ltd, Kolkata, India
e-mail: mahantasupriyo@gmail.com

L. Mandal
Institute of Engineering and Management, Kolkata, India
e-mail: mandal.lopa@gmail.com

V. E. Balas et al. (eds.), *Proceedings of International Conference on Computational Intelligence, Data Science and Cloud Computing*, Lecture Notes on Data Engineering and Communications Technologies 62, https://doi.org/10.1007/978-981-33-4968-1_31

proper grammatical rule implementation, etc. Present work considers two important languages Bengali and English for bilingual machine translation. Bengali is one of the most important languages in India and Bangladesh. The present work proposes a novel approach to convert a sentence from Bengali language to English language based on some grammatical rules and proper meaning£.

2 Literature Survey

Different approaches can be used for translation between various languages. There are some approach where syntax-based and meaning-based analysis are used for conversion. Phonetic transliteration can be used for low resourced language. Web resources are very less in Bengali language compared to English. Due to the lack of unique features of web data in low resourced language, it is difficult to prepare the training data. Many times, the stemmers, used in low resource language, cannot properly analyze the sentences [1]. One expression analysis can improve the efficiency of machine translation technique. Classification technique can be used to merge the one expression. CRF-based classifier is trained and tested for this purpose. This classification-based technique can improve the efficiency of the machine translation. It is needed to use more features for training and testing purposes [1, 2]. Text classification is one of the important things for machine translation. It can be done based on the important information of a document [3]. For repetition words, morphological rules can be implemented through Universal Networking Language. Repetition words can be found out by using the verb root and their different forms [4]. Word-level and phrase-level paraphrasing can be used to improve the translation quality. Adaption of bitexts for poor and rich resource languages can speed up the translation process [5]. In case of example-based machine translation, ontology technique can be used for low resourced languages. In this technique, chunk of strings is available in both source and target languages. Also the mapping information between source and target languages is available. This process is not suitable for unknown words. Statistical method can be used in such situations [6].

In case of Bengali language, ambiguity checking is one of the most important research fields. Vector space model can be used for this kind of job. Bengali words can be used in various perspectives. Bengali stop words are procured and removed to reduce the problems of calculation. Ambiguity checking can be done by using LSA with vector space model concept [7]. Sometimes transliteration is more useful than translation. Sometimes transliteration is needed due to inadequacy of lexical knowledge; poor grammar and insufficient phonetics. Morphological analysis is required for this [8]. Corpus-based and statistical approach can be used for Bengali to

English translation. Direct target language translation can be done using the concept of corpus-based approach. According to the linguistic rule, translated sentence can be rearranged by using the N-gram model. Most problematic thing of this Bengali to English translation is preparation of training set. This approach is not implemented for all types of sentence structures [9]. There are some sentences where some verbs are present which has different syntactic meaning. This may lead problems during the machine translation. Some procedures are there by which semantically different verb forms that can be identified during the machine translation [10]. Bidirectional long short-term memory (LSTM) can be utilized to decipher Bengali–English language. This process performed well for large dataset. Quality of the translation is increased when this technique is used [11, 12]. Neural machine translation is not much tested for Bengali to English translation. There are some linguistic challenges that may happen like agreement of subject–verb, detection of noun, rare words translation, etc. It is observed that neural machine translation can perform better than phrase-based statistical machine translation system [13]. For statistical translation system, it is problematic to use conventional decoding algorithm. Distinctive pruning technique can assist by accelerating the procedure and improve the nature of the translation. Cube pruning is one of the important approaches by which decoding speed can be increased [14]. Word2Vec and bidirectional contextual encoder can be used for this purpose. Contextually embedded deep learning-based model can be used at input level. Statistical approach can be used to calculate the cost weight at learning time [15]. Named Entity Recognition is one of the important thing at the time of translation. As Bengali is the low resource language, it is very important to do Named Entity Recognition [16].

3 Motivation and Problem Statement

It has been observed from the literature review that example-based and rule-based techniques have been widely used in machine translation. Rule-based approach can be direct or transfer-based. Corpus-based machine translation can be classified in different statistical process. Rule-based translation basically depends on knowledge sources. It can be classified in direct, transfer and interlanguage-based approach. Some important translation techniques that are used practically are knowledge-based, corpus-based, example-based and context-based machine translation [6, 12].

Corpus-based method depends on bilingual corpora. It may lead issues for closely matched words. Statistical machine translation is one of the fundamental machine translation approaches. This process performed well to select the proper lexeme. Rule-based method is required for grammatical correctness. So, performance will be increased if hybrid translation techniques can be applied.

It is also clear from the survey that due to different issues it is problematic to translate Bengali to English language. Bengali is very low resourced language for machine translation. There are so many problems due to shortage of resources, weak vocabulary, lack of analysis tools, etc. So, it is very important to analyze the sentence properly before translation. Proper language-specific grammar needs to be followed when sentences are constructed while translating.

In this context, the present work aims to propose a machine translation methodology from Bengali to English language based on a hybrid technique of rule-based and corpus-based methods.

4 Methodology

In the current work, a novel methodology has been suggested for Bengali to English machine translation. Bengali sentences are taken as input to the system, and it produces the corresponding English sentences. After taking the input of Bengali sentences, it is divided into tokens by using the parts of speech (POS) tagger. POS tagging has been done for each word. Then the words are clubbed together into three groups, i.e., subject, verb, object and some other required parts (e.g., WH- words, exclamatory expression, etc.). A few models are appeared in the accompanying figures for various sentences (Figs. 1, 2, 3, 4 and 5). For assertive sentence (Fig. 1)

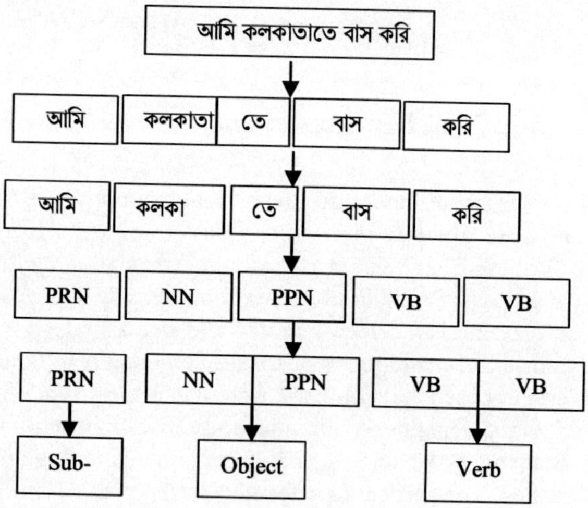

Fig. 1 Sentence analysis for assertive sentences (নির্দেশক বাক্য)

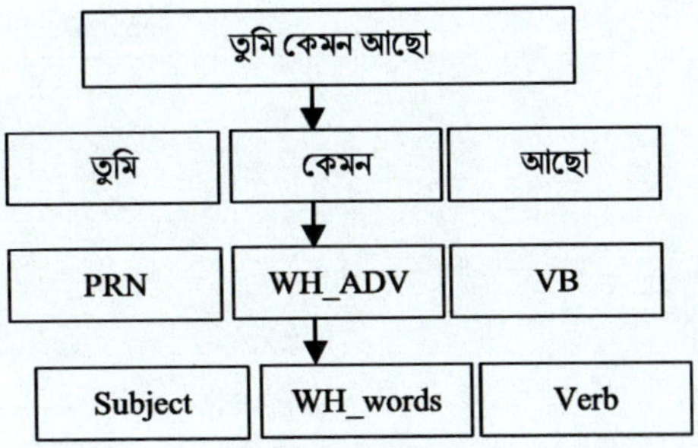

Fig. 2 Sentence analysis for interrogative sentences (প্রশ্নবাচক বাক্য)

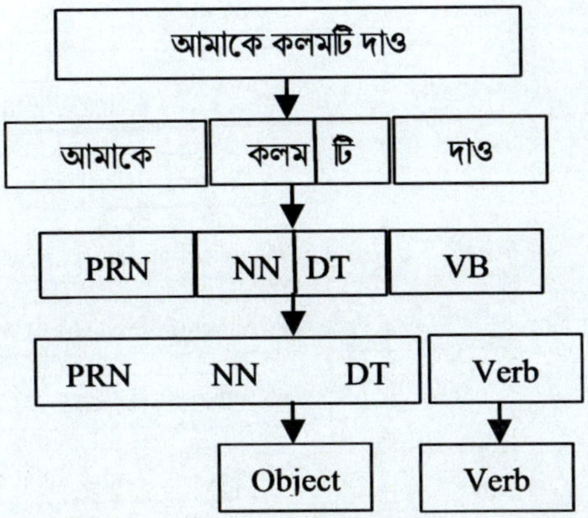

Fig. 3 Sentence analysis for imperative sentences (অনুজ্ঞাসূচক বাক্য)

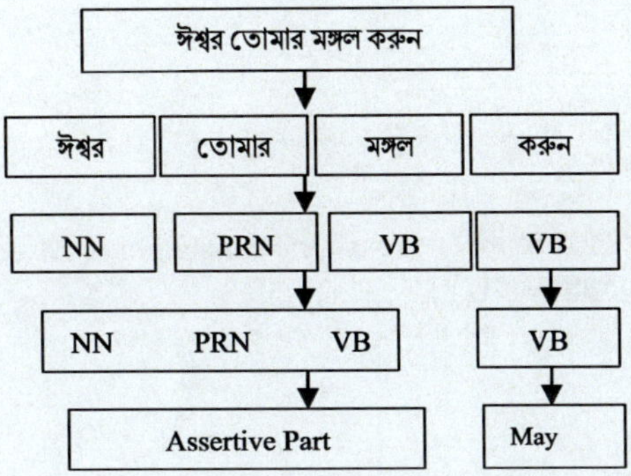

Fig. 4 Sentence analysis for optative sentences (ইচ্ছাবাচক বাক্য)

"I live in Kolkata" (আমি কলকাতাতে বাস করি) has taken. Similarly for Interrogative Sentence (Fig. 2) "How are you" (তুমি কেমন আছো), for Imperative Sentence (Fig. 3) "Give me the pen" (আমাকে কলমটি দাও) and for Optative Sentence (Fig. 4) "God bless you" (ঈশ্বর তোমার মঙ্গল করুন) have taken.

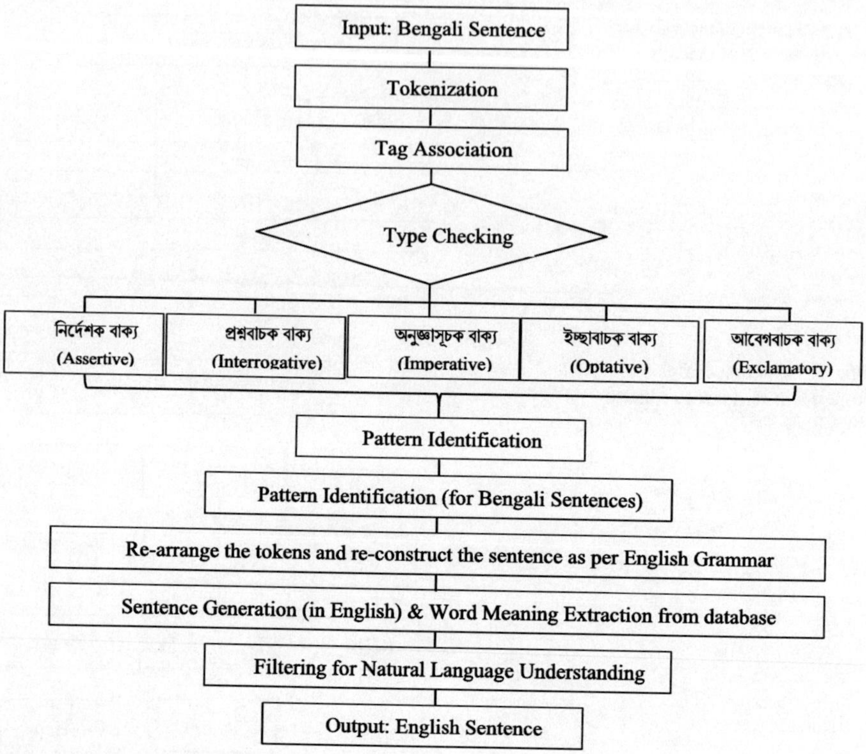

Fig. 5 Block diagram of the proposed system

5 Proposed Algorithm

The proposed algorithm translation is based on the concept of Bengali and English grammar [17].

Algorithm 1:

Procedure;
Begin:
1. Take the Bengali sentences as input.
2. Extract the words from the text and put it in a different Word Class.
3. Apply necessary stemming on each word to extract the root word for meaning retrieval.
4. Parts of Speech of each word has been marked by using the POS (Parts of Speech) Tagger.
5. Save them into different Token Class (i.e.: The set of tokens of the words of that sentence).
6. Create a parse tree based on Subject, Verb and Object and by checking the syntactic Structure.
7. If the sentence structure is "কর্তা + ক্রিয়া + কর্ম / ক্রিয়া-বিশেষণ / বিশেষণ", it will be marked as বিবৃতিমূলক বাক্য অথবা নির্দেশক বাক্য।
8. If the sentence contains 'কি', 'কেমন',.. ইত্যাদি and ended with '?', and having the structure like "কর্তা + 'কি', 'কেমন'... ইত্যাদি + ক্রিয়া", it will be marked as প্রশ্নবাচক বাক্য।
9. If the pattern is "কর্তা (অদৃশ্য) + ক্রিয়া + কর্ম", mark as অনুজ্ঞাসূচক বাক্য । subject is invisible in this case.
10. If the sentence contains মঙ্গলকর, প্রার্থনা অথবা ইচ্ছা বিষয়ক Bengali words and the pattern of the sentence is "কর্তা +ইচ্ছাবাচক/ প্রার্থনাসূচক শব্দ + ক্রিয়া", it will be marked as প্রার্থনাসূচক অথবা ইচ্ছাবাচক বাক্য।
11. Construct the pattern of the sentence using tokens and words.
12. After creating the Pattern of tokens, set the flag values (i.e.: if it is an Auxiliary verb then put AUX=True.)
13. Check the prefix verb is in correct form or not. Next it is checked that which person does the word signify?
    If it is in First Person, then put First_Per=True
14. Trigger different property of the Bengali Grammar and provided classifiers for checking of the Syntax of the Expression grammatically.
15. Then mapping has been done for Bengali to English tags.
16. Construct the pattern for each type of the sentences, and check the given sentence. Put it into that class by using a Choice Value.
17. After setting the Choice values and getting accurate sentence of words and tokens as English meaning, fetch the English meaning as per its English word from the Dictionary.
18. Use the Dictionary as database for getting the English Meaning of the Bengali words.
19. "beng" column contains the Bengali words and "eng" column contains the corresponding row's English meaning including its type (i.e- verb, adjective, noun, etc), tense (i.e- Present, Past, etc.), Person (1st, 2nd, 3rd), etc.
20. Make the Syntactic changes using some rules for better and more realistic results.
21. After fetching the meaning of English words put them into an English sentence as per Token and word classes.
22. English sentence will be the output.

End Procedure;

6 Experimentation and Result Analysis

The proposed system hasbeen developed in Python language with Natural Language Toolkit.

According to the proposed methodology, intermediate steps of the translation of an example case have been described below.

1. A Bengali text is taken as input.

   ক্লাসের পরে আমার সাথে সাক্ষাৎ করো

2. Extract the word tokens from the text.

Word Token1: ক্লাসের	Word Token2: পরে	Word Token3: আমার	Word Token4: সাথে	Word Token5: সাক্ষাৎ	Word Token6: করো

3. Apply necessary stemming on each word-token to extract the root word for meaning retrieval.

   Find the root word: ক্লাস + এর

4. Then each word will be processed through our pattern-matcher for parts of speech tagging. Apply any changes if anything required by the pattern-matcher. Example: “class” Bengali word is a proper noun. So “এই” added before it.

   এই ক্লাস এর পরে আমার সাথে সাক্ষাৎ করো

5. Then apply pos tagger to the Bengali sentence token (using B2E dictionary)

   এই/ADJ ক্লাস/NN পরে/PRPN আমার/PRN সাথে/PRPN সাক্ষাৎ করো/VB

6. After POS tagging then we have to break the sentence into three parts a. Subject b. Verb c. Object, The proposed pattern in Bengali sentence: [Obj + Sub + Verb]. Fill the parts with the POS tags:
   1. Obj list: ADJ, NN, PRPN
   2. Sub list: PRN
   3. Verb list: VB

এই ক্লাস পরে	আমার	সাথে সাক্ষাৎ করো
Object	Subject	Verb

7. Then as per English sentence pattern, we have to rearrange the Bengali sentence (one by one each of them). Proposed English pattern: Verb + Sub + Obj
8. Process each Parts until they are matching with the English pattern. rearrange the tokens as per pattern

   এই ক্লাস পরে

   ↓

পরে	এই	ক্লাস
PRPN	ADJ	NN

9. Now, the Bengali words will replaced with its corresponding English meaning fetched from our dictionary.

   সাক্ষাৎ করো আমার সাথে পরে এই ক্লাস

10. We got the Output English text

    meet me after the class

The proposed system has been compared with Google translator. Some test cases are shown here where the proposed system has performed well than Google translator (refer Figs. 6, 7, 8, 9, 10, 11, 12, 13).

Here some test cases are taken where the output of Google translator is not proper. The abovementioned figures show the Google translator output. It is examined that some of the result of Google translator is not proper and not actually matched with the standard result [18]. So, the comparative study between Google translator and proposed system is described below. Present work compared the output with the Google translator on the basis of the Bengali version used by the Bengali people of West Bengal, India. For analysis of the sequence of the output, here class "difflib.SequenceMatcher" is used in Python language. There is another class class "difflib", "Differ", by which the difference between sequences can be identified (Fig. 14; Table 1).

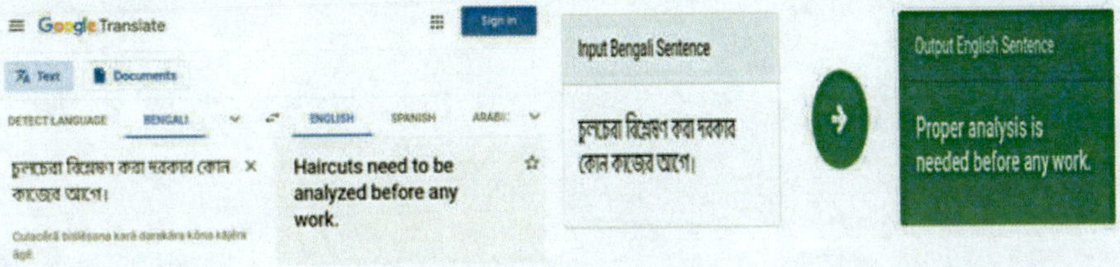

Fig. 6 Sample output of Google translator and proposed system of the sentence " চুলচেরা বিশ্লেষণ করা দরকার কোন কাজের আগে"

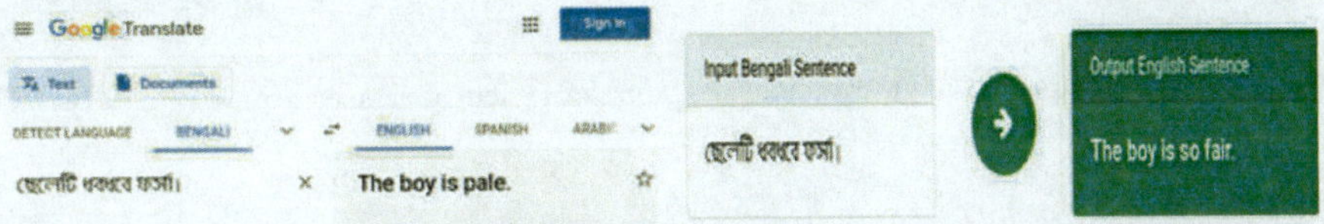

Fig. 7 Sample output of Google translator and proposed system of the sentence " ছেলেটি ধবধবে ফর্সা।"

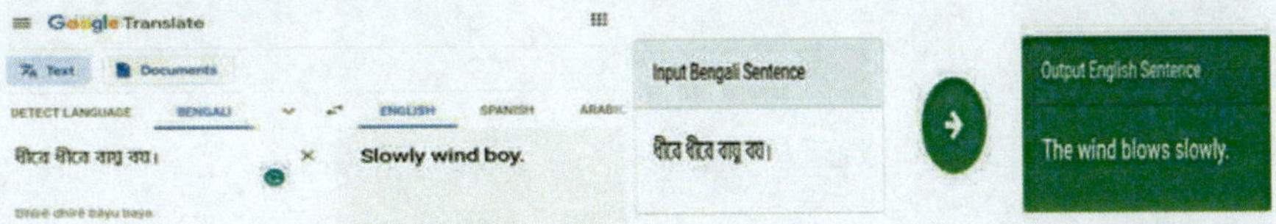

Fig. 8 Sample output of Google translator and proposed system of the sentence "ধীরে ধীরে বায়ু বয়।"

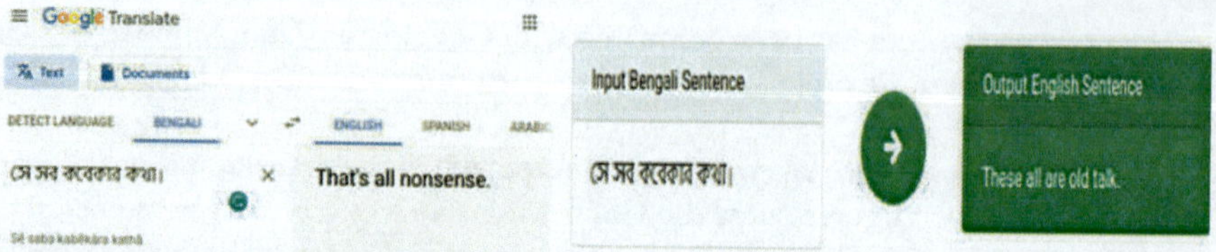

Fig. 9 Sample output of Google translator and proposed system of the sentence "সে সব কবেকার কথা।"

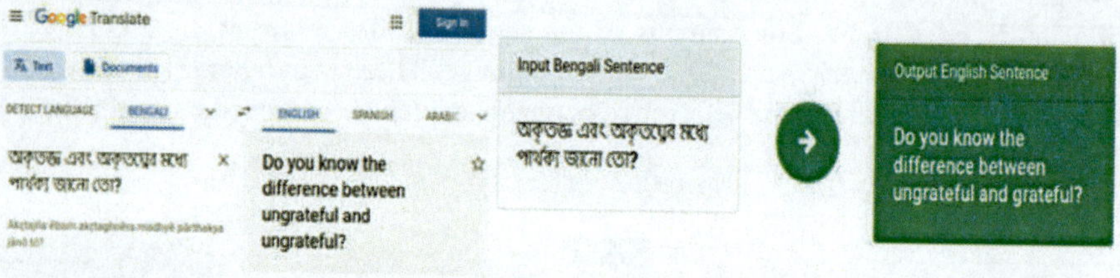

Fig. 10 Sample output of Google translator and proposed system of the sentence "অকৃতজ্ঞ এবং অকৃতজ্ঞের মধ্যে পার্থক্য জানো তো?"

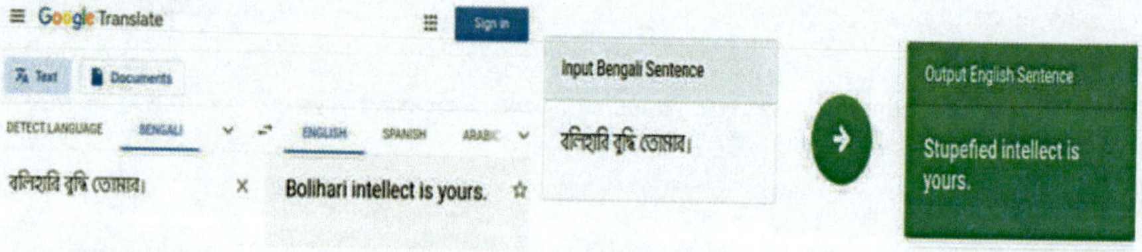

Fig. 11 Sample output of Google translator and proposed system of the sentence "বলিহারি বুদ্ধি তোমার।"

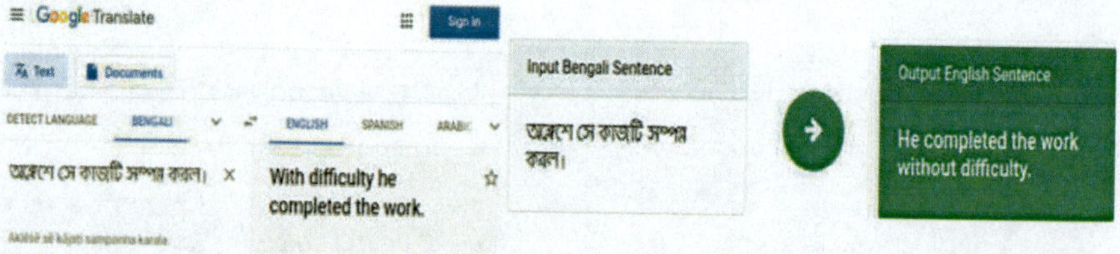

Fig. 12 Sample output of Google translator and proposed system of the sentence "অক্লেশে সে কাজটি সম্পন্ন করল।"

7 Conclusions and Future Scopes

Due to improper analysis of sentences, sometimes machine translation is not performed well. The proposed system analyzes the input Bengali sentence and

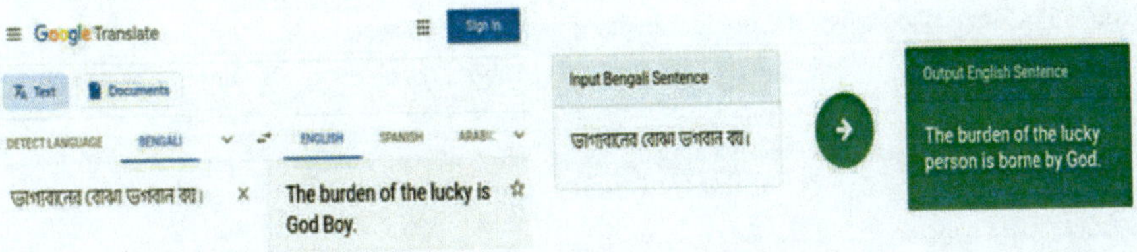

Fig. 13 Sample output of Google translator and proposed system of the sentence "ভাগ্যবানের বোঝা ভগবান বয়"

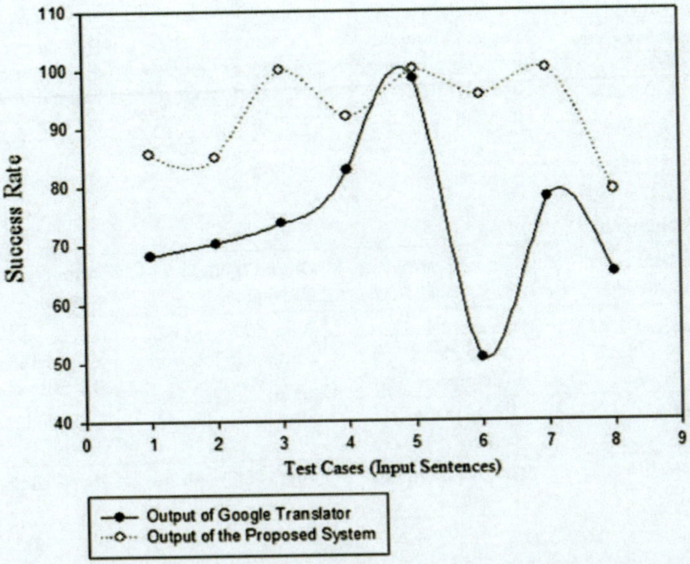

Fig. 14 Statistical result analysis between Google translator and proposed system

converts it to corresponding English sentence based on the hybrid technique of rule-based and corpus-based methods. Improved methodology for translation from Bengali to English sentence has been proposed. The work has been compared with the Google translate, and in some test cases the proposed system performed better. The work will be broadened further for different kinds of sentences in future. The resources need to be increased to get more accurate result in translation.

Table 1 Comparison table of Google translator and proposed system

Bengali sentences	Standard output	Output (google translator)	Output (proposed system)
চুলচেরা বিশ্লেষণ করা দরকার কোন কাজের আগে।	Proper analysis needs to be done before any work	Haircuts need to be analyzed before any work	Proper analysis is needed before any work
ছেলেটি ধবধবে ফর্সা।	The boy is very fair	The boy is pale	The boy is so fair
বলিহারি বুদ্ধি তোমার।	Stupefied intellect is yours	Bolihari intellect is yours	Stupefied intellect is yours
ভাগ্যবানের বোঝা ভগবান বয়।	The burden of the lucky is borne by God	The burden of the lucky is God boy	The burden of the lucky person is borne by God
অকৃতজ্ঞ এবং অকৃতঘ্নের মধ্যে পার্থক্য জানো তো?	Do you know the difference between ungrateful and grateful?	Do you know the difference between ungrateful and ungrateful?	Do you know the difference between ungrateful and grateful?

(continued)

Table 1 (continued)

Bengali sentences	Standard output	Output (google translator)	Output (proposed system)
অক্লেশে সে কাজটি সম্পন্ন করল।	He completed the work without any difficulty	With difficulty, he completed the work	He completed the work without difficulty
ধীরে ধীরে বায়ু বয়।	Slowly the wind blows	Slowly wind boy	The wind blows slowly
সে সব কবেকার কথা।	That is all old talk	That is all nonsense	These all are old talk

References

1. D. Mandal, M. Gupta, S. Dandapat, P. Banerjee, S. Sarkar, Bengali and Hindi to English CLIR evaluation, in *Advances in Multilingual and Multimodal Information Retrieval. CLEF 2007. Lecture Notes in Computer Science*, eds by C. Peters et al. vol 5152. (Springer, Berlin, Heidelberg, 2007)
2. A. Senapati, U. Garain, One-expression classification in Bengali and its role in Bengali-English machine translation. in *2014 International Conference on Asian Language Processing (IALP)*, Kuching, 2014, pp. 162–165, https://doi.org/10.1109/ialp.2014.6973489
3. Y. Ogura, I. Kobayashi, Text Classification based on the latent topics of important sentences extracted by the PageRank Algorithm, in *51st Annual Meeting of the Association for Computational Linguistics Proceedings of the Student Research Workshop, Association for Computational Linguistics*, August, 2013, pp. 46–51
4. N. Ashraf, M. Ahmad, Machine translation techniques and their comparative study. Int. J. Comput. Appl. **125**, 25–31 (2015)
5. M.N. Yousuf Ali, G. Sorwar, A. Toru, M.A. Islam, M. Shamsujjoha, Morphological rules of Bangla repetitive words for UNL based machine translation. in *Advances in Swarm and*

Computational Intelligence. ICSI 2015. Lecture Notes in Computer Science, eds. by Y. Tan, Y. Shi, F. Buarque, A. Gelbukh, S. Das, A Engelbrecht, vol 9142. (Springer, Cham, 2015)
6. P. Wang, P. Nakov, H.T. Ng, Source language adaptation approaches for resource-poor machine translation. Comput. Linguist. **42**, 277–306, 2 June (2016)
7. K.M.A. Salam, S. Yamada, N. Tetsuro, Improve example-based machine translation quality for low-resource language using ontology, in Applied Computing & Information Technology. ACIT 2017. Studies in Computational Intelligence, vol 727. (Springer, Cham, 2017)
8. A.S. Nipu, U. Pal, A machine learning approach on latent semantic analysis for ambiguity checking on Bengali literature, in *20th International Conference of Computer and Information Technology (ICCIT)*, Dhaka, 2017, pp. 1–4. https://doi.org/10.1109/iccitechn.2017.8281797
9. Md. Faruquzzaman Akan, Transliteration and translation from Bangla into english: A problem solving approach, in *British Journal of English Linguistics, European Centre for Research, Training and Development*, vol. 6, No. 6, December 2018 (UK, 2018), pp. 1–21
10. M.M. Rahman, M.F. Kabir, M.N. Huda, A corpus based n-gram hybrid approach of Bengali to english machine translation, in *21st International Conference of Computer and Information Technology (ICCIT),* (Dhaka, Bangladesh, 2018), pp. 1–6. https://doi.org/10.1109/iccitechn.2018.8631938
11. M. Haque, M. Hasan, English to Bengali machine translation: An analysis of semantically appropriate verbs, in *International Conference on Innovations in Science, Engineering and Technology (ICISET),* (Chittagong, Bangladesh, 2018), pp. 217–221. https://doi.org/10.1109/iciset.2018.8745626
12. S. Bal, S. Mahanta, L. Mandal, R. Parekh, Bilingual machine translation: english to Bengali, in *Proceedings of International Ethical Hacking Conference 2018. Advances in Intelligent Systems and Computing*, vol 811, eds. by M. Chakraborty, S. Chakrabarti, V. Balas, J. Mandal (Springer, Singapore, 2019)
13. M.A. Hasan, F. Alam, S.A. Chowdhury, N. Khan, Neural machine translation for the Bangla-English language pair. In: *2019 22nd International Conference on Computer and Information Technology (ICCIT)* (Dhaka, Bangladesh, 2019), pp. 1–6. https://doi.org/10.1109/iccit48885.2019.9038381
14. M.A. Al Mumin, M.H. Seddiqui, M.Z. Iqbal, M.J. Islam, Neural machine translation for low-resource English-Bangla. J. Comput. Sci. **15**(11), 1627–1637 (2019). https://doi.org/10.3844/jcssp.2019.1627.1637
15. D. Banik, A. Ekbal, P. Bhattacharyya, Machine learning based optimized pruning approach for decoding in statistical machine translation. IEEE Access **7**, 1736–1751 (2019). https://doi.org/10.1109/ACCESS.2018.2883738
16. I. Ashrafi et al., Banner: A cost-sensitive contextualized model for Bangla named entity recognition. IEEE Access **8**, 58206–58226 (2020). https://doi.org/10.1109/ACCESS.2020.2982427
17. M. Mitra, D. Mitra, Oxford English-English-Bengali Dictionary, Ox-ford University Press. ISBN: 9780195689648
18. Satyajit Basu, English Tutor, Class VI,chhaya prakashani pvt ltd

Review of Signboard Transliteration Using Deep Learning

Jatin Sachdeva, Mudit Gupta, Shikhar Gupta, Deepak Aggarwal, and Vedika Gupta

Abstract Extracting text from the signboard, recognizing them, and thereafter translating them are a challenging task to be done and are receiving a significant amount of attention in today's world. This fulfills applications like safe driving path planning, robot navigation, etc. We put forward and discuss the generic framework that includes deep convolutional network architectures for the purpose of signboard text detection and recognition. Systems equipped with deep learning models are known for building a scalable pipeline, since these systems are data oriented. We analyzed the sequence-to-sequence models and LSTMs to find an efficient technique, for the extensible pipeline of transliteration, by contrasting the results.

Keywords LSTM · Signboard transliteration · Deep convolutional network · Support vector machine · Neural network

1 Introduction

In India, there are 22 languages which are constitutionally recognized and are written in 13 different scripts. There is enormous amount of data present in the natural visuals. Symbols and signs are all around us in our day-to-day lives. It is next to impossible to have signboards with all 22 languages written on it in every city/state/district all over the country. Also talking about signboards, traffic sign boards must also be considered. According to a survey, there were 1,37,000 deaths took place due to road accidents in 2013.[1] Most of them were a result of being careless and being ignorant toward the traffic sign boards as well as the rules of being on road. This situation of accidents taking place in India is alarming and need to be worried of.

[1] https://sites.ndtv.com/roadsafety/important-feature-to-you-in-your-car-5/.

J. Sachdeva · M. Gupta · S. Gupta · D. Aggarwal · V. Gupta (✉)
Department of Computer Science and Engineering, Bharati Vidyapeeth's College of Engineering, New Delhi, India
e-mail: vedika.nit@gmail.com

V. E. Balas et al. (eds.), *Proceedings of International Conference on Computational Intelligence, Data Science and Cloud Computing*, Lecture Notes on Data Engineering and Communications Technologies 62, https://doi.org/10.1007/978-981-33-4968-1_32

Hence, the goal of the analysis is to get the best techniques of machine learning and deep learning in order to get an application developed using these techniques in its most efficient way. The user will point the camera toward the signboard, and the text would get transliterated via the application. In the first phases, the application will be built for up to five languages. And when achieved, the number of languages will be increased to 15. The task is divided into the following three categories

- Signboard and text region detection
- Recognition of individual characters
- Natural language processing for translation/transliteration.

The rest of the paper has been prepared in the following manner. Section 2 presents the existing work which has been proposed in the direction of signboard transliteration. Section 3 presents the generic methodology opted for the transliteration system. Section 4 suggests the techniques which can be used for the creating the transliteration system in different ways. Section 5 discusses the summaries of the work presented in the paper.

2 Existing Work

Signboard detection: Past work in signboard recognition has been dominated by SVM classifier [1–3]. There can be linear as well as non-linear SVM classifiers. Another approach is by using artificial neural networks [4]. This system proposes detection of text by neural network, SVM classifier, and HOG method. Some of the other approaches are detecting the panels and extracting the information out of it [2]. In another system, the text regions are extracted based on the edge features using contourlet transform [3]. The regions are then classified as text and non-text using GLCM features and SVM classifier [1–3]. The recognition of Gurumukhi text from images achieved accuracy over 90% [5].

Text detection and recognition: Past works discuss several techniques for detection of text and identification from Indian signboards through systems. The system first includes preprocessing of images which are achieved by conversion to gray scale and edge detection using canny edge detector, and then, SWT transformation [6] was performed. Text extraction was done using Tesseract OCR engine [6, 7]. Methods used for detection also include Maximally Stable External Regions (MSER) [8, 9] feature detectors for the detection of text regions. Non-text regions based on the stroke width variations were removed. The merged text regions were then processed for the recognition of the text in them using OCR [8–11] function technique. A common strategy can be convolutional feature fusion for multi-language text detection in natural scene images. It uses fast-RCNN [9, 12]. Recently, Zheng, et al. [13] proposed an approach for text detection using a Fully Convolutional Network (FCN). However, many have also proposed a text localization and recognition framework. It uses YOLOv2 architecture as CNN [10, 14].

Text transliteration: System captures the signboard images through a camera phone which have been proposed in [7]. The text which is recognized is then converted into the language specified by the user and is displayed as the output to the user. The steps are: acquiring sign board image; text extraction and recognition out of the captured image using tesseract OCR; and extracted text's translation into user specified language which is conducted using a translator that maintains a dictionary. The accuracy achieved was around 90%. A methodology of recognizing the text is proposed in [10] which converts into a language understandable by the user when traveling in some foreign country. A system is proposed in [11] which image is captured and sent to the central server. Another model [15] uses two types of methods—LSTM and sequence-to-sequence. The same comparative study of existing work has been represented in a tabular manner in Table 1.

3 Generic Methodology

The process of signboard translation has been represented in the form of a block diagram in Fig. 1. Signboard image has a name of place within the image that is an input, and output is the image with transliterated text. Input Image is scanned by using an optical scanner which is used to read text and translate the information into form that the computer can use. Optical scanner digitizes an image by dividing it into a grid of boxes and setting its value 0 if box is not filled and 1 if box is filled.

Images can be noisy, distorted which can create many problems that decrease the result accuracy. To avoid problem, preprocessing is done. There are many methods which are used as preprocessing tools—normalization, removing noise, binarization, and grayscale conversion.

Segmentation is a process in which partitions of the pixels of an image into groups that constitutes regions or objects in an image. The main advantage of segmentation is that the accuracy is determined by computerized analysis procedures. Segmentation algorithms are based on the basis of intensity values (Discontinuity or Similarity). Points, lines, and edges are three basic types of gray level discontinuities. Image segmentation algorithms are edge detection bases techniques, thresholding methods, region-based techniques, etc.

Feature extraction is used to get the features of images that will be useful in the classification or recognition of images. Feature extraction is a type of dimensionality reduction that represents an image as a feature vector. This approach is used when an image has a large size and is reduced in feature vectors. Feature extraction techniques are color histograms, HOG, hear wavelets, etc.

The output of feature extraction is passed through the recognition and classification block. In this block, it is verified whether text is present on the signboard. This block gives ASCII or UNICODE of the characters in the output. The output is passed to find the language of the text. The text is passed into the database of

Table 1 Tabular representation of comparative study of existing techniques

Authors	Approach	Methodology	Results	ML/DL	SBD (Signboard Detection)	Text detection and recognition	Transliteration
Yadav et al. [1]	Support Vector Machine (SVM)	Training Data—90% Testing Data—10% Working divided in three phases: • Color Segmentation • Shape Classification • Recognition	Recognized almost all traffic signs correctly Accuracy decreases when in motion	ML	Support Vector Machine (SVM)	N/A	N/A
Panhwar etg al. [4]	Neural network	• Text detection and localization • In Recognition, image text (pixel-based text) is converted into a readable and editable form	Accuracy-Signboard detection: 85% Text detection: 80% Text extraction: 71.2% Text recognition: 70.2%	ML	Text localization	Neural network	N/A
Sebastian and Priya [2]	Images represented as BOVW, SVM.	• Images represented as Bag of Visual Words (BOVW) • Classified using Support Vector Machines	Easily detects from dark, green and white regions of image	ML	Support Vector Machine (SVM)	Bag of Visual Words (BOVW)	N/A

(continued)

Table 1 (continued)

Authors	Approach	Methodology	Results	ML/DL	SBD (Signboard Detection)	Text detection and recognition	Transliteration
Kaur and Garg [5]	Vertical and Horizontal projection histograms, also Using holes, endpoints and junctions	• Segmentation using vertical and horizontal projection histograms • Feature extraction and recognition on the segmented characters—considering at least holes, endpoints and junctions	Accuracy—90%	ML	vertical and horizontal projection histograms	Using holes, endpoints and junctions	N/A
Seeri et al. [3]	Contourlet transform and SVM	• Text localized image from nature images using Contourlet transform • Classified text images using SVM	Precision rates—98.85% Recall rates—90.85% F-measure—0.95(~1) Accuracy—89.90%	ML	Support Vector Machine (SVM)	Contourlet Transform	N/A
Rahul et al. [6]	Stroke Width Transform (SWT) algorithm. Tree bagging algorithm	• Text detection—Stroke Width Transform • Language identification- tree bagging ensemble learning technique • Text extraction—off-the-shelf OCR package	Accuracy for—Kannada—64.66% English—64.91% Hindi—65.13% Average results achieved	ML	N/A	Stroke Width Transform (SWT) algorithm. Tree bagging algorithm	N/A

(continued)

Table 1 (continued)

Authors	Approach	Methodology	Results	ML/DL	SBD (Signboard Detection)	Text detection and recognition	Transliteration
Pathak, et al. [8]	MSER (Maximally Stable Extremal Regions)	Removes non-text region using Thresholds or Stroke Width	OCR function converted the text in image correctly	ML	N/A	Maximally Stable Extremal Regions (MSER)	NA
Chandio and Pickering [12]	RCNN VGG16 layers network	Weights of pre-trained model helps in small number of training images and remaining layers are trained on multilingual image text dataset	challenging to detect Arabic and Urdu text but recognize all text instances correctly	DL	N/A	Region-Convolutional neural network (RCNN)	N/A
Zhang et al. [13]	FCN and MSER	• Detection of text blocks- Fully Convolutional Network • Extraction of multi-oriented text line candidates from these text blocks by taking into account the local information	Precision −0.88 Recall −0.78 F-measure-0.83	DL	N/A	FCN and MSER	N/A
Busta, et al. [9]	RCNN and OCR	• Training of both text detection and recognition in a single end-to-end pass • RCNN preserves aspect of text & adapts its resolution to the data	F-Measure- (strong, weak, generic)-end-to-end 0.89, 0.86, 0.77 word spotting 0.92, 0.89, 0.81	DL	N/A	RCNN and OCR	N/A

(continued)

Table 1 (continued)

Authors	Approach	Methodology	Results	ML/DL	SBD (Signboard Detection)	Text detection and recognition	Transliteration
He et al. [14]	CNN and RNN	The feature of the detected text region is feed into the RNN branch for word recognition via a text alignment layer	F-Measure- (strong, weak, generic)-end-to-end 0.82, 0.77, 0.63	DL	N/A	CNN and RNN	N/A
Kavati et al. [7]	text extraction-Tesseract OCR engine text translation by using dictionary corresponding to each character	Includes 5 parts • Preprocessing • Segmentation • Feature Extraction and Classification • Postprocessing • Text Translator	precision - 93% accuracy - 90%	ML	N/A	Tesseract OCR engine	Dictionary corresponding to each character
Rohira et al. [10]	CNN with 4 layers • Convolution layer • Pooling Layer • Fully Connected Layer	CNN for dataset training and Open CV, OCR for translation	Detects text on individual frames and transliterate them	DL	N/A	CNN	OpenCV
Rajendran and Aishwarya [11]	SVM with OCR	• Signboard Translation • Text Translation • Character Recognition • Translated Signboard Images using Mobile Phone	High robustness under uneven illumination, easy to understand, implement and maintain, viable. Accuracy close to 100%	ML	N/A	SVM	OCR

(continued)

Table 1 (continued)

Authors	Approach	Methodology	Results	ML/DL	SBD (Signboard Detection)	Text detection and recognition	Transliteration
Joshi et al. [15]	Long Short-Term Memory (LSTM) and Sequence-to-Sequence models	• Two layers of bidirectional LSTM access information in forward and backward direction • Sequence-to-sequence models consist of encoder for input and decoder for generation	Seq2Seq models are better than other models in terms of performance metrics and language specific nuances. But it needs large datasets. LSTM do not need large datasets	DL	N/A	N/A	Long Short-Term Memory (LSTM) and Sequence–to-Sequence models

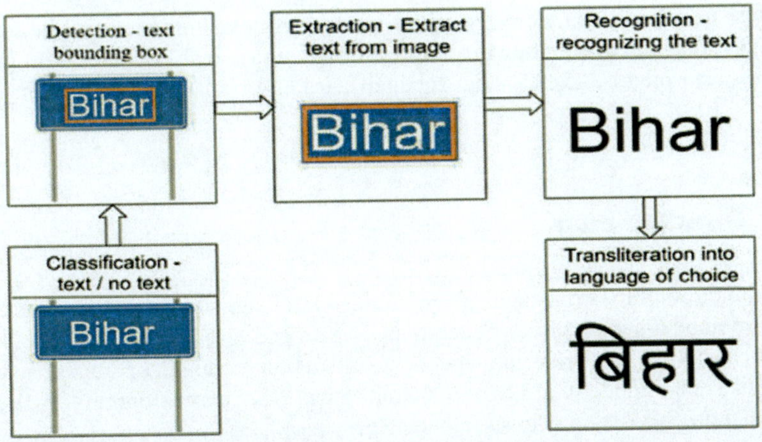

Fig. 1 Pipeline of sign board transliteration

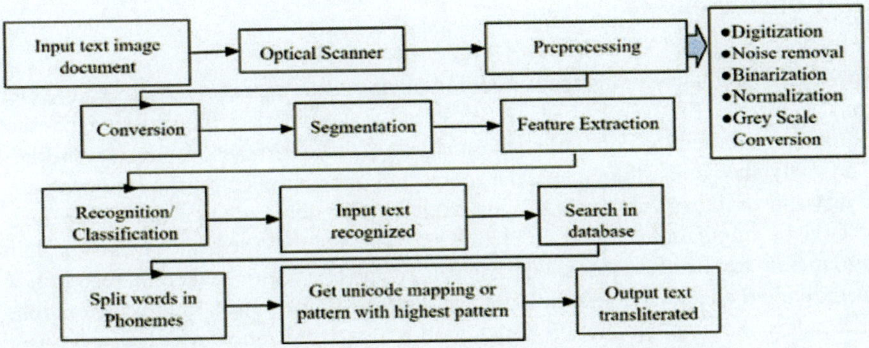

Fig. 2 Block diagram for generic methodology adopted in existing literature

the particular language in which the text needs to be transliterated. The words are converted to phonemes by splitting. The pattern with highest matching is searched in the database, and the phonemes are combined to form words, and the transliterated text is generated (Fig. 2).

4 Research Directions

The paper also identifies few prospective research directions. For the purpose of text detection of image—fully convolutional network can be used as a classifier to get the centroid of each character [16]. Single-shot text detector with regional attention can be utilized to roughly define text regions through automatically learned attentional maps [17]. Geometry-aware scene text detection may also be used to precisely localize text regions in an image of a natural scene [18]. Text recognition can be performed by CRNN with transfer learning or changing neurons in layers [19].

4.1 *Transliteration*

Transliteration of the text language from image into many languages which is easily understood by other people using semi-supervised end-to-end scene text recognition which can be helped to develop signboard translation model. Semi-supervised end-to-end is such a model that learns a spatial transformer network jointly. It can also learn to detect text regions in an image [20].

5 Conclusions

Having 22 languages recognized under Indian constitution, it is evident that a traveler may get puzzled for the varying languages used in signboards, while crossing the state borders. A detailed and comprehensive analysis of different research papers was done. The study shows that there can exist more than one approach for the said problem. Every approach has its own challenges, even with a number of algorithms, there does not exist a single unified solution, which can be labeled as the best. The task is split into mainly three different sections: Signboard and text region detection, recognition of individual characters, natural language processing for translation/transliteration.

Studies show approaches like histograms, text localization which can be used for the detection of signboard but this phase was mainly dominated by the Support Vector Machine (SVM) technique, resulting in maximum accuracy. The major stages involved are preprocessing, segmentation, feature extraction, and recognition. This was followed by the recognition of text from the detected signboard. Past works discussed several techniques for detection of text and identification from Indian signboards through systems including FCN, RCNN, MSER, etc. However, the most suitable system includes preprocessing of images which is achieved by conversion to gray scale and edge detection using canny edge detector, and then, SWT transformation was performed. The text recognized till this stage is then converted to user specified language. The system preferred for this is Optical Character Recognizers (OCR) that translates it word-by-word, looking in a 21,000—word dictionary for the desired term and convert it character by character. Accuracy achieved was close to 100%.

References

1. S. Yadav, A. Patwa, S. Rane, C. Padvekar, Indian traffic sign board recognition and driver alert system using machine learning. Int. J. Appl. Sci. Smart Technol. **1**(1) (2019)
2. S. Sebastian, S. Priya, Text detection and recognition from images as an aid to blind persons accessing unfamiliar environments. ARPN J. Eng. Appl. Sci. **10**(17), 7559–7564 (2015)
3. S.V. Seeri, J.D. Pujari, P.S. Hiremath, Text localization and character extraction in natural scene images using contourlet transform and SVM classifier. Int. J. Image, Graph. Sig. Process. **8**(5), 36 (2016)
4. M.A. Panhwar, K.A. Memon, A. Abro, D. Zhongliang, S.A. Khuhro, S. Memon, Signboard detection and text recognition using artificial neural networks. in *2019 IEEE 9th International Conference on Electronics Information and Emergency Communication (ICEIEC)*. (IEEE, 2019), pp. 16–19
5. T.P. Kaur, N. Garg, Optimized gurmukhi text recognition from signboard images captured by mobile camera using structural features. in *2015 Fifth International Conference on Advances in Computing and Communications (ICACC)*. (IEEE, 2015), pp. 412–416
6. R. Rahul, S. Bhaskaran, J. Amudha, D. Gupta, Multilingual text detection and identification from Indian signage boards, in *2018 International Conference on Advances in Computing, Communications and Informatics (ICACCI)* (IEEE), pp. 1120–1125
7. I. Kavati, G.K. Kumar, S. Kesagani, K.S. Rao, Signboard text translator: a guide to tourist. Int. J. Electr. Comput. Eng. (2088–8708), **7**(5)
8. M. Pathak, A.K. Gupta, S. Singh, A novel approach to detect and recognize text in signboard images
9. M. Busta, L. Neumann, J. Matas, Deep texts potter: An end-to-end trainable scene text localization and recognition framework. In *Proceedings of the IEEE International Conference on Computer Vision* (2017), pp. 2204–2212
10. A. Rohira, R. Shah, O. Sadarangani, M. Shinde, Therese, S. (2019). *Word Detection and Translation*. Available at SSRN 3372202
11. P.S., Rajendran, Aishwarya, M., Translation of text oriented signboard images from mobile phone camera, (2015)
12. A.A. Chandio, M. Pickering, Convolutional feature fusion for multi-language text detection in natural scene images. in *2019 2nd International Conference on Computing, Mathematics and Engineering Technologies (iCoMET)*. (IEEE, 2019), pp. 1–6
13. Z. Zhang, C. Zhang, W. Shen, C. Yao, W. Liu, X. Bai, Multi-oriented text detection with fully convolutional networks. in *Proceedings of the IEEE Conference on Computer Vision and Pattern Recognition* (2016), pp. 4159–4167
14. T. He, Z. Tian, W. Huang, C. Shen, Y. Qiao, Sun, C., An end-to-end textspotter with explicit alignment and attention. in *Proceedings of the IEEE Conference on Computer Vision and Pattern Recognition* (2018), (pp. 5020–5029)
15. A. Joshi, K. Mehta, N. Gupta, V.K. Valloli, Indian language transliteration using deep learning. In *2018 IEEE Recent Advances in Intelligent Computational Systems (RAICS)*. (IEEE, 2018), pp. 103–107
16. Z. Liu, W. Zhou, H. Li, Scene text detection with fully convolutional neural networks. Multimedia Tools Appl. **78**(13), 18205–18227 (2019)
17. C. Shiqi, Z. Ronghui, Z. Jun, Regional attention-based single shot detector for SAR ship detection. J. Eng. **2019**(21), 7381–7384 (2019)

18. F. Zhan, C. Xue, S. Lu, GA-DAN: geometry-aware domain adaptation network for scene text detection and recognition, in *Proceedings of the IEEE International Conference on Computer Vision* (2019), (pp. 9105–9115)
19. M. Liao, P. Lyu, M. He, C. Yao, W. Wu, X. Bai, Mask textspotter: An end-to-end trainable neural network for spotting text with arbitrary shapes. IEEE Trans. Pattern analysis machine Intell. (2019)
20. F. Zhan, S. Lu, Esir: End-to-end scene text recognition via iterative image rectification. In *Proceedings of the IEEE Conference on Computer Vision and Pattern Recognition* (pp. 2059–2068)

An Automated System to Identify Sentiment from Micro-Blog Texts of Tweets

Anupam Mondal, Monalisa Dey, and Darothi Sarkar

Abstract The recent trends of Internet help to produce the sentiment and emotion-based opinion at the time of conversation between human beings. During the Coronavirus disease (COVID-19) pandemic, we are spending maximum time of the day on Internet especially on social networking Web sites. The Web sites are Twitter, Facebook, and WhatsApp, which are taking a crucial role to communicate with each other in the form of messages. The messages are representing such as short-texts and micro-texts and carrying sentiment. The sentiment identification is challenging due to the length of the message. In the present paper, we are motivated to design a sentiment analysis system for Twitter micro-texts in the topic of Coronavirus disease (COVID-19). Hence, we have scrawled a dataset from Twitter using Twitter API and presented as our experimental dataset. Additionally, we have developed two state-of-the-art techniques, viz. unsupervised and supervised to build this system. The unsupervised technique helps to understand the characteristic of the dataset, whereas supervised technique assists in improving the accuracy of the system. The developed system may help to design various domain-specific applications such as annotation and emotion identification system for micro-texts in future.

1 Introduction

Sentiment analysis (SA) or opinion mining is treated as a challenging task for identifying the semantic label information from a piece of textual document in the web. We have observed that the web-based textual information contains opinion and emotion for understanding the subjective information of the corpus. In order to identify these opinions as well as emotions, sentiment analysis is taking a crucial role. The sentiment refers to three different types, namely positive, negative, and neutral, according

A. Mondal (✉) · M. Dey · D. Sarkar
Department of Computer Science and Engineering, Institute of Engineering and Management, Kolkata, India
e-mail: anupam.mondal@iemcal.com

V. E. Balas et al. (eds.), *Proceedings of International Conference on Computational Intelligence, Data Science and Cloud Computing*, Lecture Notes on Data Engineering and Communications Technologies 62, https://doi.org/10.1007/978-981-33-4968-1_33

to the polarity score. In order to assign the sentiment for the messages, the researchers have employed various natural language processing (NLP) techniques [1, 2].

On the other hand, the micro-blog Web sites present an essential information such as knowledge and opinion-oriented information on different topics like the discussion on current issues and reviews of the product which are used in the daily life [3]. Among these Web sites, Twitter is a popular micro-blogging service in which 240+ active users can create their status messages as a micro-text or "tweets". Tweets present the relevant information related to recent trends publicly by the users [4]. Hence, in this research, we have focused on extracting tweets on Coronavirus disease (COVID-19) pandemic topic due to the serious effect of our daily life. We are motivated to identify the sentiment of these extracted tweets to understand the psychological situation of the users.

In order to design this sentiment identification system, we have observed the following difficulties. (A) Twitter is presented by the short length and misspelled words. So, sentiment analysis is difficult from tweets compared to recognize the sentiment from general texts. Tweet contains maximum 140 characters in each tweet. (B) Another challenge is to build a system to detect the emotions of public and to summarize the overall sentiment of any particular issue.

To overcome these challenges, we have employed two different approaches of machine learning. First one is unsupervised learning and next is supervised learning. Both of the approaches are used for classifying the micro-texts into three different classes such as positive, negative, and neutral. The unsupervised module assists in preparing a baseline system to identify sentiments related to COVID-19 from tweets where supervised module assist in enhancing the accuracy of the system.

In addition, we have crawled micro-text tweets on the topic of Coronavirus disease pandemic and COVID-19 using a Twitter API for building both of the mentioned modules. The tweets contain hashtags, misspelled words, emojis, etc., which introduced noise in the dataset [5]. Figure 1 shows a sample tweet of our experimental dataset.

Besides, we have preprocessed the dataset after eliminating the noises such as URLs, emoticons and special characters. We have also removed the stop words from the crawled tweets to assign the initial label of the experimental dataset. The dataset has been split into two parts, namely training and test dataset. The training dataset has been applied on the naive Bayes and support vector machine (SVM) classifiers, whereas test dataset has been used for validating the output of the system. Finally, we have observed that the supervised learning provides a better performance in terms of accuracy over unsupervised learning.

The rest of the paper is organized as follows. Section 2 illustrates a background study related to sentiment analysis for micro-blogs. Section 3 and Sect. 4 discuss, how we have prepared the experimental dataset and model building part in details. Section 5 describes the result analysis and proposed modules with examples. Finally, Sect. 6 presents concluding remarks of the task and its future scope.

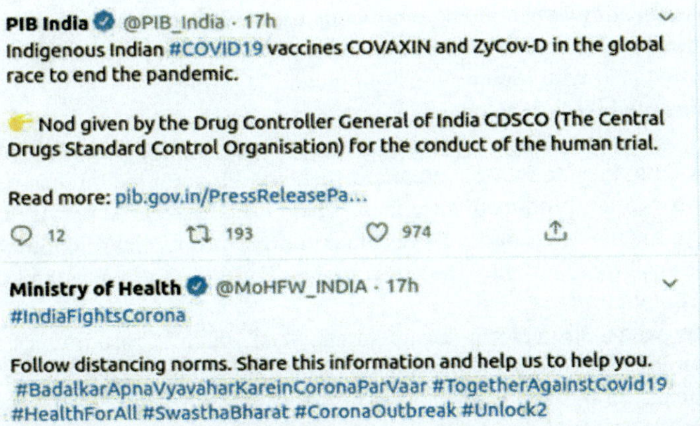

Fig. 1 Sample tweets of our experimental dataset

2 Background

In recent years, a lot of work has been done by researchers in sentiment analysis, but Twitter sentiment analysis is relatively new and difficult task. In its early stage, it was intended for bipolar classes which assigns opinions or reviews to binary classification such as positive or negative only [6]. We briefly study about related works that have been done in this field in this section.

Sentiment polarity classification is classified into document-level, sentence-level, and phrase-level classification. Document-level classification classifies document as positive, negative, or neutral [7–9]. In [10, 11], sentiment-level classification is considered and it classifies sentiment by a sentence, whereas in [12] researcher are working on sentence which are subjective or objective. There exist multiple sentiment within a single sentence. This is known as phrase-level classification. This is performed in [9, 13].

Furthermore, some researcher categorized it as the level and sentiment classes. They also considers different assumptions about sentiment sources and targets [14]. Most of the authors done the sentiment analysis by using machine learning approaches with feature vector. In [15], the researcher worked by using SVM classifier

1. to detect SA for Twitter and SMS (message-level task)
2. to detect SA for each term within the message (term-level task).

Furthermore, they have implemented the sentiment analysis on a variety of features based on surface form and lexical categories. The sentiment lexicon features was generated in both manually created and automatically. They also used n-gram features for both word and character n-grams that led to the most gain in performance.

A recent research has identified it as, an automatic collection of a corpus that can be used to train a sentiment classifier. A model to classify the tweets as positive and negative and neutral sentiments of documents proposed in [16]. They created a Twitter corpus by collecting tweets using emoticons using Twitter API that can be used to train a sentiment classifier. The classifier that this paper used is the multinomial naive Bayes method [17] with feature like N-gram and POS-tags. The unigrams, bigrams, and trigrams methods are also applied.

In [18], researcher does a sentiment analysis on movie review. Using movie reviews as data, they found that standard machine learning techniques definitively out-perform human-produced baselines. They used several supervised machine learning approaches like naive Bayes, maximum entropy classification, and support vector machines, and they show that it gives better performance than traditional topic based modeling.

One approach is to perform sentiment analysis, which is based on a function of opinion words in context. This opinion words are commonly used in case of expressing sentiments. In [4] paper, they proposed a new entity-level sentiment analysis method for Twitter. It deals with the problem of current lexicon-based and learning-based sentiment analysis approaches.

Another more related work is mentioned in the paper [19] is a novel machine learning method. This applies text-categorization techniques to the subjective portions of the document. The researcher examines the relationship between polarity classification and subjectivity detection. They apply naive Bayes polarity classifier. They also examined the relation between subjectivity detection and polarity classification, and also they show that subjectivity detection can compress reviews into much shorter extracts that still retain polarity information at a level comparable to that of the full review. In fact, for the naive Bayes polarity classifier, the subjectivity extracts are shown to be more effective input than the originating document, which suggests that they are not only shorter, but also "cleaner" representations of the intended polarity.

In [20], they present a simple rule-based model for general sentiment analysis named as VADER and then they and compare its effectiveness to eleven typical state-of-practice benchmarks including LIWC, ANEW, the General Inquirer, SentiWordNet and machine learning-oriented techniques relying on naive Bayes, maximum entropy, and support vector machine (SVM) algorithms. With the help of qualitative and quantitative methods, they first construct and genuinely validate a gold standard list of lexical features. Using Valence Aware Dictionary for sentiment Reasoning (VADER), a rule-based model to assess the sentiment of tweets, they find that VADER out-performs individual human raters (F1 classification accuracy = 0.96 and 0.84, respectively) and generalizes more favorably across contexts than any of our benchmarks.

In [21], they proposed a new entity-level sentiment analysis method for Twitter. By this method, the result will give high precision, but low recall. Thus to improve recall, lexicon-based approach is used method. Their experimental result of the proposed method dramatically improves the recall and the F-score. It out-performs the state-of-the-art baselines.

In [22], they have introduced the Twitter analytical platform (TAP), which enables developers to conduct Twitter data analytics with various analysis tools.

The background work on sentiment identification system shows a clear idea, how machine learning approaches help to identify sentiment from micro-texts corpora.

It also provides an overview related to the applied techniques for building a sentiment identification system in Tweeter. These observations motivated to develop an automated system for identifying sentiment from micro-texts.

3 Dataset Preparation

Initially, the tweets were scraped automatically using Twitter API and they were manually annotated as positive or negative in case of binary classifier and for three-way classifier as positive or negative or neutral. Initially, the dataset contains 1000 tweets in which 234 tweets are positive, 578 are negative, and remaining 188 tweets are fallen into neutral category. We reshuffle the extracted tweets every time of compilation. Thus, getting variety of result every time. Basically, we use 25% of whole dataset as test dataset and remaining 75% is for training case. We also do experiment with changing the percentage of the training and test dataset. Distributions of the experimental dataset are shown in Table 1.

Thereafter, we have performed a preprocessing step to prepare the processed experimental dataset. In the following subsection, we have discussed the preprocessing step in details.

3.1 Preprocessing Steps

We have observed that, raw micro-texts like tweets handling is to much difficult due to various types of noises. The noises are

- all caps: In Tweets, there are few words those are in upper case.
- URLs eg, http://bit.ly
- hashtags (e.g., #Coronaavirus and #Coviddisease etc.)
- Twitter special words (such as "RT") [An abbreviation for retweet, which means citation or re-posting of a message]
- @username
- Punctuation and additional white spaces
- stop words- stop word is a commonly used word (such as "the").

Table 1 A detail statistics of the experimental dataset

Dataset Statistics	Experimental	Training (75%)	Test (25%)
No. of tweets	1000	755	245
No. of hash tags	2143	1586	557
No. of positive tweets	234	181	53
No. of negative tweets	578	432	146
No. of neutral tweets	188	142	46

In order to handle the mentioned noises, we have applied the following ways on the extracted tweets.

- Lower Case—tweets are converted to lower case.
- URLs—URLs are eliminated via regular expression matching or replace with generic word URL.
- Handling #hashtags: replacing #hashtags with the exact same word without the hash since hash tags may provide some useful information. For example, #Coronaavirus disease replaced with "Coronavirus disease".
- Remove Twitter special words (such as "RT")
- @username-Remove "@username" via regex matching or replace it with generic word AT_USER.
- punctuation management: Here, we have removed punctuation at the start and ending of the tweets, e.g., "the day is beautiful!" is replaced with "the day is beautiful".
- white space removable: Replacing multiple white spaces with a single whitespace has been done.
- Removing Stop words: We also remove the stop words from tweets because these words contribute not much more emotion to a tweet.
- lengths: All Tweets must be under 140 characters.
- elongated words: the number of words with one character repeated more than two times, e.g., "soooo".
- Tokenization: We make sure that short forms such as "don't", "I'll", "she"d" will remain as one word.
- emoticons: The polarity of an emoticon was determined with a regular expression adopted from Christopher Potts' tokenized script.
  - presence or absence of positive and negative emoticons at any position in the tweet;
  - whether the last token is a positive or negative emoticon;

Thereafter, we have designed an annotation policies to annotate these preprocessed micro-text tweets. These annotated tweets assist in developing both unsupervised and supervised the sentiment identification systems.

3.2 Annotation Policy

An annotated dataset is important to design any automated system like sentiment analysis from micro-text tweets. Hence, we have designed a set of manual annotation policies that assist in annotating micro-text tweets on the topic of Coronavirus due to unavailability of structured dataset. The annotation tags, considered primarily for labeling sentiments of tweets, are namely positive (+1), negative (−1), and neutral (0). In the following part, we have pointed out a few of the annotated guidelines to annotate these tweets.

Annotated Guidelines:

1. Tweets are annotated with three fundamental sentiment classes such as positive (+1), negative (−1), and neutral (0).
2. Adjective based bag-of-words (e.g., fine, good, bad, sad, and painful) are taking important role to get the sentiment class of tweets.
3. The negation words (e.g., no, not, never, and nor) are toggled the sentiment class of tweets.
4. Depend on number of sentiment words, the annotators have decided the sentiment classes of the tweets.
5. Without any sentiment words, a tweet has been treated as a neutral sentiment class.

In the following section, we have discussed how this experimental dataset assists in developing both unsupervised and supervised versions of sentiment identification system for micro-texts.

4 Sentiment Identification Framework

Sentiment refers to the broad fundamental classes for the corpus, which helps to express a view or opinion on a specific topic. In the present research, we are motivated to understand the opinion in the form of sentiment of different group of peoples on the very emerging topic name as "Coronavirus". Coronavirus-related sentiment identification is one of the contributory tasks to understand various social and psychological effects in daily life. The effects are classified as positive, negative, and neutral of Coronavirus. The following subsections discuss, how we have developed two different modules to build the system. Figure 2 shows the overall presentation of the proposed sentiment identification system.

4.1 Unsupervised Module

In order to understand the effects of Coronavirus, primarily, we have followed the conventional sentiment-based approach in the presence of domain experts. This approach helps to identify the opinion of the experts, where sentiment-based approach assists in classifying the twitter data. We have collected a set of words (bag-of-words) related to Coronavirus and COVID-19 as a part of unsupervised module for assigning classes to the test dataset.

Additionally, we have used SentiWordNet lexicon to extract the polarity score of each of the tokens of a tweet. Thereafter, we have calculated the cumulative polarity score of tokens of a tweet and presented the average polarity score of this tweet. Finally, depend on the average polarity score of the tweet, we have set the sentiment as a positive ($0>$ to $< +1$), negative ($-1 >$ to <0), and neutral (0) for the tweet.

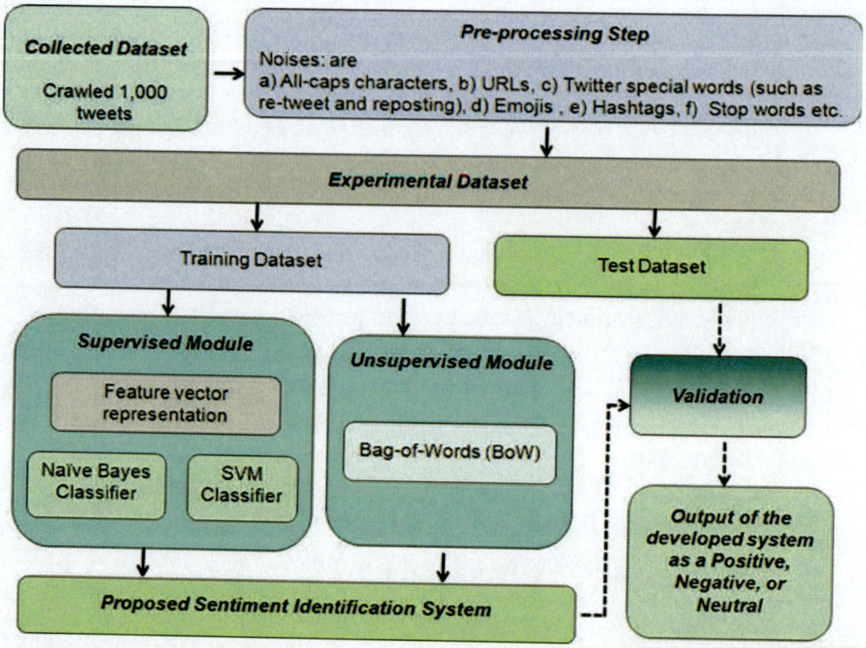

Fig. 2 A flow diagram of the developed system which identified sentiments from tweets on the topic of COVID-19

Thereafter, we have applied both of the approaches to design the classification module for the analysis of a comprehensive dataset of tweets. We have observed that the output of the proposed module depends on a specific type of keywords which are related to Coronavirus and COVID-19. So, the module is unable to reduce the error rate due to the versatile nature of the dataset. In order to overcome this difficulty, we have proposed another machine learning module using the same experimental dataset and their extracted features.

4.2 Supervised Module

In this section, we have introduced a supervised module using two different machine learning classifiers, namely naive Bayes and support vector machine (SVM). The reason for selecting these two classifiers to handle both generative and discriminant nature of the dataset. Naive Bayes classifier presents as a generative model where SVM represented as a discriminant model. Additionally, these classifiers assist in understanding the distribution of features in the form of independent or not.

Thereafter, we have extracted various linguistic as well as knowledge-based features of tweets and prepared a feature vector for each of the tweet. The extracted feature vector of training dataset has been applied on both of the classifiers individually to design the supervised module. The feature vector contains parts-of-speech (POS), n-grams, adjectives, nouns, verbs, sentiment words, emojis, all caps words, retweets, stop words, hashtags, negation words, and length of the tweet. The following algorithm is explaining the supervised module for identifying sentiment from COVID-19 related tweets in details.

Step-1: Extract the mentioned features from the training dataset and prepare a Feature vector (FV) and Label (L) it.
Step-2: Learn the Naive Bayes (NB) and SVM classifiers using FV and its corresponding Label (L) to predict the concerned classes individually.
Step-3: Thereafter, we have extracted FV of the test dataset and processed through both of the classifiers consequently and predict the output classes.
Step-4: Finally, validate the predicted output in the form of Precision, Recall, and F-Measure for the test dataset as discussed in Sect. 5.

Finally, both of the modules, viz. unsupervised and supervised, have been evaluated to justify the accuracy of the proposed sentiment identification system in the following section.

5 Result Analysis

In order to validate, both of the developed modules, we have conducted the following result analysis. The results threw up some interesting insights.

5.1 *Validation of Unsupervised Module*

In order to evaluate the unsupervised module, we have conducted an error analysis in the form of confusion matrix with the predicted output of test dataset as shown in Table 2.

Thereafter, we have presented a detailed validation of the predicted sentiment classes using standard matrices, viz. precision, recall, and F-Measure .[1] Table 3 shows the validation score in details.

We have observed that this module identified more number of negative tweets as neutral tweets. One of the reasons is the strict matching in-between bag of sentiment words with the tokens of the tweets.

[1] http://blog.exsilio.com/all/accuracy-precision-recall-f1-score-interpretation-of-performance-measures/.

Table 2 Confusion matrix for identifying sentiment from COVID-19 related tweets using unsupervised module

No. of tweets (245)		Predicted			
		Positive	Negative	Neutral	Total
Original	Positive	36	8	9	53
	Negative	17	105	24	146
	Neutral	3	6	37	46
	Total	56	119	70	

Table 3 Evaluation output of the proposed unsupervised module

	Precision	Recall	F-Measure
Positive (+1)	0.68	0.64	0.66
Negative (−1)	0.72	0.88	0.79
Neutral (0)	0.80	0.53	0.64

5.2 Validation of Supervised Module

In order to validate the supervised module, we have considered the predicted output of the mentioned test dataset into three different labels such as (A) positive (+1), (B) negative (−1), and (C) neutral (0) for the Coronavirus or COVID-19 related tweets. These labels have been evaluated through the standard matrices, viz. precision, recall, and F-Measure. Table 4 shows the validation score in details.

Additionally, we have conducted an error analysis in the form of confusion matrix with the predicted output of test dataset for both of the classifiers as shown in Table 5.

We have observed that the SVM classifier is providing better accuracy over naive Bayes classifier under the supervised module due to the selection features.

Table 4 Evaluation output of the proposed supervised module

	Precision	Recall	F-Measure
Naive Bayes			
Positive (+1)	0.74	0.72	0.73
Negative (−1)	0.79	0.91	0.85
Neutral (0)	0.85	0.61	0.71
Support vector machine			
Positive (+1)	0.79	0.78	0.78
Negative (−1)	0.87	0.93	0.90
Neutral (0)	0.87	0.73	0.79

Table 5 Confusion matrix for identifying sentiment from COVID-19 related tweets using supervised module

No. of tweets (245)		Predicted (Naive Bayes/SVM)			
		Positive	Negative	Neutral	Total
Original (Naive Bayes/SVM)	Positive	39/42	7/6	7/5	53
	Negative	12/9	116/127	18/10	146
	Neutral	3/3	4/3	39/40	46
	Total	54/54	127/136	64/55	

Finally, we can conclude that, the unsupervised module helps to prepare a baseline system for designing an automated sentiment identification system where supervised approach assists in improving the accuracy of the proposed system. This COVID-19 related sentiment identification system may help to understand the sociological and psychological condition of the users. Additionally, it assists in designing various applications related to COVID-19 like relationship extraction, recommendation, and summarization system for the users [23–25].

6 Conclusion

Nowadays, micro-blogging becomes one of the major types of the communication via Internet. Researchers have identified the large amount of information contained in micro-blogging Web sites makes them an attractive source of data for sentiment analysis. In this work, we have applied unsupervised and supervised machine learning approaches to identify the sentiment related to COVID-19 disease from the tweets. The unsupervised module helps to understand the problem, whereas supervised module assists in enhancing the accuracy of the system. Finally, both of the modules may guide to pick up other issues related to this pandemic and help to solve them.

References

1. A. Mondal, E. Cambria, D. Das et al., *Employing sentiment-based affinity and gravity scores to identify relations of medical concepts, in 2017 IEEE Symposium Series on Computational Intelligence (SSCI)* (IEEE, New York, 2017), pp. 1–7
2. A. Mondal, R. Satapathy, D. Das, et al., A hybrid approach based sentiment extraction from medical context. SAAIP@ IJCAI **1619**, 35–40 (2016)
3. A. Go, R. Bhayani, L. Huang, Twitter sentiment classification using distant supervision. CS224N Project Report. Stanford **1**(12) (2009)

4. M. Neethu, R. Rajasree, *Sentiment analysis in twitter using machine learning techniques, in 2013 Fourth International Conference on Computing, Communications and Networking Technologies (ICCCNT)* (IEEE, New York, 2013), pp. 1–5
5. E. Kouloumpis, T. Wilson, J.D. Moore, Twitter sentiment analysis: The good the bad and the omg!. Icwsm **11**(538–541), 164 (2011)
6. T. Nasukawa, J. Yi, Sentiment analysis: capturing favorability using natural language processing, in *Proceedings of the 2nd International Conference on Knowledge Capture* (ACM, 2003) , pp. 70–77
7. T. Mullen, N. Collier, Sentiment analysis using support vector machines with diverse information sources, in *Proceedings of the 2004 Conference on Empirical Methods in Natural Language Processing* (2004)
8. B. Pang, L. Lee, S. Vaithyanathan, Thumbs up?: sentiment classification using machine learning techniques, in *Proceedings of the ACL-02 Conference on Empirical Methods in Natural Language Processing-Volume 10* (Association for Computational Linguistics, 2002), pp. 79–86
9. J. Wiebe, T. Wilson, C. Cardie, Annotating expressions of opinions and emotions in language. Language Resources Eval **39**(2–3), 165–210 (2005)
10. W. Ding, X. Song, L. Guo, et al., A novel hybrid hdp-lda model for sentiment analysis, in *Proceedings of the 2013 IEEE/WIC/ACM International Joint Conferences on Web Intelligence (WI) and Intelligent Agent Technologies (IAT)*, vol. 01 (IEEE Computer Society, 2013), pp. 329–336
11. M.C. Lee, J.W. Chang, T.C. Hsieh et al., A sentence similarity metric based on semantic patterns. Adv. Inform. Sci. Service Sci. **4**(18), (2012)
12. E. Riloff, J. Wiebe, T. Wilson, Learning subjective nouns using extraction pattern bootstrapping, in *Proceedings of the Seventh Conference on Natural Language Learning at HLT-NAACL 2003*, vol. 4 (Association for Computational Linguistics, 2003), pp. 25–32
13. A. Severyn, A. Moschitti, Twitter sentiment analysis with deep convolutional neural networks, in *Proceedings of the 38th International ACM SIGIR Conference on Research and Development in Information Retrieval* (ACM, 2015), pp. 959–962
14. J. Yi, T. Nasukawa, R. Bunescu, et al., Sentiment analyzer: extracting sentiments about a given topic using natural language processing techniques, in *Third IEEE International Conference on Data Mining, 2003. ICDM 2003* (IEEE, New York, 2003), pp. 427–434
15. S.M. Mohammad, S. Kiritchenko, X. Zhu, NRC-Canada: building the state-of-the-art in sentiment analysis of tweets. arXiv preprint arXiv:1308.6242 (2013)
16. A. Pak, P. Paroubek, Twitter as a corpus for sentiment analysis and opinion mining, in *LREc*, **10**(2010) (2010)
17. A. McCallum, K. Nigam, et al., A comparison of event models for Naive Bayes text classification, in *AAAI-98 Workshop on Learning for Text Categorization*, vol. 752(1) (Citeseer, 1998), pp. 41–48
18. P.D. Turney, Thumbs up or thumbs down? Semantic orientation applied to unsupervised classification of reviews, in *Proceedings of the 40th Annual Meeting on Association for Computational Linguistics* (Association for Computational Linguistics, 2002), pp. 417–424
19. B. Pang, L. Lee, A sentimental education: sentiment analysis using subjectivity summarization based on minimum cuts, in *Proceedings of the 42nd Annual Meeting on Association for Computational Linguistics*, vol. 271 (Association for Computational Linguistics, 2004)
20. C.H.E. Gilbert, Vader: a parsimonious rule-based model for sentiment analysis of social media text, in *Eighth International Conference on Weblogs and Social Media (ICWSM-14). Available at (20/04/16)* http://comp.social.gatech.edu/papers/icwsm14.vader.hutto.pdf (2014)
21. A.Z. Khan, M. Atique, V. Thakare, Combining lexicon-based and learning-based methods for twitter sentiment analysis. Int. J. Electron. Commun. Soft Comput. Sci. Eng. (IJECSCSE) **89** (2015)
22. K. Tao, C. Hauff, G.-J. Houben, et al., Facilitating twitter data analytics: platform, language and functionality, in *2014 IEEE International Conference on Big Data (Big Data)* (IEEE, 2014), pp. 421–430

23. A. Mondal, E. Cambria, D. Das et al., Relation extraction of medical concepts using categorization and sentiment analysis. Cogn. Comput. **10**(4), 670–685 (2018)
24. A. Mondal, D. Das, S. Bandyopadhyay, A content-based recommendation system for medical concepts: disease and symptom, in *Proceedings of 15th International Conference on Natural Language Processing (ICON)* pp. 115–121 (2018)
25. M. Dey, A. Mondal, D. Das, NTCIR-12 mobileclick: sense-based ranking and summarization of english queries, in *NTCIR* (2016)

Chatbot for Query Processing in E-Learning System

Lopa Mandal and Sagnik Dutta

Abstract Chatbot is a conventional agent which is capable for communicating with users by using natural languages. Today, for every educational institution, learners need an E-Learning portal for acquiring more knowledge. This saves a lot of time and human effort, and a learner can get information regarding the institute and the courses they are interested in. In this scenario a chatbot used with an e-learning portal will be useful for learners to get quick answer for their queries relevant to that system. The present work aims to build a prototype of an intelligence e-learning chatbot to be used with an e-learning portal to facilitate query processing of a learner.

Keywords Human-chatbot interactions · Chatbot · Artificial intelligence · JSON · E-learning chatbot

1 Introduction

Chatbots are simple computer programs that interacts with users in a human like manner using natural language. The interaction can be textual or auditory depending upon the need. The applications of these chatbots are increasing in our daily life. A chatbot is often described as one of the most advanced and promising expressions of interaction between human and machine. However, from a technological point of view, a chatbot only represents the natural evolution of a Question Answering system leveraging Natural Language Processing (NLP). Chatbot applications streamline interactions between people and services, enhancing customer experience. At the same time, they offer companies new opportunities to improve the customers engagement process and operational efficiency by reducing the typical cost of customer service. There are different approaches and tools that can be used to develop a

L. Mandal (✉) · S. Dutta
Institute of Engineering & Management, Kolkata, India
e-mail: mandal.lopa@gmail.com

S. Dutta
e-mail: sagnikdutta1998@gmail.com

V. E. Balas et al. (eds.), *Proceedings of International Conference on Computational Intelligence, Data Science and Cloud Computing*, Lecture Notes on Data Engineering and Communications Technologies 62, https://doi.org/10.1007/978-981-33-4968-1_34

chatbot. There are basically three types of Chatbots. Menu/Button based chatbots are decision tree hierarchies presented to the user in the form of buttons. In this case the users need to make several selections to get the desired answer. Keyword Recognition-Based Chatbots generate responses depending on the queries users type using the keyboard. Contextual Chatbots remember interaction with a particular user and learn from the context gradually to generate appropriate answer by utilizing AI and Machine Learning techniques. The present work aims to design a web-based chatbot for query processing in E-learning system. The chatbot will be integrated in a webpage. The chatbot is supposed to help teachers and students by answering their queries. This automation in query processing is aimed to save human effort and be cost effective.

2 Literature Survey

A chatbot was implemented as a conversational agent, whose purpose was to examine the previous research on chatbots using the quantitative bibliometric analysis. This research contributed to help researchers to find the gaps in researches for the future research agenda in chatbots. It also identified deep learning as an emerging technology which can change the future direction of research in chatbot [1].

Another chatbot was made in 2017 for University Related queries. It provided a design of a chatbot to answer the queries on the basis of the dataset of frequently asked questions. Researcher used Latent Semantic Analysis (LSA) and Artificial Intelligence Markup Language (AIML) for the purpose [2].

Pardasani et al. in 2018 proposed a chatbot for differently abled people. In this work image recognition techniques were used to identify finger movement and hand motions as per American Sign Language to understand the intent of the mute person. Accordingly, the output of the chatbot was generated [3].

Another related work in 2018 introduced a platform for chatbot interaction using Python. This work analyzed a few artificial clients and introduced a web-based platform for human-chatbot interaction and a console window to take control over the interactions taking place between the user and chatbot [4].

A research in 2018 presented a Chatbot using TensorFlow for small Businesses. The chatbot proposed in the work demonstrated a method of developing chatbot which can follow the context of the conversation. This method uses TensorFlow for developing the neural network model of the chatbot and uses the natural language processing techniques to maintain the context of the conversation [5]. Recurrent Neural Network (RNN), Long Short-Term Memory (LSTM) techniques were also used in chatbots for answering frequently asked questions very effectively [6]

Albayrak et al. in 2018 presented basic concepts and general working principles of intelligent chatbots. Researchers also showed different applications of such chatbots in various sectors like banking, e-commerce, health, telecommunications etc. [7, 8].

Ramya Ravi in 2018 introduced and intelligent chatbot to be used for the purpose of web analytics. Here the authors compared widely used analytics tools and proposed an Artificial Intelligence and Machine learning based intelligent based bot to help the users to get to analytical insights regarding business by typing the query in the bot [9]

In 2019 an intelligent chatbot was introduced for the purpose of marketing. It was aimed for highly customized experiences for the customers by increasing responsiveness and solving their queries and problems [10].

In 2019 a chatbot was built by Md. Kowsher et al. for educational purpose for answering queries in Bengali language. It used Machine learning based algorithms and Bengali Natural Language Processing (BNLP) for answering users queries in Bengali language. The machine was fed with knowledge so that it can identify the appropriate sentences and make decisions to generate responses to users queries [11].

An affective chatbot was developed by Patel et al. in 2019 to be used in educational purpose. The main aim of this work was to provide mental support to students who face difficulties due to stress that leads to depression resulting poor performance. In this paper, they dealt with different emotional states like Happy, Joy, Shame, Anger, Pathetic, Sadness, Guilt, and Fear. Further, based on the emotion label, it identifies the users' mental state whether he/she is stressed or depressed using users' chat data. For emotion detection, they deployed deep learning classifiers namely, Convolutional Neural Network (CNN), Recurrent Neural Network (CNN), and Hierarchical Attention Network (HAN) [12].

Another group of researchers in 2019, also developed a chatbot that is emotionally aware. They mainly used two inputs, users' interaction with the chatbot and facial expression and used OpenCV Deep Learning for classification of different emotional states. The system was tested to track different expressions per second [13].

Another chatbot was proposed in 2019 which is an AI and Web-Based Human-Like Interactive University Chatbot (UNIBOT). The project performs in a way, where user's request is in the form of question-based message and processes it to deliver a desired response in form of message. The chatbot can provide a wide range of information as per developer's configuration. Natural Language Processing (NLP) can also be integrated to enhance the chatbot [14].

Oscar Hong et al. in 2019 introduced "Comprehensive Technology Function Product Matrix for Intelligent Chatbot Patent Mining". This research presents a text mining- based Technology Function Product Matrix (TFPM) generation approach to analyze a corpus of more than 400 chatbot technology patent documents. The analysis systematically identifies hotspot distributions across the patent landscape with respect to the key technologies, functional advances, and novel applications as current developments in chatbot technologies [15].

3 Motivation and Problem Statement

It has been found in the literature survey that Intelligent chatbot is being implemented in different facets of life for effective user interaction in a cost-effective manner. Use of such chatbots is also found in education sector for various purposes. In this context, it has been observed that, in the existing E-learning websites, users generally need to wait for discussing their queries with someone from the backend. To overcome this problem, a chatbot implemented in such educational websites can be useful to solve user's queries. The present work aims to develop an intelligent chatbot to cater this service for an E-Learning website. With the help of this chatbot a student can discuss his/her queries and a teacher can get the records of a student. Unlike the previous E-Learning websites, this one will be more handy and useful because of the implemented automation.

4 Methodology

- In the present work, at first there will be a login page, where the user needs to login. After logging in, a new site will open which will be having information regarding the courses that are available.
- There will be an option for discussing the queries with the chatbot, named as Mr. Bot.
- The bot will greet the user at first and will ask whether he/she is a student or a teacher.
- In the data pre-processing stage, given commands are divided into single words using Lancaster stemming algorithm available from 'nltk' package of python library. After that, the single words are compared to the tags present in JSON file.
- Conversational intents are needed to be defined for the implementation of chatbot framework. The present work uses a JSON file, to define chatbot intents. Each conversational intent contains: a tag, under which there are patterns and response. Depending upon the keywords, the appropriate tag is selected and then a response is given by the chatbot.
- TensorFlow library is used for fast numerical computing created on the basis of which the chatbot will generate the reply. Keras wraps TensorFlow and allows to define and train neural network models.
- In the present prototype, the chatbot will help the student by answering the queries regarding what courses are available and with the contents of the course. It also helps the teachers by updating them regarding the progress of the students in a particular course.
- The system architecture for implementing the chatbot is depicted in Fig. 1.

5 Experiment and Results

The languages used to build this chatbot prototype are HTML, CSS and Javascript. Many packages from python library were used, but mainly the following packages were used in this project:

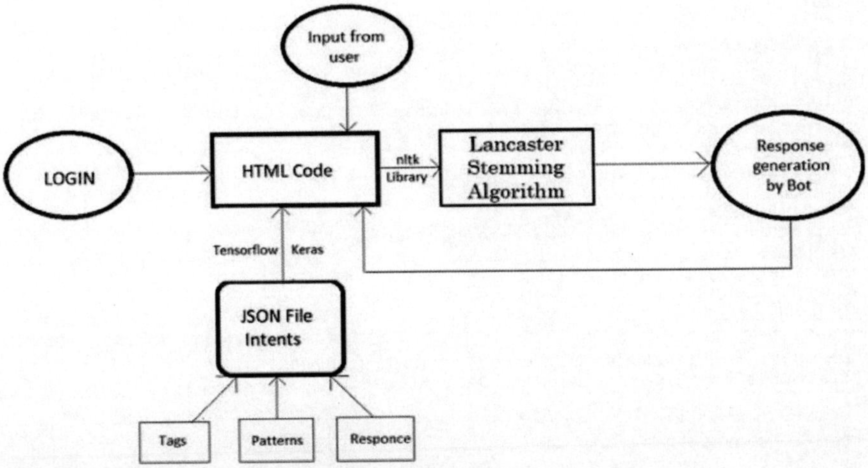

Fig. 1 Basic flow diagram of the proposed model

- nltk- for processing human language data.
- tensorflow- it is a machine learning library.
- keras- Deep learning library for theano and tensorflow

In the JSON file, the intents are defined. Some of the tags are implemented, depending upon what the teacher or student can ask the chatbot named as Mr. Bot. As Mr. Bot is a query solving chatbot, it will ask the teacher and student for their queries. This JSON file is later implemented using javascript in HTML page. The JSON file defined for the present prototype of the proposed model is shown in Fig. 2. in Sect. 5.1.

5.1 JSON File

See Fig. 2. for the JSON file defined in the present work.

```
{"intents": [
{
"tag": "greeting",
"patterns": ["Hi", "How are you", "Is anyone there?", "Hello", "Good
day"],
"responses": ["Hello, thanks for visiting", "Good to see you again", "Hi
there, how can I help?"],
"context_set": ""
},
{
"tag": "goodbye",
"patterns": ["Bye", "See you later", "Goodbye"],
"responses": ["See you later, thanks for visiting", "Have a nice day.",
"Bye! Come back again soon."]
},
{
"tag": "thanks",
"patterns": ["Thanks", "Thank you", "That's helpful"],
"responses": ["Happy to help!", "Any time!", "My pleasure"]
},
{
"tag": "courses",
"patterns": ["What are the courses available?", "What courses are
there?", "What available courses are there?"],
"responses": ["OOP", "DBMS", "C", "OS" , "DS"]
},
{
"tag": "OOP",
"patterns": ["What are the chapter in OOP?", "What are the modules of
OOP?"],
"responses": ["concepts of OOP using java", "class and object",
"exception handling", "multithreading", "applet programming"]
},
{
"tag": "DBMS",
"patterns": ["What are the chapters in DBMS?", "What are the modules of
DBMS?"],
"responses": ["basic concepts of DBMS", "normalization", "file
management", "transaction management"]
},
{
"tag": "completion",
"patterns": ["What are the past records?", "Show me the modules completed
by students!", "How many chapters are completed?"],
"responses": ["concepts of OOP using java", "class and object",
"exception handling", "multithreading",
"applet programming", "basic concepts of DBMS", "normalization", "file
management", "transaction management"]
},
]
}
```


Fig. 2 The JSON file containing the intents defined for the present prototype to validate the model

5.2 Outputs Obtained

First the user needs to go to the Login in page where they have to enter username and password to login to the E-learning portal as shown in Fig. 3. After logging in, the page will take them to the E-learning portal where they can see the courses that are available. There will be an option to chat with the chatbot named Mr. Bot with

Fig. 3 The Login page to E-Learning portal

whom the user can discuss their queries (shown in Fig. 4). The chat with Mr. Bot from the users is shown in Figs. 5a, b and 6a, b.

The present work has been compared with similar existing works. One such work is UNIBOT, an interactive university chatbot (UNIBOT) [14]. The system has been implemented using PHP for server-side programming and database for storing required data. But the present work used JSON, which is a very light weight data interchange format. Read write operations are also easier in a JSON file. In the present work, it is even easy for machines to parse and generate data in comparison to the other existing works, making it more efficient for generating quicker response and cost effective.

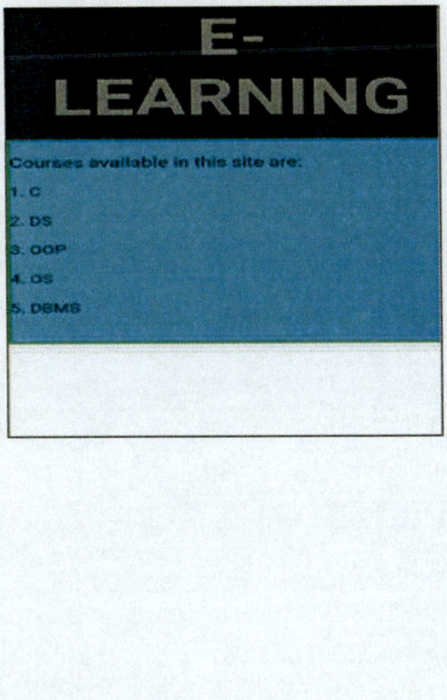

Fig. 4 The E-Learning portal showing different available courses

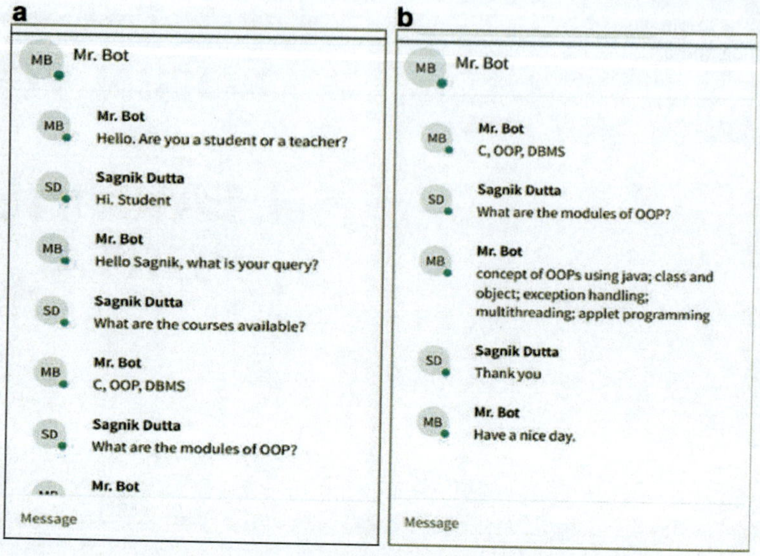

Fig. 5 **a** Chat between Student and the bot **b** Chat between Student and the bot

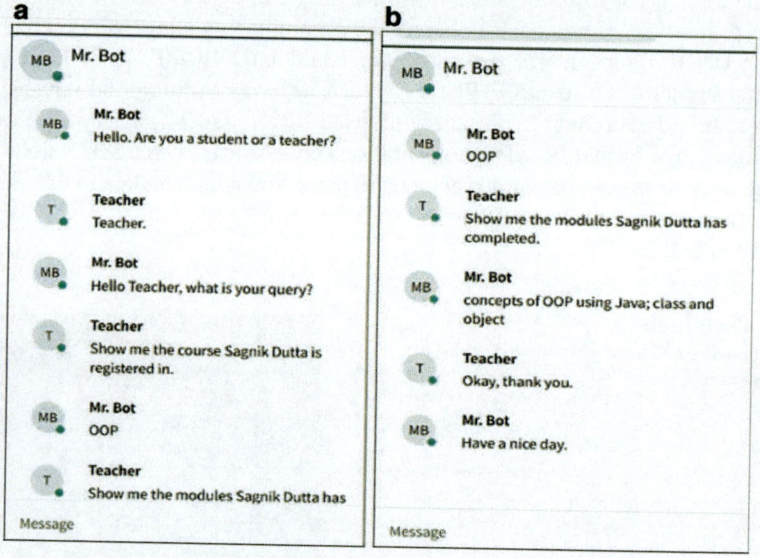

Fig. 6 **a** Chat between Teacher and the bot **b** Chat between Teacher and the bot

6 Conclusions

There are a number of E-learning websites present today, but a chatbot based website is something not very common and can be implemented for any e-learning systems. The present work has developed a prototype of a chatbot that can help the student by answering queries regarding the available course information, course contents and other related queries. The chatbot can help the teacher by showing them the progress and records of a particular student for a particular course. The main aim and benefits of this work is that, it reduces human effort, time and does not require an expensive setup. The present prototype has been compared with other similar existing systems and is found to be more efficient and cost-effective.

For future work, a huge number of intents and tags can be added so that the chatbot can have more knowledge and can converse in a better and vast way.

References

1. H.N. Io, C.B. Lee, Chatbots and conversational agents: a bibliometric analysis, in *2017 IEEE International Conference on Industrial Engineering and Engineering Management (IEEM)* (IEEE, 2017), pp. 215–219
2. B.R. Ranoliya, N. Raghuwanshi, S. Singh, Chatbot for university related FAQs, in *2017 International Conference on Advances in Computing, Communications and Informatics (ICACCI)* (IEEE, 2017), pp. 1525–1530
3. A. Pardasani, A.K. Sharma, S. Banerjee, V. Garg, D.S. Roy, Enhancing the ability to communicate by synthesizing American Sign Language using Image Recognition in a chatbot for differently Abled. In *2018 7th International Conference on Reliability, Infocom Technologies and Optimization (Trends and Future Directions) (ICRITO)* (IEEE, 2018), pp. 529–532
4. B. Kohli, T. Choudhury, S. Sharma, P. Kumar, *A Platform for Human-Chatbot Interaction Using Python*
5. R. Singh, M. Paste, N. Shinde, H. Patel, N. Mishra, Chatbot using TensorFlow for small businesses, in *2018 Second International Conference on Inventive Communication and Computational Technologies (ICICCT)* (IEEE, 2018), pp. 1614–1619
6. P. Muangkammuen, N. Intiruk, K.R. Saikaew, Automated thai-faqchatbot using rnn-lstm, *in 2018 22nd International Computer Science and Engineering Conference (ICSEC).* (IEEE, 2018), pp. 1–4
7. N. Albayrak, A. Özdemir, E. Zeydan, An overview of artificial intelligence based chatbots and an example chatbot application, in *2018 26th Signal Processing and Communications Applications Conference (SIU).* (IEEE, 2018), pp. 1–4
8. A. Bozzon, Enterprise crowd computing for human aided chatbots, in *2018 IEEE/ACM 1st International Workshop on Software Engineering for Cognitive Services (SE4COG)* (pp. 29–30). (IEEE, 2018)
9. R. Ravi, Intelligent chatbot for easy web-analytics insights. in *2018 International Conference on Advances in Computing, Communications and Informatics (ICACCI).* (IEEE, 2018), pp. 2193–2195
10. U. Arsenijevic, M. Jovic, Artificial intelligence marketing: chatbots, in *2019 International Conference on Artificial Intelligence: Applications and Innovations (IC-AIAI)* (IEEE, 2019), pp. 19–193
11. M. Kowsher, F.S. Tithi, M.A. Alam, M.N. Huda, M.M. Moheuddin, M.G. Rosul, *Doly: Bengali Chatbot for Bengali Education.* In 2019 1st International Conference on Advances in Science, Engineering and Robotics Technology (ICASERT) (pp. 1–6). IEEE

12. F. Patel, R. Thakore, I. Nandwani, S.K. Bharti, Combating depression in students using an intelligent ChatBot: a cognitive behavioral therapy, in *2019 IEEE 16th India Council International Conference (INDICON)*. (IEEE, 2019), pp. 1–4
13. K.A.L.R. Carranza, J. Manalili, N.T. Bugtai, R.G Baldovino, Expression tracking with OpenCV deep learning for a development of emotionally aware Chatbots. in *2019 7th International Conference on Robot Intelligence Technology and Applications* (RiTA). (IEEE, 2019), pp. 160–163
14. N.P. Patel, D.R. Parikh, D.A. Patel, R.R. Patel, AI and Web-based human-like interactive university chatbot (UNIBOT), in *2019 3rd International conference on Electronics, Communication and Aerospace Technology (ICECA)*. (IEEE, 2019), pp. 148–150
15. N.O. Hong, U.H. Govindarajan, Y.J.C. Chien, J.A Trappey, Comprehensive technology function product matrix for intelligent Chatbot patent mining. in *2019 IEEE International Conference on Systems, Man and Cybernetics (SMC)*. (IEEE, 2019), p. 1344–1348

Music Generation Using Deep Learning

Subham Mitra and Lopa Mandal

Abstract Machine Learning and Deep Learning has many real-world applications. Music can be considered as sequence data and over the recent past, it has become a very famous data source. It has always been the widely used signal stream. The main aim of the present work is to explore whether using Long Short-Term Memory (LSTM) in deep learning, new music can be generated automatically. In this work, simple folk tunes are used for training and generating new music (folk tunes) using deep learning. ABC notation of music is used for both training and output music generation. The dataset of music consists of mainly piano music. The generated ABC notation of music can be played in other external web players. A decent and pleasant quality music (not noise) can be generated with the help of the proposed model. The model has achieved an accuracy of 92.4% on the training data.

Keywords Deep learning · Music generation · LSTM · Recurrent neural network · ABC notation · Sequence data

1 Introduction

The act of composing music is a challenging as well as a very interesting task. Special training and creativity are required even by humans for it. So, it is expected that this task will be very challenging for machines as well. Music is considered to be the widely used signal stream sequential data. Many methods and conclusions have been attempted and proposed by researchers in this field. In 1950s the first ever music composed by computer appeared and eventually better music has continued to be generated using the computer.

S. Mitra (✉) · L. Mandal
Institute of Engineering & Management, Kolkata, India
e-mail: subham.mitra21@gmail.com

L. Mandal
e-mail: mandal.lopa@gmail.com

V. E. Balas et al. (eds.), *Proceedings of International Conference on Computational Intelligence, Data Science and Cloud Computing*, Lecture Notes on Data Engineering and Communications Technologies 62, https://doi.org/10.1007/978-981-33-4968-1_35

Since then many new and creative methods have been proposed for generating music with the help of computers. A very essential role in music generation has been played by Deep Learning. With sufficient dataset and the correct approach new music can be generated, that would sound decent enough and more importantly it would not be a noise. This report outlines an approach to music composition/generation through Recurrent Neural Network model (RNN), and in hopes of making a new piece of music it is evident that musical ideas can be produced from these algorithms. Quality music using RNN can be generated even by a non-expert. So, RNN maps a sequence of characters from ABC notations of training files to a single character and new set of ABC notations are generated, thus eventually new music is generated.

2 Literature Survey

Computational music composition or generation has always been one aspect which attracts many scholars and researchers. There are lot of research works done previously on this topic by various approaches.

Pitch, note can be generated including some random numbers using the Markov Chains [1].

Other recent attempts have been the use of machine learning and deep learning especially neural networks [2]. The results obtained through deep learning in the field of music generation has been very good. Large amount of music data can be trained and parameter tuning is done to get optimal results and to see whether it is possible to compose good music automatically [3].

All the approaches and the research works do not necessarily require one to be a music expert or to have knowledge in music theory. They have their own different problem statements and proposed solutions.

Representation of music. There are various forms in which music can be generated and so there are different forms in which a music data can be represented. The music can be generated as sequence of data. Those sequences can be encoded in some computer-understandable formats and then again decoded back into proper musical tunes. The most common form of music representation are described below.

2.1 Sheet Music

It is the most trusted form of music representation. In sheet music rhythms, pitches or chords of a song or musical instrument can be indicated using musical symbols. Figure 1 illustrates the sheet form of music notation which is in handwritten or printed format. Expert musicians initiate their music composition process by sheet music representation itself. Paper is typically the medium of sheet music especially paper or parchment. Eventually with technological advancements, sheet music can

Fig. 1 Sheet music representation (**a**)

be easily represented on the computer screens. The term 'sheet' is mainly used to differentiate it from other sound recordings format like the ones used in broadcasts or radio. It has become a basic form in which Western classical music can be noted and be used by various musicians.

2.2 *MIDI (Musical Instrument Digital Interface)*

It is a technical standard which is used to connect different varieties of musical instruments, computers for recording and playing music. Up to 16 channels of information can be carried through a MIDI cable through a single MIDI link. The sixteen channels can of various instruments as well. They carry event message, especially the data contains music instructions. As analyzed in Fig. 2 these instructions may include pitch, clock signals, note's notation. All the key presses, button presses, knob turns, slider changes are directly converted into MIDI data, when a musician plays a MIDI instrument. A MIDI instrument can be a piece of hardware (electronic keyboard, synthesizer) or just a part of software environment.

The MIDI recording could be changed to many other sounds. MIDI music can be converted to sequence of ABC format music and vice versa using python music21 library [4].

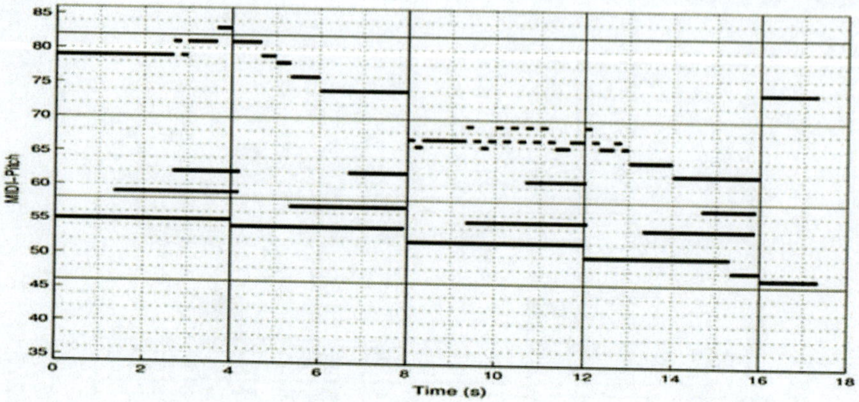

Fig. 2 Representation of music in MIDI format (**b**)

2.3 *ABC Notation*

It is the simplest form of music representation computationally. The most recent standard for ABC was released in 2011, thus solving many ambiguities, which existed in the earlier versions. In the most basic form of ABC notation, letters or characters are used to represent notes. Also other details like length of the note, key, sharpness, flatness are represented by alpha numeric characters. This format mainly began as ASCII character set code, so that music can be shared online. It was originally designed to be used for folk tunes or traditional western tunes. ABC notation can be easily edited by a text editor it is completely ASCII based. In recent times many such software packages are available that have many built-in features. Such software can be completely web based or be available for various systems like Microsoft windows, Unix/Linux, Macintosh etc.

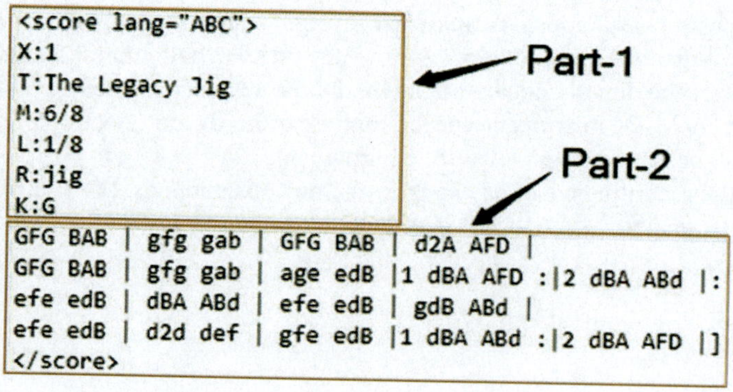

Fig. 3 Representation of music in ABC notations (**c**)

As analyzed in Fig. 3, the meta-data part is shown in part 1 which means it contains data or information about the musical tune. Part 2 mainly denotes the other essential features of a musical note [5]. It is just a sequence of characters, representing some musical notation. This example can be translated into some traditional music using some conversion tools or softwares. ABC notation of music is being extensively used in this project as it is of small size and it is easily accessible.

2.4 Deep Learning

It includes a bigger part of machine learning primarily neural networks. They consist of multiple processing layers to learn about the various representations of data. It also includes many levels of abstraction as well. It has many applications in real world projects. There are many internal parameters, tuning of which leads us to optimized results as and when required. Many different types of neural networks like Artificial neural network (ANN), Convolutional neural network(CNN), Generative adversarial network(GANs), Recurrent neural networks(RNN) etc. have been discovered. Artificial neural networks can deal with simple real world problems. It has the most simple architecture as compared to other neural networks [6]. The usage of the wide variety of neural network depends also on the type of the data present. For instance image data work best with CNNs for various classification problems, pattern recognition and computer vision [7]. GANs can work in various dataset domains and have recently been very successful in deep fakes. GANs have also attracted many researchers all around the world [8]. For many cases the input size varies (i.e. there can be sequence data) and when input size varies it should be fed to the same model architecture every time. A very dominant problem of Vanilla Neural Networks is that as they cannot work with variable sized inputs [9]. The above problems can be solved by Recurrent Neural Networks (RNNs). Music can be a considered as sequential data [10]. So this type of data can be easily given to the RNN architecture model. The core reason that RNNs are more exciting is that they allow to operate over sequences of vectors/data i.e. sequential data in the input, sequential data in the output, or sequential data in both input and output.

2.5 Architectures

As analyzed in Fig. 4, each rectangle is a vector and arrows represent functions (like matrix multiplication). Input vectors are shown in red color, output vectors are shown in blue color and green vectors show the RNN's state. One to one model or Vanilla mode of processing without RNN, from fixed-sized input to fixed-sized output like image classification. One to many model generates a sequence output. For example image captioning which takes an image and generates a sentence of words. Many to one model takes a sequence input and generates fixed-sized output. Many to many

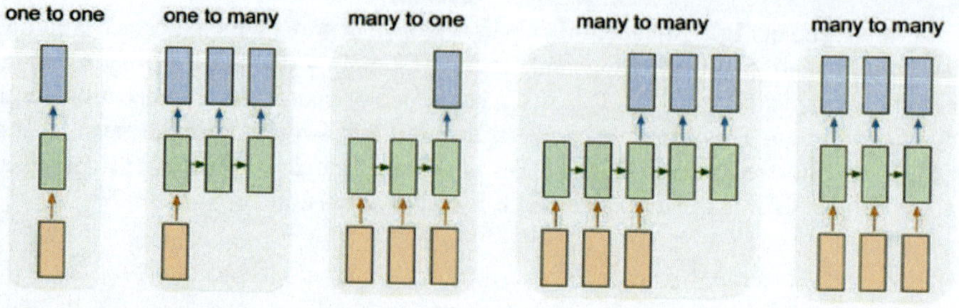

Fig. 4 Types of RNNs (**d**)

model (1st type) takes a sequence input and generates sequence output. For example, Machine Translation i.e. RNN reads a sentence in one language and then outputs a sentence in another language). Many to many model (2nd type) can be thought of as video classification in which each frame of the video has to be marked. In every case there are no pre-defined length constraints on the sequences as the recurrent transformation (green) is fixed, so they can be applied many times. As it is evident that RNNs are very powerful as compared to other fixed types of network, RNNs combine the input vector with their state vector with a fixed function to produce a new state vector. RNNs contains loops within their network which one of the reason why it makes it more and more powerful than the traditional Neural Networks.

As it can be seen from Fig. 5 that a chunk of neural network, RNN(A) looks at some input (xt) and outputs a value (ht). A loop allows information to be passed from one step of the network to the next. These loops make RNNs seem different than other traditional neural networks. A recurrent neural network can be thought of as multiple copies of the same network, each passing a message to a successor.

Fig. 5 Looped structure of RNN (**e**)

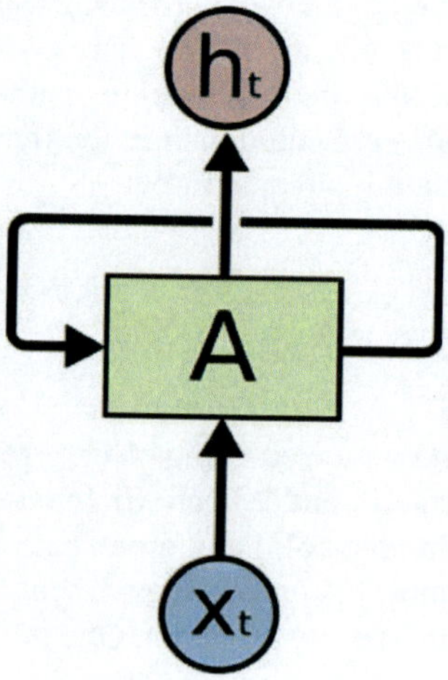

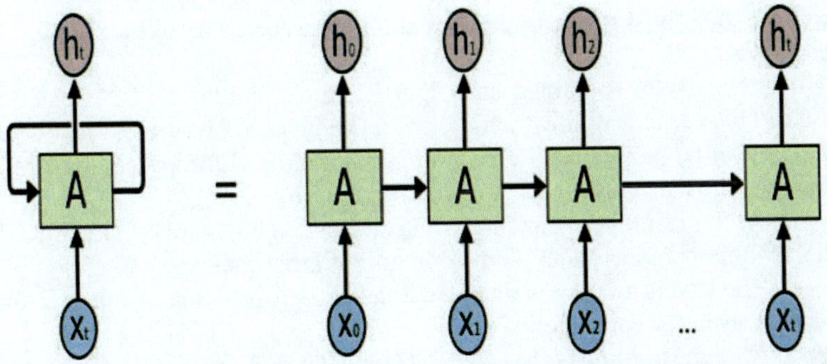

Fig. 6 Un-rolled structure of RNN (e)

Figure 6 illustrates the RNN architecture when the loop gets un-rolled. This is what simple RNNs look like. These are the architectures to be used. Although simple RNNs are not that effective because of the problem of the exploding and vanishing gradients [11]. Long Short-Term Memory (LSTM)s are special types of RNNs capable learning long term-dependencies [12]. They have very wide range of applications in real world.

Thus, LSTM are highly powerful and is a revolutionary invention. It helps us to solve lot of problems which simple RNNs cannot solve. These were primarily built for dealing with problems involving long term dependency and opens up new direction in music generation [13].

3 Problem Statement

The present work attempts to generate new music by training a variant of Recurrent Neural Network called Long Short Term Memory (LSTM) to produce a new sequence of music of ABC notation. It is accomplished by RNNs, where a trained RNN architecture maps sequence of characters to a single character from the set of all characters used in the training. It is thus a multi-class classification problem. The most important question to answer is that whether an RNN can produce music that is structured and pleasant to hear.

4 Methodology

A promising way to make machines understand sequences, is the use of Recurrent Neural Networks.

Step 1: The music is represented as [a, b, c, a, d, f, e,...]. The first character 'a' is given as an input and 'b' is expected to be generated as output from the RNN. Then in next time-step ‘b’ is given as an input and it is expected ‘c’ to be an output and it continues.

Step 2: The RNN is trained such that it should output next character in the sequence. This is the way in which the whole sequence is learnt and new sequence on its own can be generated. Here, “Categorical Cross-Entropy” is used as a loss function which is also known as “Multi-Class log-loss”.

Step 3: In the last layer, “Softmax” activations are applied [14]. Each RNN can be a LSTM that contains ‘tanh’ activation unit at input-gate.

Step 4: The RNN structure is trained using back-propagation and further “Adam” optimizer is used for convergence [15].

Step 5: After the training of the char RNN model is done, any one random character is given from the set of unique characters. This set was earlier fed to the char RNN during training time to the trained char RNN. Then, based on the sequences and patterns that the model has learnt during training phase, characters are generated automatically.

5 Experiment and Results

Data Source: ABC version of the Nottingham Music database. Programming Language used: Python. From first data-source, first two files are downloaded: Jigs (340 tunes) and Hornpipes (65 tunes). The whole experiment is done using Google Colab.

The data is fed into batches. A batch of sequences is fed at once into the RNN model. The batches are constructed with the following parameters as follows: Batch Length = 16, Sequence Length = 64.

It is found out that there are total of 155,222 characters in the data. Total number of unique characters are 87.A numerical index is assigned to each unique character.

A dictionary is made where the key is a character and the value, its index. Another dictionary is also created where key belongs to index and its value is its character.

A single RNN layer is constructed exactly like there are 256 LSTM-RNN Units in one layer of RNN. The output generated by all of the RNN units at each time step, will be an input to the next layer and also the same output will also be a input to the same RNN unit.

In the present work each RNN unit is an LSTM unit. There is a parameter called ‘return_sequences’ in Keras library. It is ‘False’ by default. But if it is True, then each character will be generated as output by each RNN unit, at each time step. The next character is expected to be generated as output by the RNN unit when given input is previous character in the sequence. There are 3 RNN layers each having 256 LSTM units. After 3 such layers of RNN, ‘Time-Distributed’ dense layers is applied with “Softmax” activations in it. There are 87 unique characters in the dataset and the output at each time-stamp is expected to be a next character in the sequence which is one of the 87 characters. Finally, 87 dimensional output is generated at each

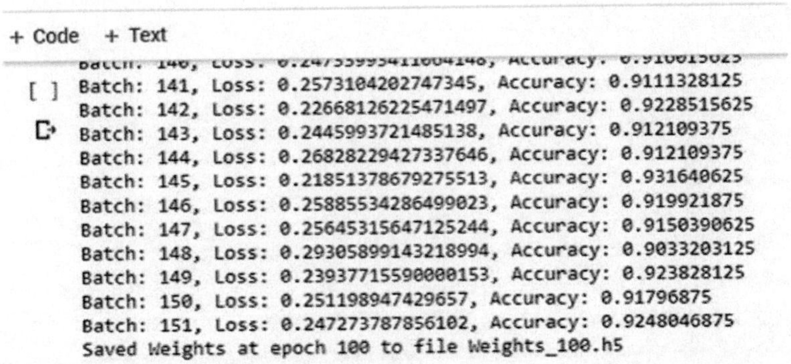

Fig. 7 Model showing 92.4% accuracy on the 100th epoch of training [using google colab]

time-stamp which will be equivalent to 87 probability values. It helps the model to maintain Many-to-Many relationship.

Now the next character is to be predicted which should be one of the 87 unique characters. So, it turns out to be a Multi-Class Classification problem. Therefore, the last layer is softmax layer with 87 activations.

The model is trained for 100 epochs each having 151 batches. After training the model, the best weights are found. The weights have been saved after every 10th epoch and the model has achieved an accuracy of 92.4% on the given dataset as shown in Fig. 7.

After that the model will be ready for making prediction. In order to make prediction, any of the 87 unique character is given as input to the model and finally 87 probability values through softmax layer will be generated. From these returned 87 probability values, the next character is chosen probabilistically and not deterministically and finally again the chosen character is fed back to the model and so on. The output characters are concatenated one after the other and music of some length is generated. This is how music will be generated/composed automatically as shown in Figs. 8 and 9.

As depicted in Fig. 10, the output text simply needs to be copy pasted in the external webpage.

That generated tune can be heard in external web page where conversion takes place.

6 Conclusions and Future Scope

The present work proposed a model using Long Short-Term Memory (LSTM) in deep learning, by which, new music can be generated automatically. The model has achieved an accuracy of 92.4% and good quality music/tune has been successfully generated using this model. There are some scopes of improvement in it as well.

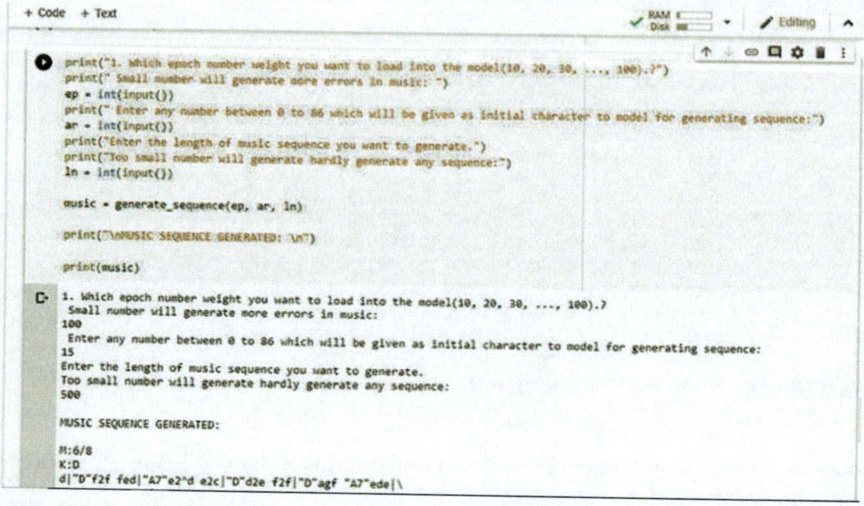

Fig. 8 Generated output by the model [using google colab]

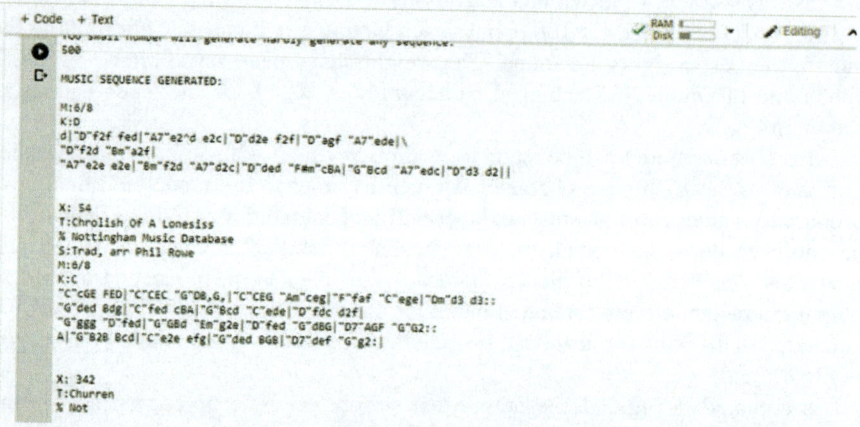

Fig. 9 Generated output by the model [using google colab]

Firstly, the generated music be made more melodious. Secondly, the model can be trained with more tunes. In the present work, model has been trained with only 405 musical tunes. By training the model with some new musical tunes, the model can learn more variety of music. Finally, the model can also be trained with multi-instrument tunes.

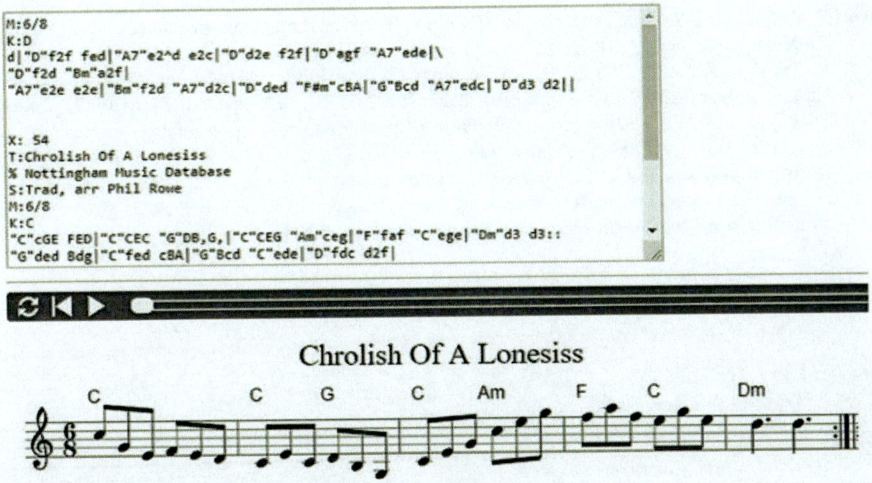

Fig. 10 Generated output by the model [using google colab]

References

1. K. Trochidis, C. Guedes, A. Anantapadmanabhan, A Klaric., CAMeL: Carnatic percussion music generation using n-gram models, in *Proceedings of 13th Sound and Music Computing Conference (SMC),* vol. 31 (Hamburg, Germany, 2016)
2. M.A. Nielsen, *Neural Networks and Deep Learning*, vol. 2018 (Determination press, San Francisco, CA, USA, 2015)
3. J.P. Briot, G. Hadjeres, F.D. Pachet, Deep learning techniques for music generation–a survey (2017). arXiv preprint arXiv:1709.01620
4. C. Raffel, *Learning-based Methods for Comparing Sequences, with Applications to Audio-to-midi Alignment and Matching* (Doctoral dissertation, Columbia University, 2016)
5. C. Walshaw, The ABC music notation. http://abcnotation.com
6. I.N. Da Silva, D.H. Spatti, R.A. Flauzino, L.H.B. Liboni, S.F. dos Reis Alves, *Artificial neural networks* (Springer International Publishing, Cham, 2017), p. 39
7. J. Gu, Z. Wang, J. Kuen, L. Ma, A. Shahroudy, B. Shuai, T. Liu, X. Wang, G. Wang, J. Cai, T. Chen, Recent advances in convolutional neural networks. Pattern Recogn. **77**, 354–377 (2018)
8. T. Kim, M. Cha, H. Kim, J.K. Lee, J. Kim, August. Learning to discover cross-domain relations with generative adversarial networks, in *Proceedings of the 34th International Conference on Machine Learning-Volume 70*. JMLR. org, pp. 1857–1865
9. T. Stérin, N. Farrugia, V. Gripon, An intrinsic difference between vanilla RNNs and GRU models. COGNTIVE **2017**, 84 (2017)
10. Z.C. Lipton, J. Berkowitz, C. Elkan, A critical review of recurrent neural networks for sequence learning. *arXiv preprint* arXiv:1506.00019
11. B. Hanin, Which neural net architectures give rise to exploding and vanishing gradients?. in *Advances in Neural Information Processing Systems* (2018), pp. 582–591
12. R. Dey, F.M. Salemt, Gate-variants of gated recurrent unit (GRU) neural networks, in *2017 IEEE 60th International Midwest Symposium on Circuits and Systems (MWSCAS)*. (IEEE, 2017), pp. 1597–1600
13. N. Agarwala, Y. Inoue, A. Sly, Music composition using recurrent neural networks, in *CS 224n: Natural Language Processing with Deep Learning* (Spring, 2014)

14. Z. Yang, Z. Dai, R. Salakhutdinov, W.W. Cohen, Breaking the softmax bottleneck: A high-rank RNN language model (2017). *arXiv preprint* arXiv:1711.03953
15. R. Tutunov, M. Li, J. Wang, H. Bou-Ammar, Compositional ADAM: an adaptive compositional solver (2020). arXiv preprint arXiv:2002.03755. a. https://en.wikipedia.org/wiki/Sheet_music. b. https://www.researchgate.net/figure/Piano-roll-representation-of-an-Aria-interpretation_fig4_36709065. c. https://en.wikipedia.org/wiki/ABC_notation. d. Karpathy, A., 2016. The unreasonable effectiveness of recurrent neural networks, 2015. *URL* http://karpathy.github.io/2015/05/21/rnn-effectiveness. e. https://heartbeat.fritz.ai/a-beginners-guide-to-implementing-long-short-term-memory-networks-lstm-eb7a2ff09a27

Med-Alert: An Android Application to Increase Adherence to Medication

Soumili Kundu and Matangini Chattopadhyay

Abstract The most dangerous problem in the modern day health-care system is missing medicines, mainly for those with complex medication schedule leading to non-adherence to medication. This is mainly due to the busy lifestyle we are leading. Life today is full of obligations and tension and people are vulnerable to various forms of diseases. Long duration of medication errors may lead to serious health issues mainly for older patients or patients with chronic illness. Since non-adherence to medication will just complicate our lives so to deal with this present day problem, to increase adherence to medication and to simplify our lives, the Med-Alert android application is introduced and discussed here in this paper. It is implemented with the objective of informing patients of their dose timings via Alarm Ringing system. The Med-Alert app discussed in this paper, is an all-in-one android application that replaces the traditional way of manually noting down date and time of taking medicines. It includes features like setting customized alarm, getting notifications on desired date and time with medicine and dosage details which also facilitate in noting down appointment details and fetching required physician contact details for booking appointments in need of emergency. Med-Alert provides provision of searching doctors based on diseases. A prototype of Med-Alert has been implemented on Android Mobile Platform. The Med-Alert app discussed in this paper increases adherence to medication and also focuses on contacting of patients with doctors.

Keywords Firebase · Real-time database · Android studio · Android application

---

S. Kundu (✉) · M. Chattopadhyay
School of Education Technology, Jadavpur University, Kolkata 700032, India
e-mail: soumilikundu0456@gmail.com

M. Chattopadhyay
e-mail: matanginic@gmail.com

V. E. Balas et al. (eds.), *Proceedings of International Conference on Computational Intelligence, Data Science and Cloud Computing*, Lecture Notes on Data Engineering and Communications Technologies 62, https://doi.org/10.1007/978-981-33-4968-1_36

1 Introduction

Digital scheduling of medicines is the core future of the medication applications. The patient group (target audience) includes all human beings, teachers, students, entrepreneurs, housewives and children. It can be difficult for a patient with a complex medication schedule to recall which pill they will take at what time, even helpful for aged/elderly [1] persons. These patients will be benefited from reminders with notifications and also provided with welcome notification [2]. An Android application is implemented here with the objective of informing patients of their dose timings via the Alarm Ringing System. The alarm can be set for several medicines and timings, including the date, time and medication summary [3]. Doctors can be searched by the patients based on diseases; patients can receive the contact details of the doctors. Users (patients) can also check out different articles related to medical fields and can also get health related tips even there is the provision to store prescription [4]. Even the app comes with admin section, which allows admin to deal with any problems faced by users while using the app, to update, delete, load records saved in database. The app provides user friendly environment, easy navigation [3]. Med-Alert can be viewed as the input-output system, where, patient login details, med-intake timings, appointment date, time, doctors' name along with specialization etc. can be treated as inputs and medical adherence, health tips, help tips, past and present news articles, notification (med-intake and appointment), list of doctors along with contact information etc. can be treated as outputs. Medical adherence primarily refers to how patients are taking their medications as prescribed, and whether they are still taking their prescribed medicines.

Problems faced are: Alarm previously set, gets dissolved when phone is shut down, Useful for only smart phone users, so persons (old persons mainly) not familiar with using smart phones, can't avail the luxury of using this kind of medical app, Ringtone/sound of the notification to be received is prefixed.

The rest of the paper is structured according to: the second section reviews some of the related work. The third section presents the proposed solution, proposed system and database used is also briefed here in this section. Implementations and results are reported in the fourth section, followed by discussing conclusion of the things studied and future work in the fifth section i.e., the final section.

2 Related Work

Park et al. [1] proposed medication reminder data-based synchronization system that transmits OMA (open mobile alliance) DS (data synchronization)-based messages containing patient medication data and device configuration data to remote manager/medical staff. This also synchronizes modified/generated data (including medication schedules) on the medication server.

Ferreira et al. [2] designed a medication assistant for elderly people. Elderly population has higher levels of non-adherence to medication thus aimed towards developing an application to aid them with medication management. The application developed following an iterative method. Along with traditional medication alerts, the application provides medication advices, and some provision for handling situations if the elder forgets to take the medication.

Deepti et al. [3] suggested an android system accompanied with automatic alarm ringing system where patients don't need to memorize the medicine dosage timings. Alarm that is used as discussed could be set i.e., time, date and medicine in-take description and accordingly the notification is sent to the patients. Easy navigation and easy-to-use user interface were the focus of the system. The developed medical recording system needs new equipment, but here they have tried to create or improve a system that is economical, time-saving and with good adherence to medicines.

Silva et al. [4] discussed about that medicine blunders are the most risky issues in cutting edge human services framework. The most common are wrong medications and organization of prescriptions obsolete. The application permits patients to enroll meds through the cell phone camera to catch the standardized identification accessible at the medicine boxes.

Wang et al. [5] Wedjat, A Mobile Phone Application has been developed that provides in-take reminders and attempts to maintain a strategic distance from medicine organization bugs. Despite the fact that mistakes can happen in each progression of medication process during medication obtainment, remedy and organization, they happen most every now and again. A considerable number of the accessible frameworks require extraordinary equipment to be purchased, which is exorbitant to bear along these lines.

Antoun et al. [6] structured a framework and the reason for the framework is to helping patients, progressively focused on seniors, so they can take as much time as necessary without the chance of missing pills, likewise diminish the danger of over or under dosing. The framework created could take care of such issues by advising and making the patients aware of take the fitting portion at the ideal time. Old patients, particularly ones with constant and intermittent medication, will profit the most for the Smart Medicine Dispenser (SMD), since it will extraordinarily increment their medication adherence.

Prasad et al. [7] discussed a med alert pro approach. This is a free App that supports up to 15 reminders. In either repeating or non-repeating alarm patterns, the user can select these. Application must generate an alarm, vibration, or LED notification at the scheduled time.

Minaam et al. [8] discussed the causes of medical mistakes or medicine non-adherence and stated that those are due to the fact that people in charge of medicine for patients or elderly people have to deal with processing vast numbers of medicines. This paper consists of designing, designing and developing a prototype pillbox intended to address this deficiency in the medical field as it has the potential to sort out the pills on its own as well as many other advanced features, with this product planned to be used by hospitals. This medication pill box focuses on patients who regularly take meds or supplements, or caregivers coping with patients. The smart pill box is programmable, enabling medical caregivers or clients to decide the quantity and timing of the pills to be taken. The pillbox will alert clients or patients to take pills using sound and light, when the med time is set and an android application owned by the patient will display the warning of pills that are scheduled.

Kim et al. [9] mainly focuses on group of patients who require complete medical attention. A computing formula is used to identify any sudden change based on body temperature, blood pressure. Many governments in the world have run medical services that visit patients to provide a more patient-oriented medical service. But due to the lack of staff, such initiatives are not so successful. U-Healthcare is a choice to overcome the threshold of chronic disease management, a remote medical monitoring service which has been of great help in solving the common issues of public healthcare. This study seeks remote monitoring of healthcare facilities for high-risk patients who require constant medical condition checks.

Park et al. [10] focuses on quality assessment of available mobile phones apps targeting medication adherence helping patients to take prescribed meds and patients' experiences with such apps. The practical finding from the study is that before selecting a suitable app, the search terms 'adherence' shows less relevant search results but 'medication', 'pills' yielded most search results.

Pérez-Jover et al. [11] estimated that a good percentage of patients have a tendency towards medical non-adherence, leading to increased morbidity and therapeutic interventions becoming unsuccessful. The goal of this study was to examine whether mobile apps are viewed as being useful to administer medication at home and if they actually lead to increasing patient adherence to treatment. In addition to smart phone applications found to facilitate adherence to medication, there are also applications for encouraging adherence to other therapeutic regimens, including reminders for leading a suitable lifestyle and tracking other health. The incorporation of such functions should be considered when developing new virtual pillbox applications. Mobile applications prevent medication from being overlooked and inappropriate administration, and thereby lead to patient safety.

There are some features which are still missing in already proposed solutions, advances are done but some developments are yet to be done. Some improvements made in the Med-Alert application, discussed in this paper, make the Android Application proposed and developed, different from many already existing applications. It helps in secure login without compromising data security, both of admin and user (patient) and also gives the provision for email verification in-case if the user forgets login password (forgot password facility); easy access of the medical background, provision for uploading prescriptions thus helps in accessing the e-prescription of a patient to the doctors, which will help them in future regarding their treatment; online system with easy record tracking of patient's medical history in real-time database. There are many apps which allow getting doctors' list based on disease along with contact details of the doctors but for that, users have to use some other application but here Med-Alert application proposed has the doctor contact list accessing facility incorporated within it, allow availability of doctors' names along with doctor contacts to patients, appointments setting facility provided, notifications allowed.

3 Proposed Solution, System and Database Used

3.1 Proposed Solution

Implementing an Android application which can overcome the problem of forgetting to take meds at proper time with proper dosage, it will not only give easy access of the medical background of a patient to the doctors, which will help them in future regarding their treatment. Basically, there are some existing ideas that are already proposed which do not provide all the facilities to replace the traditional system. The key features that make our application stand out from other proposed solutions are –

a. Direct connectivity with health app instead of providing many fields to note down health related things.
b. Allow availability of doctors' names along with doctor contacts to patients.
c. Appointments setting facility provided, notification allowed.
d. Secure login system, both of admin and user (patient) and allowing email verification in-case if the user forgets his/her login password (forgot password facility).
e. Online system allows easy record tracking of patient's medical history in real-time database.
f. Not allowing once/already registered email address, that means, each and every patient who will be registering, will be holding a unique email address thus making identification easier for a particular patient.

3.2 Proposed System

Android was designed to allow developers to build compelling mobile apps that take full advantage of what a smart phone has to offer. Android apps uses mandatory XML manifest file. This file contains important information for an Android platform to handle an application's life-cycle. Among other design and configuration properties, examples of the kinds of details contained in a manifest file are explanations of the components of the device. Components may be one of the following categories of components: Activities, Services, Broadcast Receivers and Content Providers.

User authentication the role-based access control system used in this module is used to authenticate a user. New user must register before using the device. Upon signing in, each user holds unique credentials. When the user signs in he/ she can access and edit the preceding information whenever the user wants. After registration form, comes the profile development page consisting of different personal information fields. Personal information indicates the information provided by the user during the registration process. *Database* of all patients (users) will be managed in a server running in cloud environment. The cloud service can also help to store documents, medical history and all other user information. *Alert notifications* will be sent: To the registered users (patients) at their customized time and date, to the users regarding dosage and medicine intake time.

The TWO main modules of the application are: Patient module & Admin module. Admin section is expected to perform certain functions including update, insert and delete sections/options as required.

- *Patient Module*: Patients can—Login to access the portal using correct email id and password, choose medicine name or write medicine name if in case the medicine name is not present in the list, choose date from calendar facility provided and time so that alarm can be set properly to get the time at properly scheduled time and date, choose doctor's name along with the disease name and getting contact details as well for communicating with concerned doctors. This module that has a notification sending facility consisting of Reminder sys-tem can be mainly of TWO parts: Setting Alarm and Getting notification.
  Setting Alarm module: Alarm module is helpful in remembering the medicines. User should add information about their dosing schedules. Using the given date area, one may enter the start and end dates between which he/ she was scheduled to take medicines. The time field indicates the time of dosage and the alarm will be blown on that time. The user can add the medication definition, including name, intent and other information related to it, or anything else he/ she needs. This makes everything digitalized and the records of patients available at any time & easy data handling.
  Notification module: When the alarm is activated then the warning is received by the user. The users can turn this on or off accordingly. If any individual does not need the notification then it can be switched off. A message will then be sent to the system if necessary.
- *Administrator Module*: The administrator can—Recover all registered patients' details, password reset link will be sent to the registered email-id so that only that person can have the accessibility, add new medicines, add new doctors' names to the already added list of them, view patients' medical background, update many other fields, delete some fields not needed.

Figure 1 shows the flowchart of the application which gives a pictorial idea of how the Med-Alert application is working.

Database used is real-time Firebase is much faster than HTTP. Real-time database Firebase itself offers excellent user experience. The new data is sent to Firebase the moment it is updated. When the user saves updated data, the data in the database is updated instantly. The deletion, up gradation and insertion of data are immediately reflected. That is the advantage of a database in real time. Fire-base Storage offers a convenient way to save files, most often images, but the client/ user may do little directly. Firebase Storage has its own safety rules which can also be modified to give the stored data adequate protection. It also provides authentication. Firebase has an authentication mechanism built into the email/ password. It comes in with Realtime Database and Cloud Firestore. The Sign-in method gives the provision to choose the method of sign-in. Here for the Med-Alert application, Email/Password has been enabled to serve the purpose. "Ref" and "Child" are completely different things, because "Ref" just makes a reference indicating where the data should go and "Child" is specifying the more exact location of that data. Reference represents a specific location in the database and can be used to read or write data to that place in the database, push ()

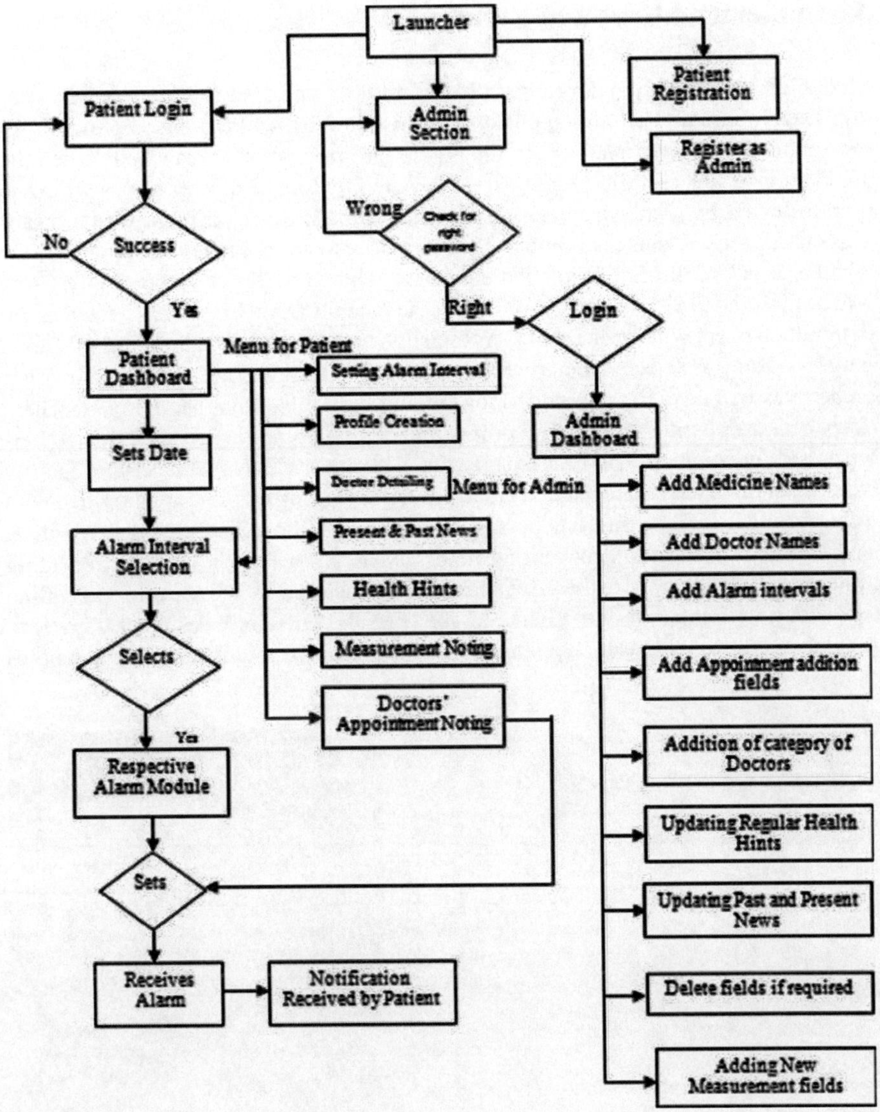

Fig. 1 Flow chart

method allows for the addition of child nodes to a list of objects in the database and assigns an auto-generated key to it. This key is unique. Once a new child has been created by the push method, it returns an initialized Database Reference object with the path to that child. Calling getKey () will return the auto-generated key on this reference.

4 Implementations and Results

Med-Alert application is developed on Android Studio, used for Android Application Development. The said application is installed in Android based phone from the application drawer available in the phone the application is launched. When the application is successfully running first the welcome screen appears with a welcome note followed by a splash screen displayed in Fig. 2. Splash screen, which disappears within seconds, is also shown. User will have to choose the type of login that is whether it is 'Admin Login' or 'Patient Login' shown in Fig. 3. Accordingly, users will be directed to different pages or activities. When 'Patient Login' is selected as the option then users will be directed to a Registration page and if 'Admin Login' (done only by admin) is selected then it will be directed to a login page where admin login credentials will have to be provided for logging in. Registration page (Fig. 4) where every new user has to register with proper and correct details in proper format, otherwise users encounter error. Login page (Fig. 5) comes with forget password facility when reset link is sent to registered email-id. After logging in, users are provided with different pages to fill in different personal details, page that is displayed here in Fig. 6. There's the provision for auto-filling fields as shown in Fig. 7. The Alarm ringing system is also shown, with the provision of selecting med-intake starting and ending date along with interval of medicine intake (Fig. 8). The clock designed (Fig. 9) to set alarms comes with user choosing facility and helps in setting alarm, which helps

Fig. 2 Welcome screen

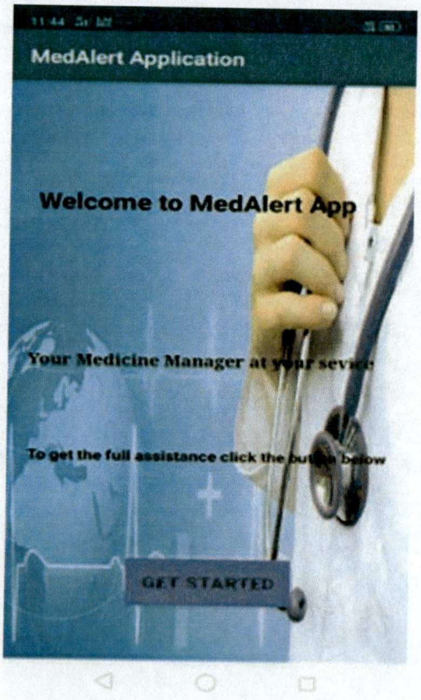

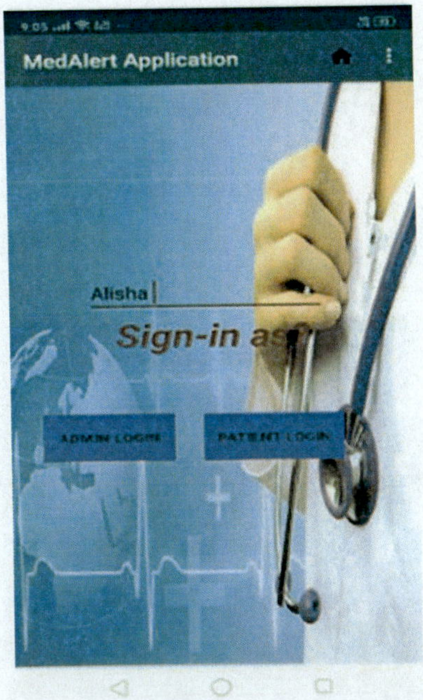

Fig. 3 Mode of entry choosing screen

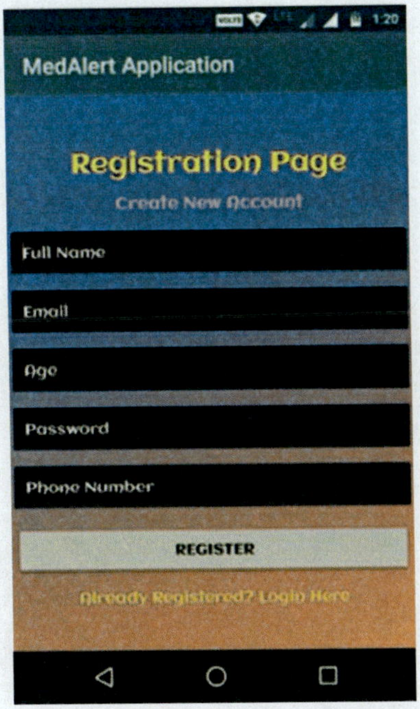

Fig. 4 Registration screen

Fig. 5 Login screen

Fig. 6 Options listed

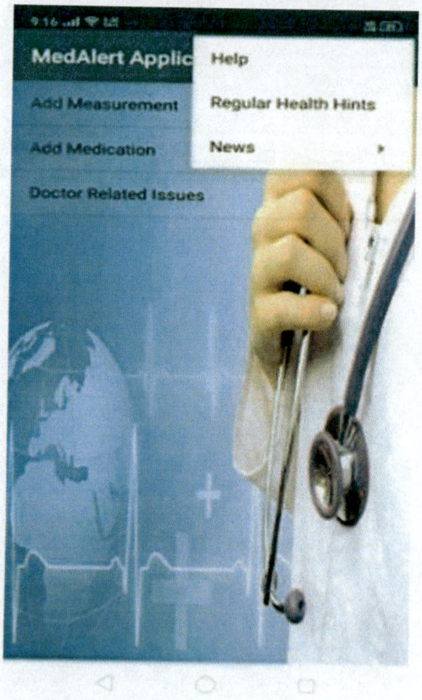

Fig. 7 Medicine list

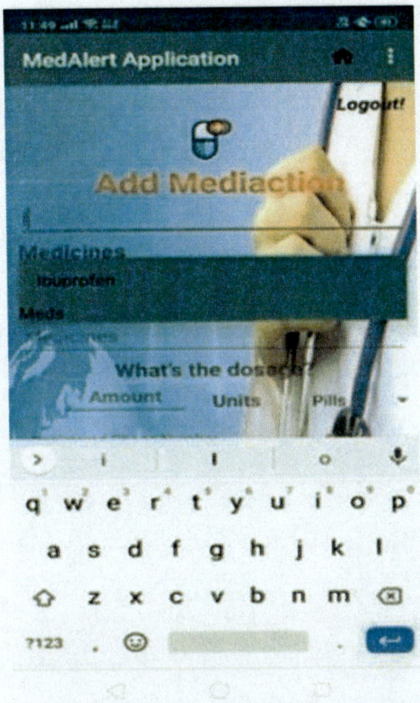

Fig. 8 Alarm setting details

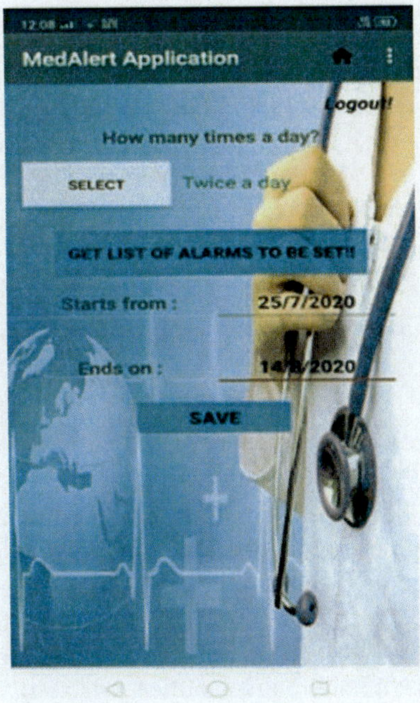

Fig. 9 Clock to set alarm

in avoiding missing meds thus increasing medical adherence. Even the facility of fixing appointments based on disease type, doctor, time availability (Fig. 10), noting it down, availability of doctor contacts, setting reminder accordingly and the provision for uploading prescription for any kind of future reference, is also incorporated into the application. Admin, with se-cure and separate login facility (Fig. 11), gets the allowance of Updating, Deletion, and Insertion of different fields (Fig. 12), and also helps in displaying inserted data. Every activity is app controlled. Even each and every application user is provided with user guidance, past and present news articles and also health tips.

5 Conclusion and Future Work

'Med-Alert' application tries to remove the drawbacks of a traditional medication procedure. A lot of research works suggested the use of smart phones in medical settings. Internet interventions can enhance medication adherence of the patients. The application has a secure and authenticated login system which makes it reliable and trustworthy for patients and even if the phone is running out of battery, system continues to give notifications until the phone is switched off.

Improvements that have been done include searching of doctors according to disease, providing the facility of scheduling appointments with the doctors and also

Fig. 10 Doctor details

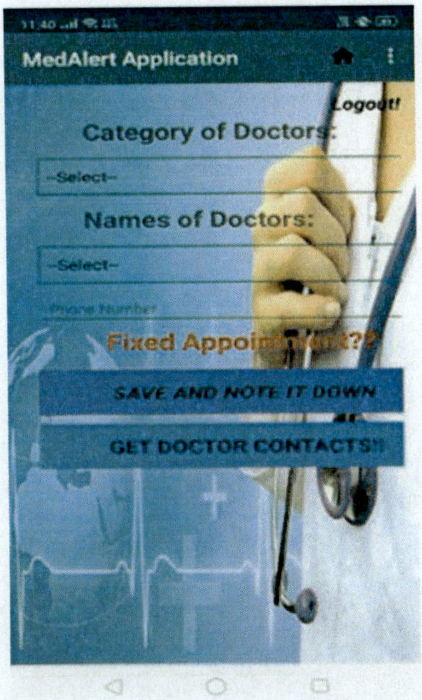

Fig. 11 Admin login screen

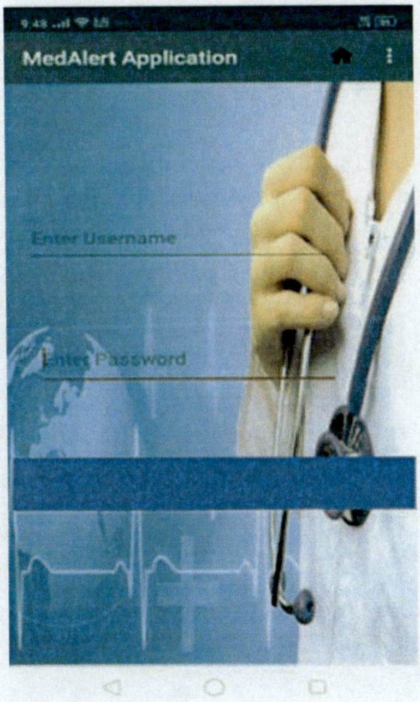

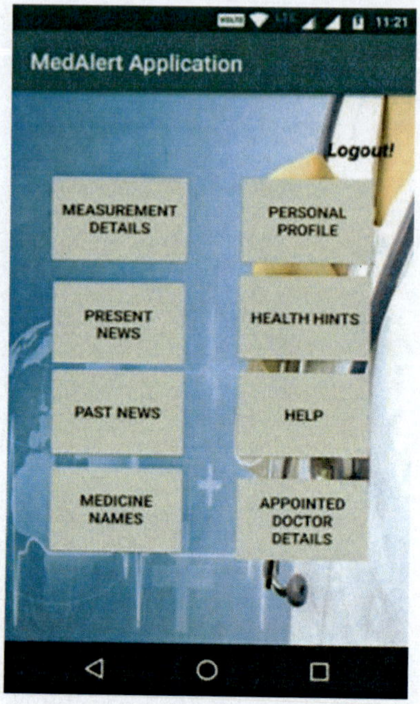

Fig. 12 Screen after admin login

helps in storing doctor's prescription. Some more features which can be included in the application and can also be suggested as the future works are as follows:

i. Improve the system so that the alarms set previously don't get switched off when the handset for any reason goes off/ switched off.
ii. Some systems having default or predefined notification tone so users cannot change, so a provision for the users to set customized alarm/ notification tone.
iii. Implementing automatic recommendation of medicine intake timings of the patients just by studying the patients' profession or by taking some characteristic inputs from the patients.
iv. Beside the secure login system more authentication tools can be implemented in the application like voice based identification of patients etc.
v. To introduce prediction based algorithm into the system so that people who have the habit of missing alerts, even after setting alerts, gets notified whenever that person launches 'Med-Alert' application and is also provided with the list of medicines that person has missed, to reduce medical non-adherence. Artificial intelligence aims at improving the working of this medical alert app and to make the app smarter.

References

1. K.H Park, S.H. Lim, Construction of a medication reminder synchronization system based on data synchronization. Int. J. Bio-Sci. Bio-Technol. **4**(4), 1–10 (2012)
2. F. Ferreira, N. Almeida, A.F. Rosa, A. Oliveira, A. Teixeira, J.C. Pereira, Multimodal and adaptable medication assistant for the elderly: a prototype for interaction and usability in smartphones, in *8th Iberian Conference on Information Systems and Technologies (CISTI)* (Lisboa, 2013), pp. 1–6.
3. D. Ameta, K. Mudaliar, P. Patel, Medication reminder and healthcare—an android application. Journal **6**(2), 39–48 (2015)
4. B.M. Silva, I.M. Lopes, M.B. Mar, J.J. P.C. Rodrigues, M.L. Proenca, A mobile health application for outpatients' medication management, In *13th IEEE International Conference on Communications (ICC)*, Budapest, 2013), pp. 4389–4393
5. M. Wang, P.H. Tsai, J.W.S. Liu, J.K. Zao, Wedjat: a mobile phone based medicine intake reminder and monitor, In *9th IEEE international conference on bioinformatics and bioengineering*, (Taichung, 2009), pp. 423–430
6. W. Antoun, A. Abdo, S. Al-Yaman, A. Kassem, M. Hamad, C. El-Moucary, Smart Medicine Dispenser (SMD), In: *4th Middle East Conference on Biomedical Engineering (MECBME)* (Tunis, 2018)
7. B. Prasad, Social media, health care, and social networking. Journal **77**(3), 492–495 (2013)
8. D.S. Abdul Minaam, M. Abd-ELfattah, Smart drugs: Improving healthcare using smart pill box for medicine reminder and monitoring system. Future Comput. Inf. J. **3**(2), 443–456 (2018)
9. Y. Kim, I. Lim, J. Lee, J. Lee, H. Namgung, J. Lee, Study on medical emergency alert algorithm based on acceleration algorithm, in *13th International Conference on Information Science and Applications (ICISA)*, (Suwon, 2013), pp. 1–4
10. J.Y.E. Park, J. Li, A. Howren, N.W. Tsao, M. De Vera, Mobile phone apps targeting medication adherence: quality assessment and content analysis of user reviews. JMIR Mhealth Uhealth **7**(1), 1–12 (2019)
11. V. Pérez-Jover, M. Sala-González, M. Guilabert, J.J. Mira, Mobile apps for increasing treatment adherence: systematic review. J. med. Internet Res. **21**(6), 1–14 (2019)

Comparison of Different Color Models for Priority Based Color Matching of Plant Parts Used in DUS Testing

Check for updates

Gopinath Bej, Tamal Dey, Sabyasachi Majumdar, Abhra Pal, Amitava Akuli, Tapas Sutradhar, Rishin Banerjee, Nabarun Bhattacharyya, and Nachiket Kotwaliwale

Abstract Distinctiveness, Uniqueness and Stability (DUS) testing is a standard practice to establish a new variety of crop. Color measurement plays a major role in DUS testing. Color of plant parts like leaf, flower, fruit and stem are very important factors to establish the uniqueness of any candidate variety. To measure the color or to describe a color objectively by some name or code, Royal Horticultural Society (RHS) color chart is used by the DUS tester. RHS color chart is available in market. In manual process of color measurement, samples are placed under the porthole of the selected color-field of the RHS color chart to find the best matched color based on the visual perception of the person being. Visual evaluation is a tricky process

G. Bej (✉) · T. Dey · S. Majumdar · A. Pal · A. Akuli · T. Sutradhar · R. Banerjee · N. Bhattacharyya
Centre for Development of Advanced Computing (C-DAC) Kolkata, Plot E-2/1, Block GP, Sector V Bidhannagar, Calcutta 700091, India
e-mail: gopinath.bej@cdac.in

T. Dey
e-mail: tamal.dey@cdac.in

S. Majumdar
e-mail: sabyasachi.majumdar@cdac.in

A. Pal
e-mail: abhra.pal@cdac.in

A. Akuli
e-mail: amitava.akuli@cdac.in

T. Sutradhar
e-mail: tapas.sutradhar@cdac.in

R. Banerjee
e-mail: rishinbanerjee2013@gmail.com

N. Bhattacharyya
e-mail: nabarun.bhattacharya@cdac.in

N. Kotwaliwale
Central Institute of Agricultural Engineering, Bhopal, India
e-mail: nachiket.kotwaliwale@icar.gov.in

V. E. Balas et al. (eds.), *Proceedings of International Conference on Computational Intelligence, Data Science and Cloud Computing*, Lecture Notes on Data Engineering and Communications Technologies 62, https://doi.org/10.1007/978-981-33-4968-1_37

as it depends on so many factors like lighting condition of the environment, gender, age, eyesight of the evaluator etc. Also, selecting of the proper color-field from the color fan of RHS color chart is fishy and it may lead to a biased result. In this paper, it is proposed to use an intelligent machine vision technology which will accurately and objectively measure color of plant parts in quick time. A database of colorimetric values of all RHS colors was used as a reference. Digital image processing and analysis algorithms were applied on the images of plant part to extract the color information using RGB, CIELab and CIELCh color models. Extracted color information was compared with the available color values in the reference database of RHS color chart using Euclidean distance, Chebyshev distance and Mahalanobis distance. These distance values were used to determine the top five matching color for each color model. Top priority was given to the color with minimum distance value. It was observed that CIELab color model using Mahalonobis distance was the most excellent model for color comparison while DUS testing. This technology may definitely help the DUS tester to accurately describe the color of plant parts in quick time.

Keywords DUS testing · RHS color chart · RGB · CIELab · CIELCh · Euclidean distance · Chebyshev distance · Mahalanobis distance · Color matching · Image processing

1 Introduction

In order to establish a new variety in the national list, the variety should be Distinct, sufficiently Uniform and Stable (DUS). This is an effective scheme for protection of plant varieties and the rights of farmers and plant breeders [1, 2]. DUS testing is mandatory to recognize the development of the new plant varieties. There is some standard DUS testing guidelines available for each crop [3–5]. In that guideline, procedures are clearly mentioned like how to prepare a field, how many replication of data is required, when to collect which plant part from field, what are the on-field measurements to be taken, what are the features of the plant part to be observed etc. Following the guidelines, a crop is grown and measures the defined parameters at defined period of growth. These defined parameters will be used in DUS testing to decide the fate of the candidate variety. Parameters defined in the DUS guidelines include color of different plant parts (i.e. leaf, stem, flower petals, fruit and seed). Almost for every crop, color of the different plant parts is the key feature according to DUS guidelines [6].

Currently, color is measured using Royal Horticulture Society (RHS) color chart which are available in market during DUS testing [7]. RHS color chart was developed in order to determine the colors manually by visual inspection. It was published for the first time in 1961 and since then it was a standard reference used by horticulturists worldwide for recording colors. In this chart each color is indicated by a code i.e. RHS color code. RHS color chart has 920 colors, divided into 4 handy color fans sorted

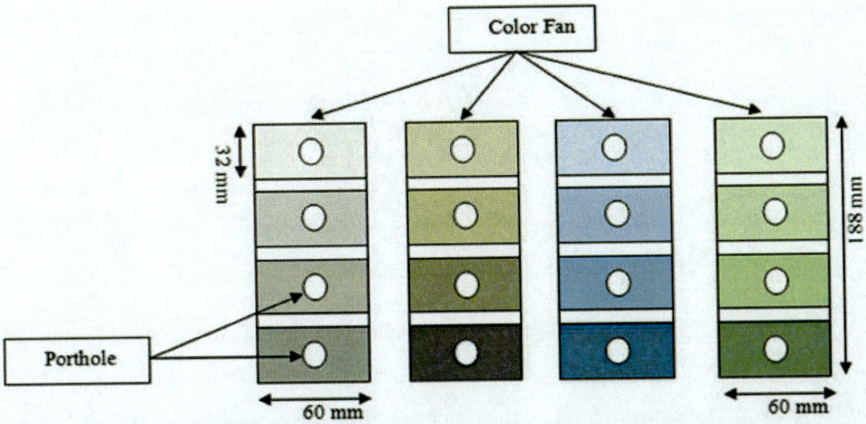

Fig. 1 Schematic diagram of RHS Color Fan

by 4 color groups (namely Yellow-Red, Purple-Blue, Turquoise-Green, and Brown-Grey) and 29 subgroups. Each color field has dimension of 60 × 32 mm whereas the dimension of one fan is 60 × 188 mm. Each color field has a porthole in the center of it as shown in Fig. 1. Sample under test is placed under the selected color field and finds the best match by viewing the sample through the porthole. This process is repeated until or unless the tester satisfied with color match within the selected color fields group. This is a manual process and purely based on visual evaluation. Moreover it is recommended to use same lighting condition for each color field. Visual color evaluation may be impaired by improper and uneven lighting condition, lack of attention to essential technical requirements, tester's age, gender, eyesight etc. Choosing the color field from the color group is fishy as the lightness and saturation differs narrowly within the same color group [8, 9].

Machine Vision technology will provide a new paradigm for objectively and accurately evaluate the color using the colorimetric knowledgebase of RHS color chart and henceforth replace the current practice of color measurement using the color fans of RHS color chart. An imaging setup with uniform lighting condition was developed and one DSLR camera was used to acquire the image of the sample collected from the field for DUS testing. All the images of plant part were captured using this setup to avoid any unintended color variance. Digital image processing and analysis algorithms were applied on the captured images to extract the color information of the plant part. Color information was extracted in RGB, CIELab and CIELCh color models. These color information was compared with the colorimetric database of RHS colors using Euclidean, Chebyshev and Mahalanobis distance. For each color model, these three types of distances were calculated while comparing the colors. Top five matching colors were determined for each distance type and each color models. The top priority was given to the color with minimum distance value. To conduct this experiment, an image processing based software was developed using Microsoft VC ++ where color matching algorithms were also implemented.

Table 1 Combination of color models and distance types

	RGB	CIELab	CIELCh
Chebyshev	Top 5 matching colors	Top 5 matching colors	Top 5 matching colors
Euclidean	Top 5 matching colors	Top 5 matching colors	Top 5 matching colors
Mahalanobis	Top 5 matching colors	Top 5 matching colors	Top 5 matching colors
Total $9 \times 5 = 45$ colors for each sample			

Software was configurable to test the color matching algorithm for each color model and each distance type. Using this developed software, color comparison result for each color model and each distance type were determined. For each sample, total nine sets of color data with five colors each were determined using combinations of color model and distance type as shown in Table 1.

The sample which are analyzed by the developed algorithms, also analyzed manually by the scientists of expert panel to find the best match color using the color fans of RHS color chart. This expert panel was formed with eight scientists of various age groups with 50% male and 50% female at NBPGR, New Delhi. Obtained results from the scientist's panel and software were cross checked and it was observed that the CIELab color model using Mahalonobis distance was giving more accurate color group and color code than any other combinations of color model and distance type.

2 Materials and Methods

2.1 Experimental Setup

A DSLR camera based imaging setup was developed to capture image of plant parts in a uniform lighting condition to avoid any unintended color variance. The DSLR camera was fixed on a height of 1.5 ft from sample. A strobe light based illumination arrangement was developed for uniform image quality. The background of the imaging region was kept white. Plant parts were placed on that white region and image was captured using the camera.

2.2 Colorimetric Database of RHS Color Chart

Standard practice used by the horticulturists globally for recording plant part color is the RHS color chart. In the first step of this experiment, the colorimetric database of RHS color chart was developed using the available colorimetric data from the website (http://rhscf.orgfree.com/) [10]. The colorimetric data was recorded by measuring every unique color in the color fields in sRGB, CIELab and CIELCh model using spectrophotometer in SCE (Specular Component Excluded) mode having illuminant/observer as D65/10°. Some of the RHS color fields are out of standard RGB model range i.e. not in the range 0–255 (e.g. color code: 6A, 9A, 9B etc.). These out of range RGB color were not included in the reference database for RGB model. All the data were included in the developed database for CIELab and CIELCh color models.

2.3 Sample Collection

Following the DUS guidelines crops were grown with replication in different locations and parameters mentioned in DUS guidelines are collected from the field for new varietal establishment. Leaf, flower, seed, fruit and stem color are important for this DUS testing. Accordingly samples are collected for color measurement.

2.4 Image Acquisition

After collecting the samples from field, image was captured using a DSLR camera. The image was captured in.jpg format with 6016 × 4000 resolutions.

2.5 Image Processing and Analysis

Color is an important parameter in DUS testing. During image processing and analysis, color information of the object was extracted using RGB, CIELab and CIELC fdh color models [11]. Digital image processing and analysis algorithms were applied on the captured images to extract the color information of the plant part [12]. Steps involved in image analysis have been described below.

```
for each pixel in the image
  if Blue(B) component of pixel > threshold_cutoff
    pixel is set to background
  else
    pixel is set to foreground
  end if
end for
```
(1)

- **Boundary Object Remove**: Segmented image had few foreground components which were touched to the boundary pixels. These noises were removed using algorithm (2).

```
for each boundary pixel
  if the pixel is foreground then
    find the 8 neighbor of the boundary pixel
    for each neighbor pixel
      if it is foreground pixel
        insert into a stack
      end if
    end for
    for each popped element from stack
      find the 8 neighbor
      for each neighbor pixel
        if it is foreground
          insert into the stack
        end if
      set pixel to background
    end for
  end if
  set boundary pixel to background
end for
```
(2)

- **Connected Component Labeling**: Connected component labeling was applied to label each connected foreground objects with unique number. Fast hybrid morphological reconstruction algorithm was used for connected component labeling. After this operation each non-touching foreground objects were uniquely labeled for further processing and analysis.
- **Noise Remove**: Each connected component has an area. Based on the area, noises were set to background pixel using an area cutoff value (3).

```
for each connected component
  if the area of component > area_cutoff value
    for each pixel of the component
      set pixel value to background                        (3)
    end for
  end if
end for
```

- **Fill Hole**: Morphological fill hole operation was applied to fill any background pixels surrounded with foreground pixel in the binary image. After this operation was performed, original color image was superimposed to get the segmented color image of plant part.
- **Color Information**: Color information from the image was extracted using RGB color space and converted to CIELab and CIELCh color spaces [13]. Average of each color component was calculated using (4) for further color matching algorithm.

$$
\begin{aligned}
&\text{Total}R = 0; \text{Total}G = 0; \text{Total}B = 0;\\
&\text{Total}L = 0; \text{Total}a = 0; \text{Total}b = 0;\\
&\text{Total}c = 0; \text{Total}h = 0;\\
&\text{count} = 0;\\
&\text{for each foreground pixel in the image}\\
&\quad RGB = \text{ Find } RGB \text{ color information of the pixel.}\\
&\quad Lab = \text{ Calculate } Lab \text{ color from } RGB\\
&\quad LCh = \text{ Calculate } LCh \text{ color from } Lab\\
&\quad \text{Total}R+ = R; \text{Total}G+ = G; \text{Total}B+ = B;\\
&\quad \text{Total}L+ = L; \text{Total}a+ = a; \text{Total}b+ = b;\\
&\quad \text{Total}c+ = C; \text{Total}h+ = h;\\
&\quad \text{count} = \text{count} + 1;\\
&\text{end for}\\
&\text{Avg}R = \text{Total}R/\text{count}; \text{Avgl}G = \text{Total}G/\text{count}; \text{Avg}B = \text{Total}B/\text{count};\\
&\text{Avg}L = \text{Total}L/\text{count}; \text{Avg}a = \text{Total}a/\text{count}; \text{Avg}b = \text{Total}b/\text{count};\\
&\text{Avg}c = \text{Total}c/\text{count}; \text{Avg}h = \text{Total}h/\text{count};
\end{aligned}
\tag{4}
$$

From the original image, color of the plant part was extracted in terms of average R, G, B values for RGB color model, average L, a, b values for CIELab color model and average L, C, h values for CIELCh color model. This color information was compared with colorimetric RHS color database and assigned a color group and code as like RHS color chart.

2.6 Inverse Covariance Matrix Calculation

After calculating the average color components of RGB, CIELab and CIELCh color models, covariance among the components is calculated for each color model using (5).

$$
\begin{aligned}
&\text{if } XYZ \text{ is the componenets of individual color model then}\\
&\mathrm{COV}_{xx} = \frac{\sum_{i=1}^{N} (X-\bar{X}) * (X-\bar{X})}{N-1}\\
&\mathrm{COV}_{YY} = \frac{\sum_{i=1}^{N} (Y-\bar{Y}) * (Y-\bar{Y})}{N-1}\\
&\mathrm{COV}_{ZZ} = \frac{\sum_{i=1}^{N} (Z-\bar{Z}) * (Z-\bar{Z})}{N-1}\\
&\mathrm{COV}_{XY} = \frac{\sum_{i=1}^{N} (X-\bar{X}) * (Y-\bar{Y})}{N-1}\\
&\mathrm{COV}_{XZ} = \frac{\sum_{i=1}^{N} (X-\bar{X}) * (Z-\bar{Z})}{N-1}\\
&\mathrm{COV}_{YZ} = \frac{\sum_{i=1}^{N} (Y-\bar{Y}) * (Z-\bar{Z})}{N-1}
\end{aligned} \tag{5}
$$

After calculating the covariance as above the covariance matrix is formed as below.

Covariance matrix	X	Y	Z
X	COV_{XX}	COV_{XY}	COV_{XZ}
Y	COV_{XY}	COV_{YY}	COV_{YZ}
Z	COV_{XZ}	COV_{YZ}	COV_{ZZ}

Inverse matrix operation is applied on above covariance matrix. This inverse matrix will be used while calculating the Mahalanobis distance. For each color model there will be one inverse covariance matrix.

2.7 Color Matching Algorithm

In this stage, the colorimetric database was ready and the color information of the plant part was extracted. The color matching algorithm was used to find the closest match of color by comparing the extracted color values with each individual color values of RHS database. This color matching algorithm is based on distance calculation. Chebyshev distance, Euclidean distance and Mahalanobis distance were calculated while comparing the colors [14–16]. The minimum distance value represents the best match. For each distance type, five best matched colors were determined using subsequent minimum distance values.

- Chebyshev distance was calculated using (6)

for each color in RHS database for *RGB* model

$$d_i^{RGB} = \max(|R_i - \text{Avg}R|, |G_i - \text{Avg}G|, |B_i - \text{Avg}B|)$$

end for

for each color in RHS database for *Lab* model

$$d_i^{Lab} == \max(|L_i - \text{Avg}L|, |a_i - \text{Avg}a|, |b_i - \text{Avg}b|)$$

end for

for each color in RHS database for *Lch* model

$$d_i^{Lch} == \max(|L_i - \text{Avg}L|, |c_i - \text{Avg}c|, |h_i - \text{Avg}h|)$$

end for (6)

- Euclidean distance was calculated using (7)

for each color in RHS database for *RGB* model

$$d_i^{RGB} = \sqrt{(R_i - \text{Avg}R)^2 + (G_i - \text{Avg}G)^2 + (B_i - \text{Avg}B)^2}$$

end for

for each color in RHS database for *Lab* model

$$d_i^{Lab} = \sqrt{(L_i - \text{Avg}L)^2 + (a_i - \text{Avg}a)^2 + (b_i - \text{Avg}b)^2}$$

end for

for each color in RHS database for *Lch* model

$$d_i^{Lch} = \sqrt{(L_i - \text{Avg}L)^2 + (c_i - \text{Avg}c)^2 + (h_i - \text{Avg}h)^2}$$

end for (7)

- Mahalanobis distance was calculated using (8)

$$
\begin{aligned}
&M_{RGB} = \left[\text{Avg}R\ \text{Avg}G\ \text{Avg}B\right] \\
&\text{for each color in RHS database for RGB model} \\
&\qquad X_i = [R_i G_i B_i] \\
&\qquad d_i = \sqrt{(X_i - M_{RGB})^T \text{COV}_{RGB}^{-1} (X_i - M_{RGB})} \\
&\text{end for} \\
&M_{Lab} = \left[\text{Avg}L\ \text{Avg}a\ \text{Avg}b\right] \\
&\text{for each color in RHS database for } Lab \text{ model} \\
&\qquad X_i = [L_i a_i b_i] \\
&\qquad d_i = \sqrt{(X_i - M_{Lab})^T \text{COV}_{Lab}^{-1} (x_i - M_{Lab})} \\
&\text{end for} \\
&M_{Lch} = \left[\text{Avg}L\ \text{Avg}c\ \text{Avg}h\right] \\
&\text{for each color in RHS database for } LCh \text{ model} \\
&\qquad X_i = [L_i C_i h_i] \\
&\qquad d_i = \sqrt{(X_i - M_{Lch})^T \text{COV}_{LCh}^{-1} (X_i - M_{Lch})} \\
&\text{end for}
\end{aligned} \tag{8}
$$

Distance values for each color in RHS database were sorted in ascending order and top five colors from the sorted list were chosen as the best matching color.

3 Results and Discussion

Applying the developed algorithms for color matching, top five colors were determined for the combinations of each individual color model and distance type. Total 9 combinations were possible (i.e. RGB & Chebysehv, RGB & Euclidean, RGB & Mahalanobis, CIELab & Chebysehv, CIELab & Euclidean, CIELab & Mahalanobis, CIELCh & Chebysehv, CIELCh& Euclidean and CIELCh & Mahalanobis). For each combination, 5 priority based colors were determined. More than 60 samples were used in this experiment. Results obtained using color matching algorithm for the sample of okra stem, leaf and fruits are shown in Tables 3, 4, 5 and 6. Overall color matching result of 60 samples are shown in Table 2. Also graphical representation of

Table 2 Matches count with expert panel's result (for 60 samples)

	RGB	CIELab	CIELCh
Chebyshev	0	**0**	0
Euclidean	23	**48**	2
Mahalanobis	35	**57**	7

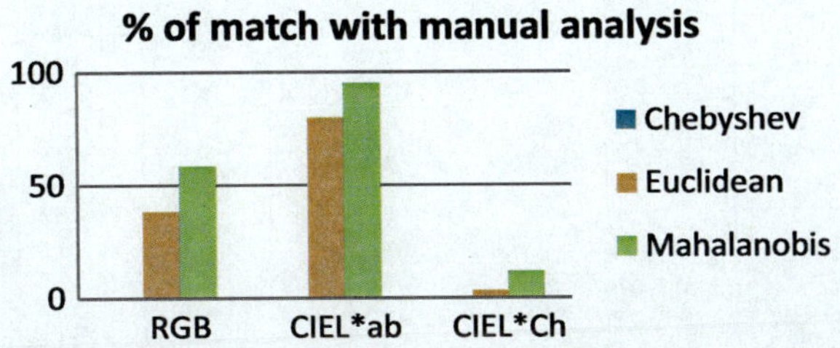

Fig. 2 Comparison of manual analysis results using different color models

% of match is shown in Fig. 2. The result depicts that the manual result of color analysis matches most of the time with the algorithm results implemented using CIELab color model and Mahalanobis distance. Though Euclidean distance also matches but sometime it deviates in comparison to manual experts' opinion.

4 Conclusion

Developed algorithm shows a promising result for color determination of plant parts using the combination of CIELab color model and Mahalanobis distance. The color matching algorithm represents the color of plant part with the defined color code of RHS. For the DUS tester, this methodology of color estimation can be well accepted as this algorithm provides the result as like RHS color charts. Having in the era of digital evaluation, the current practice of color measurement using the color fans of RHS color charts may be replaced with proposed methodology to get accurate objective results in quick time. The developed imaging setup plays an important role as any change in illumination may affect the accuracy of the developed algorithms. Moreover, this combination of CIELab color model and Mahalanobis distance can be used successfully for development of any color comparison algorithm in future.

Table 3 Color matching result of Okra stem

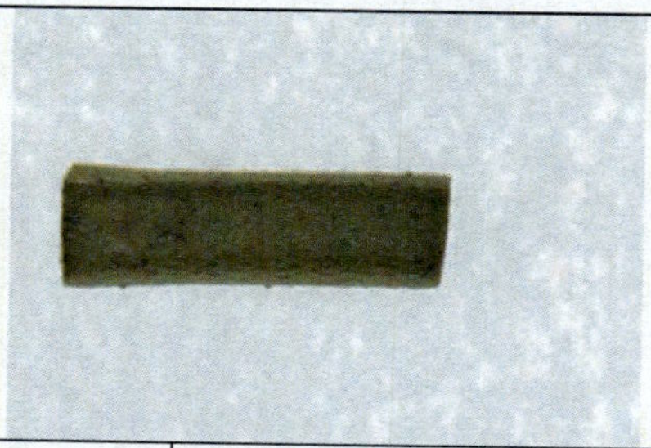

Okra Stem	Rank	RGB	CIELab	CIELCh
Chebyshev	1	25B, Orange Group	25B, Orange Group	4B, Yellow Group
	2	25B, Orange Group	25C, Orange Group	5C, Yellow Group
	3	25B, Orange Group	24A, Orange Group	3C, Yellow Group
	4	25B, Orange Group	26B, Orange Group	5B, Yellow Group
	5	25B, Orange Group	29A, Orange Group	6C, Yellow Group
Euclidean	1	148A, Yellow-Green Group	148A, Yellow-Green Group	201A, Grey Group
	2	147B, Yellow-Green Group	147B, Yellow-Green Group	201B, Grey Group
	3	137C, Green Group	148B, Yellow-Green Group	200A, Brown Group
	4	146B, Yellow-Green Group	137B, Green Group	200B, Brown Group
	5	N200B, Brown Group	197A, Greyed-Green Group	N200B, Brown Group
Mahalano-bis	1	148A, Yellow-Green Group	148A, Yellow-Green Group	N200B, Brown Group
	2	202B, Black Group	198A, Greyed-Green Group	197A, Greyed-Green Group
	3	198A, Greyed-Green Group	197C, Greyed-Green Group	N199A, Grey-Brown Group
	4	147A, Yellow-Green Group	198B, Greyed-Green Group	137B, Green Group
	5	195A, Greyed-Green Group	148B, Yellow-Green Group	147B, Yellow-Green Group
Decision taken on color observation by expert panel of scientists: 148A, Yellow-Green Group				

Table 4 Color matching result of Okra leaf

Okra Leaf	Rank	RGB	CIELab	CIELCh
Chebyshev	1	NN155D, Greyed-White Group	N25C, Orange Group	5C, Yellow Group
	2	NN155C, Greyed-White Group	25A, Orange Group	9C, Yellow Group
	3	NN155B, Greyed-White Group	N25B, Orange Group	4A, Yellow Group
	4	NN155A, Greyed-White Group	28B, Orange Group	5B, Yellow Group
	5	157D, Greyed-White Group	25B, Orange Group	6C, Yellow Group
Euclidean	1	200A, Brown Group	N137B, Green Group	200A, Brown Group
	2	N200A, Brown Group	N137A, Green Group	200B, Brown Group
	3	200B, Brown Group	147A,Yellow- Green Group	201A, Grey Group
	4	202A, Black Group	N137C, Green Group	N200B, Brown Group
	5	203B, Unknown	146A, Yellow- Green Group	166A, Greyed-Orange Group
Mahalano-bis	1	203D, Unknown	146A,Yellow- Green Group	N200A, Brown Group
	2	202A, Black Group	148A,Yellow- Green Group	N199A,Greyed- Brown Group
	3	203C, Unknown	146B,Yellow- Green Group	146A, Yellow- Green Group
	4	147A,Yellow- Green Group	N137B, Green Group	N200B, Brown Group
	5	103A, Blue Group	N137A, Green Group	141A, Green Group

Decision taken on color observation by expert panel of scientists: 146A,Yellow- Green Group

Table 5 Color matching result of okra fruit

Okra Green Fruit	Rank	RGB	CIELab	CIELCh
Chebyshev	1			3D, Yellow Group
	2	155D, White Group	17C, Yellow-Orange Group	4B, Yellow Group
	3	158D, Yellow-White Group	16B, Yellow-Orange Group	4C, Yellow Group
	4			4D, Yellow Group
	5	4D, Yellow Group	15B, Yellow-Orange Group	2C, Yellow Group
Euclidean	1			198D, Greyed-Green Group
	2	145A, Yellow-Green Group	160A, Greyed-Yellow Group	156D, Greyed-White Group
	3			156C, Greyed-White Group
	4	161A, Greyed-Yellow Group	160B, Greyed-Yellow Group	155C, White Group
	5			NN155A, Greyed-White Group
Ma-halanobis	1	153B, Yellow-Green Group	152D, Yellow-Green Group	199D, Greyed-Brown Group
	2	152D, Yellow-Green Group	160A, Greyed-Yellow Group	156C, Greyed-White Group
	3			161C, Greyed-Yellow Group
	4	195B, Greyed-Green Group	152C, Yellow -Green Group	162C, Greyed-Yellow Group
	5			163D, Greyed-Orange Group
Decision taken on color observation by expert panel of scientists: 152D, Yellow-Green Group				

Table 6 Color matching result of okra fruit

Okra Red-dish Fruit	Rank	RGB	CIELab	CIELCh
Chebyshev	1	27C, Orange Group	28B, Orange Group	5B, Yellow Group
	2	36D, Red Group	30D, Orange-Red Group	6C, Yellow Group
	3	158D, Yellow- White Group	25A, Orange Group	3B, Yellow Group
	4	158B, Yellow- White Group	N25B, Orange Group	4A, Yellow Group
	5	NN155D, Greyed- White Group	30B, Orange-Red Group	8A, Yellow Group
Euclidean	1	165A, Greyed-Orange Group	200D, Brown Group	201A, Grey Group
	2	177A, Greyed-Orange Group	165A, Greyed-Orange Group	166A, Greyed-Orange Group
	3	200D, Brown Group	177A, Greyed-Orange Group	N186C, Greyed-Purple Group
	4	199A, Grey-Brown Group	177B, Greyed-Orange Group	200A, Brown Group
	5	174A, Greyed-Orange Group	N199B, Greyed-Brown Group	178A, Greyed-Red Group
Mahalano-bis	1	200D, Brown Group	200D, Brown Group	201A, Grey Group
	2	165A, Greyed-Orange Group	165A, Greyed-Orange Group	177A, Greyed-Orange Group
	3	200C, Brown Group	177A, Greyed-Orange Group	166A, Greyed-Orange Group
	4	177A, Greyed-Orange Group	166A, Greyed-Orange Group	201B, Grey Group
	5	N199B, Greyed-Brown Group	199B, Greyed-Brown Group	165A, Greyed-Orange Group

Decision taken on color observation by expert panel of scientists: 200D, Brown Group

References

1. "Protection of plant varieties and farmers' rights authority, India", http://plantauthority.gov.in/, Accessed on 19th July 2020
2. P. Ranjan, Recent developments in India's plant variety protection, seed regulation and linkages with UPOV's proposed membership. J. World Intellect Property **12**(3), 219–243. May 2009, https://doi.org/10.1111/j.1747-1796.2009.00365.x
3. DUS Guidelines, http://plantauthority.gov.in/crop-guidelines.htm, Accessed on 19th July, 2020
4. B. Jyoti, Usha, P. Singh, Role of DUS testing in registration of plant varieties under PPV & FR Act, 2001, in *RASHTRIYA KRISHI* **10**(2) Dec 2015. e ISSN–2321–7987
5. S. Chandra, *DUS Testing* (2002). https://doi.org/10.13140/2.1.4659.3604
6. N. Kotwaliwale, K. Singh, S. K. Chakrabarty, M.A. Joshi, A. Kalne, M. Tiwari, K.K. Gangopadhyay, N. Bhattacharya, A. Akuli, D. Aggarwal, G. Bej, Machine vision for characterization of some phenomic features of plant parts in distinguishing varieties - a review. Int. J. Bio-Inspired Comput. (IJBIC), **14**(4), (2019). https://doi.org/10.1504/ijbic.2019.103960
7. The RHS Colour Chart. "https://shop.rhs.org.uk/gifts/stationery/rhs-colour-chart/rhs-large-colour-chart-sixth-revised-edition", Accessed on 19th July, 2020
8. I. Kasajima, Measuring plant colors. Plant Biotech. (Tokyo, Japan) **36**(2), 63–75 (2019). https://doi.org/10.5511/plantbiotechnology.19.0322a
9. Rob Griesbach, S. Austin, Comparison of the Munsell and royal horticultural society's color charts in describing flower color. Taxon **54**, 771 (2005). https://doi.org/10.2307/25065433
10. "Colorimetric Results of the Royal Horticultural Society Colour Charts Edition V". http://rhscf.orgfree.com/. Accessed on 19th July, 2020
11. N. Ibraheem, M. Hasan, R.Z. Khan, P. Mishra, Understanding color models: a review. ARPN J. Sci. Technol. **2** (2012)
12. Digital image processing, gonzalez rafael C
13. B. Ran, Astudyon CIELab & CIELch colour spaces, https://www.academia.edu/18929410/A_study_on_CIELab_and_CIELch_colour_spaces, Accessed on 19th July 2020
14. "DISTANCES IN CLASSIFICATION", http://www.ieee.ma/uaesb/pdf/distances-in-classification.pdf, Accessed on 19th July, 2020
15. W. Mokrzycki, M. Tatol, Color difference Delta E—A survey. Mach. Graph. Vis. **20**, 383–411 (2011)
16. G. Mclachlan, Mahalanobis distance. Resonance **4**, 20–26 (1999). https://doi.org/10.1007/BF02834632

Automated Soil Testing System to Produce Good Quality of Crops

Abdullah Al Farabe, Md. Rayhan Al Islam, Moshiur Rahman, Akash Saha, and Ghalib Ashraf

Abstract A soil test generally refers to the analysis of a soil sample to measure pH level, moisture, temperature and nutrient composition. This test will help farmers significantly to grow the good quality of crops. Most of the existing literature has worked on finding the pH level and amount of nutrients. However, they did not consider some essential measurements such as temperature and moisture. Therefore, we proposed an automated soil testing system (ASTS) from where farmers can measure the amounts of nutrients, temperature, pH level and moisture. Moreover, our device will show possibilities of which crops are suitable in our chosen area by testing that area's soil. Our experiment assumption is at least 90% accuracy for measuring pH level, moisture, temperature and nutrient composition of the soil ingredients. This proposed system will open a new chapter for farmers to work toward a future of unimagined possibilities.

Keywords pH level · Temperature · Moisture · Nutrients · Soil kit

---

A. A. Farabe (✉) · Md. R. A. Islam · M. Rahman · A. Saha · G. Ashraf
Department of Computer Science and Engineering, BRAC University, 66 Mohakhali, Dhaka 1212, Bangladesh
e-mail: abdullah.farabe@gmail.com

Md. R. A. Islam
e-mail: showrov.rayhan@gmail.com

M. Rahman
e-mail: moshiur.bracu@gmail.com

A. Saha
e-mail: sahaakash103@gmail.com

G. Ashraf
e-mail: galibashraf333@gmail.com

V. E. Balas et al. (eds.), *Proceedings of International Conference on Computational Intelligence, Data Science and Cloud Computing*, Lecture Notes on Data Engineering and Communications Technologies 62, https://doi.org/10.1007/978-981-33-4968-1_38

1 Introduction

Soil testing is a chemical analysis that provides environmentally a guideline for necessary nutrients recommendations. Different soil types, different moistures, different areas of a field could vary different fertility status of the soil. Most of the farmers do not have any scientific knowledge about how they could improve their harvest. As a result, they have to face a big amount of loss at the end of the year for various reasons. Sometimes, it also seems that the crop they are harvesting is not suitable for their land at all or does not have enough nutrients in the soil.

In the existing literature, Regalado et al. [1] have invented a process colorimetry by which it is possible to measure the amounts of nutrients (phosphorus, potassium and nitrogen) and also pH level. The soil kit has been developed in a university and it is really a good tool to measure amounts of nutrients (phosphorus, potassium and nitrogen) and pH level of the sample soil. Watching through the mixture of soil kit, water and sample soil, it is possible to measure pH level or amounts of nutrients that are given in the chart with the soil kit toolbox [2, 3]. The soil test kit used color indicators that required good vision. The color indication could be a big problem for those people who are unable to see or color blind. Their thinking about the device that could make a big help in providing strong recommendations for human errors. This study was about creating a protocol that can use colorimetry to determine the nitrogen, phosphorus and potassium content and pH of the soil to be cultivated by using the chemicals of the soil test kit in giving nutrient recommendations. By using vb.net, it processed the data and provided nutrient recommendations for every plant included in the soil test kit. The device is designed by a microcontroller unit and color sensors that measured the RGB values of the colors and data storage. Their main focus is to eliminate the lighting and distance effect by making the distance fixed from a certain origin. A big helpful project in providing standard recommendations to lessen human errors. It shortened the time analysis of soil samples by color comparison and looking for plant recommendations. The device is only a substitute in the analysis of colors to indicate the nutrient contents of the soil [1].

From another existing literature, Golicz et al. [4] have utilized an android-based cell phone application related to financially accessible Quantofix test strips to explore soil tests with an objective of giving explicit fertilizer solicitation. The app changes the smartphone into a compact reflectometer, relating the response shade of the test strips to the centralization of nutrients of soil obtainable. The combination of the smartphone and the test strip appeared to offer an agreeable screening tool for the assessment of nutrients with the possibility to bring savings on funds and decrease the supplement of nutrients to nature. Such outcomes offer impressively more accuracy than an appraisal 'by eye' compared against a coarse shading diagram. Figuring precision and accuracy of results acquired in field conditions in the sub-tropical atmosphere and the limitation for smartphone-mediated soil investigation were their motivation behind the process to screen changes in soil supplement status that were then utilized in making manure proposals. By utilizing Akvo Caddisfly, it is conceivable for farmers who may somehow effort soil testing with a straightforward support tool that can help farmers which can give data about the amount of plant accessible nitrate and phosphorus in the soil. In this article they have utilized only two elements—nitrate and phosphorus.

Our motivation is to reduce human errors and to help farmers. For color blinds as it is tough to detect those colors and measure levels of nutrients and pH. So, the device would help them to give the result directly of which crops will be the best choice. Moreover, it eliminates a huge time to analyze it. Usually, it seems that farmers have to wait for a long time to get the results from an organization for their land. But this device will let them know quickly even with the names of the crops.

In our research, the device will show the results of which crops are suitable for the user's chosen land by measuring temperature, moisture, amounts of nutrients and pH level. Therefore, after the assessment of the previous literature, a proposal has been made to make a device consisting of an Arduino, RGB color sensor, moisture sensor, temperature and humidity sensor. In this device, a moisture sensor is used to measure water level, temperature and humidity sensor, is used to detect temperature, and color sensor is used to know the amounts of three nutrients [5] and pH level from the sample soil. The device that has been made in this research could be more useful as it measures all those components which are needed before starting cultivation and the result will show in a short time. In the display, a user can see which crops (as names) are suitable for the user's chosen land before starting the cultivation. Previous researchers did not work on making this type of combination device and showing crop names on the display. In the end, by measuring the amounts of nutrients, pH level, temperature and moisture, the device shows the predicted crop names from the chart which are given in the SD card so that farmers can verify the possibilities of suitable crops for their chosen area. The whole code system is done in the Arduino Web editor.

2 Proposed System

2.1 *System Design*

Figure 1 shows the structure of the device this research is all about. A USB or 9 V battery is connected with Arduino. The breadboard is also connected with the Arduino which supplies 5 V to each of the sensors. The connected sensors are moisture sensor, color sensor, temperature and humidity sensor. An SD module is also added to access the SD card in the Arduino.

First, for the purpose of the experiment along with these sensors, soil kits have also been used for the measurement of color on which RGB color sensor will be used later. A tube/box is used where the sample soil, water and a nutrient/pH level measuring capsule (soil kit) have been mixed. The mixture will respond with a color

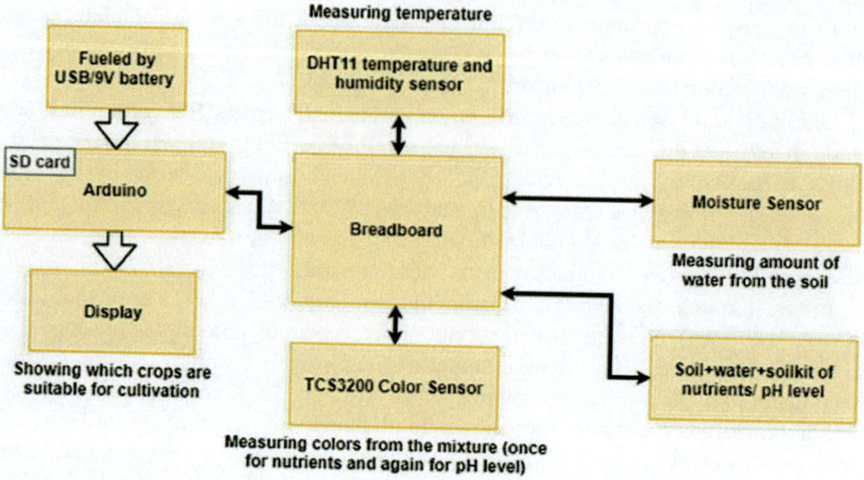

Fig. 1 Soil testing device

in the tube/box. Each color varies with different color codes and RGB values. In the proposed method, the whole project will be discussed briefly. A diagram has shown.

Figure 2 is the soil kit toolbox which shows the difference of nutrients and pH levels.

In Fig. 2, there are three different capsules for each different nutrients (nitrogen, phosphorus and potassium) and a pH level detecting capsule too. There are four boxes for three nutrients and one for pH level. Each of the nutrients has levels like—depleted, deficient, adequate, sufficient and surplus (low to high) of color levels [6]. After mixing the capsule, water and soil there will be a color on which the

Fig. 2 Soil kit

user can use the color sensor to detect the color code. It indicates the existence of nutrients, amounts of nutrients and the pH level. TCS230 color sensor is used to sense the colors generated in the tube/box. Every color stands for a specific RGB value. For example, if someone wants to detect nitrogen so the user has to mix a nitrogen capsule, sample soil and water. Then, it will show color in the tube/box and then the color sensor could be used to detect the color code.

2.2 *Hardware Part*

All the sensors have various pins and each of them has a different meaning. From them, two pins are common which are for GND and supply voltage (5 V/3.3 V). Circuits have been set from Arduino's 5 V/3.3 V and GND to the (+) and (−) row (horizontally) of the breadboard and set other sensors VCC and GND in that breadboard too to give them voltage supply. Then, other pins of the sensors have been set with the Arduino as the requirements. In Fig. 3, the circuit schematic is displayed..

Watching Fig. 3 is easier to understand the connections among the sensors and the Arduino. The Arduino Uno R3 is used as an open-source microcontroller board. The board is furnished with sets of computerized and simple info/yield (I/O) sticks that might be interfaced with different extension sheets (shields) and other circuits. It can be fueled by a USB link or by an outside 9-volt battery. However, it acknowledges voltages somewhere in the range of 7 and 20 volts.

The breadboard is fixed with the Arduino as well as connected with various sensors. The DHT11 temperature and humidity sensor are used which is basically a

Fig. 3 Circuit schematic

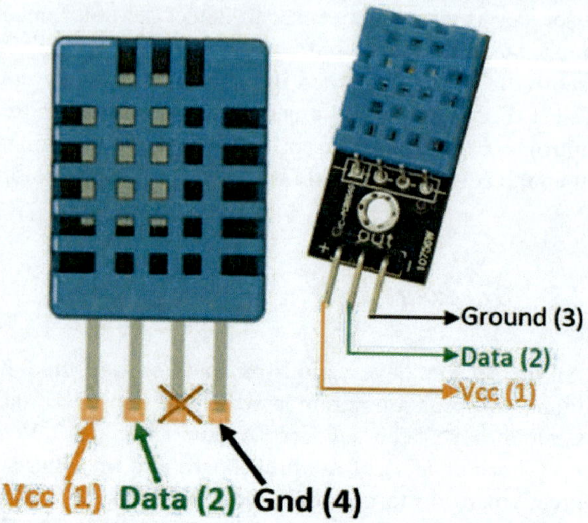

Fig. 4 DHT11 temperature and humidity sensor

thermometer attached to the breadboard. Temperature is one of the best significant contents influencing change, adsorption and the availability of soil phosphorus (P) to plants [7]. This sensor in Fig. 4 can detect temperature and humidity simultaneously of their surrounding areas of the land [8].

For now, randomly standard temperature from 18 to 45 degree Celsius has been set in Arduino. So, when the temperature sensor detects the temperature from the surrounding area, the value will be compared with the standard temperature which has been set randomly by the Arduino to show if the temperature is in control for the crops or not. If the temperature matches the fixed temperature which is given by us randomly in Arduino, then crops are suitable to harvest, and of course, the temperature will also show in the display. In addition, a moisture sensor is also used to detect the water content of soil by measuring the dielectric permittivity of the soil [9]. Figure 5 shows the moisture sensor.

The sensor in Fig. 6 helps the farmers to know the percentage of moisture of their land. Basically, for the moisture sensor, the device is just working on showing the percentage value of dryness of their soil.

Moreover, a color sensor has also been used to detect the color code as shown in Fig. 6.

After getting a mixture of sample soil, water and capsule in the tube/box the color sensor could detect the color code from the mixture. TCS3200 color sensor is a total shading identifier, including a TAOS TCS3200 RGB sensor chip and 4 white LEDs. The TCS3200 has a variety of photodetectors, each with either a red, green or blue channel, or no channel (clear) [10].

In addition, an SD card module is also playing a vital role where standard temperature, amounts of nutrients and pH level of various crops are saved as a dataset. The moisture, the temperature sensor and the color sensor parts have already been

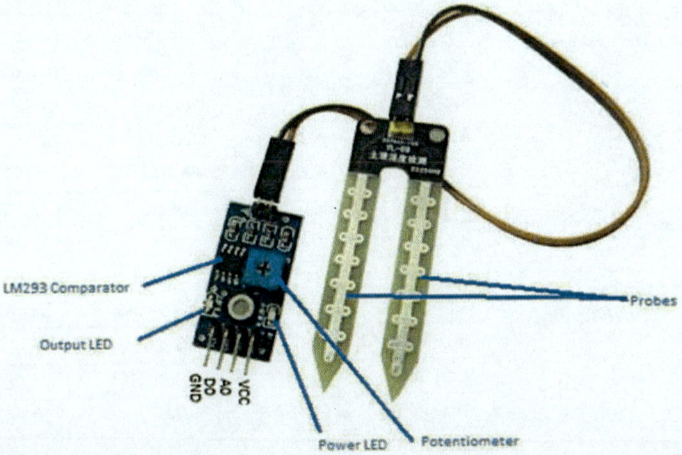

Fig. 5 Moisture sensor

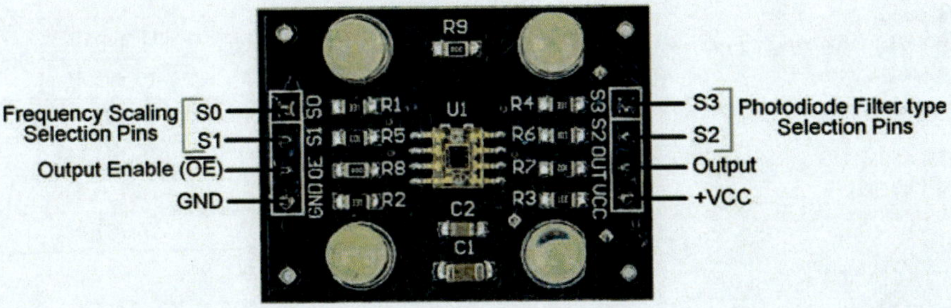

Fig. 6 TCS3200 color sensor

described. Now, this paper is going to brief about the color sensor and the reasons for which they have been used.

2.3 *Standard Soil Testing Values*

The SD card database is structured as a table carrying different nutrient values and pH levels for specific crops. This SD card has another dataset of standard color codes. In this section, it is discussed low to high color levels for nutrients. In this way, it identified the color codes and converted them to the decimal in Tables 1, 2 and 3 for nitrogen (N), phosphorus (P) and potassium (K). In each of the tables, all those hexa values have also been converted to decimal values.

Table 1 Nitrogen (converting hexa to decimal)

Amount	Colors	Color code	Decimal
Surplus	N4	660E05	6,684,723
Sufficient	N3	ffe6f0	16,770,800
Adequate	N2	Cc0066	13,369,446
Deficient	N1	99004d	10,027,085
Depleted	N0	Ffe6f2	16,770,802

Table 2 Phosphorus (converting hexa to decimal)

Amount	Colors	Color code	Decimal
Surplus	P4	800E01	128
Sufficient	P3	1E3EDE	1,982,174
Adequate	P2	3E74E7	4,093,159
Deficient	P1	79A6EC	7,972,588
Depleted	P0	B3C8E9	11,782,377

Table 3 Potassium(converting hexa to decimal)

Amount	Colors	Color code	Decimal
Surplus	K4	4D2600	13,395,456
Sufficient	K3	CC6600	13,395,456
Adequate	K2	FF8000	16,744,448
Deficient	K1	FF9933	16,750,899
Depleted	K0	Ffd9b3	16,767,411

With the help of the color sensor, it is possible to examine those color codes but it will get as hexa values. That is why it has converted to decimal to find ratios [11]. Pixie has used to get the color codes from those pictures of boxes of soil kits [12].

Below an example has given just for nitrogen—

- High value = ffe6f0 (color code) = 16,770,800 (hexadecimal to decimal convert) = 100% (converting as percentage)
- Low value = ffe6ff (color code) = 6,684,723 (hexadecimal to decimal convert) = 0% (converting as percentage).

By detecting high to low values, a ratio has measured for nitrogen 0 to 100%. For the other two nutrients (P and K) have been worked exactly in the same way. This converting process is saved for the RGB color sensor in the Arduino editor. So, basically, what has been done in Tables 1, 2 and 3 are to get ratios as a percentage of nitrogen (N), phosphorus (P) and potassium (K) for our chosen land. Then, after mixing sample soil, water and soil kit, it will give an RGB value as a percentage and let assume the value as Y. In the end, we have a dataset of nutrients values in Table 4 which are scaled in kg/ha of various crops.

Table 4 Database of crops based on nutrients [19–21]

Crops	Nitrogen (N) kg/ha	Phosphorus (P) kg/ha	Potassium (K) kg/ha
Rice(Boro)	120	25	35
Rice (Aman)	80	25	35
Wheat	120	80	80
Chilli	120	90	90
Tomato	150	100	90
Sweet potato	80	30	100
Chickpea	15	50	30
Cauliflower	110	70	80

Table 5 Database of crops based on nutrients and pH level (converting to percentage)

Crops	Nitrogen (N) %	Phosphorus (P) %	Potassium (K) %
Rice(Boro)	1.2	0.25	0.35
Rice (Aman)	0.8	0.25	0.35
Wheat	1.2	0.8	0.8
Chilli	1.2	0.9	0.9
Tomato	1.5	1	0.9
Sweet potato	0.8	0.3	1
Chickpea	0.15	0.5	0.3
Cauliflower	1.1	0.7	0.8

Table 4 is just to show in this paper for further calculations that how we have calculated Table 5 later on and Table 5 is saved in the SD card. But before that from Table 4, it seems that values have been given in kg/ha. So, in Table 5, it has to be calculated as a percentage. Equation is, 1 hectare = 10,000 square meter. In 10,000 sq m Rice (Boro) needs 120 kg N, so, 100 sq m Rice (Boro) needs.

$$X = 120 \times \frac{100}{10,000} = 1.2\%\,\mathrm{N}$$

The rest of the values of the crops for N, P and K have been calculated in the same way which is given in Table 5.

As from Table 5, this is the standard percentage value of the crops and assumes the percentage value is X which is set for Table 5 as a loop. The value which a user got from the mixture of the sample soil, water and soil kit was Y. Y percentage will try to find X from the loop. If ($X >= Y$) then in the display a user will see which crops are suitable for their chosen area from the crops that are given in the table list.

For example, —— (i)

If a user gets the RGB value for nitrogen as a percentage by the color sensor from the sample soil is $Y = 0.7\%$ then from Table 5 it seems that only Rice(Aman) and

Table 6 Measuring the RGB value of pH level (pH = 7.5–6.5)

Color	Alkaline pH 7.5	Neutral pH 7.0	Slight acid pH 6.5
Red	2.0–22.0	4.0–24.0	55.0–75.0
Green	46.0–66.0	62.0–82.0	71.0–91.0
Blue	10.0–30.0	11.0–31.0	9.0–29.0

Table 7 Measuring the RGB value of pH level (pH = 6–4.5)

Color	Acid pH 6.0	Acid pH 5.5	Very acid pH 5.0	Very acid pH 4.5
Red	102.0–122.0	146.0–166.0	137.0–157.0	143.0–153.0
Green	94.0–114.0	92.0–112.0	56.0–76.0	48.0–68.0
Blue	3.0–23.0	6.0–26.0	9.0–29.0	15.0–35.0

sweet potato are selected as their X is 0.8% which is greater than Y. In the same way, if Y for phosphorus and potassium are less than X the real standard values of Table 5 then those crops like Rice(Aman) and sweet potato will be selected based on nutrients.

The last step is about another capsule where the pH level will be measured. It is better to use another color sensor for measuring pH part to make the process easier. Again, in the same way, pH capsule, water and sample soil have to be mixed and will show various RGB colors. From the soil kit box, there are seven levels of pH and from the picture by using Pixie, RGB values have been measured. In Tables 6 and 7, RGB values have been put where alkaline (7.5) to very acid (4.5) means low to high levels of pH levels. Each pH level has seven different colors. Next RGB values have been calculated by plus 10 and minus 10 to set a range. For example, alkaline pH has been calculated for R (2.0–22), G (46–66) and B (10–30) where real values for RGB were 12, 56 and 20 for alkaline. In Tables 6 and 7, RGB values have been added and subtracted with 10 because as using a color sensor it will not give exactly one value all the time. So, a range has been set to avoid errors. Giving range might be helpful to detect if the sensor gives values within the ranges. All the values have been put in Tables 6 and 7.

In Tables 6 and 7, RGB values have given according to their pH levels. For example, after detecting the RGB values by the color sensor from the mixture of pH soil kit, sample soil and water are R = 12.0, G = 56.0 and B = 20.0 then from Table 6 it seems that it is alkaline where pH = 7.5. Tables 6 and 7 are set in a file as a dataset in the SD card for setting a range. After detecting the RGB values from the sample soil, it will be compared with the dataset of pH level directly. Table 8 is the real pH value of different foods we have got from online [13–18]. It is hard to get all nutrients and pH values of the same crops so have to find from different sources.

For example, ———— (ii)

If a user finds the RGB value for pH level by the color sensor from the sample soil is R = 150, G = 95 and B = 10 which indicates acid pH = 5.5 from Table 7.

Table 8 pH level of foods [13–18]

Crops	pH
Rice(Boro)	6.35
Rice (Aman)	6.8
Wheat	5.26–5.77
Chilli	5.20–5.93
Tomato	4.30–4.90
Sweet potato	5.30–5.60
Chickpea	5.1–8.9
Cauliflower	5.60

Fig. 7 Automated soil testing system to produce good quality of crops

Then, it will compare with Table 8 and it seems that wheat, chili, sweet potato and chickpea fell under a pH of 5.5. Down underneath, Fig. 7 shows full venture picture.

3 Results and Discussion

In the device, the program is set for the temperature between 21–45 degrees Celsius. The temperature and humidity sensor will recognize the temperature around and compare it with the standard temperature that is set. The moisture sensor will help to tell the percentage of water content.

On the other hand, the RGB color sensor is set in percentage value to ease the whole process and be more accurate. The device predicts percentages from the sample soil of pH, nitrogen, phosphorus and potassium. The percentage of nutrients and value for pH level will be compared with Tables 5 and 8 known as the standard dataset which is set in the SD card. On the display board, the user can see the crop names saved in the SD card where examples have been given in our research dataset. The temperature range is also set around 18–45 °C in the program and after detecting if it shows 35° then it is under the given range. If all of the three components—pH level, nutrients level and temperature are taken from the sample soil/area are matched with the standard value saved in the SD card then the display gives the names of the crops that are given in the tables including the percentage of moisture. In Sect. 2.3 from the examples (i) and (ii), it seemed that sweet potato and Rice (Aman) follow the contents of nutrients. Next for pH level wheat, chili, sweet potato and chickpea follow the pH rule. But for both case of nutrients and pH level, check only the sweet potato follows every step then the display will show-

Chosen crops- "sweet potato"
Moisture—65%
Temperature—35 °C.

Errors also could occur in this proposal. While checking pH level, a range has been set because taking RGB values from the sample could show different numbers which will change the color and will not match with the pH level given in Tables 6 and 7. Moreover, as it is made with different hardware components, so it should be checked if they are okay or not. But most important is the RGB color sensor. The whole project is based on the color sensor. The more good quality will be used, the more accurate results could be obtained. Our expectation is at least 90% accuracy for measuring the process.

4 Future Works

For now, we are making this device only for a few crops. In the future, more crops and data will be added. In this device, only three elements such as—nitrogen, potassium and phosphorus have calculated. More nutrients could be added. This device mainly works on color sensors as the whole project is based on RGB values. The better the color sensor will be used the better accuracy would get from the device. Moreover, to get better and accurate accuracy, the image processing could be the best procedure to be applied. Image processing could give us a better solution and will ease the complexity and more possibility to get better accuracy.

5 Conclusion

Soil testing is an experiment by which people can measure the nature of a sample soil and it is dose to increase the good quality of crops. A device with few sensors which will help users to detect components like—moisture, temperature, amount of nutrients and pH level of the soil to produce good quality crops. From the existing literature, they have worked on soil testing to measure those components by using soil kits. So, our thoughts of making a device that will show results on the display about which crops a user should choose to cultivate at their chosen land by measuring those basic components. In the end, this project will be a highly recommendable device for human errors in the world without any doubt.

References

1. R. Regalado, J. Cruz, *Soil pH and nutrient (Nitrogen, Phosphorus and Potassium) Analyzer Using Colorimetry* (2018). [online] Academia.edu. Available at: https://www.academia.edu/35375742/Soil_pH_and_Nutrient_Nitrogen_Phosphorus_and_Potassium_Analyzer_using_Colorimetry
2. Dfrobot.com., *TCS3200 RGB Color Sensor For Arduino—DFRobot* (2018). https://www.dfrobot.com/product-540.html#.U8dcIvmSxqI, [online] Accessed 3 Dec 2018
3. B.M. Quitain, R.M. David, Reliability of soil test kit [Philippines]. Supplement no. 1. 1987
4. K. Golicz, S.H. Hallett, R. Sakrabani, G. Pan, The potential for using smartphones as portable soil nutrient analyzers on suburban farms in central East China. Sci. Rep. **9**(1), 1–10 (2019)
5. Aces.edu., *Alabama Cooperative Extension System* (2018). http://www.aces.edu/pubs/docs/A/ANR-0449/ANR-0449.pdfp://www.aces.edu/pubs/docs/A/ANR-0449/ANR-0449.pdf. [online] Accessed 3 Dec 2018
6. Lusterleaf.com., *Luster Leaf Gardening Products—Soil Test Kits* (2018). http://www.lusterleaf.com/nav/soil_test.html. [online] Accessed 3 Dec 2018
7. C. Song, X. Zhang, X. Liu, Y. Chen, Effect of soil temperature and moisture on soil test P with different extractants. Can. J. Soil Sci. **92**(3), 537–542 (2012)
8. Components101.com, *DHT11 Sensor Pinout, Features, Equivalents and Datasheet* (2018). [online] https://components101.com/dht11-temperature-sensor, Accessed 3 Dec 2018
9. Vernier.com., *Soil Moisture Sensor | Vernier* (2018). https://www.vernier.com/products/sensors/sms-bta/, [online] Accessed 3 Dec 2018
10. R. Sana, Santos, D. Garcia et al., Arduino color sensor TCS230 TCS3200 (2019). https://randomnerdtutorials.com/arduino-color-sensor-tcs230-tcs3200/, Accessed 1 Dec 2018
11. W3schools.com., *HTML Color Picker* (2018), https://www.w3schools.com/colors/colors_picker.asp, [online] Accessed 3 Dec 2018
12. Pixie., in *Nattyware*. http://www.nattyware.com/pixie.php, Accessed 19 Nov 2018
13. R. Corrigan, What is the best soil ph for blackberry plants? in *Home Guides | SF Gate* (2016). Accessed 3 Dec. 2018. https://homeguides.sfgate.com/soil-ph-blackberry-plants-46235.html
14. in *pH and Acid Content of Fruits and Vegetables for Home Canning and Preserving—Alphabetized Master List*. Accessed 3 Dec 2018. https://pickyourown.org/pHofFruitsAndVegetables.htm
15. M.E. Hossain, A.N. Haque, M.E. Haque, L. Heng, Performance and productivity of bororice varieties cultivated in saline area of Satkhira. J. Biosci. Agric. Res. Accessed 3 Dec

2018. https://www.journalbinet.com/uploads/2/1/0/0/21005390/86_jbar_performance_and__productivity_of_boro_rice_varieties_cultivated_in_saline_area_of__satkhira.pdf
16. M.R. Rahman, M.A. Momin, A. Ahmed, M.M. Rashid, M.K. Alam, M.A. Islam, M.J. Alam, Performance of T. Aman rice varieties as influenced by integrated nutrient management practices. https://www.researchgate.net/publication/317512804_Performance_of_T_Aman_rice_varieties_as_influenced_by_integrated_nutrient_management_practices
17. Chillimaster, * *N Are Chillies Acidic or Alkaline*. https://originofchilli.com/are-chillies-acidic-or-alkaline/
18. (PDF) The effect of soil pH on chickpea (Cicer arietinum) genotype sensitivity to isoxaflutole. In: ResearchGate. https://www.researchgate.net/publication/226774047_The_effect_of_soil_pH_on_chickpea_Cicer_arietinum_genotype_sensitivity_to_isoxaflutole
19. En.banglapedia.org., *Banglapedia* (2018). [online] Accessed 3 Dec 2018. http://en.banglapedia.org/index.php?title=Main_Page
20. Anon, (2018) http://bbs.portal.gov.bd/sites/default/files/files/bbs.portal.gov.bd/page/1b1eb817_9325_4354_a756_3d18412203e2/AgriYearbook-2017.pdf, [online] Accessed 3 Dec 2018
21. A.Z. Moslehuddin, S. Laizoo, K. Egashira, Fertility status of Bangladesh soils-A review. J. Fac. Agric. Kyushu Univ. **1**(41), 257–267 (1997)

An Approach to Developing an Embedded System for Irregular Heart Beat Visualization and Monitoring

Partha Sarathi Paul

Abstract Heart beat measurement is a key parameter for diagnosis of various diseases. Arrythmia (irregular heart beat) is a condition which does not get captured to an adequate degree with devices like portable Blood Pressure monitor used at home. The paper presents an approach, with technology already available, to diagnose Arrythmia (irregular heart beat) with a simple and inexpensive device. The device can be coupled over Bluetooth to a smartphone for viewing results and for sending diagnostic results to medical practitioner for his intervention when necessary. This device would be useful particularly for patients in remote places or patients who may find it difficult to visit a doctor, by arming them with a basic diagnostic tool at home.

Keywords Heart beat monitor · Arrhythmia detection · Embedded devices for medical application

1 Introduction

There has been a lot of work going on lately is development of embedded systems for medical applications. Heart Beat measurement has been an important diagnostic tool since ages. There are small portable machines available with which a patient can measure his/her blood pressure and heart rate without help from any medical practitioner. The present paper is an approach to measuring heart rate along with some key parameters, which are not possible with standard portable BP monitors, with a simple and inexpensive device and sending it to medical practitioner via cell phone for his study and advice. The device is intended to work based on microphone based sensor to capture heartbeat and send key information in graphical form via cellphone. Since smart cellphones are ubiquitous these days, the device with its Bluetooth connection to cellphone is intended to make medical advice available to people without ready access to medical practitioners.

P. S. Paul (✉)
Institute of Engineering and Management, Kolkata, India
e-mail: parthasarathi.paul@iemcal.com

V. E. Balas et al. (eds.), *Proceedings of International Conference on Computational Intelligence, Data Science and Cloud Computing*, Lecture Notes on Data Engineering and Communications Technologies 62, https://doi.org/10.1007/978-981-33-4968-1_39

2 Background

Currently available devices in the market have been studied, as given in references [1–3], also recent scholarly papers on healthcare monitoring referred in [4–7] have been studied to get an idea of current IoT based health care systems but none of them are specific to detection and visualization of the problem of arrythmia or irregular heart beat. Most of them are measuring average heart rate, Blood pressure and other parameters. However, irregular heart beat is a serious medical issue as can be seen from [8, 9] and can have long term risk to the patient. The proposed device has been conceptualized to overcome this issue.

As future work enhancement, using ideas presented in paper referred in [10] has been proposed. Papers referred in [11–16] have been studied in connection with using cloud infrastructure in healthcare systems.

3 Objective

The device envisaged would serve as one of the many devices developed or being currently developed which are intended to make diagnostics easy and affordable. There are devices which can be used at home without trained personnel to measure Blood Pressure, Sugar level, Carbon Dioxide level etc. Some of them also have the facility of Bluetooth connection to cellphone and the results can be instantaneously communicated to medical practitioner, if required.

The device is envisaged to capture Arrythmia, a condition not readily detectable with existing devices. Along with heart rate it is intended to capture other key parameters like intensity and adjacent pulse period within a period of time and present the results in graphical form. Readings of heart rate, Coefficient of variation with name of patient and time stamp of the recording are also captured.

Historical data may be saved either in cellphone or shared with medical practitioner over e-mail or uploaded in cloud for future referencing.

3.1 Benefits of the Proposed System

- Inexpensive and easy to use, even by patients in remote parts of the country, from their home.
- Smart phones and internet connectivity are ubiquitous now, even in remote villages. The proposed solution builds on on smart phone and internet platform making it easier for adoption.
- Medical advice can be sought remotely from medical practitioner about irregular heart-beat issue.

- Doctors get visual analytics of the issue which would help them diagnose arrhythmia at a glance.
- The system is open and scalable, having provision of integrating other sensors and features.

The system leverages existing technology with an innovative idea for arrhythmia detection and problem visualization.

4 Methodology

The basic sensor for the device is a microphone, to capture heart beats. It can be used with an extension like what is seen in a stethoscope. This is to amplify the sound signal from the heart.

More complex sensors for heart beat monitoring are available e.g. Apple Watch uses a combination of optical sensors and detectors for heart beat measurement along with their proprietary software. However, the device envisaged would be using simple microphone for sensing purpose for reasons of cost and simplicity.

The signals from the microphone would be digitally processed in a microprocessor based embedded system. The analog signal from microphone is to be digitized and stored in the microprocessor memory for further processing. Among other things, the peaks are be identified, the time duration between peaks are recorded along with amplitude of the signal. In one measurement, say 100 pulses (this is be subject to change if required) are taken for processing.

The picture (courtesy: Firstbeat.com), Fig. 1 below, shows ECG and the interval between the R-pulses are depicted here. Corresponding to R-pulse peaks in the ECG signal, there is also a corresponding sound pulse generated which shall be captured by the proposed device and the intra-pulse intervals calculated.

The captured data may be transferred to cellphone in a tabular form for presentation in a graphical way along with readings of key parameters discussed later in results section.

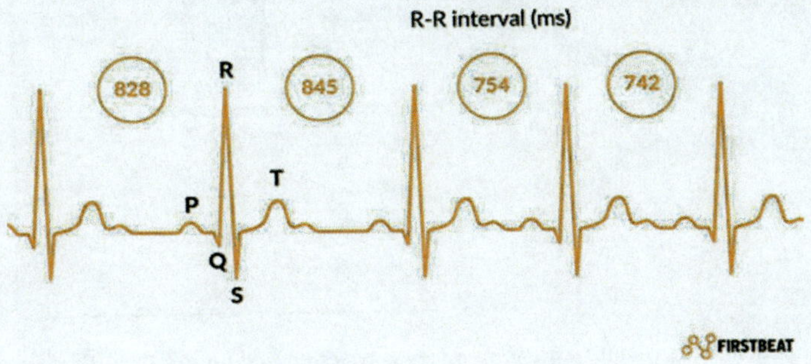

Fig. 1 General shape of a ECG signal

Technologies are already available for an application developed for the cell phone to convert the data from the embedded system to a user friendly graphical user interface. There are already available apps e.g. Graphical Calculator 3D, Quick Graph for iOS, Gnuplot which convert raw data from the embedded system to a graphical display. There are also apps from iHealthlab which are used for similar applications for various health diagnostic devices.

5 Components

Primarily the hardware and software components in the system are as follows:

Microphone with a stethoscope like extension for wiring it to processor based embedded system or alternatively, an unit with microphone with embedded hardware for Bluetooth connectivity with smartphone.

App for the device for the cellphone which takes the data from the embedded system and produce a user friendly graphical display. The app also allows electronic transfer of results to the doctor selected and allow upload of data to cloud for historical record keeping purpose, along with name of the patient and time stamp of the data.

The Fig. 2, below (courtesy reference [4]) gives a framework of the remote health monitoring system. Here the data acquisition sensing and transmission is based on microphone based sensor capturing heart beat signals, smart phone used data concentrator connecting to internet cloud where cloud processing and visualization is done for a doctor to see and advise patient.

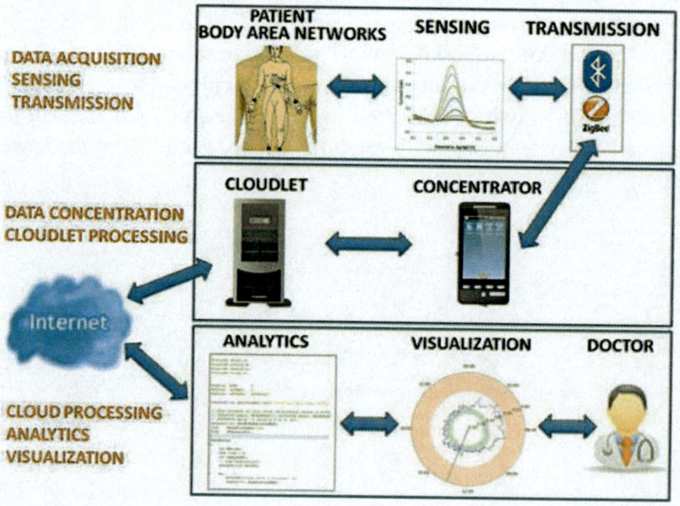

Fig. 2 Components of a remote patient monitoring system based on IoT Cloud architecture

6 Result Analysis

It is to be noted that the results presented are only indicative of the principle of operation of the device envisaged, they are not results actually obtained from patients.

A total of 100 heart beats are considered for analysis of results. This total count can be made higher as required. This would mean that the person under test has to hold the sensor close to his heart till 100 heart beats are counted for analysis.

The key parameters extracted from the sensor signal are, intra pulse interval (for adjacent pulses), Intensity and Count. X variable is the intra-pulse interval, Y variable is the intensity or amplitude of detected signal from microphone and Z variable is the count. In the table below, out of a total of 100 heart beats, 44 is of X value of 4 and Y value of 4. X values can be denominated in seconds, 0.7 to 1 s, in steps of 0.1 s in this table. The table shows 44 pulses have intra-pulse interval of 1 s. Also in a similar way 44 pulses out of 100 have an intensity level of 4. The intensity levels are divided into 7 levels with 4 being normal, 1 to 7 progressively increasing (Table 1).

In Table 1 the normal heart beat is considered $X = 4$ and $Y = 4$, the absolute levels may vary from person to person and may need to be auto adjusted in software. The intensity value is not considered important for analysis, it is merely indicative. The intra-pulse interval is more important from the physician's point of view, therefore statistical calculations are done on this measurement.

It is be useful to calculate the following from the data in the table and transfer them along with the data in the table to the cellphone over Bluetooth:

Sum of count (100 in this case)
Coefficient of variation (measures spread or standard deviation of the samples of intra-pulse interval, 12.38% in this case); greater than 15% could indicate irregular heart rate
Percentage of pulses with positive deviation from average (21% in this case, since there is a total of 7 + 7+7 = 21 out of 100 pulses); greater than 30% could flag alarm indicating a condition of having too much interval between pulses.

Table 1 Representative XYZ plot for a healthy person

		X→						
	Z↘	1	2	3	4	5	6	7
Y↓	1							
	2							
	3			7	7	7		
	4			7	44	7		
	5			7	7	7		
	6							
	7							
		0.7 s	0.8 s	0.9 s	1 s	1.1 s	1.2 s	1.3 s

The 3D rendition of the figures in the table would look as shown in Fig. 3 below (This is a typical normal heart condition):

Treatment of arrhythmia may or may not be necessary. Usually it is required only if is causing significant symptoms or of it is putting you at risk. This is all the more reason to have a basic diagnostic tool available at home so that one can decide when to see the doctor.

The tool envisaged provides diagnostics in the form of a graphical signature to classify the type of arrhythmia and help in deciding the course of treatment. The treatment can vary from using particular maneuvers like holding your breath, to using medication to restore heart rate, to using blood thinners, to restoring electrolyte balance, to using intrusive methods like ablation or implantation of Pacemaker or Bypass surgery. Visualizing the graphical signature of the heart condition helps in finding the right course of treatment.

Now the result from an abnormal situation is taken. The table changes to as shown in Table 2 (Abnormal condition).

The calculated results from Table 2 are as follows:

Sum of count (100 in this case)

Coefficient of variation (measures spread or standard deviation, of intra-pulse interval, 18.33% in this case); greater than 15% indicates irregular heart rate

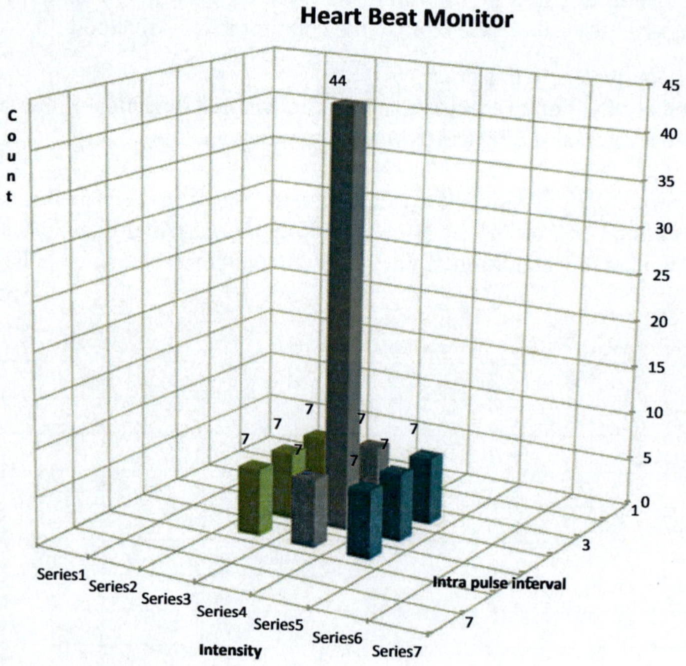

Fig. 3 Visualization of heart beat monitoring for a healthy individual

Table 2 Representative XYZ plot for a person with arrythmia

		X→						
	Z↘	1	2	3	4	5	6	7
Y↓	1						7	
	2	15		8				
	3				7			
	4		7	7	6	10		
	5			7	7		12	
	6							7
	7							
		0.7 s	0.8 s	0.9 s	1 s	1.1 s	1.2 s	1.3 s

Percentage of pulses with positive deviation from average (36% in this case, 10 + 7+12 + 7 = 36 out of 100 pulses); greater than 30% flags alarm indicating a condition of having too much interval between pulses.

The 3D rendition of the figures in the table would look as shown in Fig. 4, follows

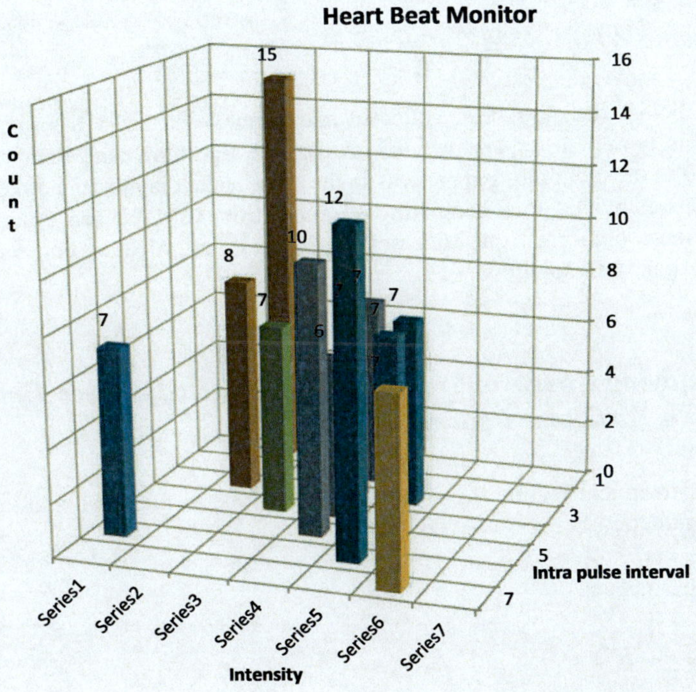

Fig. 4 Visualization of heart beat monitoring for a patient with irregular heart beat

Further detailed study has to be carried out to see if the a particular abnormal condition has a typical signature pattern in the type of 3D display, which helps in determining the line of treatment necessary for the specific patient.

Moreover, this type of display provided with time-stamp and stored over a period of time would indicate whether a patient is responding to the line of treatment.

7 Conclusion

The paper presents an approach to building a microprocessor based system for heart beat monitoring with a view to detection of arrhythmia (irregular heart beat) and help physicians determine the seriousness of the disease and possible line of treatment. It may also be used to monitor progress of the disease after a line of treatment has been determined. The ease of use by the patient from home without the help of any trained personnel should popularize the device.

8 Future Enhancements

8.1 Using AI on Data Collected from the System

After sufficient data has been collected and features extracted from the signal, from a variety of patients, an Artificial Intelligence algorithm can be employed and processed in the cloud, to find patterns in the data, make diagnosis and recommendation for treatment. Cloud computing resources from IBM-Watson, Amazon Web Server, Azure platform from Microsoft or Google platform may be used to find suitable AI and ML solution.

8.2 Studying Effects of Treatment Regime, Drugs Over Time for a Particular Patient

Effects of treatment regime may be tracked over time to find best solutions for a given set of data.

8.3 Studying Heart Rate and Its Irregularity Over Time

Variations of signal from the same patient during a 24 h period to see if the irregularity increases or decreases at certain time of the day or night depending on activities performed or stress level of the subject during a period of time.

8.4 Studying Different Patients to Find Patterns Based on

Variation of this signal over different types of patients (based on age, sex and other medical parameters like BP, Cholestorol, blood sugar and other parameters) can also be recorded in order to find patterns in the data for analysis, recommend treatment and/or giving early warning of medical emergency situation. IBM Watson platform may be used to find suitable models which yield desired accuracy.

8.5 Finding a Way for Indirect Measurement of Blood Pressure

Sometimes study can reveal a correlation between seeming unrelated parameters. As an example, Google's algorithm for retina scan revealing heart disease may be cited [10]. The Arrythmia detection device in combination infrared based clip-on sensor at the finger tip, which can give a sharp signal when the blood flows to the finger tips, can yield an indirect method of estimated blood pressure. A pattern can be found between these signals and BP through Artificial Intelligence algorithms. A large data set of labeled data has to be developed for training and testing the algorithm before adequate accuracy is achieved. If it can be done, it may result in an inexpensive and easy way to estimate BP, as it is known that Blood Pressure measurement presently is cumbersome, it requires wrapping a Velcro strap around the arm and requires expensive device.

8.6 Indirect Blood Pressure Measurement from Sensors Available in Smart Phone

If the above mentioned indirect measure of blood pressure yields satisfactory result, attempt may be made to use the sensors and actuators in the smartphone itself, to get indirect blood pressure estimate. The Microphone in the smart phone can serve as the sensor to capture heart-beat by holding the microphone close to the heart. To get the sharp peaked signal described above, the torch light in the cellphone and camera may be used to capture blood flow signal from finger tip, simultaneously with the audio signal from microphone. Current data from sensors together with prior data which have been used to train the machine learning system can yield fairly close estimate of Blood Pressure after Regression/Classification is carried out.

8.7 Integrating Other Sensors for a More Complete Diagnostic Tool at Home

The device can be made scalable enough to integrate other bio-sensors like Oxymeter, Glucose level meter, Blood Sugar measurement device and more. Some of the available features available in smart phone, like daily activities of walking, running etc. may also be integrated to the database of parameters of the patient. A more comprehensive data set for patient must be able help doctors diagnose and recommend treatment better. With a fairly well developed AI, the system may even suggest which drugs or which treatment would be best for the patient. However, the final assessment would always remain with the doctor.

References

1. Information from Apple at the website: http://www.apple.com/watch/ (Last Accessed on 03 July 2020
2. Products related to heart condition monitoring: https://ihealthlabs.com/blood-pressure-monitors/clear (Last Accessed on 03 July 2020)
3. US patent no. 3682161A Heartbeat transducer for a monitoring device. Inventor Vernon F. Alibert
4. M.M. Islam, A. Rahaman, M.R. Islam, Development of smart healthcare monitoring system in IoT Environment. SN Comput. Sci. **1**, 185 (2020). https://doi.org/10.1007/s42979-020-00195-y
5. A. Rahaman, M. Islam, M. Islam, M. Sadi, S. Nooruddin, Developing IoT based smart health monitoring systems: a review. Rev. Intell. Artif. **33**, 435–440 (2019). https://doi.org/10.18280/ria.330605
6. S. Mahmud, X. Lin, J.-H. Kim, Interface for human machine interaction for assistant devices: a review, in *2020 10th Annual Computing and Communication Workshop and Conference (CCWC)*. (IEEE, 2020). pp. 768–73
7. S.I. Ayon, M.M. Islam, M.R. Hossain, Coronary artery heart disease prediction: a comparative study of computational intelligence techniques. IETE. J. Res. 2020. https://doi.org/10.1080/03772063.2020.1713916
8. C.W. Israel et. al., Long-term risk of recurrent atrial fibrillation. https://pubmed.ncbi.nlm.nih.gov/14715182/, https://doi.org/10.1016/j.jacc.2003.08.027 (last accessed on 03 July 2020)
9. Heart rate variability, Blog by Tina Hofman: https://www.firstbeat.com/en/blog/what-is-heart-rate-variability-hrv/ (Last accessed on 03 July 2020)
10. Google's new algorithm predicts heart disease by looking at your eyes: https://www.theverge.com/2018/2/19/17027902/google-verily-ai-algorithm-eye-scan-heart-disease-cardiovascular-risk (Last accessed on 03 July 2020)
11. M. Hassanalieragh, A. Page, T. Soyata, G. Sharma, G. Mateos, B. Kantarci, S. Andreescu, *Health Monitoring and Management Using Internet-of-Things (IoT) Sensing with Cloud-Based Processing: Opportunities and Challenges* (2015). https://doi.org/10.1109/scc.2015.47

12. C. Doukas, I. Maglogiannis, Bringing IoT and cloud computing towards pervasive healthcare, in *2012 Sixth International Conference on Innovative Mobile and Internet Services in Ubiquitous Computing*, (Palermo, Italy, 2012), pp. 922–926 July 2012. View at: Publisher Site|Google Scholar
13. K. Gai, S. Li, Towards cloud computing: a literature review on cloud computing and its development trends. in *2012 Fourth International Conference on Multimedia Information Networking and Security*, vol. 146, p. 142, (Nanjing, China, November 2012). View at: Publisher Site|Google Scholar
14. D. Chappell, *The Microsoft Private Cloud, a Technology Overview*. (2011). http://www.davidchappell.com/writing/white_papers/The_Microsoft_Private_Cloud_v1.0–Chappell.pdf.View at: Google Scholar
15. M.D. Lakshmi, J.P.M. Dhas, An open source private cloud solution for rural healthcare, in *2011 International Conference on Signal Processing, Communication, Computing and Networking Technologies*, (Thuckafay, India), pp. 670–674, July 2011. View at: Publisher Site|Google Scholar
16. S. Pandey, W. Voorsluys, S. Niu, A. Khandoker, R. Buyya, An autonomic cloud environment for hosting ECG data analysis services. Future Gener. Comput. Syst. vol. 28, no. 1, pp. 147–154, 2012.View at: Publisher Site | Google Scholar

Part IV
Cloud and Security

Distributed Cloud Deep Learning Architecture for Complex Image Analysis and Run-time Prediction Tool

Shushant Kumar, Edwin Thomas, Anmol Horo, and B. Annappa

Abstract Hyperspectral imaging is a rare research tool and has been transformed into a commodity product found in a wide field. Currently, standard data processing methods that specialize in special hyperspectral accumulation structures are required. Also, with the advent of data collection and development in the field of sensory devices, it has rendered previous processing tools in vain. To manage this huge increase in the amount of data, a consistent cloud distribution method is required. Hyperspectral images (HSIs) have several spectral band channels that make the study very difficult. In this paper, an in-depth reading method of the novel with a modified autoencoder is proposed as a cloud-based use of HSI analysis, which provides a measure of lesser error rates and high accuracy of classification models. In line with this, a list of four tools has been proposed to calculate the actual number of workers, cores, and iterations required to achieve the desired accuracy for a specified amount of run-time. This will help cloud managers get a basic idea of computational needs and help them allocate resources more efficiently. The entire architecture was simulated on Spark servers and was verified experimentally by checking that the proposed architecture performs the function of efficient management and analysis of large HSI.

Keywords Apache spark server · AutoEncoder · Classification · Hyperspectral images · Regression · Spectral processing

---

S. Kumar (✉) · E. Thomas · A. Horo · B. Annappa
Department of Computer Science and Engineering, National Institute of Technology Karnataka, Mangalore, India
e-mail: shushantkmr2@gmail.com

E. Thomas
e-mail: edwinthomas444@gmail.com

A. Horo
e-mail: horoanmol@gmail.com.in

B. Annappa
e-mail: annappa@ieee.org

V. E. Balas et al. (eds.), *Proceedings of International Conference on Computational Intelligence, Data Science and Cloud Computing*, Lecture Notes on Data Engineering and Communications Technologies 62, https://doi.org/10.1007/978-981-33-4968-1_40

1 Introduction

Remote data targeting can be considered as a complex analysis of large datasets due to the high volume of data being generated per day, of various kinds including radar, visual images, sound, etc., and data production speed that needs to be processed. Concurrently, treatment with large aggregation of data is a major issue currently being dealt by various departments in the technical and analysis institutes. Cloud computing in this environment is a solution because a robust setup is provided for big computers, often found in a revenue model, and reduces users' need to buy and maintain a complex computational system. Hyperspectral images form hundreds of spectral bands consisting of innumerable wavelengths, which rises the need of effective database systems and aggregation setup [1]. Satellite and air sensing revolution have improved at such a massive scale that forces the procedural image processing to its limits. There are several approaches to this particularly in the field of machine learning. However, there are some tasks that focus on dealing with big data. An overview of machine intelligence methods for dealing with large aggregation can be found [2]; however, those methods are more sensitive to training than practice isolation. Cloud computing platforms in recent period are being used progressively standard for perception aggregation processing [3]. The cloud is now the canonic for unfocused engineering due to its advanced Web-based technology abilities, computer-assisted engineering, and efficacious computing. In the reasoning of massive amounts of hyperspectral data, cloud engineering is considered a new method [4].

In this paper, an architecture is introduced for distributed classification tasks in analyzing the hyperspectral images. Also, a tool suite is proposed which was developed using deep learning techniques that is able to give a rough estimate of the number of workers, core, and iterations that will be required in analyzing the data for given execution time, thus providing the administrators of cloud services an added edge in allocating computing and memory resources on cloud platforms. In Sect. 2, a detailed literature survey is carried out for the related works. It is followed by Sect. 3, a methodology that presents the proposed solutions. Section 4 is the experiments and results where the practical work and its outcome are mentioned. In Sect. 5, the conclusion of this research work is drawn, and this section is finally followed by References.

2 Literature Survey

Hyperspectral sensors produce hyperspectral imagery that has information in a vast number of spectral bands, and hence, it presents a challenge to the traditional data processing methods because the preprocessing like the dimension reduction is quite difficult. The techniques like minimum noise fraction [5], linear discriminate analysis [6], and principal component analysis [7] are used in the literature to reduce the dimensions of high-dimensional data to extract useful features, but it also losses nonlinearity property. Plaza et al. [8] proposes a sequential extended morphological transformation for classification and dimensionality reduction of remotely sensed HSI datasets. The orthogonal subspaces can also be applied for dimensionality reduction of HSI datasets. The dimensionality of HSI datasets can also be reduced through wavelet transformations [9], and during the signal decomposition, it preserves the low and high-frequency information. It has immense computational complexity and memory consumption but also preserves the spatial structures. The Laplacian Eigenmap preserves the relative distances among the spatial points and is used for dimensionality reduction [10]. Locally linear embedding technique is improved with a neighborhood window to reduce the dimensionality of HSI datasets [11].

A distributed computing framework provides shorter response time, data security, privacy, and system scalability [12]. Reduced time can be obtained in a face recognition application when the edge processes the photos as opposed to the cloud [13]. The distributed deep neural networks can use advantages of the distributed computing framework in a DNN application and obtain similar results. Autoencoder architectures are generally used for spacecraft's telemetry data analysis[14] and image recognition problems [15]. Luo and Nagarajan [16] propose distributed autoencoder architecture to detect anomaly present in the wireless network of sensors. The use of non-equilibrium Helmholtz free energy function in an autoencoder architecture can make it learn useful factorial feature maps [17]. DeepSpark has been proposed in [18] that provides a parallel and distributed deep learning framework on clusters using Apache Spark. It distributes parameters and workloads over active nodes automatically by utilizing Spark to support distributed parallel operations. The training results are aggregated in each iteration using an asynchronous averaging variant of the stochastic gradient descent. DeepSpark exploits the synchronized processing functionality of Spark. MapReduce was proposed by Google in [19] to support large-scale distributed data computations operations on cloud clusters. The convolutional neural networks can distribute the computation across layers effectively due to its sparsely connected structure, but the fully connected layers still need to be processed on different nodes [20]. The distributed large-scale training on nodes in a centralized cloud requires parallel data methods that can be used to scale the distributed training across various computing nodes [21].

3 Methodology

This section presents the proposed solutions. The two significant contributions in this work are an enhanced autoencoder architecture that uses fine-tuned parameters based on an empirical study and a package of four pre-training tools to estimate the crucial parameters such as workers, cores, and iterations beforehand to reduce the computational uses in trial and error setup.

3.1 Distributed Cloud Analysis Framework

The model framework of the distributed cloud [22] implementation runs on an exclusive Spark cluster, and with fault tolerance its capability to process large volumes quickly. Moreover, flexibility of the Spark cluster is defined by a master–slave framework. Spark cluster (Fig. 2) is created by using a master node, which hosts a Java virtual machine (JVM) driver on how to use and distribute resources in the cluster. The other component is the scheduler which is responsible for dividing tasks for the workers and execution nodes. These nodes that are assigned to executor nodes can be in a one-to-one or many-to-one fashion. It is the duty of worker nodes to successfully execute the tasks by generating a distributed agent executor. Worker nodes work on the actions and store the partitions of data. The architecture is shown in Fig. 1.

Distributed autoencoder is used to enforce the proposed programming model. The initial hyperspectral images data tuples are in the form of key-pair values, and the organization is based on the MapReduce model. The two key operations to be carried next are mapping and reduction. Mapping refers to the creation of key-value pairs, and reduction is to gather intermediate pairs by collecting, to give the end results.

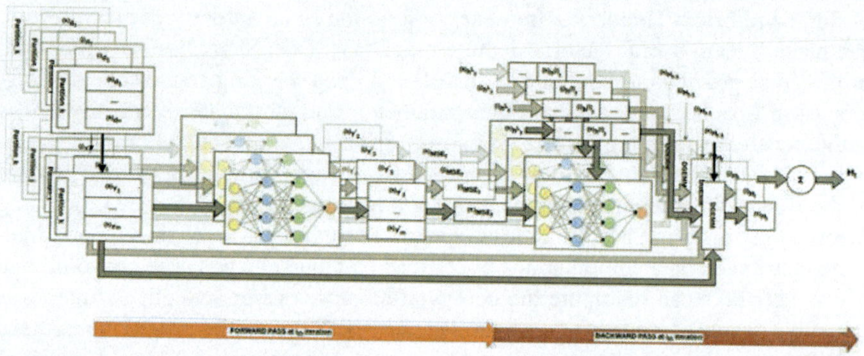

Fig. 1 Distributed architecture with the forward and backward pipelines of the training process of proposed autoencoder [22]

Fig. 2 Representation of a Generic Spark Cluster [22]

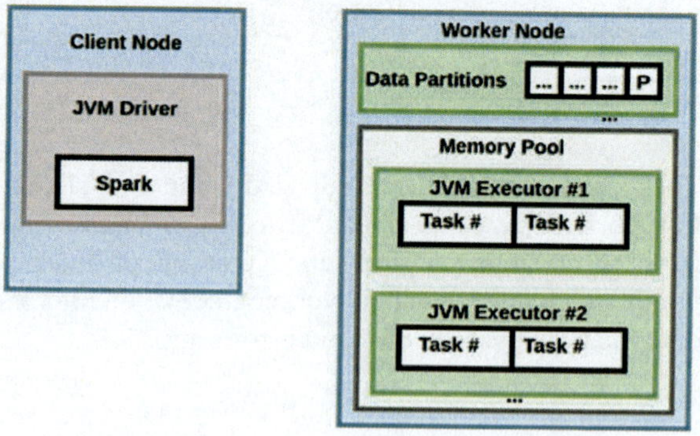

Emulating it on Spark servers, the complete data is extracted into resilient distributed dataset. The Spark handles this dataset by distributing the data when required by a particular worker node, which not only implements distributed systems but also aids parallel execution speeding up the processing of hyperspectral images. Spark drivers are directly managed by the master nodes. This driver is responsible for the partitioning of data and distributing it among the worker nodes. The executors are also created by these drivers that maintain data partitions on the worker nodes. The load distribution by the master node happens in such a way that the executor in any worker node has allocated a data partition used by the task. This is done by calling the action and the creation of a new job coordinated by the driver node. The role of the scheduler is to allocate the next set of jobs when the previous set tasks are completed successfully.

3.2 Modified Autoencoder

The autoencoder network, in general, is trained to attempt to copy its input to its output. They compress the input into a latent-space representation, which in general has dimensions restricted to force the model to learn only the principal components in the input. The input here is hyperspectral images, which have a dimension in the form $m \times n \times \omega$, where ω is the width of the HSIs. The width here refers to the number of spectral bands, and m x n is the spatial dimension of the images. In a standard autoencoder, there are three basic components encoder layers, bottleneck layer, and decoder layer. The encoder layer applies a function Ψ which represents the encoding function,

$$\theta^i = \Psi(\theta^{i-1}.w^{i-1} + b^{i-1}) \tag{1}$$

where θ^i represents the input vector at level i, w^{i-1} and b^{i-1} as weights and bias vectors at level $i - 1$, respectively. This encoding helps the model to extract essential features from the input set. Then down the model path, ω reduces. Only the prominent bands remain finally till the latent space is generated. After this, the model starts decoding where it tries to regenerate the original input, and the final output image is hence generated at the end of it. So in the bottleneck layer, a refined vector remains, which can be considered as a filtered input with essential processed spectral bands. Mean squared error function $\Xi(\Theta)$ is used to reduce the reconstruction error between the output at the end of encoding blocks θ_e and at the end of decoding blocks θ_d, which can be represented by

$$\Xi(\Theta) = min\|\Theta_e - \Theta_d\|_2 = min \sum^{m \times n} \|\theta_e - \theta_d\|_2 \tag{2}$$

In this work, a new autoencoder architecture is proposed that produces the least validation loss in the experimented architecture search. The number of hidden layer neurons of the model and monitored the model's performance based on Val loss was varied. Several combinations of neurons were used in the study whose details are presented in the results and experiments section of this paper. These combinations covered several cases in which the code size is greater, equal, or lesser than its neighboring hidden layers. Finally, the best performing model was adopted for training the tool suite.

3.3 *Modifying Workers and Cores*

To use this model on Apache Spark server, four parameters need to be assigned. First is the number of worker nodes to which the executor node distributes the processing. The second parameter that the model requires is the number of iterations. The model will be trained to collect the results. The third parameter is the number of cores to be utilized by the model in the running system. Furthermore, finally, the fourth parameter is the split of training and testing data. The fourth parameter is more of a hyperparameter that is defined better in the domain of deep learning. There has been no strict ratio of the split defined in the literature. Thus, a fixed ratio of 0.8 train to test was used by referring other research works. The remaining three parameters, however, are generic and can be exploited for the best performance on the machine. Even with massive computational power, there still exist many limitations on the allotted computation and memory resources in a cloud architecture. So, optimizing these three parameters is necessary. Hence, a Python-based tool was designed to predict the performance of the model without even training the model to get an estimate of how the model will perform on the dataset. For this reason, the simulation was executed with combinations within a range for these parameters, and the accuracy obtained was also noted. The resulting output was used as a dataset for training a regression model to predict the accuracy of any valid combination of those variables. Bayesian Ridge regression model is used for the reasons discussed in the next section. Since these parameters are restricted by the computation system so the cloud administrator can set the parameters in the proposed learned model and it will predict the classification accuracy, using this, the administrator gets a beforehand prediction that can save time and computation.

4 Experiments and Results

The complete architecture was implemented on a virtual simulation with eight virtual cores 2.5 GHz each. Eaxh node was allotted with 16 GB of RAM and 50 GB storage. The setup runs on Ubuntu 18.04 as an operating system with Spark 2.1.1, and Java open JDK version 1.8.0_252 and Scala code runner version 2.11.12. Spark MLLib was used for implementing the autoencoder and other machine learning implementations. The dataset consists of hyperspectral images obtained from two different sources. The first one is taken from the Hyperion dataset, which consists of images from six different zones collected by the Hyperion spectrometer. The other set was obtained from AVIRIS over agricultural lands. Both the datasets consisted of 220 spectral bands.

The first experiment was conducted to prove the efficacy of the new proposed autoencoder structure. For this part, the experiment conducted an empirical analysis to determine the best autoencoder architecture for the classification task. For this purpose, the number of input neurons was fixed and the output neurons (it should be noted that the output neurons represent the total number of classes or labels for the classification dataset) and varied the number of neurons in the hidden layers of the autoencoder. The architecture has five layers, including the input and output layer, which leaves three hidden layers. The first, second, and third hidden layers are represented as h1, h2, and h3, respectively. h1, h2, and h3 are varied from 128 to 32, 64 to 16, and 32 to 128, respectively, in steps of 16 neurons. Table 1. shows the combinations of h1, h2, and h3, which gave the best validation results. In this table, the referenced paper's [22] classification accuracy is also compared with the proposed method experimentally. It can be clearly analyzed that the proposed method outperforms the previous model by 8.36%.

The second set of experiments was to develop a tool suite for run-time predictions. The first Tool (Tool 1) was developed to predict the accuracy of classification given the number of iterations, obtained by using the modified autoencoder of the first experiment. The second Tool (Tool 2) is its inverse, where the number of iterations can be predicted, provided the classification accuracy of the autoencoder. Further, a Tool (Tool 3) was also developed to predict the execution run-time (total CPU time to train the modified autoencoder) of HSIs with the given combination of the number of workers, the number of allocated cores, and the maximum number of iterations. Tool 4 is an inverse of Tool 3 and was trained to predict the number of workers, cores, and iterations that would result in a run-time given as input to the tool.

For developing the tools, the experiment first simulated the setup for different combination ranges of these parameters, and the results obtained were used to prepare the dataset. Then, a simulation for creating the datasets for Tools 1–4 was conducted. For this simulation, the number of workers ranged from 2 to 32, the number of cores ranged from 2 to 32 and number of iterations ranged from 10 to 200, and the validation accuracy was monitored, and the training time of the modified autoencoder in nanoseconds (ns) using these parameters. The size of the dataset obtained from this simulation was 3132. Dataset for Tool 1 was obtained by extracting the iterations (train data) and obtained accuracy (true output) from the collected data. Dataset for Tool 2 (Tool 1 inverse) is created by simply flipping the train data and the true output before training. Furthermore, the dataset for Tool 3 is obtained by extracting workers, cores, and iterations (which together constitute train data) and the execution time (true output). Tool 4 (Tool 3 inverse) is obtained similar to Tool 2. Now, to train a suitable model for each of these tools, the Bayesian Ridge regression model was used as it is considered one of the simple yet efficient models for smaller linear datasets. Also, it has the ability to naturally solve the issue of insufficient datasets or poorly dispersed datasets. The probability distribution of data is used instead of estimating a single value.

Table 1 Best hidden layer configurations for modified autoencoder

Model	h1	h2	h3	Accuracy
Haut et al. [22]	128	64	32	0.6182
Candidate 1	64	16	48	0.7018
Candidate 2	32	16	112	0.7010

Table 2 Performance evaluation of tool suite

Tool	Mean absolute error	Means squared errors	Root mean squared error
Tool 1	0.0510	0.0044	0.0666
Tool 2	0.0862	0.0132	0.1150
Tool 3	0.0198	0.0007	0.0271
Tool 4	0.2643	0.1168	0.3417

The parameters workers, cores, and iterations serve as the input or X, and the classification accuracy is the desired output or y. The inputs are first normalized using z-score normalization and are then fed to the model. The training to testing ratio was set at 80:20. The results obtained after training the models for the four tools are presented in Table 2. It is clearly observed that all the three evaluation terms are prominently implying the efficiency of the proposed prediction tool. To get the understanding of the distribution of the classification accuracy, a univariate distribution was plotted, as shown in Fig. 3. From this figure, the mean classification accuracy is observed in the range 0.62–0.65. This distribution helps in analyzing the training data. The other two important distributions that are studied in this research work were the relation between the number of workers with respect to the classification accuracy and relation between the iterations and classification accuracy. The joint plot representation is shown in Fig. 4. From the graphs in Fig. 4a, b, it can be observed that the higher density regions between worker nodes and classification accuracy are quite evenly distributed, indicating that optimal performance due to the number of worker nodes can be obtained with lesser worker nodes if the other parameters are adjusted accordingly. Whereas from the graphs in Fig. 4c, d it can be observed that the higher density regions between iterations and classification accuracy are concentrated higher on the lower values of both, whereas with increasing iteration, better accuracy is obtained.

So, the value of iterations in general for this model should be high. The joint plot distribution of allotted cores and iterations with respect to CPU execution time is shown in Fig. 5. From the graphs in Fig. 5a, b, it is clearly visible that the correlation of allotted cores with respect to run-time is evenly distributed and has more of a linear relationship. With increasing values of cores, the run-time varies in a range indicating the other factors have a determining factor in deciding run-time along with the cores.

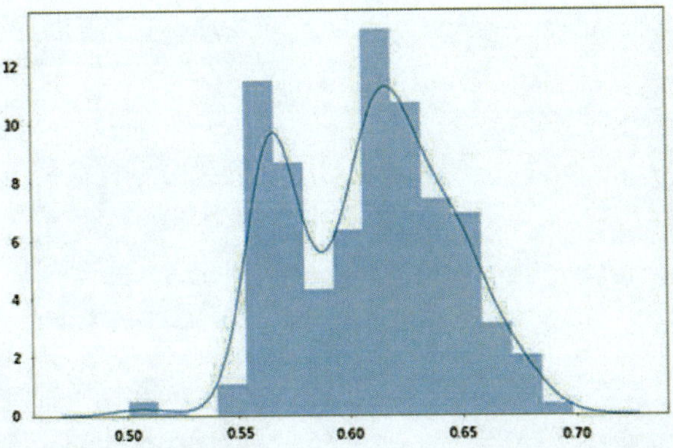

Fig. 3 Univariate distribution of classification accuracy

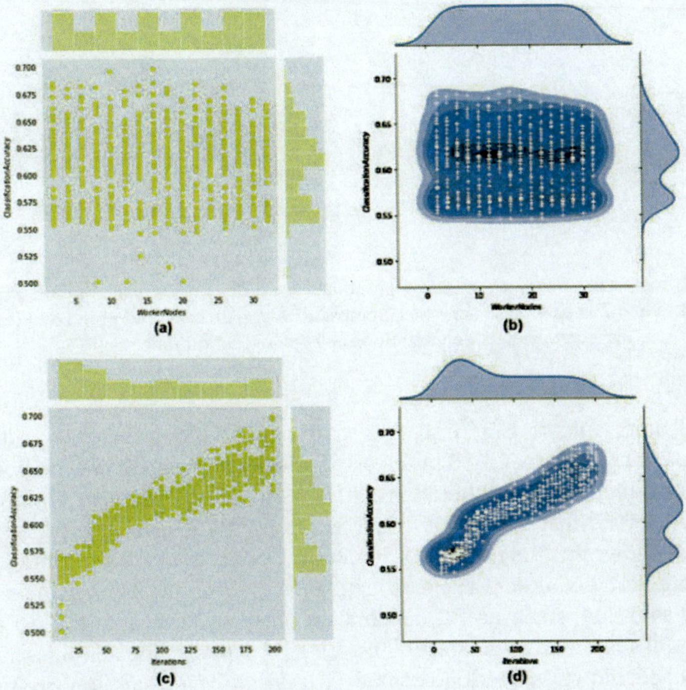

Fig. 4 **a**, **b** represents bivariate scattered graph and joint density estimates graph, respectively, between worker nodes and classification accuracy. **c**, **d** represents bivariate scattered graph and joint density estimates graph, respectively, between iterations and classification accuracy

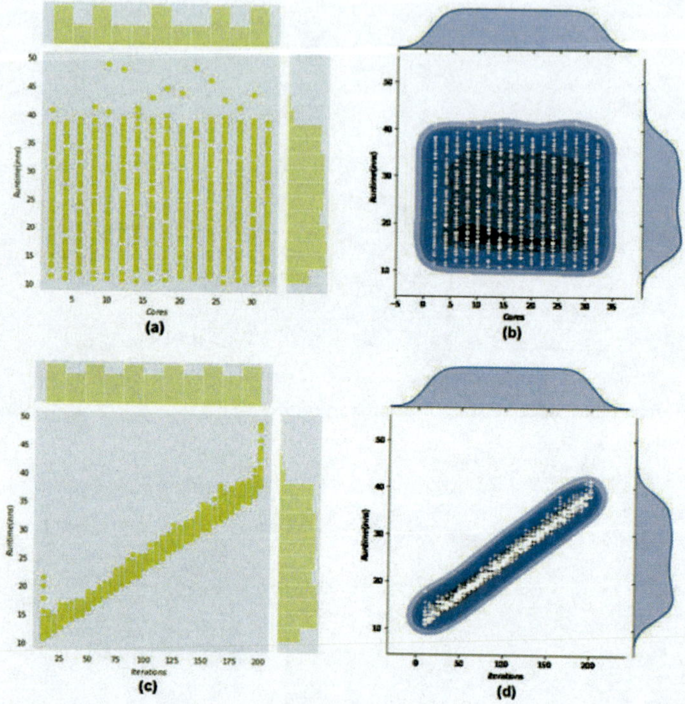

Fig. 5 **a**, **b** represents bivariate scattered graph and joint density estimates graph, respectively, between cores and cpu execution time. **c**, **d** represents bivariate scattered graph and joint density estimates graph, respectively, between iterations and cpu execution time

Whereas from graphs in Fig. 5c, d, it is visible that with an increase in iterations, the run-time increases linearly. This clearly indicates that run-time is predominantly dependent on the number of iterations. Also, comparing graphs in Fig. 4d and 5d, it can be observed that the density regions are evenly distributed in Fig. 5d. This observation shows that the number of iterations has a constant determining factor to run-time but does not affect the classification accuracy in the same manner.

Having said that, some useful insights are presented next into the usage of the proposed tool suite with an interesting use case. If suppose, the administrators need to target a specific classification accuracy within a given execution run-time. The proposed tool suite can be used in the following manner to achieve the same. Tool 2 can be used to determine approximately the number of iterations that may be required to achieve the targeted accuracy. Next, Tool 4 can be used to predict the workers, nodes, and iterations for the desired execution time. If the number of iterations given by Tool 4 and Tool 2 approximately matches, then the administrator can allocate the specified number of workers and cores to achieve the desired goal.

However, if these values differ by a large margin, then the targeted accuracy cannot be achieved within the expected run-time. Consequently, a trade-off between accuracy and computing time needs to be arrived at. This can be achieved in several ways. However, two notable ones are presented here. The first method assumes that the accuracy is less crucial as compared to run-time. In this scenario, the number of iterations can be reduced or compromised. This value is obtained from Tool 4. Simply plugging in the number of iterations as predicted by this Tool into Tool 1 gives the approximate new accuracy, which can be obtained by current allotment (compromised accuracy).

The second method assumes that the run-time is less crucial as compared to accuracy. In this aspect, it is important to keep the number of iterations (obtained from Tool 2) constant and compromise on execution time. To find the value of this run-time, either of the following methods can be performed. Using the workers (from Tool 4) and iterations (from Tool 2), a binary search on the number of cores can be performed such that the combination of these three parameters when given as input to Tool 3 produces an execution time value closest to the targeted one. Another approach is to fix the cores (from Tool 4) and iteration (from Tool 2) and perform a binary search on the number of workers on Tool 3. The administrator can then decide upon the resource allocation based on the best run-time achieved by the alternatives.

5 Conclusion

In this paper, modified autoencoder for distributed cloud-based architecture was proposed to analyze hyperspectral images and the tool suite to predict the classification accuracy given iterations (and vice versa) and to predict the execution time by the three parameters worker nodes, cores, and iterations (and vice versa). Using an autoencoder baseline model, the data gets compressed, and the model learns a non-linear solution suitable for datasets like HSIs, which are involved in nature. From the experiments carried out, it can be concluded that the proposed architecture performs better than the previous implementation of analyzing complex datasets on a cloud-based platform. The distributed cloud implementation on Spark clusters proved to be efficient in terms of scalability and reliability. Also, the prediction system can serve as an essential tool for cloud administrators to manage their resources efficiently. This will not only help them save resources but aid in the calculation of the most optimal combination of resources for different scenarios. The experiments also demonstrated that a cloud-based distributed approach would be a prominent platform for deep learning operations. This research will pave the way for developing more efficient cloud-based approaches for performing deep learning applications.

References

1. A. Plaza, J. Plaza, D. Valencia, Impact of platform heterogeneity on the design of parallel algorithms for morphological processing of high-dimensional image data. J. Supercomput. **40**(1), 81–107 (2007)
2. R. Bekkerman, M. Bilenko, J. Langford, *Scaling up Machine Learning: Parallel and Distributed Approaches* (Cambridge University Press, Cambridge, 2011)
3. Z. Wu, Y. Li, A. Plaza, J. Li, F. Xiao, Z. Wei, Parallel and distributed dimensionality reduction of hyperspectral data on cloud computing architectures. IEEE J. Sel. Topics Appl. Earth Observat. Rem. Sens. **9**(6), 2270–2278 (2016)
4. J.A. Martínez, E.M. Garzón, A. Plaza, I. García, Automatic tuning of iterative computation on heterogeneous multiprocessors with adithe. J. Supercomput. **58**(2), 151–159 (2011)
5. W. Joseph, Automated spectral analysis: A geologic example using Aviris data, north grapevine mountains, Nevada, pp. 1407–1418 (1994)
6. K. Fukunaga, *Introduction to Statistical Pattern Classification* (Academic Press, San Diego, CA, 1990)
7. I. Jolliffe, Principal component analysis. Technometrics **45**(3), 276 (2003)
8. A. Plaza, P. Martínez, J. Plaza, R. Pérez, Dimensionality reduction and classification of hyperspectral image data using sequences of extended morphological transformations. IEEE Trans. Geosci. Rem. Sens. **43**(3), 466–479 (2005)
9. S. Kaewpijit, J. Le Moigne, T. El-Ghazawi, Automatic reduction of hyperspectral imagery using wavelet spectral analysis. IEEE Trans. Geosci. Rem. Sens. **41**(4), 863–871 (2003)
10. M. Belkin, P. Niyogi, Laplacian eigenmaps for dimensionality reduction and data representation. Neural Comput. **15**(6), 1373–1396 (2003)
11. G. Chen, S.-E. Qian, Dimensionality reduction of hyperspectral imagery. J. Appl. Rem. Sens. **1**(1), 013509 (2007)
12. K. Skala, D. Davidovic, E. Afgan, I. Sovic, Z. Sojat, Scalable distributed computing hierarchy: Cloud, fog and dew computing. Open J. Cloud Comput. (OJCC) **2**(1), 16–24 (2015)
13. S. Yi, Z. Hao, Z. Qin, Q. Li, Fog computing: Platform and applications, pp. 73–78 (2015)
14. M. Sakurada, T. Yairi, Anomaly detection using autoencoders with nonlinear dimensionality reduction, pp. 4–11 (2014)
15. P. Vincent, H. Larochelle, I. Lajoie, Y. Bengio, P.-A. Manzagol, Stacked denoising autoencoders: Learning useful representations in a deep network with a local denoising criterion. J. Mach. Learn. Res. **11**, 3371–3408 (2010)
16. T. Luo, S.G. Nagarajan, Distributed anomaly detection using autoencoder neural networks in WSN for IoT, pp. 1–6 (2018)
17. G.E. Hinton, R.S. Zemel, Autoencoders, minimum description length and Helmholtz free energy (1994)
18. H. Kim, J. Park, J. Jang, S. Yoon, Deepspark: A spark-based distributed deep learning framework for commodity clusters. arXiv preprint arXiv:1602.08191 (2016)
19. J. Dean, S. Ghemawat, Mapreduce: Simplified data processing on large clusters (2004)
20. F. Seide, H. Fu, J. Droppo, G. Li, D. Yu, On parallelizability of stochastic gradient descent for speech dnns, pp. 235–239 (2014)
21. J. Dean, G. Corrado, R. Monga, K. Chen, M. Devin, M. Mao, M. Ranzato, A. Senior, P. Tucker, K. Yang, et al., Large scale distributed deep networks, pp. 1223–1231 (2012)
22. J.M. Haut, J.A. Gallardo, M.E. Paoletti, G. Cavallaro, J. Plaza, A. Plaza, M. Riedel, Cloud deep networks for hyperspectral image analysis. IEEE Trans. Geosci. Rem. Sens. **57**(12), 9832–9848 (2019)

Multilevel User Verification in Cloud Banking System

Ankur Biswas and Abhishek Roy

Abstract The recent incidents of sky high fraudulent banking transactions, ponzi monetary schemes and its corresponding corruption charges have shaken the economic backbone of India. Its impact is so high that Government has to opt for legal remedies to bring back the imposters under justice. This scenario clearly depicts that existing banking system is incapable to handle and crack down these fraudulent activities and hence it requires extensive research work to prevent similar incidents in near future. Information and communication technology (ICT)-based electronic message communication system may be utilized under controlled environment to monitor and maintain these financial transaction from its initial stage. Since cyber space is an open forum, which is easily accessible to all (including the adversaries), multilevel verification of user will help to reduce risk factors of banking transactions to significant level, if not eliminate it completely. This monitoring system will be more cohesive in nature if a single window-based integrated environment streamlines all the parties engaged in a specific banking transaction. To achieve this objective, in this paper, authors have proposed a Multipurpose Electronic Card (MEC)-based Cloud Banking System (CBS) as a part of broader Integrated Electronic Service Delivery System (IESDS), which will verify its user (i.e. Client) through multilevel architecture.

Keywords Cloud banking · User verification · Multilevel architecture

A. Biswas (✉) · A. Roy
Department of Computer Science & Engineering, Adamas University, Kolkata, West Bengal, India
e-mail: ankur2u@gmail.com

A. Roy
e-mail: dr.aroy@yahoo.com

V. E. Balas et al. (eds.), *Proceedings of International Conference on Computational Intelligence, Data Science and Cloud Computing*, Lecture Notes on Data Engineering and Communications Technologies 62, https://doi.org/10.1007/978-981-33-4968-1_41

1 Introduction

Recently, India is witnessing rapid growth of fraud banking transactions in various forms like unauthorized ATM transactions by adversaries, bank loan approvals for fraudsters, money laundering due to ponzi monetary schemes, etc., which have shaken the economic backbone of our country and caused untimely death of many victims. In most of these cases, adversary manages to conceal its activities due to its high level contacts within the system and often flew out of country to avoid legal consequences. Its impact is so high that Government has to opt for legal remedies to bring back the imposters under justice. The entire scenario clearly reveals that banking system is facing severe challenge due to its internal and external adversaries. To be precise, existing banking system is insufficient to control these fraud transactions, which calls for thorough research work to neutralize those illicit banking transactions from its initial stage. To achieve this objective, in the following sections, a multilevel user verification scheme (using single window interface, namely Multipurpose Electronic Card) has been proposed for Citizen centric Cloud Banking System (CBS).

Section 2 mentions the origin of our research work. Section 3 explains our proposed Cloud Banking System (CBS) during Citizen to Government to Bank to Citizen (C2G2B2C) type of transaction. Section 4 mentions the conclusion of our present work and explores its future possibilities.

2 Origin of Work

Information and communication technology (ICT) plays a vital role to deliver multifaceted electronic services to Citizen in cost effective and timely manner. As Internet is an open cyber space, which is equally accessible to all (including the adversaries), installation of proper security protocol will ensure detection and subsequent neutralization of illicit attempts at its initial stage. As users communicate with each other through virtual mode during electronic communication, strict verification of user's digital identity will provide significant advantage to any electronic service delivery model. Furthermore, a single window interface can be proposed to bring compactness of all Citizen centric electronic services. Though critics may raise concern about privacy issues of Citizen due to the involvement of Government during electronic transactions, as a victim of terrorism (in all possible form), we can not afford unbounded liberty of Citizen. Furthermore, the involvement of Government during these electronic transactions will also help Citizen to seek prompt assistance during any adverse situation. The block diagram of our proposed Integrated Electronic Service Delivery system (IESDS) [1–3] is shown through Fig. 1, which is also explained below:

1. Primary players of electronic transaction:
    (a) **Citizen** (i.e. Client or Service Seeker).
    (b) **Government** (i.e. State and its agencies).
    (c) **Service Provider** (e.g. **Bank** [5, 7, 8, 12, 14, 15, 19, 21], University, Hospital, Employer, Transport Services, etc.).
2. Phases of an electronic transaction:
    (a) Citizen to Government (C2G): During this phase, Citizen uses Multipurpose Electronic Card (MEC) to send **SERVICE REQUEST** (using Path—1 of Fig. 1) to Government through Public Kiosk (i.e. Public Cloud).
    (b) Government to Service Provider (G2S): During this phase, Government verifies the identity [18] of Citizen, and only in case of successful verification, **SERVICE REQUEST** is processed towards appropriate Service Provider. We have also considered privacy issues of Citizen judiciously. For this reason, Service Provider will send its **SERVICE RESPONSE** through Transaction Master, which is not under the direct control of Government.
    (c) Service Provider to Citizen (S2C): During this phase, Service Provider receives **SERVICE REQUEST** of Citizen through Government and re-verify its identity. In case of successful verification, Service Provider responds with **SERVICE RESPONSE** to Citizen (i.e. Client) using Path—2 of Fig. 1.

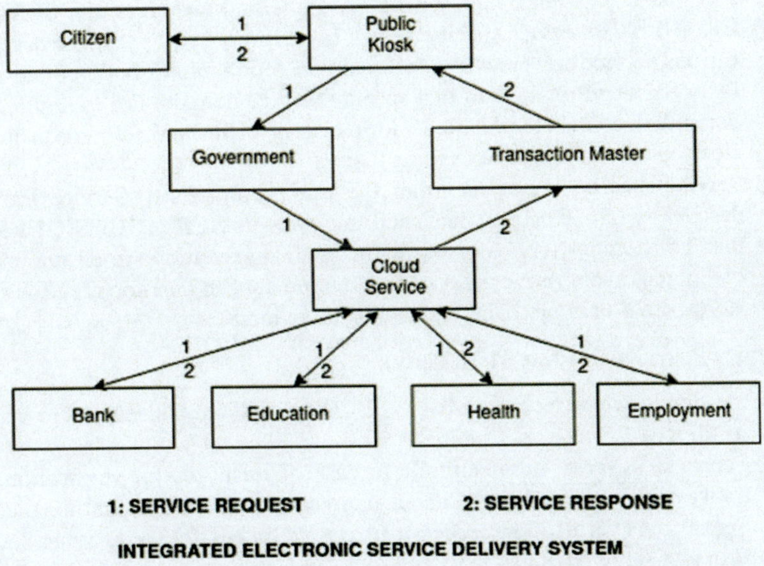

Fig. 1 Block diagram of Integrated Electronic Service Delivery System (IESDS)

As our objective is to ensure cost effective and prompt electronic service [9, 13, 20] delivery to doorstep of end user (i.e. Citizen), we have shown our conceptual diagram through Fig. 2 during Citizen to Government to Service Provider to Citizen (C2G2S2C) type of transaction, which is also described below:

1. Citizen side:
   (a) Citizen initiates electronic transaction using Multipurpose Electronic Card (MEC).
   (b) Citizen transmits its unique parameter and **SERVICE REQUEST** using Path—1 of Fig. 2 to Government through Public Kiosk (i.e. Public Cloud). The primary objective of Citizen is to avail **SERVICE RESPONSE** after successful verification of its digital identity.
2. Government side:
   (a) Cloud Governance (C-Governance) Interface receives unique parameter and **SERVICE REQUEST** of Citizen through its Firewall.
   (b) Government verifies the identity of Citizen.
   Scenario 1 (Verification Failure): In this case, transaction is aborted, and Citizen is informed through system time out.
   Scenario 2 (Verification Success): In this case, transaction proceeds towards Cloud Service (C-Service) Server for further identification of the electronic service requested by Citizen (i.e. Client).
   (c) Cloud Service (C-Service) Server performs **READ** and **WRITE** operations over the Data Centre attached its Private Cloud and transmits **SERVICE REQUEST** to service specific server. In our proposed system, we have used various service servers which will contain detailed information of all Third Party Service Providers of that specific service domain. For example, Bank Server will contain the detailed information of all the banks eligible to provide electronic banking facilities to its Client.
   (d) Service-specific server identifies the exact Third Party Service Provider desired by Citizen (i.e. Client) and transmits its **SERVICE REQUEST** to that Third Party Service Provider for its final execution. To ensure privacy of Citizen, in our proposed system, Government can monitor any **SERVICE REQUEST** only up to this phase of transaction.
3. Service Provider side (i.e. Third Party):
   (a) Service Provider receives **SERVICE REQUEST** of Citizen and re-verifies its identity.
   Scenario 3 (Verification Failure): In case of unsuccessful verification, the transaction is aborted, and Citizen is informed through system time out.
   Scenario 4 (Verification Success): In case of successful verification, Service Provider sends **SERVICE RESPONSE** to Citizen.
4. Citizen side:
   (a) Citizen access **SERVICE RESPONSE** using Path—2 of Fig. 2.

In Sect. 3, we have discussed our Cloud Banking System (CBS) using this proposed Citizen centric Multipurpose Electronic Card (MEC).

3 Proposed Cloud Banking System (CBS)

Figure 3 shows the schematic diagram of our Citizen centric Cloud Banking System (CBS) which have been offered through single window interface, i.e. Multipurpose Electronic Card (MEC). Throughout our approach, we have focused on broader Integrated Electronic Service Delivery System (IESDS), and Cloud Banking System (CBS) is a part of that system which will deliver banking facilities to Citizen through electronic medium. In order to help Government to identify illicit transactions of adversaries, we have shown multilevel user authentication [4, 6, 10, 11, 16, 17] scheme during Citizen to Government to Bank to Citizen (C2G2B2C) type of Cloud Banking transaction, which is described below:

1. Phase 1: It denotes the Cloud Governance transaction, where Government verifies the identity of Citizen (i.e. Client) and transmits **SERVICE REQUEST** to Phase 2 of transaction.
2. Phase 2: It denotes the banking transaction of Citizen (i.e. Client), which is assessed within the Cloud Governance system to identify the exact Third Party Service Provider (i.e. Bank). After identification of the exact service provider (i.e. Bank), **SERVICE REQUEST** of Citizen is transmitted to Phase 3 of transaction.
3. Phase 3: It denotes the final phase of transaction where Third Party Service Provider (i.e. Bank) interacts with Citizen to deliver the **SERVICE RESPONSE** through Public Kiosk (i.e. Public Cloud). The Transaction Master shown in Fig. 1 maintains the log of all transactions carried out by Citizen using **Transaction In Log** and **Transaction Out Log**, respectively.

The algorithm of our Cloud Banking System (CBS) is explained in Sect. 3.1.

3.1 Algorithm

The primary players of our proposed system are Client (i.e. Citizen), Government (i.e. State and its agencies) and Bank (i.e. Third Party Service Provider) which are connected through Public Kiosk (i.e. Public Cloud). The step-wise explanation of our Cloud Banking System (CBS) is mentioned below:

CLIENT Side:
Step 1: Client uses Multipurpose Electronic Card (MEC) to initiate banking transaction through Public Kiosk (i.e. Public Cloud).

PUBLIC KIOSK Side:
Step 2: As in this paper we are focusing on Cloud Banking System (CBS), we

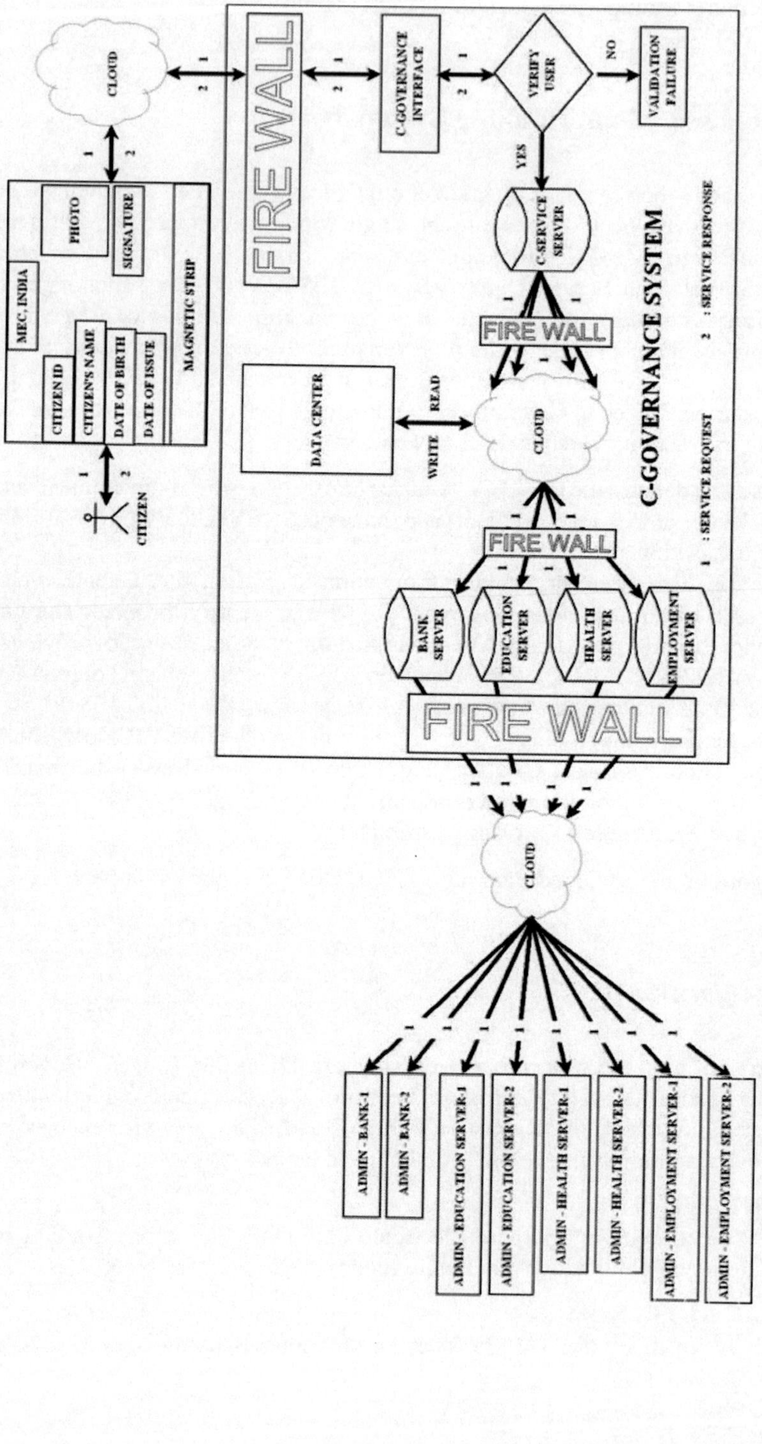

Fig. 2 Conceptual diagram of Integrated Electronic Service Delivery System (IESDS)

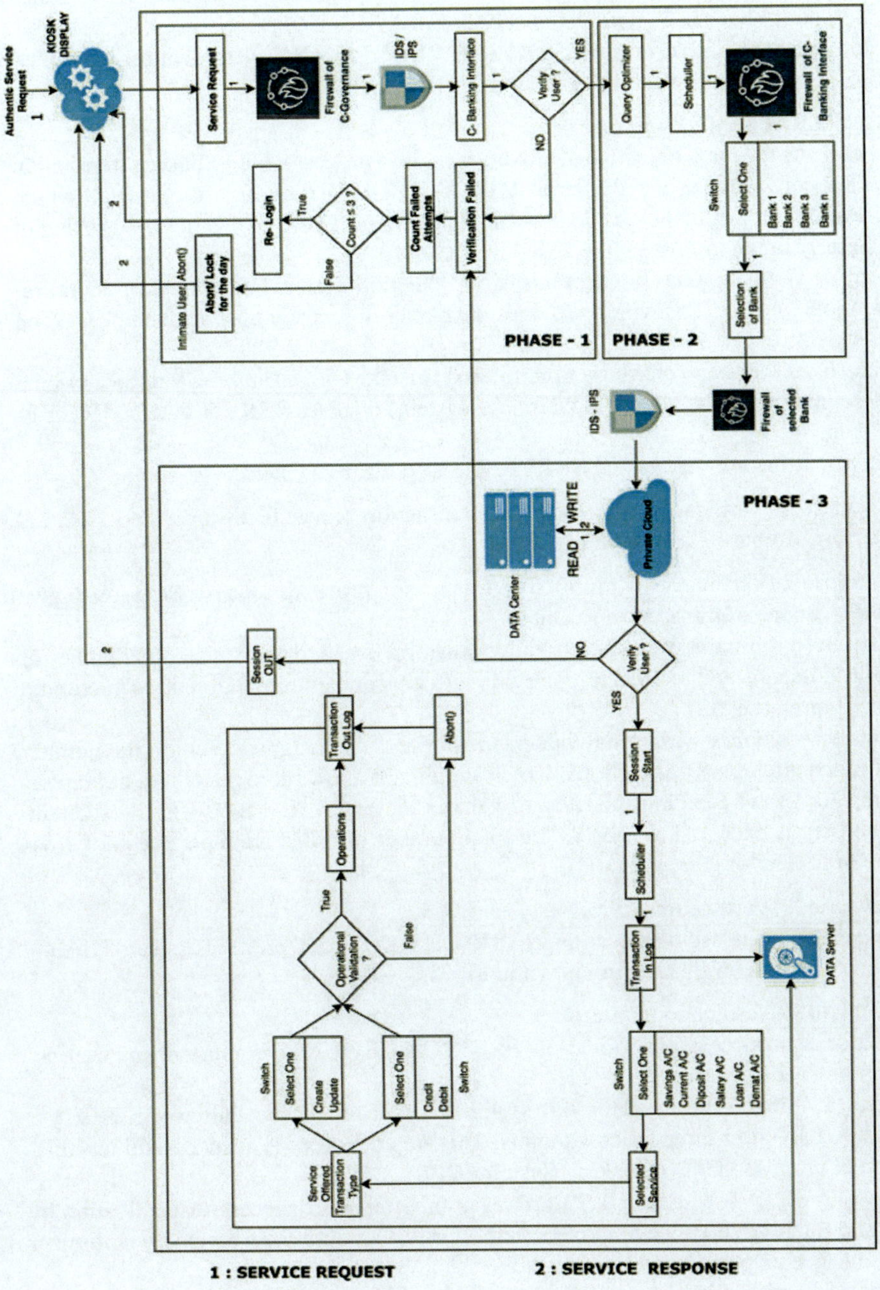

Fig. 3 Proposed Cloud Banking System (CBS)

assume that preliminary verification of Citizen using Username, Password and One Time Password (OTP), etc. (which are actually the part of core Cloud Governance transaction) have been completed successfully.
Step 3: Public Kiosk transmit **SERVICE REQUEST** to Government for identification of bank mentioned in the **SERVICE REQUEST**.

GOVERNMENT Side:

Step 4: As the main objective of Government is to process Cloud Banking transaction in a secure manner, it receives **SERVICE REQUEST** through its Firewall which prevents entry of malicious code within the system. This step denotes the starting of Phase 1 shown in Fig. 3.
Step 5: Then, it passes through Intrusion Detection System (IDS) and Intrusion Prevention System (IPS) which provides next layer of security to sensitive information of Client.
Step 6: At this stage of transaction, for first time the Cloud Banking Interface (which is mainly the part of Cloud Governance system) receives **SERVICE REQUEST** of Client.
Step 7: It further verifies the **SERVICE REQUEST** of Client.

Scenario 1 (Verification Success): The transaction proceeds towards Step—12, i.e. starting of Phase 2 as shown in Fig. 3.

Scenario 2 (Verification Failure): The transaction proceeds towards Step—8 to give further scope of transaction to Client.
Step 8: In this case, Verification Failed function is called to monitor the number of failed attempts and redirect it to provide further scope of transaction up to maximum count three (i.e. 03).
Step 9: A counter with initial value as zero (i.e. 0) is assigned to count the number of failed attempts / transactions, which should not allow more than three attempts.
Step 10: Cloud Banking Interface re-verifies the **SERVICE REQUEST** of Client. This step is executed as long as the total number of failed attempts does not cross the limit of count three (i.e. 03).

Scenario 3 (Verification Success):
In case of successful verification, **SERVICE REQUEST** proceeds towards Step—12, i.e. starting of Phase 2 as shown in Fig. 3.

Scenario 4 (Verification Failure):
In case of unsuccessful verification, **SERVICE REQUEST** is aborted, and transaction proceeds towards Step—11.
Step 11: Client is informed through system timeout about cancellation of **SERVICE REQUEST** after three failed attempts. This step denotes the unsuccessful termination of proposed Cloud Banking System (CBS).
Step 12: **SERVICE REQUEST** arrives at this step after successful verification by Cloud Banking Interface of Government. It is further optimized by Query Optimizer to enhance its performance.
Step 13: It proceeds through Scheduler for faster channelization compared to simultaneous **SERVICE REQUEST** of other Clients.

Step 14: As the primary objective of Government is to facilitate secured Cloud Banking transaction, it receives **SERVICE REQUEST** through Firewall of Cloud Banking Interface to forestall entry of malicious codes of adversary.

BANK Side:

Step 15: At this stage of transaction, **SERVICE REQUEST** identifies its specific Bank for execution at the Third Party Service Provider side.

Step 16: **SERVICE REQUEST** passes through Firewall of the selected Bank.

Step 17: To provide additional layer of security to **SERVICE REQUEST**, it further passes through Intrusion Detection System (IDS) and Intrusion Prevention System (IPS).

Step 18: As the starting of Phase 3 as shown in Fig. 3, Private Cloud of Bank receives **SERVICE REQUEST** of Client. During this phase, the Data Centre attached to Private Cloud of Bank performs **READ** and **WRITE** operations to record all banking transactions.

Step 19: Bank further verifies the identity of its Client.

Scenario 5 (Verification Success): In case of successful verification, transaction proceeds towards Step—20.

Scenario 6 (Verification Failure): In case of unsuccessful verification, transaction goes back to Step—8 to provide further scope of transaction to Client up to maximum count three (i.e. 03).

Step 20: Bank starts session to process **SERVICE REQUEST**. This step denotes the starting point of final execution of banking transaction after satisfying all security protocols of Government and Bank, respectively.

Step 21: It passes through scheduler to speed up the desired transaction.

Step 22: **Transaction In Log** is generated and stored in the Data Server.

Step 23: After generating the Transaction Log, Bank displays type of bank accounts available for the Client.

Step 24: At this stage, Client selects specific banking service from the list of available services.

Step 25: After selection of type of transaction, **SERVICE REQUEST** goes through its operational validation.

Scenario 7 (Status: SUCCESS): In case of successful validation, **SERVICE REQUEST** proceeds towards Step—26 of banking transaction.

Scenario 8 (Status: FAILURE): In case of unsuccessful validation, **SERVICE REQUEST** initiates *abort( )* function and proceeds towards Step—26 of banking transaction.

Step 26: To maintain transaction details (i.e. for both successful and unsuccessful attempts), **Transaction Out Log** is generated and stored in Data Server of selected Bank.

Step 27: BANK terminate the Session and sends **SERVICE RESPONSE** through Public Kiosk to Client.

PUBLIC KIOSK Side:

Step 28: Public Kiosk transmit **SERVICE RESPONSE** to Client.

CLIENT Side:

Step 29: Client accesses **SERVICE RESPONSE** through Public Kiosk. This step denotes successful termination of proposed Cloud Banking System (CBS).

4 Conclusion

In this paper, we have shown Cloud Banking System (CBS) within Integrated Electronic Service Delivery System (IESDS) during Citizen to Government to Bank to Citizen (C2G2BC) type of banking transaction. We have judiciously balanced between privacy concern of an individual and the necessity for state sponsored surveillance for greater benefit of society. Hence, we have sent **SERVICE RESPONSE** through Transaction Master, which is not only beyond the direct control of Government but also records all successful and unsuccessful (i.e. illicit attempts of adversaries) banking transactions of user through **Transaction In Log** and **Transaction Out Log**, respectively. For this reason, we will further explore the features of **Transaction Master** in our future work. Considering the security aspect of our nation, neither we can afford unbounded liberty of user nor we should allow absolute surveillance over user's transaction. Our objective is to provide a secured single window interface for delivery of multivariate electronic services to user and at the same time facilitate Government to respond promptly during adverse situation, so that adversaries does not get any scope to fulfil their ill intentions. However, in future, we have to explore in details all the security protocols of our proposed Cloud Banking System (CBS) to gain trust of user.

References

1. A. Roy, Object-oriented modeling of multifaceted service delivery system using connected governance, in *Automated Software Testing. ICDCIT 2019. Services and Business Process Reengineering*, ed. by A. Jena, H. Das, D. Mohapatra (Springer, Singapore, 2020), pp. 1–25. https://doi.org/10.1007/978-981-15-2455-4_1
2. A. Roy, Smart delivery of multifaceted services through connected governance, in *3rd International Conference on Computing Methodologies and Communication (ICCMC)* (IEEE, India, 2019), pp. 476–482. https://doi.org/10.1109/ICCMC.2019.8819851
3. S. Biswas, A. Roy, An intrusion detection system based secured electronic service delivery model, in *2019 3rd International Conference on Electronics, Communication and Aerospace Technology (ICECA)*, Coimbatore, India, pp. 1316–1321 (2019). https://doi.org/10.1109/ICECA.2019.8822016
4. A. Biswas, A. Roy, A study on dynamic ID based user authentication system using smart card. Asian J. Convergence Technol. (AJCT) **5**, 2 (2019). http://www.asianssr.org/index.php/ajct/article/view/871 Retrieved on 08/03/2020. ISSN 2350-1146
5. A. Mahalle, J. Yong, X. Tao, Protecting privacy in digital era on cloud architecture for banking and financial services industry, in *2019 6th International Conference on Behavioral, Economic and Socio-Cultural Computing (BESC)*, Beijing, China, pp. 1–6 (2019). https://doi.org/10.1109/BESC48373.2019.8963459

6. I. Gordin, A. Graur, A. Potorac, Two-factor authentication framework for private cloud, in *2019 23rd International Conference on System Theory, Control and Computing (ICSTCC)*, Sinaia, Romania, pp. 255–259 (2019). https://doi.org/10.1109/ICSTCC.2019.8885460
7. A. Mahalle, J. Yong, X. Tao, Ethics of IT security team for cloud architecture infrastructure in banking and financial services industry, in *2019 IEEE 23rd International Conference on Computer Supported Cooperative Work in Design (CSCWD)*, Porto, Portugal, pp. 506–511 (2019). https://doi.org/10.1109/CSCWD.2019.8791928
8. M. Singh, K.S. Tanwar, V.M. Srivastava, Cloud computing adoption challenges in the banking industry, in *2018 International Conference on Advances in Big Data, Computing and Data Communication Systems (icABCD)*, Durban, pp. 1–5 (2018). https://doi.org/10.1109/ICABCD.2018.8465412
9. X. Liao, S. Alrwais, K. Yuan et al., Cloud repository as a malicious service: challenge, identification and implication, in *Cybersecurity*, vol. 1, 14 (2018). https://doi.org/10.1186/s42400-018-0015-6
10. J. Gowthami, N. Shanthi, N. Krishnamoorthy, Secure three-factor remote user authentication for E-Governance of smart cities, in *2018 IEEE International Conference on Current Trends Toward Converging Technologies* (IEEE, India, 2018), pp. 1–8. https://doi.org/10.1109/icctct.2018.8551172
11. S. Dhal, V. Bhuwan, Cryptanalysis and improvement of a cloud based login and authentication protocol, in *2018 4th International Conference on Recent Advances in Information Technology (RAIT)* (IEEE, Dhanbad, India, 2018). https://doi.org/10.1109/RAIT.2018.8388988
12. S.P. Tripathi, A. Kumar, R. Astya, Study on secured framework for cloud based online banking, in *2017 International Conference on Computing, Communication and Automation (ICCCA)*, Greater Noida, pp. 853–858 (2017). https://doi.org/10.1109/CCAA.2017.8229915
13. N. Srilatha, G. Mrali, M. Deepthi, A framework to improve E-seva services through E-governance by using DNA cryptography, in *2017 International Conference on Algorithms, Methodology, Models and Applications in Emerging Technologies (ICAMMAET)* (IEEE, India, 2017), pp. 1–4. https://doi.org/10.1109/ICAMMAET.2017.8186748
14. S. Nagaraju, L. Parthiban, Trusted framework for online banking in public cloud using multi-factor authentication and privacy protection gateway. J. Cloud Comp. **4**, 22 (2015). https://doi.org/10.1186/s13677-015-0046-4
15. J. Park, Y. An, K. Yeom, Virtual cloud bank: an architectural approach for intermediating cloud services, in *2015 IEEE/ACIS 16th International Conference on Software Engineering, Artificial Intelligence, Networking and Parallel/Distributed Computing (SNPD)*, Takamatsu, pp. 1–6 (2015). https://doi.org/10.1109/SNPD.2015.7176235
16. N.E. Madhoun, F. Guenane, G. Pujolle, A cloud-based secure authentication protocol for contactless-NFC payment, in *2015 IEEE 4th International Conference on Cloud Networking (CloudNet)*, Niagara Falls, ON, pp. 328–330 (2015). https://doi.org/10.1109/CloudNet.2015.7335332
17. S.K.S.V.A. Kavuri, G.R. Kancherla, B.R. Bobba, Data authentication and integrity verification techniques for trusted/untrusted cloud servers, in *2014 International Conference on Advances in Computing, Communications and Informatics (ICACCI)*, New Delhi, pp. 2590–2596 (2014). https://doi.org/10.1109/ICACCI.2014.6968657
18. U. Habiba, R. Masood, M. Shibli, M. Niazi, Cloud identity management security issues and solutions: a taxonomy, in *Complex Adaptive Systems Modeling*, vol. 2, 5 (2014). https://doi.org/10.1186/s40294-014-0005-9
19. I. Sangavarapu, S. Mishra, A. Williams, G.R. Gangadharan, The Indian banking community cloud, in *IT Professional*, vol. 16, 6, pp. 25–32 (2014). https://doi.org/10.1109/MITP.2014.97
20. F. Zhu, H. Li, J. Lu, A service level agreement framework of cloud computing based on the Cloud Bank model, in *2012 IEEE International Conference on Computer Science and Automation Engineering (CSAE)*, Zhangjiajie, pp. 255–259 (2012). https://doi.org/10.1109/CSAE.2012.6272592
21. J. Jiang, D. Yang, A research on commercial bank information systems based on cloud computing, in *2011 IEEE 3rd International Conference on Communication Software and Networks*, Xi'an, pp. 363–366 (2011). https://doi.org/10.1109/ICCSN.2011.6014585

Analytical Study of Security Enhancement Methods on Diverse Cloud Computing Platforms

Moumita Roy and Monisha Singh

Abstract Cloud storage is a convenient and virtually limitless storage option for the bulk of data technology is producing in recent times. Data security in cloud is not so robust as data owners need to depend upon the service providers for the safe storage. In this paper, we have identified few broadly used cloud computing paradigms: mobile cloud, cloud-based IoT and multi-tenant cloud. Mobile cloud helps reduce the data storage overhead on the mobile device and give users access to their personal data as and when required through cloud access. Cloud-based IoT helps the network of IoT devices, which is growing exponentially, to create on-demand cloud repositories. Multi-tenant cloud platforms are cloud environment accessed by more than one user. Few recent and related research work which aims at enhanced security from all these three paradigms is discussed and analysed. Encryption and similar network securing methods are used for mobile cloud and cloud-based IoT. For multi-tenant cloud, the objective is to keep the user spaces separate to keep their resources confidential.

Keywords Cloud computing · Data security · Mobile cloud · IoT · Multi-tenant cloud

1 Introduction

The growth in use of technology in our daily lives is creating a huge bulk of data every day. Efficient and safe storage of all these data is a concern. Cloud storage service can be one convenient option. It provides worldwide accessibility, benefits of data replication and easy scalability at a much lesser operating cost than the traditional data storage mechanisms.

M. Roy (✉) · M. Singh
Computer Science Department, Christ (Deemed to be) University, Bengaluru, India
e-mail: moumita.majumder@gmail.com

Department of Computer Applications, CMR Institute of Technology, Bengaluru, India

M. Singh
e-mail: monisha.singh@christuniversity.in

V. E. Balas et al. (eds.), *Proceedings of International Conference on Computational Intelligence, Data Science and Cloud Computing*, Lecture Notes on Data Engineering and Communications Technologies 62, https://doi.org/10.1007/978-981-33-4968-1_42

Along with these benefits, cloud provides a diverse platform for usage. Cloud environment can be classified into three broad groups based on user access—the public cloud, the private cloud and the hybrid cloud. The ownership of the environment of public cloud lies with a third-party service provider and is accessible by users across the globe. The available infrastructure of such platforms can be simultaneously used by more than one user. In contrast, a group or an individual can own and manage a private cloud platform which is accessible to a group of users with special privileges. A hybrid cloud platform combines these two ideas together. The delivery model of cloud services can again be categorized into various types, and the common ones are SaaS, PaaS and IaaS. The ways in which the cloud environments are being used presently are vast. Some users use the platform as mobile cloud for the benefit of quick on-the-go accessibility. Some use cloud environment along with a bunch of IoT devices to make their lives smarter. Many multi-tenant platforms allow a group of users to access same virtual system by sharing a common cloud environment.

Cloud storage is being vastly used by individual users and organizations to manage the growth in data volume. But in a cloud platform, the owner of the data has no control on the data usage. They do not know if the data is safe from unauthorized access. To overcome this hurdle, the cloud service providers require to solve this data access control and trust issue. [1]. Extensive research is happening for quite some time on cloud data privacy protection using various encryption mechanisms and access control mechanisms [2–5]. Still, all the mechanisms that are in place currently are not flexible enough and not scalable beyond a certain level. The process to make a secure cloud has just started and needs to go a long way.

This paper aims at accumulating all the findings of reviews done involving multiple cloud platforms together and builds a stronger pathway towards protecting the data privacy in cloud environment and implement a dynamic and secure environment [6].

2 Cloud Computing Paradigms

The cloud computing paradigms discussed in this paper are the cloud storage accessed by mobile users, interact with IoT devices and the multi-tenant cloud storage which are used by more than one user.

2.1 Mobile Cloud Computing

This cloud computing paradigm combines cloud infrastructures with the mobile telecommunication environment. Users with a smart mobile handset can make use of services available on cloud to its complete strength [7, 8]. The network used to connect the platforms is mostly wireless network. Mobile cloud computing enables its users to access a plethora of applications (mail, e-commerce, chatting tools, etc.) and resources (files/documents, photographs and videos) stored on the cloud and

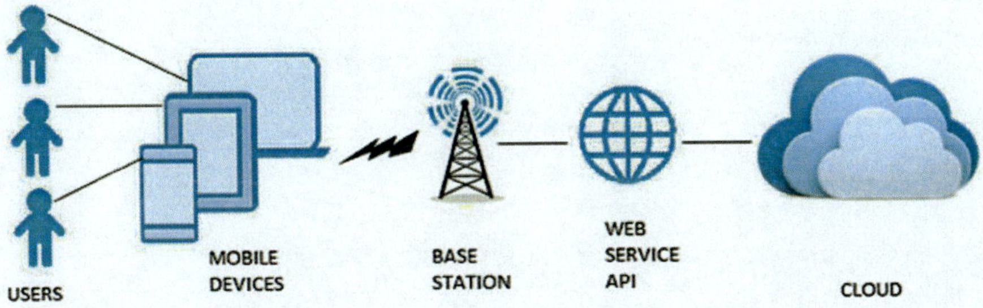

Fig. 1 Mobile Cloud Architecture

also fetch result of computations deployed on a cloud environment through their mobile handset [9 10]. Also, confidential and sensitive information (passwords/pins, information related to bank accounts, medical data) can be a part of mobile cloud storage. Advantage of using this paradigm is that the mobile device does not need to store bulk data or have a high-speed CPU for computations that can happen on cloud. The devices can be lighter in configuration and still give a good throughput. The responsibility of storage can be shifted to the centralized cloud which can be accessed as and when required using the wireless network [11]. Mobile handsets are personal devices, and due to the storage of confidential and sensitive information, securing such data stored on the cloud is a big concern. In this paper, we will discuss and analyse few methods which address the security concerns of a mobile cloud (Fig. 1).

2.2 *Cloud Storage for IoT Devices*

Internet of things (IoT) is a composite system made up of physical interconnected computing and digital machines and live individuals that are identified using unique identifiers (UIDs) that can transfer data and communicate over a network without any manual interaction. The network consists of physical objects like portable/handheld devices, vehicles, buildings, human beings, etc., that are implanted with sensors and software interfaces and connected over a network allowing these objects to collect and share data [12, 13]. The number of devices or objects joining the IoT network is increasing at an exponential rate. The count is estimated to reach 50 billion in 2020 as shown in Fig. 2 [14].

Along with the network of devices, the amount of data being created is also growing. To manage this enormous bulk of information created and improvement of the technology, cloud platforms are being used [15]. So, collaborating with a cloud platform, which has virtually limitless resources, can enhance the benefits of an IoT network by expanding its scope in real-life applications [16]. In certain cases, the cloud can form a layer connecting the top application layer and IoT at the bottom, concealing the network and functional complexities underneath. Cloud and IoT can

Fig. 2 Rate of growth of IoT devices

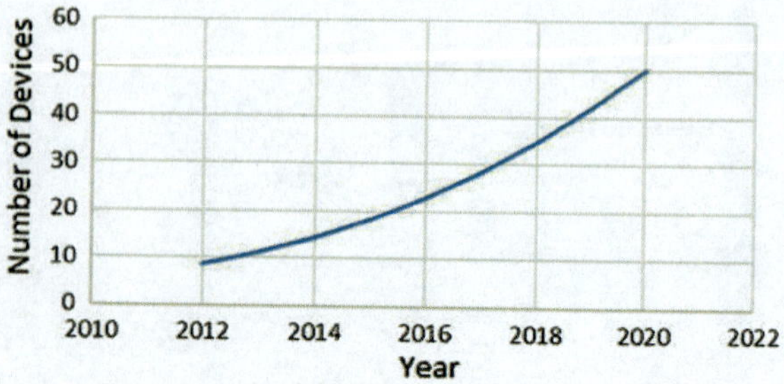

work hand in hand in real life scenarios like health care, smart cities/homes, video surveillance, smart automobiles, environmental monitoring, smart grids, etc. [17].

2.3 *Multi-Tenant Cloud*

A cloud with multi-tenant capabilities uses an architecture that enables users to store information and resources over a public or private cloud environment. Users of such cloud platforms are called tenants. The data of each tenant is kept separate and obscured from other tenants. This multi-tenant platform creation is possible through the concept of virtualization, and this enhances the scalability and extensibility of cloud platforms [18]. Thinking from economy perspective, a multi-tenant platform makes more sense to the service provider. Better utilization of resources, less infrastructural costs and less effort for maintenance are the benefits. At the same time, multi-tenant platform has its own challenges like customization (serving different requirements of each tenant) and scalability (huge number of tenants asking for the same resource) [19]. Figure 3 shows a pictorial view of single and multi-tenant platforms.

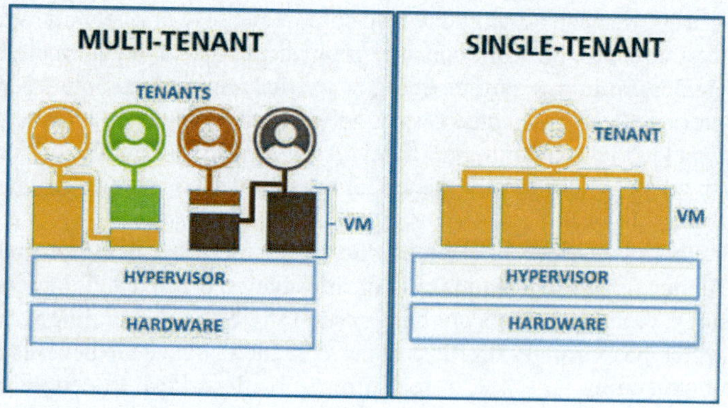

Fig. 3 Multi-tenant cloud versus single tenant cloud

3 Analytical Study

A lot of work has already happened to enhance the security of the three paradigms that we have chosen for this paper. In this section, we will discuss some of those paradigms and compare them to get an insight.

3.1 *Security Enhancement of Mobile Cloud Environment*

Devices like mobile handsets can make use of cloud storage for storing and retrieving personal data, as we have already discussed. Also, tasks involving computation can be done using cloud platforms to save mobile resources and battery [20]. Though there are so many advantages, mobile cloud computing has one serious drawback and that is security [21]. Data that is sent to cloud from a mobile app is not encrypted, and so it is extremely vulnerable.

Banerjee et al. [22] proposed CLOAK protocol which uses encryption and decryption methods that work on mobile cloud environments enhancing its security. The CLOAK encryption protocol is based on stream cipher. When data transfer is needed from one mobile device to another, the CLOAK protocol uses an external server, and a key value (CSPRN) is generated which is used to perform the encryption. Linux server, Android mobile device and AWS cloud platforms are used in this paper. The proposed protocol works well on attacks like Brute-force, Impersonation and Man-in-the-middle.

Weng and Lui [23] proposed an algorithm iForestFS for anomaly detection in mobile services. Statistical and machine learning-based concepts were used on multi-dimensional time series data collected from UCI University Machine Language Library. Three datasets (Http, SMTP and Shuttle) were collected and divided into seven different anomaly classes. The algorithm was written in Java and WEKA software is used to implement machine learning algorithms. The results of the experimental show that the algorithm successfully identifies multiple types of anomalies in mobile cloud. The proposed algorithm is also capable of identifying collective anomaly on input stream that is multidimensional, but no comparison with existing processes was done.

Elminaam et al. [24] designed a framework SMCACC that reduces the computational stress on a mobile device by implementing the concept of offloading. This framework divides a mobile application into set of methods, and the execution is shared with the mobile devices and a mobile cloud server. This method considerably reduces demands for resources like memory storage requirements, required CPU time and consumption of power in a mobile device. The experiment was done on four mobile applications, and results show average of 40% reduction in mobile device memory usage. This elastic platform also uses hybrid cryptography to improve data security on mobile cloud. The implementation of the framework along with the cryptography increases the application processing time maximum by one second.

Li et al. [25] proposed a three-tier mobile cloud platform model TALMSC that aims at trust enhancement, improved efficiency and user satisfaction. The model has three components, the JADE main container, pool of cloud resources and the transaction agents. The tree tiers or the transaction agents include mobile cloud customers, intermediators and the providers of service. Learning ability and scheduling of services are evaluated using the JADE architecture. NetLogo environment and fuzzy methods are used to evaluate the trust mechanisms. The experimental results show that when the service providers are not sure of the customer's preference of service, learning ability is of utmost importance to increase customer satisfaction. The authors also identify areas for future scope as integrating trust with other modules and adjusting market strategies based on learning abilities.

Akter et al. [26] performed quantitative and qualitative analysis on cloud storage service performance in the domain of health care. The cloud storage platforms addressed in this paper are Dropbox, Google Drive, Box and One Drive. Network packet analyser tool Wireshark is used to analyse Internet traffic packets. Seven JPEG files containing medical data are used whose size ranges from 1 MB to 1 GB. Performance evaluation of all the four storage platforms shows that Dropbox performs better than other three on the seven data files. The future scope focuses on the objective of creating a new mobile cloud service platform putting together all the positive capabilities identified in four different platforms.

Table 1 shows the comparative analysis of the experiments we discussed which aims at increased security and workability in a mobile cloud platform.

3.2 *Security Enhancement of IoT—Cloud Environment*

Cloud supported IoT is a very frequently used solution for creating smart systems. The combination of IoT and cloud can create flexible systems with improved performance at reduced cost [27]. Security is considered a major concern for this collaboration and creation of systems with increased efficiency [28]. Security lapses are frequent, and data is vulnerable during (i) transmission from cloud (ii) processing (iii) data storage in cloud [29].

Wang et al. [30] proposed a mechanism for measuring trust and reducing security related problems and also improving efficiency which is challenged by constant redundant service needs in IoT-cloud systems. The architecture uses an edge network that assists the IoT in evaluating trust by maintaining a list of trust values for each device in the network. For improved efficiency, services are segregated as local and remote services. Cloud and edge network take care of remote and local services respectively, using individual templates. The platform for simulation used is MATLAB R2016b. The network used consists of a cloud surrounded by four IoT networks of 56 devices each. Experimental output shows that the proposed architecture can improve cloud–IoT trust factor and efficiency a lot. The algorithm can be used for scheduling and pricing of IoT services in future.

Table 1 Comparative analysis for mobile cloud computing

Article references	Comparison criteria for mobile cloud platform		
	Method	Dataset and tool	Conclusion
[22]	Encryption	Android mobile, AWS Cloud	Prevents Brute-force, Impersonation and Man-in-the-middle attacks
[23]	Machine learning	Http, SMTP and shuttle datasets, WEKA	Detects network intrusion
[24]	Computation offloading, hybrid cryptography	Finger-print system, GPS Calculation system, Address Book system, Car Store system	Delivers 40% reduced usage of mobile device memory and also enhances mobile cloud security but processing time is increased by 1 s
[25]	Fuzzy comprehensive evaluation	NetLogo and JADE	Proves that learning ability of service providers enhance customer satisfaction
[26]	Quantitative and qualitative analysis	Medical report image files, Dropbox, Google Drive, Box, One Drive, WireShark	Analysis shows a comparatively better performance for Dropbox

Shen et al. [31] conceptualized an algorithm that performs phrase search and also protects the privacy of encrypted documents stored in an IoT system based on cloud. The algorithm generates secure index based on keywords and then encrypts the document. Based on user's query, a trapdoor is generated, and phrase searching works using the secure index. Alibaba cloud servers and Intel Xeon processors are used on randomly picked 2500 files from RFC dataset. Experimental results and comparisons show that the algorithm is effective but has two limitations: time taken in phrase recognition is more compared to conjunctive search and it does not support index update.

Li et al. [13] proposed a framework STRAFF that measures trust, based on reputation and security of a cloud–IoT system. For security-based assessment, several security metrics are defined, quantified for each cloud service provider and evaluated against the ideal solution. The reputation-based assessment is done in line with the feedbacks provided by the cloud service customers based on two weight factors: credibility and certainty. MATLAB R2017b and a DELL computer system with CPU with Intel i52.7 GHz, 8GBn RAM and operating system Windows 10 was used for the experiments. A real-world data set from GitHub, comprising 4500 Web services collected from 142 users across 64 different time slots, was considered. Based on attributes throughput and response time, two smaller datasets are created from the existing large pool by random selection of six services and taking feedback ratings for 100 users across ten different time slots. Results show that the framework efficiently enhances security and measures reputation of a Cloud–IoT system. A working prototype of the proposed framework is yet to be implemented.

Choi and Choi [32] designed an ontology-based service framework for providing security to power IoT–cloud environment. The proposed framework is a four-layered structure where the layers in top-down manner are application, IoT, security context inference and cloud computing. The IoT layer collects information related to device security and passes it on to security context inference layer, which then identifies security context based on data gathered from cloud and IoT layers and fixes the inference rules. Also, an analysis of attach patterns and related vulnerabilities in the cloud enabled power IoT environment was done. As inference engine, JANA, a rule engine based on Java is used, and smart metre was considered an example of a device in power system. Through the inference rules, the attack detection levels were high. To freshly detect security attacks, more security context inference rules can be added using the discussed security context reasoning method.

Thakare et al. [33] designed an access control model PARBAC, which addresses inconsistencies and operation overloads in Azure enabled IoT network, caused due to access control based on roles. The model works on priority-based authentication and is implemented on a healthcare case study. Based on the user priority, information related to assignment of user-role, and permission is generated and passed through fewer authorization and authentication levels for making the process efficient and quick without compromising on the security. Experimental outcome shows that implementation of the model proposed can produce flexible and better performing Azure IoT cloud. Organization's operational cost and human effort are reduced. In future, the proposed model can be implemented including context-based constraints like location, time etc., before access is given for a service in a health case model.

Jia et al. [34] proposed STOFDM model which transmits data between IoT network and cloud using truncated time signal which works consuming lesser bandwidth and not compromising on the amount of information passed. Privacy protected transmission for high rate of data and better spectral efficiency can be achieved in IoT network.

The algorithms discussed prove that a lot of work has happened on enhancing security, trustworthiness and workability of IoT–cloud environments. Table 2 presents a visual comparison.

3.3 Security Enhancement of Multi-Tenant Cloud Environment

One important property of the cloud environment is its multi-tenancy where more than one user can access the same resources instance simultaneously. Multiple users access cloud through the shared OS kernel but are separated through compulsory

Table 2 Comparative analysis for cloud-based IoT

Article references	Comparison criteria for cloud-based IoT		
	Method	Dataset/tool	Conclusion
[30]	Service parsing, edge network and platform	MATLAB R2016b, 4 IoT networks with 56 devices each, one cloud platform	Ensures improved trust and efficiency in cloud-IoT system
[31]	Secure indexing, encryption	Alibaba cloud, RFC Dataset	Effective in searching encrypted documents
[13]	Security based trust assessment	MATLAB R2017b, GitHub dataset	Improves security and measures reputation of cloud-IoT system
[32]	Ontology-based layered architecture	Smart meter system, JANA rule engine	Can detect all major network attacks making the system secure
[33]	Priority based authentication	Azure IoT cloud	Better performance, less manual interaction and cost
[34]	Truncated time signal	–	Improved data privacy during transmission

access control inspections at the hypervisor level [35]. But due to this sharing architecture, multi-tenancy leads to a lot of security concerns. Confidentiality and integrity violations are primary among other security issues [36].

Yu et al. [37] proposed use of network function virtualization in place of a middlebox to make the multi-tenant cloud more secure, flexible and cost effective. The proposed architecture divides the cloud into control plane and data plane. The control plane collects status related information from the hardware infrastructure, does control related decision making and passes the decision to the data plane. The data plane also communicates back about any change in status in the hardware layer. The method proposed is yet to be practically implemented.

Chen et al. [38] proposed vCNSMS, a shared system for enhancing network security which is usable in a cloud data centre serving multiple tenants. Data centre networks (DCNs) deploy security centre and peer-UTM (unified threat management). Rules are issued by security centre with the help of Web UI. Events are reported by peer-UTMs to the security centre through Web UI. The algorithm includes antivirus module, firewall module and protocol control module to enhance the network security. The test environment included hardware of Intel core 2quad CPU with 2 GB memory. Software included Ubuntu13.04 and Shield (untangle). 4-way UDP streams were used for SmartBit settings. The proposed method is capable of fulfilling security requirements of a multi-tenant cloud environment.

Ma et al. [39] have proposed tenants and user separation rules in a multi-talent cloud platform, and also, a mechanism is implemented on this rule-based model. The rules identified are tenant separation rules and access control rules. Tenant separation rules are again of two types—cloud service provider (CSP) control rules and tenant control rules. The implementation mechanism consists of three private clouds of different criticality which are connected by ethernet cables for transmission in a single direction. Three identical computers were used of configuration Intel i5 750CPU and 8 GB memory. Ethernet card used was RLT8100E100M. Results of the implementation show correctness of the model.

Thakur and Breslin [40] proposed RMM algorithm in a multi-tenant cloud using which the service provider can identify good and fraudulent users and make sure that resources are not shared between them. Reputation management is done using two methods: the informal and formal methods. In informal method, the cloud provider (CP) randomly divides users into groups of good and bad. Each group shares resources among themselves, and CP observes their behaviours and report to RMM by casting a positive or negative votes, for each user. Based on the collected votes RMM calculates a user's reputation. The formal method works according to an algorithm based on virtual network embedding. The experiment was conducted on 30 CPs and 200 users, and the result proves the correctness of the RMM.

Li et al. [41] identified a layered algorithm for progressive resource allocation that works on reducing resource fragmentation in multi-tenant clouds. The algorithm improves the tenant's QOS in the data centres and works based on concept of LP-MKP. Each data centre is represented as an information tree where the leaf nodes are the resources or physical machines. On this tree structure, the LP-MKP algorithm is applied considering the tenant requests for VMs. The experiment was conducted on four full information trees having n child nodes ($n = 2, 3, 4, 5$) and height = 5. For each leaf node, the number of VMs randomly distributed was in the range of 0–100. The experimental result of the proposed algorithm is measured against existing greedy and heuristic allocation (based on min-tree) algorithms. LP-MKP results are exceptionally great compared to greedy algorithm and also outperform min-tree algorithm. The algorithm can improve the overall quality of service for the cloud data centres for all tenants. Further studies need to be done if LP-MKP algorithm can work with open-source clouds like OpenStack and CloudStack.

Li et al. [42] proposed selection of service in a multi-tenant cloud based on services and combination of auction (CASSMT). The approach works on fulfilling quality constraints of all tenants by providing a platform for openly contesting service providers. It also accommodates comparisons among services. First, the tenants submit their service requirements (functionality, quality constraint) to the system developer. Service providers, also called bidders here, who can provide the required services and quality constraints, submit the bids. Among all the collected bids in the current round, one bid is selected as the winner, and if the exit criteria are met, the auction ends. Public dataset containing more than 2.500 Web services from the real world is used for conducting the experiment using s system with configuration of Intel i7 2.20 GHz CPU and 8 GB RAM. JDK 1.6.0 and Eclipse IDE are used, and CASSMT prototype is implemented in Java. The algorithm could significantly

Table 3 Comparative analysis for multi-tenant cloud

Article references	Comparison criteria for multi-tenant cloud		
	Method	Dataset/tool	Conclusion
[38]	Network function virtualization	–	Concept yet to be practically implemented
[39]	Deep packet inspection	Alibaba cloud, RFC dataset	Effective in searching encrypted documents
[40]	Rule based tenant and user separation	Intel i5 750 CPU 8 GB memory, RLT8100E100M ethernet card	Model works as expected
[41]	Reputation management mechanism	30 Cloud providers and 200 users	Model works as expected
[42]	Layered progressive resource allocation using multiple knapsack algorithm	4 full *n*-ary information trees (*n*-2, 3, 4, 5) of height 5	Better performance than greedy and heuristic min-tree algorithms

enhance the chances of identifying the best fitting service solution in the multi-tenant cloud environment. In future, experiments combining combinatorial auction and double auction can be tried out to improve the existing approach.

This section tells us that majorly all work on multi-tenant cloud platform aims at improving the services provided to the tenants while keeping the tenant spaces separate, thus not compromising on the security of the cloud environment. Table 3 presents a comparison of the proposed systems we discussed here.

4 Summary

Cloud computing is a necessity in present computing world. The resources available in a cloud environment are virtually limitless, very convenient and highly scalable on demand. Trust and data security issues are a major obstacle in full utilization of cloud environment benefits. In this article, we have discussed and analysed few related research works in three specific cloud architectures: the mobile cloud, cloud-based IoT and multi-tenant cloud. The methods used for mobile cloud and cloud-based IoT are mostly based on encryption and other network related privacy enhancement methods. In multi-tenant cloud, the focus is on keeping the virtualized interface of each user sealed and separate from that of other users. Also, importance is given on how service providers provide quality service to the tenants. Few gaps were identified due to the analysis which can be implemented in the future.

References

1. R.K. Kalluri, C.V.G. Rao, Addressing the security, privacy and trust challenges of cloud computing. Int. J. Comput. Sci. Inf. Technol. **5**(5), 6094–6097 (2014)
2. D. Stevenson, J. Pasek, Privacy concern, trust, and desire for content personalization, in *TPRC 43: The 43rd Research Conference on Communication, Information and Internet Policy Paper* (2015), pp. 1–30
3. R.K. Aluvalu, L. Muddana, A survey on access control models in cloud computing, in *Proceedings of the 49th Annual Convention of the Computer Society of India (CSI)*, vol. 1 (2015), pp. 653–664
4. F. Cai, N. Zhu, J. He, P. Mu, W. Li, Y. Yu, Survey of access control models and technologies for cloud computing. Cluster Comput. (to be published)
5. D.F. Ferraiolo, R. Sandhu, S. Gavrila, D.R. Kuhn, R. Chandramouli, Proposed NIST standard for role-based access control. ACM Trans. Inf. Syst. Secur. **4**, 224–274 (2001)
6. H. Takabi, Privacy aware access control for data sharing in cloud computing environments, in *Proceedings of the 2nd international workshop on Security in cloud computing* (2014), pp. 27–34
7. H.T. Dinh et al., A survey of mobile cloud computing: Architecture, applications, and approaches. Wireless Commun. Mobile Comput. **13**(18), 1587–1611 (2013)
8. N. Fernando, S.W. Loke, W. Rahayu, Mobile cloud computing: A survey. Future Gener. Comput. Syst. **29**(1), 84–106 (2013)
9. Q. Jiang, J. Ma, F. Wei, On the security of a privacy-aware authentication scheme for distributed mobile cloud computing services. IEEE Syst. J. **12**(2), 2039–2042 (2018)
10. A. Awad, A. Matthews, Y. Qiao, B. Lee, Chaotic searchable encryption for mobile cloud storage, in *IEEE Transactions on Cloud Computing*, vol. 6, no. 2 (1 Apr–June 2018), pp. 440–452
11. H.T. Dinh, C. Lee, D. Niyato, P. Wang, A survey of mobile cloud computing: Architecture, applications, and approaches. Wirel. Commun. Mob. Comput. (2013)
12. S.K. Lee, M. Bae, H. Kim, Future of IoT networks: A survey. Appl. Sci. **7**(10), 1072 (2017)
13. X. Li, Q. Wang, X. Lan, X. Chen, N. Zhang, D. Chen, Enhancing cloud-based IoT security through trustworthy cloud service: An integration of security and reputation approach. IEEE Access **7**, 9368–9383 (2019)
14. M. Burhan, R.A. Rehman, B.S. Kim, B. Khan, IoT elements, layered architectures and security issues: A comprehensive survey. Sensors (2018)
15. H.F. Atlam, A. Alenezi, A. Alharthi, R.J. Walters, G.B. Wills, Integration of cloud computing with Internet of Things: Challenges and open issues, in *2017 IEEE International Conference on Internet of Things (iThings) and IEEE Green Computing and Communications (GreenCom) and IEEE Cyber, Physical and Social Computing (CPSCom) and IEEE Smart Data (SmartData), Exeter* (2017), pp. 670–675
16. K. Lee, D. Murray, D. Hughes, W. Joosen, Extending sensor networks into the Cloud using Amazon web services, in *2010 IEEE International Conference on Networked Embedded Systems for Enterprise Applications (NESEA)* (Nov 2010), pp. 1–7
17. A. Botta, W. de Donato, V. Persico, A. Pescapé, Integration of Cloud computing and Internet of Things: A survey. Future Gener. Computer Syst. **56** (2016), 684–700. ISSN 0167-739X, https://doi.org/10.1016/j.future.2015.09.021
18. D. Gonzales, J. Kaplan, E. Saltzman, Z. Winkelman, D. Woods, Cloud-trust—A security assessment model for infrastructure as a service (IaaS) clouds. IEEE Trans. Cloud Comput. **5**, 1–1 (2015). https://doi.org/10.1109/TCC.2015.2415794
19. B.P. Rimal, M. Maier, Workflow scheduling in multi-tenant cloud computing environments, in *IEEE Transactions on Parallel and Distributed Systems*, vol. 28, no. 1 (1 Jan 2017), pp. 290–304
20. W.-T. Tang, C.-M. Hu, C.-Y. Hsu, A mobile phone based homecare management system on the cloud, in *2010 3rd International Conference on Biomedical Engineering and Informatics (BMEI)*, vol. 6. (Oct 2010), pp. 2442–2445
21. Z. Xiao, Y. Xiao, Security and privacy in cloud computing, in *IEEE Communication Surveys Tutorials*, vol. 15, no. 2, 2nd Quart. (2013), pp. 843–859

22. A. Banerjee, M. Hasan, M.A. Rahman, R. Chapagain, CLOAK: A stream cipher based encryption protocol for mobile cloud computing. IEEE Access **5**, 17678–17691 (2017)
23. Y. Weng, L. Liu, A collective anomaly detection approach for multidimensional streams in mobile service security. IEEE Access **7**, 49157–49168 (2019)
24. D.S.A. Elminaam, F.T. Alanezi, K.M. Hosny, SMCACC: Developing an efficient dynamic secure framework for mobile capabilities augmentation using cloud computing. IEEE Access **7**, 120214–120237 (2019)
25. W. Li, J. Cao, K. Hu, J. Xu, R. Buyya, A trust-based agent learning model for service composition in mobile cloud computing environments. IEEE Access **7**, 34207–34226 (2019)
26. M. Akter, A. Gani, M.O. Rahman, M.M. Hassan, A. Almogren, S. Ahmad, Performance analysis of personal cloud storage services for mobile multimedia health record management. IEEE Access **6**, 52625–52638 (2018)
27. A. Sajid, H. Abbas, K. Saleem, Cloud-assisted IoT-based SCADA systems security: A review of the state of the art and future challenges. IEEE Access **4**, 1375–1384 (2016)
28. L. Barreto, A. Celesti, M. Villari, M. Fazio, A. Puliafito, An authentication model for IoT clouds, in *2015 IEEE/ACM International Conference on Advances in Social Networks Analysis and Mining (ASONAM)*, Paris (2015) pp. 1032–1035
29. J. Singh, T. Pasquier, J. Bacon, H. Ko, D. Eyers, Twenty security considerations for cloud-supported Internet of Things. IEEE IoT J. **3**(3), 269–284 (2016)
30. T. Wang, G. Zhang, A. Liu, M.Z.A. Bhuiyan, Q. Jin, A secure IoT service architecture with an efficient balance dynamics based on cloud and edge computing. IEEE IoT J. **6**(3), 4831–4843 (2019)
31. M. Shen, B. Ma, L. Zhu, X. Du, K. Xu, Secure phrase search for intelligent processing of encrypted data in cloud-based IoT. IEEE IoT J. **6**(2), 1998–2008 (2019)
32. C. Choi, J. Choi, Ontology-based security context reasoning for power IoT-cloud security service. IEEE Access **7**, 110510–110517 (2019)
33. A. Thakare, E. Lee, A. Kumar, V.B. Nikam, Y. Kim, PARBAC: Priority-attribute-based RBAC model for azure IoT cloud. IEEE IoT J. **7**(4), 2890–2900 (2020)
34. M. Jia, Z. Yin, D. Li, Q. Guo, X. Gu, Toward improved offloading efficiency of data transmission in the IoT-cloud by leveraging secure truncating OFDM. IEEE IoT J. **6**(3), 4252–4261 (2019)
35. A. Kurmus, M. Gupta, R. Pletka, C. Cachin, R. Haas, A comparison of secure multi-tenancy architectures for filesystem storage clouds, in *Middleware 2011* ed. by F. Kon, A.M. Kermarrec. Lecture Notes in Computer Science, vol. 7049 (Springer, Berlin, Heidelberg, 2011)
36. H. Aljahdali, A. Albatli, P. Garraghan, P. Townend, L. Lau, J. Xu, Multi-tenancy in cloud computing, in *Proceedings—IEEE 8th International Symposium on Service Oriented System Engineering, SOSE 2014* (2014). https://doi.org/10.1109/SOSE.2014.50
37. R. Yu, G. Xue, V.T. Kilari, X. Zhang, Network function virtualization in the multi-tenant cloud, in *IEEE Network*, vol. 29, no. 3 (May–June 2015), pp. 42–47
38. Z. Chen, W. Dong, H. Li, P. Zhang, X. Chen, J. Cao, Collaborative network security in multi-tenant data center for cloud computing. Tsinghua Sci. Technol. **19**(1), 82–94 (2014)
39. W. Ma, Z. Han, X. Li, J. Liu, A multi-level authorization based tenant separation mechanism in cloud computing environment. China Commun. **13**(5), 162–171 (2016)
40. S. Thakur, J. Breslin, A robust reputation management mechanism in federated cloud. IEEE Trans. Cloud Comput (2017). https://doi.org/10.1109/TCC.2017.2689020
41. J. Li, D. Li, Y. Ye, Lu. Xicheng, Efficient multi-tenant virtual machine allocation in cloud data centers. Tsinghua Sci. Technol. **20**(1), 81–89 (2015)
42. X. Li et al., Quality-aware service selection for multi-tenant service oriented systems based on combinatorial auction. IEEE Access **7**, 35645–35660 (2019)

Pseudonymization of Text and Image Data to Provide Confidentiality

C. S. Manthan, K. Roopa, H. S. Bindu, B. M. Apoorva, and Mala D. Madar

Abstract It is important to maintain secrecy of one's valuable data when it is loaded in the Internet. Unintended parties can view the data, resulting in breach of confidentiality. Another serious concern is the possibility of misuse of data to the advantage of an unintended party. To overcome these problems, this paper presents an algorithm developed and implemented to make the data that is uploaded to the servers, pseudo anonymous (pseudonymous). Typically, user-sensitive information exists as either texts or images. Hence, such data is used as input to the developed algorithm. From the textual data, the algorithm can detect keywords called tags, such as names, places, financials, and replace the original tag with newly created false tags. Detection of the tags is done by the existing natural language processing (NLP) algorithms. For image input, the algorithm detects face and replaces it with a falsified one. This is done by using generative adversarial networks (GANs). Further, the original or true image is hidden in the falsified image using SteganoGAN. This image will be sent over the network. At the receiver, both anonymized text and image data are de-anonymized as per the developed algorithm.

Keywords Confidentiality · Pseudonymous · Text · Image · Natural language processing · Generative adversarial networks

C. S. Manthan · K. Roopa (✉) · H. S. Bindu · B. M. Apoorva · M. D. Madar
Department of Electronics and Communication Engineering, The National Institute of Engineering, Mysuru, Karnataka 570008, India
e-mail: roopakodnad@gmail.com

C. S. Manthan
e-mail: mnthn.ai@gmail.com

H. S. Bindu
e-mail: binduhs2016@gmail.com

B. M. Apoorva
e-mail: bmapoorva.6@gmail.com

M. D. Madar
e-mail: malajnv@gmail.com

V. E. Balas et al. (eds.), *Proceedings of International Conference on Computational Intelligence, Data Science and Cloud Computing*, Lecture Notes on Data Engineering and Communications Technologies 62, https://doi.org/10.1007/978-981-33-4968-1_43

1 Introduction

There has been a rise in the number of private data breaches in online systems and services [1]. Actually, there are strict rules on personal data breach notifications [2]. Some of the major data breaches of recent times [3] are either intentional or otherwise. But solution is inevitable. Several users have opted for new methods such as data anonymization to protect their information. Data anonymization is a process by which personal data is irreversibly altered. The anonymized output will denote some meaning but with the sensitives replaced by different data. Even if this modified data is used for training, development, testing and analytics, it will not reveal the identity of the actual person or source. According to Balaji Raghunathan [4], data anonymization is a technique to de-identify sensitive data while at the same time preserve its format and data type. The masked data may be real or random data. Anonymization algorithms have to transform the data into a form that provides a privacy guarantee with the minimum possible distortion of the original data. Reference [5] describes techniques of data anonymization by generalizing the data and by adding noise to data. A few other techniques are discussed in references [6, 7].

De-anonymization is the reverse process in which anonymous data is cross-referenced with other data sources to re-identify the anonymous data source. The process of obscuring data with the ability to re-identify it later is also called pseudonymization [8, 9].

Recent anonymization and pseudonymization techniques use generative adversarial networks (GANs) [10]. GAN is used in image identification, generation and classification. GANs are basically made up of a system of two competing neural network models which compete with each other and are able to analyse, capture and copy the variations within a data set. In GANs, there is a generator and a discriminator. The generator generates fake samples of data while the discriminator, on the other hand, tries to distinguish between the real and fake samples.

This paper discusses new algorithms formulated using existing concepts to provide pseudonymization of text and face image data, giving confidentiality for those involved. To achieve this, different forms of GANs are used. Python programming language [11] is used for algorithm implementation. A few GitHub resources are also used with appropriate referencing.

The motivation for the present approaches stems from the following: (i) Natural language processing (NLP) bridges the gap between human language and machines by text classification using pretrained deep learning models. (ii) SpaCy statistical model is used here as it is an open-source library for NLP and can handle large volumes of text. (iii) Text pseudonymization algorithm is based on an idea of representing text into an image. (iv) Steganography offers a feasible alternative or a supplement to encryption. Use of GAN-based steganography helps in powerful construction of generators and discriminators. Moreover, GANs are a relatively new concept in machine learning, introduced for the first time in 2014.

2 Related Work

Hukkelas et al. [12, 13] discussed a model called deep privacy making use of a conditional generative adversarial network (GAN). They evaluated their model by anonymizing the validation set of the WIDER Face data set [14] and then performed face detection of the anonymized images to measure the impact of anonymization. The conditional GAN they used generated images based on the surrounding of the face and sparse pose information. They used seven key points to describe the pose of the face; left/right eye, left/right ear, left/right shoulder and nose.

Karras et al. [15] proposed a generator architecture for GANs leading to an automatically learned and unsupervised separation of high-level attributes.

Anonymization of text in particular has practical applications in maintaining patient records in electronic form which are useful for research. Anonymization or de-identification allows research uses of clinical data [16–19]. Neubauer and Heurix [20] discuss a solution which provides a guaranteed data privacy allowing primary and secondary use of the data at the same time. Rahul et al. [21] present a review of various natural language processing (NLP)-based machine learning approaches for summarizing the text data.

In the work reported in this paper, text as well as image data are used as input to the pseudonymizer. It is called so because, at the receiver, de-anonymization is done to retrieve the true information by the intended user.

NLP is used to identify text tag words. Additionally, an algorithm is written for text anonymization. For image anonymization, face images are used. To anonymize face, a fake face has to be generated which would mask the original face. This is done by using GAN [10], StyleGAN [22] or StyleGAN2 [23–25].

Though the terms anonymization and pseudonymization are used interchangeably to describe the implemented work for ease of use of terminologies, it is actually the pseudonymization that has been carried out since the anonymized data is also de-anonymized to get back their true forms as shown in Fig. 1.

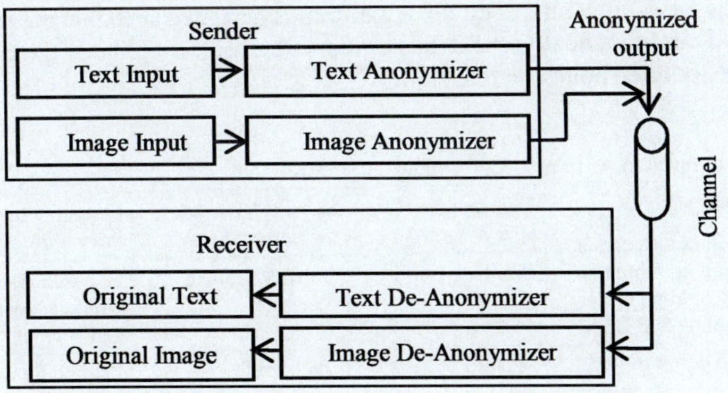

Fig.1 Overall diagram of the anonymizer

3 Methodology

3.1 Implementation of Text Anonymization

Textual data can contain a lot of user-related information such as names, places and contact details. Such user-sensitive data can be identified using NLP [26–28] and categorized according to names, locations and so on. In this work, the main focus for text input is on anonymization of location-based tags. A library called SpaCy [29] is available in Python to carry out different NLP tasks.

The steps in text anonymization are as follows:

(i) Identify all the tags that are related to location using SpaCy and NLP.
(ii) Save all identified tags (original tags) in 'detected tags' array.
(iii) Substitute the original tags with other fake tags randomly from a table containing all the city names in India [30].
(iv) In the text data input, replace all detected tags with fake tags.
(v) Perform encryption [31] of all the tags. The result turns out to be an image with concentric circles.

Two fictitious examples are shown in Table 1, with the identified tag, substituted tag and the output text with the replaced tags (keywords).

To encrypt a tag, each of its character is used. Before the actual encryption, a blank space is added between the detected tag (original tag) and the replaced tag and saved in a variable, **'joint_tag'** as,

$$\text{joint\_tag} = \text{original tag} + \text{“ ”} + \text{replaced tag} \tag{1}$$

Inclusion of a space in between the original and replaced tag helps in retrieving the original text during de-anonymization.

An algorithm to encrypt **'joint_tag'** is shown in Fig. 2. A circle is drawn whose radius is equal to the length of the **'joint_tag'**. Each character in the **'joint_tag'** is to be plotted as a point on this circle. Algorithm is designed such that the number of points on a circle is equal to the radius of that circle, e.g. if radius of the circle is 3, then there are three points on that circle.

Table 1 Examples for tag word identification and substitution

Sl No	Input	Tags detected by spacy	Substituted tag for the detected tag	Output with replaced keywords
1	Bob told Alice that he would go to Bombay and Delhi	Bombay, Delhi	Kolkata, Bhopal	Bob told Alice that he would go to Kolkata and Bhopal
2	I am going to Mysuru this weekend	Mysuru	Bhubaneshwar	I am going to Bhubaneshwar this weekend

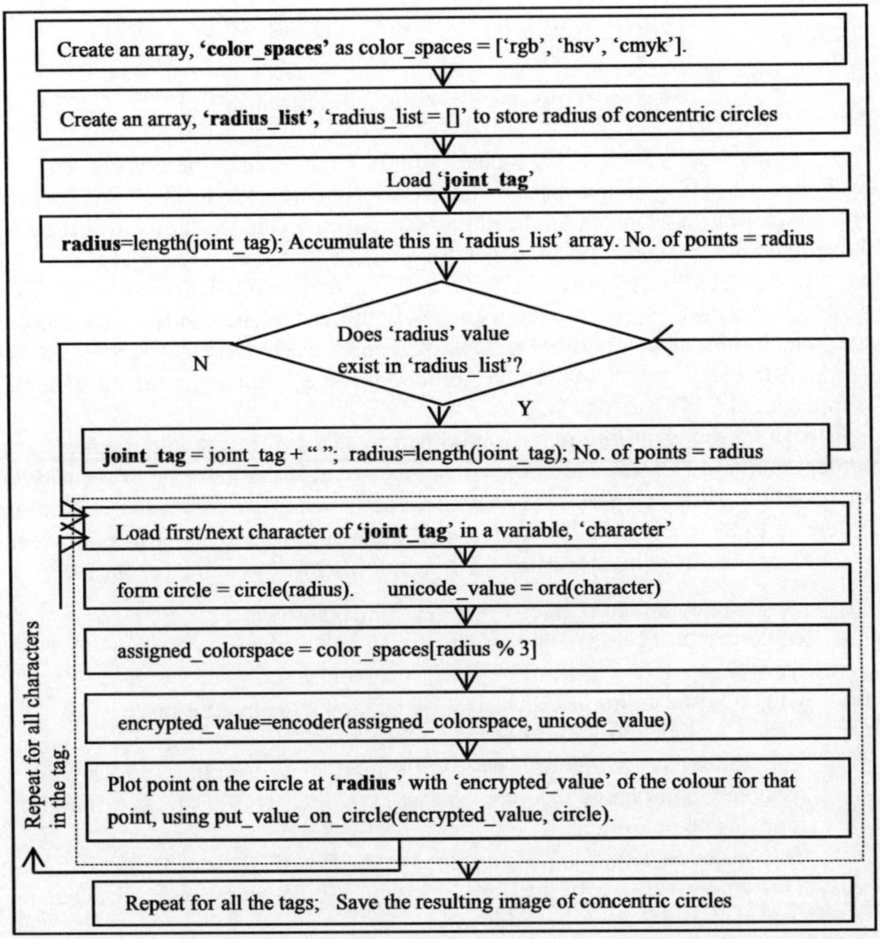

Fig. 2 Flow diagram to encrypt **'joint_tag'**

While anonymizing multiple tags, there might be a chance that a new **'joint_tag'** will have the same length as that of a previously generated **'joint_tag'**. This will lead to an overlap of data on the circle as the radii become the same. Whenever such cases arise, **'joint_tag'** is padded with another blank space at the end, and then radius value is incremented. These extra spaces are removed while decrypting [30] to get the original data back.

The colour of the point depends on the Unicode value (**'unicode_value'**, in Fig. 2) returned by the **'ord'** method in Python for a character in the **'joint_ tag'**. This value is added to a colour component in one of the three colour spaces. The colour spaces are declared in the array as shown in (2). The assigned colour space to a character depends on radius value as shown in (3),

$$\text{color\_spaces} = [\text{'RGB'}, \text{'HSV'}, \text{'CMYK'}] \tag{2}$$

$$\text{assigned\_colorspace} = \text{color\_spaces}[\text{radius}\%\,3] \tag{3}$$

If 'radius% 3' in RHS of (3) evaluates to 0, 'RGB' colour space is chosen, if it evaluates to 1, 'HSV' colour space is chosen, else, if it evaluates to 2, 'CMYK' colour space is chosen, in which particular colour component in the '**assigned_colorspace**' added with the '**unicode_value**' depends on which character in 'joint_tag' is being encrypted, e.g., if first character in 'joint_tag' is to be encrypted, first colour component in the '**assigned_colorspace**' is added with the '**unicode_value**'. As another example, if third character is being encrypted and the selected colour space [calculated as per (3)] is say, 'CMYK', the '**unicode_value**' is added to the third colour component of CMYK which is Y′.

If desired, the algorithm may be modified to select different colour spaces and colour components. Use of multiple colour spaces adds another level of security as all the values of the string cannot be found properly in a single colour space. Plotting is done in Python code with the help of a package known as **matplotlib** [32].

The steps involved in encryption are explained below through an example:

(i) Original tag: Mysuru, Substituted tag: Bhubaneshwar.
(ii) **joint_tag** = Mysuru Bhubaneshwar (Note the space in between the city names).
(iii) length(Mysuru Bhubaneshwar) = 19, including the middle space.
(iv) Hence, a circle of radius 19 is to be plotted with 19 points on it.
(v) The index for '**color_spaces**' is 19 modulo 3 which is 1. Therefore, **assigned_colorspace** = color_spaces [1] which is 'HSV' since the array indexing starts from zero.
(vi) The '**unicode_value**' of 'M' is added to the Hue (H) part of HSV.
(vii) This colour value is plotted on the circle. Similar steps are carried out for subsequent characters in '**joint_tag**'.
(viii) The points are plotted on the circle in a clockwise direction, with the initial point being plotted on the positive Y-axis.
(ix) The process is repeated for every character and for all tags. Thus, concentric circles of tags are generated with a common centre at the origin (0, 0). The output of text anonymization is shown in the 'results' section.

3.2 Implementation of Text De-Anonymization at the Receiver

The input to the decrypter is a substituted message and an image consisting of concentric-dotted circles having dots of different colours. For text de-anonymization, the following steps are used:

(i) Detect the origin or centre of the circles.
(ii) Group all points that are equidistant from the centre. Number of groups formed will be the number of tags that were replaced by fake data in the string.
(iii) The point that lies on the Y-axis is the starting point and continues clockwise. The initial colour space is found by finding the modulo of the total number of points or equivalently, the length at which the points were found.
(iv) The '**unicode_value**' of the character that was encoded in the point is obtained by decrypting the point in the colour space.
(v) The '**unicode_value**' is converted back to the letter.
(vi) The process is repeated for all the points, and the decrypted letters are pushed into an array. The elements of the array are joined to get a decrypted string which has the original tag and the replaced tag with a space in between.
(vii) The decrypted string is split at the space to get the original tag and the replaced tag. The replaced tag is found in the received message and replaced by the original tag to get the original message back.

3.3 Working of Anonymizer for Pictorial Input Data

In this work, the focus is on anonymizing images that contain faces. Original face (true image, in Fig. 3) is anonymized using a fake face. With the help of a StyleGAN, highly realistic photos of anonymized faces can be generated.

Figure 3 shows the block diagram of the image anonymizer which involves the following steps:

(i) The StyleGAN-generated fake image is given to the anonymizer.
(ii) The anonymizer identifies the convex hull of the faces using the '**dlib**' package provided by Python, in both the original image and StyleGAN-generated image by triangulating over the faces by a method called Delaunay triangulation [33].
(iii) The coordinates where the convex hull of only the face is found in the entire true input image are saved. Various other details (other than face) which may be present in the true input image are not considered for processing. They remain as it is. This step is shown in Fig. 4a. In the figure, the convex hull is indicated with a rectangle bounding box. The image shown is representative and not of original size for the actual coordinates of (X, Y, W, H) = [143, 601, 130, 150]. It is also possible to take any other information in the image

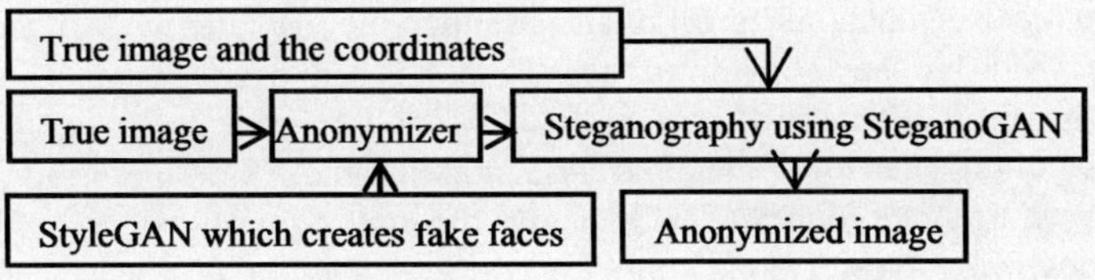

Fig. 3 Block diagram of the image anonymizer

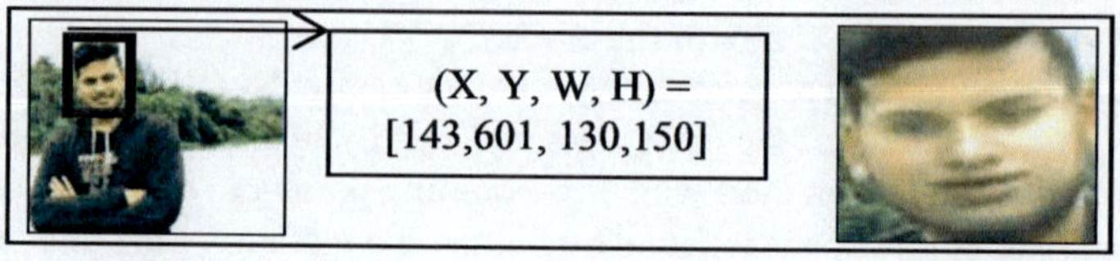

Fig. 4 **a** Original input image with face and other details. Also shown is the convex hull identified for face at (X, Y, W, H) = [143, 601, 130, 150]. **b** Original face anonymized by overlapping with a StyleGAN-generated image

as input, instead of face. In this work, face of the input image is considered for anonymization as it is a prime feature.

(iv) The true face image alone as shown by the bounding box in Fig. 4a and its convex hull coordinates [143, 601, 130, 150] are input to steganography using SteganoGAN. The other input to SteganoGAN is the anonymizer output as shown in Fig. 3, and its image is shown in Fig. 4b.

(v) In the convex hull, the triangles of one face are imposed on another by matching the vertices. This produces a hard output of the face. This makes sure that the expression of the original face is also preserved.

(vi) The hard imposition after detecting the convex hull of both the faces is smoothened for brightness and contrast using OpenCV's seamless clone algorithm [34] which gives a realistic fake face.

(vii) The original convex hull [35] is hidden inside the smoothed image using steganographic technique, SteganoGAN [36, 37]. Image steganography is used to hide messages inside pictures. In this work, true image (face) is hidden in the anonymized image. SteganoGAN is a tool for creating steganographic images using adversarial training. Optionally, the convex hull can be encrypted with other cryptographic algorithms such as Advanced Encryption Standard (AES) and then sent for steganography at the cost of additional computations. This process generates the final anonymized image. The output of image anonymization is shown in the 'results' section.

3.4 *Retrieval of Original Face from the Pseudonymized Image at the Receiver*

The image received contains the original face and the coordinates where it was hidden (steganographed) in the anonymized face. Thus, it is sent to the SteganoGAN which retrieves the original face and coordinates.

The coordinates retrieved are an array containing the X, Y points and the width and height of the face found. (X, Y) specify the starting point of the original face in the anonymized face. This is simply used to retrieve the original face from the

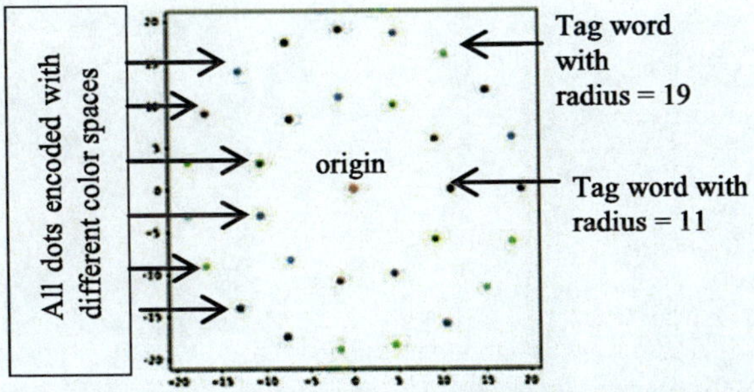

Fig. 5 Results of text anonymization

anonymized image according to the width and height it was originally steganographed in. This will output the original image as it was, before anonymization, and the retrieval is complete.

4 Results

The algorithm was tested for a total of 50 text input data with single and multiple real city names. It was also tested for a total of 50 images data set generated by the authors. The system worked well for both data sets in terms of anonymization and de-anonymization. The programs were executed in Apple MacBook Air system with 1.8 GHz Dual-Core Intel i5, 8 GB, 1600 MHz DDR3. The average execution times were 656 ms and 1453 ms for anonymization of text and image, respectively. The average execution times were 224 ms and 183 ms for de-anonymization of text and image, respectively. The execution times are within the acceptable limits for general applications.

An example of an anonymized text output in the form of an image with two concentric circles with radii 19 units and 11 units is shown in Fig. 5. Figure 6 shows the results of face anonymization. Figure 6a shows the original face image input to the anonymizer. Its triangulated image is shown in Fig. 6b. Figure 6c, e, g and i are taken from StyleGAN-generated images which are input to the anonymizer. Figure 6d, f, h and j are the respective anonymized output images from SteganoGAN. De-anonymization will result in the true image of Fig. 6a.

5 Conclusion

This paper discussed a pseudonymizer for text and image input data. This software system will be useful in applications like sending medical data of a patient over public network or sharing confidential data over social networking sites. The system worked well for a total of 100 representative inputs randomly taken inclusive of text

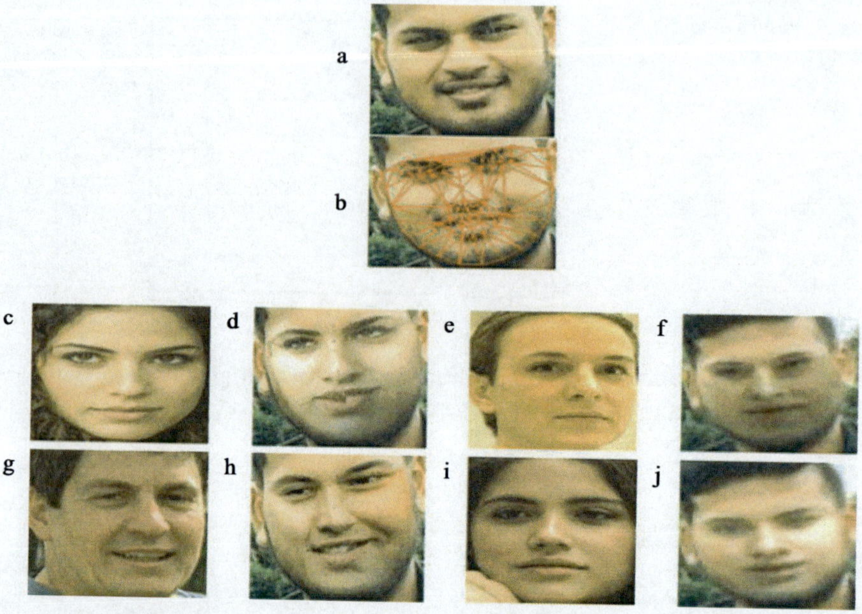

Fig. 6 Results of face anonymization **a** input face image **b** face triangulated **c, e, g, i** are taken from StyleGAN-generated images which are input to the anonymizer; **d, f, h, j** are the respective anonymized output images from SteganoGAN

and image. However, the limitations are that, as the textual algorithm uses an NLP model based on probability to detect tags, there is a possibility that the model might miss out some tags when a different, test data set is used and in the case of image input data, the posture of the face is not specifically considered here for validation. Hence, for posture-variant image data, the algorithmic performance evaluation has to be carried out separately.

As future scope, the authors suggest the following:

(i) Provide option to select multiple types of tags, e.g. city, personal names and numbers.
(ii) Provide option to choose entire face, only eyes, only hair style or background, etc.
(iii) Provide flexibility in the extent to which anonymization is to be performed. A suggestive measure for this would be, in triangulation, the precision must be controlled by providing a variable parameter so that the user can preview and select the % anonymization which is satisfactory for a particular application.
(iv) Evaluate the quantitative performance parameters showing the difference between the original and the anonymized image.

References

1. https://www.enisa.europa.eu/topics/data-protection/personal-data-breaches, accessed June 2020.
2. https://www.i-scoop.eu/gdpr/personal-data-breach-notification/, accessed June 2020.
3. https://www.csoonline.com/article/2130877/the-biggest-data-breaches-of-the-21st-century.html, accessed June 2020.
4. B. Raghunathan, *The Complete Book of Data Anonymization: From Planning to Implementation* (Infosys Press and CRC Press, 2013)
5. https://policies.google.com/technologies/anonymization?hl=en-US, accessed June 2020.
6. https://www.imperva.com/learn/data-security/anonymization/, accessed June 2020.
7. https://iapp.org/media/pdf/resource_center/Guide_to_Anonymisation.pdf, accessed June 2020.
8. R. Schwartmann, S. Weiß (eds.), *White Paper on Pseudonymization*. Drafted by the Data Protection Focus Group for the Safety, Protection, and Trust Platform for Society and Businesses in Connection with the 2017 Digital Summit (2017)
9. Pseudonymisation Techniques and Best Practices, November 2019, Available on https://www.enisa.europa.eu/publications/pseudonymisation-techniques-and-best-practices, accessed June 2020.
10. https://developers.google.com/machine-learning/gan, accessed June 2020.
11. https://www.python.org/, accessed June 2020.
12. H. Hukkelas, R. Mester, F. Lindseth, DeepPrivacy: A generative adversarial network for face anonymization, in *Advances in Visual Computing: 14th International Symposium on Visual Computing, ISVC 2019* (Springer International Publishing, 2019), pp. 565–578. Available on https://arxiv.org/pdf/1909.04538.pdf, accessed June 2020.
13. www.github.com/hukkelas/DeepPrivacy, accessed June 2020.
14. S. Yang, P. Luo, C.C Loy, X. Tang, WIDER FACE: A face detection benchmark, in *2016 IEEE Conference on Computer Vision and Pattern Recognition (CVPR)*. Las Vegas, NV (2016), pp. 5525–5533. https://doi.org/10.1109/CVPR.2016.596
15. T. Karras, S. Laine, T. Aila, A style-based generator architecture for generative adversarial networks, in *2019 IEEE/CVF Conference on Computer Vision and Pattern Recognition (CVPR)*. Long Beach, CA, USA (2019), pp. 4396–4405. https://doi.org/10.1109/CVPR.2019.00453
16. R.N. Cardinal, Clinical records anonymisation and text extraction (CRATE): An open-source software system. BMC Med. Inform. Decis. Making **17**(1), 50 (2017). https://doi.org/10.1186/s12911-017-0437-1, available on https://www.ncbi.nlm.nih.gov/pmc/articles/PMC5405523/, accessed June 2020.
17. https://crateanon.readthedocs.io/en/latest/nlp/index.html, accessed June 2020.
18. N. Mamede, J. Baptista, F. Dias, Automated anonymization of text documents, in *2016 IEEE Congress on Evolutionary Computation (CEC)*. Vancouver, BC (2016), pp. 1287–1294. https://doi.org/10.1109/CEC.2016.7743936
19. S.L. Ribeiro, E.T. Nakamura, Privacy protection with pseudonymization and anonymization in a health IoT system: Results from OCARIoT, in *2019 IEEE 19th International Conference on Bioinformatics and Bioengineering (BIBE)*. Athens, Greece (2019), pp. 904–908. https://doi.org/10.1109/BIBE.2019.00169
20. T. Neubauer, J. Heurix, A methodology for the pseudonymization of medical data. Int. J. Med. Inf. **80**(3), 190–204 (2011)
21. Rahul, S. Adhikari, Monika, NLP based machine learning approaches for text summarization, in *2020 Fourth International Conference on Computing Methodologies and Communication (ICCMC)*. Erode, India, pp. 535–538 (2020). https://doi.org/10.1109/ICCMC48092.2020.ICCMC-00099
22. https://machinelearningmastery.com/introduction-to-style-generative-adversarial-network-stylegan/, accessed June 2020
23. https://medium.com/analytics-vidhya/from-gan-basic-to-stylegan2-680add7abe82, accessed June 2020

24. https://towardsdatascience.com/stylegan2-ace6d3da405d, accessed June 2020
25. https://neurohive.io/en/news/stylegan2-new-improved-stylegan-is-the-new-state-of-the-art-model/, accessed June 2020
26. https://machinelearningmastery.com/natural-language-processing/, accessed June 2020
27. https://towardsdatascience.com/your-guide-to-natural-language-processing-nlp-48ea2511f6e1, accessed June 2020
28. https://www.wonderflow.co/blog/natural-language-processing-examples, accessed June 2020
29. https://spacy.io/, accessed June 2020
30. https://simplemaps.com/data/in-cities, accessed July 2020
31. W. Stallings, *Cryptography and Network Security Principles and Practice*, 6th edn. Pearson (2014)
32. https://matplotlib.org/, accessed July 2020
33. https://in.mathworks.com/help/matlab/math/delaunay-triangulation.html, accessed June 2020
34. https://www.learnopencv.com/seamless-cloning-using-opencv-python-cpp/, accessed June 2020
35. https://www.cs.auckland.ac.nz/software/AlgAnim/convex_hull.html, accessed June 2020
36. https://arxiv.org/abs/1901.03892, accessed July 2020
37. https://github.com/DAI-Lab/SteganoGAN, accessed July 2020

Secure Data Management Using IPFS and Ethereum

P. Poornima Devi, Srinivasan Ananthanarayanan Bragadeesh, and Arumugam Umamakeswari

Abstract Blockchain technology is receiving increased attention over the past few years. Blockchain is essentially a decentralized, distributed digital ledger in which data is organized into blocks. The data stored in Blockchain is secured using cryptographic properties. Storage is the main concern when the volume of transactions increases. Handling data in huge amount may lead to increased overheads as data can only be appended to it, and the previous data cannot be deleted. So, it led to the development of the InterPlanetary File System (IPFS). IPFS is a distributed database with peer-to-peer file system and uses content-addressable hash value which is used to verify the data. Using IPFS, one can upload the data in the form of text, images as well as videos. While uploading the data in IPFS, the protocol returns a unique hash value for the data via gateway. This hash value is securely stored in Blockchain. Based on the hash value, the data can be retrieved. In IPFS, data is distributed among the network participants locally. The proposed work combines both IPFS and Blockchain technologies. This scheme uses IPFS for storing large volume of data, removes duplication, share files, and hosts them. The proposed work minimizes the central server storage space, improves data security by using unique content-addressable hashes for data verification and also reduces computing power.

Keywords Blockchain · Ethereum · Hash value · IPFS · Sharing data · Smart contracts

---

P. Poornima Devi · S. A. Bragadeesh · A. Umamakeswari (✉)
School of Computing, SASTRA Deemed to be University, Thanjavur, Tamil Nadu, India
e-mail: a_umamakeswari@yahoo.com

P. Poornima Devi
e-mail: bragadeesh.prithvi@gmail.com

S. A. Bragadeesh
e-mail: sa_bragadeesh@yahoo.co.in

V. E. Balas et al. (eds.), *Proceedings of International Conference on Computational Intelligence, Data Science and Cloud Computing*, Lecture Notes on Data Engineering and Communications Technologies 62, https://doi.org/10.1007/978-981-33-4968-1_44

1 Introduction

Due to tremendous increase in the amount of data produced by IoT devices such as sensors, smart homes, and smart-phones, etc., it is not possible to store the huge data in a single local machine [1]. To address the storage requirements, cloud computing infrastructures are used, but there may be security issues, unauthorized modification of data, requirement of a third party to establish trust, and ensure privacy. Nakamoto [2] introduced the concept of bitcoin in 2008 which is mainly used to store transaction details based on crypto-currency without the specific need for a central authority. Blockchain technology is the fundamental concept behind bitcoin, which is a distributed ledger. It improves security and there is no possibility of single point of failure as all the nodes in the network has the copy of data. It contains data, hash, and previous hash of the data. Over time, the data is increasing which leads to drastic increase in computing power to store and process the data. The data in Blockchain can only be added and cannot be modified or deleted permanently. When the number of transactions increases, then data storage will be a big concern.

Conventionally, data communication is enabled via a centralized server. In the centralized server, the data is communicated between client and server, and the speed of the data transmission will be very slow. The main pitfalls of the centralized server are increased bandwidth utilization, low efficiency, requires additional networking capabilities, increased energy consumption, and latency. There is also high possibility of DDoS attacks and single point of failure. In a decentralized server, the data will be distributed across the network participants. So, attackers or unauthorized users are unable to tamper the entire data. It is also possible to achieve faster data transfer rates. Many applications now utilize decentralized data storage.

In the proposed work, a decentralized application is created by combining Blockchain and IPFS. Blockchain is mainly used to secure the data from unauthorized user. Blockchain is a growing list of blocks that are connected by cryptographic algorithms. It is based on distributed ledger technology (DLT) which is a mechanism for documenting digital transactions in a distributed storage with no centralized file systems. IPFS is a hypermedia P2P protocol and shared file system that can be used to replace traditional client server infrastructures. The data is stored using Merkle tree-based directed acyclic graph (DAG) [3] in which the data in root node is a function of its leaf nodes. This distributed way of organizing the data eliminates single point of failure. HTTP has many drawbacks such as inefficiency, lack of historical versioning, and centralization which can be overcome by using IPFS.

The aim of this paper is to overcome the shortcomings of the existing centralized storage systems by using Blockchain technology which is tamperproof and decentralized. It also ensures data traceability to enhance the trustworthiness and accountability of cloud solutions. The major contributions are outlined as follows:

(a) Blockchain is used for the secure storage of data.
(b) The data stored in IPFS is hashed, and the corresponding hash value is stored in Blockchain.
(c) A user interface is created to store, retrieve, and display data in a secure manner.

The rest of the paper is organized as follows. Section 2 presents the background and current related works existing in the literature, Sect. 3 provides information about the architecture of our proposed solution. Section 4 explains the experimental setup which is followed by performance evaluation in Sect. 5. Finally, the concluding remarks and some possible future directions are presented in Sect. 6.

2 Background and Related Works

In this section, the fundamental technologies used in the proposed work and the existing state of the art works are presented for understanding what is done currently.

A. *Blockchain*
Blockchain is a P2P decentralized network and which contains original data, unique hash, and previous hash. The main characteristics of Blockchain are confidentiality, no need for 3rdparty vendors, and the data can be easily accessible to all the nodes in the distributed network. There are many forms of Blockchain, either private or public and both the bitcoin and Ethereum are popular but they differ in several ways. Any data recorded in the block cannot be modified [4]. Blockchain networks are commonly used in today's crypto-currency. Blockchain technology is fast evolving and its suitability for other applications have been explored widely.

B. *Smart Contract*
In 1997, Nick Szabo first introduced the idea of smart contracts and discussed about its role for applications that involve collateral, contract bonds, real estate, etc.[5]. It is a self-executable computer program that allows conditional transfer of information by enabling transactions. For example, it may be content, property, share, and money. The contract uses the Ethereum platform. Especially, it will be used for the supply chain management. It was written in solidity programming language. Used to build decentralized exchanges, tokenized assets, games, etc.

C. *IPFS*
IPFS is a shared file system with the purpose of linking all computer devices to the same file system. Blockchain is very similar to IPFS but in IPFS, one can store large volume of data for low cost. The drawback of Blockchain is inability to keep a vast amount of data on it. Using IPFS, data will be more secure, high-throughput, content-addressed [6].

D. *Ethereum*
Ethereum is one of the most popular Blockchain systems in the world [7]. It is a decentralized platform that allows smart contracts to run on the Ethereum platform. Smart contracts has a very flexible programming language. Here, Ethereum allows the smart contract to perform computation on a distributed ledger. In Ethereum, the miners will validate and approve the Blockchain transaction. In

order to add a new block to the chain, the miner solves a computationally complex problem [8].

Blockchain offers a distributed ledger, owned by all network members based on the consensus process [9]. Different kinds of applications are used in Blockchain technology. Pitbull et al. [10] examined the mechanism for secure transmission. Du et al. [11] suggested a private encoding, storage for preservation, and data protection scheme for video surveillance. Recent years have seen the rapid development of a modular data security framework. Zhu et al. [12] used Blockchain to handle cloud data on the basis of higher voting approval. Liu et al. [13] employed Blockchain and smart contracts in IoT environments for data carrier applications. Facebook's data leakage problem to Cambridge Analytical took social media security issues to a bigger level. The challenges faced by social media networks today include are presented in [14]. An Ethereum platform-based applications is proposed for data recording and use IPFS [15]. It minimizes the central server storage space, improves data security by using unique content-addressable hashes for data verification, and also reduces computing power. Akasha is a social media framework built on dedicated Blockchain using Ethereum and IPFS. This project offers an instruction on how to create social media apps with the benefits of the Blockchain system. They explained in detail how to create a decentralized application using the Ethereum framework, web3.js library, and ReactJs as the front-end. Guidance and tips on building up a private Blockchain and making the whole DApp work-ready is provided.

3 Proposed Methodology

In this proposed work, the IPFS is mainly used to store the data in a distributed manner among the network participants locally. It minimizes the central server storage space, reduces computing power, improve security, and reduce cost. To improve data security, IPFS uses unique content-addressable hashes [16] for data verification if we store our content in Blockchain, it will be secure but it can handle only a small amount of data and the cost will be more.

IPFS is peer-to-peer storage; User can get the exact data from anyone in the globe. So, it is much quicker to transfer it. Using a unique hash value, one can verify that the data that is received is correct or not.

In this project, Blockchain is used for storing transaction hash securely. By creating a decentralized application, it is possible to share data with people whom we want to. We can upload the file in the form of text, images, and videos.

A. *Overall Framework*

Figure 1 explains the overall framework. First, the data will be stored in the IPFS which provides access to the correct data using hash code [17]. The data will be encrypted, and the hash code will be stored in Blockchain. Using smart contract, the

Fig. 1 Overall framework

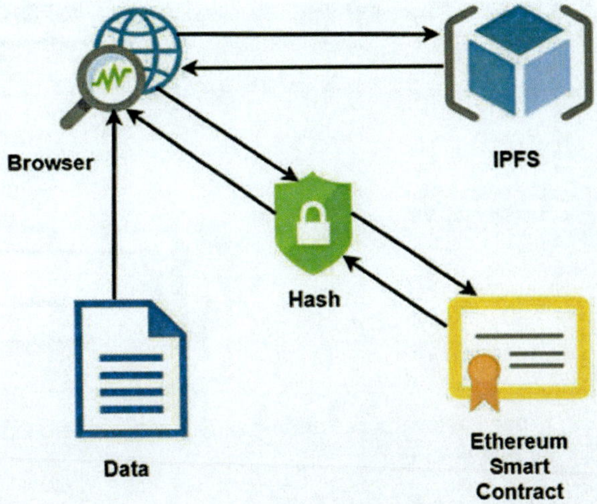

data can only be read. Before uploading the data, MetaMask extension will ask for the confirmation. It will check whether the persons are authorized user or not.

The proposed model works in **four steps**.

i. *Selecting the data*: The data can be either in the form of images or videos. We have to enter the details about the data in the description box.
ii. *Upload the data*: After selecting the data, click the submit button and it will be uploaded to the IPFS.
iii. *Storing the data*: After uploading, the hash value will be generated which will be stored into Blockchain. A public key and transaction ID will be associated with the hash value.
iv. *View the Transaction Details*: Only authorized users can view the file creation date, file creation time, original hash value, file access date, file access time, and hash modified file.

For implementing the smart contract, solidity language is used. Such smart contracts can also be configured to record real-world properties that are changeless. Without the need for a third party, the data can be traded securely. Every transaction that occurs in the network is recorded in to the blocks. Then, mining process is carried out in which the nodes in the network come to a consensus and a block is created. The new block created is then updated in all the nodes of the network.

B. *Distribution of data*

Every user in the system can send or receive data among them. Nevertheless, the data is not transmitted directly between users (e.g., A and B) but via the following steps.

i. First, send the data and it will store it in the IPFS as shown in Fig. 2 anyone can access the encipher data according to the data's hash value, but cannot modify it. They can only read the data.

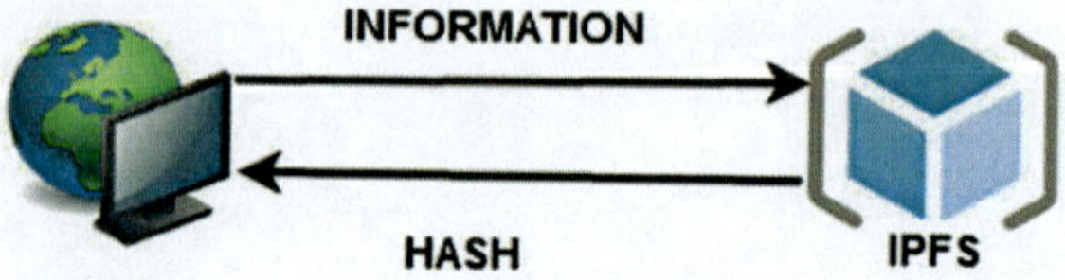

Fig. 2 Uploading the data to IPFS

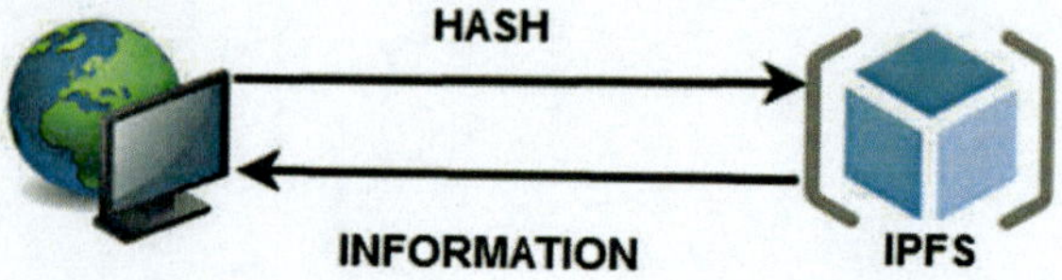

Fig. 3 Fetching the data to IPFS

ii. Using hash value, an authorized person can fetch, download, and modify the data as shown in Fig. 3.

A consensus protocol is chosen to ensure that the recorded data are not repudiated. The IPFS can also record a new transaction in about 3–6 s to enable nearly real-time delivery of data. Data distribution system based on Blockchain will ensure that person A has tried to send the data to the person B; if person B reads the details, the smart contract may monitor them further.

In Fig. 4, if the person wants to add a new data to the application, the admin ensures that the request is only from the authorizer user. The smart contract contains a list of address and only those addresses are allowed to add a new instance of the image to the contract. If the application does not have a similar image, it will be stored in IPFS. The transaction results and unique hash value will be added to the smart contract. If the image is already stored in IPFS, it will close the transactions. The admin and user will connect peer-to-peer to achieve an authorization agreement. Both the user and admin can communicate in two directions. The public key and the message to be sent will be received initially when sending a message. The message is encrypted with the receiver's public key, and then the packet is generated using this encrypted message. The packet is being transmitted on the network. Using the private key, the file can be decrypted. The position of the sender and receiver is interchanged between admin and user at the same time. After deciding on the amount, the user will send to the admin, the data will be made payable by the admin. It makes it possible for the customer to give the negotiated amount to the proprietor. After successfully sending value to the admin, a new contract to record this transaction is established. When a new contract is successfully created, the recipient will obtain the original data.

C. *Access to data via smart contract*

It can be implemented in a traceable and permanent way without third parties. In 1997, the smart contract was proposed by Nick Szabo. The Blockchain 2.0 functionality is also known as the smart contract. Using a contract, one can monitor the data, read the data, and verify the data ensuring transparency. To create a record, every user must

Fig. 4 Flowchart for finding the duplicate image

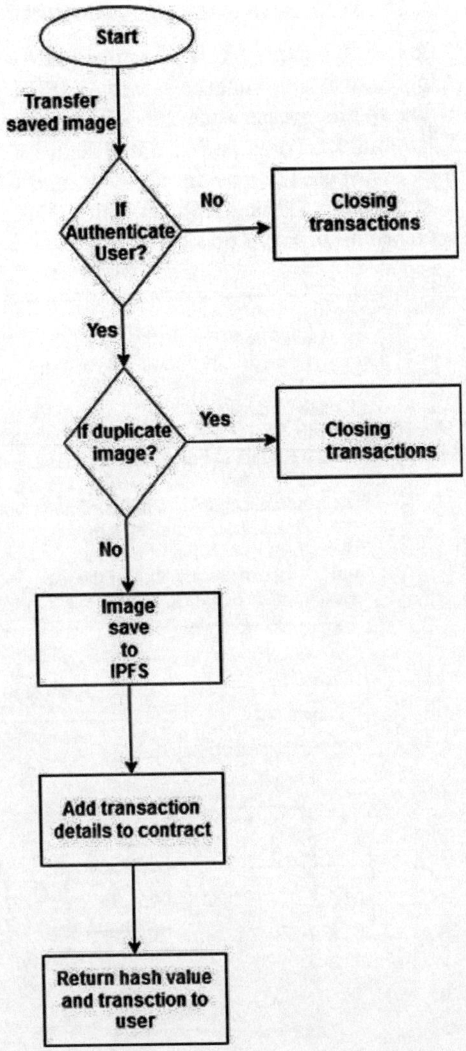

sign the smart contract, so that the data will be read by the user. In Blockchain, the contract will be registered and distributed to everyone in the network. When person B receives person A data and key, to read the data person B executes the contract. The contract offers both encryption and decryption services. For reading the data, the person must know the key value. When the person wants to view the information, it can be viewed only after decrypting the data, and the transaction will be written in the Blockchain. The user's identity and request for data decryption which is stored in the Blockchain helps us to know who has accessed it.

D. *Model for demonstrating the functionalities of IPFS*

Algorithm 1 explains the pseudocode for the sequence of steps to explain what happens when the data is stored in IPFS. Whenever a file is submitted by a user if the size is greater than 256 kB, it is divided in to data chunks of 256 kB each. Each chunk of data is associated with its own unique hash value. The chunks are then distributed across the network, and the chunk with corresponding hash value is stored in a distributed hash table (DHT). A sample for values stored in DHT is represented in Fig. 5 and Table 1.

```
Algorithm 1:
Input: Data File
Output: Content Addressable Hash Value

    Create root hash Hr
    for File Size > 256 kB
        split data D into 256kB chunks D1, D2... Dn
        assign hash value for each chunk h1, h2, ... hn and link it to Hr
    distribute the chunks to network nodes and update DHT
    create final Hash value Hr(h1 h2..... hn)
    store Hr(h1 h2..... hn) in blockchain
    link Hr with private key Kpriv of the user
    share Hr with network
```

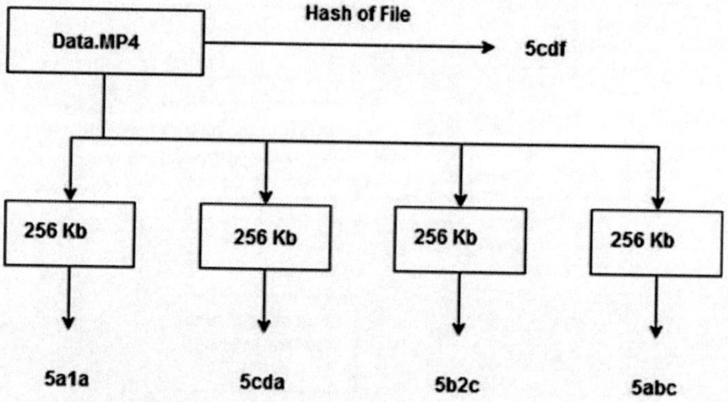

Fig. 5 Data is split into chunks of 256 kB each

Table 1 Sample contents of DHT

CHUNK	PEER ID	Hash value
D_1	6def (NODE 2)	h1
D_2	6def (NODE 2)	h2
D_3	8c2d (NODE 4)	h3
D_4	5abc (NODE 1)	h4
D_5	ON YOUR NODE	h5

Request for data access: Use Public key K_{pub} along with H_r to retrieve stored data in IPFS.

Share Data with other users: Provide H_r to the user.

Hash Value Key Generation: $H_r(h1\ h2 \ldots h_n)$ is the hash Key generated for Data D.

4 Experimental Setup

Configuring and managing the Blockchain-based data distribution framework is implemented using a web-based system. We can upload data into the system and download the data through the created interface.

A. *UI Design*

As the front-end framework, we use a truffle suite to call specific Web3 APIs to communicate with the confidential Blockchain. Decentralized application created called as Flick O' Post can be used for data hosting and sharing of data. After uploading, the data will be encrypted and stored in IPFS. Only the unique hash value will be stored in the Blockchain. By using the gateway and unique hash value, the person can retrieve the data. All the transaction details will be stored in the Blockchain.

A contract address will be shown upon successful login or registration. After the data is stored successfully, the MetaMask will pop-up the dialog box indicating that the transaction is confirmed. A person can view the details in Ropsten testnet network. By using the Ropsten testnet network, we can overview the balance of the ether. Only the authorized person can view transaction hash, transaction fee, block size, the value of ether, gas used, gas limit, average gas limit, and reward as shown in Figs. 6 and 7.

In Ropsten testnet network, authorized persons can view the incoming transactions, outgoing transactions, pending transactions, failed transactions, and completed

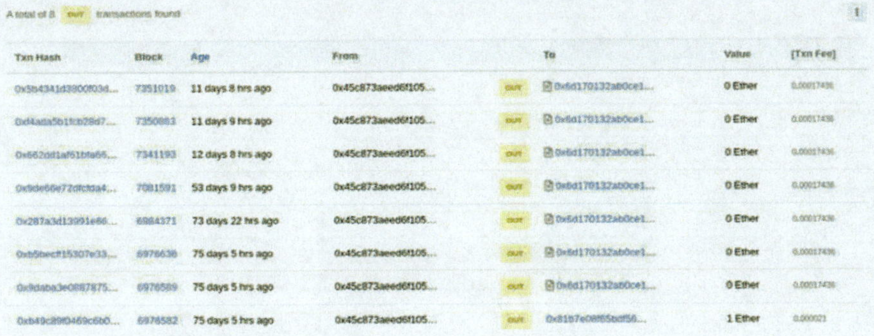

Fig. 6 Total number of transactions found

Block	Age	Txn	Uncles	Miner	Gas Used	Gas Limit	Avg.Gas Price	Reward
7318520	38 secs ago	16	0	0x635b4764d1939df...	1,669,092 (20.86%)	8,000,029	9.44 Gwei	2.01575 Ether
7318519	47 secs ago	1	0	0xe216ebdbc12adb...	21,000 (0.26%)	8,000,029	1.00 Gwei	2.00002 Ether
7318518	50 secs ago	13	0	0x635b4764d1939df...	1,414,102 (17.68%)	8,000,029	5.47 Gwei	2.00773 Ether
7318517	1 min ago	0	0	0x635b4764d1939df...	0 (0.00%)	8,000,029	-	2 Ether
7318516	1 min ago	12	0	0x635b4764d1939df...	1,350,462 (16.88%)	8,000,029	9.52 Gwei	2.01285 Ether
7318515	1 min ago	11	1	0x635b4764d1939df...	7,831,071 (97.89%)	8,000,029	11.64 Gwei	2.15366 Ether
7318514	2 mins ago	21	0	0x635b4764d1939df...	5,009,197 (62.61%)	8,000,029	5.31 Gwei	2.02661 Ether
7318513	2 mins ago	1	0	0x635b4764d1939df...	22,603 (0.28%)	8,000,029	2.20 Gwei	2.00004 Ether
7318512	2 mins ago	8	0	0x635b4764d1939df...	1,252,600 (15.66%)	8,000,029	1.51 Gwei	2.00189 Ether

Fig. 7 Data access record in the system

transactions. The authorized person can download (export) their transaction data independently. The legitimate person can view the latest blocks, latest transactions, miner details, and each transaction time. When the number of blocks increases simultaneously the Eth cost will increase but when the number of transaction increases, the cost will remain same. In Fig. 8, it shows the transaction details for only one data. It contains whether the transaction is success, transaction fee, transaction hash and timestamp, etc.

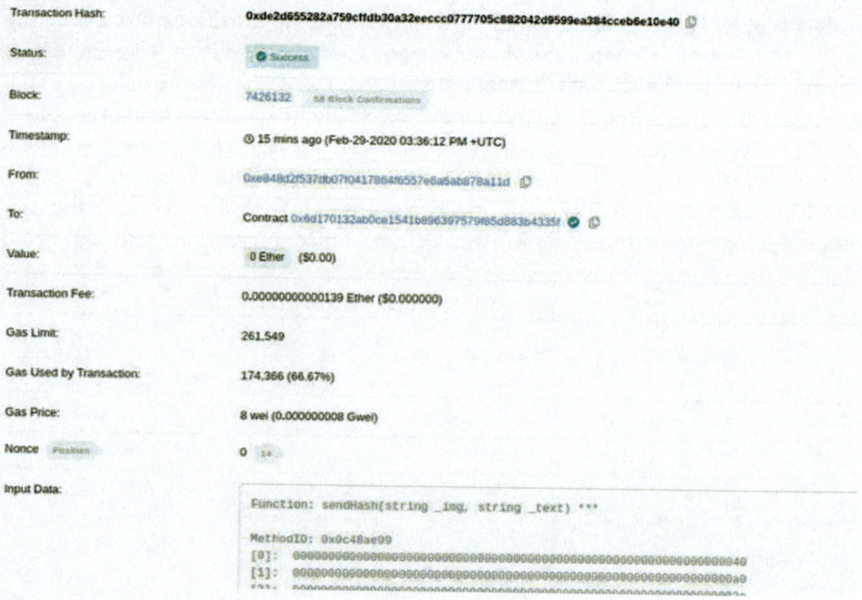

Fig. 8 Transaction details for only one data

B. *Simulation Setup*

Implementation configuration requirements are: Intel(R) Pentium(R) CPU A1018 @2.10 GHz, 2100 MHz, 2 Core(s), 2 Logical Processor(s). To write a smart contract, solidity programming language is used. JavaScript is used for user-interactive forms, and using Bootstrap 4.0, GUI is developed by front-end. The key methods used to build this program are given below:

i. NPM—NPM stands for node package manager personal BC will look like a real Blockchain that can be connected to the public (anyone can run on PC)
ii. VSC—VSC stands for visual studio code. It is a lightweight platform mainly used for editing the code. It is efficiently used for highlighting the syntax, refactoring the code, completing the code, and variety of tools are used for debugging the code.
iii. Ganache—It will be running on both servers and computers. Ganache is used to deploy the smart contract, run test, and to run script opposed to the network.
iv. MetaMask Extension—It is a kind of browser, able to connect to the network and personal account. In MetaMask, one can add token, account details, confirm transactions, import, or remove account.

5 Performance Evaluation

Gas price is set to 2 Gwei to run the corresponding smart contracts. 1 Gwei is equal to 10^9 Gwei [18]. The costs for smart contracts assessed as shown in Figs. 9 and 10. Using a contract, one can know the consumption values of gas. After uploading the data, the value of gas used, transaction fee, and contract storage space are registered. The transaction fee is the amount of data sent to the IPFS, and the execution price is used to execute the task. When a user sends data to the IPFS, the transaction fees will

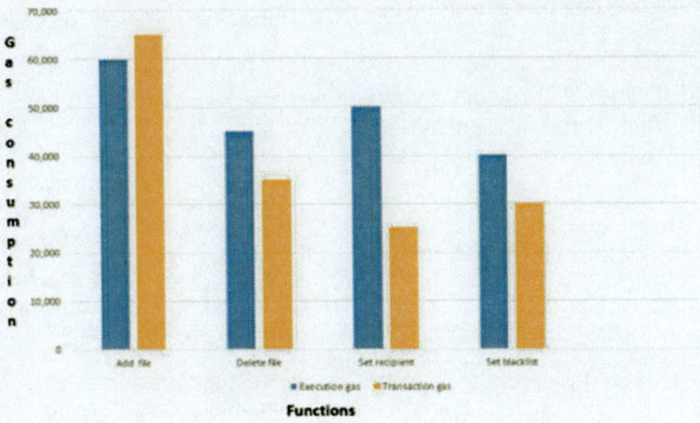

Fig. 9 Gas consumption for admin

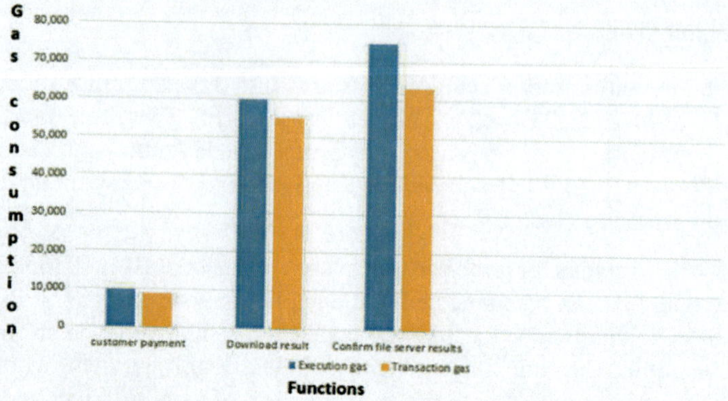

Fig. 10 Gas consumption for user

be approximately 0.00017436. Likewise, the delete data feature is invoked when any file is removed from the device. Then, the event confirm file deletion is activated. The actual cost of performing this role is $0.0066 USD. When a new person is authenticated, then the person will be a member. As a result, the set blacklist function has a recorded cost of $0.0034 USD. Here, the set blacklist function will also be recorded. As a result, gas consumption will be recorded for adding the file, deleting the file, adding the recipient, and set blacklist as shown in Fig. 9. Two contracts with Ethereum address were checked 0 × 4b4d29c4e53b6d9d7e920513fdc7c5410897b9c4e536b and 0 × 58376bd6afabb30214f025140225031d113 are checked. The other two contract roles are download results and confirm results are calculated as shown in Fig. 10. Later downloading the file from the IPFS, the user can later check the download results.

To evaluate the performance, we performed data creation, modification, and deletion. Figure 11 shows the average response time for different sizes. The key performance metrics for the proposed systems are average response time, the gas consumption rates for smart contract execution and other functions. Based on the experimental

Fig. 11 Average response time with different file size

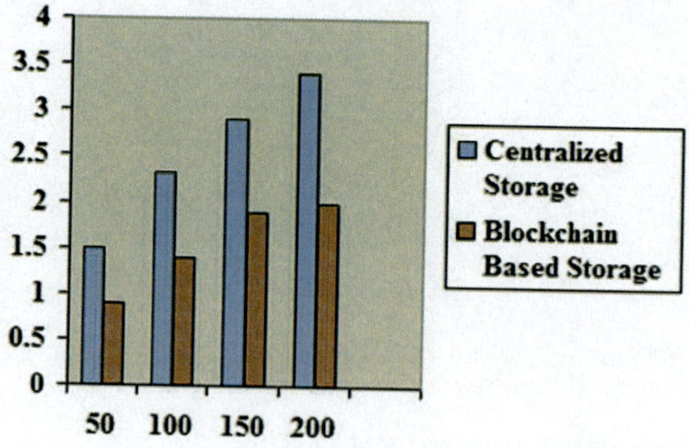

results, it is seen that the time taken by the traditional approach, and Blockchain is very high whereas IPFS takes less time. These results are obtained by evaluating the random file names and content for 100 repetitions with size of 2000 MB. The data is stored in IPFS, and only the hash value will be stored in Blockchain.

In this paper, we are focusing on the storage system. When compared to the traditional and Blockchain approach, the IPFS takes less storage space and cost. Storing large data in Blockchain incurs high cost and is not an efficient. IPFS provides content addressable, scalable, distributed storage which is hash-based. The transactions of the system are recorded using Blockchain for ensuring security and accountability. If the number of file size increases, the time for storing the data takes less time. For example, if a user stores a file size as 500 MB, it takes 20 ms time to store data in IPFS. Using IPFS, to store, to share, to edit the data will be easier.

6 Conclusion and Future Scope

This paper has presented how data can be stored in IPFS. In Blockchain, one can store a sensitive data but cost will be more expensive. Instead of storing in Blockchain, one can store a sensitive data efficiently in IPFS. A Blockchain-based communication framework is created to store the hash value securely. IPFS is used to address the storage problem of Blockchain. Tests were performed based on IPFS. This framework is used to minimize the central server storage space, improve data security by using unique content-addressable hashes for data verification, and also reduce computing power. It can be used for any kind of IPFS transaction, as it has relatively good distributed storage efficiency. This work can be enhanced by implementing the performance testing model for distributed networks and improve the handling of web3 errors.

Acknowledgements The authors wish to express their sincere thanks to the Department of Science & Technology, New Delhi, India (Project ID: SR/FST/ETI-371/2014) and express their sincere thanks to the INSPIRE fellowship (DST/INSPIRE Fellowship/2016/IF160837) for their financial support. The authors also thank SASTRA Deemed to be University, Thanjavur, India for extending the infrastructural support to carry out this work.

References

1. P.K. Sharma, M.Y. Chen, J.H. Park, A software defined fog node based distributed blockchain cloud architecture for IoT. IEEE Access **6** (c), 115–124 (2018)
2. S. Nakamoto, *Bitcoin : A Peer-to-Peer Electronic Cash System*, pp. 1–9
3. A. Rajalakshmi, K.V. Lakshmy, M. Sindhu, P.P. Amritha, A Blockchain and IPFS based framework for secure Research record keeping. Int. J. Pure Appl. Math. **119**(15), 1437–1442 (2018)

4. R.A. Saritekin, E. Karabacak, Z. Durğay, E. Karaarslan, Blockchain based secure communication application proposal: Cryptouch, in *2018 6th International Symposium on Digital Forensic and Security (ISDFS)*, pp. 1–4 (2018)
5. N. Rifi, E. Rachkidi, N. Agoulmine, N. C. Taher, Towards using blockchain technology for eHealth data access management, in *2017 Fourth International Conference on Advances in Biomedical Engineering (ICABME)*, pp. 0–3 (Oct 2017)
6. W. Huang, *A Blockchain-Based Framework for Secure Log Storage*, pp. 96–100 (2020)
7. Q. Xu, Z. Song, R.S.M. Goh, Y. Li, Building an ethereum and IPFS-based decentralized social network system, in *2018 IEEE 24th International Conference on Parallel and Distributed Systems (ICPADS)*, pp. 986–991 (Dec 2019)
8. L. Chen, W. Lee, C. Chang, K.R. Choo, Blockchain based searchable encryption for electronic health record sharing. Futur. Gener. Comput. Syst. **95**, 420–429 (2019)
9. N. Nizamuddin, K. Salah, M. Ajmal Azad, J. Arshad, M.H. Rehman, Decentralized document version control using ethereum blockchain and IPFS. Comput. Electr. Eng. **76**, 183–197 (2019)
10. S. Pirbhulal, O.W. Samuel, W. Wu, A.K. Sangaiah, G. Li, A joint resource-aware and medical data security framework for wearable healthcare systems. Futur. Gener. Comput. Syst. **95**, 382–391 (2019)
11. L. Du, W. Zhang, H. Fu, W. Ren, X. Zhang, An efficient privacy protection scheme for data security in video surveillance. J. Vis. Commun. Image Represent. **59**, 347–362 (2019)
12. L. Zhu, Y. Wu, K. Gai, K.K.R. Choo, Controllable and trustworthy blockchain-based cloud data management. Futur. Gener. Comput. Syst. **91**, 527–535 (2019)
13. X. Liu, K. Muhammad, J. Lloret, Y.W. Chen, S.M. Yuan, Elastic and cost-effective data carrier architecture for smart contract in blockchain. Futur. Gener. Comput. Syst. **100**, 590–599 (2019)
14. Y. Li, Z. Zhang, Y. Peng, H. Yin, Q. Xu, Matching user accounts based on user generated content across social networks. Futur. Gener. Comput. Syst. **83**, 104–115 (2018)
15. Y. Ranka, J. Bagrecha, K. Gandhi, B. Sarvaria, M. Chawan, U.G. Student, A survey on file storage and retrieval using blockchain technology, in Int. Res. J. Eng. Technol. 763 (2008)
16. M.S. Ali, K. Dolui, F. Antonelli, IoT data privacy via blockchains and IPFS. ACM Int. Conf. Proceeding Ser. (2017)
17. V. Jacynycz, *Towards a Decentralized Publication System : A Proposal Using Blockchain and P2P Technologies* (2018)
18. M. Naz et al., A secure data sharing platform using blockchain and interplanetary file system. Sustainability **11**(24), 1–24 (2019)

Implementation of Audio Image Steganography—A Technique to Secure Data

Satvik Yadav, Shreya Dewan, Sumit Agrahari, Dheeraj Singh, and Tania Gupta

Abstract The modern era comes with a very crucial job of encrypting the important information before sharing it in order to avoid its misuse. Steganography is the most efficient way of encrypting the data and hiding it in a carrier audio or image file, with minimal increase in the size of the file, and sharing the data securely between two parties. The objective of this paper is to implement audio and image steganography, i.e., concealing of data within an audio file or an image file. The aim is to develop an application for encoding and decoding. Various cryptography techniques like Caesar cipher have been used to achieve steganography. MATLAB tool has been used to achieve the same.

Keywords Audio · Image · LSB · Steganography

1 Introduction

Information security is the most crucial factor which is observed when a confidential information is been shared between two parties [1]. The involvement of third party applications in the process of sharing the data can some time lead to not only loss of confidential information, but also misuse of confidential information. Thus, encrypting the confidential data before sharing is a necessity.

S. Yadav (✉) · S. Dewan · S. Agrahari · D. Singh · T. Gupta
ABES Engineering College, Ghaziabad, India
e-mail: satvik.16bec1049@abes.ac.in

S. Dewan
e-mail: shreya.16bec1028@abes.ac.in

S. Agrahari
e-mail: sumit.16bec1140@abes.ac.in

D. Singh
e-mail: drajiitr@gmail.com

T. Gupta
e-mail: taniagupta2409@gmail.com

V. E. Balas et al. (eds.), *Proceedings of International Conference on Computational Intelligence, Data Science and Cloud Computing*, Lecture Notes on Data Engineering and Communications Technologies 62, https://doi.org/10.1007/978-981-33-4968-1_45

For sharing the encrypted data, the two techniques Steganography and Cryptography are being used. Cryptography is used to send the data in encrypted form using a key (public or private). In Cryptography, if the key is found, the data can easily be decrypted. Also, the third party knows that a confidential information is been shared between two parties [1].

The Steganography is better approach as it not only encrypts the data but also hides the encrypted data inside an audio or image file, so that any third part is not able to recognize that the hidden data is shared. In this paper, we are hiding the confidential information using suitable methods via Steganography technique.

Steganography hides the secret information in a cover objects or carriers such as audio or image files [2]. The cover objects that is hiding the data is called the stego object. The secret information can be text, image, or audio. In this paper, the cover is an audio or image file, and the secret data is hidden in the cover files. In this paper, we are hiding the data such that the size of the cover file is minimally increased. The capacity of the technique is the amount of the data that the technique can hide with minimal increase in the size and without disrupting the quality of the cover objects. Along with the capacity of the data hiding schemes, it is also required to see the robustness of the scheme. The robustness is the measure of how much the hidden data can withstand the attacks while sharing hidden data in the stego objects. The attacks may include the addition of white noise, rescaling, distortion, resizing, compression, etc.

In the audio Steganography, the confidential information is hidden into the digitally processed audio signal which can slightly alter the binary sequence of the audio file. The methods used are LSB coding, amplitude modulation, etc. [3].

In image Steganography [4], the algorithm to hide the data inside the image cover can be a slightly different based on the cover file format. Most popularly the used algorithms for image Steganography is LSB insertion, masking and filtering. Using these algorithms, the message signal is shared with utmost security with only minimal increase in the size and almost no loss.

1.1 Steganography

Steganography is made up of two Greek words *steganos* and *grapehin* which mean "concealed" and "writing," respectively. It is the practice of concealing a secret message within an ordinary message and extracting the same at the destination end. It takes cryptography a step further by using various other techniques which will be discussed further in this paper.

This paper focuses on **Audio Image Steganography** which discusses about how data can be concealed within a cover image or an audio file.

Audio signals are analog in nature. To digitally process the audio signals, it is required to sample those signals periodically with time [5]. Audio files are present in various formats. The easy format is WAV file. WAVs store samples such that there is no need of any preprocessing. The most popular format for audio files is MP3.

The MP3 is first processed and converted into the samples that are periodic to time. Then, the samples are ready to be used for further Steganography process.

Most popularly the cover used for hiding the secret data is an Image file [6]. The cover images may be gray images or colored. Color Images have way more opportunity to store the confidential data than the gray images. Thus, color image Steganography is preferred than gray image Steganography. Colors can be of different formats like Red Green Blue (RGB), hue, saturation, value (HVS), YUV, YIQ, etc. The color image Steganography can be done in any of the above color patterns [2].

An application for decoding has been developed by us for the same. Using the application, we first retrieve the encoded media, then using various algorithms, we retrieve our encoded message from within the received media, and display it to the user.

2 Modeling

There are two inputs available to us at the sending end, which are the cover object (image or audio) and the secret data. The embedding of secret data is done via Steganography algorithms. The **stego object** is obtained.

This flow is shown in the following block diagram (Fig. 1).

When the stego object reaches its receiving end, the extraction of secret data takes place using detecting nature of steganography algorithms used. The cover object (image or audio) are separated from the secret data, and the data is displayed to the user. This has been illustrated in the following block diagram (Fig. 2).

Fig. 1 Block diagram for flow of steganography technique at sender's end

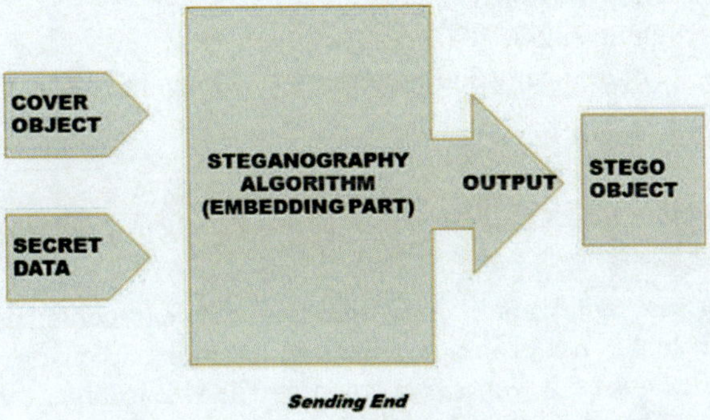

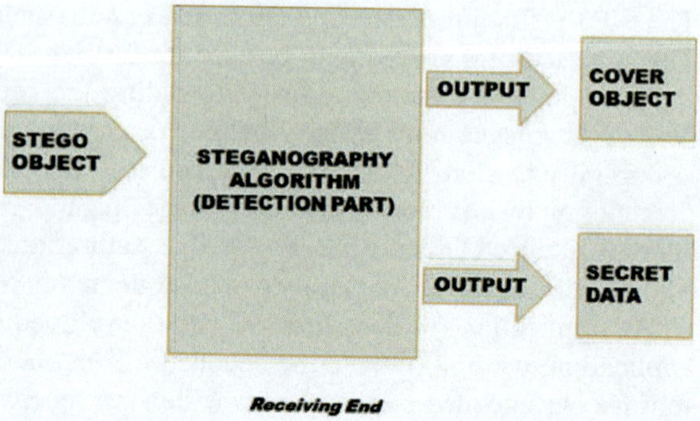

Fig. 2 Block diagram for flow of steganography technique at receiver's end

3 Image Steganography

3.1 Image Component

An image file consists of little dots called pixel, where each pixel is represented by 3 bytes. In the color wheel, there are 3 basic colors which mix to form all the other existing colors. These are Red, Green, and Blue. The 3 bytes in a pixel are made up of these colors. Each byte is interpreted as an integer, which tells how much of that color is used to form the final color of the pixel. The range of these pixel values is from 0 to 255 for a grayscale image. Below is an example of how "orange" color is formed.

Red: **11,111,000**—248.
Green: **11,001,001**—201.
Blue: **00,000,011**—3.
Orange: **248 201 3** (RGB value).
For orange, value of Red is 248, Green is 201, and Blue is 3.

3.2 LSB-Based Data Embedding for Image

Least significant bit algorithm is an efficient method used for image Steganography. It makes use of image's pixel information and changes it according to the message being sent. It works best when the file size is larger than the message being sent.

The concept of LSB embedding is to change the last bit value of a pixel. The result is a minor distortion in the cover image.

Mathematical Formula:

$s(i, j) = c(i, j) - 1$, if LSB = 1 & SM = 0.
$s(i, j) = c(i, j) + 1$, if LSB = 0 & SM = 1.
$s(i, j) = c(i, j)$, if LSB = SM;

where,

c(i, j) SB of cover image;
s(i, j) Stego image;
SM next message bit to be embedded.

As it can be seen from the example below, only a few pixels of the cover image are affected, and the result is only a minor difference in the shade of the pixel, which goes unnoticed to all outside entities except the receiver which has the tool to extract the hidden data from the image (Fig. 3).

We have taken the following image as our input sample (Fig. 4).

After LSB-based embedding of data "*message*," the output image we receive is (Fig. 5).

Thus, we can conclude that data can be successfully hidden without affecting the original image to a noticeable extent.

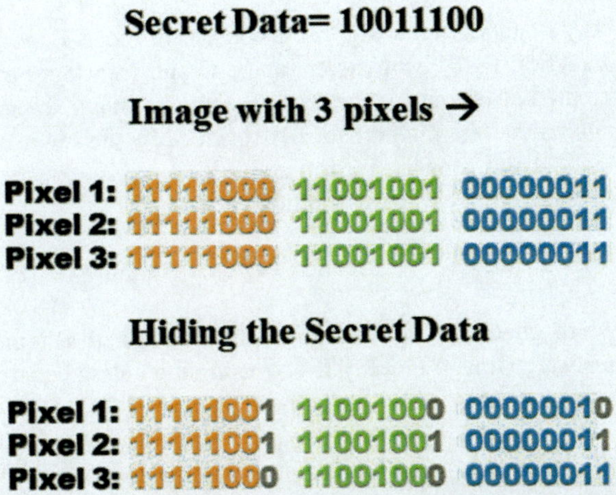

Fig. 3 Hiding the secret data in image by replacing the last bit using LSB algorithm

Fig. 4 The input cover image in which we will hide the data

Fig. 5 The stego image containing the hidden data

3.3 Mathematical Modeling

Any steganographic algorithm consists of Stego-Function F as well as inverse Stego-Function F-1. Using cover image C and information I as input, Stego-Image S is created as output. At the receiver end, the Stego-Image S is fed into decoding algorithms that are mathematically reversed by the inverse Stego-Function (represented as F-1) and generate information I.

Mathematical representation:

$$\mathrm{S} = \mathrm{F}(\mathrm{C},\ \mathrm{I})\ \mathrm{and}\ \mathrm{I} = \mathrm{F} - 1(\mathrm{S})\ \mathrm{and}\ \Psi = \{\mathrm{F},\ \mathrm{F} - 1,\ \mathrm{C},\ \mathrm{S},\ \mathrm{I}\}.$$

To check the robustness of the mentioned scheme, we can check the fidelity measure (Image Fidelity IF) by using the below equation:

$$\mathrm{IF} = 1 - \frac{\sum_{I,J}(P(i,\ j) - S(i,\ j))^2}{\sum_{I,J} P(i,\ j)xs(i,\ j)}$$

3.4 Application

We have used MATLAB tool to achieve audio steganography.

- When we run our script for encoding, following dialog box appears.
- This requires an image to be selected for encryption. A secret text has to be entered in the black box.
- By selecting Crypt-it, the file is encrypted and can be transferred to the destination required to be sent to.
- When we run our script for decoding, following dialog box appears.
- This requires to select the encrypted file to be decrypted.

- When we select De-Crypt it, the secret text hidden within the file is displayed played in the bottom text box (Figs. 6 and 7).

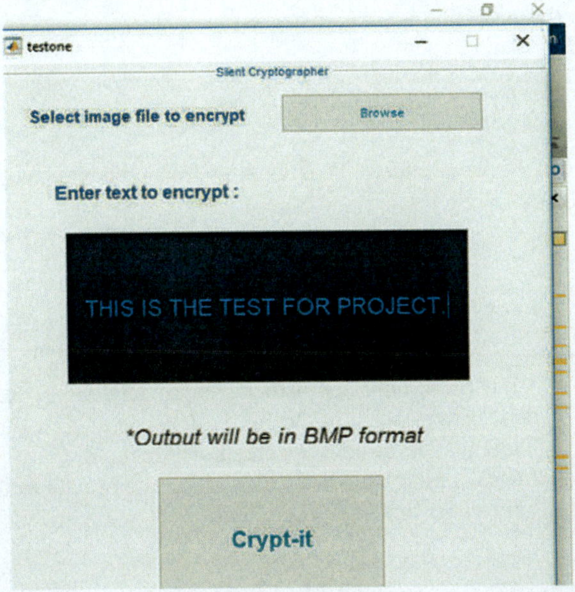

Fig. 6 The encrypt user interface, used for encrypting the data in cover object

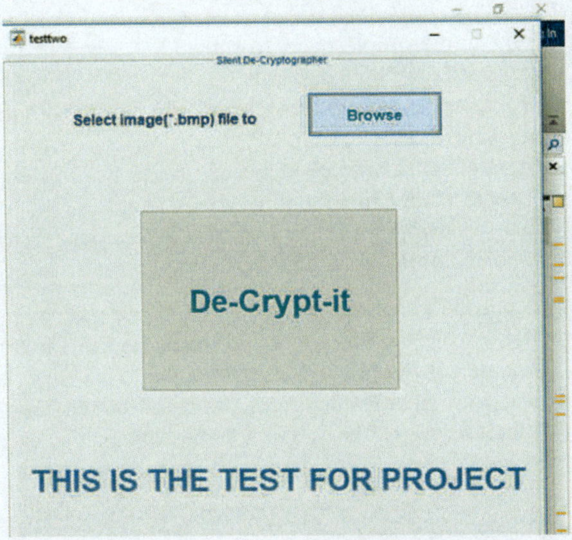

Fig. 7 The decrypt user interface, used for decrypt the stego object and retrieve the hidden data

4 Audio Steganography

To embed secret data onto a digital audio file, there are a few coding techniques introduced:

- LSB coding
- Phase coding
- Parity coding.

We have chosen LSB Coding for audio Steganography over phase and parity coding.

4.1 *Why LSB Coding?*

- Human ear can often detect even the slightest noise introduced into a sound file, as it is very sensitive to noise. This poses as a major disadvantage with parity coding, as it introduces some noise in the file.
- Phase coding has low transmission rate of data as the secret message is encoded only in the first signal segment.

4.2 *LSB-Based Data Embedding for Audio*

It is by far the simplest way to embed a message in a digital audio file, offering with it a considerably high data rate.

We replace the LSB of each sampling point with the given binary information in encryption.

In decoder, the receiver needs access to the sequence of samples used in the process of embedding the binary message.

Let's consider an example where a message "Yes" is to be encoded in a 16-bit sample audio file using LSB embedding.

- Here, the given secret data is "Yes" and the cover file is the given 16-bit audio file. "Yes" is to be encoded in the audio file. Firstly, "Yes" and the sample audio file are converted into ***bit streams***.
- The least significant column of bits of the sample is replaced by the bit stream of secret message "Yes". The file then obtained is called stego-file or object.

4.3 *Application*

Following script is run in MATLAB for hiding the secret data. Here, "chirp.wav" is the input audio file stored in attribute in_file.

The secret message hidden is "sumit."

Following script is run for recovering the message "sumit" from out_file chirp_out.wav.

The following waveforms are obtained for in_file and out_file that is the original and the stego file (Figs. 8, 9, 10).

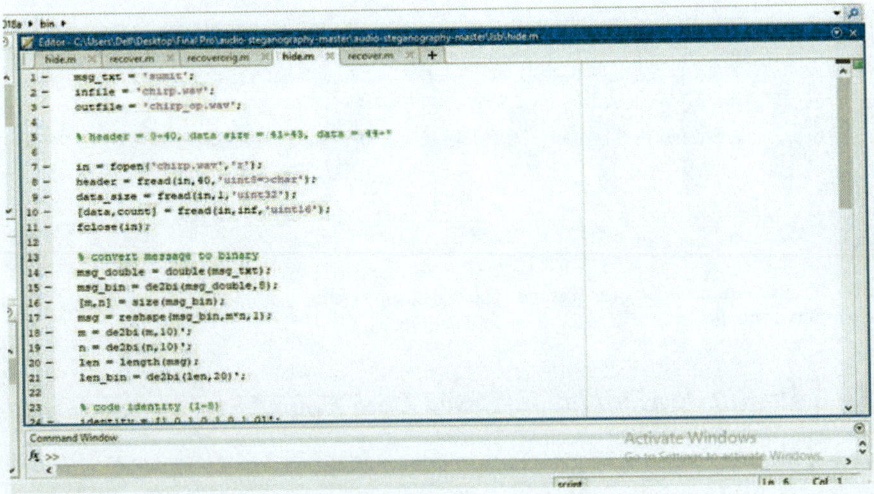

Fig. 8 Code used for hiding the data in audio taken as cover object

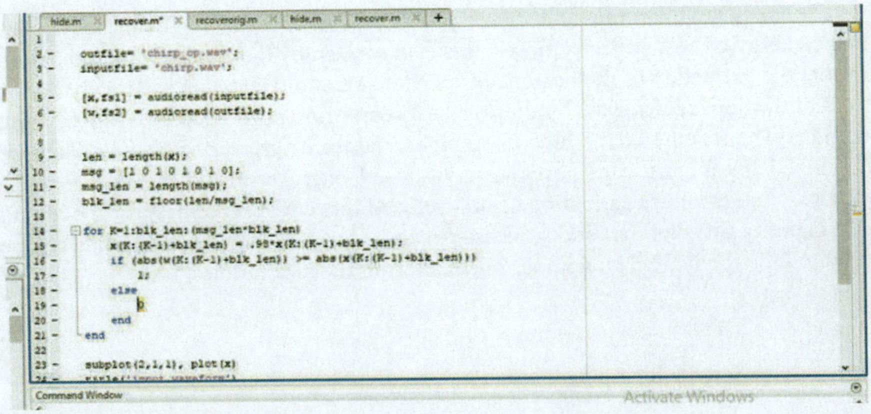

Fig. 9 Code used for decrypting the data hidden in audio stego object

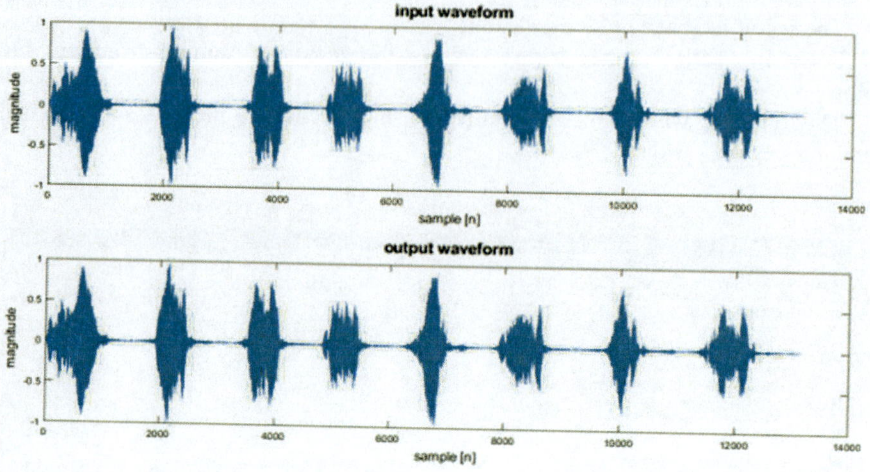

Fig. 10 Amplitude modulation method. The waveforms shows that there is almost no loss in the cover object quality even after encryption and decryption

4.4 Amplitude Modulation-Based Data Embedding for Audio

This technique involves division of audio signal in numbers of non-overlap portions/pieces, the average of energy of all the portions/pieces are inspected, and the portions/pieces having the greatest energies are taken as the better hosts for the undisclosed bit.

The change in the coefficient values, produced by technique, might cause changes in the energy of host portions/pieces, this is prevented by the algebraic manipulation so that the extraction of the host pieces/portions of audio signal is accurately done.

Bit insertion is carried out by altering the selected host portions/pieces into discrete domain. Then, some of the high coefficients values of greater frequency band are quantized for them use to host the undisclosed bits. The result is given that the flawless, and errorless extraction of undisclosed bits could be achieved. The hiding rate is pretty tiny, but the hidden message can be large in the size. Thus, this can be used to hide the large data without much altering the size (Table 1).

5 Conclusion

In this paper, we have shown the application of audio and image steganography through the explained methodologies. Any secret message can be embedded into an audio or image file and can be sent with minimal changes in size and type of file. Maximum efficiency is provided via LSB algorithm. This reduces the chances of detection of the embedded message.

Table 1 Comparison with the existing technique

Parameters	Implementation of audio image steganography	Qadir M.A. "A Review Paper On Cryptography" Software Engineering Department, Firat University
Concept	Using encryption and decryption algorithm	Using public and private keys
Algorithms	Least significant bit (LSB) method, parity check, amplitude modulation	Caesar cipher, simple substitution ciphers, transposition ciphers
Security	High	Low
Time taken for encryption and decryption	More	Less
Third party interference	Very low	High
Efficiency	High	Low
Reliability	High	Low

The Steganography technique is widely an advanced version of a technique to send hidden data called as Cryptography. In Cryptography technique, the data is been encrypted on the sender side via a key (public or private key) and is decrypted by the receiver side using the same key it was encrypted.

So, the data is sent securely to the receiver without being misused by the man-in-the-middle or a hacker. In today's advanced technology world, the key can also be stolen by a third party and the data can be misused.

However, in the Steganography technique, the data is encrypted inside the image or audio file(a carrier), with a very minimal increase in the file size, the information is securely sent to the receiver without any third party recognizing that the encrypted information is hidden inside the image or audio file. Cryptography confuses the information, but it discloses the existence of hidden information. Steganography hides the existence of hidden information so no party other than the sender and receiver can recognize that the confidential information is been hidden.

Thus, Steganography is a better technique than Cryptography.

We have picked the best possible methods and tools at our disposal. Although, these methods have few limitations such as limited bit rate and are only specific to certain type of file. These may be improved with further research and methods like modified LSB and PVD for audio steganography.

However, in the Steganography technique, the data is encrypted inside the image or audio file(a carrier), with a very minimal increase in the file size, the information is securely sent to the receiver without any third party recognizing that the encrypted information is hidden inside the image or audio file. Cryptography confuses the information, but it discloses the existence of hidden information. Steganography hides the existence of hidden information so no party other than the sender and receiver can recognize that the confidential information is been hidden.

Thus, Steganography is a better technique than Cryptography.

We have picked the best possible methods and tools at our disposal. Although, these methods have few limitations such as limited bit rate and are only specific to certain type of file. These may be improved with further research and methods like modified LSB and PVD for audio steganography.

References

1. T. Aiswarya, S. Mansi, Aishwarya Talekar, Pallavi Raut, Steganographic technique for hiding secret audio in an image. Int. J. Res. Eng. Appl. Manage. (IJREAM), **03** (Issue 04) (May 2017)
2. S. Choudhary, C. Panwar, Key based image steganography using DWT and Chaotic Map. Int. J. Eng. Manage. Res. (IJMER) **4** (4), pp. 94–97 (2014)
3. F. Djebbar, B. Ayady, H. Hamamz, K. Abed-Meraimx, A view on the latest audio steganography techniques, in *International Conference on the Innovations in Technology* (2011)
4. T Morkel, J.H.P Eloff, M.S Olivier, An overview of image steganography, in *Proceedings of the Fifth Annual Information Security South Africa Conference (ISSA2005)* (2005)
5. P. Dutta, D. Bhattacharyya, T.H. Kim, Data hiding in audio signal. Int. J. Database Theory Appl. (2009)
6. R. Chandramouli, M. Kharrazi, N. Memon, Image steganography and steganalysis: Concepts and practice, in *IWDW 2003, LNCS 2939* ed. By T. Kalker et al. (Springer-Verlag Berlin Heidelberg, 2004), pp. 35–49

A Unique Approach for Detection and Removal of Key Loggers

Mukund Kr Kedia and Baisakhi Das

Abstract Key loggers are a significant threat to cyber security and privacy worldwide. Though they are useful in some scenarios, they do more harm than good. Companies put key loggers in their systems, to keep an eye on the working of their employees. Most often, attackers send key loggers to victims to get sensitive information like the id, password, credit card details, etc. The antivirus identifies these key loggers in most scenarios, but new key loggers are implemented every day and modern techniques to escape antivirus. In this paper, some methods have been discussed to detect a key logger if installed in the system and also to render the key logger useless even if present in the system. Also, it shows a novel technique of detection and removal of key logger. The approach is to filter and sort out those processes from the Windows Process Explorer, which shows maximum tendencies of being malicious. Then, most malicious application/processes execution is stopped, and their list is presented to user. User has the choice to let them remain in the system or discard them.

Keywords Key logger · Security · Process explorer · PowerShell · Task manager

1 Introduction

Key loggers have become a significant threat to the privacy and security of the user. It can log every chat, every credential, and information that may pass on to the computer through the keyboard. Though keyboards are the prime source of information gathering for a key logger, attackers have advanced them even to take screenshots at regular intervals to capture even the onscreen keyboard, which a person may use as a security measure. Also, key loggers are not like the legacy viruses and worms

M. K. Kedia (✉) · B. Das
Institute of Engineering and Management (IEM), Kolkata, West Bengal, India
e-mail: mukundkedia3@gmail.com

B. Das
e-mail: baisakhi.das@iemcal.com

V. E. Balas et al. (eds.), *Proceedings of International Conference on Computational Intelligence, Data Science and Cloud Computing*, Lecture Notes on Data Engineering and Communications Technologies 62, https://doi.org/10.1007/978-981-33-4968-1_46

that we are familiar with. It is tough to detect a key logger because key logging does not entirely fall in the blacklist category. Many legitimate software and applications require this functionality, and thus, attackers take advantage of this privilege provided [1]. Between January 2003 and July 2006, the number of key loggers has risen by 500% posing serious threat to cyber space. Meanwhile, key loggers are becoming more diverse, sophisticated, deceptive, and increasingly difficult to detect by antivirus software and anti-key loggers based on the signature analysis.

The key logger has become a real issue in today's world. Lots of data are leaked daily, and there is no account of that. Standard security measures are just not sufficient to curb this rising threat. Key loggers are regularly developed with new styles and techniques, and automated tools can only detect them up to a certain extent. Though tools are being developed and updated daily, there is still a lot of space for improvement in implementing security measures against them. Key loggers have not always been the black hat hacker's tool. It started as a security measure by big companies who wished to track their employees' activities. This led to software development that would be able to log keystrokes from the user's keyboard. Attackers saw it as an opportunity to exploit the new feature and use it for their benefit. Key logger is a kind of spyware that spies/monitors the victim's activity and sends the report to the attacker. This is very dangerous for systems that are accessed daily to initiate primary functions like fund transfer and governmental issues.

Key logging is one of the significant password attacks being done today. Other significant attacks include shoulder surfing, dictionary attack, phishing attacks. Many security features and techniques have been evolved to counter these attacks, such as keystroke dynamics, graphical passwords, biometrics, and time signature. These and many more techniques are developed to prevent key logging attacks. The input of passwords and essential credentials through the keyboard is quite prevalent, so there is a severe need to make key logger detection schemes and techniques [2].

Section 2 of the paper contains the literature review about the previous works that have been done on key logger, their removal and prevention from data theft through it. Section 3 gives some detail about the key logger and its working. Section 4 contains a novel approach of detecting key logger if installed in computer, and Sect. 5 concludes the paper.

2 Literature Review

A Windows hook is the core of key loggers. A hook is a point in the mechanism of system handling where procedure is installed to intercept the message traffic before reaching the target windows procedure [3]. There are majorly two types of hooks: system-wide and application-specific. System-wide hook is used to collect information from all the applications, whereas application-specific hooks filter and log the keystrokes corresponding to a specific thread [4]. For the same reason, anti-hook techniques are being implemented to detect every process, executable, DLLs, which may be using hooks to log the keystrokes. The current technique can intercept events before they reach an application. Functions can modify, discard, or act on events. Those functions which receive events are filter functions. Hooks use these filter functions to modify, record mouse and keyboard events.

A key logger can also be hardware-based. It may be in the form of a small USB drive or a small ps2 connector [5]. Keyboards use ps2 connector to connect to the motherboard, and these key loggers are attached between the port and keyboard wire, thus gaining access to every keystroke sent by the keyboard [6]. Another method is to install a key logger circuit into a standard keyboard. It has the advantage of no physical evidence of user monitoring.

Kernel-level key loggers require all the administrator privileges from the user, and once they get those privileges, it is almost an impossible task detecting them. Therefore, kernel-level key loggers are also challenging to create and need high expertise in this field. Whereas the user-space key loggers work at the application/user level, and they do not require any privileges as such [7].

Though general applications work at ring 3 level and key logger works at ring 0 level [8]. Therefore, more than 90% of key loggers encountered are user-space key loggers. The key logger starts running automatically as we open the file with which it comes attached, and also, they hide are not easily visible. This is the reason that users do not get a hint of key loggers being installed on their computers.

Dynamic detection technique must be used for the detection as some of the static methods used in the past are no longer trustworthy for key logger detection [5]. Binary signature alone was the criteria to filter the legitimate versus illegitimate applications. Many black box detection techniques have been implemented, though it is challenging to accurately determine the key loggers from the vast number of applications running simultaneously. Key loggers also have evading algorithms implemented with them to escape detection or bypass any filters used upon them. Signature-based detection problem was resolved with the use of TAKD algorithm. It incorporated anomaly and log-based detection. Taint data analysis is also gaining popularity; it consists of tracking data in a program at runtime. It uses a host-based intrusion detection system (IDS); to taint, monitor, and examine the keyboard data at the keyboard device driver level. Previous works [5] have been based on the detection of applications using the Windows Keyboard Hook method, Keyboard State Table method, and kernel-based Windows Filter Driver method. In contrast to many other malicious programs, a key logger has a very well-defined behaviour that is easy to model [9]. The general function of key logger is to eavesdrops and log each keystroke and then log the collected data into a file. In this scenario [4], the events triggering the malicious activities are always known in advance and could be reproduced and controlled.

As it is challenging to detect key loggers most of the time accurately, some techniques have also been developed to let the key loggers prevail in the system, and by making it useless for the attackers. This implementation is by using multiple keyboard layouts. When the key is pressed, it is transformed into other characters based on the different key-layout set [10]. Thus, the key logger logs the wrong keyword for every keystroke and cannot get any meaningful information out of the system [11]. Research is going on using virtual keyboard to counter the screen capturing technique of key loggers. To protect personal data, virtual keyboards with irregular arrangement of characters come out to turn the tables. They present a visual interface on screen for users to click instead of tapping on traditional keyboards so that they bypass the keystroke input method and secure the credentials from key loggers.

With increasing security measures against key logging activity, also there has been an enormous increase in new key logging techniques. By instructing the GPU to carefully monitor via DMA the physical page where the keyboard buffer resides, a GPU-based key logger can record all user keystrokes and store them in the memory space of the GPU. An essential property of key logger is that it does not rely on any kernel modifications besides altering the page table. It also uses a small code snippet that needs to run just once from the kernel context to acquire the physical address of the keyboard buffer. Fortunately, some measures can be effective against these GPU-based key loggers. GPU code analysis and runtime detection are the methods to countermeasure these key loggers, as most of the prior techniques were based only on CPU-based key loggers [12].

Drop zones are publically writable directories on the Internet. These are mostly used for malicious purposes. These drop zones usually have a large amount of data exchange taking place. Taking use of this, hackers try to use it to collect their logged files sent from key loggers. Key loggers are designed so that the recorded data by key loggers are sent to these drop zones instead of mailing them. The attacker knows the data of key loggers and can easily take away their file. Also, the detection of logged files from drop zones is difficult because it has a vast amount of random data to filter [13].

Usually, key loggers send the data as soon as they are logged, or sometimes they store it in a file before sending it. But security implementations have been made to check the behaviour of applications for constant uploads, and also, system files are reviewed regularly to detect the log file if present. To evade this, attackers have implemented to store the logged keystrokes in RAM for the time being and then send the data at once [14].

An algorithm has been implemented inspired by the immune system and dendritic cells, which are the natural intrusion detection of the human body. This algorithm is named the dendritic cell algorithm. It works by correlating different behaviours of different applications and, according to the result, predicts the suspicious processes. The three main events monitored and analysed by this algorithm are key logging activity, file access, and network communication. Applications using Windows hooks or any means for logging keystrokes are marked, applications without legitimate binary signature accessing files from the system are marked, and applications using network communication in any of the ways (SMTP, FTP, HTTP) are marked [1].

The novel approach that is to be shown in this paper is that the list of processes will be taken from the Windows Process Explorer. This list will be then narrowed down to only a few processes which have a high probability of being a key logger. This narrow down will take place, based on some filters which would be provided to the list. Score would be generated based on the filters for each suspicious process. High scored process list, then at last, would be presented to the user, who could take action accordingly.

3 Basics of Key Logger

Key loggers are the software that logs the keystrokes from keyboard and stores those in a file. The main objective of such software is to send the recorded keystrokes from the victim's computer to the attacker. The transfer of the log file can be in several ways. Maybe through email, or FTP port, or the attacker can use a Website to dump the logged data, from where they can extract it, something like drop zones. Key logger can be both hardware and software and in software–user space and kernel. Figure 1 shows software key loggers of different types and their distribution percentages. The kernel-based key logger requires user/administrator permission and is generally hard to build as it requires high expertise. Hook-based key logger uses Windows hooks to get the logged keystrokes. The other type of key logger uses DLL's such as WindowsAPIGetKeyState. The key logger's general function is to get in between any two links in a chain of events, from typing on the keyboard to displaying on the monitor. It can intercept at different levels depending on its type. Key logger can intercept the wirings of the computer itself, or the input/output. It can substitute the keyboard driver and intercept the kernel functions or DLL functions. The key loggers may get installed by opening an attachment to a malicious mail, or through open-access directory, via Web page scripts written by some hacker, or through already installed malicious software (Fig. 2).

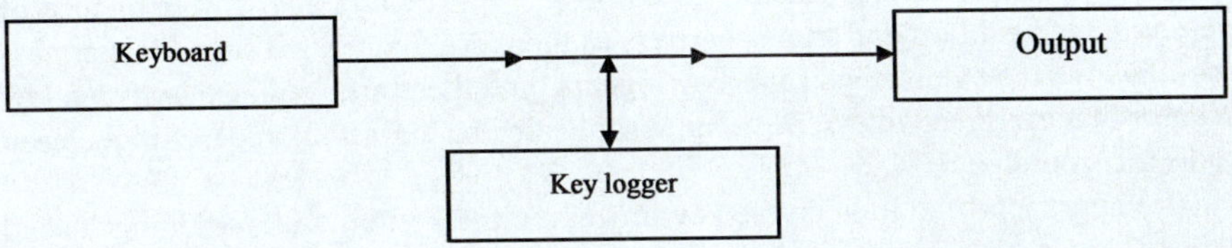

Fig. 1 Basic key logger layout

Fig. 2 Distribution of different software key loggers

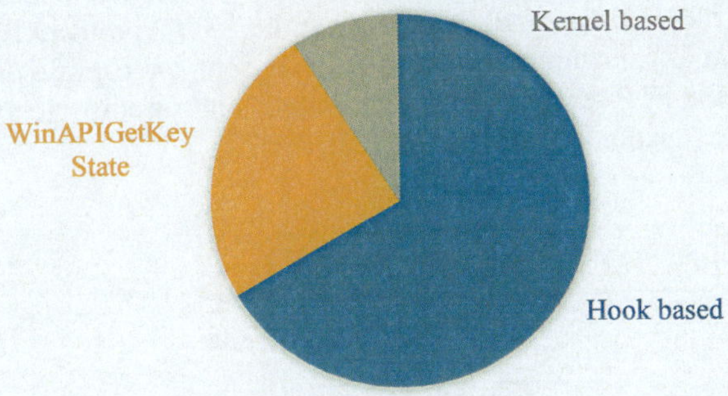

4 Approach Used to Implement Key Loggers

There are two ways to implement the detection process. First, we will write a code to segregate all the applications that use Windows hook to hook the getKeyboardState and getAsyncState. These processes will log the keystrokes from the path between the keyboard and the processor. Then, on those selected applications, we will add a filter to get a high probability of being key loggers. Autostart is the main feature of key loggers. If an application starts at the start of the system, it comes under the category, and if the application logs keystrokes with other applications. Such as if a word pad and an application are simultaneously logging keystrokes, there is something fishy. Most legitimate applications will have their signatures with them, so those are not in the suspect list. Another approach is that we call windows service explorer and on that place filter, as discussed above. We can keep scores for the applications as they pass the filters to be more close to key logger traits. We then can prompt the user of ten10 applications or so, which have the highest score in resembling key loggers. Users can then take appropriate action for all those applications (Fig. 3).

Anti-hook technique has been discussed in some previous works, and also, it will be a part of the approach. Hooking prevents function call or messages which pass through various software components. This technique an scan all the processes, static executables, Dynamic Link Libraries (DLLs) and detect suspicious processes or files that use hooks. On the whole, this technique collects the complete details about that process or file which uses hooks [3, 4]. There is also a high chance of getting false positives and true negatives, as there are many legal services that pass the traits we have described for a key loggers, and there are some well-written key loggers that evade the filters. So, a more sophisticated algorithm and technique need to be incorporated as well.

The approach used to keep the key loggers in the system without letting it send relevant credentials is to monitor the SMTP protocol and the FTP protocol constantly. Key loggers are useful only if they send the logged keystrokes to the people/attackers behind it. They may either send a mail from the victim's computer with an attachment or create an FTP server on the victim server, which the attacker can log on to. We will monitor the sender's address at the SMTP protocol for every mail that is sent, and if any unauthorized email is being sent, we capture that request and analyse it. For the FTP server, we keep it disabled and set it to such that it needs admin privileges to enable FTP on the machine (Fig. 4).

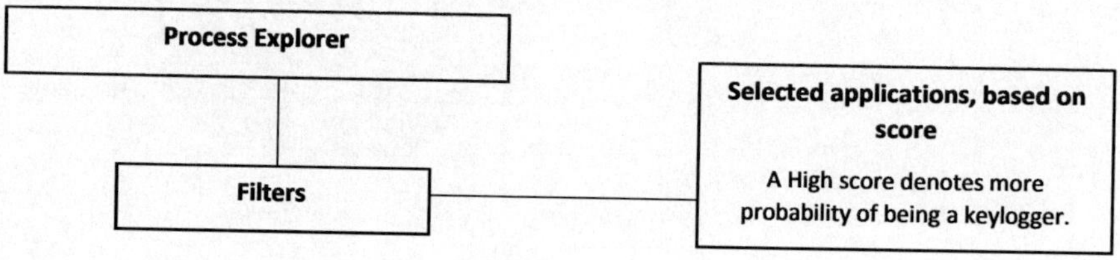

Fig. 3 The novel approach based on separating with highly probability of vulnerable features

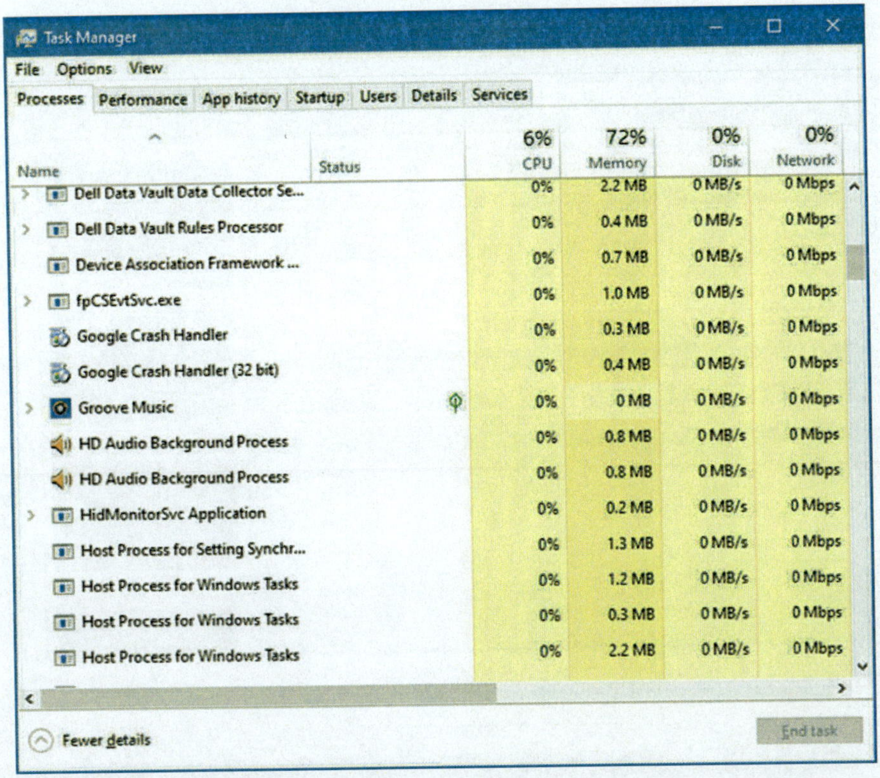

Fig. 4 Windows currently running processes

This approach has used command-line shell scripting. Used Windows PowerShell, to extract the processes which show malicious behaviour Fig. 1.

At first, all the processes using Internet are separated. This is done by capturing their PID. Then, those PIDs are to be searched in the list of all PIDs (all the running processes). Also, we make a list of PIDs of processes with top active running time. The intersection of those PIDs narrows us down to the processes with highest running times using Internet. Thus, it will remove necessary system processes. Further, we would get the processes PIDs of those which have auto run enabled Fig. 5.

Get-Process | Sort CPU -descending | Select -first 5 | Out-File <file location path>

This command is to get high running time processes and also store them in a file (Fig. 6).

Netstat –o | Out-File <file location path>

This command gives us the list of processes using Internet and stores them in file (Fig. 7).

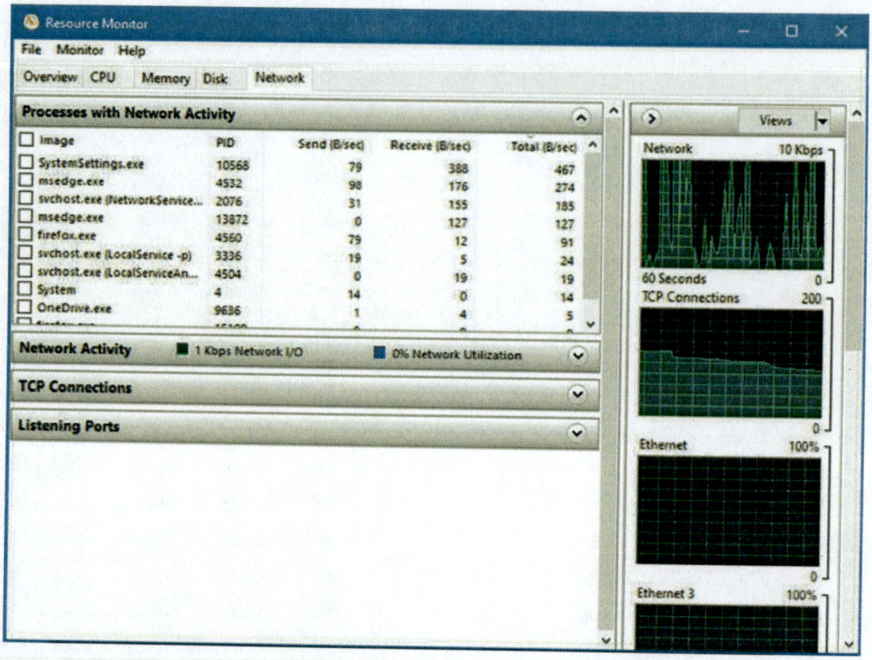

Fig. 5 Processes connected to ethernet

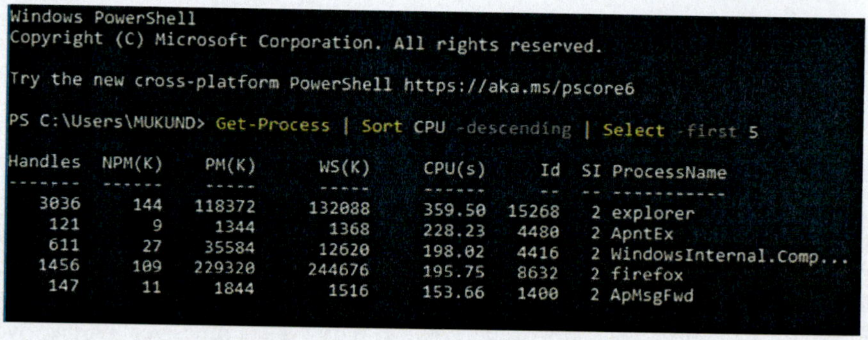

Fig. 6 Power shell command to get processes with maximum running time

5 Conclusion

Key loggers are generally very easy to implement; the Windows hook feature makes it a straightforward task to develop an application that captures keystrokes. Only the technical part is to hide the key logger and sending the recorded information to the attacker. These tasks are also not very hard to achieve. Therefore, detection of an installed key loggers is a much harder task than to create a key logger. The method

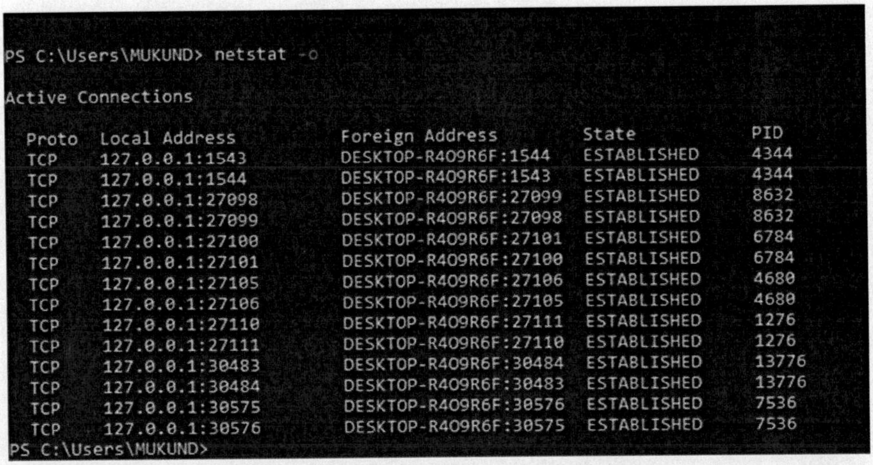

```
PS C:\Users\MUKUND> netstat -o

Active Connections

  Proto  Local Address          Foreign Address         State           PID
  TCP    127.0.0.1:1543         DESKTOP-R4O9R6F:1544    ESTABLISHED     4344
  TCP    127.0.0.1:1544         DESKTOP-R4O9R6F:1543    ESTABLISHED     4344
  TCP    127.0.0.1:27098        DESKTOP-R4O9R6F:27099   ESTABLISHED     8632
  TCP    127.0.0.1:27099        DESKTOP-R4O9R6F:27098   ESTABLISHED     8632
  TCP    127.0.0.1:27100        DESKTOP-R4O9R6F:27101   ESTABLISHED     6784
  TCP    127.0.0.1:27101        DESKTOP-R4O9R6F:27100   ESTABLISHED     6784
  TCP    127.0.0.1:27105        DESKTOP-R4O9R6F:27106   ESTABLISHED     4680
  TCP    127.0.0.1:27106        DESKTOP-R4O9R6F:27105   ESTABLISHED     4680
  TCP    127.0.0.1:27110        DESKTOP-R4O9R6F:27111   ESTABLISHED     1276
  TCP    127.0.0.1:27111        DESKTOP-R4O9R6F:27110   ESTABLISHED     1276
  TCP    127.0.0.1:30483        DESKTOP-R4O9R6F:30484   ESTABLISHED     13776
  TCP    127.0.0.1:30484        DESKTOP-R4O9R6F:30483   ESTABLISHED     13776
  TCP    127.0.0.1:30575        DESKTOP-R4O9R6F:30576   ESTABLISHED     7536
  TCP    127.0.0.1:30576        DESKTOP-R4O9R6F:30575   ESTABLISHED     7536
PS C:\Users\MUKUND>
```

Fig. 7 Command to get the process ID (PID) of processes using internet

discussed above is not the full proof method; they do have some flaws and might not detect all the types of key loggers. A key loggers may not use SMTP or FTP to share the logged data file; then, the approach will fail. If the key loggers do not need the admin rights to create an SMTP server or send mail, plan will fail again. Also, this approach is not valid for GPU-based key loggers as getting processes used in GPU and applying filters on that is an entirely new study. It can be taken as a reference for future work in this field. Also, some work is needed to detect applications logging keystrokes and select them as key loggers, as many applications get listed if not filtered correctly. I have used PowerShell commands for my approach, which is available in all Windows systems. The commands are not at all complex. Users will be able to see the processes involved on their own. This approach checks different parameters to short list the malicious looking applications, which has not been tried. Approaches have been made by insertion of keys that could lure the key logger to give its identity. This approach will find out all the malicious looking software and present it to the user for their verification.

Reference

1. J. Fu, Y. Liang, C. Tan, X. Xiong, Detecting software key loggers with the dendritic cell algorithm, in *International Conference on Communications and Mobile Computing* (2010)
2. M. Raza, M. Iqbal, M. Sharif, W. Haider, A survey of password attacks and comparative analysis on methods for secure authentication. World Appl. Sci. J. (Jan 2012)
3. M. Aslam, R.N. Idrees, M.M. Baig, M.A. Arshad, Anti-hook shield against the software key loggers, in *National Conference on Emerging Technologies* (2004)
4. A. Solairaj, S.C. Prabanand, J. Mathalairaj, C. Prathap, L.S. Vignesh, *Key Loggers Software Detection Techniques* (2016)

5. Y.A. Ahmed, M.A. Maarof, F.M. Hassan, M. Muse, Survey of key logger technologies (Feb 2014)
6. T. Olzak, Keystroke logging (Keylogging) (May 2008)
7. S. Ortolani, C. Giuffrida, B. Crispo, Unprivileged black-box detection of user-space key loggers. IEEE Trans. Dependable Secure Comput. **10** (1) (Jan/Feb 2013)
8. A. Wajahat, A. Imran, J. Latif, A. Nazir, A. Bilal, A novel approach of unprivileged key logger detection, *in International Conference on Computing, Mathematics and Engineering Technologies—iCoMET* (2019)
9. S. Ortolani, C. Giuffrida, B. Crispo, *Bait your Hook: a Novel Detection Technique for Key loggers* (2010)
10. T.O.M. Ali, O.S.A. Awadelseed, A.E.W. Eldewahi, Random multiple layouts key logger prevention technique, in *Conference of Basic Sciences and Engineering Studies (SGCAC)* (2016)
11. S.C. Hsu, C.T. Huang, C.Y. Weng, S.J. Wang, A logging-in scheme in virtual keyboard protection from shoulder surfing and screenshot capturing, in *IEEE 4th International Conference on Computer and Communication Systems* (2019)
12. E. Ladakis, L. Koromilas, G. Vasiliadis, M. Polychronakis, S. Ioannidis, You can type, but you can't hide: A stealthy GPU-based key loggers, in *Proceedings of the Annual Computer Security Applications Conference (ACSAC)* (2006)
13. T. Holz, M. Engelberth, F. Freiling, Learning more about the underground economy: A case-study of key loggers and dropzones. *Conference Paper* (Sept 2009)
14. M.B. Ahmed, M. Shoikot, J. Hossain, A. Rahman, Key logger detection using memory forensic and network monitoring. Int. J. Comput. Appl. (0975–8887) **177** (11) (Oct 2019)

Part V
Data and Security

A Black Widow Optimization Algorithm (BWOA) for Node Capture Attack to Enhance the Wireless Sensor Network Protection

Ankur Khare, Rajendra Gupta, Piyush Kumar Shukla, Rajdeep Chowdhury, and Pushan Kumar Datta

Abstract The resource exploitation is inhibited in the huge region of WSN (Wireless Sensor Network) and various attacks are introduced in vast vulnerable network. The intruder seizes the sensor vertices (nodes) and pilfers the secret data to control laudatory WSN under the node capture attack. We develop a Black Widow Optimization Algorithm (BWOA) to discover the sensor vertices having superior possibility of compromising by attack. BWOA is applied on multi objective function combining the various factors such as link key utilization, link stability time, link participation and link energy expenses of vertices and experiment results illustrate the superior performance of BWOA in terms of energy expenses cost, rounds of attack, compromised portion of traffic and complexity against MA, FGA, MREA, ACO and PSO.

Keywords Black widow · Compromised · Energy expenses · Link stability · Participation

A. Khare (✉)
Computer Science & Engineering Department, Rabindranath Tagore University, Bhopal 464993, India
e-mail: khareankur94@gmail.com

R. Gupta
Computer Science Department, Rabindranath Tagore University, Bhopal 464993, India
e-mail: rajendragupta1@yahoo.com

P. K. Shukla
Computer Science & Engineering Department, University Institute of Technology, RGPV, Bhopal 462023, India
e-mail: pphdwss@gmail.com

R. Chowdhury
Computer Society of India, Mumbai, India
e-mail: dujon18@yahoo.com

P. K. Datta
School of Engineering and Technology, Amity University, Kolkata, India
e-mail: pkdutta@kol.amity.edu

V. E. Balas et al. (eds.), *Proceedings of International Conference on Computational Intelligence, Data Science and Cloud Computing*, Lecture Notes on Data Engineering and Communications Technologies 62, https://doi.org/10.1007/978-981-33-4968-1_47

1 Introduction

A huge amount of research work is performed in the field of WSN such as armed forces, health and organizational region in recent years [1]. This quick growth in WSN also enhances the vulnerabilities allowing the intruder to demolish the WSN by domineering the laudatory network [2, 3]. Some sensor vertices are seized by intruder to cooperate entire network under the node capture attack [4, 5]. Various methodologies are implemented with utmost competence for pilfering the confidential data of sensor vertices. Unified Modelling Language (UML) [6], imaginary, Epidemic theory [7], receptiveness, and probabilistic approaches [8] are frequently developed for node capture attack [9].

Sensor vertices are elected erratically and keys and confidential data are filched in Random Attack (RA) [10]. In Maximum Key Attack (MKA), the intruder is seized the sensor vertices based on maximum keys for cryptosystem [11]. The keys in unused path are also utilized in MKA to enhance the overhead. In Maximum Link Attack (MLA), the intruder is seized the sensor vertices based on maximum links among nodes [12]. The links in unused path is also utilized in MLA to increase the overhead. The sensor vertices are seized by intruder on the basis of maximum traffic in Maximum Traffic Attack (MTA). The sensor vertices-path association is not utilized in MTA [11]. The superior vulnerabilities are measured in Greedy Node capture Approximation using Vulnerability Evaluation (GNAVE) [10]. The unused sensor vertices are also utilized to enhance the overhead in network. This issue is resolved by applying Matrix Algorithm (MA) having the node-path association with least energy [10]. The power expenses and competence of attack is not covered in MA. The performance of MA is enhanced by using Full Graph Attack (FGA) to get better efficiency of node capture attack by decreasing the compromising sensor vertices [11]. The power expenses are a limitation of FGA which is further resolved by utilizing Minimum Resource Expenditure node capture Attack (MREA) with key utilization approach [12].

All the above described approaches are developed on any one object among several objects such as cost, links, traffic, vulnerability, and node-path association. Therefore we applied two well known optimization approaches [13, 14] ACO (Ant Colony Optimization) and PSO (Particle Swarm Optimization) on multi objective function [14, 15] to get better performance of node capture attack. ACO is an optimization approach based on the behaviour of ants and PSO is approach based on particle swarm with local and global best location. We implemented a Black Widow Optimization Approach (BWOA) to get better results which is described in our work.

2 Proposed Black Widow Optimization Approach (BWOA) for Node Capture Attack

A network $N^n = (V^n, L^n)$ of sensor vertices (nodes) V^n and links L^n is created with single and multipath approaches of routing for information transmission. Firstly a packet of information is encrypted by using a shared key $K_{p,q} = K_p \cap K_q$ and transmitted by a link $L_{p,q}$ between vertices V_p and V_q in transmission range T_R. Various packets are encrypted by using various keys $K_p \subset K^n$ and transmitted by a vertex $V_p \subset V^n$. The attacker model intricate a procedure for vertex capturing and taking away the useful information from vertices like keys for encryption, routing methodology and plain text. We implemented a BWOA to generate optimal vertices to capture for destroying the whole network. The capturing of a link $L_{p,q} \in L^n$ is directly related to the capturing of single key $K_{p,q} \in K^c$ used for that link. The capturing of a path P_p is directly related to the capturing of any link for that path $P_p \in P$. The capturing of a route $R_{s,d}$ is directly related to the capturing of any path for that route $R_{s,d} \in R$.

The key objective of intruder is to restrain minimum vertices for destroying the entire WSN based on maximum link key utilization, maximum link stability time, maximum link participation and minimum link energy expenses of vertices. Few general symbols and their illustrations are represented in Table 1.

Table 1 Symbols and illustrations

Symbols	Illustrations	Complexity
V^n	The number of sensor vertices (nodes) in WSN	O(N)
V_p	The pth vertex	O(1)
K^n	The total keys	O(1)
K_p	The key set to allocate to V_p	O(1)
L^n	The total links in WSN	O(N^2)
	The link between connected vertices V_p and V_q	O(1)
T_R	Vertices transmission range	O(1)
V^c	Total captured vertices	O(1)
K^c	Total captured keys	O(N)
S^{SV}	Source sensor vertices	O(1)
D^{SV}	Destination sensor vertices	O(1)
$R_{s,d}$	A route from source to destination sensor vertex	O(1)
P	Total Paths in WSN	O(N)
R	Total routs in WSN	O(1)
Rp	Repetitions	O(N)
Pv	Population values of black widow spiders	O(N)

2.1 Link Key Utilization

A single packet of information is encrypted by a single key which is transmitted by a vertex to another through a link. Several packets are transferred by a link after the encryption through several keys between two neighboring vertices. It means a link has various keys which are generated as Link Key Utilization (LKU) by evaluating Number of Keys of a Link (NKL) using Eq. 1. A vertex is selected to capture having maximum links with maximum values of LKU.

$$\mathrm{LKU}_{p,q} = \begin{cases} \sum_{\mathrm{key}=1}^{K^n} \mathrm{NKL}_{p,q} & \text{If} \exists \text{keys} \in V_p, V_q \\ 0 & \text{Otherwise} \end{cases} \tag{1}$$

Here,

$\mathrm{LKU}_{p,q}$ Link Key Utilization between two vertices V_p and V_q.
$\mathrm{NKL}_{p,q}$ Number of Keys of a Link between two vertices V_p and V_q.
K^n Total number of keys.

2.2 Link Stability Time

Link stability is depended on the mobility of vertices in WSN. The location of vertices is obtained from GPS (Global Positioning System). The Link Stability Time (LST) is evaluated by using Eq. 2. A vertex is selected to capture having maximum links with maximum value of LST.

$$\mathrm{LST}_{p,q} = \frac{-(ef + gh) + \sqrt{(e^2 + g^2)T_R^2 - (eh - fg)^2}}{e^2 + g^2} \tag{2}$$

Here,

$\mathrm{LST}_{p,q}$ Link Stability Time of a link between V_p and V_q.

$$e = v_p \cos \Phi_p - v_q \cos \Phi_q \tag{3}$$

$$g = v_p \sin \Phi_p - v_q \sin \Phi_q \tag{4}$$

$$f = x_p - x_q \tag{5}$$

$$h = y_p - y_q \tag{6}$$

v_p and v_q velocity of two vertices V_p and V_q.
$(x_p, y_p)\&(x_q, y_q)$ Position of two vertices V_p and V_q.

2.3 *Link Participation*

A vertex is to be captured with maximum participation in information transmission. Every vertex has been sent the HELLO packets to the all other vertices and received the acknowledgement from the neighbouring vertices in their transmission ranges T_R to obtain the Number of Neighbours (NN) as Vertex Participation (VP). VP is evaluated by using Eq. 7.

$$\mathrm{VP}_p = |\mathrm{NN}(p)| = \sum_{p \neq q} \{\mathrm{dis}(p, q) < T_R\} \tag{7}$$

Here,

$\mathrm{dis}(p, q)$ distance between two vertices V_p and V_q.
$|\mathrm{NN}(p)|$ Number of Neighbours of vertex V_p.
VP_p Vertex Participation of vertex V_p.

After that 1 is divided by VP to allocate equivalent distribution of vertex participation to each vertex. Therefore, Link Participation (LP) percentage is evaluated by using Eq. 8.

$$\mathrm{LP}_{p,q} = \frac{1}{\mathrm{VP}_p} \tag{8}$$

Here,

$\mathrm{LP}_{p,q}$ Link Participation of a link between two vertices vertex V_p.

2.4 *Link Energy Expenses*

The Link Energy Expenses (LEE) is evaluated as energy expenditure of every link to transmit the information packets. LEE is calculated by using Eq. 9. A vertex is selected to capture having maximum links with minimum value of LEE.

$$\mathrm{LLE}_{p,q} = \frac{\mathrm{LP}_{p,q}}{\mathrm{CC}_{p,q}} \tag{9}$$

Here,

$\mathrm{LLE}_{p,q}$ Link Energy Expenses of a link between of vertex V_p and V_q.
$\mathrm{CC}_{p,q}$ Capturing Cost of a link between V_p and V_q.

2.5 Final Link Contribution

Final Link Contribution (FLC) is obtained for every link by utilizing four factors ($LKU_{p,q}$, $LST_{p,q}$, $LP_{p,q}$ and $LEE_{p,q}$) by Eq. 10.

$$\text{Maximum}\quad FLC_{p,q} = (wt_1 \times LKU_{p,q}) + (wt_2 \times LST_{p,q}) + (wt_3 \times LP_{p,q}) + \left(wt_4 \times \frac{1}{LEE_{p,q}}\right) \tag{10}$$

Here,

wt_1, wt_2, wt_3 and wt_4	Weight Factor with $wt_1 + wt_2 + wt_3 + wt_4 = 1$.
$FLC_{p,q}$	Final Link Contribution of a link between vertex V_p and V_q.

2.6 Black Widow Optimization Approach (BWOA)

After obtaining the Eq. 10 for each and every links in WSN, BWOA is initiated to generate maximum value of multi objective function FLC and to obtain links with optimal result from available links. BWOA initiates with primary spiders population, therefore every spider shows a prospective solution. These primary spiders seek to imitate the next creation. In the phase of mating, female eats the male black widow and she takes away the provided sperms in her sperm thecae and liberates them into egg sacs. After some times, spider lings appeared from egg sacs. They then abscond through being taken away on the wind. The flowchart of BWOA is represented in Fig. 1.

Population Initialization. The prospective solution to every problem has been measured as a Black widow spider representing the problems variables standards in BWOA. A black widow is represented by using Eq. 11.

$$\text{Widow} = [x_1, x_2, x_3, \ldots, x_{N_v}] \tag{11}$$

Every widow is floating point number and fitness function is calculated by Eq. 12.

$$\text{Fitness} = f(\text{Widow}) = f(x_1, x_2, x_3, \ldots, x_{N_v}) \tag{12}$$

where,

N_v	Number of variables (dimensional optimization problem).
x_1, x_2, x_3	Values of variables.

An aspirant widow matrix ($N_p \times N_v$) is obtained through primary spider's population (N_p). After that couple of parents arbitrary is picked to execute procreating action by mating.

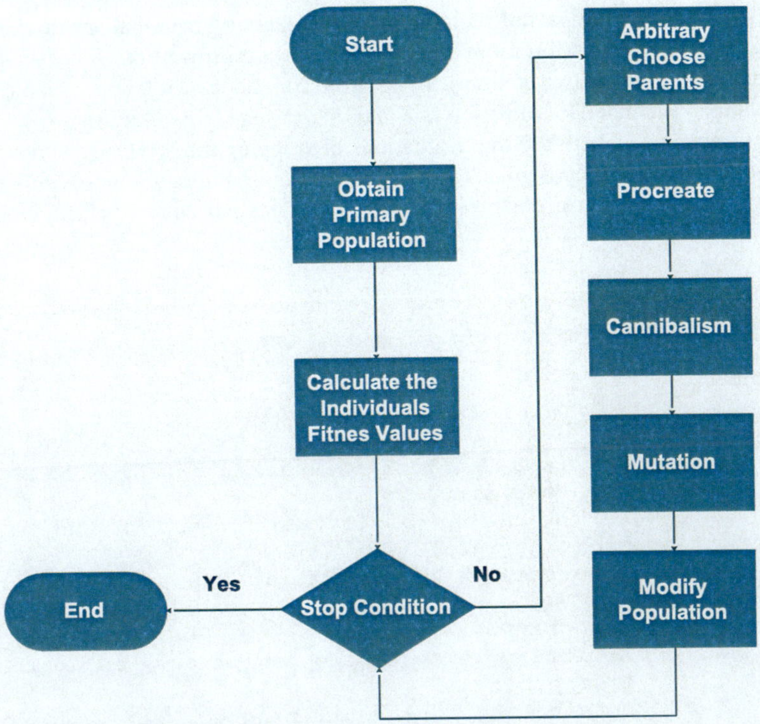

Fig. 1 Flow chart of Black Widow Optimization Approach (BWOA)

Procreate Action. Every couple of parents are performed mating separately and imitated the new creation in parallel. Strong children of spiders are survived form numerous number of eggs. An alpha array is obtained and offspring is generated by utilizing alpha (Eq. 13).

$$\begin{cases} y_1 = \alpha \times x_1 + (1 - \alpha) \times x_2 \\ y_2 = \alpha \times x_2 + (1 - \alpha) \times x_1 \end{cases} \quad (13)$$

where,

x_1, x_2 Parents spiders.
y_1, y_2 Offspring.

This procedure is iterated for $N_v/2$ times. At last an array is created by mother and children utilizing the fitness values based on cannibalism marking. This procedure is initiated to all couples.

Cannibalism and Mutation. There are three types of cannibalism; (1) in sexual cannibalism; female eats the male black widow in the phase of mating and female-male recognizes with fitness values, (2) in sibling cannibalism, sturdy spider lings eat their weaker siblings and remaining are indicated by cannibalism marking, (3) in this type, child eats the mom spider. All types are utilized the fitness values to decide the sturdy and weak siblings. We arbitrary choose Mutepop variables from populations evaluated by mutation rate and interchange two values in array.

Convergence. There are three situations of stopping the BWOA; (1) previously provided number of repetitions, (2) adherence of no modification in optimal widow fitness value for various repetitions, (3) achieving the particular level of precision.

```
BWOA Algorithm
Input: Maximum repetitions, procreating rate, cannibalism
rate, and mutation rate
Output: the best objective function value
Start
Provide the primary population values of black widow
spiders
      Every pop is a D-dimensional array of chromosomes
WHILE (until the end situation)
      Evaluate the number of imitation (ni) on the basis
      of procreating rate;
      Choose the optimal ni results in pop and store them
      in pop 1;
      FOR i=1 to ni DO
            Arbitrary choose two solutions as parents
            from pop1;
            Obtain D children utilizing eq. 11;
            Obliterate father;
            Obliterate few children on the basis of
            cannibalism rate (new provided
            solutions);
            Store the remain solutions into pop2;
      END FOR;
      Evaluate number of mutation children "mn" on the
      basis of mutation rate;
      FOR j=1 to mn DO
            Choose a solution from pop1;
            Mutate arbitrary one chromosome of the
            solution and obtain a new solution;
            Store the new one into pop3;
      END FOR
      Modify pop=pop2+pop3;
      Inveterate the optimal solution;
      Inveterate the optimal solution from pop;
END WHILE
Stop
```

2.7 Final Vertex Contribution

After applying the BWOA, the optimal values of FLC for each and every links are generated.

Then a matrix FLC_p of optimal FLC values for each and every link of vertex V_p is obtained by using Eq. 14.

$$\mathrm{FLC}_p = \left[\mathrm{FLC}_{p,q}\right]_{1\times \mathrm{VP}_p} \tag{14}$$

Here,

FLC_p Final Link Contribution with optimal values of all links of vertex V_p.

After that Final Vertex Contribution (FVC) with optimal values is evaluated by using Eq. 15.

$$\mathrm{FVC}_p = \frac{\sum_{q=1}^{\mathrm{VP}_p} \mathrm{FLC}_{p,q}}{\mathrm{VP}_p} \tag{15}$$

Here,

FVC_p Final Vertex Contribution with optimal value of vertex V_p.

Similarly we are evaluated FVC optimal values of each and every vertices of WSN. The vertices V^c with maximum FVC optimal values have been selected due to have greater chance to capture for compromising the whole WSN.

3 Result and Analysis

The analysis of BWOA is performed on MATLAB 2019a environment with windows 8.1, Intel(R) Core(TM) i3-3110 M, 2.40 GHz, 4 GB RAM and results are calculated independently more than 30 runs for 350 repetitions against MA, FGA, MREA, ACO and PSO on the basis of few circumstances over WSN provided in Table 2. The sensor vertices (350) in several simulation scripts are equivalently deployed in 100 m × 100 m network area in which 25 source sensor vertices and 7 destination sensor vertices are dynamically selected. The various keys are arbitrarily distributed to every sensor vertices from total keys ($K^n = 350$).

Table 2 Simulation circumstances

Circumstances	Values
Number of sensor vertices (nodes) (V^n)	350
Network area	100 m * 100 m
Source sensor vertices (S^{sv})	25
Transmission range ()	35
Destination sensor vertices (D^{sv})	7
Total Keys (K^n)	350
Population values of black widow spiders (pv)	350
Repetitions (Rp)	350

3.1 Compromised Portion of Traffic

It is a part of compromised paths (traffics) to whole paths (traffics), generating in single and multipath approaches of routing in WSN. Figures 2 and 3 illustrates the superior attacking competence of BWOA in terms of compromised portion of traffic with capturing minimum number of sensor vertices to generate 15.38% and 11.76% better efficiency against the PSO, 21.43% and 16.67% better efficiency against ACO,

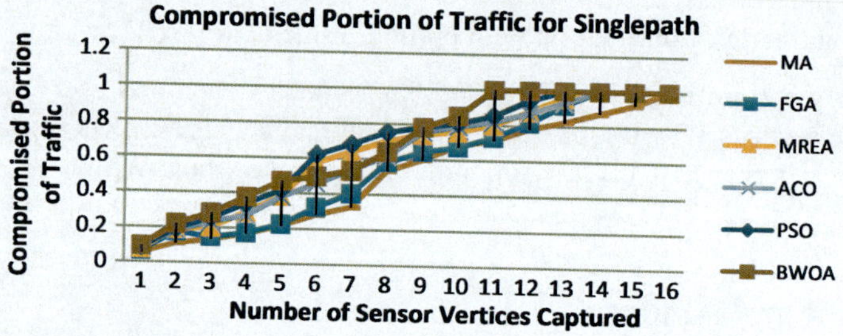

Fig. 2 Compromised portion of traffic (Single path approach for routing)

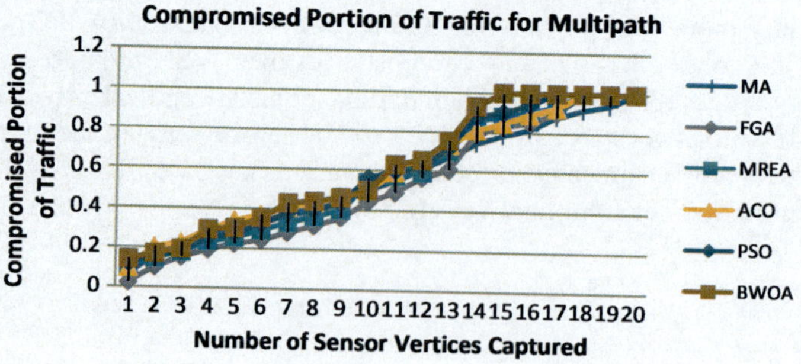

Fig. 3 Compromised portion of traffic (Multipath approach for routing)

MREA, and FGA, and 31.25% and 25% better efficiency against MA for single and multipath approaches for routing, respectively to control entire WSN.

3.2 *Energy Expenses Cost*

The Energy Expenses Cost is the cost which is utilized for spending the energy for capturing the sensor vertices to compromise entire WSN. It is enhanced with enhancing the number of keys and captured sensor vertices. Figures 4 and 5 illustrates the superior performance of BWOA in terms of energy expenses cost with respect to number of keys through capturing minimum number of sensor vertices to obtain 6.38 and 8.36% better efficiency against the PSO, 16.41 and 16.08% better efficiency against ACO, 34.07% and 29.36% better efficiency against MREA, 38.12% and 26.05% better efficiency against FGA, and 54.14% and 38.70% better efficiency against MA for single and multipath approaches for routing, respectively to control entire WSN.

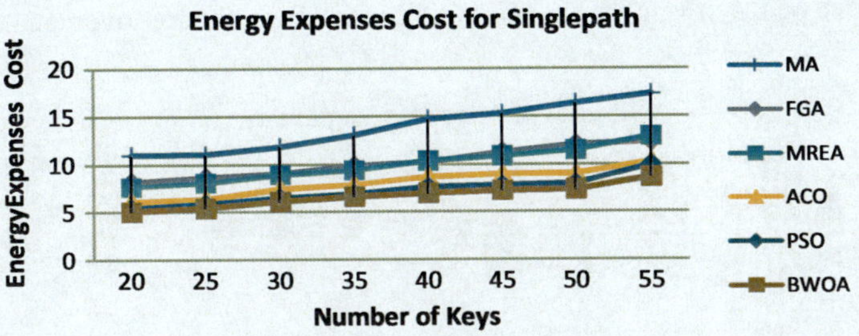

Fig. 4 Energy expenses cost (Single path approach for routing)

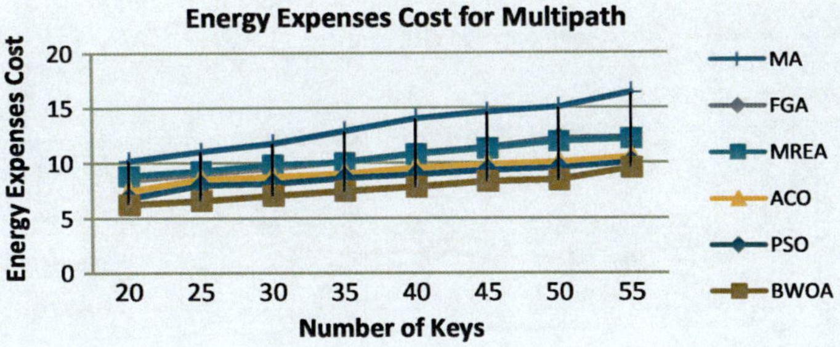

Fig. 5 Energy expenses cost (Multipath approach for routing)

3.3 Rounds of Attack

The rounds of attack are defined as the number of rounds to attack the WSN for capturing the least sensor vertices to compromise entire WSN. It is directly dependent on compromised portion of traffic. Therefore Figs. 6 and 7 illustrates the superior results of BWOA in terms of rounds of attack with respect to number of keys through capturing minimum number of sensor vertices to provide 15.38% and 11.76% better results against the PSO, 21.43% and 16.67% better results against ACO, MREA, and FGA, and 31.25% and 25% better results against MA for single and multipath approaches for routing, respectively to control entire WSN.

The overall results analysis is indicated the minimum efficiency of MA among all approaches due to not consider power expenses and any indirect relationship among sensor vertices and paths. The indirect relationship is considered in FGA, so it obtains better results than MA. The power expenses are a limitation of FGA which is further resolved by MREA with key utilization approach, so MREA obtains better results than FGA. The performance of node capture attack approach is again enhanced by using ACO which provides more optimal sensor vertices than MREA. The slow convergence of ACO is further resolved by using PSO; so PSO generates some better quality results than ACO. At last the computation overhead of PSO is

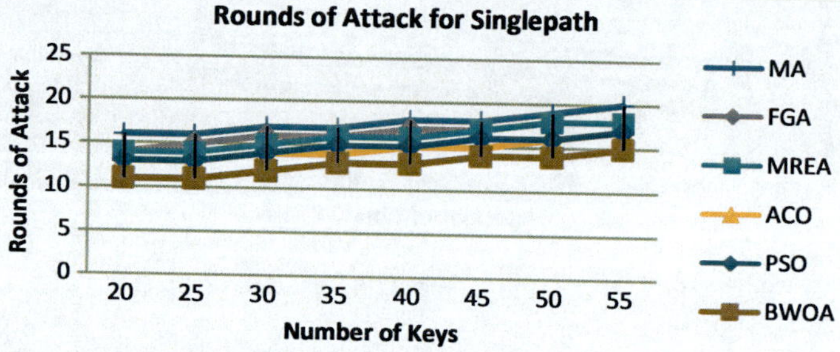

Fig. 6 Rounds of attack (Single path approach for routing)

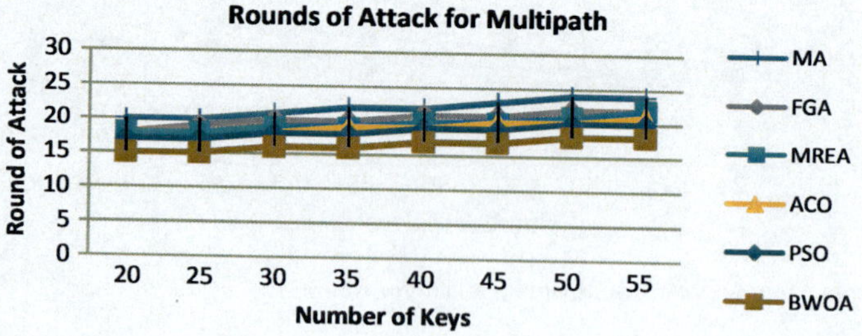

Fig. 7 Rounds of attack (Multipath approach for routing)

resolved by utilizing BWOA; so BWOA generates better results than PSO, ACO, MREA, FGA and MA.

3.4 Complexity of Approaches

The complexity of approaches is calculated using the information about complexities available in Table 1 and shown in Table 3.

In MA, the attacker initially obtains the relationship between nodes and paths directly and indirectly and therefore calculates the cost of capturing of nodes to control whole WSN. Consequently, the complexity of MA is calculated as $O(V^n K^n P) = O(N \times 1 \times N) = O(N^2)$. In FGA, the attacker obtains the total links of every node to form the full graph. Consequently, the complexity of FGA is calculated as $O(V^n P) + O(V^n K^n P) = O(N \times N) + O(N \times 1 \times N) = O(N^2) + O(N^2) = O(N^2)$. In MREA, the attacker obtains the least cost nodes having highest total paths with direct and indirect association. Consequently, the complexity of MREA is calculated as $O(V^n K^n P) = O(N \times 1 \times N) = O(N^2)$.

In ACO, the attacker initially generates the four factors of sensor vertices link key utilization ($O(K^n L^n) = O(1 \times N^2) = O(N^2)$), link stability time ($O(V^n L^n) = O(N \times N^2) = O(N^3)$), link participation ($O(V^n L^n) = O(N \times N^2) = O(N^3)$) and link energy expenses ($O(V^n L^n) = O(N \times N^2) = O(N^3)$). Final Link Contribution for every link is evaluated as $O(K^n L^n) + O(V^n L^n) + O(V^n L^n) + O(V^n L^n) = O(N^2) + O(N^3) + O(N^3) + O(N^3) = O(N^3)$. After that ACO is applied on final link contribution utilizing few parameters population value and repetitions in $O(P_v R_p) = O(N^2)$. Therefore final complexity of ACO is $O(N^3) + O(N^2) = O(N^3)$. Similarly, the PSO is applied on final link contribution utilizing same parameters as ACO. Therefore final complexity of PSO is $O(N^3)$. Similarly, the BWOA is applied on final link contribution utilizing same parameters as ACO and PSO. Therefore final complexity of BWOA is $O(N^3)$. It means all the approaches MA, FGA, MREA, ACO, PSO and BWOA are solvable in polynomial time.

Table 3 Complexity of approaches

Name of approach	Values
MA	$O(N^3)$
FGA	$O(N^3)$
MREA	$O(N^3)$s
ACO	$O(N^3)$
PSO	$O(N^3)$
BWOA	$O(N^3)$

4 Conclusion

The resource management is repressed in the massive province of WSN and a variety of attacks are deputized in immense vulnerable WSN. The intruder seizes the nodes and filches the surreptitious data to manage sycophantic WSN under the node capture attack. We implement a Black Widow Optimization Algorithm (BWOA) to identify the nodes with greater probability of competence by attack. BWOA is functionalized on multi objective function combining the several factors such as key utilization, stability time, participation and energy expenses of nodes links and experiment outputs describe the better-quality performance of BWOA in terms of energy expenses cost, rounds of attack, compromised portion of traffic and complexity against MA, FGA, MREA, ACO and PSO.

References

1. H.R. Shaukat, F. Hashim, M.A. Shaukat, K.A. Alezabi, Hybrid multi-level detection and mitigation of clone attacks in Mobile Wireless Sensor Network (MWSN). Sensors, MDPI **20**, 1–23 (2020)
2. Xiong C., Li S., Liu L., Li R. and Jin Y.: A Hybrid Key Pre-distribution Scheme for Wireless Sensor Networks. IOP Conf. Series, Journal of Physics, pp. 1–18, (2019). (doi:https://doi.org/10.1088/1742-6596/1229/1/012066)
3. I.Q. Kolagar, H.H.S. Javadi, M. Anzani, Hypercube bivariate-based key management for wireless sensor networks. J. Sci., Islamic Repub. Iran, Univ. Tehran. **28**(3), 273–285 (2017)
4. B. Butani, P.K. Shukla, S. Silakari, An exhaustive survey on physical node capture attack in WSN. Int. J. Comput. Appl. **95**(3), 32–39 (2014)
5. G. Wu, X. Chen, M.S. Obaidat, A high efficient node capture attack algorithm in wireless sensor network based on route minimum key set. Secur. Commun. Netw. **6**, 230–238 (2013)
6. S. Hong, S. Lim, Analysis of attack models via unified modeling language in wireless sensor networks: A survey study. IEEE, 692–696 (2010)
7. P. De, Y. Liu, S.K. Das, Deployment aware modeling of node compromise spread in wireless sensor networks using epidemic theory. ACM, 1–29 (2008)
8. A. Albakri, L. Harn, S. Song, Hierarchical Key Management Scheme with Probabilistic Security in a Wireless Sensor Network (WSN). Secur. Commun. Netw., Hindawi, 1–11 (2019). (https://doi.org/https://doi.org/10.1155/2019/3950129)
9. J.W. Ho, Distributed detection of node capture attacks in wireless sensor networks. Intechopen, 345–363 (2010)
10. C. Lin, G. Wu, Enhancing the attacking efficiency of the node capture attack in WSN: A matrix approach. J. Supercomput. (Springer Science & Business Media), 1–19 (2013). https://doi.org/10.1007/s11227-013-0965-0
11. C. Lin, G. Wu, C.W. Yu, L. Yao, Maximizing destructiveness of node capture attack in wireless sensor networks. J. Supercomput. (Springer Science & Business Media) **71**, 3181–3212 (2015). https://doi.org/10.1007/s11227-015-1435-7
12. C. Lin, T. Qiu, M.S. Obaidat, C.W. Yu, L. Yao, G. Wu, MREA: A minimum resource expenditure node capture attack in wireless sensor networks. Secur. Commun. Netw. (Wiley Online Library) **9**, 5502–5517 (2016). https://doi.org/10.1002/sec.1713
13. D. Molina, J. Poyatos, J.D. Ser, S. Garcia, A. Hussain, F. Herrera, *Comprehensive Taxonomies of Nature- and Bio-inspired Optimization: Inspiration versus Algorithmic Behavior, Critical Analysis and Recommendations*. cs.AI, pp. 1–76, (2020). (arXiv:2002.08136v2)

14. R. Bhatt, P. Maheshwary, P. Shukla, P. Shukla, M. Shrivastava, S. Changlani, Implementation of Fruit Fly Optimization Algorithm (FFOA) to escalate the attacking efficiency of node capture attack in Wireless Sensor Networks (WSN). Comput.Commun. (Elsevier) **149**, 134–145 (2020)
15. P.K. Shukla, S. Goyal, R. Wadhvani, M.A. Rizvi, P. Sharma, N. Tantubay, *Finding Robust Assailant Using Optimization Functions (FiRAO-PG) in Wireless Sensor Network* (Hindawi Publishing Corporation, Mathematical Problems in Engineering, 2015), pp. 1–8. (http://doi.org/https://doi.org/10.1155/2015/594345)

Proposing an Architecture for IoT Based on WSN

Fadi Abusafat, Tiago Pereira, and Henrique Santos

Abstract Internet of things (IoT) is considered the future of communication and information technology due to the countless benefits it provides through its use in several fields. It is associated with several communication technologies, communication protocols, wide applications and many devices that communicate, collect and share data through connecting by wireless sensor network technology (WSN). Therefore, it is important to improve all these components to improve IoT itself. However, IoT has many characteristics based on its nature, connected devices and operations inside IoT. These characteristics imposed several challenges as well as future trends. In this paper, we proposed an architecture for IoT to provide a clear and comprehensive image of components, communication technologies, operations and challenges of IoT. This image helps to provide a clear understanding of IoT. Besides and based on the revenue of IoT, security issues are considered one of the most critical challenges of IoT, and it should be addressed as well as WSN.

Keywords IoT challenges · IoT functionality · IoT characteristics · IoT security · WSN

1 Introduction

The Internet of things (IoT) is considered as one of the most important fields in information technology in the twenty-first century. It has very wide scope into most industrial and technological fields. The core importance of IoT is to make everything accessible by connecting to the Internet [1]. Therefore, it is defined as a computing system that communicates and shares data between mechanical, technological and digital devices while every device has a unique identifier (IP). These devices are diverse starting from computers, mobiles, street sensor, monitors camera, health devices, etc. [2, 3]. Figure 1 shows a summarized framework of several IoT devices.

F. Abusafat (✉) · T. Pereira · H. Santos
Algorithm Center, University of Minho, Guimarães 4710-570, Portugal
e-mail: fabusafat1987@hotmail.com

V. E. Balas et al. (eds.), *Proceedings of International Conference on Computational Intelligence, Data Science and Cloud Computing*, Lecture Notes on Data Engineering and Communications Technologies 62, https://doi.org/10.1007/978-981-33-4968-1_48

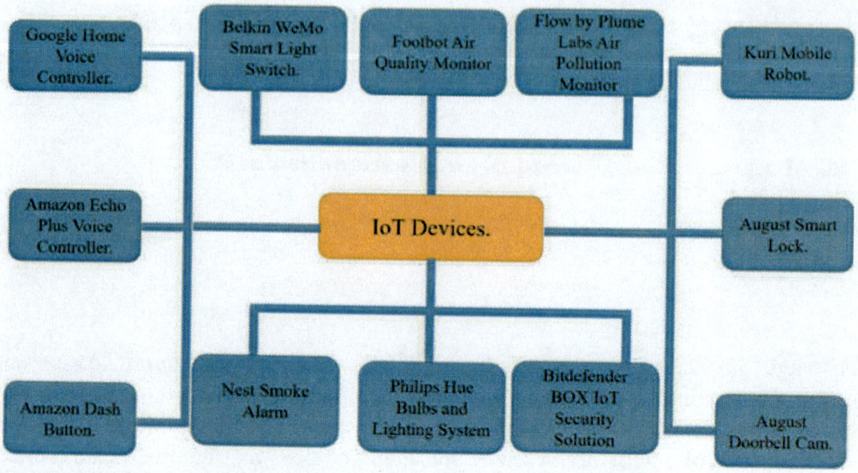

Fig. 1 IoT devices. Adapted from: [2, 3]

Communications between these devices can be achieved either by wireless or cable connection. There are several applications associated with IoT such as educations, surveillance, entertainment, traffic monitor, environment monitor, industrial, transportation, smart building, smart grid and health care [1, 4, 5]. Figure 2 shows a summarized framework of several IoT applications.

Several kinds of research were conducted to forecast the number of connected devices to the Internet, such as a research conducted by Cisco in 2011, showing the number of connected devices to the Internet was around 500 million units in 2003. While the number of connected devices was 12.5 billion devices in 2010, in 2015 and

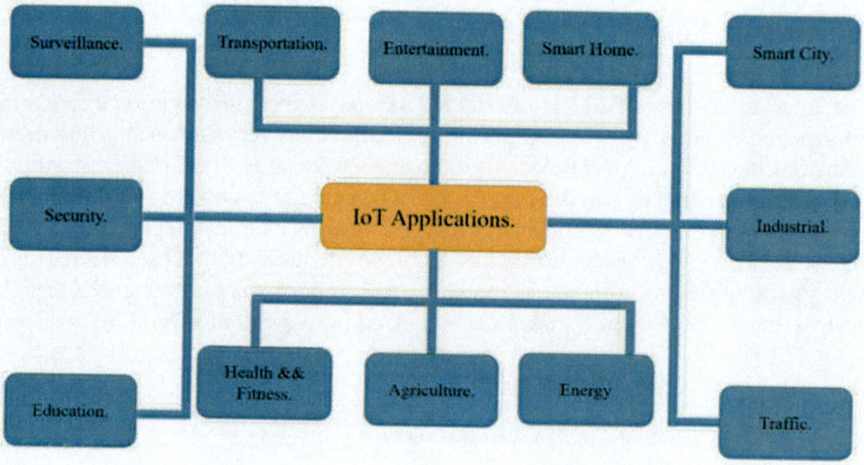

Fig. 2 IoT applications. Adapted from: [1, 4, 5]

2020 the number of connected devices was 25 and 50 billion devices, respectively [6]. With regards to a research conducted by Statista, number of connected devices in 2025 will be around 75.44 billion devices [7]. While a study conducted by IoT analytics shows the number of IoT devices in 2018 was around 7 billion, and it is expected to be 21.5 billion devices in 2025 [8]. A study conducted by IBM claims the number of connected devices in 2015 was around 1 trillion devices, while Gartner's estimated number of connected devices in 2016 is around 6.4 billion devices. International Data Corporation expected 9 billion devices and HIS's estimated 17.6 billion devices in 2016 [9].

Several studies show enormous revenue from adapting IoT, for instance, a study conducted by IoT-NOW in 2020 shows revenue will be around 1.5 trillion USD from adapting IoT [10]. Another study conducted by PWC in 2018 shows benefits were around 1710.4 billion USD [11]. Another study conducted by Transforma Insights which is a specialized research firm in digital transformations shows the global market of the global IoT will have around 24.1 billion devices into 2030 and annual revenue around 1.5 trillion USD [12]. Another study conducted by Markets and Markets shows the impact of IoT into bank and financial sector will be the growth from 249.4 million USD in 2018 to 2,0030.1 million USD in 2023 [13]. Another study conducted by Gartner shows revenue from using end points electronics will be totalled 389 billion USD in 2020 [14]. Another study conducted by Mckinsey & Company shows the impact of IoT could reach 11.1 trillion USD by 2025 [15]. Therefore, we can see the enormous revenue of adapting IoT into several life sectors. Figure 3 shows a summarized framework of IoT revenue based on several studies.

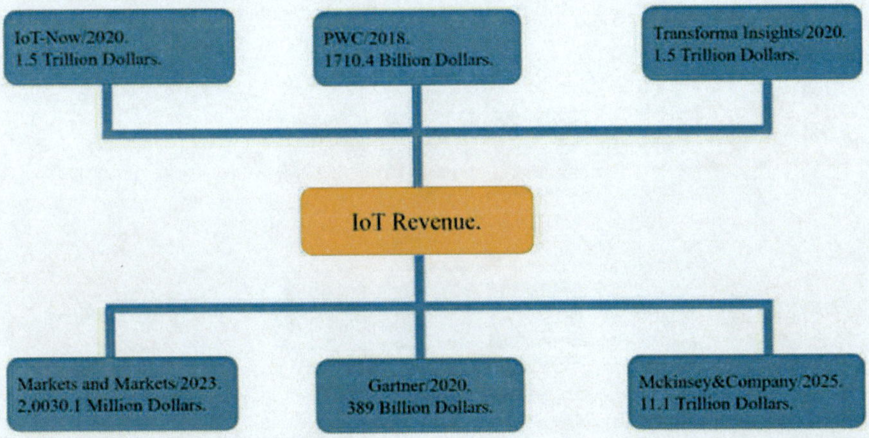

Fig. 3 IoT revenue. Adapted from: [10–15]

2 Overview of Communications Technologies in IoT

In this chapter, we are going to investigate several issues related to communication technologies in IoT such as communications model, communication protocols, wireless sensor network (WSN) and characteristics of WSN.

2.1 Wireless Sensor Network (WSN)

WSN can be defined as scattered technological nodes that are used to share data between nodes and servers as well as shares it between nodes. Therefore, it is considered heterogeneous due to the ability to communicate between several types of nodes such as mobile devices, vehicles and environment. It consists of data collection/sensing unit, data distributed/transceiver unit and main server/processing unit. Data collection or sensing unit is responsible for sensing data from nodes and passes it for the main unit to access it. Distribution unit is responsible to send data between nodes and servers. All of these units are connected to the power source to do their functionalities [16, 17]. Figure 4 shows WSN components.

There are various communication technologies in IoT such as Bluetooth, near-field communication (NFC), Wi-Fi, cellular 3G, 4G, 5G LTE, ZigBee, SigFox, narrowband IoT (NB-IoT) and LoRaWAN. These technologies are various in terms of range, bandwidth, reliability, latency, power consumption, data rates and costs [17]. The range for IoT devices can be extended from centimeters to kilometers [18]. Bandwidth is an important issue for IoT devices as it affects the transmission of data

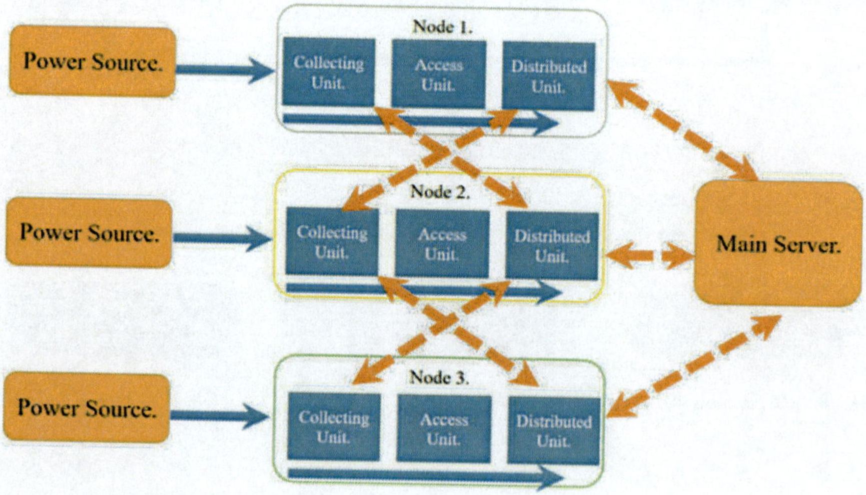

Fig. 4 WSN components. Adapted from: [16, 17]

between devices. For example, in fire detection devices, an alarm will be activated once a fire is conducted while other devices such as camera need to be in constant communication. Therefore, increasing bandwidth will affect the transmission of data [19]. Reliability reflects the efficiency of IoT systems. For instance, IoT devices that are used into the health sector to monitor patient's health should transmit data for a destination on time as well as it should be accurate. Therefore, reliability is an important issue for the IoT systems [20].

Latency is defined as the required time to transmit data from source to destination. This issue is considered one of the most important issues in the IoT systems. For instance, in the health sector, in case sensors are delayed in transmission of data related to patient's health, it could cost patient life. So, it is very critical into IoT systems [21]. Energy consumption is considered one of the main attributes for IoT devices. The core of IoT nodes to work autonomously is through forwarding data between nodes. Therefore, the life span of IoT devices is an issue. Despite power consumption of IoT devices that have been improved sharply due to several improvements of functionality of IoT devices such as eliminating full-duplex, half-duplex and reducing the amount of time needed to transmit data, but considering other issues such as quality, operating system, security and capacity, it keeps power consumption under developing stages. Also, there is a climate factor that affects IoT devices such as street sensors and traffic lights, which are in nature all the time [22]. Coverage can be defined as a length of the area that wireless signals can be covered [23]. Figure 5 shows wireless categories and clarifications. It divides into five categories:

1. Wireless Body Area Network (WBAN): This network is designed to collect/connect personal parameters such as patient health, temperature, blood pressure and heart rate [24].

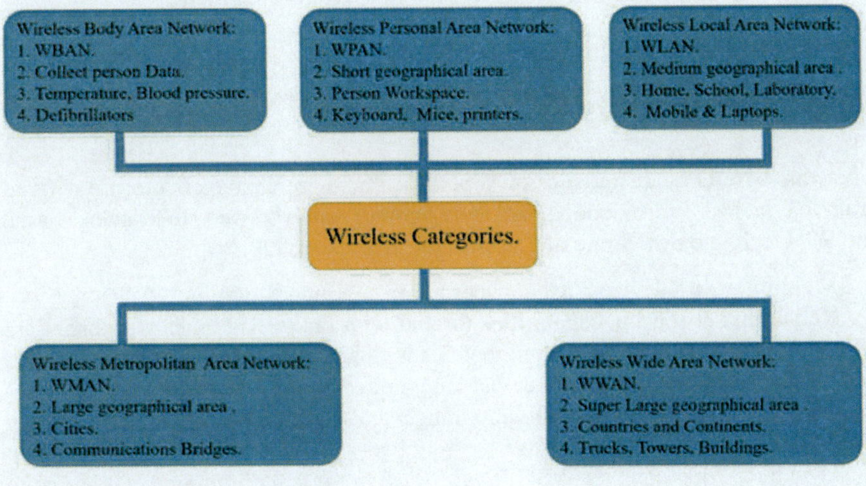

Fig. 5 Wireless categories. Adapted from: [23, 24]

2. Wireless Personal Area Network (WPAN): This network is designed to cover/connect for a short geographical area such as headphones [23].
3. Wireless Local Area Network (WLAN): This network is designed to cover/connect for a medium geographical area such as home and school [23].
4. Wireless Metropolitan Area Network (WMAN): This network is designed to cover/connect for a large geographical area such as city [23].
5. Wireless Wide Area Network (WWAN): This network is designed to cover/connect super large geographical areas such as countries and continents [23].

Cost/price is variant for wireless communication technologies starting from 2 to 5 USD for Bluetooth devices while 60 USD for SigFox devices. Bluetooth is developed by Special Interest Group (SIG), and it is used for a short distance. It operates into 2.4 GHz and low in terms of energy consumption. There are several versions of Bluetooth but since 2011 version 4 has been introduced, it has low time for install as well as low-power consumption. Also, it supports an unlimited number of nodes in a star topology. The best example of this technology is blood pressure measurement [25]. NFC technology, developed by Sony and Philips, works based on transfer data into two way once two devices are sitting side by side. Range of NFC is up to 10 cm. It operates at 13.56 MHz. An example of this technology is contactless payment cards [26]. ZigBee is introduced earlier in twenty-first century in the USA as a standardized by IEEE. Range of it can be up to 100 m, and it is used to transfer low data rate [27]. NB-IoT is based on a 4G mobile network, and it belongs to LET-IoT. The range of this technology is large, and it operates from 450 MHz to 3.5 GHz. Also, it can operate on 2G and 3G as well. It operates based on half-duplex, time division duplex TDD and frequency division duplex FDD [28]. LoWaWAN technology belongs to low-power wide area network (LPWAN), and it covers a long range [28]. Table 1 shows a comparison between wireless communication technologies.

2.2 Characteristics of WSN

There are several characteristics of WSN. Ideally, these characteristics are derived from IoT devices, IoT functionalities and communication between IoT devices based on WSN technologies. Some of these characteristics are [29, 30]:

1. Mobility of IoT devices: IoT devices are in continuous motion [29, 30].
2. Reliability: It reflects capabilities to deal with faults. This means, sensors are able to maintain network functionalities with an interruption [29, 30].
3. Limitation for resources: IoT devices are limited for power consumption, computational capabilities, transmission range and data, reliability and immunity [29, 30].
4. Dynamic: Topologies of WSN in an IoT environment are unpredictable due to nodes communicating with each other. Therefore, it self-organizes itself. This will introduce several issues such as latency, capacity and routing data [29, 30].

Table 1 Wireless communication technologies adapted from: [17, 25–28]

Wireless communication technology	Range	Reliability	Latency	Coverage	Power consumption	Bandwidth	Example	Date rate
Bluetooth	<10 m	Medium	<10 ms	Small	Low	<2.4 GHz	Wireless keyboards	Up to 25 Mbps
ZigBee	<100 m	High	<20 ms	Small	Low	<2.4 GHz	Street lights	20–250 Mbps
Wi-Fi	<100 m	High	<10 ms	Small	Average	2.4–5 GHz	Computers	2–54Mbps
NFC	<10 cm	High	100–250 ms	Short	Low	13.56 MHz	Contactless payment card	106–848 Kbit/s
LoRAWAN	<5–20 km	Medium	1–10 s	Large	Low	500 kHz–900 MHz	Smart city	27 Kbps
NB-IOT	>10 km	High	1.5–10 s	High	Low	450 MHz–3.5 GHz	Smart parking	26–66 Kbps
SigFox	10–50 km	Medium	1–30 s	Large	Low	100 Hz–900 MHz	Smart home	600 Bits
3G/4G	>30 km	High	100 ms–1 s	Large	Low	200 kHz–900 MHz	Fire detection	8–50 Mbps

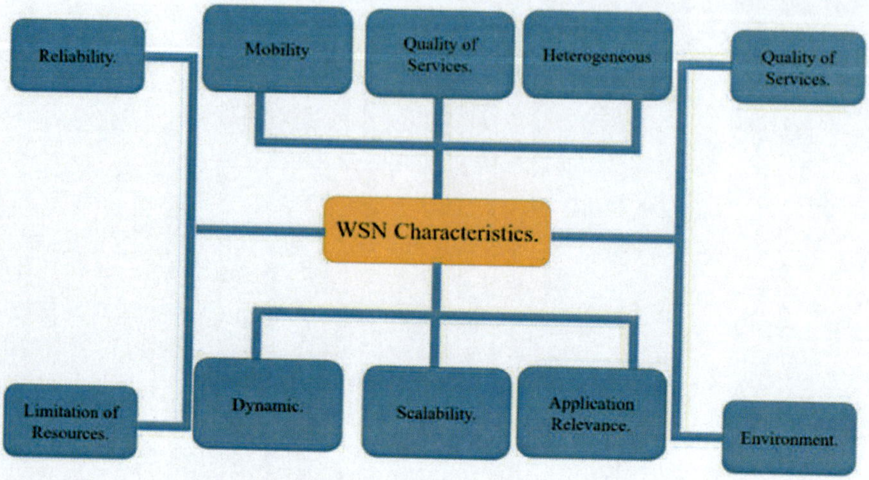

Fig. 6 Characteristics of WSN. Adapted from: [29, 30]

5. Scalability: In IoT environment, there are several types of devices communicating with each other to collect and share data. This massive number of devices is known by scalability issue. Ideally, scalability affects reliability and accuracy due to a massive number of devices [29, 30].
6. Application relevance: The main principle of WSN is to collect and share data between IoT devices. Therefore, it needs applications to achieve that such as Uber application for picking a car. This means the application itself can affect WSN [29, 30].
7. Environment: IoT devices operate in open areas such as traffic lights [29, 30].
8. Heterogeneous: Several types of IoT devices are communicating with each other [29, 30].
9. Quality of Services: Some application in IoT is dealing with sensitive or important fields such as the health sector. Therefore, IoT devices with a WSN-based communication should be capable for data delivery accurately on time. Therefore, low latency and high accuracy are very important [29, 30] (Fig. 6).

2.3 Communication Model and Protocols

Communication in IoT environment consists of sensors that connect to IoT gateways and gateway into connection with a further network using WSN. Gateway is responsible for establishing a connection between sensors and the Internet while sensors are responsible for collecting data from environment and nodes. However, there are several communication models in IoT such as three-layered IoT model, open system

OSI Communication Model.

OSI Layer	Responsibility.	Protocols.
Layer 7- Application.	Responsible to provide access for applications that are connect to network through an Interface.	FTP, IPS, SSH, SNMP XMPP, Telnet,
Layer 6- Presentation.	Responsible to present data into translated format from application to one that used into network. Also, it deals with encryption and decryption, compression and decompression.	SSL, ARP, DTP, STP
Layer 5- Session.	Responsible for Synchronization process of shared data on separated devices. Also, it responsible for create, open and close sessions.	HTTP, TCP, UDP, MQTT, CoAP, DDS
Layer 4- Transport.	Responsible for provide reliable data delivery between sending and receiving through provides communication services.	TCP and UDP
Layer 3- Network	Responsible for identified destination, address and routed data through different networks.	IPv4, IPv6, ARP, RARP, IGMP,OSPF
Layer 2- Data Link.	Responsible for error detection and correction, Reliable delivery, Framing and link access, Half and Duplex. It has two sub layers: A. Logical Link Control (LLC): responsible to be as an interact layer between DAC and Network layer. B. Media Access Control (DAC): responsible for interaction with transmission media.	SDLC, HDLC, SLIP, PPP, LCP, NCP
Layer 1- Physical.	Responsible to define physical network and communication topology.	Ethernet, Token Ring, FDDI.

Fig. 7 Communication layers and protocols. Adapted from: [31–34]

interconnection model OSI and TCP/IP model. Ideally, these models have differences in terms of communication layers. For instance, three-layered model is considered a basic model and consists of three layers known as application, network and perception layers, while OSI consists of application, presentation, session, transport, network, data link and physical layers. TCP/IP model consists of application, transport, network and data link layers [31–34]. Figure 7 clarified layers functionalities as well as protocols used at each layer.

2.4 *Challenges of IoT and WSN*

The main reasons for the proposed challenges of IoT and WSN are IoT devices and functionalities of IoT devices. These challenges are as follows:

1. Interoperability.
   It reflects the heterogeneously on IoT devices that need to communicate with each other to share and collect data. Therefore, IoT devices should be interoperable to generate data. There are several challenges of interoperability such as technical, syntactical, semantic and organizational. Technical refers to communication protocols and infrastructure that are used between devices while syntactical refers to formats of data that are used in communication. Semantic refers to the capability of a computing system to exchange data while organizational means to establish an automatic link among processes [35].

Hacking & Data Breaches.

Company	Year/ Impact
Adobe	2013 / 153 Million user records.
Adult Friend Finder	2016/ 412.2 million accounts
Canva	2019/ 137 Million user accounts.
eBay	2014/ 145 Million users
Equifx	2019/ 147.9 Million users.
LinkedIn	2012 and 2016 / 165 million user accounts.
Zynga	2019 / 218 Million user accounts.

Fig. 8 Hacking and data breaches in the twenty-first century. Adapted from: [36]

2. Security.
   Security is considered the main concern for any field in information technology. In IoT, devices are open to many threats, vulnerabilities and attacks. Therefore, it needs a security mechanism and tools to secure communication between devices, data and operations [30]. Figure 8 shows the impact of several security breaches conducted in the twenty-first century [36].
3. Enterprise.
   It reflects massive numbers of connected IoT devices. This introduced the availability issue and has an impact on real-time business [35].
4. Storage Management.
   In an IoT environment, IoT devices are constantly generating data as well as interacting to access data. Therefore, the massive sizes of data will be generated, and this will need massive storage such as data warehouse [35].
5. Server Technologies.
   It requires high-speed computational power to provide services for an enormous number of users [35].
6. Data Center Technologies.
   It reflects the requirement of an effectively designed data center to process generated data by IoT devices. In fact, there are three data centers known by center of big data, center of distributed data and center of micro data [35].
7. Shared Infrastructures.
   IoT includes several sensors, softwares and processes. In terms of software, there are several open sources which help to reduce the price, but in an IoT environment, there are few open-source tools [35].
8. Common Standards.
   IoT devices are using their own protocols to communicate and share data as there is no standard in IoT. This proposed a challenge for IoT developers. For instance, Eclipse foundation has introduced an open source to deal with all

main IoT standards besides framework to design IoT gateway and smart home [30, 35].

9. Power Consumption.
   IoT devices are limited for power consumption, and there is a need for long-life batteries [30]. There are several proposals to deal with power consumption.
   1. Keep device in an idle state when not used for a while [30].
   2. Keep device in halt and use when devices are in a specific location or not in use [30].
   3. Body-attached devices can be re-charged from body temperature. Heartbeat can be also used for charging wearable devices users [30].
10. Data Sharing and Data Privacy.
    Many of IoT devices are presenting data such as location while the user would not like to present it. Therefore, this is considered critical toward privacy issue. Also, corporations use customer's personal data for several issues such as survey or advertisements. This is considered a breach of customer privacy [30, 37].
11. IPv6.
    Every device into IoT has a unique IP that identified it. However, due to a massive number of IoT devices, there is a need for a new IP supplied due to which IPv4 will not be sufficient [30, 37].

3 The Functionality of IoT

Communicate, collect and share data into IoT environment conducting between enormous heterogeneous devices. Therefore, there are several requirements to establish these functions:

1. Machines/Devices:
   IoT devices or machines are responsible for sense, collect, access and share data between devices and servers [30].
2. Interactions/Communications:
   IoT devices are collect and share data based on several communication protocols. This highlights on the importance of IoT communication protocols due to which it is different rather than normal communication protocols in terms of data size, length, checksum, etc. [37].
3. Operation Services:
   It means IoT devices are capable of using services of IoT system such as device control, discover service, downloading and publishing data. [37].
4. Big Data, Centralization Management and Access Interface.
   Central management means data in IoT system should be visualized through readable format for all users through various communication protocols. This readable/structured data are accessed through an interface/application. Ideally, data generated from IoT system are known as big data, and it has several characteristics [38] such as

    1. Volume: Size of data ranged from terabyte to zettabytes [38].
    2. Variety: Several formats for generated data need to be accessed to present it into a useful format [38].
    3. Velocity: Data in IoT applications are accessed to achieve into high speed [38].

5. Security:
   IoT applications and environment should be secure to prevent any unauthorized access or theft of user's credits or information. There are various sophisticated attacks, threats and vulnerabilities toward IoT systems that should be covered. Besides, there are several security mechanisms such as blockchain and intrusion detection mechanisms (IDs) should be developed to cover these attacks [30, 39].
6. Clouding:
   The main aim of adapting clouding is to make services, resources and data access available through the Internet. Clouding characteristics are wide such as elasticity, resource pooling, service and ubiquitous access. Clouding features became an important issue to develop in order to be compatible with IoT vision. Also, cloud can provide several important issues such as sharing resources, access data without having geographical or time limitation. Ideally, these issues are also the main ideas of IoT. Therefore, it is important to consider cloud besides IoT as both of them will provide a great opportunity in information technology area [40].
7. 5G:
   Generating massive traffic by connecting to the Internet leads to developed communication technology from 3G and 4G to be in 5G. 5G is considered the future of wireless communication technology. However, there are several requirements that should be addressed through 5G technology such as [41–43]

    1. Latency should be reduced.
    2. Power consumption should be addressed.
    3. The massive number of connected devices.
    4. High data rates and data bandwidth.
    5. Being compatible with artificial intelligent (AI).
    6. World Wide Wireless Web (WWWW).
    7. Heterogeneous network.

4 Conclusion

Benefits of adapting IoT into information and communication technology are countless as it is associated with most life fields. However, in order to obtain maximum benefits, we have to improve it as much as possible. Therefore, we have to dig deeply into IoT in order to get the best understanding for IoT environment, IoT devices, IoT operations, IoT characteristics and communication technologies into IoT. These issues are considered very important due to which they are playing the main role in identifying IoT challenges. Based on analyzed IoT challenges as well as considering benefits/revenue from IoT, security is considered one of the most critical and important challenges. Therefore, there is a need to understand the security challenge in comprehensive in order to keep the benefits of IoT raised.

5 Future Works

This work is produced to identify a comprehensive understanding of IoT environment and challenges. The security challenge is considered critical and needs to develop. This work is considered the starting point of improving security into IoT. At present, the author of this paper is working on defining a security architecture for smart city based on intrusion detection mechanism (IDs) to detect major attacks. This architecture considers threats, vulnerabilities and attacks associated with IoT and smart city. Also, it considers popular IoT devices. Therefore, it is important to establish a concrete foundation that provided a clear image of IoT in order to overcome IoT challenges.

References

1. I. Lee, K. Lee, The internet of things (IoT): Applications, investments, and challenges for enterprises. Bus. Horiz. **58**, 431–440 (2015). https://doi.org/10.1016/j.bushor.2015.03.008
2. Kenneth, *Top 10 Most Popular IoT Devices In No Coding, Excel-like Database No Coding* (2019), pp. 1–18. https://robots.net/tech-reviews/top-iot-devices/ (accessed 6 Aug 2020)
3. Software Testing Help, 18 Most Popular IoT Devices in 2019 (Only Noteworthy IoT Products). Softw. Test. Help. 2020 (2019). https://www.softwaretestinghelp.com/iot-devices/ (accessed 6 Aug 2020)
4. D. Miorandi, S. Sicari, F. De Pellegrini, I. Chlamtac, Internet of things: Vision, applications and research challenges. Ad Hoc Netw. **10**, 1497–1516 (2012). https://doi.org/10.1016/j.adhoc.2012.02.016
5. N. Sathiyanatha, P. Selvaprasanth, A brief study on IoT applications. Int. J. Trend Sci. Res. Dev. **4**, 2456–6470 (2020). https://www.researchgate.net/publication/338423030_A_Brief_Study_on_IoT_Applications
6. D. Evans, The Internet of Things—How the next evolution of the internet is changing everything. CISCO White Pap. 1–11 (2011). https://doi.org/10.1109/IEEESTD.2007.373646
7. M. Greenfield, Internet of Things (IoT) connected devices installed base worldwide from 2015 to 2025. Stat. Res. Dep. **8**. (2016). https://www.statista.com/statistics/471264/iot-number-of-connected-devices-worldwide/ (accessed 4 Aug 2020)
8. K.L. Lueth, *State of the IoT 2018: Number of IoT devices now at 7B-Market accelerating*. Market Update Global number of Connected Devices: 17B, pp. 5–9 (2018). https://iot-analytics.com/state-of-the-iot-update-q1-q2-2018-number-of-iot-devices-now-7b/ (accessed 2 Aug 2020)
9. A. Nordrum, Popular Internet of Things forecast of 50 billion devices by 2020 is outdated—warning: All projections for the Internet of Things are subject to change. IEEE Spectr. 2019–2020 (2016). https://spectrum.ieee.org/tech-talk/telecom/internet/popular-internet-of-things-forecast-of-50-billion-devices-by-2020-is-outdated

10. A. D'mello, Global smart city revenue to reach $1.7 trillion by, 2019–2020 (2020). https://www.iot-now.com/2020/05/20/102937-global-iot-market-to-grow-to-1-5trn-annual-revenue-by-2030/ (accessed 2 Aug 2020)
11. M.D. Fernandes, *Internet of Things New digital technologies* (2018). www.pwc.pt/iot (accessed 4 Aug 2020)
12. Transforma Insights, Global IoT market will grow to 24.1 billion devices in 2030, generating $1.5 trillion annual revenue (2020). https://transformainsights.com/news/iot-market-24-billion-usd15-trillion-revenue-2030 (accessed 6 Aug 2020)
13. F.S. Market, IoT in Banking and Financial Services Market, 2023 (2020) 1–6. https://www.marketsandmarkets.com/Market-Reports/iot-banking-financial-services-market-172304505.html (accessed 4 Aug 2020)
14. Gartner, Newsroom, (2019) 2019–2021. https://www.gartner.com/en/newsroom/press-releases/2019-08-29-gartner-says-5-8-billion-enterprise-and-automotive-io (accessed 4 Aug 2020)
15. D. Macarrone, *10 relevant IoT trends in 2020*. Loriot. (2020) 1–8. https://www.loriot.io/blog/IoT-trends-2020.html (accessed 6 Aug 2020)
16. A. Salleh, K. Mamat, M.Y. Darus, Integration of wireless sensor network and web of things: Security perspective, in *2017 IEEE 8th Control System Graduate Research Colloquium, ICSGRC 2017* (2017), pp. 138–143. https://doi.org/10.1109/ICSGRC.2017.8070583
17. M. Sikimic, M. Amovic, V. Vujovic, B. Suknovic, D. Manjak, An overview of wireless technologies for IoT network, in *2020 19th International Symposium INFOTEH-JAHORINA, INFOTEH 2020* (2020), pp. 18–20. https://doi.org/10.1109/INFOTEH48170.2020.9066337
18. W. Anani, A. Ouda, A. Hamou, A survey of wireless communications for IoT echo-systems, in *2019 IEEE Canadian Conference of Electrical and Computer Engineering* (2019), pp. 1–6. https://doi.org/10.1109/CCECE.2019.8861764
19. K. Shafique, B.A. Khawaja, F. Sabir, S. Qazi, M. Mustaqim, Internet of things (IoT) for next-generation smart systems: A review of current challenges, future trends and prospects for emerging 5G-IoT scenarios. IEEE Access **8**, 23022–23040 (2020). https://doi.org/10.1109/ACCESS.2020.2970118
20. V. Kharchenko, M. Kolisnyk, I. Piskachova, N. Bardis, Reliability & security issues for IoT-based smart business center: Architecture & Markov model, in *2016 Third International Conference on Mathematics and Computers in Sciences and in Industry (MCSI)* (2017), pp. 313–318. https://doi.org/10.1109/MCSI.2016.60
21. S. Bhandari, S.K. Sharma, X. Wang, Latency minimization in wireless IoT using prioritized channel access and data aggregation, in *GLOBECOM 2017—2017 IEEE Global Communications Conference* (2017), pp. 1–6. https://doi.org/10.1109/GLOCOM.2017.8255038
22. A.A. Gendreau, M. Moorman, Survey of intrusion detection systems towards an end to end secure internet of things, in *2016 IEEE 4th International Conference on Future Internet of Things and Cloud (FiCloud)* (2016). https://doi.org/10.1109/FiCloud.2016.20
23. K. Sharma, N. Dhir, A study of wireless networks: WLANs, WPANs, WMANs, and WWANs with Comparison. IJCSIT Int J. Comput. Sci. Inf. Technol. **5**, 7810–7813 (2014)
24. H.K. Lau, High-speed short-range systems for wireless personal area networks, in *2009 Wireless Telecommunications Symposium WTS 2009* (2009) 2006–2009. https://doi.org/10.1109/WTS.2009.5068981
25. C. Gomez, J. Oller, J. Paradells, Overview and evaluation of bluetooth low energy: An emerging low-power wireless technology. Sensors (Switzerland) **12**, 11734–11753 (2012). https://doi.org/10.3390/s120911734
26. K. Curran, A. Millar, C. Mc Garvey, Near field communication. Int. J. Electr. Comput. Eng. **2** (2012). https://doi.org/10.11591/ijece.v2i3.234
27. S. Long, F. Miao, Research on ZigBee wireless communication technology and its application, in *2019 IEEE 4th Advanced Information Technology, Electronic and Automation Control Conference (IAEAC)*. (2019) 1830–1834. https://doi.org/10.1109/IAEAC47372.2019.8997928

28. K.K. Nair, A.M. Abu-Mahfouz, S. Lefophane, Analysis of the narrow band internet of things (NB-IoT) technology, in *2019 Conference on Information Communications Technology and Society (ICTAS)* (2019). https://doi.org/10.1109/ICTAS.2019.8703630
29. S. Chen, H. Xu, D. Liu, B. Hu, H. Wang, A vision of IoT: Applications, challenges, and opportunities with China perspective. IEEE Internet Things J. **1**, 349–359 (2014). https://doi.org/10.1109/JIOT.2014.2337336
30. B.B. Zarpelão, R.S. Miani, C.T. Kawakani, S.C. de Alvarenga, A survey of intrusion detection in Internet of Things. J. Netw. Comput. Appl. **84**, 25–37 (2017). https://doi.org/10.1016/j.jnca.2017.02.009
31. M. Bagga, P. Thakral, T. Bagga, A study on IoT: Model, communication protocols, security hazards countermeasures, in *2018 Fifth International Conference on Parallel, Distributed and Grid Computing (PDGC)* (2018), pp. 591–598. https://doi.org/10.1109/PDGC.2018.8745984
32. M. Centenaro, L. Vangelista, A. Zanella, M. Zorzi, Long-range communications in unlicensed bands: The rising stars in the IoT and smart city scenarios. IEEE Wirel. Commun. **23**, 60–67 (2016). https://doi.org/10.1109/MWC.2016.7721743
33. Y. Li, D. Li, W. Cui, R. Zhang, Research based on OSI model, in 2011 IEEE 3rd International Conference on Communication Software and Networks (2011), pp. 554–557. https://doi.org/10.1109/ICCSN.2011.6014631
34. H. Mechanical, E. Polytechnic, H. Changsha, *Research and Application of TCP/IP Protocol in Embedded System* (2011), pp. 584–587
35. G.S. Matharu, P. Upadhyay, L. Chaudhary, The Internet of Things: Challenges & security issues, in *2014 International Conference on Emerging Technologies (ICET)* (2014), pp. 54–59. https://doi.org/10.1109/ICET.2014.7021016
36. B.D. Swinhoe, *Data breaches affecting millions of users are far too common. Here are some of the biggest, baddest breaches in recent memory* (2020). https://www.csoonline.com/article/2130877/the-biggest-data-breaches-of-the-21st-century.html (accessed 8 Aug 2020)
37. S.R. Jino Ramson, D. Jackuline Moni, Applications of wireless sensor networks—A survey, in *2017 International Conference on Innovations in Electrical, Electronics, Instrumentation and Media Technology (ICEEIMT)* (2017), pp. 325–329. https://doi.org/10.1109/.2005.1467103
38. E. Bertino, Data security and privacy: Concepts, approaches, and research directions, in *2016 IEEE 40th Annual Computer Software and Applications Conference (COMPSAC)* 1 (2016), pp. 400–407. https://doi.org/10.1109/COMPSAC.2016.89
39. S. Neogy, Security management in Wireless Sensor Networks, in *2015 International Conference on Cyber Situational Awareness, Data Analytics and Assessment (CyberSA)* (2015), pp. 9–12. https://doi.org/10.1109/CyberSA.2015.7166112
40. A.R. Biswas, R. Giaffreda, IoT and cloud convergence: Opportunities and challenges, in *2014 IEEE World Forum Internet Things, WF-IoT 2014* (2014), pp. 375–376. https://doi.org/10.1109/WF-IoT.2014.6803194
41. U.B. Shukurillaevich, 5G Technology, 5G Technol. (2020). https://doi.org/10.1002/9781119236306
42. A.K. Jain, R. Acharya, S. Jakhar, T. Mishra, Fifth Generation (5G) Wireless Technology "Revolution in Telecommunication," in *2018 Second International Conference on Inventive Communication and Computational Technologies (ICICCT)* (2018), pp. 1867–1872. https://doi.org/10.1109/ICICCT.2018.8473011
43. S. Kumar, G. Gupta, K.R. Singh, 5G: Revolution of future communication technology, in *2015 International Conference on Green Computing and Internet of Things (ICGCIoT)* (2016), pp. 143–147. https://doi.org/10.1109/ICGCIoT.2015.7380446

Banking Transaction Considerations on Standardized Browser Architecture

Indraneel Mukhopadhyay and Abir Ghosh

Abstract Web applications and Mobile apps have their own importance and needfulness. Almost all banks prefer to have their separate web application and mobile app for banking transactions of their customers. Mobile apps offer customization compared to web applications. Progressive Web Applications (PWA) try to maintain a middle path however there are some disadvantages like lacking stricter security control which becomes very critical in case of banking transaction. Creating a web browser which works for only banking transactions adds this extra layer of security. It will have benefits of PWA as well as in-built security of browser which will make it more user-friendly. Customer won't have to download mobile app frequently from store and won't have to switch between browsers to do banking transactions in most secured manner. Thus it will have a standard approach of taking into account design considerations into browser side which almost every latest banking web app targets to achieve. This standardization approach of having banking transaction specific browser will also save lot of development effort and cost. Initial resistance might be there from industry and customers but considering the benefits and comforts it has a great prospect of wide acceptance from industry and customers.

Keywords Cyber security · Banking transaction · Web browser security · Progressive web application · Web application · Browser architecture standardization

---

I. Mukhopadhyay (✉)
Institute of Engineering & Management, Kolkata, India
e-mail: imukhopadhyay@gmail.com

A. Ghosh
University of Engineering & Management, Kolkata, India

V. E. Balas et al. (eds.), *Proceedings of International Conference on Computational Intelligence, Data Science and Cloud Computing*, Lecture Notes on Data Engineering and Communications Technologies 62, https://doi.org/10.1007/978-981-33-4968-1_49

1 Introduction

Modern banking applications need to be aligned on multiple factors to meet the global standards. Growing security concerns, enriched customer experience and time to market are just some factors which banking applications need to factor in their design considerations. Many standardization approaches are already started and gaining the required attention, in this context banking transaction specific browser makes a strong case. Not only it will help standardization of the customer experience of doing banking transactions but also it will help the development efforts since many design aspects will be there as in-built browser feature. Otherwise these features would be required to be developed and globally many such banks and technology companies are spending such effort.

1.1 Issues Faced by Banking Sites

Security is a growing concern faced by almost all the banks globally. Fraudsters are coming up with innovative way to take control of customer account and siphon money out. Amidst this, it takes lot of effort to make banking sites secure to prevent fraud. Not just security standards, banking applications need to consider many other aspects like responsive design, easy navigation, multi-media content etc. Apart from functioning properly, these aspects also decide whether any banking application will be liked by customers or not. There is also another factor which has become critical is browser compatibility. It is expected that banking application will work perfectly fine in all browsers. Time to market is becoming more and more critical for banks nowadays considering the stiff challenges faced by banks from Financial Technology (fintech) and Mobile Network Operator (MNO) wallet providers. Keeping the above factors in mind, one standardized browser can be built only for banking transactions as it will take away some of the critical design factors of a banking web application.

1.2 Existing Alternatives

Banks take care many such aspects discussed above for securing application, responsiveness etc. while designing internet banking application. Different browser compatibility is considered in design aspects. There are tools and technology advances to support it however the significant cost and effort involved can't be ruled out. Banks prefer now mobile app as customers are more mobile friendly now a day. Mobile app also provides the required customization features which banks prefer to connect with their customer. However mobile apps also don't solve all the problems. Mobile app developments are time intensive effort. Also customers face challenges as they repeatedly get alerts for latest version of app download from store. Not only does it can consume large amount data but also raises question in customer mind about authenticity of app.

1.3 Standardization

Efforts are already in place to address these issues faced by banking applications. In the architecture side, Banking Industry Architecture Network (BIAN) tries to take a standardized approach how the banking service can be offered to the customers. Adopting BIAN standards is high among banks and that also proves the acceptance of standardization. Not just the service offering, customer experience can also be standardized by creating a financial transaction specific browser which specifically focuses on security and customer comfort doing financial transactions.

2 Banking Web Applications and Security Perspective

Though internet banking applications provide great facility to banking customers there are various other factors which are great concern for banks across world. It is convenience, ease, time flexibility which attracts customer to use online banking but simultaneously customers are worried about security and about their compromised information which prevent them from using it [1]. Customers are largely concerned about identity theft and Phishing like attacks though they are aware about the conveniences offered by online banking [2]. The sophistication of phishing, vishing and spoofing attacks are also identified and confirmed by 76% of the bank's executives. The phishing, the falsification, the hacking and the identity theft online are some of the main challenges for banks [3]. Symantec saw a number of groups previously involved in spreading ransomware move to delivering other malware such as banking Trojans and information stealers [4].

It is a matter of concern for banks that key element of any user's trust is all about user's perception of security. Banks are more vocal about their transmission encryption than their storage encryption [5]. It is also shown in a Deloitte survey that globally customers are more likely to use online banking more frequently if security is increased, real-time problem resolution provided more and regular banking transactions are allowed more to be complete online [6]. The security of private browsing is undermined by many popular browser extensions and plug-ins [7].

Table 1 Comparative analysis between PWA, mobile app and specific purpose browser

	PWA	Mobile App	Specific Purpose Browser
Instant Use Opportunity	Yes	No	Yes
Compatibility	High	Low	High
Implementation Effort	Low	High	Low
App Store Dependency	No	Yes	No
Speed	Lower than Mobile app	Higher than PWA	Lower than PWA and Mobile app
Personalization	Not Easy implemntation	Can be done easily	Not easy implementation
Network	Online	Online & Offline	Online
Security	Relatively lower	Higher	Can be made robustly secure

3 Mobile App Versus Progressive Web App Versus Specific Purpose Browser

Current browsing implementations provide privacy against some local and web attackers but it stands weak against some determined attackers [7]. Support for service workers in Apple's iOS Safari browser is a requirement if we consider PWA to have same potential as cross-platform app development. A PWA-enabled web app wouldn't deliver a consistent experience across browsers and platforms without such support [8]. Whether mobile app or PWA or banking specific browser, which one is best suited to support secure banking transaction? There is no clear answer to it. Every tool has its own benefits and shortcomings. Below list shows a high level a comparative analysis: (Table 1)

Browsers can adopt the following techniques to minimize the danger faced by users:

1. The attack severity can be reduced by applying the principle of least privilege in the browser architecture. By applying this technique the damage could be minimized when any attacker exploits vulnerability.
2. Ensure that updates are developed and deployed as quickly as possible which will reduce the window of vulnerability. It will minimize the number of vulnerable browsers which can be targeted by an attacker.
3. Filter out known malicious content to reduce how often users are exposed to attacks. This will protect users during the vulnerable time windows.

The Google Chrome team has considered each of these factors and focused into it to provide a secure browser without compromising compatibility with existing web content [9].

4 Browser Architecture

Tremendous progress happened on browser development work in the recent past. From monolithic to multi-process browser architecture, it has changed extensively

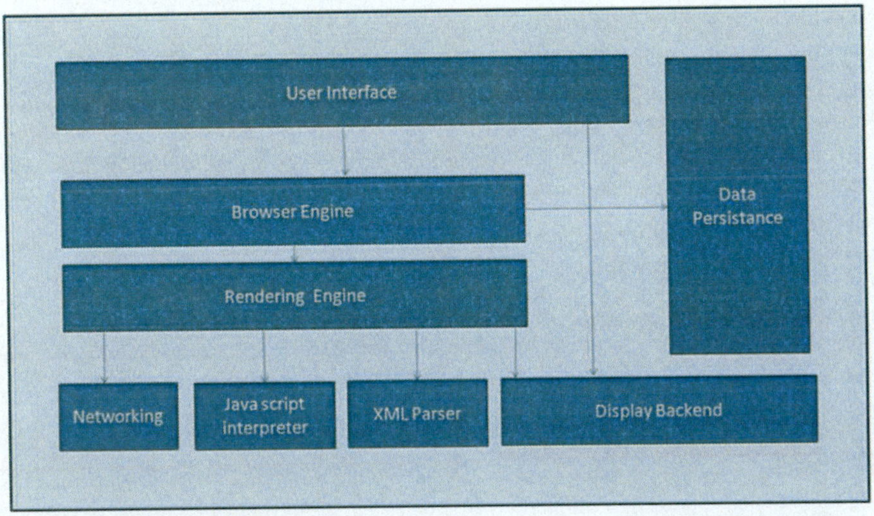

Fig. 1 Browser architecture

to support more and more features [10]. Below diagram shows the components of typical browser architecture: (Fig. 1)

- The User Interface: This is user's interaction with the browser. Address bar, back button, next button, refresh, home button and stop, bookmark option, etc. all are part of user interface.
- The Browser Engine: This works as bridge between UI and rendering engine. Based on user inputs, it interacts with rendering engine.
- The Rendering Engine: The rendering engine renders the web page on browser screen. It interprets the HTML, XML and images and generates layout for displaying in user interface.
- Networking: This component retrieves the URLs and handles all internet communication aspects and securityy
- JavaScript Interpreter: This component does the interpretation and execution of java script code which any web site uses. The interpretation results are sent to rendering engine for display purpose.
- UI Backend: UI backend is used for drawing basic widgets.
- Data Persistence/Storage: This is a persistence layer which is like a small database created on the local drive. It manages cookies, cache, bookmarks and user preferences.

5 How Specific Purpose Browser Can Help

Banking transactions differ from other browsing activities in the sense that customer doesn't want to cache the data, wants to avoid auto-filling, wants to use data encryption and data masking as much as possible. Below are some desirable features with respect to banking transactions which no single browser provides today:

1. No Browser History maintenance
2. No Browser Data caching
3. Every customer data input handling like password
4. No Auto-filling capability
5. Automated Log-out
6. In-built Biometric authentication
7. In-built spyware detection
8. Effective Password Management.

Though some of these features are developed by application developers to provide the security to customer, it will greatly reduce build time when such features are standardized and available as in-built features in the browser.

6 Related Work

Lot of work has already been done to enhance browser security and user privacy. Below are some features which are provided by privacy browsers [11]: (Table 2)

Table 2 Privacy browsers

Privacy Browsers	What they provide
TOR	1.Block Trackers 2.Defend against surveillance 3.Resist fingerprinting 4.Multi-layered encryption
Brave	1.Automatic HTTPS connection upgrades 2.Ability to block ads and scripts, manage cookies 3.Native password manager. 4.Customize shield settings on a "per-site or browser-wide basis."
Safari	1. Uses machine learning to prevent the tracking of personal data, browsing history. 2.Sandboxing 3.Warnings for unsafe websites, 4.DuckDuckGo default search engine

Browser like Brave and Tor provides the much needed privacy required for banking Transactions however these are not built for specifically banking transactions [12]. Similarly some other browsers are there which specifically addresses some concerns which other normal browsers don't handle effectively. Specific purpose browser brings some additional unique benefits which are really critical for banking transactions.

7 Key Consideration for Creating Banking Transaction Purpose Browser

Various types of attacks are potential threat for banking web application which is listed below [13]:

1. Social Engineering
2. Port Scanners
3. Packet Sniffers
4. Password Cracking
5. Trojans
6. Denial of Service Attacks
7. Server Bugs
8. Super User Exploits.

Intrusion Prevention System (IPS) can be seen as an improvement upon firewall technologies which makes access control decisions based on application content. Whereas in case of traditional firewalls it is based on IP address or ports. Intrusion Detection and Prevention Systems (IDPS) is capable of detecting and preventing such attacks from happening [14–16]. Private browsing modes also don't protect against such threat models [7]. There is clearly a need to redesign the browser architecture to address such concerns. Not every component of browser architecture needs to re-haul to create banking transaction specific browser. Below table suggests some design considerations to achieve the desired features: (Table 3)

8 Issues and Challenges

There are different categories of challenges lying ahead before financial transaction specific browser can become a reality. Some of these challenges are technical, some are banking industry specific and some are related to customer acceptance:

Technical Challenges: It is quite complex technical development work to create a financial transaction specific browser. It needs changes in almost all the component layers like user interface, networking, User Interface (UI) backend and data persistence layer. To highlight the complexity around networking layer changes, it will require participation from banking organizations so that filtered routing is possible.

Table 3 Key design consideration

Features	How it is managed now	Key Design considerations how it can be managed in banking transaction specific browser
No Browser History maintenance	Caching policy defines what can be cached or not. The various Cache-Control directives are no-cache, no-store etc.	The caching policy which is handled mostly by web applications, need to be handled similarly inside the browser components
No Browser Data caching		
Every customer data input handling like password	Currently not all data fields are handled as securely as password field in the case of banking applications. There are similar other sensitive data like ID Number, Pin etc. which needs similar treatment	All the fields whether sensitive or not, needs to be encrypted and managed through browser Normally banking transactions are not very heavy with respect to data handling
No Auto-filling capability	Managed through autocomplete field. If it is set as "off" then browser won't save data input done by customer. Also it doesn't allow browser to cache data	The same handling needs to be in-built in browser and auto complete should be defaulted to "off"
Automated Log-out	This is managed by the web applications to make the session inactive.	By default, browser needs to track non-activity and force user to login again.
In-built Biometric authentication	It works by comparing two sets of data: preset biometric data of owner and same set of data of device visitor. In case of identical data, access is allowed	The similar design aspects to be considered which are considered by web application to build the bio-metric authentication
In-built spyware detection	Google Chrome has this feature in-built however most other latest popular browser doesn't have this feature	This is a must feature and already implemented in one of the popular browser. However issues and concerns need to be addressed
Effective Password Management	Most browser gives option to save password and when user saves it, it is stored in the registry	The similar design aspects to be considered which are considered by web application to manage password effectively to have ease of operation by customer and to prevent fraud
Allowing only Banking sites	Banking sites restrict illicit users to access site though userid and password validation. Currently there is no such browser which allows only banking sites to be accessed	This requires maintaining a registry of allowed site of global banks. Lots of phishing attempts can be prevented if browser can control it

Acceptance from Industry: This can be useful if industry wide acceptance is there and necessary support is there. Support from regulators will be an added benefit in this regard.

Customer Acceptance: Customer needs to see the tangible benefits so that acceptance is there. Benefits will be visible only when banks start promoting it. In this regard reference of Technology Acceptance Model (TAM) is important.

Technology Acceptance Model: The TAM is an information systems theory that models user's response to accept and use a technology. TAM suggests that when users are presented with a new technology, there are various factors which influence their decision about how and when they will use the new technology [10]: Perceived usefulness (PU)—"The degree to which a person believes that using a particular system would enhance his or her job performance"—as defined by Fred Davis.

Perceived ease-of-use (PEOU)—"The degree to which a person believes that using a particular system would be free from effort"—as defined by Davis.

9 Conclusion

Creating banking transaction specific browser may give banking customers the required security, comfort and unique experience which are demanded from banking transaction applications. It will help fight frauds worldwide and will also remove the need of downloading data intrinsic app from store for banking purpose. Browser will provide the in-built biometric support and responsiveness which provides great ease of operation to banking customers. Banking transaction specific browser will also bring in the required standardization which will help in reducing the application development effort and cost substantially. Browser architecture needs to be redesigned to achieve some of the objectives which also demand more research work. Whether banking transaction specific browser becomes a reality or not that depends on multiple factors like wider acceptance from industry but it highlights the need to consider some critical factors into browser design from suitability to banking transaction perspective.

References

1. Analyzing online banking from banker's perspective Mukta Sharma1, Dr. R.B. Garg2, Richa Joshi,"http://tmu.ac.in/college-of-computing-sciences-and-it/wp-content/uploads/sites/17/2016/10/CCSIT202.pdf
2. N.A. Sharaaf, M.N. Haamid, S.S. Samarawickrama, C.N. Gunawardhane, K.R.S.C.B. Kuragala, D. Dhammearatchi, Improved E-Banking System With Advanced Encryption Standards And Security Models, https://www.ijstr.org/final-print/oct2016/Improved-E-banking-System-With-Advanced-Encryption-Standards-And-Security-Models-.pdf
3. A. Bamrara, Evaluating database security and cyber attacks: a relational approach. J. Internet Bank. Commer. 20. https://doi.org/10.4172/1204-5357.1000115

4. Internet Security Threat Report, vol 24, https://docs.broadcom.com/doc/istr-24-2019-enFeb 2019
5. D.R. Moscato, S. Altschuller, International perceptions of online banking security concerns, https://core.ac.uk/download/pdf/55330369.pdf
6. V. Srinivas, R. Wadhwani, The value of online banking channels in a mobile-centric world, https://www2.deloitte.com/us/en/insights/industry/financial-services/online-banking-usage-in-mobile-centric-world.html
7. G. Aggarwal, E. Bursztein, C. Jackson, D. Boneh, An analysis of private browsing modes in modern browsers, https://crypto.stanford.edu/~dabo/pubs/papers/pri-vatebrowsing.pdf
8. A. Biørn-Hansen, T.A. Majchrzak, T.M. Grønli, Progressive web apps: the possible web-native unifier for mobile development, https://pdfs.semanticscholar.org/a4ab/d85fa63cecf41f1941f4b3947d3474610b81.pdf
9. C. Reis, A. Barth, C. Piza, Browser security: lessons from google chrome, no, https://www.researchgate.net/publication/220310104_Browser_Security_Lessons_from_Google_Chrome
10. C. Reis, S.D. Gribble, Isolating web programs in modern browser architectures, https://static.googleusercontent.com/media/research.google.com/en//pubs/archive/34924.pdf
11. A. George, The best browsers for privacy, https://www.digitaltrends.com/computing/best-browsers-for-privacy/, 9 May 2020
12. Browse Privately, Explore Freely, https://www.torproject.org/
13. Z. Omariba, N.B. Masese, G.W. Wanyembi, Security and privacy of electronic banking, https://www.researchgate.net/publication/268060917_Security_and_Privacy_of_Electronic_Banking
14. I. Mukhopadhyay, M. Chakraborty, S. Chakrabarti, A comparative study of related technologies of intrusion detection & prevention systems. J. Inf. Secur. **2**, 28–38 (2011)
15. I. Mukhopadhyay, M. Chakraborty, "Hardware realization of artificial neural network based intrusion detection & prevention system", J. Inf. Secur. **05**(04), (2014), doi:10.4236/jis.2014.54015
16. A.S. Al-Ajam, K.M. Nor, Influencing factors on behavioural intention to adopt internet banking service, World Appl. Sci. J. **22**(11), 1652–1656, (2013), ISSN 1818–4952

Intrusion Detection System for Securing the SCADA Industrial Control System

Shobica Sivakumar, Bragadeesh Srinivasan Ananthanarayanan, and Umamakeswari Arumugam

Abstract Traditional industrial control system (ICS) is a system that consists of interconnected components in order to manage and control the hardware components in the industries. Supervisory control and data acquisition (SCADA) control system is one such control system that is used for collecting information and managing the critical infrastructures. Due to the reliable nature of the SCADA systems for enabling wider connectivity, they are connected to the corporate networks and to the Internet for better performance and productivity. The interconnectivity of the control systems with other networks leads to many security concerns. Attackers can exploit and take hold of the control systems even if the vulnerability exists in any one of these systems. As a result, it leads to possible hardware failure, low system performance, functional losses, casualties, environmental pollution, etc. Intrusion detection is one of the major processes that is been done for ensuring safety of the connected systems. An intrusion detection system (IDS) is a means of security that detects and provides prevention against variety of attacks. It acts as a second line of defence for network security in addition to the firewall. The objective of this work is to build a network intrusion detection system using random forest with adaptive boosting algorithm and to improve the detection rate for providing better accuracy and low false positive rates. In order to evaluate the proposed system, gas pipeline dataset is being used. In this work, RI Attacks (response injection), CI Attacks (command injection), DoS Attacks (denial of service) and RA (reconnaissance attacks) are successfully detected and an accuracy of 98%, 97%, 93% and 91%Analysis (PCA) is one of the greatly used is achieved, respectively.

Keywords Intrusion detection · Adaptive boosting · SCADA · Cyber-attacks

S. Sivakumar · B. S. Ananthanarayanan · U. Arumugam (✉)
School of Computing, SASTRA Deemed to Be University, Tirumalaisamudram, Thanjavur, Tamilnadu 613401, India
e-mail: a_umamakeswari@yahoo.com

S. Sivakumar
e-mail: bragadeesh.prithvi@gmail.com

B. S. Ananthanarayanan
e-mail: sa_bragadeesh@yahoo.co.in

V. E. Balas et al. (eds.), *Proceedings of International Conference on Computational Intelligence, Data Science and Cloud Computing*, Lecture Notes on Data Engineering and Communications Technologies 62, https://doi.org/10.1007/978-981-33-4968-1_50

1 Introduction

SCADA systems control and provide a perception of critical infrastructure frameworks. These systems usually consist of four major components. Actuators and sensors form the first set of components. The sensors include pressure controllers, laser sensors and water level gauges that gather information about the system. The condition of the system is maintained by the actuators like pumps and motors. The programmable logic controller (PLC) forms the second set of components. They control and gather data that decides the condition of the system. Such controllers are usually known as remote terminal units (RTUs). The RTUs integrate with the first set of components of the SCADA system. The sensor information in the predetermined registers is stored here. The supervisory control forms the third set of components in the SCADA system. The supervisory control is managed by the master terminal unit (MTU). The RTUs are communicated with the help of MTUs. For example, the RTU in a gas pump system receives commands from the MTU to get the pump turned on. The RTU registers that comprises the present pressure measurement can be read by the MTU by sending a read query. There are variety of protocols that enable this kind of communication like Profibus, Fieldbus, Modbus and Distributed Network Protocol Version 3 (DNP3). The human–machine interface (HMI) form the fourth set of components, where the operator views the sensor information gathered by the MTU. The HMI generally comprises of visualization of the system and the working of the subsystems. It also helps to alter the features and states inside the SCADA system by communicating through the MTU.

The gas pipeline management system deploys a gas pipeline that is used to distribute natural gas and other petroleum products to the market. Cyber-attacks such as intrusions into these systems that monitor and manage gas pipelines can lead to loss of data and loss of control of gas pipelines. In either of the cases, it may cause physical damage to the gas pipeline or may affect the people in the region of gas pipeline at the time of occurrence of incident. In 1991, a Bellingham Washington process control system got inert and the pressure relief valve was left closed. This failure caused the pipeline to rupture and spilled more than 25,000 gallons of gasoline into two closes by streams. The spilled gas later got blasted, killing the lives of three people and harming eight other people. The exact reason for this failure remains unknown due to the unavailability of forensic evidence. However, Abrams and Weiss analysed that the human–machine interface (HMI) became inert during the damage of the pipelines which would otherwise update the screen data for every five to seven seconds. This made the administrators unaware of the current state of the system and unable to protect the system from the damage. To avoid such failures and to intimate the operators whenever something goes wrong intrusion detection systems play a major role. Figure 1 shows the diagram of the SCADA gas pipeline control system.

An intrusion detection system (IDS) is a security mechanism that monitors and raises alerts whenever intrusions take place. IDS can be generally split into two categories: Host-based IDS and Network-based IDS. Host-based IDS makes use of

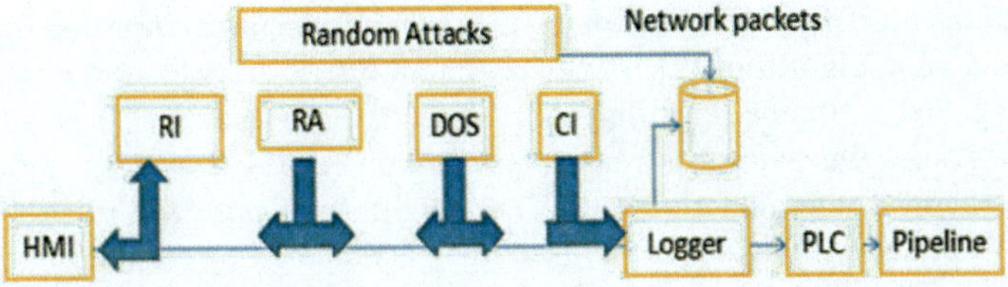

Fig. 1 SCADA gas pipeline control system

several review information of the host machine. Network-based IDS collects information from the IP packets that are obtained through the network hardware such as switches and routers. Despite the fact that this kind of information are not hugely available as that of host-based IDS, network-based IDS is used predominantly to detect various distributed intrusions among the entire network and reduces the work of all separate host machines. The objective of this paper is to build a network intrusion detection system for securing the SCADA gas pipeline management systems using machine learning technique random forest with adaptive boosting. It is observed that our results show improvements in the accuracy and give low false positive rates than the other traditional machine learning techniques.

2 Related Works

In order to understand the possible set of cyber-attacks that can occur in the industrial control systems, the network traffic logs are gathered from the gas pipeline. These captured gas pipeline data logs describe the normal operation as well as the attack scenarios [1]. Many SCADA researchers have deployed intrusion detection systems to check and prevent cyber intrusions. As a solution to address the severity of low detection rates on several cyber-attacks in the industrial control systems, an adaptive coupled method that integrates both white list filtering and machine learning algorithm was proposed [2]. Various relevant machine learning algorithms were deployed like the support vector machine (SVM) [3] and random forest modelling [4]. An anomaly-based IDS was introduced using the genetic algorithm (GA) and SVM to improvise the efficiency of the intrusion detection process [5]. An intrusion detection technique that enforces neural network algorithm including adaptive mutation and white list filtering was proposed. The neural network technique helps to identify the attack scenarios that have never happened before [6].

An IDS that enhances the existing backpropagation neural network algorithm was developed using the PCA-BP method (Principal Component Analysis-Back Propagation) [7]. Another intrusion detection technique which combines both extreme learning machine (ELM) and k-nearest neighbour (KNN) algorithm where the classifier helps to identify various kinds of cyber-attacks was proposed in [8]. A multi-level hybrid intrusion detection technique that is used to majorly minimize the training duration of the classifiers was built using the support vector machine and the extreme learning machine algorithm [9]. A new efficient intrusion detection system which brought together different algorithms like support vector machine, k-nearest neighbour (KNN) and the particle swarm optimization (PSO) showed a good amount of improvement in the detection rate. The weights are optimized using PSO algorithm, and a better intrusion detection accuracy is obtained by combining both SVM and KNN classification methods [10]. An intrusion detection method using particle swarm optimization—support vector machine (PSO-SVM) was developed where the penalty parameters and the kernel function of the support vector machines are optimized by the particle swarm optimization and it also helps to enhance the classifying efficiency of support vector machine [11]. Various intrusion detection techniques using genetic algorithm (GA) [12], extreme learning machine [13], backpropagation neural networks (BPNN) [14] and K-means clustering [15] are developed to ensure security of the ICS. An efficient intrusion detection was approached using enhanced random forest techniques [16]. However, the techniques used in these methods were not efficient enough in detecting the all kinds of cyber intrusions.

As a response to rectify the above issue, this paper proposes the usage of random forest (RF) along with adaptive boosting algorithm to enhance the accuracy and to minimize the false positive rate (FPR) than the other traditional machine learning algorithms.

3 Network Framework

As presented in Fig. 2, the components that comprise the SCADA gas control systems are interconnected with the Internet. The attack detection process commences by gathering the network packets stored in the HMI registers or the PLCs. There are various features like the response address and command address that help us to identify whether it is a normal scenario or an attack scenario. Based on a certain set of features, decision trees are made using the random forest algorithm. Then on each of these decision trees adaptive boosting is carried out sequentially where a number of weak classifiers are identified and developed into a strong classifier. Finally, the predictions are made and the system identifies the type of attack.

4 Methodology

A. *PCA data pre-processing*

Principal component analysis (PCA) is one of the greatly used and a standard data analysis technique that helps for dimensionality reduction. The PCA algorithm is applied on the features extracted from the gas control system, and the dimensionality

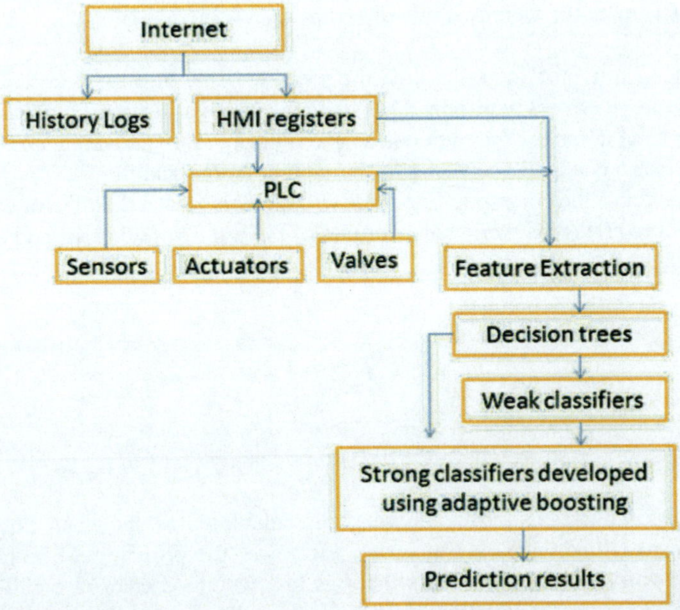

Fig. 2 Network framework

is reduced based on multiple features such as response memory, response address and command address.

B. *Decision tree using Random Forest algorithm*

Random forest (RF) is an ensemble technique that helps to enhance the accuracy of the trained model. RF comprises of several decision trees based on which voting is carried out to make the final decision. The reduced features of the gas pipeline data are used to form n number of random subsets from which n number of decision trees are trained.

C. *Adaptive boosting algorithm*

Adaptive boosting is a boosting ensemble technique. The basic key of this model is that it learns from its previous mistakes. So, the algorithm works by selecting and gathering a group of weak classifiers and constructing them into a strong classifier.

Step 1: Initialize the weights

First of all, the weights are initialized based upon the number of data points in the decision tree. Provided there are (X1, Y1)……. (Xn, Yn) total set of data points, the weights are initialized with available number of data points as 1 as given in Eq. (1).

$$D(i) = 1/n(i = 1, 2, \ldots n) \tag{1}$$

Step 2: Evaluate the weighted rate of errors

The decision tree is trained and then the weighted rate of error is calculated. The weighted rate of error is determined based on the number of wrong predictions made out of the total number of predictions as given in Eq. (2). The weights of the wrong predictions are changed according to the weight of its data points. The higher the weight, the more the corresponding error will be weighted at the time of calculation of $e(t)$; where $e(t)$ is the weighted error rate of the wrong predictions and $D(i)$ is the initial weight of the data points.

For $t = 1 \ldots T$

$$e(t) = \sum_{i=0}^{n} D(i) + e(t) \tag{2}$$

Step 3: Calculate the weight of the decision trees

The corresponding decision tree's weight is calculated as shown in Eq. (3). The higher the weighted error rate of a tree, the lower the priority will be given to it during the voting. Hence, the trees that has low weighted error rate will be given more priority during the voting.

$$\alpha(T) = 0.5 * \log((1 - e(t))/e(t)) \tag{3}$$

where $\alpha(T)$ is the decision tree's weight in the ensemble and 0.5 is chosen as the balanced learning rate.

Step 4: Update the weights

The weights of the wrongly classified data points are updated using Eq. (4). If the model has correctly classified the data points, the weights of such data points remain the same. If the model has wrongly classified the data points, then the new weights are updated accordingly.

$$U(i) = \frac{D(i)\exp(-\alpha_t y_i h_t(x_i))}{Z_t} \tag{4}$$

where Z_t is the normalization factor for keeping $D(i)$ as a distribution.

Step 5: Repeat the steps

All the previous steps are repeated until the maximum number of decision trees that is set to train is reached.

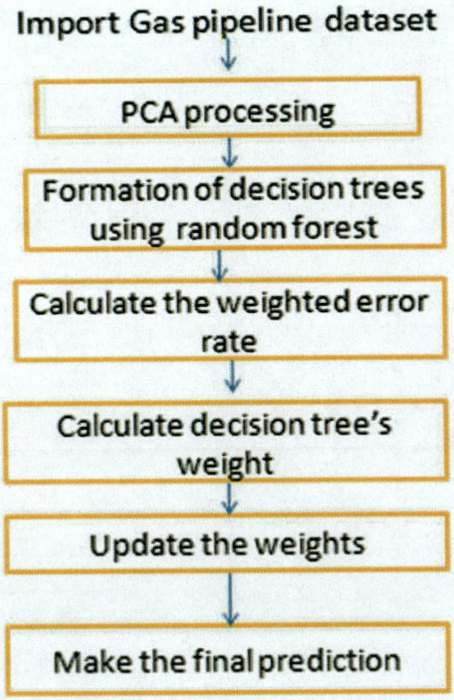

Fig. 3 Flowchart of RF with adaptive boosting

Step 6: Final prediction

Final prediction is made by voting. The tree with higher weight has more priority in the prediction of results. Figure 3 shows the flowchart of the random forest with adaptive boosting algorithm.

5 Dataset Description

The algorithm used in this paper is evaluated using the gas pipeline dataset. The gas pipeline dataset is an industrial network dataset from the industrial control systems simulated at a laboratory environment through the MODBUS application layer protocol. A network logger that controls and captures the Modbus traffic information through RS232C connection is used to collect the dataset. The list of attacks that is contained in the gas pipeline dataset is listed in Table 1.

The seven major categories of attacks in the dataset are divided into four major categories such as CI attacks (command injection), RI attacks (response injection), DoS attacks (denial of service) and RA (reconnaissance attack). This paper focuses upon these four types of attacks. The statistical information regarding the gas pipeline dataset is given in Table 2.

Table 1 Attack categorization of gas pipeline dataset

Type of attack	Abbreviation	Type of threat
Normal	Normal (0)	N/A
Naive Malicious Response Injection	NMRI (1)	Tampering
Complex Malicious Response Injection	CMRI (2)	Tampering
Malicious State Command Injection	MSCI (3)	Tampering
Malicious Parameter Command Injection	MPCI (4)	Tampering
Malicious Function Code Injection	MFCI (5)	Tampering
Denial of Service	DoS (6)	Interruption
Reconnaissance	Reconn (7)	Interception

Table 2 Gas pipeline data

Attack type	Training data	Testing data
Normal	30,541	15,045
RI	10,245	5165
CI	3267	2160
DOS	1247	543
RA	3589	1593

6 Evaluation Metrics

Evaluation metrics helps us to figure out how efficient the intrusion detection system is working by calculating the performance of the system and comparing with the results gathered from the datasets. Generally, the efficiency of an intrusion detection system is assessed using the five major evaluation metrics: Accuracy, false positive rates (FPR), false negative rates (FNR), Precision, Recall and F-Score of which the most important metric is the accuracy and the FPR. Normally, IDS consists of four scenarios: true positives (TP), true negatives (TN), false positives (FP) and false negatives (FN). Figure 4 explains all the four scenarios. A well-performing intrusion detection system is the one which has a high accuracy and a very low false positive rate. In this paper, we have calculated all the essential metrics required for an IDS

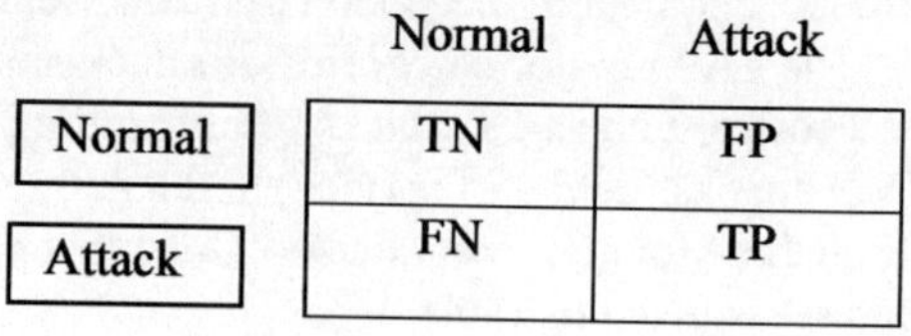

Fig. 4 Confusion matrix representation

and compared those metrics with the IDS developed using other machine learning techniques such as PSO-SVM, GSA and random forest.

A. *Accuracy*

The accuracy is obtained by calculating the total number of correct predictions as the true cases to the total amount of data that are present as presented in Eq. (5). The true positives and the true negatives are referred as the true cases, and the false positives and the false negatives are referred as the false cases. The accuracy metric will be maximum when the numbers of false cases are less since they do not affect the accuracy much.

B. *FPR (False Positive Rate)*

False positive rate is referred as the probability that the intrusion detection system will raise false alerts; i.e. a false alert will be raised when the true value is negative. The FPR can be calculated using Eq. (6). This metric is very essential in evaluating the performance of the IDS. When the false positive rate is very less, we can say that the performance of the IDS is great.

C. *FNR (False Negative Rate)*

False negative rate is referred to the probability that the intrusion detection system will not raise any alert when there is an actual attack; i.e. no alerts will be raised when the true value is positive. The FNR is calculated using Eq. (7). This metric is also important in evaluating the efficiency of the intrusion detection systems.

$$\text{Accuracy} = \frac{\text{TP} + \text{TN}}{\text{TN} + \text{FP} + \text{FN} + \text{TP}} \tag{5}$$

$$\text{FPR} = \frac{\text{FP}}{\text{FP} + \text{TN}} \tag{6}$$

$$\text{FNR} = \frac{\text{FN}}{\text{FN} + \text{TP}} \tag{7}$$

$$\text{Precision} = \frac{\text{TP}}{\text{TP} + \text{FP}} \tag{8}$$

$$\text{Recall} = \frac{\text{TP}}{\text{FN} + \text{TP}} \tag{9}$$

$$\text{F} - \text{Score} = 2.\frac{\text{Precision} \times \text{Recall}}{\text{Precision} + \text{Recall}} \tag{10}$$

D. *Precision*

Precision can be obtained by calculating the totally available data points that are exactly identified as true value to the total amount of data points that are predicted as true. Equation (8) gives the formula for calculating the precision value.

E. *Recall*

Recall can also be referred to as true positive rate or sensitivity. This metric can be obtained by calculating the total number of data points that are correctly identified as true value to the total number of true values that are present according to Eq. (9). Precision and recall both are essential for evaluating the efficiency of an IDS. They are used for calculating the F-Score. Individually, they do not serve of much purpose in evaluating the performance of IDS.

F. *F-Score*

The F-Score metric is obtained by taking into consideration of both precision and recall equally. It refers to the harmonic mean of the precision and recall as shown in Eq. (10). This metric gives more of a realistic way to evaluate the performance of the IDS. It brings about a balance between both precision and recall.

7 Experimental Results

The hardware used for the evaluation has the following specification: Processor of Intel(R) Core (TM) i5-4210U CPU at 1.72 GHz, 16 GB of RAM running 64-bit Windows 8 operating system. The dataset is trained and tested using MATLAB 2016a, and the corresponding results achieved are presented below.

The information given in the gas pipeline dataset consists of 26 dimensionality features. The standard dimensionality reduction algorithm principal component analysis (PCA) is applied on these features, and they are finally reduced to about 10 features. The selected data features of the gas pipeline dataset are shown in Table 3.

Table 3 Selected features of gas pipeline dataset

Principal component	Description
Response_address	Response device address
Response_memory	Response memory
Comm._read_function	Read command function code
Resp_write_fun	Write response function code
Comm._write_fun	Write command function code
Response_memory_count	Response store count
Command_memory_count	Command store count
Command_address	Command device address
Resp_length	Response packet length
Resp_read_fun	Read response function code

The accuracy comparison shows that the proposed model has an improved accuracy than the other machine learning techniques. For the DoS attack type, the proposed model has achieved an accuracy of 93% and a false positive rate of 6.6%. For the response injection attack, an accuracy of 98% and FPR of 5% is achieved. An accuracy of 97% and FPR of 6.4% is obtained for the command injection attack. For the reconnaissance attack, an accuracy of about 91% and FPR of 8.8% is obtained. The experimental results show that random forest with adaptive boosting has achieved better accuracy and very low false positive rates than the other machine learning techniques such as particle swarm optimization–support vector machine (PSO-SVM), genetic search algorithm (GSA) and random forest. Figures 5 and 6 show the comparison of accuracies of experimental results.

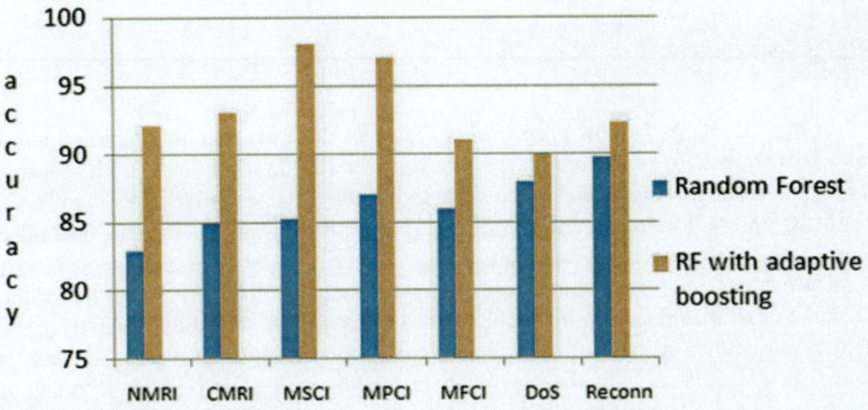

Fig. 5 Accuracy comparison

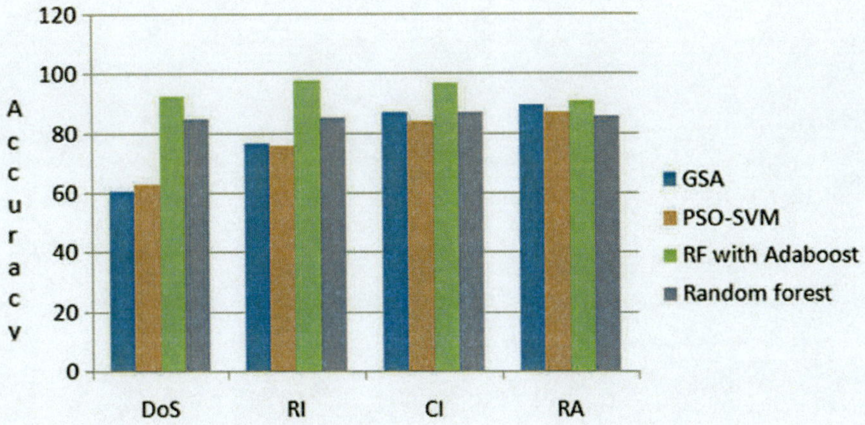

Fig. 6 Comparison of accuracies of obtained results

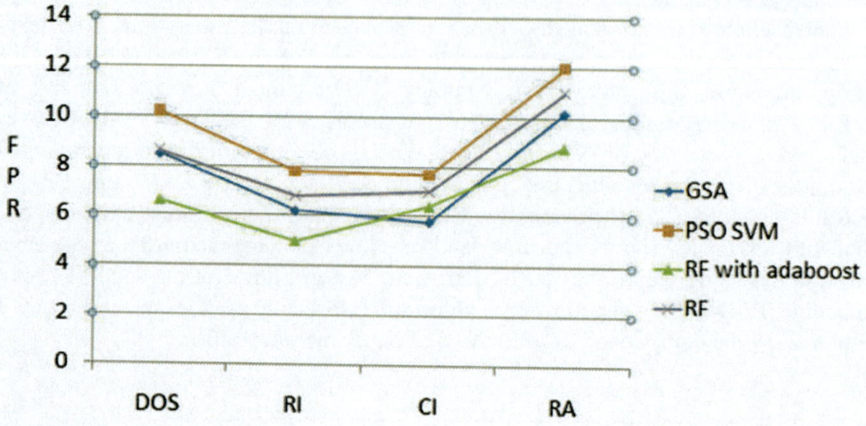

Fig. 7 Comparison of FPRs

In Fig. 7, comparison of false positive rates of different machine learning techniques is shown. The experimental results show that the random forest with adaptive boosting algorithm has achieved comparatively less false negative rates than PSO-SVM, GSA and random forest algorithms. Figure 8 shows the comparison of false negative rates of experimental results.

The evaluation of F-measure shows that proposed algorithm achieves higher value than the other machine learning techniques. The experimental results comparing the F-measure values are shown in Fig. 9. In the proposed algorithm, each tree learns from the mistakes made by the previously trained trees. It optimizes the accuracy of detection since it learns from its weak classifiers.

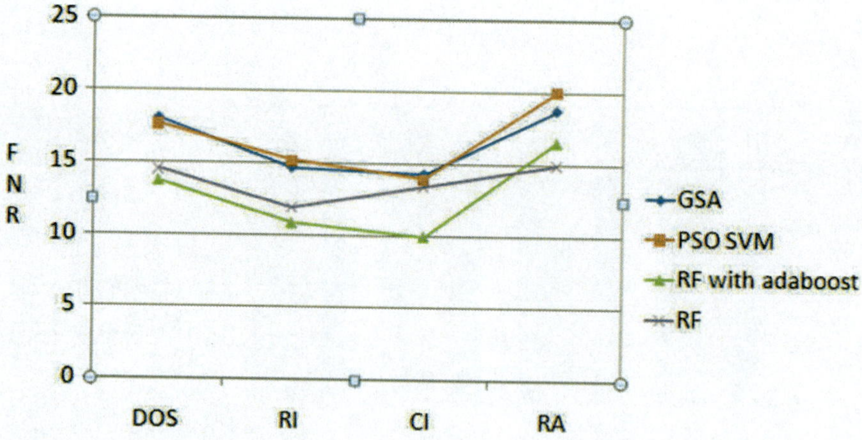

Fig. 8 Comparison of FNRs

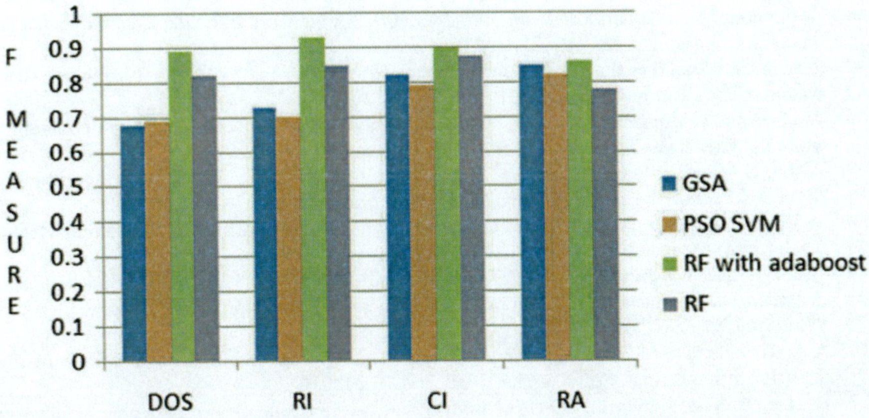

Fig. 9 Comparison of F-measure

8 Conclusion

In order to ensure the security of the SCADA gas control systems, an intrusion detection system was constructed using the random forest with adaptive boosting technique to bring an improvement in the accuracies and the false positive rate of the IDS. Comparison is also done with other machine learning techniques such as PSO-SVM, GSA and random forest. From the experimental results with the gas pipeline dataset, it is seen that the proposed technique has achieved preferably more accuracy and a very low false positive rate than the other techniques providing an assurance of SCADA gas control system safety.

In the future works, the detection rates and the false positive rates of the IDS can still be enhanced by adjusting the learning rate of the adaptive boosting algorithm.

Acknowledgements The authors wish to express their sincere thanks to the Department of Science & Technology, New Delhi, India (Project ID: SR/FST/ETI-371/2014), and express their sincere thanks to the INSPIRE fellowship (DST/INSPIRE Fellowship/2016/IF160837) for their financial support. The authors also thank SASTRA Deemed to be University, Thanjavur, India, for extending the infrastructural support to carry out this work.

References

1. D.J. Kang, J.J. Lee, S.J. Kim, J.H. Park, "Analysis on cyber threats to SCADA systems," Transm. Distrib. Conf. Expo. Asia Pacific, T D Asia 2009, pp. 1–4, (2009)
2. W. Chen, T. Liu, Y. Tang, D. Xu, Multi-level adaptive coupled method for industrial control networks safety based on machine learning. Saf. Sci. **120**(June), 268–275 (2019)
3. W. Shang, P. Zeng, M. Wan, L. Li, P. An, "Intrusion detection algorithm based on OCSVM in industrial control system," Secur. Commun. Networks. **9**(June 2011), 1040–1049, (2016)

4. N. Farnaaz, M.A. Jabbar, Random Forest Modeling for Network Intrusion Detection System. Procedia Comput. Sci. **89**, 213–217 (2016)
5. B.M. Aslahi-Shahri et al., A hybrid method consisting of GA and SVM for intrusion detection system. Neural Comput. Appl. **27**(6), 1669–1676 (2016)
6. W. Chen and L. Dongzhe, "Intrusion detection method in industrial control network combining white list filtering and neural network," *J. Comput. Appl.*, vol. 2, 2018
7. W. Jia, D. Zhao, T. Shen, S. Ding, Y. Zhao, C. Hu, An optimized classification algorithm by BP neural network based on PLS and HCA. Appl. Intell. **43**(1), 176–191 (2015)
8. M. Latah and L. Toker, "An Efficient Flow-based Multi-level Hybrid Intrusion Detection System for Software-Defined Networks," pp. 20–27 (2018)
9. W.L. Al-Yaseen, Z.A. Othman, M.Z.A. Nazri, Multi-level hybrid support vector machine and extreme learning machine based on modified K-means for intrusion detection system. Expert Syst. Appl. **67**, 296–303 (2017)
10. A.A. Aburomman, M. Bin Ibne Reaz, A novel SVM-kNN-PSO ensemble method for intrusion detection system. Appl. Soft Comput. J. **38**, 360–372 (2016)
11. B. Han, X. Bian, A hybrid PSO-SVM-based model for determination of oil recovery factor in the low-permeability reservoir. Petroleum **4**(1), 43–49 (2018)
12. S.N. Pawar, R.S. Bichkar, Genetic algorithm with variable length chromosomes for network intrusion detection. Int. J. Autom. Comput. **12**(3), 337–342 (2015)
13. Y. Li, R. Qiu, S. Jing, Intrusion detection system using online sequence extreme learning machine (OSELM) in advanced metering infrastructure of smart grid. PLoS ONE **13**(2), 1–16 (2018)
14. H.I. Ahmed, N.A. Elfeshawy, S.F. Elzoghdy, H.S. El-sayed, O.S. Faragallah, A neural network-based learning algorithm for intrusion detection systems. Wirel. Pers. Commun. **97**(2), 3097–3112 (2017)
15. U. Ravale, N. Marathe, P. Padiya, Feature selection based hybrid anomaly intrusion detection system using K Means and RBF kernel function. Procedia Comput. Sci. **45**(1), 428–435 (2015)
16. K. Park, Y. Song, Y. G. Cheong, "Classification of attack types for intrusion detection systems using a machine learning algorithm," Proc.—IEEE 4th Int. Conf. Big Data Comput. Serv. Appl. BigDataService 2018, pp. 282–286, (2018)

Cyber Threats of Modern Era

Shaurya and Moutushi Singh

Abstract Todays world is called the age of information technology where it is impossible to circumvent computers. Information is now floating everywhere. But then again there is a negative facet of this. Due to the increased usage of cyber space, the dimensions of cyber threats have also shoot up highly. This risk is principally higher in the developed countries due to the high use of automation techniques in almost all activity areas. Artificial intelligence is the booming area of Information and Communication Technology. This paper explains the role of artificial intelligence in cyber world and establishes different statistics of security breaches that have happened in recent past. Billions were invested for starting up the business organizations who are specialized in artificial intelligence and machine learning. These approaches in cybersecurity domain boost up big data intelligence and analytics by nearly hundred billion dollars around 2021. Business industries, who are exclusively vulnerable to cyber-attacks, are strongly interested in using machine learning and artificial intelligence for threat detection by analysis of data and applying legal procedures for protecting both organizational and personal data. The importance of cyber-attacks and its consequent financial damage must not be taken lightly. When cyber-attacks involve serious setup, they hover international security which leads to cyber terrorism. Recently, it has been noticed that due to casual security concerns, huge amount of money and personal data are sacrificed. Many security breaches are also explained in this paper to establish the fact of establishment of a secure framework.

Keywords Cybercrime · Threat · IoT · Social engineering attack · Network attack · Artificial intelligence · Intrusion detection and prevention system

---

Shaurya (✉) · M. Singh
Institute of Engineering & Management, Kolkata, India
e-mail: shaurya1337@gmail.com

M. Singh
e-mail: moutushi.singh@iemcal.com

V. E. Balas et al. (eds.), *Proceedings of International Conference on Computational Intelligence, Data Science and Cloud Computing*, Lecture Notes on Data Engineering and Communications Technologies 62, https://doi.org/10.1007/978-981-33-4968-1_51

1 Introduction

Cybercrime is the one of the greatest global threats that are faced by an organization for its survival. The activities of cyber criminals is one of the biggest challenges that humankind will face in coming two decades [1].The extent of threat can be understood by a cybercrime report of 2017 by Steve Morgan, Editor-in-Chief of Cybersecurity Ventures which says "Cybercrime damages will cost the world $6 trillion annually by 2021." This signifies the greatest transfer of economic prosperity in history, risking the incentives for invention and speculation and will be more gainful for cyber criminals than global trade of all major illegal drugs combined after the massive COVID-19 pandemics [2].

Additionally, the threat of cybercrimes is not only limited to organizations but the clients are also at a great extent. Several data breaches of credit cards have already revealed the threats and received concerns of security experts. One of the largest credit card breaches in US includes, Capital One revealing a breach on July 29, 2019, that exposed highly personal details of nearly 106 million customers [3]. In Oct 2019, it is spotted that 1.3 million credit and debit card details from Indian banks were sold on Darknet. Group-IB a Singapore-based cyber security organization established the fact that the information was sold for $100 for every card [4].

Apart from the stealing and cracking the card details, the healthcare services of India were hit by cyber criminals in Feb 2019. FireEye, a US-based cyber security organization revealed the fact that security breaches happened into a leading Indian healthcare service Web site, and nearly 6.8 million records were stolen that contains patient information, personally identifiable information (PII), doctor information as well as similar credentials [5]. Without divulging the name of the Web site, the organization stated that the cyber criminals were mostly from a typical country, who directly sells data stolen from healthcare organizations globally including India. The importance of healthcare information can be understood by the concern that this data can be misused to plan a Global Bio-warfare strategy.

According to a white paper by CISCO, there will be roughly 50 billion connected IoT devices at the end of 2020 [6]. The number of IoT devices will increase by manifolds in the coming future as vast developments will be made in the field of IoT. This rapid cumulative number of IoT devices in world also promotes the concern of safety of these devices. The danger of corruption in IoT devices will become one of the most concerned issues of near future. The breach in IoT devices impacts numerous actors at the same time. An attack on IoT device used in domestic usage results in breach of PII of clients. Exposure of any such attack reduces the popularity and sale of devices resulting in loss to the company selling that product. The risk of ransomwares produces huge damage to clients, firms, businesses and government organizations. The risk of IoT ransomwares will be palpable and may result in irretrievable damage to IoT devices in an explicit time limit.

There was a study on cyber security workforce by ISC^2. According to them, there is severe shortage of cyber security professionals in recent days. The study estimated the current number of skilled cyber security professionals around the world is 2.8 million but the current demand requires 4.07 million more cyber security professionals to defend the organizations [7]. This raises the concern for cyber security, as the number of cybercrimes committed every day is increasing rapidly but the shortage of cyber security professionals is severe.

There are a few domains of technology where innovations and development are being made rapidly to combat the adversary effects raised due to cyber-attacks. The domains are artificial intelligence (AI), machine learning (ML), and deep learning (DL). The development of AI is successfully decreasing the human dependency in several domains. The incorporation of AI has also reduced the chances of human error. Although cybersecurity and AI are two separate domains, when we observe the current scenario of cyber threats and the shortage of cyber security professionals around the world, the integration of cyber security and AI becomes an appropriate solution to counter the arising problems. The incorporation of AI in cyber security will not only reduce the workload and human dependency but will also reduce the time required to enable counter measures against an attack. In Table 1, the attractive valuable information of business organizations is given. This information are the target areas of cyber criminals.

The paper is organized in the following way: In Sect. 2, different statistics for cyber security threats are given. Section 3 surveys different literatures related to this area. Section 4 discusses security breaching and a few security attacks. Finally, the paper is concluded in Sect. 5 by illustrating the necessity of building secure framework.

Table 1 Most valuable information to cyber criminals according to 21st EY Global Information Security Survey (GISS)

Sl. No.	Top 10 most valuable information to cyber criminals	Percentage (%)
1	Customer information	17
2	Financial information	12
3	Strategic plans	12
4	Board member's information	11
5	Customer password	11
6	R & D information	9
7	M & A information	8
8	Intellectual property	6
9	Non-patented IP	5
10	Supplier information	5
11	Others	4

2 Statistics for Cyber Security Threats

To elaborate the cyber security threats, the statistical data accumulated over various research journals, cyber security organizations, survey reports are given in the following section, accordingly with their subcategories.

2.1 Cybercrime Statistics According to the Type of Attack

- In 2019 DBIR, 94% of malwares were delivered through email. (Verizon) [8].
- IoT devices experience an average of 5,200 attacks/month. (Symantec) [9].
- 51% of businesses experienced denial of service (DoS) attacks in 2018. (Cybint Solutions) [10].
- 71% of breaches were financially motivated, and 25% were motivated by espionage. (Verizon) [10].
- 52% of breaches highlighted hacking, 28% had malware, and 32–33% had phishing or social engineering attacks, respectively. (Verizon) [10].
- The highest malicious email attachment types are .doc and .dot which make up 37% of the whole. The very next one is .exe at 19.5%. (Symantec) [10].
- By 2020, the assessed number of passwords used by humans as well as machines worldwide will raise to 300 billion. (Cyber Security Media) [10].
- 100,000 groups in at least 150 countries and more than 400,000 machines were infected by the WannaCry virus in 2017, at a total cost of around $4 billion. (Technology Inquirer) [11, 12].

2.2 Industry Specific Cyber Security Statistics

- Financial services had 352,771 exposed sensitive files on an average, while Healthcare, Pharma and Biotech have 113,491 files on average—the highest when industries are compared. (Varonis) [9].
- 15% of the breaches are tangled in healthcare organizations, 10% are in the financial industries, and 16% are found in the public sectors. (Verizon) [10].
- Banking sectors suffered the most cybercrime costs in 2018 at $18.3 million (Accenture) [11].
- The estimated losses in 2019 for the healthcare industry were $25 billion. (SafeAtLast) [13].
- WannaCry ransomware attack cost the National Health Service (NHS) over $100 million. (Datto) [14, 15].

2.3 Security Spending and Cost Statistics

- The average cost of a malware attack on a company is $2.6 million. (Accenture) [16].
- Health Care is having the highest data breach costs at $429/record. (IBM) [17].
- The average annual security spending per employee is doubled, from $584 in 2012 to $1,178 in 2018. (Gartner) [18].
- Turnover of customers, amplified customer acquisition activities, reputation losses and moderated goodwill, the cost of business loss globally is highest for US companies at $4.13 million/company. (Ponemon Institute's Cost of Data Breach Study) [19].
- Ransomware damage costs have risen up to $11.5 billion in 2019, and a business becomes victim to a ransomware attack every 14 s. (Cybersecurity Ventures) [20] (Fig. 1).

3 Literature Survey

The authors of [22, 23] have explained the applications of integration of AI with cyber security. Across the industries, applications using ML as well as AI are broadly being used for all types of services like data collection, process development, output generation as well as for outcome analysis. At the same time, dependency and inefficiency of human beings to handle huge amount of data can easily

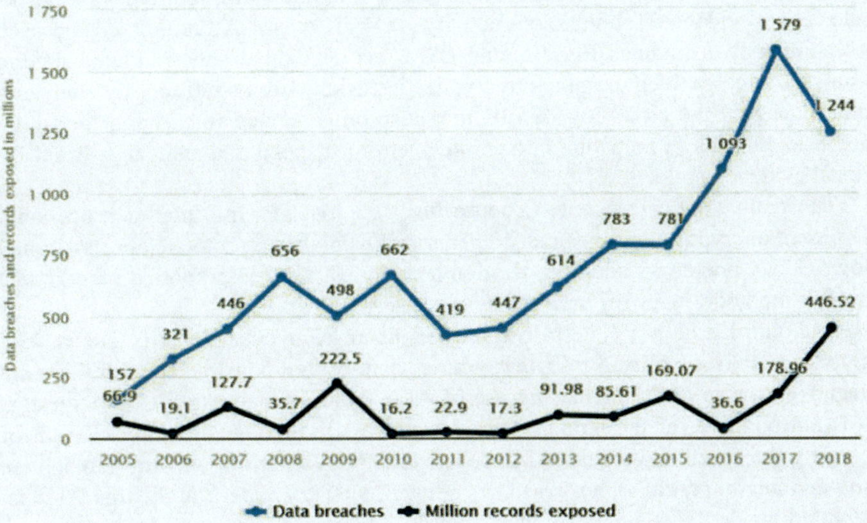

Fig. 1 Number of data breaches and the number of records exposed increased at an extraordinary rate from 2016 to 2018. [21]

be restored and processed using AI and ML. Hence, the capacity to handle cyber security threats is also increased. According to them, six threat management techniques are used broadly in industries, namely Expert Systems, Neural Nets, Intelligent Agents, Search, Learning, and Constraint Tackling. Expert systems are one of the most popular AI tools. It is a software used for discovering answers to a problem, displayed either by a user or by a product. Databases of expert knowledge are utilized to make a decision or offer advice. For medical diagnosis and in accounting, they are used vastly. Neural nets or DL is a propelled branch of AI. It functions like a human mind. Artificial neurons or "perceptrons" consolidate with each other to solve problems. So, in cyber security, it may be able to distinguish any malicious or legitimate data without the intervention of human beings. Neural nets will also be crucial in recognition of new cyber security threats arising from malwares or any vulnerability present in frame work of an organization [24]. The intelligent agent (IA) is a sovereign entity that accomplishes its task by taking inputs via sensors. It performs real-time learning by communicating with the environment. It also has memory-based model storage capacity and recovery measurements. Supervised and unsupervised learning are the two ways of ML. Both of them are the AI-based techniques used for minimizing the problem-solving efforts. Apart from them, self-organizing maps (SOM) are another AO-based techniques used today.

In [25], the author stated that security is an indispensable characteristic which is needed for an improved utility service. In recent years, cloud is one of the popular efficacies framework which is prone to several malicious activities and attacks. These attacks not only affect user data but also lower the cloud performance. Forensic investigation helps to improve the security framework in the cloud domain to detect and collect evidences of criminal activities. It is also to prevent future attacks. The author projected a strong forensic architecture to monitor different types of attacks and collect evidences for malicious activity in cloud domain. An automated approach combining two modules FESNA and ESMA is developed and simulated for log analysis. It helps to envisage and recognize intrusion with its intruder by analyzing and managing the cloud log. A FIR may also be generated for driving the future forensic analysis. It helps the law enforcement to suspend and analyze warrant for detail evidence collection.

The authors of [26] presented a neuro-fuzzy approach for information processing. Different encryption outlines can be generated using keys, protocols, propositional logic, fuzzy operators, and optimization techniques. They also applied several data mining methodologies for multi-party cryptosystem.

Ljubomir Lazic in [27] described the benefits of AI in cyber security. He has also given the statistics related to cyber security data of the industry. Though there are negative aspects of the increasing use of AI in cyber domain, still it has its usage. It is not possible for the human beings to process these huge amounts of data for keeping the speed and safety. AI does it without any interruption. But humans are still an integral part of cyber world. As a result, industry experts argue that AI must be fitted inside the business's Cyber Security Operation Center (CSOC), to ensure an appropriate AI enabled cyber security solutions. Analysis techniques are applied after detection of security measures has been done.

4 Security Breaching

4.1 A Recent Incident

On July 15, 2020 between 20:00–22:00 UTC, several Twitter accounts of World's high profile personalities counting, Bill Gates(Entrepreneur and Founder of Microsoft), Barak Obama (former United States President), Elon Musk(Entrepreneur, CEO of Tesla Motors, SpaceX), Jeff Bezos(Entrepreneur and Founder of Amazon), as well as World's topmost firms, including Apple (Hardware and Software Giant), Uber (Car Aggregator service), Binance (Cryptocurrency Exchange Company), and many more were hacked [28].

All the hacked twitter accounts were used for promoting a Bitcoin Scam in where 415 transactions amounting to $117,457.58 were made in Bitcoin equivalent [29]. After enquiring, Twitter found that the Scammers actually used the tools which were accessible only to internal support teams for targeting 130 Twitter accounts. For 45 of those accounts, the attackers initiated a password reset, which prohibited the owner to log into their accounts [30]. Twitter remarked that the attackers actually initiated coordinated social engineering attacks using phone spear phishing methodologies on some employees and were able to access internal systems and tools [31]. On July 31, 2020, three individuals were charged for the major twitter hack, according to US Department of Justice [32, 33].

4.2 Breaching Methodologies

There are two methods in which a breach can occur within an organization:

1. *Social Engineering Attacks:* Hackers or configured bots will spam the employees, usually via emails or as concealed traffic on a Web site. They will try to persuade top employees to enter the passwords in fake logins through emails which is the idea behind spear phishing, to click a button or open a file or access materials that pretend to be something secure and vital to the company. If a hacker can get a direct entry point into the network, then the vital part of security may be compromised [34].
2. *Network Attacks:* A skilled cracker can research in network defenses, such as weak passwords, outdating operating systems or even ineffective anti-virus software. The tunnel through such weak spots can be used to implant a backdoor or worm in the system. The worst network attacks can pose indefinite data extraction risks that will take a complete system renovation to patch [35].

Twitter breach incident is one of the best examples that used social engineering to gain access of an organization's internal services.

4.3 Stages of Deploying a Social Engineering Attack

There are three stages involved in deployment of a social engineering attack:

1. *Research:* The attacker collects information about the target like organizational structures, roles and activities related to the employees. The attackers gather this data via Web sites, social media platforms, or even in person visits.
2. *Planning:* The stages include the selection of the targets and selection of techniques of attacks along with a strategy to exploit the target individual's weakness.
3. *Execution:* The attacker attacks the target individuals through specific techniques. If the target individual falls in the trap laid by attackers, then the attackers are successfully able to gain access of employee's internal services, hence resulting in a breach.

4.4 Social Engineering Techniques

There are several social engineering techniques used by the hackers. Some of them are explained below:

(a) *Phishing*: Phishing attacks are executed in a multitude of ways, but their main purpose is the same as having an actual user to share confidential information. Business-critical details like network passwords or personal data such as emails, passwords, credit card numbers, bank accounts, and medical history may be the information at risk. Often phishing operations are more effective than other forms of data stealing methods as they use a human input to get what they need. Phishing hackers also send messages and frame data as if it had originated from a trusted source. They can also look like emails from established individuals or services, such as a co-worker or bank, including links that redirect to a non-intrusive Web site. By gaining the access with the credentials obtained in phishing, the hacker is able to gain access of internal tools and services of an organization or implant malware and backdoors onto the server, while still pretending to be authentic (Fig. 2).

(b) *Watering Hole*: A watering hole attack has two phases. In the first phase, the attacker is able to gain access to a legitimate service of an organization by exploiting a vulnerability. The second phase involves adding of a backdoor trojan to a downloadable file that has to be downloaded by the targets. Once the backdoor is installed in victim's device, the attacker can easily use the trojan to monitor, spy and control the victim's devices. As an example, an organization's blogging Web site can be accessed by exploiting a vulnerability. A trojan backdoor is added to the files, which may be downloaded from the Web site by an employee of that organization. The compromised site helps in installing a backdoor in the devices of employees that can be remotely monitored or even controlled by the attackers.

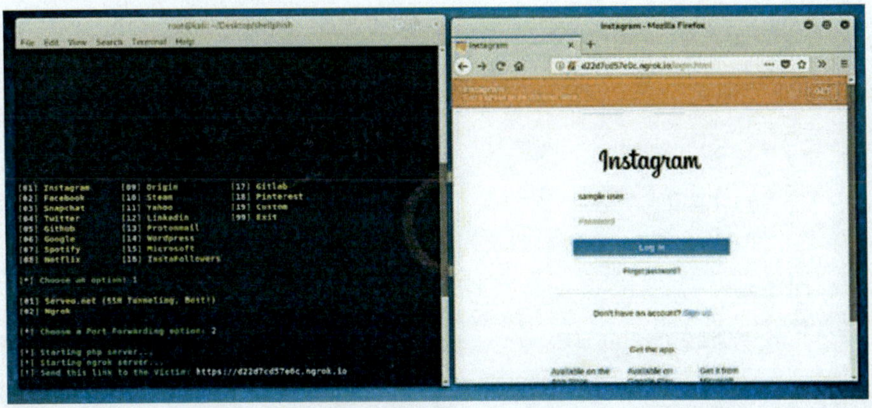

Fig. 2 Illustration of a phishing tool launching and phishing page for Instagram

Water hole attacks are performed by skilled attackers who discovered Zero-Day Exploitation.

(c) *Spear Phishing Attacks*: Spear phishing are type of phishing attacks on highly specified targets which have privileged access to a system or have some highly sensitive information. A spear phishing attack could be much more complicated than a regular phishing attack. Techniques like email spoofing is often used in spear phishing attacks in which a spoofed email is sent to a victim and the mail seems to come from an email address with which the victim is already familiar (Fig. 3).

Apart from the abovementioned techniques, several other techniques of social engineering are also established nowadays like Pretexting, Baiting, Vishing, Scareware, Diversion theft, Honey trap, and Tailgating.

5 Conclusion: Building a Secure Framework

Building a secure framework includes both hardware components that are capable of biometric monitoring and software components which include induction of AI in cyber security domain.

The hardware components include biometric monitoring technologies that are able to detect face or fingerprint of the user/employee on a regular basis. That hardware may include algorithms to perform infrared mapping and 3D facial mapping. The devices like mouse and keyboards can be equipped with capacitive fingerprint sensors that will be able to continuously intake the fingerprints of the user for authentication.

The software components include AI, ML, intrusion detection and prevention systems (IDPS) to monitor and authenticate the user/employee's biometric data at

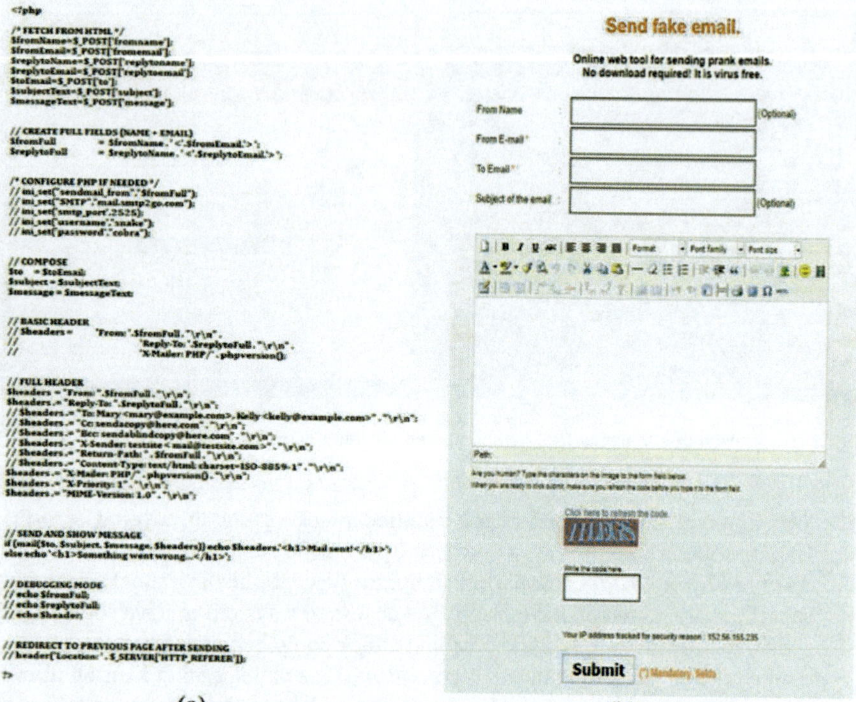

(a) (b)

Fig. 3 **a**. PHP script that allows an attacker to send spoofed emails. **b**. email spoofing service with advanced text formatting options

regular intervals along with the monitoring of user and entities behavioral analytics continuously for finding malfunctioning.

To eliminate chances of social engineering attacks, AI is vastly utilized nowadays to trace phishing and spoofed emails. Modern intrusion detection system (IDS) are designed specifically for detecting any malicious attempts to destroy, steal, alter, resource or gain unauthorized access. To block attacks, an Intrusion Prevention System (IPS) scans the whole network traffic to identify the malformed data. The system does this by analyzing vast amounts of available data on possible threats, to monitor and block the progression of suspected files and malware throughout the network to prevent outbreaks in the devices connected to the network. Hence, IDS and IPS working together with powered by AI will be capable of handling emerging cyber threats.

Deep learning or neural networks are also used in cyber security to analyze the vulnerabilities present in a system while also analyzing the new attack vectors that emerge every day.

The number of cyber-attacks is increasing continuously and is expected to grow manifolds. Amid the scarcity of cyber security researchers, AI can be integrated with cyber security to counter the new emerging threats. Integration of AI techniques in cyber security domain is used for defense mechanisms. When a human challenger with an inimitable circumvention goal attacks the system, the framework may fail. So, to mitigate this effect, AI requires human alliance and training. Though the threats are arising day by day, the amalgamation of both the entities has been proven worthy for smooth operation of the secured cyber world.

References

1. https://cybersecurityventures.com/hackerpocalypse-cybercrime-report-2016
2. https://www.prnewswire.com/news-releases/cybercrime-damage-costs-may-double-due-to-coronavirus-covid-19-outbreak-301027007.html
3. https://edition.cnn.com/2019/07/29/business/capital-one-data-breach/index.html
4. https://economictimes.indiatimes.com/industry/tech/8-biggest-data-leaks-of-2019-that-hit-indian-users-hard
5. https://www.databreaches.net/hackers-attack-indian-healthcare-website-steal-68/
6. https://www.cisco.com/c/dam/en_us/about/ac79/docs/innov/IoT_IBSG_0411FINAL.pdf
7. https://www.cpomagazine.com/cyber-security/cybersecurity-workforce-shortage-contin-ues-to-grow/
8. https://www.phishingbox.com/news/phishing-news/verizon-data-breach-investigations-re-port-dbir-2019
9. https://symantec-enterprise-blogs.security.com/blogs/expert-perspectives/istr-2019-inter-net-things-cyber-attacks-grow-more-diverse
10. https://www.varonis.com/blog/cybersecurity-statistics/
11. https://enterprise.verizon.com/resources/reports/dbir/
12. https://www.reuters.com/article/us-cyber-attack-europol/cyber-attack-hits-200000-in-at-least-150-countries-europol-idUSKCN18A0FX
13. https://www.accenture.com/_acnmedia/PDF-96/Accenture-2019-Cost-of-Cybercrime-Study-Final.pdf#zoom=50
14. https://www.datto.com/blog/ransomware-news-wannacry-attack-costs-nhs-over-100-million
15. https://safeatlast.co/blog/ransomware-statistics/
16. https://www.accenture.com/_acnmedia/pdf-96/accenture-2019-cost-of-cybercrime-study-final.pdf
17. https://healthitsecurity.com/news/data-breaches-cost-healthcare-6.5m-or-429-per-patient-record
18. https://ocd-tech.com/2019/04/22/cybersecurity-spend/
19. https://www.ibm.com/downloads/cas/ZYKLN2E3
20. https://cybersecurityventures.com/ransomware-damage-report-2017-part-2/
21. https://www.statista.com/statistics/273550/data-breaches-recorded-in-the-united-states-by-number-of-breaches-and-records-exposed/
22. Dr. S. Bhutada, P. Bhutada, "Applications of artificial intelligence in cyber security", Int. J. Eng. Res. Comput. Sci. Eng. **5**(4), (Apr 2018)
23. B.H. Rashmi, "Impact of artificial intelligence in cyber security", Int. J. Comput. Sci. Eng. **6**(12), (Dec 2018)
24. A. Panimalar, G. Pai, "AI techniques for cyber security", IJRET. **05**, 122–124
25. P. Santra, "An expert forensic investigation system for detecting malicious attacks and identifying attackers in cloud environment", Int. J. Sci. Res. Network Secur. Commun. (IJSRNSC). **6**(5), (Oct 2018)
26. P. Chakrabarti, "Information security: an artificial intelligence and data mining based approach", IACSIT Int. J. Eng. Technol. 1(5), (Dec 2009), ISSN: 1793–8236

27. L. Lazic, "Benefit from AI in cyber security", In *The 11th International Conference on Business Information Security (BISEC-2019)*, 18th October 2019, Belgrade, Serbia
28. https://www.inverse.com/culture/a-complete-list-of-every-twitter-account-hacked-on-July-15
29. https://www.theverge.com/2020/7/15/21326200/elon-musk-bill-gates-twitter-hack-bitcoin-scam-compromised
30. https://twitter.com/TwitterSupport/status/1284331129416200192
31. https://twitter.com/TwitterSupport/status/1289000138300563457
32. https://securereading.com/florida-teen-arrested-twitter-hack-celebrity-accounts/
33. https://www.weny.com/story/42438476/three-people-charged-in-twitter-hack-that-hit-biden-and-obama
34. F. Salahdine, N. Kaabouch, "Social engineering attacks: a survey", Future Internet. **11**(89), (April 2019), https://doi.org/10.3390/fi11040089
35. M. Pawar, J. Anuradha, "Network security & types of attacks in network", Proc. Comput. Sci. **48**, https://doi.org/10.1016/j.procs.2015.04.126, May 2015

Cyber Security Issues in the COVID-19 Times

Moutushi Singh and Indraneel Mukhopadhyay

Abstract The term cyber security defines a widespread range of questions and solutions related to protecting communications service providers (CSPs), residential subscribers and business customers from malevolent Internet activity and detrimental content. We must be prepared for a COVID-like global cyber pandemic that may blowout faster and further compared to a biological virus, with an equal or may be greater economic influence. The coronavirus catastrophe provides intuitions into how business organizations can prepare recovering for such cyber risks. In this pandemic time, cyber security got even more importance, as the situation is just right for cyber criminals to airstrike. This paper scrutinizes the cyber security threats during the COVID-19 pandemic. This presents a snap of the current scenario and how the pandemic due to COVID influenced the cyber world so far. Society has seen an enormous intensification on cyber threats during this pandemic, and this paper aims to explore that. This paper delivers all the current cyber security attacks during the pandemic. The impact of COVID-19 on business organization perspective is also provided and what are the key factors to secure the organizations are also discussed.

Keywords Cyber-attack · Cyber security · COVID-19 pandemic · Phishing attack · Malwares · Data breaches

1 Introduction

COVID-19 times has brought a sea change in the way the employers and employee are working. Everyone is afraid of the virus; they are worried of the health of the family members, and employers are afraid of the rising phishing attacks as more and more people are using unsecured computer to access secured systems. Apart from

M. Singh (✉) · I. Mukhopadhyay
Institute of Engineering & Management, Kolkata, India
e-mail: moutushi.singh@iemcal.com

I. Mukhopadhyay
e-mail: imukhopadhyay@iemcal.com

V. E. Balas et al. (eds.), *Proceedings of International Conference on Computational Intelligence, Data Science and Cloud Computing*, Lecture Notes on Data Engineering and Communications Technologies 62, https://doi.org/10.1007/978-981-33-4968-1_52

phishing attacks other cyber security scams are happening like ransomware threats to government, healthcare facility and even pandemic research institutes [1].

In the wake of work from home, many organizations weren't prepared to handle the new norms of how to manage the unsecure to secure network. Beyond ensuring employees and customers don't fall for phishing scams and ransomware threats, the cyber security teams are working day and night under COVID conditions so that the virtual private network (VPN) is able to withstand the load of the applications. Many organizations have given their employees hardware lock via which they are able to log into the secure systems of the organization like Deloitte, Infosys, and the like. Confidentiality, integrity, and availability (CIA) triad is rushed to the top of security team's to-do lists [2]. Overnight, investing in new policies, systems and technologies for endpoint data protection has become very urgent in a country where only 15–20% of the budget is spent on security.

Social Engineering attacks are also in the rise as it is also a sophisticated attack that exploits the human nature of trust. Whenever cyber crime happens, we must remember that it happens between a human being and its device. TRUST is the most important aspect. For example, one can get an email with the PayPal logo and click on the link because he/she trust PayPal and then find out hackers have abused his/her trust in PayPal to make him/her download malware.

In this paper, an overview of threats in cyberspace is given and also on corona virus phishing scams and review best practices which are needed now is elaborated [3]. There are many predictions done by cyber security analysts and Chief Security Officer (CSO) on ransomware attacks. This paper will prove invaluable as it navigates the turbulent cyber security effects of COVID-19 and helps to have better preparation for a future pandemic.

2 Key Cyber Security Considerations

Before the start of the fateful year 2020, an industry white paper was published by the title "Cyber Security: Considerations and Techniques for Network-Based Protection" by Sandvine. Business strategies are changing with the fourth industrial revolution. Data has become the main issue of every organization. The future will be created by blending new business models, new technologies, and new partnerships. In twenty-first century, there are some merciless capitalists, called cyber criminals, sitting on the wrong side of the law, are making money in this new economy [4]. It poses a new challenge to organizations for thinking about how to protect their competitive advantage and develop new security models to save their secure information. Cyber security professionals must exhibit that they can protect this newly transformed business domain with their skills, which is at par with the cybercriminals. The cyber resilience requires strategic alliance of cyber line of attack with incident response, business continuity, and disaster recovery. Technology is an opportunity, not a threat. Cyber security is becoming a key factor for organizations in the present scenario.

There are six key considerations that must be maintained in this crucial era. They are:

1. Aligning business goals with security needs
2. Digital trust and consumer authentication
3. The evolving security teams
4. The next wave of regulation
5. Cloud transformation and resilience
6. Automating the security function.

"Cyber security" has been formally defined as "the preservation of confidentiality, integrity, and availability of information in the cyberspace" by the ISO/IEC which also well-known as CIA rules (confidentiality, integrity, and availability). For better understanding, we described some fundamental security concepts below [5–7].

Confidentiality: It makes sure that the message is received by the authenticated entity. Restricts the illegitimate entities and grant access to the legitimate ones.

Integrity: It ensures that the information can be change, modify, or alter by the legitimated entities only. Protects data in a constant, accurate, and trustworthy manner.

Authentication: It guarantees on receiver and senders are authenticated with each other before transmitting information to each other.

With the CIA triad, there are other security concepts which we should know.

Availability: The services which are mentioned for security must be present when the system needed. Various systems have dissimilar availability requirements; they must be accessed easily whenever they needed.

Accountability: Accountability itself cannot stop networks to be attacked but supportive in making sure other security mechanisms are working appropriately or not. In addition, it adds liability and confident procedures to improvise security.

Information Security: It completely depends on the defense of secrecy, reliability, and accessibility of data records to supply the desires of the correct information abuser.

No-repudiation: The intention of non-repudiation states that it delivers certain confirmations in some cases where the abuser cannot reject an act.

Network Security: This notified with the design, execution, and action of a set of connections to attain the use of information security on networks within, among and between associations and abusers.

Internet Security: It relates to defending internet-related services as well as linked ICT schemes along with the system as an expansion of system safety in associations and at home, to accomplish the intention of protection.

Critical information infrastructure protection: It guarantees to facilitate the methods, and networks are privileged and tough not in favor of information safety threats, network protection threats, Internet protection threats along with cyber security hazards.

Auditing: It is a well-organized progression of measuring security how well it established for a particular entity. It concludes any vulnerable faults that place data in danger.

Cyber Crime: It describes the illegal action where works or functions in the cyberspace utilized on behalf of or else the objective of a fault, or where the cyberspace is the cause, tool, objective, or consign of a fault.

Cyber Safety: It exploits the circumstance of being confined not in favor of physical, communal, spiritual, economic, political, poignant, work-related, emotional, educational or other kinds or consequences of malfunction, harm, mistake, disaster, damage or any other incident in the cyberspace which might be measured non-desirable.

2.1 Aligning Business Goals with Security Needs

Business organizations are spending huge amount of money especially on cyber security, both on tooling and personnel [8]. To manage costs and to ensure that business and security priorities are aligned, they are automating their cyber functionalities by applying digitized cyber risk management methodologies for benefitting organization's operational and business strategies. The concept of business-driven risk scenarios is still lagging far behind. Business needs must align with the perspective of concept of cyber security. Forthcoming risk scenarios should be identified by the business organizations. The entire enterprise must be analyzed in terms of risk factors. IT risks are now expressed in terms of business risks. Larger organizations have spent billions of dollars on IT security over the last 10–15 years. But now they are realizing that a new model must be developed by lowering the costs through an automated approach. Hence, right people must be placed in the right roles. Most companies are now engaged in a digital transformation and automating their cyber and risk management processes.

2.2 Digital Trust and Consumer Authentication

During the COVID times, we have become habituated with doing online transactions. We have started both online banking and shopping as we thought it was safe and we were doing the same from the machine which we were using to connect to our office servers. Is it safe? Many of our systems do not have antivirus or firewall, many teenagers used the same system to play online games and hence the vulnerability of those systems kept on increasing. Many organizations use second-level authorization which was not very easy for people who are not of the millennium generations. Many modern companies have started using modern technologies like machine learning (ML), artificial intelligence (AI), and deep learning (DP) so that many clients have unique experience and these companies can use the modern technologies to also store client's unique characteristics and behavioral patterns. These patterns can be later used to the organizational benefit. Digital companies must work to connect the data, authentication, and fraud teams systematically and programmatically. Understanding of the governance requirements like type of pulled data, its owner, source and destination of data must be controlled. From there a better experience for the customers must be derived. Hence, they are being asked questions to authenticate themselves and to do atypical transactions. The interactions must be as easy and painless as possible, but a little friction is incorporated to make sense algorithmically upon common behaviors. The privacy and data concerns around how, and by whom, the cloud data is going to be used is of utmost priority. Hence, encryption and protection of data are needed. It can be solved technically, but is based on the business's desire for customers to have a better knowledge across every way they interact with the company digitally.

2.3 *The Ever-Evolving Security Team*

The Chief Information Security Officers (CISO) are working to adjust the changing dynamics of the COVID times of the business, and they have become a trusted and important voice at the strategy table [9]. CISO along with the security teams were able to visualize the organization's exact operational priorities and map it with internal business needs and subsequently implement them in the business recovery plan. They assure regulatory requirements from both time and cost perspectives in an efficient way. Security teams listen to different perspectives and communicate with the business heads about the organization's need. The cyber security team of digitally transforming company should look for a strategic perspective and present themselves as the connection between the business, digital, and security. Identification of the data, which is going to be placed on the cloud, must be identified. Interactions are required between the development and production environments. After that, mapping of expectation is done within the security plan. The team must work very closely with the corporate, who are intimately involved with customer experience. Messaging strategies are formulated if a worst-case scenario materializes to inculcate the trust in consumers. The team must establish the role of AI to handle and what truly requires the shade of human thought. The role of cyber security is established in the organization's environmental, social, and governance (ESG) agenda.

2.4 *The Next Wave of Regulation*

Companies understand the need for additional investment in tooling and process development for configuration and maintenance in terms of cyber talent for cyber governance and risk strategy analysis. Many companies hired IT professionals who lack cyber security perspective to understand the new paradigm.

There are three areas on which cyber regulations are focused:

i. the underlying operational technology;
ii. outsourcing for data that is processed via third parties; and
iii. resiliency of the company to detect, respond to and recover from cyber-attacks.

The role of CISO must be embedded in the first line to link these tasks to annual performance targets. The second line of defense is the security task force which helps to implement the business continuity plan. The third line review and assess the work of the first two lines of defense. This makes the cyber security needs including regulatory compliance across the entire organization.

2.5 *Cloud Transformation and Resilience*

IT has been primarily focused on the challenges of the ground before the commencement of cloud. It was the responsibility of the security team to scan that infrastructure for vulnerabilities but they often do not know what to scan due to the disconnection with IT on an updated threat list. For cloud, the priority is information protection. The data that is being deployed in the cloud is often not resilient. There are not only multiple availability zones, but there is a major breaching area for recovery of data. Modern business organizations must develop it with a hard code to measure security breaches without the security team's involvement, which the cloud can facilitate [10]. Organizations must understand and communicate with all the connections between business enablement, resilience, and information protection.

2.6 *Automatically Securing Customer Security*

Under the COVID conditions, it has become very difficult to securely collect know your customer (KYC) data which is required for many financial and health applications. The organizations are currently trying to understand their digital customers, via new methodologies of AI, ML but also acquiring deeper information. Security professionals are also using various third-party tools like IBM security product which has gained high prominence due to the current scenario [11].

Finally, to conclude, organizations cannot be held back by any cyber risk. Nowadays, the security professionals understand that cyber security is about risk management—not risk elimination.

3 Data Breaches in India Q1 2020

If we compare first quarter (Q1) 2020 to that of last quarter (Q4) 2019, there was an increase of 37% due to cyber-attacks. The Kaspersky Security Network (KSN) report showed that its products detected and blocked 52,820,874 local cyber threats in India between January and March 2020. In the data released it showed that India ranked 27th globally in the increase if the number of web threats detected by different organizations. Protection against such attacks not only requires an antivirus but also a firewall capable of defending the data breaches. Under the present scenario, it is a must for all users to also use anti-malware and anti-rootkit functionality tools to save ourselves from such attacks.

The mobile smartphone users are also being targeted due to mass consumption and increased digitalization. Risks like data leakage, connection to unsecured wi-fi networks, phishing attacks, spyware, applications with weak encryption (also known as broken cryptography) are some of the common mobile threats that Android users face [12, 13].

In order to mitigate some of the major risks like data breaches, targeted ransomware attacks, and large-scale distributed denial-of-service (DDoS) attacks, organizations need to allocate their budgets correctly to build a stronger security infrastructure.

A comparison can take on world perspective regarding cyber breaches (Fig. 1).

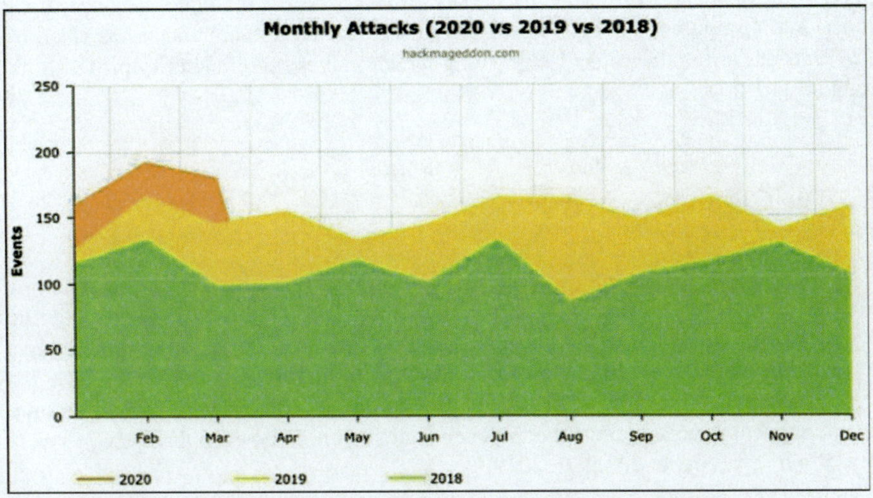

Fig. 1 Comparison of monthly cyber-attacks in the year 2018, 2019 and 2020 [12]

4 Cyber Security Attacks

The nine key cyber security statistics are presented at a glance [14]:

I. 94% of malware is delivered via email
II. Phishing attacks account for more than 80% of reported security incidents
III. $17,700 is lost every minute due to phishing attacks
IV. 60% of breaches involved vulnerabilities for which a patch was available but not applied
V. 63% of companies said their data was potentially compromised within the last twelve months due to a hardware- or silicon-level security breach
VI. Attacks on IoT devices tripled in the first half of 2019.
VII. File less attacks grew by 256 percent over the first half of 2019
VIII. Data breaches cost enterprises an average of $3.92 million
IX. 40% of IT leaders say cyber security jobs are the most difficult to fill.

From the above figure, it was found 94% of the malware infected the machines via email. 80% of all social engineering attacks happen via phishing, so to save our device from such attacks we need to install real-time protection tools like Malware Protection Live. But again, being the user to the system we should not become to complaisant that we have all the tools to stop malware we have to use our common sense to protect our very own device.

If we compare the number of malwares, it has always been in the rise from 2018 to 2019 by 14% and from 2019 to 2020 by 20%. Currently, malware does not attack individual computers but they attack organizational computers and it has risen by 224%. Mutating malware is also growing which makes life of security professional tougher [15–17].

5 The Cost of Security Failures

At present, cost of loss every minute globally is around $17,700 due to phishing attacks only. Nearly 71% of all attacks are financially motivated. With potential losses looming, organizations are realizing that they have to increase their budget in securing themselves. It has been seen that many big organizations, i.e., 18% list cyber security as their lowest priority. This attitude has driven by a certain amount of complacency: 66% believe that a cyber-attack is unlikely, even though 67% were actually hit by a cyber-attack in 2019.

6 Conclusion

COVID-19 is not the only risk which speedily and exponentially disrupts the way of human beings living. This disaster shows that the world is far more susceptible to disturbances occurred due to pandemics, cyber-attacks or environmental listing points than history signposts. The "New Normal" is not COVID-19 only—it is COVID-like occurrences. A cyber pandemic is perhaps as evident as a forthcoming pandemic caused by diseases. The financial impact of an extensive digital shutdown would be of the same degree or may be even greater than the current situation. If cyber-COVID arises, then 30% of the infected systems may become asymptomatic and spread the computer virus although half may continue with degraded functionalities. In the interim, 15% may be "rubbed" with total data loss, resulting in a complete system reinstallation. To conclude, 5% of the devices will result in inoperability [17].

As a result, millions of devices may be taken disconnected in a single day and business organizations will face a complete shutdown. The solitary solution to stop these exponential propagations of cyber-attack is to fully detach all susceptible devices from the Internet to avoid chances of infection. The entire world may face cyber lockdown and will wait for a digital vaccine. Business transactions and transmission of data may become impassable. Research showed that a single day without Internet facility results in more than $50 billion loss and just a 21 day global cyber lockdown will cost more than $1 trillion. This paper focused on some cyber security considerations for business organizations. Adopting to those considerations will help the business giants to continue their normal operations.

References

1. https://www.modernhealthcare.com/cybersecurity/hackers-taking-advantage-covid-19-spread-malware
2. https://www.forcepoint.com/cyber-edu/cia-triad
3. Francois Mouton, Arno de Coning, "COVID—19: Impact on the Cyber Security Threat Landscape", https://www.researchgate.net/publication/340066124_COVID-19_Impact_on_the_Cyber_Security_Threat_Landscape
4. https://venturebeat.com/2018/04/21/the-web-of-profit-a-look-at-the-cybercrime-economy/
5. D. Kwon, H. Kim, "A survey of deep learning-based network anomaly detection", Cluster Comput. 6(22), 949–961
6. V. Vapnik, "Estimation of dependences based on empirical data", Springer, Berlin (1982)
7. H. Drucker, C. Burges, L. Kaufman, A. Smola, V. Vapnik, "Support vector regression machines", MIT Press, Cambridge, (1997)
8. https://resources.infosecinstitute.com/six-things-you-should-be-spending-your-cybersecurity-budget-on/#gref
9. https://home.kpmg/be/en/home/insights/2020/03/ta-the-role-of-cio-and-cisos-in-times-of-covid-19.html
10. https://www.infoq.com/articles/improving-cloud-security/
11. "India Tribune News Article on 23rd May 2020" http://indiatribune.com/37-increase-in-cyber-attacks-in-india-in-q1-2020-report/
12. https://www.hackmageddon.com/

13. https://cisomag.eccouncil.org/india-reports-37-rise-in-cyberattacks-in-the-first-quarter-of-2020/
14. Top cybersecurity facts, figures and statistics for 2020 by Josh Fruhlinger, March 2020
15. https://www.csoonline.com/article/3153707/top-cybersecurity-facts-figures-and-statistics.html
16. https://www.weforum.org/agenda/2020/06/covid-19-pandemic-teaches-us-about-cybersecurity-cyberattack-cyber-pandemic-risk-virus/
17. https://cio.economictimes.indiatimes.com/news/digital-security/cyber-security-amidst-the-covid-19-outbreak/75172086

SQL Injection Attack Detection, Evidence Collection, and Notifying System Using Standard Intrusion Detection System in Network Forensics

Sonam Bhardwaj and Mayank Dave

Abstract Structured query language (SQL) injection is an attack method that explores the functional and storage vulnerabilities of web applications that have data stored in a database. The attacker is capable of affecting the security by intentionally deciding the content that will be forwarded to the database for information retrieval. The attacker gets the benefit by exploiting the syntax and storage vulnerabilities that are responsible for weak points generated in the DBMS security system. This study makes use of the Snort intrusion detection system log files that contain information affiliated to attackers and can provide timed attack notifications via digital notification systems, like emails. In this research, a web server-based network system is initialized using the Snort intrusion detection system (IDS) to detect various methods of SQL injection attacks possible. The method used is based on NIST standards which are based on major risk assessment phases. This is a five phase-based research that performs exploit site testing, simulating attack circumstances, configuring IDS, collecting data and final phase of performing analysis. This study contributes to a web server-based IDS snort system that is capable of detecting a significant number of SQL injection attacks and real-time response notifying system via digital notifications.

Keywords Network forensics · Privacy · SQL injection attack · Cyber attacks · Intrusion detection · Cyber security

S. Bhardwaj (✉) · M. Dave
Department of Computer Science and Engineering, National Institute of Technology, Kurukshetra 136119, India
e-mail: sonam_6180082@nitkkr.ac.in

M. Dave
e-mail: mdave@nitkkr.ac.in

V. E. Balas et al. (eds.), *Proceedings of International Conference on Computational Intelligence, Data Science and Cloud Computing*, Lecture Notes on Data Engineering and Communications Technologies 62, https://doi.org/10.1007/978-981-33-4968-1_53

1 Introduction

The Internet has single-handedly revolutionized technology, connecting countless devices to the world of sharing and exploring information digitally. Internet-connected networking devices come with lots of exploitable vulnerabilities making the Internet the most attack prone medium of sharing information and extracting data. Whenever data is transferred from one computing device to another, it goes through numerous intermediate network devices which means greater threats and attack vulnerabilities to the data. Internet security is compromised almost every hour in one form or another, which comes under illegal acts of cyber exploitation under cyber laws. Computer network security and data privacy are of utmost importance as far as cyber security is considered. Maintaining confidentiality, integrity, and availability of data before and after data sharing is a challenge itself. To detect and take preventive measures for security purposes, one way is to opt for an intrusion detection system. IDS detects various ways of intrusion and risk assessment capabilities. Most of the IDS systems work on the basic integration of sniffing and network mapping tools to carefully observe the flow of data traffic in a network and report any abnormal behavior after careful analysis of data traffic logs and track files. Based on the report, the network administrator can alter the network security features to prevent an intrusion or halt any ongoing intrusion. There are various techniques via which security can be compromised and network-related vulnerabilities can be exploited. SQL injection is one such technique that harnesses vulnerabilities of web applications that use a basic database for storing data. The attacker utilizes a certain combination of literals, constants, and SQL recognizable statements that are initially accepted as a query but later initiate an injective attack on the database directly exploiting data (interest specific to the attacker).

Identifying SQL injection attacks on a web server can be done using a forensic approach model for gathering attack information, analyzing it, and reporting to the administrator. On the other hand, handling such intrusion risks is done using forensic methods like NIST SP 800-30, ISO, or tools like GNU Octave. These methods aid network forensic investigation that includes the collection of digital pieces of evidence, preserving evidence, obtaining legally valid information by analyzing the data traffic. There are preexisting network security components like network firewalls, anti-malware, anti-spyware that can easily be bypassed by an expert attacker. Digital evidence can be retrieved from web server log files containing frequently accessed pages and browsing activities [1]. IDS can monitor inbound and outbound data traffic in a network system, exhibit analysis over data being observed, and recognize evidence or traces that help in identifying the intrusion activities [2]. Snort IDS shows remarkable performance for real-time data traffic analysis, observing IP addresses, perform checks on networking ports, and also identify outsider attacks like buffer overflows, probing, digital fingerprint forgery, and auto-logging [3].

To execute SQL injection attacks, the attacker exploits inserting query commands in such a way that it grants the attacker access for back-end-database. Once an attacker breaches security, privileges are misused for purposes like data retrieval, unethical modifications, tampering data integrity, or simply destroying data structures holding a huge amount of data which results in data loss and denial of service attacks. If military level confidential information gets compromised then a simple SQL injection intrusion can become a major threat to national security [4, 5].

2 Procedure and Experimental Setup

The local server is used with the URL as http://192.168.231.2/SQLinjection/ as an illustration of the SQLi exploits attack. It is designed using primitive PHP supported by the MySQL database [6]. A simple web page containing information gathering "Form" for input data so that after successful login, access to database information is feasible. It also supports a simple message transfer mechanism through which multiple users can send and receive messages. A web server using MS Windows 10 operating system with 8 GB RAM, 2 TB HDD with an additional 256 GB SSD integrated with web applet under PHP 5.6.31, apache, and supporting database of MySQL version 5.0. The intrusion detecting system of Snort executes all its majority of functions in Linux OS. For supporting network forensic procedures, the server utilizes 192.168.231.2 as the static IP address. In the initial stages of configuration, ground rules need to be set up regarding the proper detection of SQLi attacks by the Snort IDS [7], set this ground rule as a default rule for all the SQL injection attacks being tested for intrusion detection.

3 Standard NIST Method for SQL Injection Detection and Analysis

SQL injection attacks are difficult to mitigate and in real-time scenarios like dynamic nodes interacting with a web server, changing clusters frequently results in frequent changes in IP addresses as devices undergo reconnecting IP address switches [8]. Scenarios are not perfect but attacker waits for the perfect opportunity to exploit a certain vulnerability. Functioning and implementation details are given ahead.

3.1 Standard NIST 800-86 and NIST 800-30 Standard Methods

(i) **Standard NIST 800-30**: It is a documented article developed by the National Institute of Standards and Technology(NIST), which performs all procedural and legal activities under the CSA Act 1987(Computer security Act) and the IMR 1996. It performs both assessment and mitigation of risks at various levels

[9]. Following are the major nine phases of risk assessment and mitigation under this standard.

Phase 1: ***System Categorization***: Functional boundaries of information and computer systems are defined.

Phase 2: ***Detecting Threats and Vulnerabilities***: Identifying various threats and vulnerability possibilities.

Phase 3: ***Data traffic Analysis***: Analyzing network-related attacks and malicious data within traffic response.

Phase 4: ***Likelihood Estimation***: Ranking and estimation of a vulnerability depending on the control measures.

Phase 5: ***Evaluating Impact Factor***: It deals with the degree of negative impact due to vulnerability exploitation.

Phase 6: ***Determining Risk***: Assessing risk level and threat level of that risk is identified.

Phase 7: ***Measures to control Risk***: This phase contains collective measures against a certain type of threat.

Phase 8: ***Intrusion Avoidance***: Checks if minor alterations in control measures are capable enough to avoid threats.

Phase 9: ***Final Documentation***: A risk assessment and risk-mitigating report is documented in this phase. All the further forensic investigatory activities and risk measures are operationally based on this report.

(ii) **Standard NIST 800-86**: The National Institute of Standards and Technology (NIST) developed NIST standard 800-86 to adhere to its responsible context under the FISMA Act of 2002, Public Law of 107-347 [10]. It is oriented toward guiding and helping digital forensic investigators in investigating security compromising events in IT sectors and troubleshooting the cause of cybercrime, withholding cybercriminals based on digital pieces of evidence collected, and procedural prosecution within the jurisdiction of cyber laws. NIST 800-86 guidance gives standardization in holding computer and network forensic investigations. The process for performing digital forensics comprises the basic phases of collection, examination, reporting, and action [11]. A broad diagram of network forensic data flow is shown in Fig. 1, Table 1.

3.2 SQL Injection (Vulnerability Exploitation) Scenario

By standard procedure, the following blocks are explained as below:

Block 1 Clustering Information: Using cmd (command-line interface) and systematic series of commands, an attacker is capable of attacking the vulnerabilities of the security system and gain control over the web server-based application. Using sequences of valid injecting commands, information like database names, structure, tables, server information, and metrics of the operating system which is operational currently can be retrieved.

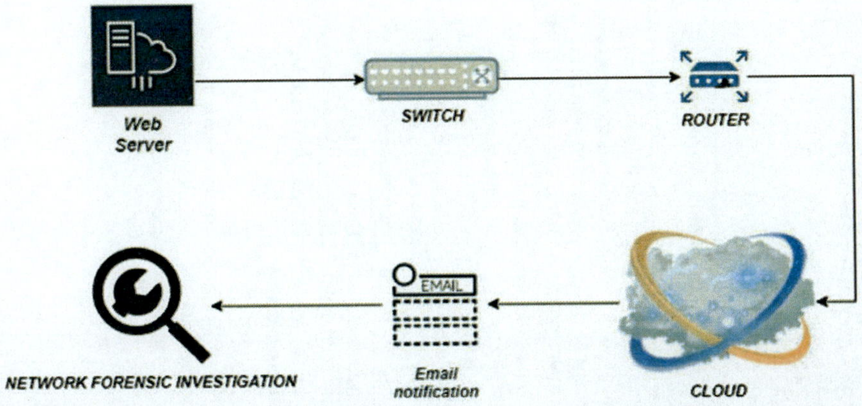

Fig. 1 Flow of information from a web server to the network forensic investigator along with basic alert notification via email

Block 2 Database information leakage (private to public): For an attacker to harness vulnerabilities in a web server-based application, server info is of vital importance. SQL command given to extract server information is "sqlmap.py –u localhost/sonambhardwaj/index.php?id = 1 –dbs". This command returns metadata about the database being used for storing data.

Block 3 Exploiting relational data in Database: Databases that might contain confidential information affiliated directly or indirectly to (say) debit cards, forex cards, login credentials are found at higher risks from an attacker's point of view. Data retrieved from

"sqlmap.py –u localhost/sonambhardwaj/index.php?id = 1 –D db –tables" command is shown in Fig. 2.

Block 4 Drop-Dumping attribute values: The target of the attacker is to exploit private information by bypassing irrelevant information. A user-defined relation in a database has attributes designated to hold login credentials like "username" or "password". The SQLmap command to drop-dump the contents of any specific or bunch of attributes is "sqlmap.py –u localhost/sonambhardwaj/index.php?id = 1 –D db –T user –C Uname, Pword, creditpoints –dump". The command above is capable of drop-dumping the attribute values of the relational table in the target database. After this, required credentials are disclosed to the attacker as well as the attacker is granted access to the web server and its back-end database with "administrative privileges". Figure 3 shows the dumping of attribute values.

3.3 Snort IDS Initialization and Setup

Any type of demand request in the form of pull-based demand or SQL injection attacks is required to be identified at the earliest. Figure 4 illustrates the SQLi attack

Table 1 Comparison of different standard approaches adopted for attack detection

Standard method	Methodology	Information Gathering/Communication	Human resources	Software tools	Documentation
NIST 800 30	Allows for third-party execution,	Questionnaires, interviews and document reviews	Does not address human resources as a possible organizational asset	Relies on role definition to determine use for testing purposes	Security Requirements Checklists for the security areas of management, operational and technical
OCTAVE	Self-directed	Workshop-based approach	Identify human resources that may be a "mission-critical" asset with respect to IT issues	Uses software tools specifically for previously identified vulnerabilities.	Relies upon the creation of three catalogs of information: catalog of practices, threat profile and catalog of vulnerabilities
ISO 27005	Geared toward higher level, management practices	Observation of processes mentioned in organization policies	Specifically covers human resource security which include employees, contractors and third—party users	Uses system and network audit tools for technical compliance checking	Covers all security controls clauses defined in ISO 27002 standard
NIST 800-86	Allows for forensic investigation and methods	Forensics tools to acquire data from hosts that are being redeployed or retired	Specifically involves IT professionals, incident handlers and forensic investigators	Uses Network Monitoring tools, IDS, Investigatory tools	Covers the type of incident, source of attack and preventive measures for future

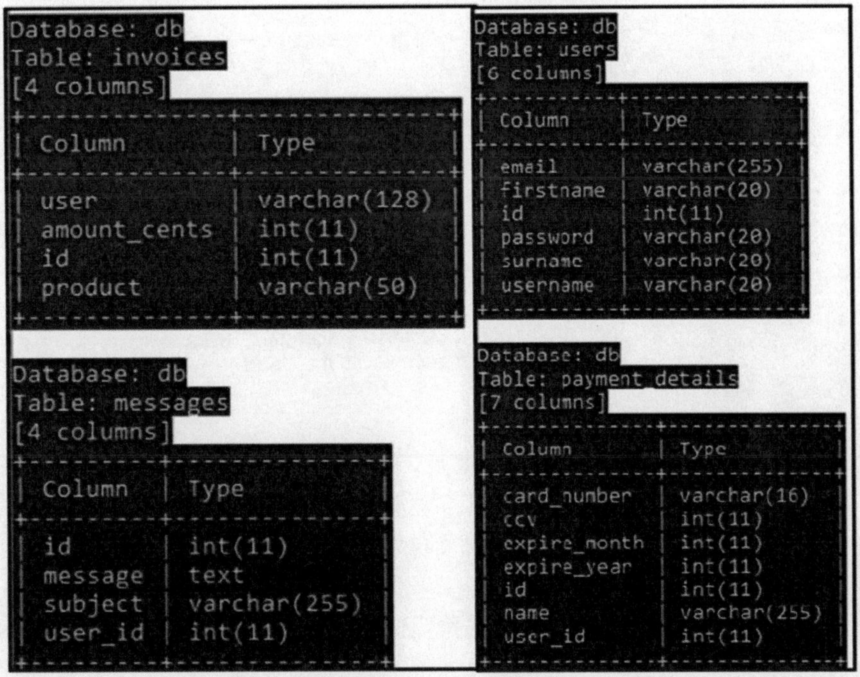

Database: db
Table: invoices
[4 columns]

Column	Type
user	varchar(128)
amount_cents	int(11)
id	int(11)
product	varchar(50)

Database: db
Table: messages
[4 columns]

Column	Type
id	int(11)
message	text
subject	varchar(255)
user_id	int(11)

Database: db
Table: users
[6 columns]

Column	Type
email	varchar(255)
firstname	varchar(20)
id	int(11)
password	varchar(20)
surname	varchar(20)
username	varchar(20)

Database: db
Table: payment_details
[7 columns]

Column	Type
card_number	varchar(16)
ccv	int(11)
expire_month	int(11)
expire_year	int(11)
id	int(11)
name	varchar(255)
user_id	int(11)

Fig. 2 Data retrieved after successful execution of command above

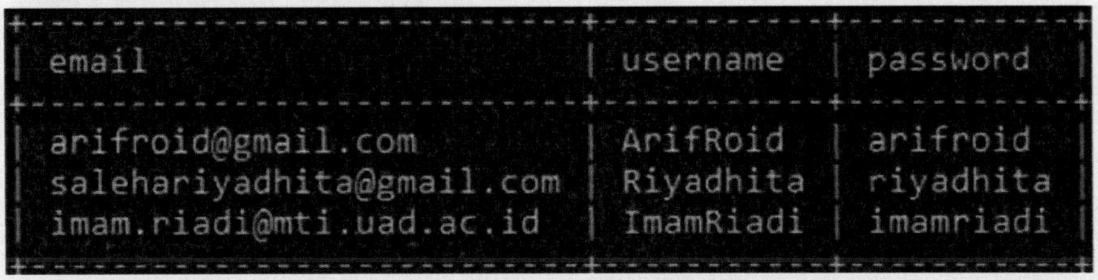

email	username	password
arifroid@gmail.com	ArifRoid	arifroid
salehariyadhita@gmail.com	Riyadhita	riyadhita
imam.riadi@mti.uad.ac.id	ImamRiadi	imamriadi

Fig. 3 Drop-dumped attribute values from the database record

detection process flow in snort and initiating an alerting system via email notifications to the administrator when a network is compromised

The procedure for detecting SQLi attacks [12] is described below:

- The system is fed with SQL injecting packets.
- SQLi attack library is identified by the detection setup system and preprocessed by the attack detection engine (ADE).
- The suspected attack packet is fetched and compared with the rules defined in snort IDS to check if it contains malicious attack content or not. If found, proceed to the next step otherwise terminate.

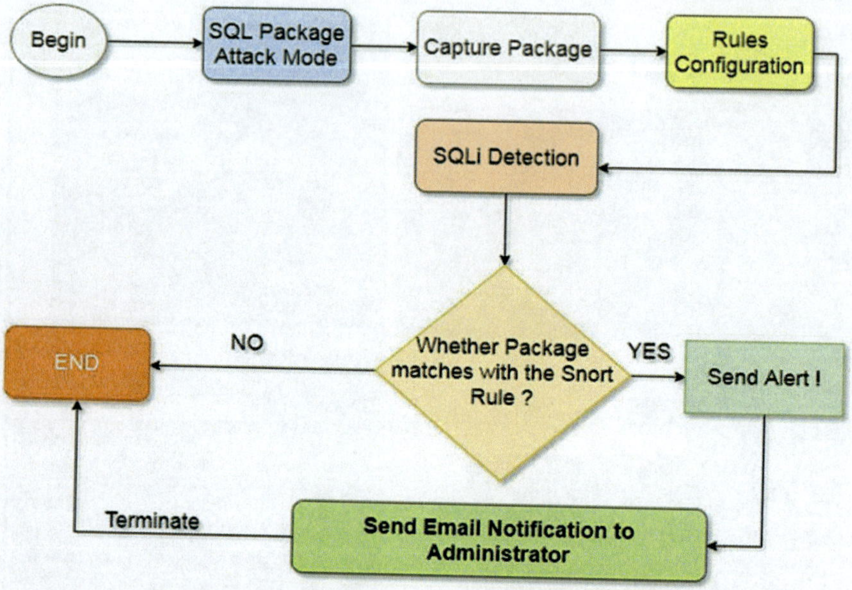

Fig. 4 SQLi attack detection process flow in snort and initiating an alert via email notifications to the administrator

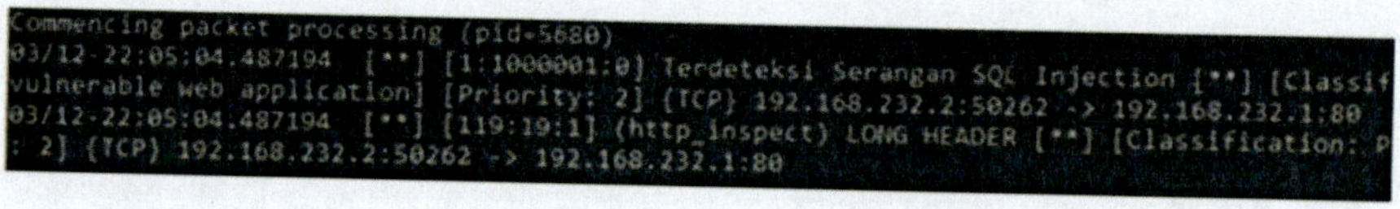

Fig. 5 Logged files processed by Snort IDS for one of the various SQLi attacks tested

- Network traffic packets detected as attacks are stored as network evidential logs.
- These are then processed ahead using "Folder_Watchdog" which pushes email notifications to the administrator for alerting purposes.

Overall we designed a web server-based system in which data is stored and secured in the ASCII coded form. Snort IDS is assigned for SQLi attack detection purposes. Figure 5 shows the logged files processed by IDS for one of the various SQLi attacks. Tracing of the log files is done using Wdog_2 which triggers the email notifications to the admin as a part of the alerting system.

3.4 *Collecting Traffic Data and Trace Log Files*

Digital evidence is collected using snapshots recorded as a part of network traffic observation using pre-configured IDS. The procedure of data and log reconstruction initiates as soon as the server pushes emails to the administrator. Logs are stored and observed under the Wdog_2 directory folder. Logs are visible under directory "C:/snort/log". The database is online, and later the switch is integrated to the IDS server in case of any maliciousness if identified, an automatic alert will be sent to the administrator in a timed manner [13].

3.5 SQL Injection Attack Mitigation and Risk Assessment

Risk assessment and mitigation include basic efforts to reduce the impact possibility of an SQLi attack. Reducing the risk levels is a top priority in this. Wireshark and Network miner are the required tools to collect possible digital evidence in the form of logs and trace files as network forensic evidence [14, 15]. This procedure is accompanied by alerting the administrator whenever there are new attacks detected even during a parallel examination of such pieces of evidence. As mobile devices are the quickest medium to reach an individual account, notifying admin via emails is found to be quicker and easier instead of other time lagging OTP verification [16] and alerting systems [17] comparatively. To send a notification to the destination email, destination email is pre-configured to the snort IDS in the early initiation stages [18]. Figure 6 illustrates the notification mechanism in which there is a log file modified in the snot IDS logs named Snort.log.6116546969 on 8th September 2019 at 1:03. The speed by which the email is sent to the network admin is a factor dependent on the Internet connection speed provided ample weather conditions. Further analysis is carried forward using Wireshark and Network Miner for digital evidence investigation [19].

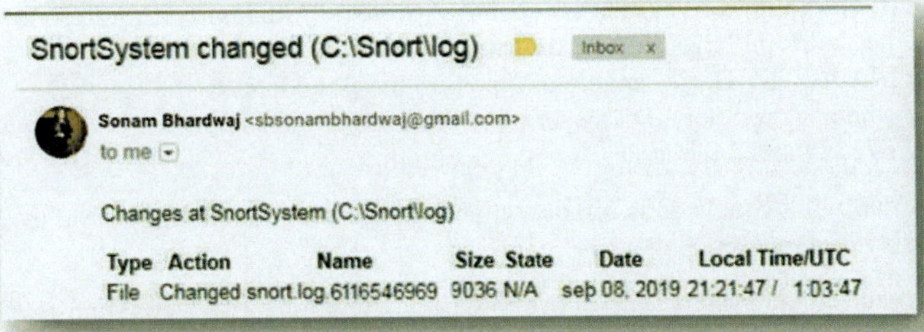

Fig. 6 Email notification received after successful detection of SQL injection attack

4 Testing, Verification and Proof Generation

User agreement verification Test (UAVT) and User Acceptance Test (UAT) [20–22]: It is a standard procedure aimed at producing a proof digital document that the user agrees with and accepts the validity of the document. It is generated by executing several SQL injection attacks in a common simulated environment and analyzes the mitigating capability of system setup (IDS). Ten major rule-based checklist questionnaires built for testing include:

- Deny (TS)
- Intermediate approval (KS)
- Approve (S)
- Strong Approve (SS)

Following are the testing checklist in the Questionnaire:

- Test whether intrusion detection works when a procedure or program call is made.
- Test whether IDS is capable of monitoring the data traffic going through an under suspicion IP address.
- Test whether the implemented intrusion detection system can detect the SQLi requests.
- Test whether the Snort intrusion detection system can collect and hold logs when SQLi attacks occur.
- Test whether system setup can put on display the real-time alerts on system commands when SQL injection attacks occur.
- Test whether IP Address under suspicion can be a source of exploitable information from the attacker.
- Test whether the "Log_port_source" can be a source of information from the SQLi attacker.
- Test whether the Snort system is capable of protecting and mitigating from severe risk factors from SQL injection attacks.
- Test whether the IDS is capable of blocking the IP address of the attacker.
- Test whether the Snort IDS system can send email alert notifications in a given time slot to the Administrator's email account.

The following conclusions are drawn after successful completion of the testing and observation phase:

- The user of the system who chose to deny (TS) before risk mitigation (BRM) received 39% and those who chose after risk mitigation (ARM) received 1% of the attack success rate.
- The user of the system who chose the intermediate approval (KS) option before risk mitigation (BRM) received 41% and those who chose after risk mitigation (ARM) received 17%.

- The user of the system who chose approve (S) right before risk mitigation (BRM) received a 21% score and those who chose after mitigation fired up a score of 51%.
- The user of the system who chose the strong approve (SS) option before risk mitigation received 1% and those who chose after risk mitigation were able to fetch a score of 49%.

5 Conclusion and Future Scope

This study concludes that implementation of standard intrusion detection techniques on a web server-based application environment integrated with a back-store database can be used to identify certain SQL injection attacks and also provide SQLi attack detection information in the form of log files capable of being investigated using network tools like Wireshark and Network Miner. It can be forensically investigated by following simple NIST standards like NIST 800-86 and NIST 800-30 by utilizing pre-configured SQL injection rules that are implemented by the Snort IDS system. Log files are securely moved from IDS data store to the secure storage device or on web services for analyzing illegal malicious activities that took place in the real-time web server-based network environment. This approach of SQLi attack mitigation can be applied to broader networks with more integrated tools for better SQLi detection, risk assessment, evidence collection, and improved network monitoring.

References

1. M. Babiker, E. Karaarslan, Y. Hoscan, Web application attack detection and forensics: a survey, in *6th International Symposium on Digital Forensic and Security (ISDFS)*. IEEE (2018), pp. 1–6
2. H.J. Liao, C.H.R. Lin, Y.C. Lin, K.Y. Tung, Intrusion detection system: a comprehensive review. J. Network Comput. Appl. **36**(1), 16–24 (2013)
3. R. Hunt, S. Zeadally, Network forensics: an analysis of techniques, tools, and trends. Computer **45**(12), 36–43 (2012)
4. F. Alanazi, M. Sarrab, The history of web application security risks. Int. J. Comput. Sci. Inf. Secur. **9**(6), 40 (2011)
5. D.A. Kindy, A.S.K. Pathan, A survey on SQL injection: vulnerabilities, attacks, and prevention techniques, in *2011 IEEE 15th International Symposium on Consumer Electronics (ISCE)*, IEEE (2011), pp. 468–471
6. L. Welling, L. Thomson, PHP and MySQL Web development. 2nd edn. (Sams Publishing, 2003)
7. R.U. Rehman, *Intrusion detection systems with Snort: advanced IDS techniques using Snort, Apache* (PHP, and ACID. Prentice Hall Professional, MySQL, 2003)
8. D. Kar, S. Panigrahi, S. Sundararajan, SQLiDDS: SQL injection detection using query transformation and document similarity, in *International Conference on Distributed Computing and Internet Technology*, (Springer, Cham, 2015), pp. 377–390
9. A. Syalim, Y.Hori, K. Sakurai, Comparison of risk analysis methods: mehari, magerit, NIST800-30 and microsoft's security management guide. in 2009 International Conference on Availability, Reliability and Security, IEEE, pp. 726–731 (2009)

10. K. Kent, S. Chevalier, T. Grance, Guide to integrating forensic techniques into incident. NIST Pubs, Tech. Rep. pp. 800–86 (2006)
11. A. Aminnezhad, A. Dehghantanha, M.T. Abdullah, A survey on privacy issues in digital forensics. Int. J. Cyber-Secur. Digit. Forensics **1**(4), 311–324 (2012)
12. S. Roy, A.K. Singh, A.S. Sairam, Detecting and defeating SQL injection attacks. Int. J. Inf. Electron. Eng. **1**(1), 38 (2011)
13. S. Li, B. Cui, Research on association analysis technology of network attack trace based on web log, in *International Conference on Innovative Mobile and Internet Services in Ubiquitous Computing*, (Springer, Cham, 2020), pp. 33–43
14. S. Senthivel, I. Ahmed, V. Roussev, SCADA network forensics of the PCCC protocol. Digit. Invest. **22**, S57–S65 (2017)
15. M. Hikmatyar, Y. Prayudi, I. Riadi, Network forensics framework development using interactive planning approach. Int. J. Comput. Appl. **161**(10), 41–48 (2017)
16. J. Zhang, H. Duan, W. Liu, X. Yao, How to notify a vulnerability to the right person? case study: in an ISP scope, in *GLOBECOM 2017–2017 IEEE Global Communications Conference IEEE*, (2017), pp. 1–7
17. M. Khera, Think like a hacker: insights on the latest attack vectors (and security controls) for medical device applications. J. Diabetes Sci. Technol. **11**(2), 207–212 (2017)
18. R. Gaddam, M. Nandhini, An analysis of various snort based techniques to detect and prevent intrusions in networks proposal with code refactoring snort tool in Kali Linux environment, in *2017 International Conference on Inventive Communication and Computational Technologies (ICICCT)*, IEEE, (2017), pp. 10–15
19. A. Kumar, J.B. Yadav, Comparison: wireshark on different parameters. Int. J. Eng. Comput. Sci. **5**(3), (2016)
20. H. Croitoru, Developing and testing solutions, in Agile Office 365 (Apress, Berkeley, CA, 2018), pp. 119–138
21. A.S. Dewi, H. Setiawan, Implementation of SHA-256 and AES-256 for securing digital Al Quran verification system, in *2019 Fourth International Conference on Informatics and Computing (ICIC)*, IEEE, 2019, October, pp. 1–8
22. S.K. Shivakumar, Digital workplace testing, in *Build a Next-Generation Digital Workplace* (Apress, Berkeley, CA, 2020), pp. 143–171

Part VI
IoT and Blockchain

Dual-Mode Enabled IoT-Based Smart Balcony Gardening in Urban Area

Subhalaxmi Chakraborty and Abhinav

Abstract Presently, wide adaptation of IoT in different areas of applications has gained extreme exposure. In the field of agriculture, the role of IoT is remarkably increasing in recent days. In this paper, a novel smart farming technique is introduced. Here, few crops, like tomato, mint, onion, and chickpea are considered for the experimental purpose, can be grown irrespective all seasons. In this paper, sensors are embedded in the lawn exploiting the data related to different environmental parameters and depending on the status of the obtained data it is operated in two modes manual mode and automatic mode, respectively. The uniqueness of this prototype is with inside the dual mode of operation of the device in concurrence with the cost analysis.

Keywords IoT · Sensors · Balcony gardening · DHT11 · ESP32 · Dual-mode

1 Introduction

In recent world of information and communication technologies, automation enabled smart agriculture is a potential area for the application of Internet of Things [1]. Nowadays, due to high density of human habitation in urban area, balcony gardening has become an emerging area of exploration of nature. IoT-based automation and controlling of gardening environment and exchange of sensor-based obtained data to the cloud through the device layer are obtained [2]. Moreover, increased crop productivity with minimized water consumption for automatic watering system is showcased [3]. Further, associated works in the discipline of modernized watering system in agriculture are based on wireless networks [4], IoT-based green house control system with data analysis [5], and automated plant watering system [6]. Moreover,

S. Chakraborty (✉) · Abhinav
Department of CSE, University of Engineering and Management, Kolkata, India
e-mail: subhalaxmi2008@gmail.com

Abhinav
e-mail: abhinav33303@gmail.com

V. E. Balas et al. (eds.), *Proceedings of International Conference on Computational Intelligence, Data Science and Cloud Computing*, Lecture Notes on Data Engineering and Communications Technologies 62, https://doi.org/10.1007/978-981-33-4968-1_54

literatures related to smart irrigation using Raspberry Pi [7] study on photovoltaic schemes for smart agriculture [8] which are promising area for exploration. Application of cloud computing and stick-based live temperature and moisture monitoring is implemented in the paper [9]. Development of Arduino microcontroller-based smart irrigation system with DC pump is illustrated in [10] as well as the concept of plant phenol-typing is proposed in the presence of IoT-cloud gateway in [11]. Study of crane-based various control strategy on smart agriculture is presented in [12]. Further, WSN-based smart crop monitoring is also showcased in [13]. In this paper, a smart balcony-based gardening system is represented utilizing the advantages of IoT. The intension of this design implementation is to monitor the moisture and temperature with the ability to control and maintain those parameters automatically. The circumstance of the lawn can be supervised remotely via the advanced digicam device in conjunction with the quantity of water to pour relying on the considered parameters. These system parameters can also be maintained and controlled remotely. This feature can create an edge over other existing IoT-based balcony gardening system. So, this prototype system can be considered as a smart gadget that can efficiently operative both in the manual and automated mode; thus, it can be termed as dual mode of operation. For this farming, some vegetables are selected that can be grown in any season. For this system, tomato, mint, onion, and chickpea are chosen which are found in all seasons in the north and eastern parts of India. The different parameters required for such type of farming of those vegetables are shown in Fig. 1 with proper figure of chosen vegetables Fig. 1a–d.

2 Principle of Operation

The recent progress in the area of IoT-based communication system involves smart balcony gardening. Such system can be implemented by employing various sensor devices [9, 14] like light, humidity, temperature, and soil moisture for monitoring and automatic controlling those parameters for nourishing the indoor plants. The IoT-based smart balcony gardening monitoring system is presented by Fig. 2.

An IoT-based system is implemented using input–output interface for sensors, memory for storage, audio and video, and Internet connectivity for operation of the monitoring applications. The requisite component for implementation of smart balcony garden monitoring system is described below:

(i) The incorporated DHT11 sensor in this prototype acquires the required humidity and ambient temperature for various crops. This sensor works relentlessly to update the temperature and details of humidity to the microcontroller.
(ii) Soil moisture sensor is used to detect the moisture of the soil as the irrigation of soil is based on the initial moisture sensor.

Fig. 1 Parameters require for particular vegetables, **a** Vegetable Name—Tomato (Temperature required: 18–35 °C, Soil type: any, Water required: 1.3 L per/ day), **b** Vegetable Name—Mint (Temperature required: 20–40 °C, Soil type: any, Water required: 1.5 L per/day), **c** Vegetable Name—Onion (Temperature required: 16°C to 30°C, Soil type: any, Water required:2 L per/ day), **d** Vegetable Name—Chickpea (Temperature required: 24–30 °C, Soil type: any, Water required: 1 L per/day)

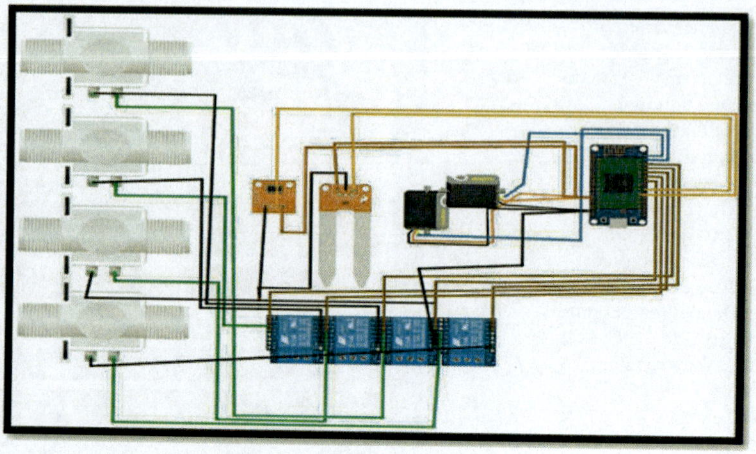

Fig. 2 Proposed model for Smart Balcony Gardening

(iii) In this system, the servo motor is used for the two-axis movement by an angle of 45° in four directions of the ESP32 cam platform, consisting of a microcontroller and camera module.
(iv) Esp32 camera is used as a microcontroller as well as an IP Camera. All sensors are connected to this module and all data will be uploaded to the cloud through this module only.
(v) There is a solenoid valve that is largely used for controlling the glide of water like a tap.
(vi) The necessity of a water flow sensor is to degree the extent of water in the pots in terms of 500 ml or 1,000 ml.
(vii) SMPS (5 V, 20 Amps): SMPS is used to power all associated circuits. From microcontrollers to all valves, this SMPS is connected to 220 V main supply.

The proposed IoT-based smart balcony garden monitoring system can operate in dual mode [15]. As per situation demand, it can operate in manual mode as well as automatics mode using android-based application. The following sections describe the details of the each of modes.

(i) Automatic mode: The principle of operation of automatic mode is shown in Fig. 2. In this mode, all the actuators work according to the given inputs obtained from sensors, i.e., temperature, humidity, and soil moisture sensors [14]. Initially, the android application in this model is fed with the information of the four different pot and their corresponds crop types values (Fig. 3). Then the system use to automatically keep track of temperature, humidity, and soil moisture for each crop in those pots and maintain watering accordingly. The water flow sensor measures the amount of water pouring to pots from valves and when it reaches the limit, valves are closed automatically. In every 10 min,

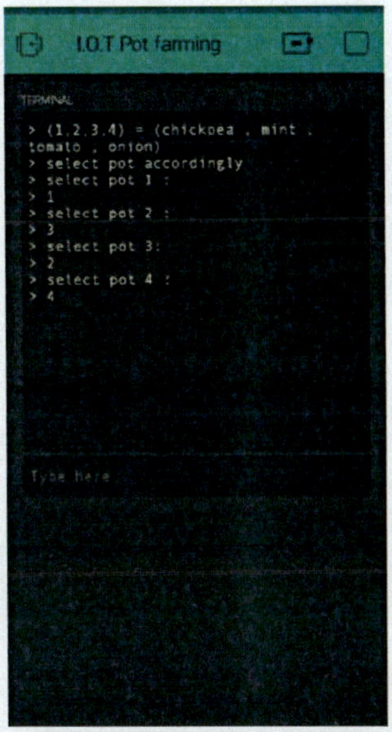

Fig. 3 Parameter variation in automatic mode. Initialization of input parameter

this system collects the data from sensors for 1-minute means remains in inactive mode for 1 min and then takes the actions if required and goes to sleep mode again. This system will update the condition of all the pots on the android application every 10 min (Fig. 4).

(ii) Manual mode: The operation of manual mode is shown in Fig. 3. In manual mode, there are buttons for operation of each valve for each pot, and below it, there is a small display that will display the amount of water going to the particular pot (Fig. 5). There is a camera also which will give the live feed of balcony so that anyone can easily monitor the pots and by their decision can be taken to water more or less to any particular pot (Fig. 6).

3 Analysis of Power Consumption and Cost in Dual Modes

The implemented model shows smarter performance in terms of consumption of power. Here, ESP32 and SMPS are used. Here, connection is done of all sensors as well as ESP32 camera Vin pin with 5 V pin of SMPS. Max operating voltage of the sensors is 5 V. The power consumption operating current for every component is presented in Table 1. The power consumption operating current for automatic mode is presented in Table 2. The power consumption operating current for manual mode is presented in Table 3.

From this comparative analysis of power consumption in between two modes, it is observed that automatic mode is much more cost-effective than manual mode of operation. Depending on the need of the situation, this system can be operated in

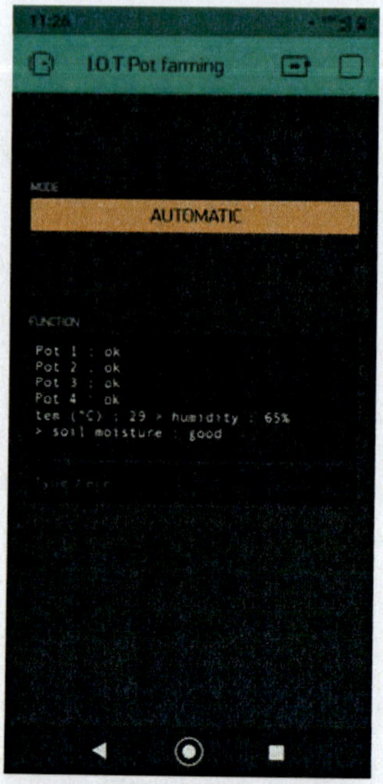

Fig. 4 Parameter variation in automatic mode. Updated condition details

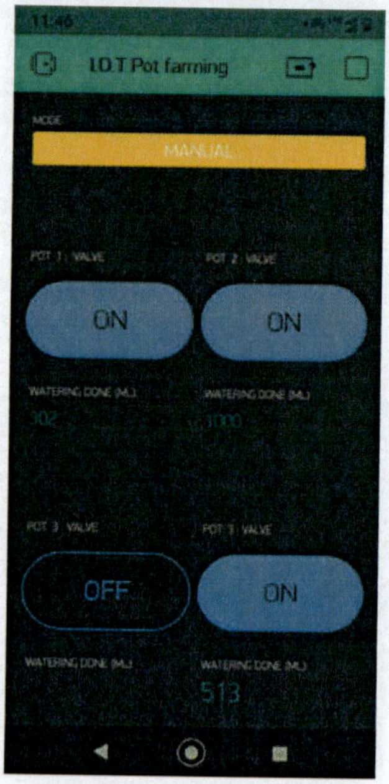

Fig. 5 Parameter variation in manual mode. Initialization of manual mode

Fig. 6 Parameter variation in manual mode. Visual mode for taking decision

Table 1 Power consumption by the components

Sl. No.	Component	Operating Current
1.	DHT11	1.5 mA
2.	Soil moisture	5 mA
3.	Solenoid valve	6 mA (aprox)
4.	Water flow sensor	15 mA
5.	Servo motors (2)	10 mA

both the modes simultaneously throughout the month. In that scenario, electricity cost becomes on an average Rs. 6 in a month. From this perspective, this system can be considered as cost-effective system.

4 Conclusion

In this paper, the hardware and materials used to develop the prototype, allowed to make an accurate and cost-effective system. Moreover, the proposed model is quite economical as well as easily installable for users. In this model, dual mode of operation helps user to maintain balcony garden smartly. Thus, it can be summarized that this prototype system is beneficial, easy to use, and easily deployable in pot farming and lawn farming to effectively monitor crops with a user-friendly application and other alert functions.

Table 2 Power consumption in automatic mode

Calculation of power consumption per hour for sleep mode, (1 min active, 9 min sleep mode)		
Sl. No.	Description	Operating Current
1.	Current consumption in Active mode (when not watering)	(48 + 1.5 + 5) mA = 54.5 mA
2.	Current consumption in Active mode (when watering)	(48 + 1.5 + 5+6 + 15) mA = 75.5 mA
3.	Current consumption in Sleep mode	(3.5 + 1.5 + 5+6 + 15) mA = 31 mA
4.	Current consumption in 10 min (when watering)	(Current consumption in Active mode) + (Current consumption in Sleep mode) = (75.5 + 31) mA = 1.65 A
5.	Current consumption in 1 h (when watering)	(Current consumption in 10 min) × 6 = (1.65 × 6) A = 9.9 A
6.	Current consumption in 10 min (when not watering)	(Current consumption in Active mode(when not watering)) + (Current consumption in Sleep mode) = (54.5 + 31)mA = 0.86A
7.	Current consumption in 1 h (when not watering)	(Current consumption in 10 min (when not watering)) × 6 = (0.86 × 6) A = 5.16 A
8.	Current consumption in 1 day	Current consumption in 1 h (when watering) + Current consumption in 1 h (when not watering) × 23 = 9.9 + (5.16 × 23) A = 128.58 A
9.	Power consumption in 1 day	Power = Voltage x Current = 5v × 128.58 A = 642.9 W
10.	Consumption of electric units per days	Kilowatt per hour = (0.6429/24) = 0.02 units
11.	Average cost of per unit electricity is Rs. 8 (CESC). So, electricity bill to be paid for system	0.02 Units × 30 days x Rs. 8 = Rs 4.8 (approx)

Table 3 Power consumption in manual mode

Manual mode for 5 h in a day in rest 19 h Automatic mode is on		
Sl. No.	Description	Operating current
1.	Current consumption in Sleep mode	= 31 Ma
2.	Current consumption in Active mode(with camera on)	= 62.5 mA
3.	Current consumption in one day	(62.5 × 5) +(31 × 19) = 9.01 A
4.	Power consumption in a day	(Power consumption of automatic mode in a day) + (0.79 × power consumption in automatic mode in a day) = {[(9.01A × 5v) x 5 h] + (0.79 × 642.9)} W = (225.25 + 507.9) W = 733.2 W
5.	Consumption of electric units per days	Kilowatt per hour = (0.7332/24) = 0.03 units
6.	Average cost of per unit electricity is Rs. 8 (CESC). So, electricity bill to be paid for system	0.03 Units × 30 days x Rs. 8 = Rs 7.2 (approx)

References

1. D.L. Yang, F. Liu, Y.D. Liang, "A survey of the internet of things," in *Proceedings of the 1st International Conference on E-Business Intelligence (ICEBI2010)* (Atlantis Press, 2010), pp. 524–532
2. G. Suciu, S. Halunga, A. Vulpe, V. Suciu, "Generic platform for IoT and cloud computing interoperability study," in *International Symposium on Signals, Circuits and Systems ISSCS2013*, IEEE, (2013), pp. 1–4
3. R.N. Rao, B. Sridhar, "IoT based smart crop-field monitoring and automation irrigation system," in *2018 2nd International Conference on Inventive Systems and Control (ICISC)*, IEEE, (2018), pp. 478–483
4. A.H. Abbas, M.M. Mohammed, G.M. Ahmed, E.A. Ahmed, R.A.A.A.A. Seoud, "Smart watering system for gardens using wireless sensor networks," in *2014 International Conference on Engineering and Technology (ICET)*, IEEE, (2014), pp. 1–5
5. Y.-J. Chen, H.-Y. Chien, "IoT-based green house system with splunk data analysis," in *2017 IEEE 8th International Conference on Awareness Science and Technology (iCAST)*, IEEE, (2017), pp. 260–263
6. D. Divani, P. Patil, S.K. Punjabi, "Automated plant watering system," in *2016 International Conference on Computation of Power, Energy Information and Commuincation (ICCPEIC)*, IEEE, (2016), pp. 180–182
7. T. Sahu, A. Verma, "Automated smart irrigation system using raspberry pi", Int. J. Comput. Appl. **172**(6) (2017)
8. A. Shamma et al., "Photovoltaic energy conversion smart irrigation system-dubai case study (Goodbye overwatering & waste energy, hello water & energy saving)," in *2018 IEEE 7th World Conference on Photovoltaic Energy Conversion (WCPEC)(A Joint Conference of 45th IEEE PVSC, 28th PVSEC & 34th EU PVSEC)*, IEEE, (2018), pp. 2395–2398

9. V.P. Anand Nayyar, V. Puri, "Smart farming: IoT based smart sensors agriculture stick for live temperature and moisture monitoring using arduino, cloud computing & solar technology," in *The International Conference on Communication and Computing Systems (ICCCS-2016)* (Gurgaon, India, 2016), pp. 9781315364094–121
10. O.K. Ogidan, A.E. Onile, O.G. Adegboro, Smart irrigation system: a water management procedure. Agri. Sci. **10**(01), 25 (2019)
11. J. Hadabas, M. Hovari, I. Vass, A. Kertész, "IoLT smart pot: an IoT-cloud solution for monitoring plant growth in greenhouses," (2019), pp. 144–152
12. S. Jagannathan, R. Priyatharshini, "Smart farming system using sensors for agricultural task automation," in *2015 IEEE Technological Innovation in ICT for Agriculture and Rural Development (TIAR)*, IEEE, (2015), pp. 459–53
13. B.V. Ashwini, A study on smart irrigation system using IoT for surveillance of crop-field. Int. J. Eng. Technol. **7**(4.5), 370–373 (2018)
14. P. Zhang, X. Li, Y. Liu, X. Kang, Y. Liu, SDU: state-based dual-mode sensor search mechanism toward internet of things. IEEE Access **7**, 147962–147974 (2019)
15. H. Omar et al., "IoT-based interactive dual mode smart home automation", in *2019 IEEE International Conference on Consumer Electronics (ICCE)*, IEEE, (2019), pp. 1–2

A Study of Device Fingerprinting Methods

Abhishek Sinha and Mamta Ramteke

Abstract Security and privacy are the two most important issues in today's world of Internet and communication. Most of us might think of privacy as an area of concern for those with highly sensitive data but actually, each one of us gets tracked the moment we open up our browsers. This paper tries to explore most of the major unconventional approaches of online tracking with their suitable classification into different categories. Also, it discusses how the extensions claiming of privacy enhancement themselves become a victim of the process. Further, the paper highlights the enhancement in the accuracy of the fingerprint due to the addition of the hardware and network-based techniques and the methods that can be opted to achieve an optimum level of privacy for the users in general.

Keywords Browser · Cache · Device · Fingerprinting · Privacy · Tor

1 Introduction

Talking of the unconventional approaches, many contemporary methods of distinctive information gathering have evolved targeting the modern desktop and handheld devices. The primary focus of such information collection is to segregate particular devices from the rest over the Internet for tracking. This approach is known as device fingerprinting. The process can leverage both the browser-dependent and independent features like user agent information, list of available plugins, supported fonts, current time zone, locale, screen resolution, cookie-enabled/disabled, etc. [1, 2] which are the examples of the properties directly accessible through the browser-based APIs whereas the features like ingrained biased behavior and inaccurate readings by the sensors are the examples of the features independent of the browser [3, 4] prevailing

A. Sinha (✉)
Sri Sukhmani Institute of Engineering and Technology, Dera Bassi, Punjab, India
e-mail: sinha@live.in

M. Ramteke
National Institute of Technical Teachers Training & Research, Chandigarh, India
e-mail: mamtapj@yahoo.com

V. E. Balas et al. (eds.), *Proceedings of International Conference on Computational Intelligence, Data Science and Cloud Computing*, Lecture Notes on Data Engineering and Communications Technologies 62, https://doi.org/10.1007/978-981-33-4968-1_55

as a result of the intrinsic properties of the material used in manufacturing and the production tolerance of the instrument. Extending the latter part, the behavior of the network protocol and its associated attributes are leverageable to extract distinctive information from a device. Targeting only a specific browser on a device for tracking is known as single-browser fingerprinting, while the ability to track a user across multiple browsers is known as cross-browser fingerprinting. Keeping these in mind, an understanding of the defense mechanism is very essential to safeguard our right to privacy.

2 Fundamental Idea

Previously, cookies were considered to be good enough for maintaining the state of the user over the stateless protocols but over the time, it was observed that the modern browsers focusing on privacy, started rolling out new features like disabling cookies, private mode browsing, etc. [5], hence, severely challenging the reliability of the cookies. Therefore, techniques targeting the browser and hardware attributes were proposed to be used as fingerprints [1] [6]. The uniqueness of a fingerprint is estimated via Shannon's entropy equation [7]:

$$H(F) = -\sum_{i=1}^{n} p_i \times \log_2 p_i \tag{1}$$

where H is the calculated entropy, expressed in bits, F is any given feature, n is the number of fingerprint samples collected and p_i represents the probability of the occurrence of the ith value. A higher value of entropy suggests a higher degree of uniqueness and hence easy distinction. Studies [1, 8] suggest that just about 15–20 bits of this information is enough to precisely spot a browser over the internet, where attributes like 'Plugins' can alone have an accuracy of up to 82% due to their higher value of entropy [9]. But all the data that we collect for this purpose are susceptible to change due to the interaction between the user and the device making entropy less reliable when evaluated alone. Therefore, the second parameter is the stability which is the rate of change of fingerprint over time [1].

3 Fingerprinting Techniques

3.1 Browser-Specific Fingerprinting

This method of fingerprinting mainly focuses on exploiting the features of the browsing environment including properties like plugins, font size/family, userAgent, etc., that are accessible via JavaScript, Flash, or Java [1], [3] though Java not being considered anymore for commercial fingerprinting due to a huge number of vulnerabilities within its plugins [10]. The various methods involved are:

JavaScript Attributes Fingerprinting. This approach targets the *navigator* and the *screen* objects available in the browser where the former mainly consists of the browsing environment properties accessible via *'window.navigator'*. The fixed order of the enumeration of this object across the browsers except for Google Chrome makes it suitable for the categorization of the browser [10]. The exceptional behavior of Google Chrome (GC) is not just limited to the version but is also influenced by the underlying platform; i.e., the different versions of GC have a different order of enumeration on the same operating system (OS), and the same version of it has a different order of enumeration on different OS, hence reveling both browser and OS identity.

Similarly, the screen object represents the screen of the device and can be accessed via *'window.screen'*. The value of its properties varies across the browsers and can have an entropy of about 7.41 bits [4]. Also, properties like screen resolution have a dependency on zoom level in the case of Mozilla Firefox and Internet Explorer requiring it to be considered when dealing with such browsers. Additionally, *window.name* can be used as a cookieless session variable due to its reluctance to page reload and allowance of cross-domain access [11].

JavaScript API-Based OS Detection. Platform dependency or inaccuracy in the reimplementation of the built-in APIs lead to the variation in the output across platforms which reveal information about the underlying OS and the architecture [10, 12]. A few such APIs are:

Battery API Fingerprinting. Browsers allow access of battery information via a JavaScript API, *navigator.getBattery()* which in the case of Mozilla Firefox running on a Linux displayed a double-precision battery level while a Windows, Mac, or an Android-based systems displayed only two significant digits for the same API [13].

Math Object Analysis. *'Math'* is a JavaScript object that provides static functions for mathematical computations. The output of these functions has a variation not only across the OS but also across different architectures for the same OS [12]. This abnormality has a greater stability due to the fixed nature of the underlying architecture.

Font Detection. Fonts are fingerprintable due to the difference in their attributes accessible via the APIs provided by ActionScript or through side-channel inference [10]. In the latter, an unknown font is applied to any given text and due to the fallback mechanism of the browser, the default font gets applied this is utilized to obtain the dimensions of the supported default font. Further keeping the font size as constant, a set of predefined font-families is applied to the given text, and the dimensions hence obtained are compared with the default values obtained earlier. A mismatch in the two proves the existence of the font on the browser or otherwise. This way almost the entire list of fonts can be obtained. Also, the API returns the list in an unsorted order which makes it the second-largest contributor to the overall entropy [1, 9].

NoScript Whitelist Fingerprinting. NoScript is a prevalent Firefox plugin focused on providing a defense mechanism against the execution of the baleful JavaScript code available online by blocking them all except for the one been whitelisted by the user. Because this list contains all the preferred websites of the user, it can reveal a substantial amount of information related to them [14]. The method emphasizes on loading a JavaScript code from a potentially whitelisted domain and further checking for the existence of a predefined object in that code. The existence of the object confirms the presence of the domain in the whitelist or otherwise.

Flash Fingerprinting. Flash is a proprietary plugin that has two major classes called *capabilities* and *font* that can be used to extract fingerprintable information from the browser. The *capabilities* class provides information associated with the underlying systems such as the name of the OS and information related to running applications, whereas the *font* class can reveal information related to the installed fonts on the system [8, 15]. The Flash APIs accessible via JavaScript are sometimes the re-implementation of the actual browser-based functions with an output different from their browser-based counterparts [10]. This allows the malicious servers to identify the browser, its underlying architecture, and the exact kernel version. For example, while trying to identify the underlying platform for a Firefox browser running on a 64-bit Linux OS would return 'Linux x86_64' when queried via browser-based API, whereas it would return 'Linux 3.2.0-26-generic' when queried via Flash-based implementation. The value returned by the latter is the full version of the underlying kernel which is undesirable from the view of system security. Another example of such difference in the output can be noticed in the case of the API that returns the resolution of the screen, where the Flash-based API returns the resolution as the sum of the two individual screens for a dual monitor setup, unlike the actual. This approach might not work in the future as the Flash is becoming obsolete due to performance and security concerns [7, 15].

Canvas Fingerprinting. This process involves the rendering of 2D graphics on the canvas element of HTML5 followed by an invocation of the *toDataURL()* method on its object, to obtain a PNG image with Base64 encoding that holds the content of the canvas used as fingerprint [3].

Browser History Sniffing

Interactive History Sniffing. In this attack, a series of puzzles with hyperlinks concealed under a transparent image is presented to the users who interact with them to get access to the website. Each interaction with the puzzles involved an event like keypress or mouse click to be triggered that resulted in changing of the color of the word such that if the URL was visited then the color would become black otherwise it would remain as the color of the background, hence creating a distinction between the visited and unvisited links [16]. A single word in the puzzle revealed information about more than one visited website.

CSS Timing Attack. Most of the current browsers, process CSS from right to left and any CSS rule with the:*visited* selector on the rightmost side in the stylesheet, as in *[class* = "className"]:visited{... }* would result in a delay in loading the page if a link on the page was visited earlier. This delay can be analyzed and compared with a standard value to sniff the browsing history [16].

Cache Timing Attack. This attack exploits the fact that a page loaded locally will be faster as compared to one loaded remotely. This requires a threshold timing to be determined in advance which is done by initially loading a resource from a website with caching enabled and recording the time taken for the process to complete. Any subsequent access to the same resource would probably be served from the cache which can be compared with the threshold to determine the source [17].

Vendor-Specific Fingerprinting. The specific implementation of objects and properties by some of the vendors can be used to reveal information about a browser and associated environment [10]. For example, the *typeof* operator in JavaScript can be used to check for the existence of a specific attribute for identifying the browser's family.

if (typeof navigator.webkitGetUserMedia){console.log("Chrome")}

3.2 Hardware Fingerprinting

The degree of error in every hardware component can be exploited to generate a fingerprint that apart from being unique would also be stable due to its reluctance toward tampering hence making them suitable for cross-browser fingerprinting [4]. The methods involved in this process are:

CPU Virtual Core Estimation. *window.navigator.hardwareConcurrency* is a browser API that provides the number of logical processors available on an individual's computer to run threads [18]. For cross-browser fingerprinting, this feature has shown to have an entropy of 0.59 and a stability of 100%. This API might not be available in some of the older browsers but the value can still be estimated through a timing attack that requires the execution of simultaneous web workers [4].

GPU Fingerprinting. This method of fingerprinting measures the GPU clock skew with CPU clock as the reference which involves the rendering of complex 3D graphics on an HTML5 canvas using WebGL and measuring the number of frames rendered in a given time [19]. The time of the rendering of the frames is measured using the CPU clock. The maximum entropy yield in this method is known to be 5.14 for a sample space of 130 devices, but this value can be as huge as 9 [20].

Camera-Based Fingerprinting. The pixel non-uniformity noise associated with the sensor of every digital camera has a linear relationship with the intensity of light hitting the pixel [21]. The noise is multiplicative and is unique to each sensor that gets embedded in the pictures taken by the camera. The noise due to its deterministic pattern serves as a fingerprint of the sensor that can be exploited to recognize a camera with a precision of up to 90.8% [22]. The noise is caused by the minor dimensional and manufacturing imperfections associated with the CCD and CMOS sensors of the camera that causes a variation in the pixel output on receiving a uniform light on the sensor. The other type of noise that can be leveraged for this task includes the fixed pattern noise (FPN) (aka dark current noise) [21, 23] that exists as a result of insufficient illumination of the camera sensor at the time of photography, unlike the PNRU that is due to the abundance of light. This method is insusceptible to the operations like cropping, scaling, or compression [24, 25]. Pixel imperfections like hot pixels, dead pixels & pixel trap can also be used for fingerprinting [26].

Audio Fingerprinting. The exploitation of the subtle variations in the processed audio signals to generate identifiable information is called audio fingerprinting. It involves:

Audio Device Detection. For any given speaker or microphone, the frequency response must ideally remain the same for all the frequencies in the given range [6] but [19] suggests that the existence of variation in the response curve of these devices over a given frequency range can be leveraged to generate a unique identity for the machine on which the device is installed. The process would involve simultaneous playing and recording of the audio using a pair of speakers and a microphone on the given device. The existence of variation in the output is usually due to various manufacturing imperfections and production tolerance which is permissible up to ± 2db.

Cuccovillo's work [27] in this field relies on identification of the disparities in the audio signal captured and processed by a typical microphone, where the captured signal S_2 is processed by the microphone to produce a signal S_3 acting as an input to the analog to digital (A/D) converter which samples and quantizes the received signal to produce a digital output S_3 [n]. The frequency response of the microphone obtained in this process is examined with the help of a radial basis function (RBF) kernel SVM with parameters (C_{RBF}, γ_{RBF}), where C_{RBF} and γ_{RBf} are the cost variable for the given SVM and the gamma parameter for the given RBF kernel respectively. The SVM is trained with the feature matrix designed by summing up the feature vectors created using Eq. (3) for different devices. Thermal noise produced in this process is considered to have a negligible impact on the final output and is hence ignored. The process can mathematically be expressed as:

$$S_2(f) = S_1(f) \times F_{echo}(f) + N_{env}(f) \quad (2)$$

where S_1 is the source signal exposed to the echo and the environmental noise represented by F_{echo} and N_{env}, respectively.

$$S_3(f) = \int_{spectrum} F_{mic}(f) S_2(f) df + N_{mic}(f) + N_{ENF}(f) \quad (3)$$

F_{mic} is the frequency response function of the microphone, N_{mic} is the thermal noise generated by the mic and N_{ENF} is the electric network frequency influence.

$$S_3[n] = \text{quantization}\ (b, \text{sampling}(\text{fs}, S3(t))) \tag{4}$$

S_3 [n] is quantized with b bits per sample applied to the values considered by S_3 (t) at a given time, $t = n \times T$, S_3 (t) represents the inverse Fourier transformation of S_3 (f) and f_s is the sampling frequency in Hertz. Further S_3 $[n]$ is encoded and stored on the machine. The method works for both desktop & mobile devices.

AudioContext Fingerprinting. Another study [28] suggests the use of AudioContext interface [29] and OscillatorNode interface which provides a wide range of functionalities for audio signal processing through the audio stack and sound card available on a computer. The signal generated using OscillatorNode is processed via DynamicsCompressorNode and further fed into AnalyserNode to obtain a set of frequencies to fingerprint. The amplitude remains constant for the peak frequencies during the process while it shows unique values for the remaining when tested across multiple browsers running on the same computer, hence making the former a suitable candidate for cross-browser fingerprinting while the latter can be used for individual-browser fingerprinting [4]. The overall entropy observed in the experiment was 5.4 for a sample space of 18,500. A demo of this technique is available at [30].

Accelerometer Fingerprinting. The accelerometer is an instrument that measures the acceleration of a body in motion along all the three axes of rotation. However, these instruments suffer from linear bias given by,

$$V_m = V_t \times S_a + O_a \tag{5}$$

where v_m is the measured value of acceleration, v_t is the true or actual value of acceleration, S_a and O_a are the sensitivity and offset of the instrument, respectively, along a given axis 'a'. S_a and O_a are also known as the calibration parameters for the sensor having ideal values, 1 and 0, respectively. An accelerometer usually reports -1 in case the phone is facing upwards and 1 in case it is facing downwards. A fingerprint can be obtained for mobile gadgets by measuring the variation in the instrument's calibration parameters under the condition of zero acceleration. It requires JavaScript code to be embedded in a web application which when visited by the user extracts the required data. The bias in the obtained data acts as a unique signature for the device. The entropy of this data can be up to 7.502 with an accuracy of device identification of up to 58.7% [6]. Using machine learning accuracy can be increased up to 95% [31].

Mouse Fingerprinting. Even the safest considered browsers such as Tor are susceptible to a mouse interaction-based attacks where the acceleration on a page during a scroll varies across machines due to its dependence on the algorithm used by individual OS [32]. This data can not only be collected for wired, but also for the wireless mouse revealing not just the trackable information but also the graphically generated passwords [33]. e.g.: *document.getElementById('Id').addEventListener("wheel. function(e){console.log(e.deltaY)});*

Also, the angle based features of a mouse can be leveraged as a behavioral parameter to track an individual accurately up to 98.7% [34].

3.3 Network Attribute Fingerprinting

Network Traffic Summary Fingerprinting. The flow log maintains a record of the inbound and the outbound IP traffic within a typical network which can be used to determine the running application software on a host even without having access to the actual network payload. The process involves the training of a support vector machine (SVM) on a given dataset comprising of the observed flow of the traffic for a given number of websites over a significant number of days that can be used to obtain a probability of the belongingness of a given instance of a browser to a particular type under the condition of it being at a minimum distance from the separating hyperplane determined by the SVM. The method has an accuracy of about 71% [35].

Origin Bound Certificate (OBC) Analysis. OBC also known as TLS ChannelID is the stable self-signed certificates generated by the browsers to authenticate themselves on an HTTPS server during a handshake. The certificates are associated with the cookie and validated by the server with each request from the client before beginning the actual data transfer [36]. Since the signature on the certificate is unique to a browser, it can serve as trackable information. Also, the use of the same OBC across multiple sessions makes it susceptible to third-party tracking [37].

Cache Metadata Fingerprinting

E-Tag-based Identification. E-Tag is a web cache validation mechanism of HTTP protocol that facilitates the server to save version information of a web resource on the client which is again sent back through *If-None-Match* header whenever a new request is made for the same resource. In case of a match for this value on the server, results in an HTTP code 304 being forwarded to the client, which means the resource on the server has not been modified, and the client has the liberty to use the cached version of the same. Because the method allows the storage of data on the client machine in a similar way to that of a cookie, it can be exploited to identify a device [38].

Last-Modified- based Identification. Another mechanism to ensure the existence of the updated resource in the client cache is the *Last-Modified* HTTP header that allows the server to store date and time on the client with a minimum length of 32 bits. This data is communicated back to the server along with every request from the client to access the resource where the server in return responds with an XHR code 200 if the resource has been modified or 304 otherwise. Since the mechanism allows the storage of information on the client-side, it can be used to save identifiable information [38].

DNS Cache History Sniffing. A DNS cache is a storage location for the results of the recent DNS lookups, maintained by an OS [39]. Since lookups to the server are usually considered expensive in terms of processing, hence any subsequent request would result in the response being served from the cache which is considerably faster as compared to the actual lookup. Since the difference in the timing from both the sources is significant, it can be used for the distinction between the two, hence leading to the conclusion about the URL being accessed previously [16].

Tor Detection. TorDNSEL is a DNS-based record of the exit nodes of the Tor network that can be queried to verify for a requesting IP address belonging to one of the Tor exit nodes [40]. The query results in *127.0.0.2* in case of the existence of a valid Tor node or *NXDOMAIN* otherwise. The request can either be automatic or manual.

Browserprint project demonstrates the use of this feature to distinguish between the requests from Tor and non-Tor based clients [41].

HTTP Strict Transport Security (HSTS) Tracking. HSTS is a web server directive that mandates the use of HTTPS from source to destination to mitigate man in the middle attack. This is achieved by forwarding an HSTS header by the server instructing the client to use HTTPS for subsequent communications resulting in the storage of this information as a single bit identifier on the client-side. A malicious server can exploit this feature to create a trackable binary string by including URLs from multiple domains in their web page which upon receiving the request would respond with HSTS header being randomly forwarded to the client. Further, any subsequent request to the server would result in a connection being established to the embedded URLs depending upon their HTTPS preference set by the HSTS header, which can be analyzed to generate a binary sequence and hence use it for tracking [42]. Also, any attempt to connect to port 443 of a server using HTTP would result in an error in the browser not having HSTS pinned for that server. In case the server was visited previously, then HSTS will enforce the use HTTPS making it a valid request. Therefore, the former would trigger *an onerror* event while the latter would trigger onload in JavaScript which can be exploited for history sniffing [43].

Intermediate CA Certificate Fingerprinting. This approach relies on the incapability of the misconfigured HTTPS server to send intermediate CA certificates to the client. Since the browsers to enhance its performance stores these certificate in its cache from the previously accessed websites, it can be exploited to generate a cookie-like string by loading a resource from a misconfigured server because usually the resource would not be loaded in the absence of the intermediate certificate but since it is available in the cache same would be utilized for the purpose. A successful invocation of the *onload* event would conclude the existence of the required certificate at the client, simultaneously validating the client's visit to one of the user website of the certificate. The method has shown to provide information related to the geographical location and browsing preferences of the individual [44].

4 Analysis

Looking at the built-in properties of the browser, we found that the list of plugins and fonts had a contribution of almost 27% and 25%, respectively, to the overall entropy of the fingerprint, hence making them the highest contributor of the trackable information supported by the features like order of enumeration of fonts, vendor-specific API prefixes and non-uniform precision in battery API help in mounting browser-specific attacks. Also, the inaccuracies in the implementation of built-in APIs help in choosing a kernel-specific attack based on information revealed by them. Despite its ease of implementation, browser-based fingerprinting seems to be quite unstable due to the changing value of the properties which makes it less reliable in the case of identically configured devices [45]. Nonetheless, a simple heuristic can be applied to track the evolution of the fingerprint [1]. Further, we discovered that generally browser plugins and extensions fail to protect against fingerprinting due to their unrealistic behavior including the 'NoScript' plugin [10, 14] contradicting the study made earlier [1]. Even the CSS attacks are still possible through social engineering and screen interaction.

We also discovered that the hardware-based approach alone was accurate enough up to 99.24% and was effective in distinguishing the users across different machines even when concealed behind Tor which was not the case for the browser-based methods which had the best accuracy of 90.84%. This contradicts the study [1] suggesting mobile devices being less fingerprintable which actually can be traced with up to 95% accuracy [31], [6].

Almost 70% of Tor browsers before version 5.5 are susceptible to fingerprints [46] but the issues like the canvas fingerprinting, CPU virtual core analysis, screen orientation analysis, sensor detection, Audio API fingerprinting, cache timing analysis, etc., in its versions ranging from 5.5 to 7.0. HSTS considers security over privacy as the former seems crucial while the latter is just theoretical.

5 Countermeasures

5.1 *Browser-Specific Countermeasures*

Since browser-based JavaScript objects are one of the most crucial agents for revealing the fingerprintable information disabling them by default and allowing only the trusted sources could be a good approach [1]. Also, having a uniform behavior of the JavaScript objects & APIs including the uniform order of enumeration of the fonts and uniformity in names is desirable which might be a bit difficult to achieve due to conflict of opinion among the vendors [10]. The timing attacks relying on cache and CSS can be mitigated by turning off the browser cache and clearing out the browser history each time the browser closes while the timing attacks leveraging CPU & GPU can be countered by setting up a fixed amount of time interval for each request at the OS-level [47]. Further, media players like Flash, RealPlayer, and QuickTime should be disabled as they may not only leak OS-related data but also the IP address of the machine. Other generic approaches at the user level might include avoiding the customization of browser settings which might otherwise add to the entropy as well as disabling the WebRTC which might leak the IP address even on a good VPN. Although it has been recommended previously [46] to use Tor version 5.5 and above, we based upon our analysis recommend the use of version 7.0 and above because it not only addresses the issues covered in the former but also addresses the latest issues such as Audio API fingerprinting and hardware concurrency analysis.

5.2 *Hardware Fingerprinting Countermeasures*

To prevent clock skew attacks for GPU fingerprinting, the value of time can be monotonically increased rather than the clock tick of the device to avoid estimation of the drift [48]. The PNRU-based digital camera fingerprinting can be mitigated by applying the noise removal strength obtained from multiple images taken by the given camera, expressed as:

$$J' = J(1 - \beta K_1) \tag{6}$$

where *J'* is the final image free from PNRU, *J* is the original image containing PNRU, K_1 is the estimated PNRU, β is the removing strength such that $\beta \approx 1/q$, q being quality of K_1. Also, noise filters can be designed using the approach suggested in [49] that can be used to eliminate sensor-based noise from the photographs.

Trackpads on the laptop have a varying delta value; therefore, the use of a conventional mouse is recommended which shows uniform behavior eventually reducing the entropy. A more generic approach toward mitigating hardware fingerprinting can be achieved by calibrating devices after being manufactured or by adding a constant random value to its output at the OS level [6].

5.3 *Network Fingerprinting Countermeasures*

OBC-based third-party tracking can be mitigated by using the encrypted client certificates extension [50], which will ensure the transfer of OBC only after a secure channel is established via transport layer security. Since E-Tag and last-modified are the parts of the browser cache validation mechanism, disabling or clearing the cache would result in the clearing of these values along with the actual data hence prohibiting the attack. Also, private mode browsing can be used to partially avoid the tracking, where the user will remain trackable only until the browser is open [51]. The DNS cache fingerprinting can be mitigated by disabling the DNS prefetching on the browser. The intermediate CA certificate fingerprinting can be countered with regular cleaning of the profile and blocking the third party requests [44].

6 Conclusion

While comparing the accuracy of the three major categories discussed in the paper, we found the hardware-based methods to be most accurate with a mean accuracy of 94.83% while the maximum achievable accuracy was 99.24% followed by the browser-specific approach and then comes the network-based approach.

It will be incomplete to evaluate accuracy-based only on entropy and neglecting the stability for the reasons discussed in section-2. Since the hardware-based entropy is material ingrained, therefore its modification is difficult ensuring higher stability which we also observed in our study having a value of up to 100%. Also, the browser-based approach was found to be the least stable.

Based on the above factors and the analysis done in section-4, we consider the hardware-based method as the winner among the three approaches. Further, we conclude that a combination of various hardware-based methods can help achieve near 100% accuracy and stability.

Also, the use of most of the additional plugins and extensions to protect against fingerprinting is not recommended as they increase the overall entropy of the fingerprint. A significant focus is required on countering the network-based approaches as not many of the options are available to counter these methods even though its accuracy is fairly high. Further, Tor seems to have fixed some of the latest issues, such as audio API fingerprinting but still, techniques like mouse fingerprinting throw its claim of considerable anonymity into question. Nonetheless, for a common user, the optimum level of anonymity can be attained by using a Tor version, greater than 7.

References

1. P. Eckersley, *How Unique is Your Web Browser?*. LNCS 6205, (Springer, Berlin, Heidelberg, 2010), pp. 1–18
2. C. Blakemore, J. Redol, M. Correia, Fingerprinting for web applications: from devices to related groups, in IEEE TrustCom/BigDataSE/ISPA, IEEE, (2016), pp. 144–151
3. R. Upathilake, Y. Li, A. Matrawy, A classification of web browser fingerprinting techniques, in IEEE 7th international conference on new technologies, mobility, and security. IEEE, (2015)
4. Y. Cao, S. Li, E. Wijmans, (Cross-)browser fingerprinting via OS and hardware level features, in NDSS Symphonium, 2017
5. X. Zhang, Q. Zhou, C. Gu, L. Han, The location privacy preserving of social network based on RCCAM access control. IETE Tech. Rev. (2018)
6. H. Bojinov, D. Boneh, Y. Michalevsky, Mobile device identification via sensor fingerprinting. arXiv: 1408.1416v1 [cs.CR] (2014)

7. P. Laperdrix, W. Rudametkin, B. Baudry, Beauty and the beast: diverting modern web browsers to build unique browser fingerprints, in IEEE symposium on security and privacy, IEEE, (2016), pp. 879–894
8. A.F. Khademi, M. Zulkernine, K. Weldemariam, An empirical evaluation of web-based fingerprinting, in IEEE Software 32, IEEE (2015), pp. 46–52
9. P. Laperdrix, W. Rudametkin, B. Baudry, Mitigating browser fingerprint tracking: multi-level reconfiguration and diversification, in 10th international symposium on software engineering for adaptive and self-managing systems, (2015), pp. 98–108
10. N. Nikiforakis, A. Kapravelos, W. Joosen, C. Kruegel, F. Piessens, G. Vigna, Cookieless monster: exploring the ecosystem of web-based device fingerprinting, in IEEE symposium on security and privacy, IEEE, (2013), pp. 541–555
11. T. Bujlow, V. Carela-Español, J. Solé-Pareta, P. Barlet-Ros, A survey on web tracking: mechanisms, implications & defenses. Proc. IEEE **105**(8), 1476–1510 (2017)
12. P. Laperdrix, B. Baudry, V. Mishra, FPRandom: randomizing core browser objects to break advanced device fingerprinting techniques. LNCS 10379, (2017), 97–114
13. L. Olejnik, G. Acar, C. Castelluccia, C. Diaz, The leaking battery a privacy analysis of the HTML5 battery status API. LNCS 9481, (Springer, Cham, 2016), pp. 254–263
14. K. Mowery, D. Bogenreif, S. Yilek, H. Shacham, Fingerprinting information in JavaScript implementations, in W2SP ed. by H. Wang, IEEE Computer Society, (2011)
15. Al-Fannah, N.M., Li, W.: Not All Browsers are Created Equal: Comparing Web Browser Fingerprintability. In: IWSEC 2017, LNCS 10418, pp. 105–120. Springer, (2017)
16. Z. Weinberg, E.Y. Chen, P.R. Jayaraman, C. Jackson, I still know what you visited last summer leaking browsing history via user interaction and side channel attacks, in 2011 IEEE Symposium on Security and Privacy, IEEE, (2011), pp. 147–161
17. W. Felten, M.A. Schneider, Timing attacks on web privacy, in 7th ACM Conference on Computer and Communications Security, pp. 25–32, (2000)
18. M. Schwarz, F. Lackner, D. Gruss, JavaScript template attacks: automatically inferring host information for targeted exploits, in Network and Distributed Systems Security Symposium 2019, (2019), pp. 1–15
19. G. Nakibly, G. Shelef, S. Yudilevich, Hardware Fingerprinting Using HTML5. arXiv:1503.01408v3 [cs.CR], 2015
20. K. Mowery, H. Shacham, Pixel perfect: fingerprinting canvas in HTML5, in web 2.0 security & privacy 2014. (2012)
21. J. Lukas, J. Fridrich, M. Goljan, Digital camera identification from sensor pattern noise. IEEE Trans. Forensics Secur. **1**(2), 205–214 (2006)
22. T. Filler, J. Fridrich, M. Goljan, Using sensor pattern for camera model identification, in 15th IEEE International Conference on Image Processing, (2008), pp. 1296–1299
23. M. Goljan, Digital camera identification from images-estimating false acceptance probability. LNCS 5450, (Springer, Berlin, Heidelberg, 2009), pp. 454–468
24. M. Goljan, J. Fridrich, Camera identification from cropped and scaled images, in Electronic Imaging, Security, Forensics, Steganography, and Watermarking of Multimedia Contents X, SPIE, 68190E. (2008)
25. M. Goljan, M. Chen, P. Comesana, J. Fridrich, Effect of compression on sensor-fingerprint based camera identification, in Electronic Imaging Media Watermarking, Security, and Forensics, (2016), pp. 1–10
26. Z.J. Geradts, J. Bijhold, M. Kieft, K. Kurosawa, K. Kuroki, N. Saitoh, Methods for identification of images acquired with Digital cameras. Enabling Technol. Law Enforcement Secur. **4232**, 505–512 (2001)
27. L. Cuccovillo, S. Mann, M. Tagliasacchi, P. Aichroth, Audio tampering detection via microphone classification, in IEEE 15th International Workshop on Multimedia Signal Processing, (2013), pp. 177–182
28. S. Englehardt, A. Narayanan, Online tracking: a 1-million-site measurement and analysis, in ACM SIGSAC Conference on Computer and Communications Security, (2016), pp. 1388–1401
29. B. Smus, *Web Audio API* (O'Reilly Media, Inc., Publisher, 2013)

30. A. Narayanan, D. Reisman, The princeton web transparency and accountability project. Transparent data mining for big and small data, studies in big Data, vol 32, (Springer, Cham. 2017), pp. 45–67
31. Z. Yang, R. Zhao, C. Yue, effective mobile web user fingerprinting via motion sensors, in 17th IEEE international conference on trust, security and privacy in computing and communications/ 12th IEEE international conference on big data science and engineering, IEEE, (2018), pp. 1398–1405
32. Q. Xiao, M.K. Reiter, Y. Zhang, Personalized pseudonyms for servers in the cloud. Proc. Priv. Enhancing Technol. **2017**(4), 191–209 (2017)
33. X. Pan, Z. Ling, A. Pingley, W. Yu, K. Ren, N. Zhang, X. Fu, Password extraction via reconstructed wireless mouse trajectory. IEEE Trans. Dependable Secure Comput. **13**(4), 461–473 (2016)
34. N. Zheng, A. Paloski, H. Wang, An efficient user verification system via mouse movements, in ACM conference on Computer and communications security, 2011
35. T.F. Yen, X. Huang, F. Monrose, M.K. Reiter, Browser fingerprinting from coarse traffic summaries: techniques and implications. LNCS 5587, (Springer, 2009), pp. 157–175
36. M. Dietz, A. Czeskis, D. Balfanz, D.S. Wallach, Origin-bound certificates: a fresh approach to strong client authentication for the web, in 21st USENIX Conference on Security Symposium, (2012), pp. 317–332
37. M. Dietz, A. Czeskis, D. Balfanz, D.S. Wallach, Origin-bound certificates: a fresh approach to strong client authentication for the web, in 21st USENIX Security Symposium, (2012), pp. 317–331
38. C. Sy Burkert, H. Federrath, M. Fischer, Tracking users across the web via TLS session resumption, in Annual Computer Security Applications Conference (2018)
39. B.N. Silva, M. Khan, K. Han, Internet of things: a comprehensive review of enabling technologies, architecture, and challenges. IETE Tech. Rev. **35**(2), (2017)
40. S. Khattak, D. Fifield, S. Afroz, M. Javed, S. Sundaresan, V. Paxson, S.J. Murdoch, D. McCoy, Do You See What I See? differential treatment of anonymous users, in Network and Distributed System Security Symposium, (2016), pp. 1–15
41. M.S. Grebenyuk, I.V. Styopochkina, Attacks on technologies of ensuring privacy on the Internet, in XVI All-Ukrainian Scientific and Practical Conference on Theoretical and Applied Problems of Physics, Mathematics and Informatics, (2018), pp. 88–91
42. B. Eriksson, A. Sabelfeld, AutoNav: evaluation and automatization of web navigation policies, in The Web Conference 2020, (2020), pp. 1320–1331
43. A. Huang, C. Zhu, D. Wu, Y. Xie, X. Luo, An adaptive method for cross-platform browser history sniffing, in Measurements, Attacks, and Defenses for the Web Workshop 2020 (San Diego, 2020), pp. 1–7
44. A. Klink, Fingerprinting Firefox users with cached intermediate CA certificates (#fiprinca), https://shiftordie.de/blog/2017/02/21/fingerprinting-firefox-users-with-cached-intermediate-ca-certificates-fiprinca/, last accessed 2020/09/07
45. K. Boda, A.M. Foldes, G.G. Gulyas, S. Imre, User tracking on the web via cross-browser fingerprinting, in information security technology for applications (Springer Berlin Heidelberg, 2012), pp. 31–46
46. T. Saito, K. Takahashi, K. Yasuda, K. Tanabe, M. Taneoka, R. Hosoya, Tor fingerprinting: tor browser can mitigate browser fingerprinting?. LNDECT 7, (Springer, 2017), pp. 504–517
47. A. Brumley, D. Boneh, Remote timing attacks are practical, in 12th USENIX Security Symposium, (2003), pp. 1–13
48. V.K. Khanna, Remote fingerprinting of mobile phones. IEEE Wireless Commun. **22**(6), 106–113 (2015)
49. K. Irie, A.E. McKinnon, K. Unsworth, I.M. Woodhead, A technique for evaluation of CCD video-camera noise. IEEE Trans. Circuits Syst. Video Technol. **18**(2), 280–284 (2008)

50. A. Parsovs, Practical issues with TLS client certificate authentication, in 2014 Network and Distributed System Security Symposium, (2014), pp. 1–13
51. X. Lin, C.-T. Li, Preprocessing reference sensor pattern noise via spectrum equalization. IEEE Trans. Inf. Forensics Secur. **11**(1), (2016)

Air Quality Index Prediction Using Azure IoT & Machine Learning for Smart Cities

Shaon Hossain Sani, Md. Shopon, and Sakhawat Hossen Rakib

Abstract In this paper, we are concerned about the air pollution & Air Quality Index (AQI) level. Today, every major city around the world is facing a worsened situation about public health & environmental disturbance. So, we have proposed an IoT-based solution that evolved with cloud computing platform likes Microsoft Azure, Azure IoT Hub and Azure machine learning (AML). These enhanced the scalability and life cycle of the system. The system classifies the health effects of human life from the air quality index level and gives interactive real-time data visualization results and push notification through the system. We have worked on Dhaka's air quality dataset. Among major criteria pollutants, we have focused on $PM_{2.5}$, PM_{10}, CO, SO_2, NO_2 that can help us to classify the air quality index level. Multiple classification algorithm is used to train the dataset. Among them, we have acquired the highest accuracy 99.89% with multiclass decision forest.

Keywords Air quality index · Azure machine learning · Multiclass classification · IoT Hub · Cloud computing

1 Introduction

For every living being, natural resources like air, water is essential to sustain their existence. Around 70% of the earth's surface is covered with water. The Earth atmosphere is covered with air, and air contains many gases and dust particles. All living

S. H. Sani · Md. Shopon (✉)
Department of Computer Science and Engineering, University of Asia Pacific, Dhaka, Bangladesh
e-mail: shopon.uap@gmail.com

S. H. Sani
e-mail: shaon.sani.ss1@gmail.com

S. H. Rakib
Department of Electrical & Electronic Engineering, University of Asia Pacific, Dhaka, Bangladesh
e-mail: rakib.h.eee@uap-bd.edu

V. E. Balas et al. (eds.), *Proceedings of International Conference on Computational Intelligence, Data Science and Cloud Computing*, Lecture Notes on Data Engineering and Communications Technologies 62, https://doi.org/10.1007/978-981-33-4968-1_56

being including animals, plants depend on air. Nowadays, the sustainable quality of air and water has been decreasing dramatically. Every developing country has lots of demand to construct mills, factories, industries and residential household for their people. So, the industrialization in the urban and semi-urban area has been affecting our earth's atmosphere most. Environmental contamination has occurred through industrial pollution. They have to process various materials and emits not only numerous gases like CO_2, CO, SO_2, nitrogen gas but also they disperse solid particles($PM_{2.5}$, PM_{10}). For instance, cities like Dhaka and Delhi are considered the most polluted city among other polluted cities. Every smart city uses different types of electronic devices attached to sensors to collect data to solve different problems. In order to the lack of proper city planning, people have suffered a lot from environmental pollution, especially air pollution. The people of those cities are being affected by serious health trouble. Moreover, people are not aware of the AQI and its veritable effects on their health [1]. So in this situation, an Internet of things(IoT) [2] would bring a solution to the problem which involves not only monitoring the real-time air quality but also acquainted about long- and short-terms effects on the human body and given suggestions to take necessary step based on AQI Level. The IoT in smart cities has a great impact on technology advancements. Consequently, the smart city's aim is to provide a smart solution with the help of public networks and the latest technology. So, in this paper, we contribute to making an IoT-based air quality monitoring system that deploys on the Azure cloud server, together with the Raspberry Pi, the whole system has been designed and constructed. Specific sensor nodes have been set up with Raspberry Pi to collect SO_2, NO_2, CO, $PM_{2.5}$, PM_{10} value. In recent times, cloud computing is one of the popular platforms for storing data and performing data analysis without any user interaction. Among them, Amazon Web service (AWS), IBM, Microsoft Azure, Google Cloud platform are mentionable. But Microsoft has an IoT working platform named Azure IoT Hub where millions of devices are deployed and run simultaneously. Using the advancement of artificial intelligence, we have used the machine learning technique to classify the AQI Level. Besides, construct meaningful correlation between AQI level and adverse health effects on human respiratory system. Recently, the WHO reported that the main cause of respiratory health risks is outdoor air pollution. Ambient air pollution caused globally 4.2 million premature death. Children are extensively suffered from acute respiratory infections. The adults are mainly affected by lung cancer, chronic obstructive pulmonary disease, heart attack, stroke and heart disease. So, it is important to classify the air pollution level so that we can identify the risks of health issues. Moreover, along with the IoT system, an advanced machine learning classification technique would help to get rid of this problem. Azure ML Studio has an identical feature drag and drops functionality that conveys fewer efforts to build a classification model and after that deploy the predictive model as Web services. We have developed an ASP.NET Web app to deploy the model for real-time monitoring. Later, published the Web app to the Azure platform. In addition, the sensor data pass through Azure IoT Hub gateway and stored in Blob storage after that raw data sent to Web app as a Web service input parameter. Then, classify real-time AQI to generate necessary suggestion for current pollution level. Both end users and Gov't would be benefitted from the advancement of the system.

This paper, we have made the following contributions:

- Build an IoT device which can monitor pollutants and air quality index using Raspberry Pi.
- Build a high accuracy classification model that classifies the AQI level perfectly.
- Deploy the model in the cloud as Web service and interact with Azure IoT Hub and end devices.

The rest of the paper is organized as follows. In Sect. 2, we discuss about background. We discuss about related works which are mentioned in Sect. 3. Then, in Sect. 4, we present the methodology of this paper. In Sect. 5, we present the result analysis. Finally, Sect. 6 concludes the works. This paper proposed a machine learning technique to classify the AQI level using Azure IoT Hub and Raspberry Pi.

2 Background

2.1 *Air Quality Index (AQI)*

The AQI is a term which reflects the daily air quality reporting. It indicates our atmospheric condition, how clean or polluted our air and associated with the health effects [3]. The AQI concerns mainly with health effects that occurred or experienced within short-term periods of time through inhaling the contaminated air. The AQI value from 0 to 50 indicates the air quality is good and no health effects after respiration of the air. The value from 51 to 100 exhibits that the air quality is moderate and sensitive people might be affected by inhaling the polluted air. AQI value 101 to 150 indicates the air is unhealthy for a sensitive group of people who are suffering from various respiratory diseases. The value ranges from 151 to 200 implied that we are living in an unhealthy atmosphere which affects almost everyone. Everyone might have faced immensely serious health issues from AQI value 201 to 300 which indicates a very unhealthy air quality. And lastly, from 301 to 500, the air is in hazardous level and emergency condition for the entire community of that area (Table 1).

2.2 *Azure IoT Hub*

Azure IoT Hub is a scalable cloud platform that communicates with IoT applications and IoT devices and works as a service (PaaS) cloud computing model. Azure IoT Hub conducting individual IoT device registry service includes IoT devices data storage platform and data security service to devices from the objectionable requests of unauthorized users. Azure IoT Hub supports bidirectional communications with

Table 1 Air quality index (AQI) level

AQI level	Level of health concern
0–50	Good
51–100	Moderate
101–150	Unhealthy for sensitive
151–200	Unhealthy
201–300	Very unhealthy
301–500	Hazardous

the end devices and the cloud. Azure IoT Hub would simultaneously working with Azure ML Studio and Power BI tools that can be used for data visualization.

2.3 *Microsoft Azure ML Studio Classic*

Microsoft Azure Machine Learning Studio is a cloud computing platform, simplifying traditional machine learning model construction procedure where predictive analysis, data science concept, cloud resources merged. Azure ML Studio is a drag and drop tool where anyone can design a machine learning solution effortlessly. Building a model, testing model, evaluating and scoring a model can be done efficiently. After scoring several models, the best-valued model can be deployed as Web services (Fig. 1).

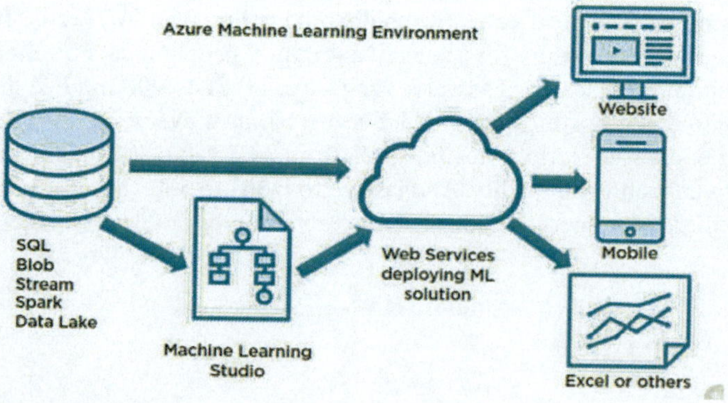

Fig. 1 Azure ML Studio classic

2.4 Raspberry Pi

We have used Raspberry Pi 3(model B) which is a single-board computer. It holds Cortex-A53 (ARMv8) and 1 GB of RAM along with 1.4 GHz. It has built-2.4 GHz and 5 GHz IEEE 802.11 wireless LAN. Pi 3 has also an Extended 40 GPIO header pinout. One HDMI port, four USB 2.0 ports, one CSI camera port, Micro SD slot for inserting External MicroSD card as a storage. For connecting PI touch screen display, it has DSI display port also. It works with 5 volt and 2.5amp power supply.

2.5 Sensor Nodes

To monitoring the air quality level, the sensor nodes make up with five sensors. The sensor nodes are functionally used for various air pollutants measuring purposes. For $PM_{2.5}$ and PM_{10}, we have used Sensirion Particulate Matter Sensor SPS30 which is measuring particles range from 1.0 μm to 10.0 μm. SPS30 is a high-quality optical particle sensor which uses laser scattering. To obtaining particulate matter measurement SPS30 has ingenious contamination resistance terminology. For sensing SO_2, we have used a $4SO_2$-20 sensor. It has a nominal range of 0–20 ppm and a maximum range of up to 150 ppm. Along with SO_2 gas, it may also respond to various gases like hydrogen sulfide, nitric oxide, nitrogen dioxide, carbon monoxide. MQ-7 has been used for sensing carbon monoxide gas. It is a semiconductor-based sensor that has a long lifetime. Carbon monoxide has been detected when the temperature base is lower (voltage around 1.5 V). MiCS-4514 is used for measuring NO_2 gas. The MiCS-4514 has two completely distinct sensing components in a knot. It is a MOS sensor type including compact design. Detectable NO_2 gas measuring range is 0.05 – 10ppm.

2.6 Classification Algorithm

Classification is a machine learning technique, where a state is mapped to one of many labels. The decision forest model is a classification algorithm that works similarly to the ensemble learning method. The model performs by constructing multiple decision trees, and the most popular class would be voted as output class. Multiclass decision jungle is an extended version of the decision forest which is based on ensembling decision directed acyclic graphs (DAGs). Decision jungle is different from traditional decision trees that one consent path to every node, and decision jungle along with a DAG allows multiple paths from the root to distinct leaf node. Among various classification models, logistic regression is extensively used in machine learning. Logistic regression is not a conventional regression model rather than it is a probabilistic model that classifies different classes. Neural networks are a set of algorithms designed to recognize the pattern, mimic of the human brain. Neural networks work for both regression and classification problems. Multilayers neural networks classify complex patterns.

3 Related Works

Gore et al. [1] mainly concern classification of AQI level using Naive Bayes and Decision Tree J48 algorithm. They have got 91.99% accuracy with Naive Bayes and 86.663% with J48. Classifying air pollution level affects health in a various way. They categorized health effects on five terms very unhealthy, unhealthy, unhealthy for sensitive, moderate and good. The author would not mention anything about how to get rid of health risks by applying or taking any actions with the help of advanced technology. Jo et al. [2] have proposed an Internet of things (IoT)-based solution running over Microsoft Azure Machine Learning Studio platform along with principal component analysis and artificial neural networks which provides R^2 value 0.6654 and RMSE value 0.2104, respectively. The author did not mention any simplified way to construct such infrastructure with low cost and one centered monitoring device. Jo-et al. [4] discussed an intelligent system that was proposed for underground coal mines station. With Arduino and gas sensors, they integrate the whole system with cloud for monitoring toxic gases and give necessary alerts from real-time data. They have worked with spatiotemporal data analysis and probabilistic theory and got sensitivity 87.50% and positive predictive value 77.70%. Author mainly focused on monitoring the gases in coal mine rather analyze the data. Ivanova et al. [5] introduced an intelligent system with Raspberry Pi that monitors air quality in the different city of Bulgaria. They have used MLP regression model for predicting different air pollutants like ozone, nitrogen oxide, nitric oxide and PM_{10}. The Raspberry Pi collects temperature, humidity and barometric pressure from the environment and fits into as neural network model inputs and gets expected pollutant concentrations. Their model has R2 scored 0.65. Author bound only with static devices and local area networks. BARAN [6] conducted a study on air quality index prediction using ELM algorithm. He mainly worked with six important parameters that were influential to air pollution. He classified the AQI level from 1 to 6 by standard AQI Indexing. Ten fold cross-validation applied for getting better learning from the dataset which was taken from March, June, September and December of 2017. He got the highest accuracy of 66.67% with Hardlimit activation function. Biondi et al. [7] build a system that can detect air pollution, and using Raspberry Pi and Azure Cloud Framework, the whole system was integrated. With the help of the Microsoft Azure Logic App, the system can also alert through mail to the users. They have worked with few sensors to analyze the result. Gore et al. [8] used random forest and multiclass classifier to predict health risks from NO_2, O3, CO, and SO_2 air quality data. They have got 94.61% accuracy with the multiclass classifier algorithm and 93.46% accuracy with the help of random forest algorithm. The author did not give any solution to overcome such kind issues.

4 Methodology

4.1 Dataset

Data Source We have collected dataset from a project running named [9] Clean Air & Sustainable Environment (CASE) under the Ministry of Environment & Forest of Bangladesh. They have collected real-time continuous air quality data. We have worked with from 2016 to 2019 time period data.

Data Attributes The dataset comes from two stations of Dhaka city, which contain SO_2, NO_2, NO, NO_x, CO, $PM_{2.5}$, PM_{10}, O3, Solar Radiate, Barometric Pressure, Temperature, Humidity, Rain details. But, as we have worked for the air quality index level, we mainly concern with SO_2, NO_2, CO, $PM_{2.5}$, PM_{10}.

Data Preprocessing Data in the real world are seldom clean and lucid. Our dataset carried numerous missing data fields. So, we have applied various statistical techniques. In this regard, the ANN model helps significantly for filling up missing data fields. We split the date column into Year, Month, Week number, Day and add those columns into our dataset and train the ANN model. After training the ANN model with two hidden layers with 150 hidden nodes, we have successfully construct the model which fill the missing data. Finally, we predict the missing value with the trained model and filled those up.

4.2 System Architecture

Our proposed system Fig. 2 basically works in four steps. In the first step, all the sensors data will fetch by Raspberry Pi [10]. In the second step, establish a gateway between Raspberry Pi and Azure IoT Hub [11] in order to pass data from end device to cloud. Third step, from IoT Hub, data went into stream analytics, and there data are being processed for visualization through Power BI visualization apps. On the other hand, data are being stored in Azure Blob storage for further operations. Fourth step, data from the Blob storage go to Web app (ASP.NET) deployed through Azure ML Web service API as a Web service input parameter. A pre-trained classification model which was trained in Azure ML Studio platform with drag and drops provides the Web service API. Azure provides more flexibility to train any machine learning model without any hard coding. A model can be tuned with user requirements and analyze the model with powerful functionality. The model classifies real-time health effects from the current AQI level. End user can monitor real-time air quality level and its corresponding health effects via Web application and Power BI application that can help to take necessary steps in any alarming situation.

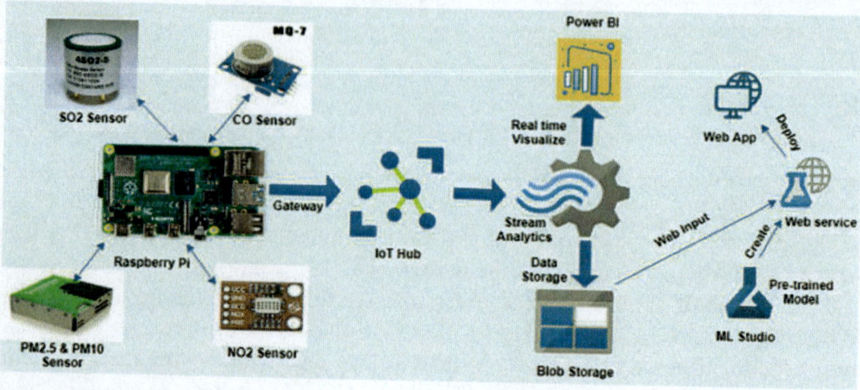

Fig. 2 System architecture

4.3 Model Design

The four models Fig. 3 were trained in the Azure ML Studio platform which is a cloud-based integrated development environment used for data science analysis. During the training, 75% data have been used for training the model, and 25% data have been allocated for testing purpose. In the first model, multiclass decision forest (MDF) has been tuned such as the number of decisions tree (16), and the maximum depth of decision tree's (32), the number of random splits parameter per node (128), the minimum number of samples per leaf node was tuned (1). For multiclass decision jungle (MDJ) has been tuned with the number of decision DAGs (16), the maximum depth of decision DAGs (32), the maximum width of decision DAGs (128) and the number of optimization steps per decision DAGs layer (2048). Multiclass logistics regression (MLR) has been tuned with L1 regularization weight (0.005), L2 regular-

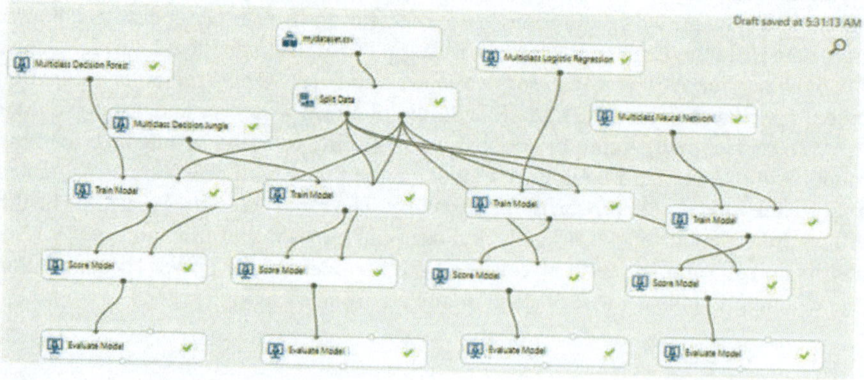

Fig. 3 Four classification models on AML studio

ization weight(0.001) and memory size for L-BFGS (15). Multiclass neural networks have been built with (150) hidden nodes, (0.001) learning rate, 1000 iterations, initial learning weights diameter (0.1) and Min-Max normalizer.

4.4 *Web Service and Web Application*

After evaluating Fig. 4 the four models, the best model has been chosen for the predictive model and deployed as a Web service Fig. 4a. In Azure ML Studio, we create a predictive model with selected best model and the predictive model Web service input and output nodes that take input when it deployed as a Web service. Azure ML Studio provides us a REST API key, through the key, we build a Web application Fig. 4b using ASP.NET. ASP.NET offers some flexibility to build the Web application. Azure ML Studio and ASP.NET are from Microsoft, so it is always preferred by Azure to build the Web app using ASP.NET rather than Django or other frameworks. ASP.NET gives us more comfort to work with the API and the whole application deployment process. And finally, the Web application which is deployed as a RESTFUL API works fine in Azure platform along with the Web application.

4.5 *Flowchart*

In Fig. 5 initially, the Raspberry Pi fetches the SO_2, NO_2, CO, PM_{10}, $PM_{2.5}$ raw data from particular sensor node with the time between an interval of every hour. After getting all data in a row, Raspberry Pi sends the raw data toward Azure IoT

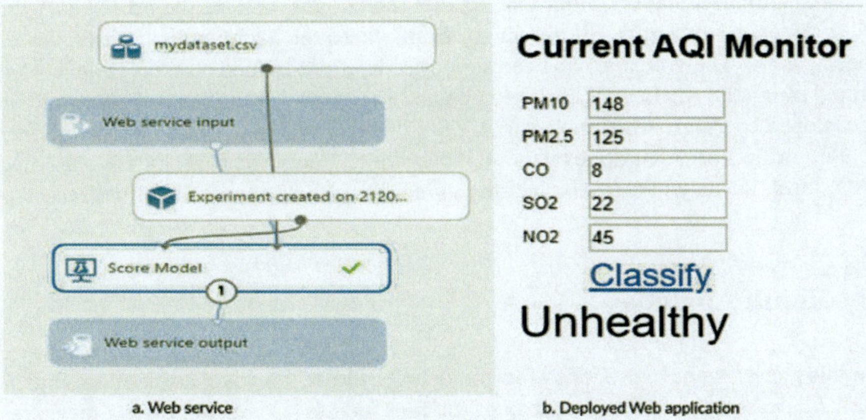

Fig. 4 Web service model & Deployed ASP.NET Web application

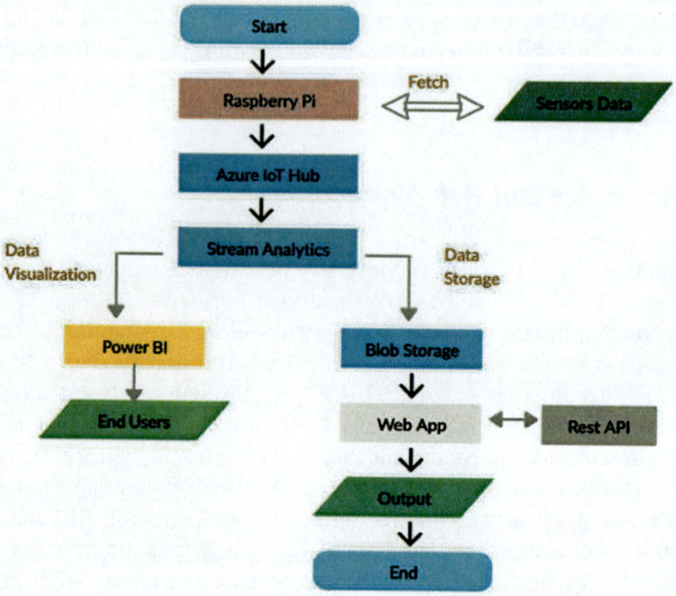

Fig. 5 Flowchart

Hub through MQTT protocol [12]. The process considers as the data transmission process. After that, in IoT Hub platform, an event processing engine named stream analytics processes the rest of the data. In order to visualize the data, it move forward the sensors data to an analytical service tool named Power BI where the data representation is done in an intelligent manner. Power BI has very attractive user interface where data were visualized iteratively and user can easily understand from the visual representation rather than textual representation. One the other hand, the data deposited in Azure Blob storage, from where the Web application input has been passed. The Web app has been supported by a pre-trained classification model, and through the REST API, the Web app accesses the Web service facility, which is developed by Azure Machine Learning Studio. The Web app gives us real-time air pollution alert, which is done by the AQI classifier model. And based on the classified AQI level, we may take necessary step to reduce the risk of many health diseases.

5 Result Analysis

In this paper, we have proposed four models for the AQI level classification. Figs. 6 and 7 have shown confusion matrix of the four models. Confusion matrix measures the performance of the machine learning model, where the diagonal line determines the true positive and true negative value of the classification model. In the confusion

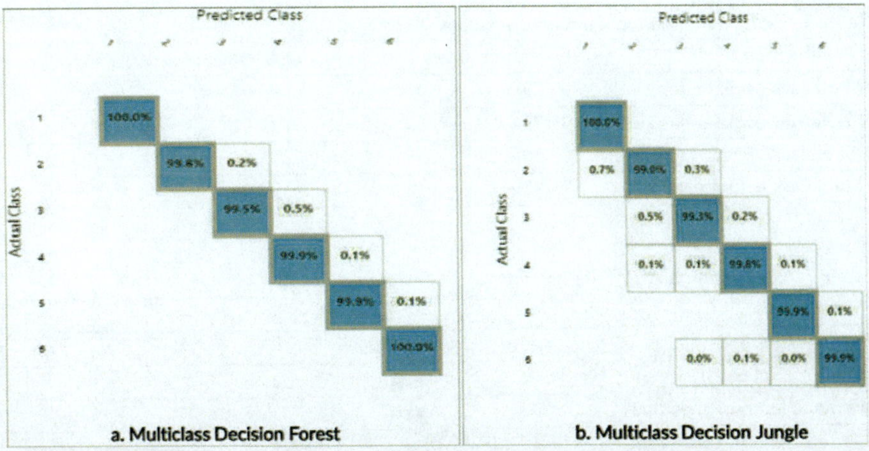

Fig. 6 Confusion matrix of multiclass decision forest and multiclass decision jungle

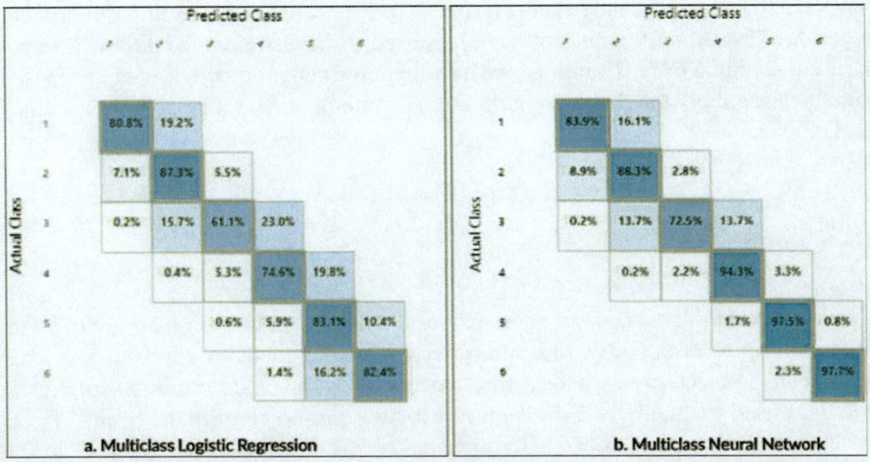

Fig. 7 Confusion matrix of multiclass logistic regression and multiclass neural network

matrix, class-1 labeled for good air quality index, class-2 indicates moderate AQI, class-3 labeled unhealthy for sensitive groups, class-4 labeled for unhealthy AQI level, class-5 labeled for very unhealthy situation and class-6 for hazardous air quality and extreme condition for people. From Table 2, multiclass decision forest (MDF) has achieved the highest performance on both overall accuracy and average accuracy. Though multiclass decision jungle and multiclass neural network are so close to the MDF model. In comparison with existing works, Table 3 showed our proposed multiclass decision forest has a higher accuracy of 99.89% than paper [1] which

Table 2 Different classification model accuracy

Classification model	Overall accuracy (%)	Average accuracy (%)
Multiclass decision forest	99.89	99.96
Multiclass decision jungle	99.73	99.91
Multiclass logistic regression	80.60	93.53
Multiclass neural networks	93.83	97.94

Table 3 Past result on AQI classification

Model	Accuracy (%)
Naive Bayes and	91.99
Multiclass classifier	94.61
Extreme learning machine	66.67

have 91.99% accuracy, and paper [8] have 94.61% accuracy with multiclass classifier algorithm. In addition, paper [6] have proposed ELM technique, and they have got an accuracy of 66.67%. Therefore, we have acquired better outcomes to our proposed classification model and able to build an IoT system which relies on both accuracy and efficiency.

6 Conclusion

The purpose of this work is to build a system that can get real-time air quality data along with predict the adverse health effects from the air quality index level. This system has connectivity with local area networks [13] and cloud computing platform. We have used the multiclass decision forest algorithm to classify the health effects with an accuracy of 99.89%. Microsoft Azure IoT Hub has been backed up the whole system with its advanced Cloud Hosted Solution and Applications. Real-time data visualization tools and air pollution monitoring notification would be given through Power BI and our deployed Web application. Our research work opens a new dimension to integrate a system with powerful cloud architecture and alteration of heavy physical air pollution monitoring infrastructure. Our future work includes setting up the device in the areas where the pollution levels are alarming with Govt Collaboration. With the help of this device, the Government can take necessary steps to reduce the air contamination in the critical areas.

References

1. R.W. Gore, D.S. Deshpande, An approach for classification of health risks based on air quality levels, in *1st International Conference on Intelligent Systems and Information Management (ICISIM)* (2017). https://doi.org/10.1109/ICISIM.2017.8122148
2. B.W. Jo, R. Muhammad, An internet of things system for underground mine air quality pollutant prediction based on Azure. Mach. Learn. **18**(4), 930 (2018). https://doi.org/10.3390/s18040930
3. S. Lemeš, Air Quality Index (AQI)—comparative study and assesment of an appropriate model For B&H, in *2th Scientific/Research Symposium with International Participation 'Metallic And Nonmetallic Materials'* (2018)
4. B.W. Jo, R.M.A. Khan, Arduino-based intelligent gases monitoring and information sharing Internet-of-Things system for underground coal mines. J. Ambient Intell. Smart Environ. **11**(2), 183–194 (2019). https://doi.org/10.3233/AIS-190518
5. D. Ivanova, A. Elenkov, Intelligent system for air quality monitoring assessment using the Raspberry Pi Platform, in *IEEE International Conference on Information Technologies* (2019). https://doi.org/10.1109/InfoTech.2019.8860883
6. B. Baran, Prediction of air quality index by extreme learning machines, in *International Conference on Artificial Intelligence and Data Processing* (2019). https://doi.org/10.1109/IDAP.2019.8875910
7. K. Biondi, E. Al-Masri, Air pollution detection system using edge computing. Sensors **18**(9), 3021 (2018). https://doi.org/10.3390/s18093021
8. R. Gore, D. Deshpande, Air data analysis for predicting health risks. IJCSN - Int. J. Comput. Sci. Network **7**(1), 36–39 (2018)
9. CASE Project Homepage. http://case.doe.gov.bd/
10. S. Kumar, A. Jasuja, Air quality monitoring system based on IoT using Raspberry Pi, in *International Conference on Computing, Communication and Automation* (2017). https://doi.org/10.1109/CCAA.2017.8230005
11. S. Forsström, U. Jennehag, A performance and cost evaluation of combining OPC-UA and microsoft Azure IoT Hub into an industrial internet-of-things system, in *Global Internet of Things Summit* (2017). https://doi.org/10.1109/GIOTS.2017.8016265
12. W. Choi, D. Hwang, Fine dust monitoring system based on Internet of Things, in *International Conference on Information and Communication Technology Robotics (ICT-ROBOT)* (2018). https://doi.org/10.1109/ICT-ROBOT.2018.8549878
13. S.J. Johnston, P.J. Basford, City scale particulate matter monitoring using LoRaWAN based air quality IoT devices. Sensors **19**(1), 209 (2019). https://doi.org/10.3390/s19010209

A Review of Risks and Threats on IoT Layers

Subir Panja, Arup Kumar Chattopadhyay, and Amitava Nag

Abstract In the present scenario, the significance of the Internet of Things (IoT) in different application domains is not easy to avoid. Human beings are receiving services very effectively and efficiently from the IoT in smart decision-making through machine-to-machine (M2M) communication from different locations. The IoT field sensors play an essential function in sensing the smart object/things and making an impact decision. The IoT applications are very much demanding in all the sectors. But until now, all the privacy and security attributes for IoT applications are not adequately implemented. Due to the use of unsecured applications, users may face different types of difficulties and can even face life-threatening situations. Thus, implementation of privacy and security for IoT applications is one of the major concerns of the researchers. In this paper, we discuss about the risks and the threats of IoT. There are different security techniques in the TCP/IP network model, but they are not perfectly suitable in IoT architecture. Therefore, to explain this, we provide a layer-wise comparison of the different technologies or protocols used in the TCP/IP network model and IoT architecture. Here, we emphasize that the security solution based on Internet protocol is not sufficient for IoT enabled applications. Here, we will also discuss different threats and attacks at different layers of IoT stack.

Keywords Internet of things · IoT security · TCP/IP · Risks · Threats

S. Panja (✉) · A. Nag
Central Institute of Technology Kokrajhar, Assam, India
e-mail: panja.subir@gmail.com

A. Nag
e-mail: amitavanag.09@gmail.com

A. K. Chattopadhyay
Institute of Engineering and Management, Kolkata, WB, India
e-mail: ardent.arup@gmail.com

V. E. Balas et al. (eds.), *Proceedings of International Conference on Computational Intelligence, Data Science and Cloud Computing*, Lecture Notes on Data Engineering and Communications Technologies 62, https://doi.org/10.1007/978-981-33-4968-1_57

1 Introduction

"Internet of Things (IoT)" acts as an umbrella that covers a variety of features such as the expansion of internet usage for extensive embedded scattered devices, transfer, and the actuation abilities [1], etc. The fast progress of the variety of digital devices, starting with simple sensors to extremely difficult cloud servers, form the Internet of Things, including smart locks, smart bulbs, IP cameras, different electronic appliances, alarm clocks, and many more. The ability to communicate over the Internet is one of the main objectives of all the IoT devices. It is not possible to remote access without the internetwork connectivity of objects in the IoT environment. These IoT objects are smart by using different essential technologies such as ubiquitous computing, pervasive computing, data communication, internet protocols, etc. As a result, we are getting more absorption of applications for IoT. As defined by Derhamy et al. [2], the Internet of Things (IoT) is the network connected with real things that have implanted technology to interconnect and sense and act in an organized way in their inner states or the exterior environment. The connected things to the Internet are expected to cross more than 20 billion very shortly. The Internet of Things (IoT) is suitable for the largest established application domains. With the introduction of IoT technologies, our technology dependency index is showing a drastic upward shift.

The faultless study of the accumulated data provides the capability to create an enhanced prospect for future generations and finally make the perfect decision. IoT is a network that has many physical objects such as sensors, attractors, smart devices, etc. IoT is now considered as an extended version of the wireless sensor networks (WSN), machine-to- machine (M2M), or cyber-physical systems (CPS). Thus, security is a major challenge for IoT deployment. It is also necessary to protect the privacy of the architecture and maintain the integrity and confidentiality of data present in the public network. Hence, security is the major issue related to the IoT network, which must be improved. Many challenges are there to secure IoT devices and ensuring node-to-node security within connected devices of IoT. In recent years, the researchers' outstanding effort has been noticed in the area of security and privacy for IoT.

2 Application Domains

IoT is massively utilized in the every aspects of our daily life. As presented in [3], IoT will transform our daily lives and technology. Many companies are trying to capture this almost open market by an impressive way to give surprises us by developing different IoT devices and applications. Major domains identified are: smart agriculture, smart home, smart healthcare, smart grid, etc. The market share of IoT distributed across these domains are as displayed in Fig. 1.

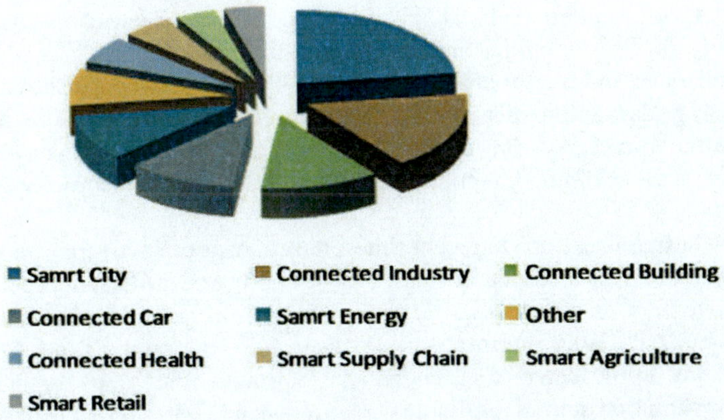

Fig. 1 Global share of IoT project [3]

Smart Agriculture: As described by Rossi et al. [4] in the present scenario, food production and agriculture are expected to grow. So, to improve agriculture production, it requires the growing usage of the latest technologies of computer science and electronics. That's why, a decision support system was developed to help farmers deciding the field of agriculture. Depending on the weather and type of soil, the IoT system suggests the farmers, which crops and vegetables can be produced more in a similar environment. If any fertilizers are needed to be added to the soil, that will also be suggested by IoT system depending on previous data stored. So, farmers will be benefitted more as they will also avoid losing time, money, and effort.

Smart Home: Smart home is becoming very well known to our modern civilization. In the Smart home, as explained by Jacobsson and Davidsson, [5] the idea has been included with embedded of brain-power to daily usable objects, like as home appliances, door locks, closed- circuit cameras, the door of the garage, locks used on doors, water motor pump, toaster, chapatti making machine, and so on. These objects are accessible from a remote area using a mobile phone, smartphone, tablet, or laptop using the Internet or private network. This technological facility added more assistance to improved human social lives, such as improved convenience, robust security, and efficient use of natural assets. For example, we can easily save energy, which is very much essential for our environment, using IoT applications in our home.

Smart Health Care: Smart health care is essential nowadays. It increases the efficiency level of the health care system as well as reduces the cost of healthcare [6]. At present, in medical setup, with the help of a smart healthcare system, healthcare experts can easily observe patients from a distance using the Internet or a private network. They can also collect data from the attached devices and decide who needs more attention. So, without being physically present, depending on collecting medical data from different sensors, senior doctors can instruct the junior medical experts as needed. Healthcare professionals can take the necessary steps before the situation going out of control.

Smart Grid: Mahmood et al. [7] have defined it as an electrical framework that consists of different functional and energy measuring devices, such as smart meter, smart appliances, and energy-efficient resources. The main concept in the background of the smart grids is automatically collecting data and estimating the behavior of electricity customers and providing efficient electricity use. So, depending on the requirement in an area, optimum electricity will be supplied by the electricity generation house.

Smart Transportation: In recent times, the concept of smart transportation has become popular. It is a branch of smart city infrastructure. Different problems such as car movement blocking, road safety, automatic toll collection facilities can be provided by IoT. Using optimized route suggestion, the public can be benefitted by estimating traffic congestion, suggest an optimum route, and for that reason, energy consumption, and car emission can be reduced [8]. It is possible to suggest an optimum route from the sensors which are attached to the vehicles or mobile devices and other different devices. Using this, we can be get benefited by getting parking reservations, preventing accidents, and drive autonomously.

Smart City: Nowadays, citizens can take the facilities of the smart city by using their smartphones and mobile devices, as well as from devices connected in cars and homes. The different areas within the smart city are intelligent. They can improve energy distribution, collect garbage, minimize traffic jams, improve air quality, and enhance security (based on surveillance, biometry, etc.). A smart city must have the following properties: smart energy (usage- based on smart meters, smart grid, intelligent energy storage, etc.), smart building (based on automation, smart lighting equipment), and even smart government and smart education based on IoT applications. Alaba et al. [9] presented different security issues for different application areas of IoT.

In Table 1 we compared these IoT applications with respect to many important parameters such as network size, bandwidth requirement, security requirements, and protocols.

3 Architecture of Iot

End-to-End IoT Architecture (Horizontal Direction)

A whole IoT system includes four different major components. These are (i) sensors or devices, (ii) IoT gateway or connectivity, iii) cloud server or data processing and iv) end-user interface as shown in Fig. 2. A brief description is given for basic functionalities of those components.

Table 1 Requirement of different security properties in IoT application

Characteristics	Smart home/office	Smart city	Smart agriculture	Smart grid	Smart transportation	Smart healthcare
Network size	Small	Medium	Medium/Large	Large	Large	Small
Data management	Local server	Shared server/cloud server	Local server and cloud server	Cloud server	Shared server/cloud server	Local server and cloud server
Bandwidth requirement	Small	Large	Medium	Medium/large	Medium/large	Medium/large
Security Requirement	A, B, D, E, F, G, H, I, J	A, B, C, D, E, F, G, H, I, J	A, B, C, D, E, F, H, I, J	A, B, C, D, E, F, G, H, I, J	A, B, C, D, E, F, G, H, I, J	A, B, C, D, E, F, G, H, I, J
Technology and protocols used	WiFi, 6LoW-PAN, Bluetooth LE(BLE), ZigBee, Z—Wave	IEEE 802.15.4, 6LoWP AN, ZigBee, MANET, UAANET, WiFi, Ethernet	IEEE 802.15.4, 6LoWPAN, ZigBee, MANET, UAANET, MAVLink	IEEE 2030, IEEE 802.15.4, 6LoWPAN, IES 62056, ZigBee, Ethernet	IEEE 802.15.4, IEEE 1609, 6LoWPAN, Bluetooth LE (BLE) ZigBee, Ethernet, VANET, UAANET	IEEE 802.15.4, IEEE 802.15.6, 6LoWPAN, Bluetooth, ZigBee, Ethernet

(where A: accountability, B: authenticity, C: authorization, D: availability, E: confidentiality, F: integrity, G: non-repudiation, H: physical security, I: privacy, J:reliability)

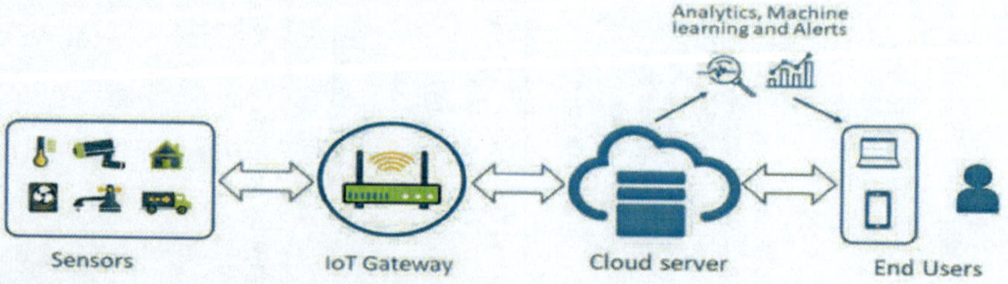

Fig. 2 IoT Architecture in Horizontal direction

Sensors/Devices

At first, it automatically senses and collects data from the environment using these sensors depending on the purpose of the IoT devices. Temperature sensor, proximity sensors, pressure sensors, smoke sensors are the list of commonly used sensors in different IoT applications. These sensors make smart device from an ordinary device. It may be embedded with a device or can standalone. This could be very simple, like a temperature reading for smart home applications or analyzing a real-time video feed to control the traffic in Smart cities.

IoT Gateway/Connectivity

The sensors or devices can be connected to the cloud in different ways, such as satellite, low—power wide-area networks (LPWAN), Bluetooth, WiFi, or directly connected to the Internet. A gateway placed in between sensors and devices to create bridge with the cloud. Gateways are the responsible to minimize data transmission, data processing, and requirements of data storage using sort out the data which is generated by sensors/devices.

Cloud Server/Data Processing

Once data is accepted and gathered into the cloud, the software performs some kind of processing on it. Mainly, data analysis, comparison with previously stored standard data sets, and machine learning activities are done during this process and depend on these process outcomes; an alert can be sent, if needed, to the decision-maker.

A simple example is to check the temperature within a suitable range. But what necessary action should be taken if the temperature is too high? In this case, user interaction is needed.

End-user Interface

At last, the information is crucial to the end-user. The information can be an alert to the user's interacting devices or as any kind of notification. Also, a user has an interface dashboard that allows them to access the system, such as the user may want to check the CCTV footage of a particular camera of the house with the help of a web browser or phone app. However, this is not permanently a one-way communication.

Depending on the IoT application, the user may take action on the system. For example, the user may adjust the temperature of their room by an app.

Layer-wise Architecture (Vertical Direction)

The layers of IoT architecture are briefed as follows:

Perception Layer

This layer consists of intelligence implanted technology and sensor technology. Recognizing single objects is the main objective of this layer. This layer is accountable for collecting information from the real world. Initial process of data and transmitting is also done in this layer. The physical IoT devices are fitted into this layer. The fitted IoT devices sense the analog data from the real world. For the purpose of communication, this analog data pass through the digitization process. Different sensors [10] are used in this layer, such as RFID, sensor gateways, GPS receivers, and sensor nodes.

Network Layer

It contains wireless sensor networks and communication channels of a different network from perception layers. Network layer works as linker in between perception and application layer. This layer's main task is to collect information from different variety of sensors/devices and send them to an information processing system (IPS). The network layer provides information security, network transmission, and delivers an access environment to the perception layer. Mobile devices, cloud computing, and the Internet, WiFi, Bluetooth, LTE, ZigBee, Z-Wave, etc., are the main components in this network layer [11].

Application Layer

In the application layer, the realistic and appropriate applications are expanded based on the need of user requirements or scope of the industry. User access the IoT applications via smart phone and computer or tablets, using the interface of application layer. Different applications such as smart home, smart traffic, smart agriculture, smart transportation, etc., are represented in this layer. Based on the offered services, this layer can be present in different ways. This layer has the ability to be seen by the user. This layer works as the interface for end-users. Main different protocols that are used

- Hypertext transfer protocol (HTTP)
- Constrained application protocol (CoAP) [12],
- Message queue telemetry transport (MQTT) [13],
- Extensible messaging and presence protocol (XMPP) [14],
- Advanced messaging queuing protocol (AMQP) [15],
- Data distribution service (DDS) [16].

4 Threats and Attacks

Anything can be a threat that may interrupt the process of functioning, reliability, or accessibility of a system or network. In security of computer, threat is representing probable negative activity that outcomes in an undesired impression to application of computer or compute system. A threat may be diverse intensity of negative effect on IoT applications. A high intensity of threat probably damages any asset of IoT application. A mid level threat intensity may damage partially of IoT application and in low threat intensity may disclose the information or just eavesdrop.

In contrast, attacks stand for illegal action against different types of concern such as government concern, IT firm, private or public concern. Here, actions can be many destructive procedures such as stealing the data, updating the data, or destroying the data.

Numerous scholarly and technological challenges require improvement and implementation of different IoT applications. Any device connected to a public network (Internet) is always vulnerable because the data collected by IoT devices always catch the attackers' attention. We present a layer-wise comparison between TCP/IP and IoT stacks in Fig. 3. The most important security challenges are segregated layer-wise in IoT architecture and are shown in Fig. 4.

Eavesdropping: Eavesdropping is listening to a conversation secretly. It is a very old security problem. In this kind of attack, the attacker passively listens to network communications to achieve access to secret information, such as sensitive data of the application, the identification numbers of the node, or updated information about routing [17]. The attacker can use this secret information to understand nodes in the network and disrupt routing or degrade the application's performance. The attacker may obtain secret data if the data packets are either not strongly secure or not secured at all.

Node Capture: On a physical layer of wireless sensor networks, when an attacker can physically access the nodes and manage the nodes or the whole network, this type of attacks are the most familiar and essential category of threats that take place

IoT Layer	Protocol / Technology used	TCP/IP Layer	Protocol / Technology used
Application layer	CoAP, MQTT, XMPP, AMQP, DDS etc.	Application layer	HTTP, SMTP, SNMP, FTP, TELNET, DNS etc.
Network layer	IPv6, WiFi, NFC, Bluetooth, Zigbee, Z-Wave, LTE etc.	Transport layer	TCP, UDP
		Network layer	IPv4
Perception layer / Physical layer	Sensor, RFID tag, Camera, WSN etc.	Physical layer	Ethernet, Token Ring, Other Link Layer

Fig. 3 Layer-wise comparison between TCP/IP and IoT stacks

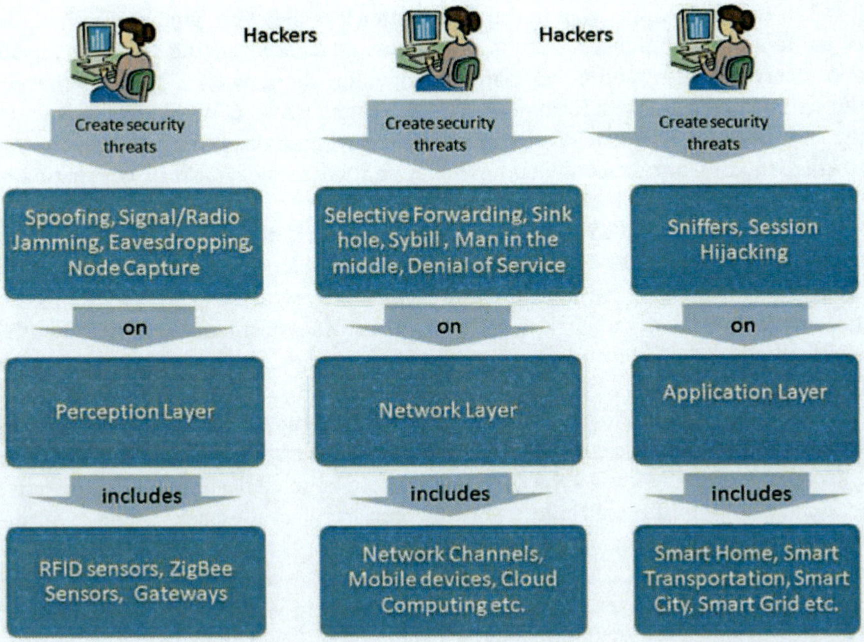

Fig. 4 Attackers activity on IoT layers

and is identified as node capture [18, 19]. In this way of attack, the attacker directly takes control of an intention node. Using this node, the attacker can pull out the encryption key or get the expected received information from another sender node. A captured node can be removed by an attacker or can transform this attacked node to a malicious node.

Spoofing: At first, the attacker act as a listener does not convey any signal. When the valid node discontinue to sending signal to the legitimate receiver, spoofer acts as a sender for sending the untrustworthy signal [20]. During this type of attacks, the security issues of IoT devices have detected a systems problem. Generally, for generating an incorrect message, the routing loop has been created by an attacker.

Selective forwarding: Selective forwarding attack is the harmful attacks which can affect the whole network communication of IoT devices. In this kind of attack, malicious nodes may decline to pass the messages and just drop them. They try to establish that those messages are not transmitted anymore. So, affected nodes will determine that the message has failed to transmit and to resolve try to find any other available route.

Sybil Attack: In a wireless network, to degrade the IoT functionality, Sybil attacks are initiated by malicious Sybil nodes. These Sybil nodes are using false identities. On the physical layer, a Sybil node may use random copied physical network addresses to act as a different device while targeting at users of network resources. As an effect, the authentic nodes may be not accessible.

Man in the Middle: Man in the middle attack is the cyber attack, which is one of the familiar and oldest forms. In this attack, attacker sitting between the sender and receiver parties, and either notice or operates the network [21] This may be through involvement with legal networks or creating a fake network, which is under the control of the attacker. The attacked network is then disclosed of all encryption to keep, modify, or change the communication route to the attacker's destination of choice.

Denial of Service: DOS attacks attempts to prevent authentic users from accessing some eligible services. A denial-of-service (DoS) attack occurs when a system maliciously gets overflowed with traffic or information by which attackers try to make a system crash or creates a block to authentic users. The most familiar way is a buffer overflow attack, which sends more traffic to a network address than it can handle [21].

Sink Hole: The sinkhole attack is referred to as the most destructive attack among all wireless routing attacks [22, 23]. In this type of attack, the attacker declares to its adjacent nodes that it knows the shortest path to a targeted destination. Its objective is to attract the traffic load into a specific area to reject the packets and spoil the network communication.

Session Hijacking: During authentication and session management, session Hijacking attack discloses private identities by exploiting IoT security's weakness. This type of attack is very familiar and may have remarkable effects on the legitimate user. An attacker can do anything with the identity of an authentic user [24].

5 Security Feature

The principal goals of network security can be illustrated using major three fundamental security properties: confidentiality, integrity, and availability, generally termed as CIA triad. The above security properties have been broadly classified day by day, such as authenticity, authorization, non-repudiation, accountability, reliability, privacy, and physical security. Below, we provide brief descriptions of these security properties:

Authenticity. It protects from illegal access. Only authentic users are right to use the system or users information. It assures that the user actually communicate with the partner with whom user wants.

Authorization. A user or device has the permission to access resources, such as systems, programs, files, devices, etc. So, we can say authorized user/device have the authenticity to use the resources. Depends on permission level, authorized users have the types of accessibility such as read, write or anything else.

Confidentiality. Information should not be disclosed to any unauthorized entity. Confidentiality helps to protect the transmitted data between connected devices.

Integrity. Related information should not be corrupted, and only authorized users can update or access.

Availability. The targeted information or resources should be available in a perfect location in a proper manner.

Reliability. It means guaranteed information, as well as a guaranteed system that users can access.

Privacy. It refers to the data collected from the sensors or actuators of IoT from the user's daily life, which must not be disclosed to any others. This deals with the right of an individual to control personal information.

Non-repudiation. This makes the protection to deny who/where a message came from, also confirming the message authenticity.

Physical Security. It refers to prevent the physical damage of IoT devices.

Accountability. It relates to the traceability of any action of any IoT devices.

In the IoT ecosystem, a large number of heterogeneous devices/objects participate and cooperate to perform different tasks. In such a dense environment where devices irrationally discover other devices, trust management becomes a very important issue. Authors in [25] surveyed several trust management techniques for the IoT systems and also provided a comprehensive analysis of these trust management schemes. As discussed by Hassija et al. [26], edge computing in IoT ecosystem plays a very important role to overcome the following security threats:

1. **Data Breaches**: In fog computing, attackers may wait to take the benefit of data movement in between fog layer and devices. But as no transmission of data between different local networks or different devices occurs in edge computing technology, the risk of data breaches can be prevented using edge computing technology.
2. **Data Compliance Issues**: Using edge computing technology, it is possible to restrict the data movement within a particular geographical region.
3. **Safety Issues**: We know time is precious in all aspects. Similarly, in safety issues, time—delayed responses may conduct different safety issues.
4. **Bandwidth Issues**: Huge amount of data generated by IoT application may face security issues and require high bandwidth to send all those generated data to cloud. But using edge computing, only the summarized data and/or essentials are to be sent to the cloud, if required.

Software defined networking (SDN) and network function virtualization (NFV) are the potential augmentations of IoT ecosystem. Farris et al. [27] described that the strategy formed with SDN and NFV to monitor, protect, and react to IoT security threats could revolutionize the IoT cyber security system. Zeadally and Tsikerdekis [28] presented the potential of artificial intelligence and machine learning techniques in enhancing the security of IoT devices. Applications are based on different important parameters in Table 1. We know IoT devices are in danger of exposing user's private data. As per Lv et al. [29], to use permissioned Blockchain, this can secure devices of IoT. To protect data manipulation, we can also use Blockchain. In [30], Yu et al. proposed a solution in which systems will reject any kind of data for device updates. However, we can also use the fog layer in IoT. The fog layer others different security features for applications of IoT. But the data computation and transmission to the fog layer generate different vulnerabilities. Here, fog plays an important role in between cloud and end-users. Since all threats must be accepted through the fog.

It is clear that there is huge quantity of IoT devices deploying all over the world for making it smart and generating a bulk amount of data which is related to surroundings and users. A lot of confidential information may open to everyone, and it is one of the causes of threat. As a effect, most important improvements in the modern structure of IoT application are desired to make it trustworthy, secure, and vigorous. In this observation, we can say the following security enhancement is needed [30–33].

i. IoT security using Blockchain.
ii. IoT security using fog computing
iii. IoT security using machine learning and
iv. IoT security using edge computing.

6 Conclusion

The IoT has recently appeared as a significant topic of interest for researchers. IoT is a perfect blend of different technologies used to sense and communicate with other objects over the Internet without human interaction. In this paper, we have discussed the growth of IoT applications in different sectors. The layers of IoT network in both horizontal and vertical directions have been described. We have presented the challenges for development of secure IoT applications. We have compared between IoT protocol stack and TCP protocol stack, and corresponding technologies. We have also compared different security requirements and protocols available for different applications of IoT.

References

1. D. Miorandi, S. Sicari, F. De Pellegrini, I. Chlamtac, "Internet of things: vision, applications and research challenges." Ad hoc Networks. **10**(7), 1497–1516 (2012)
2. H. Derhamy, J. Eliasson, J. Delsing, P. Priller, "A survey of commercial frameworks for the internet of things," in *2015 IEEE 20th Conference on Emerging Technologies & Factory Automation (ETFA)*, IEEE, 2015, pp. 1–8
3. L. Columbus, "charts that will challenge your perspective of IoT's growth." Forbes. June 6 (10), 2018
4. V. Rossi, F. Salinari, Stefano Poni, Tito Caffi, Tiziano Bettati, Addressing the implementation problem in agricultural decision support systems: the example of vite. net®. Comput. Electron. Agric. **100**, 88–99 (2014)
5. A. Jacobsson, P. Davidsson, "Towards a model of privacy and security for smart homes, " in *2015 IEEE 2nd World Forum on Internet of Things (WF-IoT)*, IEEE, 2015, pp. 727–732
6. G. Leroy, H. Chen, T.C. Rindflesch, Smart and connected health [Guest editors' introduction]. IEEE Intell. Syst. **29**(3), 2–5 (2014)
7. K. Mahmood, S.A. Chaudhry, H. Naqvi, T. Shon, H.F. Ahmad, "A lightweight message authentication scheme for Smart Grid communications in power sector." Comput. Electr. Eng. **52**, 114–124 (2016)

8. A. Al-Dweik, R. Muresan, M. Mayhew, M. Lieberman, "IoT-based multifunctional scalable real-time enhanced road side unit for intelligent transportation systems," in *2017 IEEE 30th Canadian Conference on Electrical and Computer Engineering (CCECE)*, IEEE, 2017, pp. 1–6
9. F.A. Alaba, M. Othman, I.A.T. Hashem, F. Alotaibi, "Internet of Things security: a survey." J. Network Comput. Appl. **88**, 10–28, (2017)
10. H. Suo, J. Wan, C. Zou, J. Liu, "Security in the internet of things: a review," in *2012 International Conference on Computer Science and Electronics Engineering*, vol 3, IEEE, 2012, pp. 648–651
11. P. Pongle, G. Chavan, "A survey: attacks on RPL and 6LoWPAN in IoT," in *2015 International Conference on Pervasive Computing (ICPC)*, IEEE, 2015, pp. 1–6
12. Z. Shelby, K. Hartke, C. Bormann, "The constrained application protocol (CoAP)." (2014)
13. D. Locke, "Mq telemetry transport (mqtt) v3. 1 protocol specification." IBM developerWorks Technical Library 15 (2010)
14. P. Saint-Andre, "Extensible messaging and presence protocol (XMPP): Core." (2004)
15. S. Vinoski, Advanced message queuing protocol. IEEE Internet Comput. **10**(6), 87–89 (2006)
16. M. Ammar, G. Russello, B. Crispo, Internet of things: a survey on the security of IoT frameworks. J. Inf. Secur. Appl. **38**, 8–27 (2018)
17. M. Nawir, A. Amir, N. Yaakob, O.B. Lynn, "Internet of Things (IoT): taxonomy of security attacks," in *2016 3rd International Conference on Electronic Design (ICED)*, IEEE, 2016, pp. 321–326
18. S.F. Bahari, "Security of wireless sensor networks in the presence of captured nodes." Ph.D. diss., 2008
19. R. Bahrampour, R.E. Atani, Y. Deldjoo, "A survey of node capture attacks in wireless sensor networks"
20. M.H. Yılmaz, H. Arslan, "A survey: spoofing attacks in physical layer security," in *2015 IEEE 40th Local Computer Networks Conference Workshops (LCN Workshops)*, IEEE, 2015, pp. 812–817
21. Engin Leloglu, A review of security concerns in Internet of Things. J. of Comput. Commun. **5**(1), 121–136 (2016)
22. L. Wallgren, S. Raza, T. Voigt, Routing attacks and countermeasures in the RPL-based internet of things. Int. J. Distrib. Sens. Netw. **9**(8), 794326 (2013)
23. C. Karlof, D. Wagner, Secure routing in wireless sensor networks: Attacks and countermeasures. Ad Hoc Netw. **1**(2–3), 293–315 (2003)
24. F.A. Alaba, M. Othman, I.A.T. Hashem, F. Alotaibi, Internet of Things security: a survey. J. Network Comput. Appl. **88**, 10–28 (2017)
25. I.U. Din, M. Guizani, B.S. Kim, S. Hassan, M.K. Khan, Trust management techniques for the Internet of Things: a survey. IEEE Access **7**, 29763–29787 (2018)
26. V. Hassija, V. Chamola, V. Saxena, D. Jain, P. Goyal, B. Sikdar, A survey on IoT security: application areas, security threats, and solution architectures. IEEE Access **7**, 82721–82743 (2019)
27. Ivan Farris, Tarik Taleb, Yacine Khettab, Jaeseung Song, A survey on emerging SDN and NFV security mechanisms for IoT systems. IEEE Commun. Surv. Tutorials **21**(1), 812–837 (2018)
28. S. Zeadally, M. Tsikerdekis, Securing Internet of Things (IoT) with machine learning. Int. J. Commun Syst **33**(1), e4169 (2020)
29. P. Lv, L. Wang, H. Zhu, W. Deng, G. Lize, An IoT-oriented privacy-preserving publish/subscribe model over blockchains. IEEE Access **7**, 41309–41314 (2019)
30. Y. Yu, Y. Li, J. Tian, Jianwei Liu, Blockchain-based solutions to security and privacy issues in the Internet of Things. IEEE Wireless Commun. **25**(6), 12–18 (2018)
31. Y. Jararweh, S. Otoum, A.I. Ridhawi, "Trustworthy and sustainable smart city services at the edge." Sustain. Cities Soc., p. 102394, (2020)
32. W. Rafique, L. Qi, I. Yaqoob, M. Imran, R. ur Rasool, W. Dou, "Complementing IoT services through software defined networking and edge computing: a comprehensive survey." IEEE Commun. Surv. Tutorials (2020)
33. J. Kharel, H.T. Reda, S.Y. Shin, Fog computing-based smart health monitoring system deploying lora wireless communication. IETE Tech. Rev. **36**(1), 69–82 (2019)

IoT-Blockchain Integration: The Way Ahead

Prajna Paramita Ray, Bhargavi, Baisakhi Das, and Abhishek Das

Abstract IoT allows communication to take place globally between human to human, human to things, things to things. As data are shared among different participants in a distributed network, there are lots of security issues that are to be handled properly to secure the data. Blockchain technology which is an emerging technology can play a vital role to solve different security issues in IoT devices, the data they retrieve, produce, and process. Apart from security issues, the blockchain has different perspectives. Smart contracts can be created using the blockchain technology to make data immutable in a distributed peer-to-peer networking environment apart from cryptocurrencies like "BitCoin." This paper focuses on the study of the Internet of Things (IoT), different perspectives of blockchain technology, how integration of IoT with blockchain can achieve the trustworthiness of data in a distributed network, different distributed IoT applications are discussed which are implemented through blockchain methodology.

Keywords Internet of things · Blockchain · Security · Smart contract · IoT-blockchain integration

1 Introduction

Both of Internet of Things (IoT) and blockchain are the emerging concepts and technology nowadays that can create new possibilities by transforming the existing concepts [1]. Applications can be created by merging them. The decentralized nature

P. P. Ray (✉)
GNIT, Kolkata, India
e-mail: prajna.ray@hotmail.com

Bhargavi · B. Das
IEM, Kolkata, India
e-mail: baisakhi.das@iemcal.com

A. Das
Aliah University, Kolkata, India
e-mail: adas@aliah.ac.in

V. E. Balas et al. (eds.), *Proceedings of International Conference on Computational Intelligence, Data Science and Cloud Computing*, Lecture Notes on Data Engineering and Communications Technologies 62, https://doi.org/10.1007/978-981-33-4968-1_58

of blockchain can serve as a benefit for the IoT. Apart from computers, other gadgets like laptops, TVs, fridges, stoves, electrical appliances, cars, and smartphones are examples of different equipment that can be connected to communication networks [2, 3]. Smart cities, smart healthcare, and smart home are several types of applications within the IoT domain. Due to the increasing number of IoT applications, the unseen, impenetrable, common collection, processing, and distribution of data in the people's private lives are open to serious security and privacy threats [4, 5]. Here, blockchain holds the security and privacy in IoT [4]. Standard mechanisms and protocols are needed to support the huge expansion of IoT to reduce the existing heterogeneity in the field. The adoption of the IoT has been reduced due to this heterogeneity problem [6]. However, apart from heterogeneity and integration challenges trustworthiness is another important factor that needed to be addressed [7]. Nowadays, the information provided by the government and financial entities trusted by the public. But, we cannot guarantee that the information provided by the external entities such as the IoT companies [7, 8]. Information can be altered by the untrusted entities according to their interest, hence completely reliable information might not be provided by them. So, it is a must to verify whether the information has been modified or not. All the participants must guarantee that the IoT data remains immutable through a distributed service trusted by all the participants. The trustworthiness can be achieved if all the participants can guarantee that the data does not tamper if there is a proper verification mechanism. Moreover, the information provided to citizens by the government should have a system that guarantees data reliability [7, 8].

Blockchain is the technology that can be leveraged for a peer to peer distributed network to achieve the above purpose [9].

The paper has contributed mainly on followings like:

- Study of blockchain, a different perspective of blockchain technology.
- Analysis of IoT, security issues in IoT, and how security issues can be resolved through blockchain.
- Study of integration of IoT with blockchain, related applications, and case studies of IoT and blockchain integration.

Section 2 describes the preliminaries of blockchain technology, a brief description of different terminologies of blockchain. Section 3 explains the background of IoT and analytics of IoT. Section 4 describes the different applications of IoT-Blockchain integration. Section 5 concludes the paper.

2 Blockchain

A blockchain is nothing but an immutable ledger that replaces the concept of the traditional ledger by creating blocks and linking them cryptographically to form an almost unbreakable chain. For example, if a person wants to buy a new house and got the authority of that house, then the purchase is a valid transaction, and a block will be generated and linked with the previous blocks of other people who have purchased

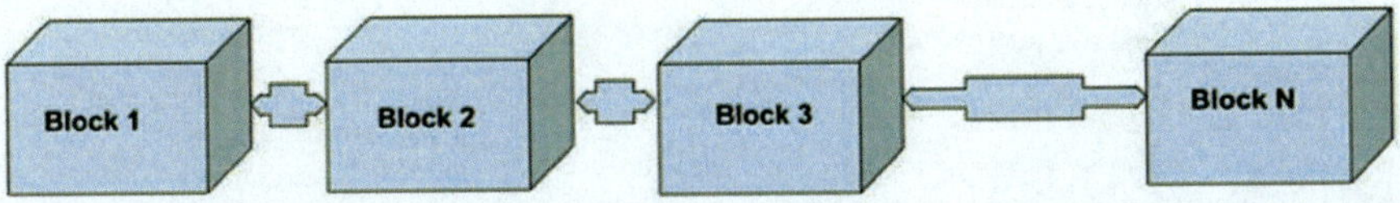

Fig. 1 Representation of Blockchain

houses earlier. Blockchain is a solution to all the participating nodes in a distributed network to maintain a continuously growing list of data records. A public ledger records all the data, including information after the completion of every transaction. No third party organization is needed in the middle by blockchain. All nodes in the blockchain maintain every transaction ever completed in blockchain and share among themselves. All anonymous nodes in blockchain make it more secure for other nodes to confirm the transactions. The first application was "Bitcoin" that introduced Blockchain technology. [10] In a decentralized marketplace, the participants can make transactions to buy and sell products through Bitcoin [11].

Figure 1 is a diagrammatic representation of blockchain, where *N* blocks are connected to form the chain. Lakhani [12] in 2017 defined blockchain as "A blockchain is an open, distributed ledger that can record transactions between two parties efficiently and in a verifiable and permanent way."

Other related definitions of blockchain are as follows:

Jesus et al. [2] defined blockchain as "Blockchain is a data structure that stores transactions in an ordered way and linked to the previous block, serving as a distributed system of records. This structure is divided into two parts, header and transactions, and stores detailed information about the transactions it contains."

Blockchain is the mechanism that allows transactions to be verified by a group of unreliable actors. It provides a distributed, immutable, transparent, secure, and auditable ledger. The blockchain can be consulted openly and fully, allowing access to all transactions that have occurred since the first transaction of the system and can be verified and collated by any entity at any time [8].

The evolution of blockchain can be categorized into three generations. Bitcoin was the first application of blockchain and termed as Blockchain 1.0. Blockchain 2.0 was used for financial services and contacts. Blockchain 3.0 was introduced to implement applications for the public sector, healthcare and is used for more general purposes.

2.1 Blockchain Structure

The blocks are tamperproof. The structure of the blockchain has two portions. They are blockchain header and transactions. The blockchain header has (a) hash value of the block, (b) the hash value of the previous block, (c) nonce, and (d) Merkle root. Another is the transactions, i.e., the data stored in the block. Figure 2 shows a

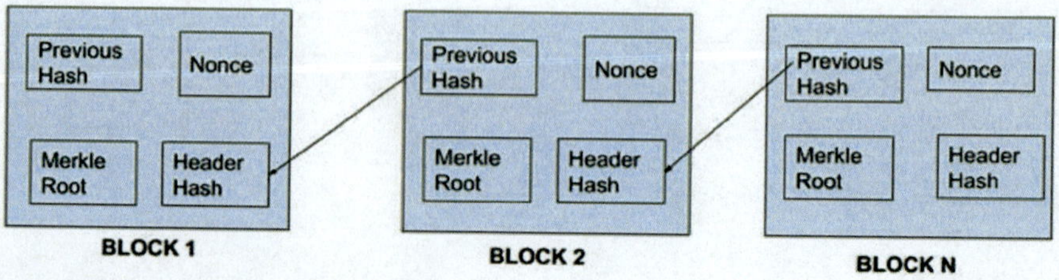

Fig. 2 Blockchain header

diagram of the blockchain header. Each block starting from Block 1 to Block N all have the four fields in their header. These are as follows:

A hash value of the header: The principal block identifier is the hash value of the header. Each complete node computes after receiving the new block. To identify a block unambiguously, the hash is used.

Previous Hash value: This field holds the hash value of the previous block. It is used to create the chain by linking the blocks with each other.

Nonce: A nonce is a number that is used and validated before generating the header hash value of the current block.

Merkle root: This concept of the Merkle tree was introduced by Ralph Merkle. Also known as a hash tree.

- Each leaf node of the Merkle tree is labeled with the hash of a data block.
- Each non-leaf node of the Merkle tree is labeled with the cryptographic hash of the labels of its child. Merkle tree root is one of the fundamental data structures behind blockchain. Construct a Merkle tree with all set of transactions, and if someone wants to change any transaction, then the Merkle root will be changed.

2.2 *Consensus*

The miners need to perform the proof of work (or mining) to generate the hash which will connect the block in the chain. This proof of work is termed as a consensus algorithm. Some of the consensus algorithms are described below:

- **Proof of work (PoW)**: Mining is also termed as proof of work which performs critical calculations and generates blocks. In blockchain to mine, one new hash for a block takes 10 min. Power consumption is very good. PoW consensus is open and entirely decentralized. The scalability is excellent in terms of number of nodes, and throughput is limited. The power consumption is very poor and has high latency. It is used in permissionless blockchain and vulnerable to Sybil attack [13].
- **Proof of stake (PoS)**: **PoS** concept states that a person can mine or validate block transactions according to how many coins he or she holds.

- **Byzantine fault tolerance (BFT)**: In BFT, tens of thousands per transaction can be possible but cannot accommodate more nodes. Power consumption is very poor. BFT is for permissioned blockchain. Beyond cryptocurrency, the blockchain is used by a smart contract. A smart contract is based on the state-machine replication protocol where a smart contract is a state machine and across multiple nodes, in a distributed environment its consistent execution is achieved using state-machine replication. The state-machine replication protocol belongs to the family of Byzantine fault tolerant (BFT) [13, 14].
- **Delegated Byzantine fault tolerance (DBFT)**: A variant of standard BFT is termed as delegated Byzantine fault tolerance (DBFT). NEO whitepaper describes within a peer-to-peer network, there are two types of clients. They are bookkeepers and ordinary nodes. The bookkeepers take place in the consensus process, and they are elected by the ordinary nodes [15, 16].
- **Proof of burn (PoB)**: This consensus algorithm combines both the PoW and PoS and partially overcomes their shortcomings and the miners must prove that they have burned some coins. PoB has no energy costs involved, and the cryptocurrency for PoB is slim coin [16].
- **Proof of elapsed time (PoET)**: Intel has developed this consensus protocol which is a lottery-based consensus protocol for Sawtooth Lake, a blockchain-based distributed application platform [3]. Efficiency and fairness are the two approaches that have made this consensus more advantageous than consensus like PoW. This algorithm allows the miner who has the least waiting time to mine the next block [17].
- **Proof of activity (PoA)**: It is a hybrid version of PoW and PoS. Shortcomings of both the algorithms are overcome with the higher computing power to mine a new block. PoW and PoS are vulnerable to 51% attack [18, 19], double-spend attack [20]. PoA prevents the chance of the above-mentioned attacks.
- **Raft-based Consensus**: Raft is a consensus algorithm that is designed to be easy to understand. It is equivalent to Paxos in fault-tolerance and performance. The difference is that it is decomposed into relatively independent subproblems, and it cleanly addresses all major pieces needed for practical systems [14].

2.3 *Smart Contracts*

A contract between two parties without a third party involved in a decentralized networking environment or platform. It is faster and cheaper. A smart contract is one of the applications of the blockchain. It is termed as Blockchain 2.0. Suppose supporters will fund a product team for a project. Supporters and product team will be agreed upon on some protocols. Then, those protocols will be written which will be put in a blockchain platform. Once the rules are fixed, no one will be able to tamper the contract. Suppose in the code, it is written that after 1 month if the product team does not meet the criteria, the supporters will get back their refund from the product team. And, as the contract is already written, no one can change the rule. The features of a smart contract can be as follows:

- It is distributed
- Code written within the contract will be executed
- Based on blockchain, if the blockchain is stable and everyone can see the transactions.
- Here "Code is the law"
- Secure peer-to-peer communication
- Security caused attacks on Ethereum smart contract ("The DAO attack")
- Quorum is a permission version of Ethereum.

Different platforms are there to run smart contracts like ethereum, hyper ledger fabric, quorum, etc. Table 1 compares different blockchain platforms to run smart contracts. Ethereumis a platform that supports smart contracts. It is a blockchain framework where different applications for the IoT devices can be developed using this platform, where the applications are known as DApps. The built-in language to write codes is known as solidity. The EVM is the Ethereum virtual machine, and the cryptocurrency is known as ether. Coins are obsolete [21]. The network type is public or private for ethereum. PoW and PoS are the consensuses supported by ethereum. The industry focus is cross-industry. Hyper LedgerFabric is a platform where various projects of blockchain-based on smart contracts are developed. Bitcoin as a cryptocurrency is used here. IBM's Bluemix platform, which is a hyper ledger fabric eases the integration of IoT by offering it as a service [22]. The network type is private for hyper ledger fabric. Network accessibility is permissioned. BFT and PBFT are the different consensuses for it. Dapps are developed on this platform. Cross-industry is the industry focus. Quorum is a permission-based platform where smart contracts can be developed to support only financial industries [22]. Raft-based

Table 1 Different blockchain platforms to run smart contracts

Platform name	Network type	Network access	Consensus	Suitable for IoT	Industry focus
Ethereum	Public(or private)	Permissionless	PoW,PoS	Yes	Cross-industry
Hyper ledger fabric	Private	Permissioned	BFT,PBFT	Yes	Cross-industry
Quorum	Special version of Ethereum	Permissioned	Raft-based Consensus,Istanbul BFT, Clique POA Consensus	Yes	Financial service
Corda	open-source distributed ledger platform	Permissioned	Several consensus algorithms can be run using notary nodes	Yes	Cross-industry

consensus, Istanbul BFT, Clique POA are the consensus for a quorum. Corda [23] is a permissioned blockchain and open-source distributed ledger platform mainly for financial industries where the data is not kept publicly in the blockchain rather it is kept by the financial organizations. Several consensus algorithms can run using notary nodes.

3 Internet of Things (IoT) Background

IoT visualizes the whole world as a distributed peer-to-peer network where all the nodes can store and share data globally. Examples of IoT applications are smart home, smart car, smart water, smart grid, wearable, industrial internet, smart retail, smart supply chain, smart farming, etc. IoT allows communication over the globe where the IoT nodes are not only any smart gadgets but also real-life objects and things [24–26].

The 2018 ranking of Top IoT Segments [27] states that according to the latest IoT analytics by 2018, it has been shown that most of the IoT projects identified are on smart city (367 projects), followed by industrial settings (265) and connected building IoT projects (193). The USA ranked top by making up maximum IoT projects (45%), followed by Europe (35%), and Asia (16%). When looking at individual IoT segments and regions, there are large differences that can be seen. Europe (45%) ranked top in the smart city projects, while the USA is an expert in connected health (55%) and connected car (54%). Smart agriculture projects (31%) are explored maximum in the Asia Pacific region [28].

4 Problems of IoT Without Blockchain

IoT will grow drastically in the nearer future and will become an enormous information source. Cisco Inc. predicted to have 50 billion connected devices by 2020. But can we ensure that the data provided by the IoT devices to the various public and financial sector are not altered or changed? [8] As the deployment of IoT devices might be in the remote places and left without monitoring, it might happen a malicious attacker attacks any IoT node, extract the cryptographic secrets, and alter the programs or data. However, most of the IoT devices are vulnerable to different security attacks [29]. Security issues need to be addressed at different levels as the range of the IoT devices varies from small embedded processing chips to large high-end servers. Security issues can be classified into (i) low-level security issues, (ii) intermediate-level security issues, (iii) high-level security issues.

Low-level Sybil and spoofing attacks [30], insecure physical interface [31] are the examples of low-level security attacks. Buffer-Reservation attack [32], RPL Routing attack [33],insecure neighbor discovery, Sinkhole, and wormhole attacks, replay or duplication attacks due to fragmentation, Sybil attacks on intermediate layers, authentication and secure communication, transport level end-to-end security [33], session establishment and resumption, privacy violation on cloud-based IoT [35] are the examples of intermediate level security attacks. CoAP security with Internet [36], insecure interface [34] and middleware security [35], insecure software/firmware are the examples of high-level security attacks.

4.1 How Integration of IoT with Blockchain Will Solve the Problem

In a distributed network, if all the participants can assure that the data provided by them are immutable and they have means to verify that the data have not tampered, then trustworthiness can be achieved. Blockchain can be integrated with IoT in this context to achieve this trustworthiness. Several projects are running by integrating these two technologies and creating a huge scope for research scholars as there are lots of challenges that are untraced, need to be addressed.

For example, the ability to sense and actuate over the Internet in many areas is provided by IoT. Food packaging can be done safely and securely without compromising with the quality and price by integrating IoT and Blockchain. A smart contract is deployed and executed. The firing of an event would be there if the measurement exceeded a predefined threshold. These events will be listened to by management applications, and notifications will be sent to the shipping companies, retailers, manufacturers, and clients. If no event is raised, the shipment is carried out in optimal conditions as guaranteed by the blockchain. A secure and reliable processing engine for IoT would be provided by smart contracts. Recording and managing all the interactions will be done by a smart contract. The result of reliable and secure processing is an action. The application logic of IoT applications can be securely modeled by a smart contract. A lot of challenges will be there in this kind of integration. Oracles are special entities that provide real-world data in a trusted manner following the rules mentioned in a smart contract. Validation compromise may be there for a smart contract due to the instability of IoT. Moreover, these contracts can be overloaded when multiple data sources are accessed. Smart contracts are distributed and decentralized but resource sharing, task distribution, and addressing a large amount of computation are not done by these smart contracts. In other words, a single node is responsible for the execution of the smart contract whereas, simultaneously the code executions are done by multiple nodes. For the validation process, only the distribution is done; instead of doing to distribute tasks. Distributed capabilities of cloud computing and big data have been leveraged by the IoT to increase the processing power. Addressing the IoT data as a whole would be done by the data mining techniques. A better understanding of the IoT is also enabled by data mining techniques. For example, cloud computing enlarges the processing power. The processing of large amounts of data simultaneously is enabled by big data. The extraction of knowledge from large datasets is also allowed by big data. Integrating blockchain with IoT, the distributed nature of smart contracts should be leveraged. IoT consists of constraints and heterogeneity and these should be taken into account by smart contracts. To enable the application of IoT depending on the context and requirement should be complemented by a smart contract. The enablement of device inclusion on the fly is done by the discovery mechanism which eventually makes this application more powerful. To enable faster reactions with IoT, the actuation mechanism from a smart contract is responsible [13].

5 Different Applications of Integrating Blockchain with IoT

Different IoT-based front end devices run different applications where blockchain will be there in the backend and serve as a distributed ledger to ensure the trustworthiness in a distributed peer-to-peer network.

Different smart contract case studies, platforms, and the description are discussed and listed in Table 2.

Fast payment protocol [36]: is a fast payment protocol that is implemented through smart contracts on the ethereum platform to prevent the double-spending attack. Digital payment is done through IoT-based front end devices, and the blockchain in the back end ensures the validity of the payments across the system. However, the blockchain is vulnerable to double-spending attacks.

Traceability system for agriculture product [37]: is a traceability system for agriculture products based on IoT and blockchain technology which analyzes food quality problems. IoT technology is more reliable than the traditional, manual recording of data, and the blockchain is more credible than the existing traditional database.

Architectures for IoT-based application using blockchain [38]: it proposes 3 different architectures for IoT frontend blockchain applications. The applications are developed on the ethereum platform as bitcoin is obsolete for micropayments due to time constraints to mine a block.

Trust-list [39]: is a trust-list to a blockchain network. It represents the distribution of trust among IoT-related stakeholders like service providers, developers and network operators, and autonomous enforcement of IoT traffic provided by software defined networking (SDN) and blockchains.

MioT framework [40]: is a general framework and solutions have been proposed to find out the origin of ownership for medical IoT devices. MioT plays an important role in the healthcare sector. It is essential to trace out whether the MioT device is original or counterfeited as one device can be used by several patient parties during its lifetime. The framework is implemented on the Ethereum platform where two smart codes are executed. One is a manufacturer smart contract, and another is IoT smart contract. The codes will be executed without any third party involvement or central authority control.

Implementation of smart contracts for blockchain-based on IoT applications [41]: An application for buying and selling data (measurements) of IoT sensors

Table 2 Different case studies based on Blockchain and IoT integration are listed

Sl.No.	Smart contract case studies	Description	Platform
1.	Fast payment protocol [36]	Fast payment protocol which will be implemented through the smart contract	Ethereum
2	Traceability system for agriculture product [37]	A system for agriculture products based on IoT and blockchain technology that analyzes food quality problems	Ethereum
3	Architectures for IoT-based application using blockchain [38]	3 different architectures for IoT-based application using blockchain in the background is being proposed	Ethereum
4	Trust-list [39]	It is a trust-list to a blockchain network. It represents the distribution of trust among IoT-related stakeholders	Ethereum
5	MioT framework [40]	Tracking out the owner of the medical IoT device where one device can be used by several patient parties during its lifetime	Ethereum
6	Implementation of smart contracts for blockchain-based on IoT applications [41]	An IoT application is developed which is based on sensing–as-a-service (s^2aas) business model and implemented through blockchain	Ethereum
7	Safe Farming [42]	IoT-based prevention system implemented through blockchain is built to protect the crop from the animal's attack	Ethereum
8	Privacy in blockchain-enabled IoT devices [43]	An artificial guideline for blockchain-enabled IoT devices used to develop not only financial applications, but also other decentralized applications	Ethereum
9	Auto pay [44]	An auto-pay service system which provides security and trusts by embodying an authorized human being	Ethereum
10	Food supply chain traceability system [45]	A system in a distributed network for real-time food tracing which is based on hazard analysis and critical control paths (HACCP), blockchain, and IoT	BigChainDB
11	Emergency service for smart home [46]	A smart home service is a collection of IoT sensors integrated with blockchain here	Ethereum

(continued)

Table 2 (continued)

Sl.No.	Smart contract case studies	Description	Platform
12	IoT-based healthCare [47]	Blockchain and IoT are integrated here to design a decentralized IoT-based e-healthcare system	Ethereum

related to weather. The huge amount of data collected by IoT sensors can be shared to achieve economic scalability and to reduce costs. Here, an IoT application is developed which is based on sensing–as-a-service (s^2aas) business model and implemented through blockchain. Two Dapps, i.e., two smart contracts are deployed and interacted over ethereum blockchain. There is a front end or the interface between the users and the blockchain which displays the lists of sensor data, the time interval for which to buy the selected data, etc.

Safe farming [42]: IoT-based prevention system is built to protect the crop from the animal's attack. IoT sensors report any disturbance to a repelling and notifying system (RNS) in the field, which produces ultrasonic waves for animals thus they can leave the field. Farmer management system (FMS) maintains a blockchain that generates a new block wherever an attack happens, and this blockchain shares all the data related to different attacks in the field.

Privacy in blockchain-enabled IoT devices [43]: This is an artificial guideline for blockchain-enabled IoT devices is being proposed. Applications are built on the ethereum platform. The applications are not only financial applications but also other decentralized applications. Here, the authors have described their work by illustrating an example of CCTV, where the CCTV camera is controlled by the smart contract. The smart contract is created by the manufacturer, and the contract will run when the owner rents the camera to the renter for a particular duration of time. When the renting duration will get over, the contract will stop running, and the owner will get some benefits.

Auto pay [44]: is an auto-pay service system that provides security and trusts by embodying an authorized human being. This personification is done through smart contracts and its blockchain interface. Auto-pay service is acting as an autonomous payment device. Auto-pay service can initiate several features in an embedded system, for example, in a car's display. It can be integrated with an application called 'Journey Planner.' Whenever a person enters his/her destination data in the journey planner, it checks the vehicle's essential data and proposes the best route to the destination. If the car is low on fuel, it will search a route where the person can find a gas station on the way. The payment for fuel can also be done through the smart contract feature of the auto-pay service. It can also be integrated into a person's smart phone's AI application (like Apple's Siri) to get information about parking availability at the destination. It can also integrate a smart grocery management application where ordering, payment, and delivery can be coordinated through this auto-pay service block chain interface. Access to the system can be given to another person with limited privileges. For example, the authorized person can give his/her car's auto-pay service to his/her children but auto-pay service may not work in some specified cases. The authorized person can also quickly find out the unauthorized transactions, tried by his/her children in the immutable transaction history on auto-pay's interface to the car's blockchain ledger.

Food supply chain traceability system [45]: is a system in a distributed network for real-time food tracing which is based on hazard analysis and critical control paths (HACCP), blockchain, and IoT. This system provides a distributed platform where all the supply chain members communicate with each other with openness, transparency, security, and reliability. As scalability is one of the main disadvantages of blockchain, a new concept BigChainDB is being proposed which combines the key characteristics of distributed databases like high throughput, low latency, and high capacity with the key characteristics of blockchain-like decentralized control, immutability, and creation & movement of digital assets.

Emergency service for smart home [46]: The smart home system (SHS) is composed of IoT sensors that collect the environmental data and send it to the home service providers (HSPs) as an emergency call whenever the values of sensor data are more than the predefined threshold. This total communication takes place in a decentralized network and executed through a smart contract among all the participants like HSP, SHS, etc.

IoT-based healthCare [47]: IoBHealth data flow architecture is proposed where all the e-healthcare related sensitive data are managed and accessed in a trusted, secure, and efficient manner in a distributed environment. Blockchain and IoT are integrated here to design a decentralized IoT-based e-healthcare system.

6 Conclusion

Internet of Things (IoT) devices have become smart and autonomous, and they can communicate over a distributed decentralized network. IoT applications are generating huge amount of data, and securing those IoT data is a concern. Here, blockchain technology plays an important role. As blockchain achieves the trustworthiness among the participants in a decentralized distributed network by providing the means of verifying the immutability of the data, it can be integrated with IoT to secure the IoT data. In this paper, we discussed the different applications and integration of blockchain and IoT. Several security attacks on IoT and that how they can be solved using blockchain technology.

References

1. G.G. Dagher, J. Mohler, M. Milojkovic, P.B. Marella, Ancile: Privacy-preserving framework for access control and interoperability of electronic health records using blockchain technology. Sustainable cities and society **39**, 283–297 (2018)
2. E.F. Jesus, V.R. Chicarino, C.V. de Albuquerque, A.A.D.A. Rocha, A survey of how to use blockchain to secure the internet of things and the stalker attack. Secur. Commun. Networks, (2018)
3. N. Fotiou, G.C. Polyzos, Smart contracts for the internet of things: opportunities and challenges, in *2018 European Conference on Networks and Communications (EuCNC)*, IEEE, (June 2018), pp. 256–260
4. A. Dorri, S.S. Kanhere, R. Jurdak, Towards an optimized blockchain for IoT, in *Proceedings of the Second International Conference on Internet-of-Things Design and Implementation*. ACM, (April 2017), pp. 173–178
5. A. Panarello, N. Tapas, G. Merlino, F. Longo, A. Puliafito, Blockchain and iot integration: a systematic survey. Sensors **18**(8), 2575 (2018)
6. D. Evans, The internet of things: how the next evolution of the internet is changing everything. CISCO white paper **1**(2011), 1–11 (2011)
7. L. Atzori, A. Iera, G. Morabito, The internet of things: a survey. Comput. Networks **54**(15), 2787–2805 (2010)
8. A. Reyna, C. Martín, J. Chen, E. Soler, M. Díaz, On blockchain and its integration with IoT. Challenges and opportunities. Future Generation Computer Systems **88**, 173–190 (2018)
9. M. Conoscenti, A. Vetro, J.C. De Martin, Peer to Peer for privacy and decentralization in the internet of things, in *2017 IEEE/ACM 39th International Conference on Software Engineering Companion (ICSE-C)*. IEEE, (May 2017), pp. 288–290
10. S. Nakamoto, Bitcoin: a peer-to-peer electronic cash system, (2008)
11. J. Yli-Huumo, D. Ko, S. Choi, S. Park, K. Smolander, Where is current research on blockchain technology?—a systematic review. PLoS ONE **11**(10), e0163477 (2016)
12. M. Iansiti, K.R. Lakhani, The truth about blockchain. Harvard Bus. Rev. **95**(1), 118–127 (2017)
13. N. Kshetri, Can blockchain strengthen the internet of things? IT Prof. **19**(4), 68–72 (2017)
14. M. Castro, B. Liskov, Practical byzantine fault tolerance, in *OSDI*, vol. 99, 173–186 (Feb 1999)
15. L.M. Bach, B. Mihaljevic, M. Zagar, Comparative analysis of blockchain consensus algorithms, in *2018 41st International Convention on Information and Communication Technology, Electronics, and Microelectronics (MIPRO)*. IEEE, (May 2018), pp. 1545–1550
16. I. Makhdoom, M. Abolhasan, H. Abbas, W. Ni, Blockchain's adoption in IoT: the challenges, and a way forward. J. Network Comput. Appl. (2018)
17. L. Chen, L. Xu, N. Shah, Z. Gao, Y. Lu, W. Shi, On security analysis of proof-of-elapsed-time (poet), in *International Symposium on Stabilization, Safety, and Security of Distributed Systems* (Springer, Cham, Nov 2017), pp. 282–297
18. M. Bastiaan, Preventing the 51%-attack: a stochastic analysis of two phase proof of work in bitcoin, in *Availab le at* http://referaat.cs.utwente.nl/conference/22/paper/7473/preventingthe-51-attack-astochastic-analysis-of-two-phase-proof-of-work-in-bitcoin.pdf. (Jan 2015)
19. X. Li, P. Jiang, T. Chen, X. Luo, Q. Wen, A survey on the security of blockchain systems. Future Gener. Comput. Syst. (2017)
20. M. Rosenfeld, Analysis of hashrate-based double-spending. arXiv preprint arXiv:1402.2009. (2014)
21. A.M. Antonopoulos, G. Wood, Mastering ethereum: building smart contracts and dapps. O'reilly Media. E. Androulaki, A. Barger, V. Bortnikov, C. Cachin, K. Christidis, A. De Caro, D. Enyeart, C. Ferris, G. Laventman, Y. Manevich, S. Muralidharan, 1801. Hyperledger fabric: a distributed operating system for permissioned blockchains (2018). arXiv preprint arXiv:1801.10228. (2018)
22. E. Androulaki, A. Barger, V. Bortnikov, C. Cachin, K. Christidis, De Caro, A., D. Enyeart, C. Ferris, G. Laventman, Y. Manevich, S. Muralidharan, 1801. Hyperledger fabric: a distributed operating system for permissioned blockchains (2018). *arXiv preprint* arXiv:1801.10228

23. R.G. Brown, J. Carlyle, I. Grigg, M. Hearn, Corda: an introduction. *R3 CEV, August*. (2016) [13] Z. Hao, R. Ji, Q. Li, FastPay: a secure fast payment method for edge-IoT platforms using blockchain, in *2018 IEEE/ACM Symposium on Edge Computing (SEC)*. IEEE, (Oct 2018), pp. 410–415
24. E.A. Kosmatos, N.D. Tselikas, A.C. Boucouvalas, Integrating RFIDs and smart objects into a UnifiedInternet of Things architecture. Adv. Internet Things **1**(01), 5 (2011)
25. S. Madakam, R. Ramaswamy, S. Tripathi, Internet of Things (IoT): a literature review. J. Comput. Commun. **3**(05), 164 (2015)
26. R. Aggarwal, M.L. Das, RFID security in the context of the internet of things, in *Proceedings of the First International Conference on Security of Internet of Things*. ACM, Aug 2012, pp. 51–56
27. The Top 10 IoT Segments in 2018 – based on 1,600 real IoT projects. Available online: https://iot-analytics.com/top-10-iot-segments-2018-real-iot-projects/
28. A. Dawod, D. Georgakopoulos, P.P. Jayaraman, A. Nirmalathas, "Advancements towards global IoT device discovery and integration," *2019 IEEE International Congress on Internet of Things (ICIOT)*, (Milan, Italy, 2019), pp. 147–155, https://doi.org/10.1109/iciot.2019.00034. M.M. Hossain, M. Fotouhi, R. Hasan, Towards an analysis of security issues, challenges, and open problems in the internet of things, in *2015 IEEE World Congress on Services*. IEEE, Jun 2015, pp. 21–28
29. M.M. Hossain, M. Fotouhi, R. Hasan, Towards an analysis of security issues, challenges, and open problems in the internet of things, in *2015 IEEE World Congress on Services*. IEEE, Jun 2015, pp. 21–28
30. K. Zhang, X. Liang, R. Lu, X. Shen, Sybil attacks and their defenses in the internet of things. IEEE Internet of Things J. **1**(5), 372–383 (2014)
31. M.A. Khan, K. Salah, IoT security: review, blockchain solutions, and open challenges. Future Gener. Comput. Syst. **82**, 395–411 (2018)
32. R. Hummen, J. Hiller, H. Wirtz, M. Henze, H. Shafagh, K. Wehrle, 6LoWPAN fragmentation attacks and mitigation mechanisms, in *Proceedings of the sixth ACM Conference on Security and privacy in wireless and mobile networks*. ACM, Apr 2013, pp. 55–66
33. A. Dvir, L. Buttyan, VeRA-version number and rank authentication in rpl, in *2011 IEEE Eighth International Conference on Mobile Ad-Hoc and Sensor Systems*. IEEE, Oct 2011, pp. 709-714
34. E. Kosenko, Industrial IoT development for condition monitoring purposes, 2018
35. D. Conzon, T. Bolognesi, P. Brizzi, A. Lotito, R. Tomasi, M.A. Spirito, The virtus middleware: an xmpp based architecture for secure IoT communications, in *2012 21st International Conference on Computer Communications and Networks (ICCCN)*. IEEE, Jul 2012, pp. 1–6
36. W. Hong, Y. Cai, Z. Yu, X. Yu, An agri-product traceability system based on IoT and blockchain technology, in *2018 1st IEEE International Conference on Hot Information-Centric Networking (HotICN)*. IEEE, Aug 2018, pp. 254–255
37. M. Pustišek, A. Kos, Approaches to front-end IoT application development for the ethereum blockchain. Proc. Comput. Sci. **129**, 410–419 (2018)
38. M. Alblooshi, K. Salah, Y. Alhammadi, Blockchain-based ownership management for medical IoT (MIoT) devices, in *2018 International Conference on Innovations in Information Technology (IIT)*. IEEE, Nov 2018, pp. 151–156
39. K. Kataoka, S. Gangwar, P. Podili, Trust list: internet-wide and distributed IoT traffic management using blockchain and SDN, in *2018 IEEE 4th World Forum on the Internet of Things (WF-IoT)*. IEEE, Feb 2018, pp. 296–301
40. M. Alblooshi, K. Salah, Y. Alhammadi, Blockchain-based ownership management for medical IoT (MIoT) devices, in *2018 International Conference on Innovations In Information Technology (IIT)*, IEEE, pp. 151–156 (Nov 2018)
41. G. Papadodimas, G. Palaiokrasas, A. Litke, T. Varvarigou, Implementation of smart contracts for blockchain-based IoT applications, in *2018 9th International Conference on the Network of the Future (NOF)*. IEEE, (Nov 2018), pp. 60–67
42. R. Iqbal, T.A. Butt, Safe farming as a service of blockchain-based supply chain management for improved transparency. Cluster Comput. pp. 1–12 (2020)

43. A. Pouraghily, M.N. Islam, S. Kundu, T. Wolf, Privacy in blockchain-enabled IoT devices, in *2018 IEEE/ACM Third International Conference on Internet-of-Things Design and Implementation (IoTDI)*. IEEE, (Apr 2018), pp. 292–293
44. S. Huckle, R. Bhattacharya, M. White, N. Beloff, Internet of things, blockchain, and shared economy applications. Proc. Comput. Sci. **98**, 461–466 (2016)
45. F. Tian, A supply chain traceability system for food safety based on HACCP, Blockchain & Internet of things, in *2017 International Conference on Service Systems and Service Management*. IEEE, Jun 2017, pp. 1–6
46. T. Tantidham, Y.N. Aung, Emergency service for a smart home system using Ethereum blockchain: system and architecture, in *2019 IEEE International Conference on Pervasive Computing and Communications Workshops (PerCom Workshops)*. IEEE, Mar 2019, pp. 888–893
47. P.P. Ray, D. Dash, K. Salah, N. Kumar, Blockchain for IoT-based healthcare: background, consensus, platforms, and use cases. IEEE Syst. J. (2020)

Secure Blockchain: Assessing Specific Security Threats

Ganguly Ananya, Das Priyanjali, Das Baisakhi, and Das Abhishek

Abstract Blockchain technology originates from the Bitcoin system, and due to its great success it can draw awareness to other fields because of its global acceptable features like decentralization, anonymity, and audibility. Other than finance, blockchain technology has evolved in various applications like legal, supply chain, health care, etc. Though blockchain technology is a bit difficult to maintain and monitor, still a blockchain network is more secure than other technologies. In this work, we want to give some avenues of the blockchain security system and also highlight some specific security threats. The issues related to mining or fork problems have also been discussed. It is found that several unexpected fields contribute to security concerns rather than decentralized computing and any cryptographic issues.

Keywords Blockchain · Security · Vulnerability · Smart contract · Threats

1 Introduction

The first application of blockchain technology was the Bitcoin, which is treated as a kind of digital currency or cryptocurrency in the real world. Bitcoin turned out to be a great success. People tried to utilize blockchain technology in different fields and services, such as the financial market, IoT, supply chain, voting, medical treatment,

G. Ananya
Acharya Jagadish Chandra Bose College, Kolkata, India
e-mail: a.ganguly.it@gmail.com

D. Priyanjali · D. Baisakhi
IEM, Kolkata, India
e-mail: priyanjali.das2000@gmail.com

D. Baisakhi
e-mail: baisakhi.das@iemcal.com

D. Abhishek (✉)
Aliah University, Kolkata, India
e-mail: adas@aliah.ac.in

V. E. Balas et al. (eds.), *Proceedings of International Conference on Computational Intelligence, Data Science and Cloud Computing*, Lecture Notes on Data Engineering and Communications Technologies 62, https://doi.org/10.1007/978-981-33-4968-1_59

document savings, and storage. Blockchain technology has distributed architecture which helps to make communication in a peer-to-peer network of miners where the transactions are transparent among the involved miners. So here the fundamental question comes to know about the blockchain's security, privacy, and limitations. As we are using these technologies, tools, or services in our daily life, cybercriminals also get the opportunity to engage in cybercrime [1, 2].

As per the architectural model, a blockchain stores its database across the decentralized network of computers. Computers which are connected for any transaction in networks called a node in the network, and every node in the network fetches an identical copy of the required documents of blockchain that automatically gets downloaded but the main copy of the document stored in the main blockchain makes the blockchain transparent and decentralized and visible to every node in that blockchain. Here all the transactions are maintained with proper cryptographic measures as well as with several security measures. This security is computed by several mathematical calculations related to one another. The complex mathematical calculation of blockchain makes this technology more secure, and it is almost impossible to find out or guesses the security codes, making it tougher for the transactions to be cracked which secures the blockchain technology properly to make the blockchain technology database as a "World Wide Ledger" where all data and nodes concerning all new transactions as well as old transactions are propagated to all nodes.

In this paper, we considered the main security challenges as well as privacy issues with regard to the blockchain technology along with its possible preventing measures and its future trends. Although blockchain is secure, several threats are still existing which may create disputes to the transparency and decentralization of blockchain. Our work discusses the numerous fields which are used in blockchain and how security issues and privacy issues can be challenging in those applications [3]. Every application is using blockchain technology to have a different approach to the usability and authenticity of blockchain, hence different security and privacy concerns arise. Here we discuss various fields of usage of blockchain along with its related security and privacy issues. Furthermore, we also tried to propose some possible precautions that can be used to achieve a more secure blockchain environment. We consider the earlier research work and try to explore their related security and privacy challenges and probable precautions.

The basic concept of blockchain is discussed in Sect. 2, Sect. 3 focuses on different applications of blockchain, in Sect. 4 we tried to cover about the security measures of several blockchain limitations issues and how to prevent those challenges to overcome security issues, here the cause or the background of mining and fork problem, and in Sect. 5, Sect. 6 concludes the paper with the prospects of those challenges.

So, at first, we should make clear some basic concepts of blockchain which we'll discuss in Sect. 2.

2 The Concept of Blockchain

Blockchain technology is not a single technique or mechanism, rather it is a combination of several techniques like cryptography, mathematics, networking, algorithm, and economic model, combining with peer-to-peer networks and also distributed consensus algorithm which is used to solve the traditional distributed database problem, so we can say, it is an integrated multi-field infrastructure [3–5]. Blockchain technology is a combination of six key elements.

Decentralized: This is the basic feature of blockchain, and it means that blockchain does not have to rely on a centralized node anymore, when the data can be recorded, stored, and updated over the distributed environment.

Transparent: The data recorded by the blockchain system must be transparent to each node, and it should also be transparent on updating the process of data, that is why blockchain is trusted.

Open Source: Most of the blockchain system is open to everyone, all the records can be checked publicly, and people can also use blockchain technologies to create an application as they want.

Autonomy: Due to the basic structure of the consensus mechanism, each node on the blockchain technology-based system can transfer or update data very securely and safely, which make the whole system more secure and trustworthy where it is transparent to all rather than a single person, and no one can intervene it.

Immutable: In the case of blockchain when a record is recorded, then any record will be reserved and stored forever and cannot be changed unless someone can take more than 51% control at the same time.

Anonymity: Blockchain technology is able to recover the confidence problem between every node, so data transfer or even transaction can be nameless, and only needs to know the person's blockchain address that means the hash value of the previous node, hash of that block, timestamp of that node, and of course the data.

2.1 Blockchain Working Technique

The main working processes of blockchain follow six basic techniques which are as follows:

1. Generally, when the new node comes to the blockchain, then it sends the records of new data and broadcasts that records to the network.
2. The receiving node must check the authenticity and correctness of that message which it received new, and if it is satisfied, then add that data to a block.
3. All the new receiving node in a single-specific network must execute that nodes proof of work (PoW) or proof of stake (PoS) algorithm to the block for verification and authenticity purpose.

4. The blocks are following the checking process by executing the consensus algorithm of every node, and those nodes are honest in the network to admit as block and will continuously extend the chain base on this block.

2.2 The Architecture of Blockchain

Generally, a single block contains the actual main data, the hash of the previous block, the hash of the current block, timestamp, and if required other relevant information. Figure 1 shows the structure of the block where all required data are recorded in a block and they are connected to each case in a chaining way.

Main data: Depending on the field of that specified blockchain area and service, the relevant data are stored in that block, for example, bank balance, debit credit records, contract record, IoT data record.

Hash: When a specified transaction is executed, then it generates a unique code which is termed as the hash code of that broadcast, it contains thousands of transaction records in each node's block, and the blockchain used the Merkle tree function to generate a final hash value of each block, which is also called Merkle tree root. Here this final hash value must be recorded in the header part of the block (hash of current block), by using the Merkle tree function, which also reduced the data transmission and resource computing.

Timestamp: Time of block generated.

Other Information: Like signature of the block, nonce value, or other data that the user defines and relevant to that blockchain.

2.3 Consensus Generating Process

Consensus: Consensus is a mathematical expression or mechanism by which it is determined that which block will add by pitching the nodes against others in

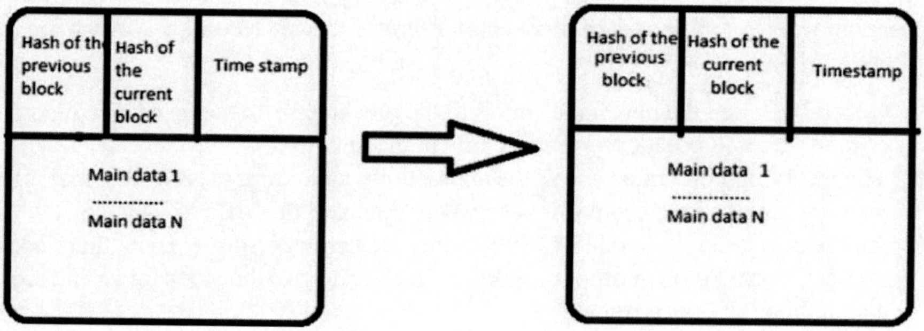

Fig. 1 Structure of the blockchain

the blockchain cryptographic race. Though blockchain is a decentralized system, then here the third party need not become trustworthy, but the data must be reliable and consistent. For this reason, blockchain generally adopts the decentralized consensus mechanism [6]. There are several existing blockchain consensus mechanisms [7]: proof of work (PoW), proof of stake (PoS), proof of importance (PoI), practical Byzantine fault tolerance (PBFT), and delegated proof of stake (DPoS). Other consensus mechanisms, such as proof of bandwidth (PoB) [8], proof of elapsed time (PoET) [6], proof of authority (PoA) [9, 10], PoSpace, proof of importance (PoI), and minimum block hash (MBH), measure of trust (MoT) are also used in some blockchain systems. The two most popular blockchain systems are Bitcoin and Ethereum, and both follow the PoW mechanism. Ethereum also used the PoA mechanism. The consensus function is a special and specified mechanism that makes all blockchain nodes must have agreed with them in that corresponding message, and it also can make sure that the latest block has been added to the chain correctly, along with that, guaranteed to that the message that stored by the node was the identic one and will not happen "fork attack", even can prevent thee malicious attacks (Table 1).

2.4 *Different Types of Blockchain Based on Needs and Requirements*

To minimize the manipulative activities, to provide transparency as well as traceability of the block records, to ensure the immutability, and to increase security levels of the blockchain can be described as different types [11]:

Public Blockchain: To ensure the access more transparent and public, the public blockchain or permissionless blockchain is the most desirable ones. Here all public protocols rely on the proof of work consensus algorithms. Here the communities can access easily them and participate in the public blockchain [11, 12].

Private Blockchain: To enhance internal business management and keeping all internal information private, basically the private blockchain or permission-based blockchain is used. It provides the centralized sources of authority and read-only permissions to the all nodes connected to that blockchain's specified and restricted viewers [12, 13].

Federated or Consortium Blockchain: Federated or consortium blockchain is mostly managed by a group of individuals or organizations. There is no possibility of free access to any other users [12, 13]. They can access or view a transaction only after the verification is done by the authority. It is also known as the semi-decentralized, permission-based blockchain.

Hybrid Blockchain: It combines the best features of three blockchains and provides a new hybrid one. The hybrid blockchain solves security issues and prevents various types of attacks and security threats [11, 13]. It was specifically able to determine what information should stay private and what information is made public.

Table 1 Comparative study of different mining techniques

Mining Techniques	Implementing-Platform	Resource	Using smart contract	Rewards miners	Advantages	Disadvantages
PoW [7]	Bitcoin/Ethereum	High computational power	Yes	Yes	On the basis of users or nodes, it becomes scalable and it is most popular	Its time complexity is very high, and computation is also expensive
PoS [7]	Ethereum	Power of stake	Yes	No	No computation needed	Essential to get all the stake of required mining
PBFT [6, 7]	Hyperledger	Nil	Yes	No	Faster than all because of no mining	More than 20 nodes scalability power is not there
DPoS [7, 8]	Bitcoin	Democratic system	No	No	It is more efficient and effective. Here the selection of block producers allows transaction validation in seconds	It is a forging technique and should know all stake of selected mining
PoB [8]	TorCoin	Rewards relay process	Yes	No	The proportional amount of bandwidth is required	Due to malicious relay, not able to give accurate consensus
PoET [6]	Permissioned blockchain in different cryptocurrency	Mining right	Yes	Yes	Used for randomly selected waiting time by the winning participant. Running code is also trusted and control consensus cost	Submit trades are virtual so the risk of transfer is there

(continued)

Table 1 (continued)

Mining Techniques	Implementing-Platform	Resource	Using smart contract	Rewards miners	Advantages	Disadvantages
0PoA [9, 10]	Permissioned and permissionless	Power of authority	Yes	Yes	Allows non-consecutive block approval so the cost identity is lesser	It's centralized soothe risk of serious damage is there
PoSpace [9, 10]	Permacoin, SpaceMint, and Burstcoin	High money	No	Yes	Fairer and greener because of general-purpose storage	Problems in irretrievability and storage functions
PoI [9, 10]	NEM	Node significant	No	Yes	Due to the amount of crash, there is no scope for the rich to keep getting richer on the platform	Use of dummy transaction rewards sending back and forth in the cheat algorithm
MBH [10]	Bitcoin extension	None	No	Yes	Miners can choose randomly based resources and also save energy	The probability of miner's selection and approach of implementation is very hard
MoT [10]	Not implemented	Trustworthiness	Yes	Yes	Rewards miners Selection makes pool creation complexity lesser	Risk of trustworthiness

Further decentralization about primarily a centralized private blockchain can be achieved in various ways.

No matter what types of blockchain are, it has an advantage. Sometimes we need public blockchain because of its convenience, but sometimes we may need private control like consortium blockchain or private blockchain, depending on what service we over or what place we use it.

After knowing the concept, we should know the main application of blockchain, and it is discussed in Sect. 3.

3 Different Applications of Blockchain Technologies

Blockchain technologies can be used in many areas, not only in financial applications but also in other industries.

3.1 *Digital Currency: Bitcoin*

The data structure and transaction of the bitcoin system was built by blockchain technology which makes Bitcoin as a digital currency and online payment system. By using this encrypted technique, funds transfer can be processed and does not need to rely on any centralized bank. Bitcoin also used the public key encryption to send the address and receiving the cryptocurrency and recorded the transaction details, and the personal ID was nameless. The transaction confirmation process needs other user's computing power to get the consensus and then records the transaction into that network.

3.2 *Smart Contract: Ethereum*

A smart contract is one such digital contract that deals and controls all the user's digital assets, evaluating the participant's right and obligation, and also will automatically execute by the computer network system. It is not only just a computer network procedure but also is seen as a contract participant and will respond to the message what it receives and stores as data, and it can also send message or value to the outside. A smart contract is treated as a person whom you can be trusted, can hold your assets temporarily, and will treat as an authentic order which has already been proposed in the program [14]. Ethereum is an open-source blockchain platform which combines smart contract, to offer a decentralized virtual machine which can able to handle the contract, by using its digital currency called ETH, by which people can create many different services, applications, or contracts on that platform [15, 16].

3.3 Hyperledger

Another open-source blockchain platform is hyperledger, which is started by the Linux Foundation, to support the blockchain-based distributed open ledgers. Its main focus is on ledgers designed by which it is able to support global business transactions, including technological, financial, banking, and other supply chain companies, to improve several aspects of performance and also enhance the reliability. It is an open protocol that brings many independent and individual efforts and standards together to provide a modular framework that supports different components for different uses. This would be able to include several and various blockchain with their consensus and storage models, as well as services for identity, access control, and contracts.

3.4 Other Applications

Still, there have been many use cases regarding blockchain technologies, like protection of intellectual property, traceability in the supply chain, voting, banking, performance reporting, identity certification, insurance, international payments, IoT, patient's details along with their privacy for medical treatment, or prediction market asset [17, 18].

Several unexpected fields contribute to the security concerns other than the decentralized computing issues or any cryptography issues which will be discussed in Sect. 4.

4 Blockchain Security Issues and Challenges

Here are some major blockchain security issues to look out for in the recent environment.

4.1 More than 50% Attacks or Majority Attacks

With the help of proof of work, the probability of mining a block is depending on the work done by the miner. Due to this mechanism, miners will try to joint mining more blocks and become "mining pools" and place the holding request to get computing power. When it holds 51% computing power, it is able to take control of that specified blockchain. It causes security issues [19, 20]. If someone is able to get than 51% computing power, then that miners can find more nonce value quicker than others, and it means that miners are able to get the authority to decide which block is permissible.

What it can do is:

1. The authority gaining miner is able to modify the transaction data, and it may cause a double-spending attack [21, 22].
2. The miner is able to stop the block verifying the transaction.
3. The miner may also stop miner mining of any available block.

A majority attack was maximumly feasible in the past when most transactions were more significant than the block reward and when the network hash rate was much lower to reorganization about new miners [5, 23].

51% attacks are mostly recognized blockchain security threats. In a 51% attack, one, or more than one, malicious entities can achieve or gain the majority control of that specified blockchain's hash rate. By using this major hash rate, the miners are able to reverse the transactions, and it promotes to perform double-spends attacks and prevent other honest blocks from coming and confirming blocks.

In 2018, several cryptocurrency and blockchain such as ZenCash, Verge, and Ethereum Classic fell victim to 51% attacks [24, 25].

If a blockchain utilizes a proof of work (PoW) consensus mechanism, then that blockchain needs to have checked the security measures in place to prevent a 51% attack [24, 26].

4.2 *Exchange Hacks*

One of the most expensive blockchain security issues is not a problem with blockchain technology itself. When blockchain generates and creates one by one blocks that time, it comes as a lucrative honeypot for hackers due to various reasons or poor security practices. Many exchange platforms are making a co-relation or bonding between major blocks and act as innately centralized, so they make the decentralized benefits of blockchain fully obsolete [27].

4.3 *Social Engineering*

Another blockchain security issue that blockchain users should be aware of is social engineering. Social engineering may come to miners in many forms, but the main motto is always the same: to obtain the private keys, login information, or more directly the details of blockchain [28, 29].

Another very famous part of social engineering is phishing. In a phishing attempt, some or one malicious actor sends victims some email, message, or even sets up a Web site or social media account imitating a company brand which someone may trust. Often, the say to share victims to their credentials under the pretense of a giveaway or critical issue to force a sense of urgency. After giving information, there is a little chance to resist them and none can do to stop them from clearing out victims' assets [29–31].

The prevention of social engineering is a relatively straightforward matter. Never send anyone your login credentials or private keys to anybody or anywhere. Creating social awareness is the only solution to this problem.

4.4 *Software Flaws*

In most of the security issues, the biggest one is this one. Blockchain (Bitcoin, Ethereum) has tried to overcome rather has been able to overcome all types of attacks. But several apps are built on top of the actual blockchain and are still very susceptible to bugs. As per our survey, the maximum bug and damages come to blockchain from this issue.

It should be mandatory or important that any software using blockchain technology under the hood must undergo rigorous testing and review. This process must include code reviews, penetration testing, and smart contract audits. Additionally, several reputable applications should follow some redundant security measures in place. A bug or two will inevitably slip through the cracks, so you want to be prepared when they do.

For security purposes, the user must check before using any blockchain-based apps that it must go through a third-party security audit. Ideally, the code is also open source so that anyone can go in and review it [30–32].

4.5 *Malware*

This blockchain security issue ranges anywhere from the malicious cryptomining software to code that may be able to shut down a company's server.

Cryptojacking is the most dangerous type of malware that is mostly associated with blockchain and cryptocurrency. Normally, we can say cryptojacking is an unauthorized and also unnoticeable takeover of the computer's resources to mine the cryptocurrency. In this case, the cryptojackers do not directly steal any information from the victims, and they incorporate some malware or inject some malware causes to the victim's node which creates problems for the performance issues, increases electricity usage, and opens the door for other hostile details.

Because of cryptojacking and other malware's secretive fictitious nature, the yearly cost to victims is largely unknown. The only way to avoid this issue is blockchain needs more and more cautiousness to check and avoid unwanted links [31, 32].

4.6 Mining Pools Security Threats

Mining means miners, and pools mean work together. So, we can say mining pools attack means different attack vectors exploit the vulnerabilities in the pool-based mining approach [33–35]. Here both types, i.e., external and internal attackers attack the come in mining pools. In internal attacks, several dishonest miners collect more than their fair shares and create an interruption to the honest miner in the pool and in external attacks on the pools in such a way that the dishonest miners increase their hash power and perform some double-spending attacks.

4.7 Block Withholding Attack

Here the block withholding attack is very similar to selfish mining, but here the blocks are discarded by the dishonest miners [36, 37]. There are two forms of block withholding found based on the types of users or miners in a pool. One of the users is the pool manager, and another one is the regular pool users. The first form is *sabotage*. An attacker in a mining pool holds the blocks and can harm that pool where the attacker also belongs as shown in Fig. 2 (where block color is showing dishonest miners, and gray color is showing that honest miners and dishonest miners try to hamper that sabotage where they hold). However, it is not profitable to attackers. The second form is *lie in wait* where an attacker postponed submitting blocks to increase his reward. Contrary to *sabotage*, it is treated as profitable to the attacker [21]. In Rosenfeld's idea, it is more evolved. Unlike developed, it is a type of attack in between a mining pool and in a mining pool rather than any individual [21, 36]. As a mining pool manager can share his or her reward by partial proof of work (pPoW),

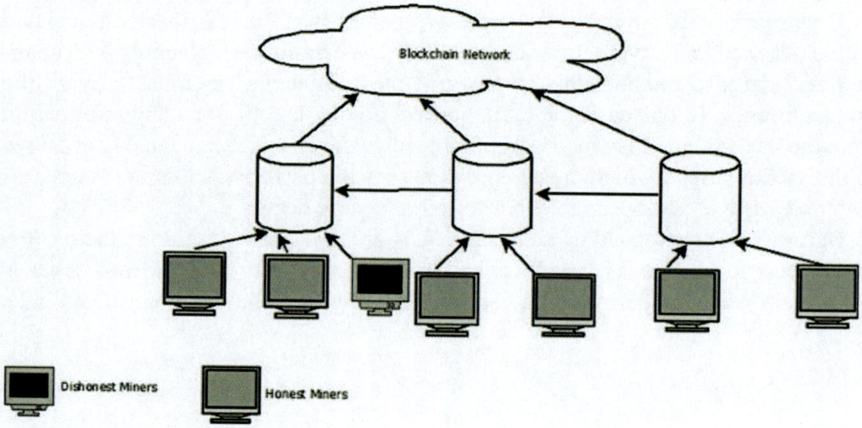

Fig. 2 Block holding attack

the honest miners can be cheated for their contribution. The pool can get the reward of block generation by the publication of full proof of work (fPoW). Now if miners do not submit fPoW on some specific purpose, then they can get a share though they do not contribute to the pool. In other cases, the attacker comes as a regular miner in a pool, and by using its reward it can discard other miners. In block withholding attack, the reward of the attacker is evaluated based on the assumption of the total mining power of the network [33, 38, 39]. To avoid this attack, only recognized, well-known, and trustworthy miners only should allow the pool.

4.8 Fork Problems

Another security threat is the fork problem. The fork problem is generally related to a decentralized node, where the agreement is done when the software is upgraded. It is a very important issue because it involves a huge area in the blockchain.

Types of Forks: With help of the new advanced consensus mechanism, blockchain can be divided into two types: the new nodes and the old nodes. This type of security threats can face four situations as follows:

1. The new nodes agree with the transaction which is sent by the old nodes.
2. The new nodes do not agree with the transaction which is sent by the old nodes.
3. The old nodes agree with the transaction which is sent by the new nodes.
4. The old nodes do not agree with the transaction which is sent by the new nodes.

Due to these four reasons, the fork problems can be divided into two types: the *hard fork* and the *soft fork*.

Hard Fork: Hard fork occurs when the system gets some new version or agreement, which may not be compatible with the previous version and the old nodes may not agree with the new nodes mining, then the one chain became divided into two chains, though the new nodes computing power may be stronger than the older one till the old miners continue with the older one.

Soft Fork: Soft fork generally happens when the system gets a new version or new agreement and as similar as hard fork here also the newer version is not compatible with the previous version, then the new nodes could not agree to mine with the old nodes. Because the computing power of new nodes is stronger than old nodes, so the blocks which are approved by the older one are not accepted by the newer ones, but they may not separate and continue with the same blockchain.

4.9 Scale of Blockchain

When one by one node is added to the blockchain, then it becomes longer and longer which may create trouble to store and compute, as a result, difficult for users to synchronize data, whereas at that time also the data are continuously increasing [40]. It also creates trouble, and the cybercrime may enter in that chaotic situation.

Now, out of these issues on some threats background, we are able to find and that will be discussed in the next phase.

In Sect. 4, we can able to several security issues and their measures. Now we tried to find out some issues root or background in Sect. 5.

5 Background of Security Threats of Blockchain

The several consensus mechanisms like PoW comprise two parts in between , one is the probabilistic consensus mechanism, which is based on solving (computational solving) technique (depending on a hash function) to generating a new valid block, which one is broadcasting the block via the peer-to-peer networking. Here the miners execute the consensus protocol and consequently registered the new specific block on the blockchain to generate reward. The consensus protocol is generated by the use of race condition and only those miners who are able to solve the consensus mechanism to broadcast the corresponding block in the earliest round; once that happens, the other miners start their new round of mining by updating the chain when the new block comes. While the second one consensus mechanism is designed to be computationally fair (based on distribution technique), here the miners experience high variance when finding the corresponding blocks and winning rewards. Here, a miner is computing the global-scale group of other miners, and the mining difference is adjusted accordingly, to find the block is very scattered and bursty.

To generate a more stable stream of reward, miners form a pool used to combine their computational power and share the power as well as the mining reward winnings to the blockchain. To increase their pooling power, they try to enhance their computational power by pooling them together and generating the winning a block whose reward gets split across the pool. To estimate each of the miner's contributions, the mining pool generally finds their consensus solutions which depend on their shared blocks in that blockchain, which usually compares the correlation between the same block as termed as block header of the block but with some easier difficulty; a block solution is a share but a share means not compulsorily a block. The reward of a pool using such shares is also prepared in such a way which is able to compute easily. To manage the mining pool, the pool manager generally keeps the track of the individual miner and their contributions along with their broadcasts record.

The forking problem generates when two or more blocks give collided propagation results. This means some nodes receive one block earlier than the others, whereas the other nodes already received some other block first, which creates a partition between the miners regarding the matter that which block to use for its impending round of mining [21, 41, 42]. Blockchain systems provide a partial reward for uncle blocks to increase fairness between the miners, i.e., reduce the effect of the networking discrepancy.

6 Conclusion

There is no scope to deny that blockchain is one of the hottest and promising issues in recent years or in future years too. After the success of Bitcoin, people now are trying to utilize blockchain technology in many other fields and services, such as digitized financial market, IoT, supply chain, voting, medical treatment, performance report generating, and storage. Several security threats are not yet preventive or recovered. We have to implement such preventing tools that bring secure impactor to the current system of blockchain. In this paper, we tried to explore a survey on blockchain security threats. We initially try to generate an outline of blockchain technology and the characteristics of blockchain along with some challenges and issues that may attack blockchain development. Also, we tried to summarize their background cause by which we can prevent those attacks in near future.

References

1. E.U. Opara, O.A. Soluade, Straddling the next cyber frontier: The empirical analysis on network security, exploits, and vulnerabilities. Int. J. Electron. Inf. Eng. **3**(1), 10–18 (2015)
2. J. Singh, Cyber-attacks in cloud computing: A case study. Int. J. Electron. Inf. Eng. **1**(2), 78–87 (2014)
3. S. Nakamoto, *Bitcoin: A Peer-to-Peer Electronic Cash System* (2013). (http://bitcoin.org/bitcoin.pdf)
4. J. Garay, A. Kiayias, N. Leonardos, *The Bit-coin Backbone Protocol: Analysis and Applications* (Springer, Berlin, Heidelberg, 2015), pp. 281–310
5. A. Gervais, G.O. Karame, V. Capkun, S. Capkun, Is bitcoin a decentralized currency? IEEE Secur. Priv. **12**, 54–60 (2014)
6. X. Li, P. Jiang, T. Chen, X. Luo, Q. Wen, A survey on the security of blockchain systems. Future Gener. Comput. Syst. (2017)
7. Z. Liu, N.C. Luong, W. Wang, D. Niyato, P. Wang, Y.C. Liang, D.I. Kim, A survey on blockchain: A game theoretical perspective. IEEE Access **7**, 47615–47643 (2019)
8. M. Ghosh, M. Richardson, B. Ford, R. Jansen. *A TorPath to TorCoin: Proof-of-bandwidth altcoins for compensating relays* (Naval Research Lab, Washington DC, 2014)
9. K. Toyoda, P.T. Mathiopoulos, I. Sasase, T. Ohtsuki, A novel blockchain-based product ownership management system (POMS) for anti-counterfeits in the post supply chain. IEEE Access **5**, 17465–17477 (2017)
10. X. Xu, I. Weber, M. Staples, L. Zhu, J. Bosch, L. Bass, C. Pautasso, P. Rimba, A taxonomy of blockchain-based systems for architecture design. in *2017 IEEE International Conference on Software Architecture (ICSA)* (IEEE, 2017), pp. 243–252
11. Lloyd's London presents a report called "Emerging Risk Report 2015"
12. Gervais et al., Novel quantitative framework to analyze the security and performance implications of various consensus and network parameters of Proof of Work (PoW) blockchains, Retrieved 27 Feb 2018
13. Apostolaki, Zohar, Vanbever, Bitcoin's Hijacking (2017)
14. A. Kosba, A. Miller, E. Shi, Z. Wen, C. Papamanthou, Hawk: The blockchain model of cryptography and privacy-preserving smart contracts, in *2016 IEEE Symposium on Security and Privacy (SP'16)* (May 2016), pp. 839–858
15. H. Watanabe, S. Fujimura, A. Nakadaira, Y. Miyazaki, A. Akutsu, J. Kishigami, Blockchain contract: Securing a blockchain applied to smart contracts, in *IEEE International Conference on Consumer Electronics (ICCE'16)* (2016), pp. 467–468

16. E. Heilman, A. Kendler, A. Zohar, S. Goldberg, Eclipse attacks on bitcoin's peer-to-peer network, in *24th USENIX Security Symposium, Washington, D.C.* (2015), pp. 129–144
17. L. Luu, V. Narayanan, C. Zheng, K. Baweja, S. Gilbert, P. Saxena, A secure sharding protocol for open blockchains, in *Proceedings of ACM SIGSAC Conference on Computer and Communications Security (CCS'16), New York, NY, USA* (2016), pp. 17–30
18. W.T. Tsai, R. Blower, Y. Zhu, L. Yu, A system view of_nancialblockchains, in *IEEE Symposium on Service-Oriented System Engineering (SOSE'16)* (2016), pp. 450–457
19. N.T. Courtois, L. Bahack, On subversive miner strategies and block withholding attack in bitcoin digital currency, CoRR, arXiv:1402.1718 (2014)
20. I. Eyal, E.G. Sirer, Majority is not enough: Bitcoin mining is vulnerable, CoRR, arXiv:1311.0243 (2013)
21. J. Bonneau, A. Miller, J. Clark, A. Narayanan, J.A. Kroll, E.W. Felten, Sok: Research perspectives and challenges for bitcoin and cryptocurrencies, in *IEEE Symposium on Security and Privacy* (2015), pp. 104–121
22. Y. Kwon, D. Kim, Y. Son, E. Vasserman, Y. Kim, Be selfish and avoiddilemmas: Fork after withholding (faw) attacks on bitcoin, in *Proceedings of the2017 ACM SIGSAC Conference on Computer and Communications Security, ser. CCS '17* (ACM, New York, NY, USA, 2017), pp. 195–209. Available: http://doi.acm.org/10.1145/3133956.3134019
23. L. Luu, R. Saha, I. Parameshwaran, P. Saxena, A. Hobor, On power splittinggames in distributed computation: The case of bitcoin pooled mining, in *2015 IEEE 28th Computer Security Foundations Symposium* (2015), pp. 397–411
24. I. Eyal, E.G. Sirer, Majority is not enough: Bitcoin mining is vulnerable. Commun. ACM **61**(7), 95–102 (2018)
25. I. Eyal, *The Miner's Dilemma*, in the IEEE Symposium on Security and Privacy (Cornel University, Oakland, 2015), Archived from the original on 09 Aug 2017. Retrieved 23 May 2017
26. C. Schlötterer, R. Tobler, R. Kofler, V. Nolte, Sequencing pools of individuals—mining genome-wide polymorphism data without big funding. Nat. Rev. Genet. **15**(11), 749 (2014)
27. E. Morvai, *Block-Build: A Decentralized Project Delivery Method*. Available at https://ssrn.com/abstract=3233798 (2018)
28. T. Bamert, C. Decker, L. Elsen, R. Wattenhofer, S. Welten. Have a snack, pay with bitcoins, in *13-th IEEE International Conference on Peer-to-Peer Computing* (2013)
29. E. Heilman, A. Kendler, A. Zohar, S. Goldberg, Eclipse attacks on bitcoin's peer-to-peer network, in *24th USENIX Security Symposium, USENIX Security 15, Washington, D.C., USA, August 12–14, 2015.*, ed by J. Jung, T. Holz (USENIX Association, 2015), pp. 129–144
30. J. Golosova, A. Romanovs, The advantages and disadvantages of blockchain technology, in *2018 IEEE 6th Workshop on Advances in Information, Electronic and Electrical Engineering (AIEEE)* (IEEE, 2018), pp. 1–6
31. J.H. Huh, K. Seo, Blockchain-based mobile fingerprint verification and automatic log-in platform for future computing. J. Supercomput. **75**(6), 3123–3139 (2019)
32. H. Treiblmaier, The impact of the blockchain on the supply chain: a theory-based research framework and a call for action. Supply Chain Manag. Int. J. **23**(6), 545–559 (2018)
33. J.H. Mosakheil, *Security Threats Classification in Blockchains* (2018)
34. A. Miller, A. Kosba, J. Katz, E. Shi, Nonoutsourceable scratch-off puzzles to discourage bitcoin mining coalitions, in *Proceedings of the 22nd ACM SIGSAC Conference on Computer and Communications Security* (ACM, 2015), pp. 680–691
35. J. Garay, A. Kiayias, N. Leonardos, The bitcoin backbone protocol: Analysis and applications, in *Annual International Conference on the Theory and Applications of Cryptographic Techniques* (Springer, Berlin, Heidelberg, 2015), pp. 281–310
36. M. Rosenfeld. Analysis of bitcoin pooled mining reward systems, *arXiv preprint* arXiv:1112.4980 (2011)
37. K. Liao, J. Katz, Incentivizing double-spend collusion in bitcoin, in *Financial Cryptography Bitcoin Workshop* (2017)
38. S. Lee, S. Kim, *Countering Block Withholding Attack Efficiently* (2018)

39. P. McCorry, S.F. Shahandashti, F. Hao, A smart contract for boardroom voting with maximum voter privacy, in *International Conference on Financial Cryptography and Data Security* (Springer, Cham, 2017), pp. 357–375
40. G. Karame, On the security and scalability of bitcoin's blockchain, in *Proceedings of ACM SIGSAC Conference on Computer and Communications Security (CCS'16), New York, NY, USA* (2016), pp. 1861–1862
41. V. Buterin, Ethereum: A next-generation smart contract and decentralized application platform (2014), Available https://github.com/ethereum/wiki/wiki/White-Paper. Accessed 22 Aug 2016
42. F. Piersigilli, V. Bhandari, Biomarkers in neonatology: the new "omics" of bronchopulmonary dysplasia. J. Matern. Fetal Neonatal Med. **29**(11), 1758–1764 (2016)

An Analysis of Energy Consumption of Blockchain Mining and Techniques to Overcome It

Ghosh Eshani, Das Rajdeep, Raj Shubhankar, and Das Baisakhi

Abstract Blockchain is found to be an evolving technology that is playing a major role in storing information securely. The architecture of blockchain is such that it provides security at the cost of an enormous amount of energy. The proof of work mechanism that it uses to mine block consumes a lot of energy. This protocol as a whole is responsible for causing pollution. It also increases the energy cost associated with it which overshadows the benefits that one can get from blockchain. The hash rate and the difficulty level, while mining of blocks, are the factors responsible for the consumption of the maximum amount of energy. This paper here presents a relationship between the hash rate and difficulty level and also a predictive model of the difficulty level for mining blocks. It also explores the solution for reducing energy consumption by looking for efficient mining hardware and a mining hardware recommender system which can help in finding more similar hardware.

Keywords Blockchain · Energy consumption · Proof of work · Mining hardware · Regression

G. Eshani (✉) · D. Rajdeep · R. Shubhankar · D. Baisakhi
Department of Information Technology, Institute of Engineering and Management, Salt Lake Electronics Complex, Sector V, Kolkata, India
e-mail: ghosh.eshani34@gmail.com

D. Rajdeep
e-mail: dasrajdeep450@gmail.com

R. Shubhankar
e-mail: shubhankarraj81@gmail.com

D. Baisakhi
e-mail: baisakhi.das@iemcal.com

V. E. Balas et al. (eds.), *Proceedings of International Conference on Computational Intelligence, Data Science and Cloud Computing*, Lecture Notes on Data Engineering and Communications Technologies 62, https://doi.org/10.1007/978-981-33-4968-1_60

1 Introduction

Blockchain technology was first introduced as Bitcoin in 2009 by Satoshi Nakamoto [1]. It evolved to be a very efficient method for storing information securely. The chain of blocks in a blockchain is connected by cryptographic links. Each block has an index, a data field, a nonce field and a previous hash value and uses the proof of work protocol to get mined. It is also a peer-to-peer network-based structure. Any change which is made to an existing block gets nullified, thus, making it immutable. The protocol for mining new blocks requires a cryptographic puzzle to be solved. Being a heavy computational work, several miners take part in this process. There is a particular range for the hash value of the new block which needs to be considered while generating a hash value for the block. But, only one or a few miners come up with the result and share the solution with others. So, the huge computational power that is used by all the participating miners is a complete wastage [2]. This is an alarming situation as most of the power is generated from non-renewable sources of energy which is responsible for increasing carbon footprint. The non-renewable sources of energy are exhaustive which makes the situation more dismaying.

The process of mining in blockchain requires a hardware implementation that can perform computational work by providing the necessary resources and power in terms of electricity. The hardware is then put in a pool of hardware which helps in increasing the hash rate of the blockchain because it is difficult to find the perfect block in solo mining. But the problem aroused when the electricity consumption increased for all the mining hardware working simultaneously. Mining pools turned out to be a major consumer of electricity [3]. So, industries try to place the pools in places where the electricity cost is low. This resulted in the concentration of mining pools only in few places. But the benefits of blockchain can be seen majorly if its usage can be made available to all regions. So, instead of going for mining protocols, the mining hardware can be changed. There is hardware with different internal structures and implementations which consume a different amount of electricity. Using efficient mining hardware can increase the hash rate and at the same time consume less power.

In this paper, the energy consumption issue has been discussed and the technique that can be used to overcome it is mentioned. Section 2 describes the related work, and block mining in the blockchain is discussed in this section. Section 3 gives a prediction analysis using the Ethereum dataset, Sect. 4 explains the mining efficiency of blockchain, Sect. 5 shows a mining hardware recommender system, and Sect. 6 concludes the paper.

2 Related Work

In an article named Socialism and Blockchain, we get an estimate that Bitcoin mining's annual energy consumption is 3.38 terawatt hours (TWh) [4]. It further discusses the CO_2 that was released in the environment due to Bitcoin mining which was equal to 13 tones of carbon, in the first half of 2018. The effect of emission is such that it can contribute to an overall global temperature rise of 2 °C, as approximated by a team in Hawaii. This is a frightful situation. There were several alternatives suggested by many researchers all over the world. For example, some proposed to switch to the proof of stake protocol [5], proof of authority protocol [6], etc., for mining of blocks. But the expenses associated with the change in protocol make this alternative unaffordable. Switching to renewable sources of energy was thought of as an alternative to non-renewable sources of energy which is being used for the mining of blocks. But, due to the uncertainties of weather conditions related to the renewable sources of energy, this could not be adopted as a solution for reducing energy consumption.

So, instead of going for other alternatives if the hardware is analyzed in terms of its efficiency and power consumption, then it can help in reducing the energy consumption of blockchain. The most efficient hardware needs to be found which has less energy consumption but high performance.

3 Prediction Analysis

An analysis has been done on the Ethereum dataset which shows the relationship between the hash rate and difficulty level. Further, linear regression techniques have been used for making a predictive model. For prediction analysis, the Ethereum dataset has been used which contains data about various factors associated with mining. The various data columns found in the database are as follows:

Eth_tx: This refers to the transaction fee required for a transaction.

Eth_address: It refers to the custom Ethereum tokens [7].

Eth_marketcap: It refers to the current market price of Ethereum.

Eth_blocks: It refers to the number of blocks present in Ethereum-based blockchain at that instance.

Eth_uncles: It refers to the orphan blocks in the Ethereum-based blockchain [8].

Eth_blocksize: It refers to the space required in bytes for storing the blocks [9].

Eth_blocktime: It refers to the average time for mining a new block.

Eth_gaslimit: It refers to the maximum amount of computational work [10].

Eth_ethersupply: Amount of ether required for performing on Ethereum [8].

Eth_difficulty: It refers to the number of nodes involved in the computational work.

Eth_hashrate: It refers to the number of blocks mined per unit time [11].

Eth_difficulty_predicted: It is the value of difficulty predicted from the analysis.

Using the given data in the dataset, the prediction model has been made using simple linear regression.

3.1 *Difficulty Level Prediction—Simple Linear Regression*

In Table 1, we get the predicted values of the Ethereum difficulty level by applying the simple linear regression technique. It is seen that the values which are predicted are almost similar to the values which are present in the actual difficulty level. This shows that the system is predicting the values right. From this analysis, we also get to know that as the value of the hash rate is getting increased, the difficulty level is also getting increased.

Results of simple linear regression analysis

- The test size is taken to be 40% of the given data.
- The linear_regression model from the sklearn library is used for the prediction.
- The train_test_split model is applied to the dataset.
- The X_train and y_train values obtained from the train_test_split are used to train the model.
- The y_pred is calculated from the values of y_test.
- The r_2_score is calculated from the y_pred and y_test, and the value is found to be 0.98.

The prediction analysis using simple linear regression was found to be good. It predicted the values almost correctly.

3.2 *Relation of Ethereum Hash Rate and Ethereum Difficulty*

From Fig. 1, it is found that the hash rate of Ethereum is increasing with the hash rate. It is found that the hash rate of Ethereum increases with an increase in the difficulty level. For increasing the performance of the system, the difficulty level is increased. But that in turn uses an enormous amount of energy. A study of mining efficiencies is done in the later sections to better understand the mining process and to find out efficient hardware.

4 Mining Efficiency of Blockchain

The mining of blocks in blockchain is the most time-consuming process in its entire architecture. An analysis of the Ethereum dataset has shown that the hash rate increases with the increase in power consumption. But going for different mining

Table 1 Difficulty level prediction using simple linear regression

eth_tx	eth_address	eth_marketcap	eth_hashrate	eth_blocks	eth_uncles	eth_blocksize	eth_blocktime	eth_gaslimit	eth_ethersupply	eth_difficulty	Eth_difficulty_predicted
4156	18298	85.89	343.12	5003	303	669	17.27	3141656	26137.96	5.58	6.11
4487	18089	92.41	325.24	5097	335	680	16.95	3141658	26738.09	5.17	10.17
953429	23937049	101889.94	203454.77	5910	1119	26059	14.29	7998293	20401.03	2765.4	2450.7
1008759	24828260	99840.470	221004.0257	5778	1348	26328	14.59	7997985	20542.12	2685.9	2666.2
704758	26638444	91549.38	243106.15	5958	1064	20711	14.22	7996526	20372.25	3115.0	2991.2

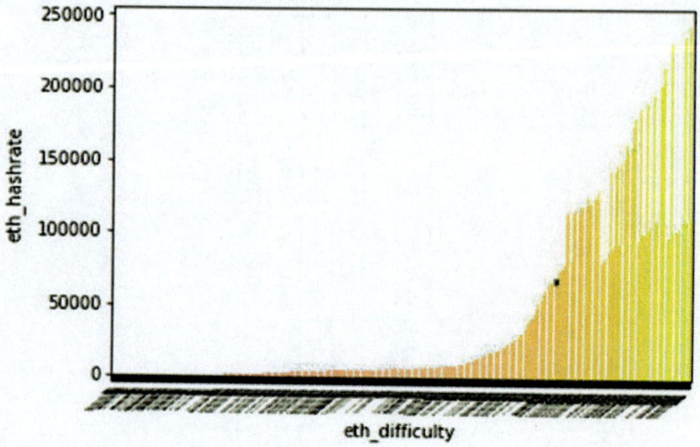

Fig. 1 Graph showing Ethereum mining hash rate versus Ethereum mining difficulty

hardware can be a better way of reducing energy consumption. That mining hardware which uses less power but at the same time produces good hash rate can be used. An analysis of the Ethereum hardware dataset [12] has been done. The mining hardware dataset consists of the following data:

- **Number**: It states the index number of the dataset.
- **Type**: It states whether it a CPU type or GPU type.
- **Release date/date of original information retrieval**: It states the release date of the product or the date of the original information about the product was obtained.
- **Source**: It states the source of information and the Web sites from where the information about the product was obtained.
- **Accessed on**: It states when the product was accessed for use.
- **Manufacturer**: It is the name of the manufacturer of the product.
- **Product**: It is the name of the actual product and the model number.
- **Hash rate in Mh/s**: Hash rate refers to the number of the hash of blocks generated per unit time [13].
- **Power in W**: It is the amount of power consumed by the hardware model used [14].
- **Efficiency in Mh/J**: It refers to the efficiency of each of the mining hardware considering the power consumption and hash rate [14].

4.1 Efficiency of Each Product

An analysis of the dataset in Fig. 2 shows that the efficiency of GTX Titan V is the maximum of all. Here in the given graph X-axis denotes the product names and the Y-axis denotes the efficiency in Mh per Joules. The efficiency of Radeon HD 7870

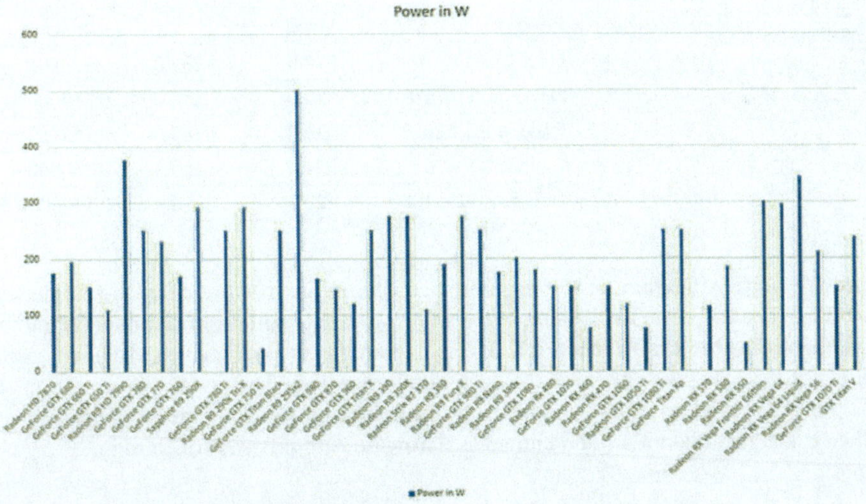

Fig. 2 Efficiency of mining product hardware

is found to be the least. There are other mining hardware models like Radeon RX 570 is having the second-highest efficiency.

4.2 *Power Versus Product*

In Fig. 3, it is seen that the hardware with maximum efficiency is not having the minimum power consumption. So, another hardware needs to be observed which can

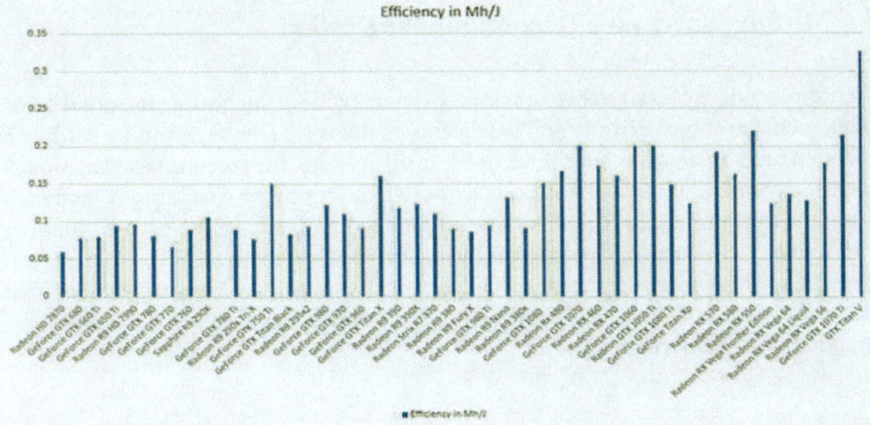

Fig. 3 Power versus product

Table 2 List of some efficient hardware

Product	Hash rate	Power	Efficiency
GTX Titan V	77.0	237.0	0.324895
GeForce GTX 1070 Ti	30.0	150.0	0.200000
Radeon RX 580	30.0	185.0	0.162162
Radeon RX 470	24.0	150.0	0.160000
GeForce GTX 1080	27.0	180.0	0.150000

provide better efficiency and lower energy consumption. It is found that the efficiency of GTX Titan V is the maximum, and also its power consumption is moderately low. Though it is seen that Radeon RX 550 is having the second-highest efficiency, still it is not a good choice because the hash rate is very low. The power consumption of GeForce GTX Titan Black is found to be the maximum, and its efficiency is also less than 0.10. So, it is not a good choice as hardware is to be used for mining.

4.3 *Mining Hardware with Better Efficiencies*

Some of the efficient hardware were found from the analysis.

In Table 2, the products having higher efficiencies are taken. It has been found that the product with the highest efficiency out of the given list is having a maximum power consumption but at the same time the highest hash rate of all. But still, the power consumption of GTX Titan V is found to be less as compared to all other hardware in the given dataset. There are other products with very low power consumption but their efficiencies are not found to be good. To make this process of finding efficient hardware, a mining hardware recommender model is being developed.

5 Mining Hardware Recommender System

A recommender system refers to a system that is capable of suggesting users the items of their choice based on some previously chosen items [15]. Here, a mining hardware recommender system has been developed. In this system, the content-based approach of recommendation has been used because we are trying to find similar hardware based on the dataset, and there is no need for another dataset to be used for finding similarities [16]. Each name of the hardware provided by the user is taken as the pivot. The other hardware which is near to it is taken as the suggestion. The suggestions give the index values of the hardware which are recommended. Using the index values, their names are determined.

5.1 Analysis Result of Recommendation

In Table 3, the name of some efficient hardware models is shown, and the corresponding recommended models are shown. It does the recommendation based on the value of the efficiencies that have been provided. The analysis of the hash rate and difficulty level shows that to increase the hash rate, the power consumption has to be increased in terms of difficulty level. But, choosing better hardware can increase efficiency by providing a better hash rate and an acceptable amount of power consumption [17]. Using the recommender system, one can also find the hardware which is similar to the efficient ones.

6 Conclusion

The analysis of the given Ethereum dataset here shows that the hash rate increases when the difficulty level is increased. Increasing the difficulty level is an alarming situation because it increases the overall energy consumption. Simple linear regression has been used to predict the difficulty level for a given value of other factors. Using the mining hardware dataset, the efficiencies and power consumption of each of the models are compared. The efficient hardware is found. Using the given dataset, a mining hardware recommender system has been developed which is recommending similar types of the model as per the model asked by the user. This system can be used to determine the efficient hardware from a given set of models if only one efficient model is known. So, it has been found that there exists hardware which can consume less energy and has efficiency as compared to others. Thus, switching to efficient hardware can help in reducing the energy consumption of blockchain.

Table 3 Recommended hardware for the given model

Model name	Recommendations
GeForce GTX 660 Ti	'Radeon RX 470', 'Radeon Rx 480', 'Radeon HD 7870', 'GeForce GTX 770', 'Radeon R9 290x Tri-X'
GTX Titan V	'Radeon HD 7870', 'GeForce GTX 770', 'Radeon R9 290x Tri-X', 'GeForce GTX 680', 'GeForce GTX 660 Ti'
GeForce GTX 1070 Ti	'GeForce GTX 1070', 'Radeon Rx 480', 'Radeon RX 470', 'GeForce GTX 660 Ti', 'Radeon HD 7870'
Radeon RX 580	'Radeon HD 7870', 'GeForce GTX 770', 'Radeon R9 290x Tri-X', 'GeForce GTX 680', 'GeForce GTX 660 Ti'
Radeon RX 470	'Radeon RX 580', 'Radeon HD 7870', 'GeForce GTX 770', 'Radeon R9 290x Tri-X'

References

1. S. Nakamoto, *Bitcoin: A Peer-to-Peer Electronic Cash System.* Available online: https://bitcoin.org/bitcoin.pdf
2. H. Zhao, *Bitcoin and Blockchain Consume an Exorbitant Amount of Energy* (2018). Available online: https://www.cnbc.com/2018/02/23/bitcoin-blockchain-consumes-a-lot-of-energy-engineers-changing-that.html
3. J Tuwiner, *Buy Bitcoin World* (2015). Available Online: https://www.buybitcoinworldwide.com/mining/
4. S. Huckle, *Bitcoins Energy Consumption is a Concern But it May be a Price Worth Paying* (2018). Available Online: http://theconversation.com/bitcoins-high-energy-consumption-is-a-concern-but-it-may-be-a-price-worth-paying-106282
5. Z. Zheng, S. Xie, H.N. Dai, X. Chen, H. Wang, *Blockchain Challenges and Opportunities: A Survey* (2018)
6. S. De Angelis, L. Aniello, R. Baldoni, F. Lombardi, A. Margheri, V. Sassone, *Golden. Proof of Authority.* Available Online: https://golden.com/wiki/Proof_of_authority
7. Blockgeeks. *How to Setup an Ethereum Wallet and Buy a Custom Token* (2017). Available online: https://blockgeeks.com/how-to-setup-anethereum-wallet-and-buy-a-custom-token-a-guide-for-coinbase-users/
8. Investopedia, Uncle block (2017). Available online: https://www.investopedia.com/terms/u/uncle-block-cryptocurrency.asp
9. Bit on Blocks. *Gentle Introduction to Ethereum* (2016). Available online: https://bitsonblocks.net/2016/10/02/gentle-introduction-ethereum/
10. Masterthecrypto. Guide to Ethereum: what is Gas, gas-limit, gas-price. Available online: https://masterthecrypto.com/ethereum-whatis-gas-gas-limit-gas-price/
11. Bitcoin (2009). Available Online: https://bitcoin.org/en/vocabulary
12. J. Michel, Bitcoin and Ethereum Hardware. *Data Mendeley* (2018). Available online: https://data.mendeley.com/datasets/4dw6j3pxz5/1
13. W. Suberg, Cointelegraph (2020). Available online: https://cointelegraph.com/news/its-never-been-harder-to-mine-bitcoin-ashash-rate-hits-new-highs
14. S. Küfeoğlu, M. Özkuran, *Energy Consumption of Bitcoin Mining* (University of Cambridge, England, 2019)
15. R. Mall, Recommender system. *Towards Data Science* (2019). Available Online: https://towardsdatascience.com/recommender-system
16. S. Luo, Introduction to recommender system. *Medium* (2018). Available online: https://towardsdatascience.com/intro-to-recommender-system-collaborativefiltering-64a238194a26
17. J.O.D. Karl, D. Malone, Bitcoin. *Bitcoin Mining and its Energy Footprint* (Hamilton Institute National University of Ireland Maynooth, Ireland)

Author Index

V. E. Balas et al. (eds.), *Proceedings of International Conference on Computational Intelligence, Data Science and Cloud Computing*, Lecture Notes on Data Engineering and Communications Technologies 62, https://doi.org/10.1007/978-981-33-4968-1